A Color Atlas of the Chinese Mecoptera

Ji-Shen Wang • Bao-Zhen Hua

A Color Atlas of the Chinese Mecoptera

天龙昆虫学基金
Tianlong Entomological Foundation

河南科学技术出版社
HENAN SCIENCE AND TECHNOLOGY PRESS

Springer

Ji-Shen Wang
College of Agronomy and Biosciences
Dali University
Dali, Yunnan, China

Bao-Zhen Hua
College of Plant Protection
Northwest A&F University
Yangling, Shaanxi, China

ISBN 978-981-16-9557-5   ISBN 978-981-16-9558-2 (eBook)
https://doi.org/10.1007/978-981-16-9558-2

Jointly published with Henan Science and Technology Press
The print edition is not for sale in China (Mainland). Customers from China (Mainland) please order the print book from: Henan Science and Technology Press.

This Springer imprint is published by the registered company Springer Nature Singapore Pte Ltd.
The registered company address is: 152 Beach Road, #21-01/04 Gateway East, Singapore 189721, Singapore

Foreword (First Edition)

Speaking of insects, some common groups immediately appear in our minds: butterflies, dragonflies, bees, mosquitos, flies, and cockroaches. These insects are closely related to the daily life of humans and frequently occur in the screen and other cultural works. Few people know, however, that among over a million species of extant insects, there are a group of peculiar members with elongated rostrum in their heads; some of them have enlarged and recurved male genitalia resembling the tail of scorpions, hence the common name "scorpionflies"; their external morphology and biological characteristics are highly varied; and they play a significant role in the study of the evolution and phylogeny of the order Insecta. This group of insects has been assigned in the order Mecoptera by entomologists. Because of the lack of specific literatures, the identification of Mecoptera is relatively difficult in a long period for the general entomological researchers and hobbyists.

The extant Mecoptera, a small relic order in the class Insecta, consists of over 700 species in nine families and ca. 40 genera. By contrast, the fossil Mecoptera includes ca. 620 species in more than 40 families and 210 genera; a great number of new taxa (families, genera, and species) have unceasingly been found. Among them, more than three-fourths of families were extinct. Mecopterans have a long history of origin and evolution, with the earliest fossil records in the Early Permian (ca. 270 mya). The Mesozoic era (252–65 mya) was an important period which links the past with the present days for the evolving Mecoptera. The mecopterans reached their peak of diversity in this era and held a large proportion in the Mesozoic fossil records of Holometabola. Nine extant families in Mecoptera have been found in the Jurassic-Cretaceous period. Many extinct families (e.g., Cimbrophlebiidae, Orthophlebiidae, and Choristopsychidae) not only enrich our knowledge of the diversity of Mecoptera, but also provide precious and solid evidence for the study of the origin and evolution of this order.

The mecopterans have their mouthparts, wings, and genitalia highly varied among different groups, corresponding to their diverse biological characteristics. Frequently, by comparing ancient, extinct mecopterans with extant ones, we are so amazed at their highly varied biological characteristics. For instance, members of the families Aneuretopsychidae, Mesopsychidae, and Pseudopolycentropodidae were ancient pollinators sharing a complex co-evolution with plants, and their special, slender siphoning mouthparts were able to feed the pollination drops of the gymnosperms; *Juracimbrophlebia ginkgofolia*, belonging to the family Cimbrophlebiidae, has a complex mimicry and symbiotic relationship with the plant *Ginkgo*, the relationship between them lasted over a hundred million years; some male insects in the family Holcorpidae have exaggeratedly elongated abdomen and genitalia, special structures not only utilized to forcedly fight with other males, but also an sexual display to the females. There are enormous number of unknown structures, biological characteristics, and evolutionary laws await our further discovery and research.

For the past many years, the research team of Mecoptera, Northwest A&F University, leading by Professor Bao-Zhen Hua, dedicated to the study of morphology, behaviors, taxonomy, and phylogeny of Mecoptera and made a lot of remarkable achievements. They published some excellent works in many Chinese and abroad journals, making Northwest A&F University one of the global centers of the study on Mecoptera. Meanwhile, professor Bao-Zhen Hua also dedicated to teaching works and trained a group of young entomologists. His college courses such as scientific writing, general entomology, insect morphology, and insect taxonomy are popular among students. A title of "Prominent Educator of Shaanxi" was granted for his unique teaching idea and rigorous working attitude.

A Color Atlas of the Chinese Mecoptera is a valuable and unmatched work, in which all the aspects of knowledge of Mecoptera are shown by distinct elegant pictures and accurate scientific texts. It is a great original work that gathered scientific, artistic, and enjoyable qualities, suitable for the general entomological researchers and hobbyists, and a practical reference for the forefront workers in agriculture and forestry. I am honored to recommend this atlas to the readers. Meanwhile, I also expect more excellent discoveries from the research team of Mecoptera leading by professor Bao-Zhen Hua.

Capital Normal University, Beijing, China Dong Ren
September 2018

Preface (Second Edition)

Mecoptera is a small order of holometabolous insects and consists of over 700 known species worldwide, including scorpionflies and hangingflies, etc. They are important indicators of ecological environments and are often ideal materials for behavioral studies due to their interesting courtship and mating behaviors. They are ancient members in Holometabola and play a significant role in the evolutionary research of Insecta.

China is rich in Mecoptera all over the world. More than 300 species in 11 genera of three families have been recorded in China hitherto. However, literature regarding the identification of the Chinese Mecoptera is very scarce. Hence, we spent three years to compile the first edition of this color atlas based on the abundant collections from the Entomological Museum, Northwest A&F University, and our intensive field investigations in recent years, to introduce the common species of the Chinese Mecoptera. In the second edition, we list and illustrate 242 species (updated from 170 species in the previous edition) of the Chinese Mecoptera. It is often difficult to identify to the species-level based exclusively on live photographs without careful examinations under a microscope. Therefore, we also provide many pictures of pinned or alcohol-preserved specimens. For most species, we present pictures of the adults, and the male and female genitalia. A geographical distribution map is also provided for each species.

This work cannot be finished without the generous helps from numerous friends and colleagues during our writing and editing courses. For the identifications of species, we mostly benefit from the late Professor Io Chou, who had collected specimens and literature and initiated the studies of the Chinese fauna of Mecoptera decades ago. We also profited from the postgraduate students of the research team of Mecoptera, Northwest A&F University, including Peng-Ying Huang, Jiang-Li Tan, Jing Chen, Li-Jun Cai, Na Ma, Wen Zhong, Lu Jiang, Qiong-Hua Gao, Meng Wang, Gui-Lin Hu, Ying Miao, Sha Xie, Tao Peng, Xiao-Yan Hou, Shu-Yu Liu, Gang Yan, Hao Xu, Xue Li, Qiang Fu, Jun-Xia Zhang, Chao Gao, Wei Du, and Kai Gao.

Our sincere thanks also go to Yan-Dong Chen, Zhi-Wei Dong, Kai-Wen Gao, You-Jing Gong, Li He, Zhu-Qing He, Zheng-Kun Hu, Gui-Qiang Huang, Si-Yao Huang, Ri-Xin Jiang, Shi-Xiang Jiang, Zhuo-Heng Jiang, Qi-Long Lei, Ze-Jian Li, Ye-Jie Lin, Lin Lyu, Jian-Yue Qiu, Lu Qiu, Hong-Jian Wang, Yong Wang, Chao Wu, Hao Xu, Qi-Cheng Yang, Qiao-Zhi Yang, Mao Ye, Kun Yu, Cheng-Hui Zhan, Jie Zhang, Ming-Zhi Zhao, Yu-Chen Zheng, Chao Zhou, De-Yao Zhou, Dan-Chen Zhu, and Wen-I Chou (Taiwan). They collected many valuable specimens and made it feasible to show those species to the readers.

We thank Jun Chen and Hong Liu (IZAS), Li Chen and Zhu Li (ISWU), Pavel Chvojka and David Král (NMCZ), Xing-Yue Liu and Yu-Chen Zheng (ECAU), Toshiya Hirowatari and Satoshi Kamitani (KYSU), Li-Zhen Li and Zi-Wei Yin (SHNU), Hong Pang and Bing-Lan Zhang (SYSU), Dong Ren (CANU), Min Wang and Xing-Min Wang (SCAU), Mei-Cai Wei (CUFT), Jian-Yue Qiu, Cheng-Bin Wang, and Hao Xu (MYNU), Shinji Yano (OMGM), and the late George W. Byers (EDKU) for arranging the access to examine and loan of the specimens.

Our recent field works were greatly supported by Yan-Qing Hu, Shuang Xue, Yuan Hua, Finks Bakrie (Indonesia), and Sepni Juhansah (Indonesia). We would also like to express our thanks to Dalton S. Amorim (Brasil), Cristiano Lopes-Andrade (Brasil), Tomoya Suzuki

(Japan), and Jason G.H. Londt (South Africa) for fervidly sending precious literatures. We also thank John Horstman, Lu Jiang, Bo Lei, Ye-Chan Lin (Taiwan), Wei-Liang Xie, Han Xu, Wei-Wei Zhang, and Wen Zhong for allowing the use of their beautiful photographs. A special thank is due to Messrs Ben-Qing Zhou and Yi-Kun Li from the Henan Science and Technology Press. Under their encouragement, this atlas began to compile. The publication of the current form is also due to their excellent editorial work.

This work was financially supported by the Tianlong Entomological Foundation, the National Natural Science Foundation of China (Grant Nos. 30070101, 30370179, 30670255, 30970386, 31172125, 31301898, and 31672341), the 2018 National Publication Foundation of China, and the Starting Foundation for the High-level Talents, Dali University (Grant No. KY2096124040).

Due to our limited knowledge, there might be some mistakes and flaws in this atlas. We expect criticism and correction from experts and readers.

Dali, China
October 2021

Ji-Shen Wang

Contents

Part 1

Introduction

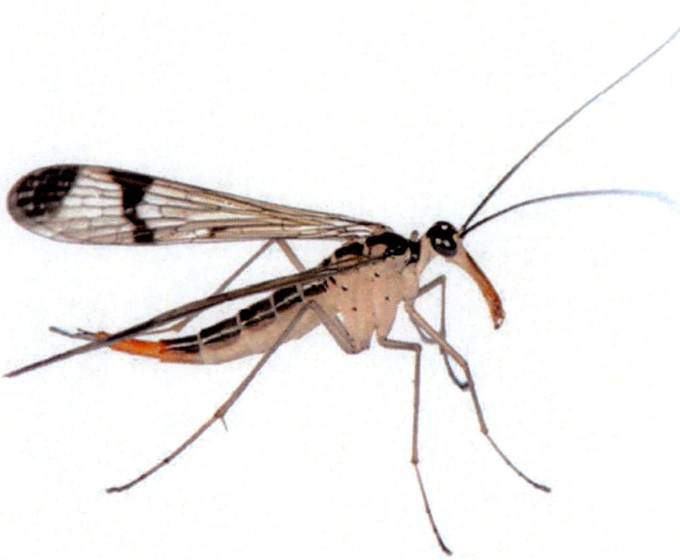

Fig. 1 Female adult of ***Neopanorpa triangulata***, lateral view

Systematic Status

1

The term Mecoptera is derived from the ancient Greek μέκο (= meco, long) and πτερόν (= pteron, wing), referring to the elongated wings of these insects.

According to the fossil records (Figs. 1.1 and 1.2), mecopterans probably originated in the Lower Permian (ca. 270 mya), reached their peak of diversity in Mesozoic (250–65 mya), and once occupied various ecological niches (Grimaldi and Engel 2005). Nearly 400 fossil species in 87 genera and ca. 36 families were recorded (Dunford and Somma 2008; Byers 2009; Soszyńska-Maj et al. 2022). Some long proboscis-bearing mecopterans may have been the earliest pollinators and shared a close co-evolution with the gymnosperms, long before the present-day pollinating insects such as flies in Diptera and bees in Hymenoptera (Ren et al. 2009). In addition, an extinct hangingfly belonged to the family †Cimbrophlebiidae showed a case of leaf-mimicry to the ginkgo leaves (Wang et al. 2012). Some mecopterans bear greatly reduced hindwings, which superficially resembled the halteres of Diptera (Grimaldi et al. 2005). After of a series of drastic climatic changes and several mass extinctions (Erwin 1990; Benton 1995), many mecopterans failed to survive, and the rest of them became a relict group.

Extant members of Mecoptera are a small group of the holometabolous insects, comprising approximately 700 species in nine extant families (Penny and Byers 1979; Bicha 2018). They are one of the most ancient groups of holometabolous insects (Kristensen 1981; Dunford and Somma 2008) and constitute the superorder Antliophora (means "pump-bearers" literally owing to their males bearing a sperm pump in genitalia) together with Diptera and Siphonaptera (Figs. 1.3 and 1.4).

Mecopterans play a significant role in the study of the evolution of Insecta. Diptera, Lepidoptera, and Trichoptera may originated from ancestors that were closely related to Mecoptera (Grimaldi and Engel 2005). Mecoptera, Siphonaptera, and Diptera are grouped into the Antliophora (from Latin *antlia* + *-phora*, "pump-bearers"), a superorder proposed for having a sperm pump in their male genitalia (Hennig 1969). The monophyly of Antliophora was supported by morphological (Friedrich and Beutel 2009) and molecular evidence (Whiting et al. 1997; Misof et al. 2014).

Fig. 1.1 Fossil mecopterans (I). Sedimentary rock, Nannochoristidae: †*Itaphlebia* sp. (donated by Dong Ren)

The systematic position of Mecoptera in Holometabola is open to debate for considerably a long time (Dunford and Somma 2008; Friedrich et al. 2013). Previously, Hennig (1969) regarded Mecoptera as a sister group to Diptera, though Kristensen (1999) considered the order to be more related to Siphonaptera. Subsequently, phylogenetic research based on concatenated data of multi-genes (Whiting 2002) points to the close relationship between Mecoptera and Siphonaptera, and Siphonaptera is sister to the snow fleas Boreidae. There are two synapomorphies shared by Siphonaptera and snow fleas: (1) the possession of resilin (an elastomeric protein found in many arthropods) within their bodies and (2) the panoistic ovariole (ovarioles that have no specialized nurse cells) in their female reproductive systems. Therefore, Mecoptera were regarded as a paraphyletic group. This standpoint of view, however, was in a lot of doubts. After investigating the ultrastructure of sperm, Dallai et al. (2003) discovered a significant difference between Boreidae and Siphonaptera, and reconfirmed the

J.-S. Wang, B.-Z. Hua, *A Color Atlas of the Chinese Mecoptera*, https://doi.org/10.1007/978-981-16-9558-2_1

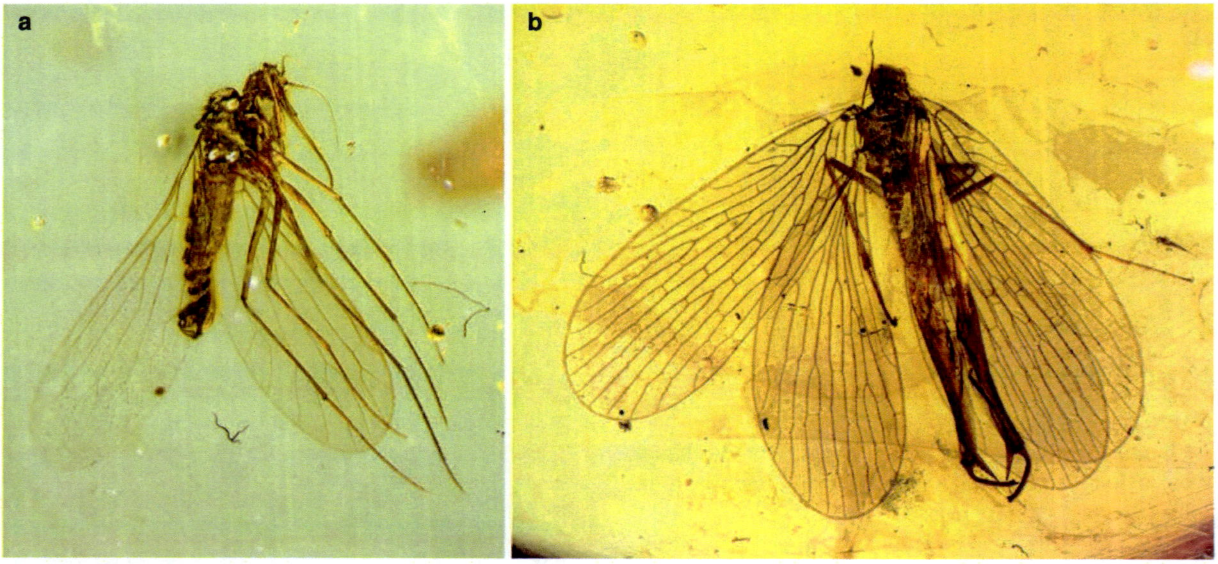

Fig. 1.2 Fossil mecopterans (II). Burmese amber, (**a**) †Pseudopolycentropodidae: †*Parapolycentropus paraburmiticus*; (**b**) Meropeidae: †*Burmomerope eureka*

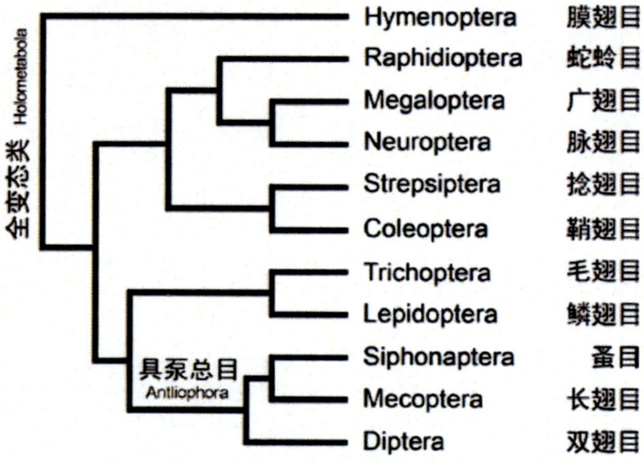

Fig. 1.3 Phylogenetic relationships among Holometabola (after Misof et al. 2014)

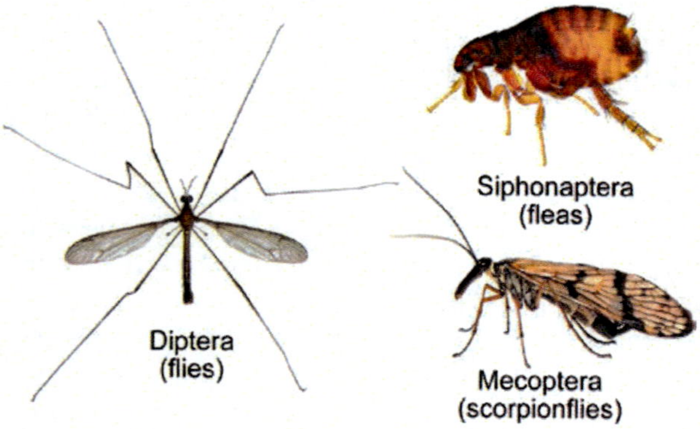

Fig. 1.4 Three orders in Antliophora and their representative members

monophyly of Mecoptera. Misof et al. (2014) also supported the monophyly of Mecoptera based on their phylogenetic analysis of Insecta from 1,478 protein-coding genes and amino acid sequences. Therefore, the phylogenetic relationship within Antliophora should be as follows: Diptera + (M ecoptera + Siphonaptera). However, recent molecular studies suggest that Siphonaptera is closely related to Nannochoristidae, and probably should be treated as an infraorder within Mecoptera (Meusemann et al. 2020; Tihelka et al. 2020) in the future.

2 Historical Review

As early as the Ancient Rome, Pliny the Elder (AD 23–79), a famous natural historian, had mentioned that "some of which really have wings" in the chapter of scorpions in his *Naturalis Historia* (The Nature History). Later, the paleontologist George Cuvier claimed that the creatures Pliny described might be scorpionflies, belonging to the order Mecoptera (Bostock and Riley 1855).

Carolus Linnaeus (AD 1707–1778) described two species of scorpionflies, *Panorpa communis* and *Panorpa germanica*, which were assigned to Neuroptera in his famous *Systema Naturae* (10th edition) published in 1758 (Fig. 2.1). Meanwhile, another species, *Panorpa coa* was described along with them, but this species is a real neuropteran, which belongs to Nemopteridae, the spoonwings or thread-winged-antlions. Since then, its name has been modified into *Nemoptera coa* (Linnaeus, 1758). Linnaeus's original descriptions of the genus *Panorpa* was in Latin: "*Rostrum corneum, Cylindrium: palpis... Antennae thorace longiores. Cauda maris chelata,*" meaning "rostrum horn-like; palpi (including maxillary and labial palpus) cylindrical; antennae longer than thorax; and tail claw-like." This clearly depicts a general image of the scorpionflies.

The early researchers of Mecoptera include Latreille (1802), MacLachlan (1875, 1894), Handlirsch (1906), Navás (1908, 1911, 1936), and Lieftink (1933, 1936), etc. Esben-Petersen's (1921) *Mecoptera*: *Monographic Revision* was the earliest fundamental work at the global scale. Over 90 species in five families were included in this monograph.

Issiki (1927, 1929a, b, 1931, 1933) and Miyaké (1908, 1913) recorded and described the mecopterans from Japan and adjoining areas. Tjeder (1936, 1950, 1973) recorded some mecopterans from China and Southeast Asia. Carpenter published a series of papers dealing with the Chinese (1938, 1945, 1948) and North American (1931, 1935, 1936) faunas, respectively. Riek (1954, 1973) studied the Australian fauna. Byers's work included the Mecoptera faunas of Southeast Asia (1965, 1966, 1999), China (1970), North America (1958, 1962, 1973, 1993, 1996), South America (1989), and Africa (1975). Cheng published a series of papers (1949, 1952, 1957a), and made a revision on the Chinese fauna (Cheng 1957b). Rust and Byers (1976), Chau and Byers (1978), and Cheung et al. (2006) recorded the Indian, Indonesian, and Canadian faunas. Penny and Byers (1979) published a checklist of the extant Mecoptera species of the world. Bicha recorded some mecopterans from North America (1983, 2006, 2007, 2015a) and Southeast Asia (Bicha, 2010, 2011, 2015b; Bicha et al. 2017).

Chou et al. (1981, 1987, 1988) investigated the mecopterous fauna in Shaanxi and Hunan Provinces. Afterwards, Hua (1997, 1998, 1999, 2004), Hua and Chou (1997, 1998a, b, 1999), and Hua et al. (2001) recorded a series of mecopterans from Jilin, Liaoning, Henan, Fujian, Hainan, Sichuan, and Tibet. Tan and Hua (2008a, b, c, 2009a, b), Cai et al. (2008), Ma and Hua (2011), Zhong and Hua (2013a), and Gao (2016) established and described several Chinese genera and species. Zhou (2000, 2001, 2003, 2005, 2006) also published a series of research articles regarding the Chinese mecopterans. Wang and Hua (2016, 2017a, 2019a, b) published a series of Chinese genera and species and conducted a fundamental phylogenetic analysis of Panorpidae based on morphology (Wang and Hua 2020).

The morphology of Mecoptera also draw great interests from scholars. Ferris and Rees (1939) conducted a comprehensive anatomical study of Panorpidae. Mickoleit (1976) summarized the female reproductive systems of Mecoptera. Kaltenbach (1978) reviewed the morphological studies of the whole order. Biological researchers of Mecoptera include Miyaké (1912), Yie (1951), Byers (1963), Gassner (1963), etc. Constrained by poor technical conditions, these studies mainly focused on the external morphology and larval chaetotaxy. The scorpionfly *Panorpa qinlingensis* was described and illustrated in detail for its immature stages by Cai and Hua (2009b) using electron microscopy. A series of systematic work of larval morphology were conducted by Jiang and Hua (2013, 2014, 2015). Some groups of Mecoptera, especially Panorpidae and Bittacidae, exhibit an interesting "nuptial feeding" behavior, which draws great attentions from entomologists (Zhong and Hua 2013; Zhong et al. 2015a, b).

J.-S. Wang, B.-Z. Hua, *A Color Atlas of the Chinese Mecoptera*, https://doi.org/10.1007/978-981-16-9558-2_2

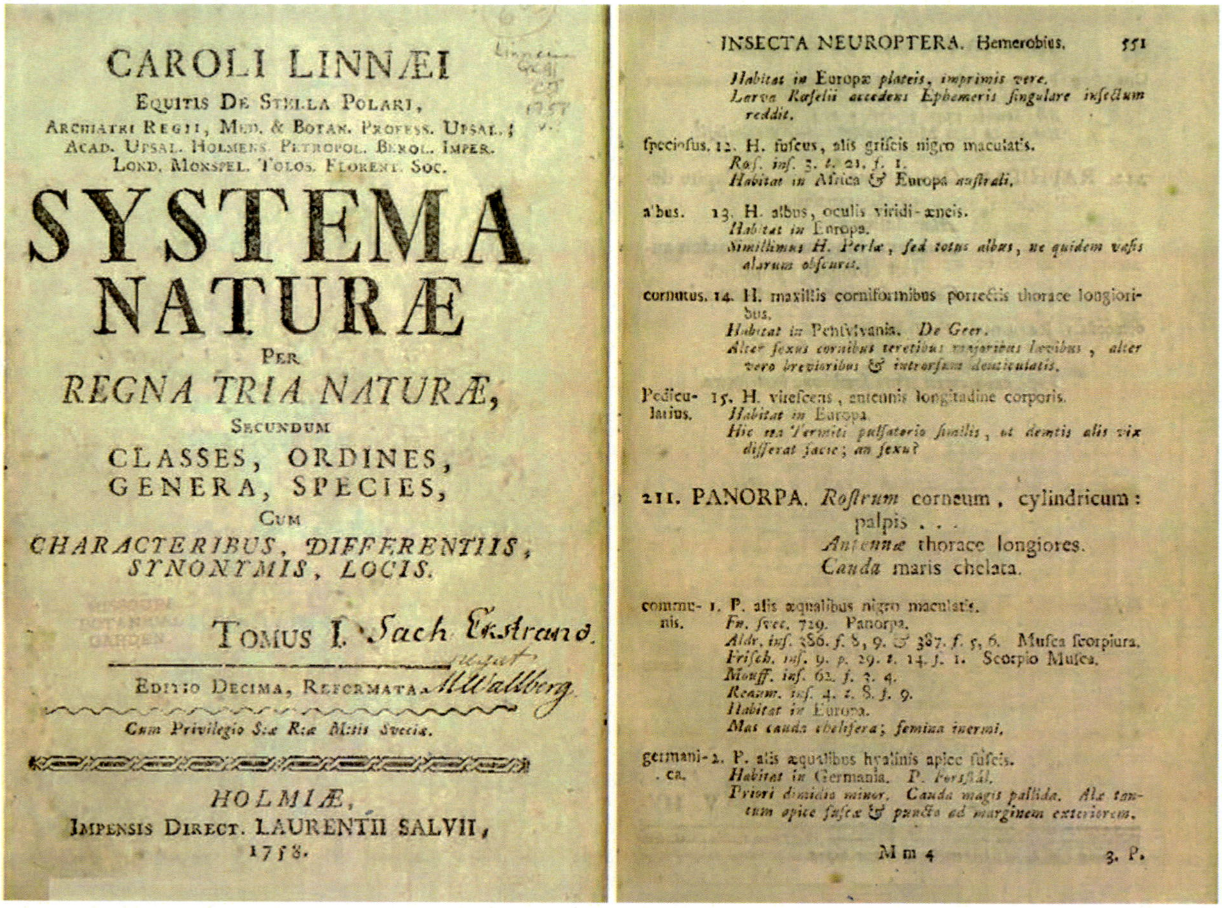

Fig. 2.1 Linnaeus's *Systema Naturae*, 10th edition. Cover and descriptions of *Panorpa*

In addition, studies regarding other fields are also constantly emerged in recent years, including phylogeny (Ma et al. 2012; Hu et al. 2015), structure of visual organs (Chen and Hua 2016), embryonic development (Du et al. 2009; Yue and Hua 2010; Kou and Hua 2016), and cytotaxonomy (Miao et al. 2017, 2019).

Classification System

3

Panorpida is a group consisting of the orders Mecoptera (scorpionflies), Siphonaptera (fleas), Diptera (true flies), Trichoptera (caddis flies), and Lepidoptera (moths and butterflies) (Grimaldi and Engel 2005) (Fig. 3.1). The extant Mecoptera consists of nine families, i.e. Nannochoristidae, Bittacidae, Boreidae, Meropeidae, Eomeropidae, Apteropanorpidae, Choristidae, Panorpodidae, and Panorpidae (Figs. 3.2 and 3.3). Based on the width of wings, Tillyard

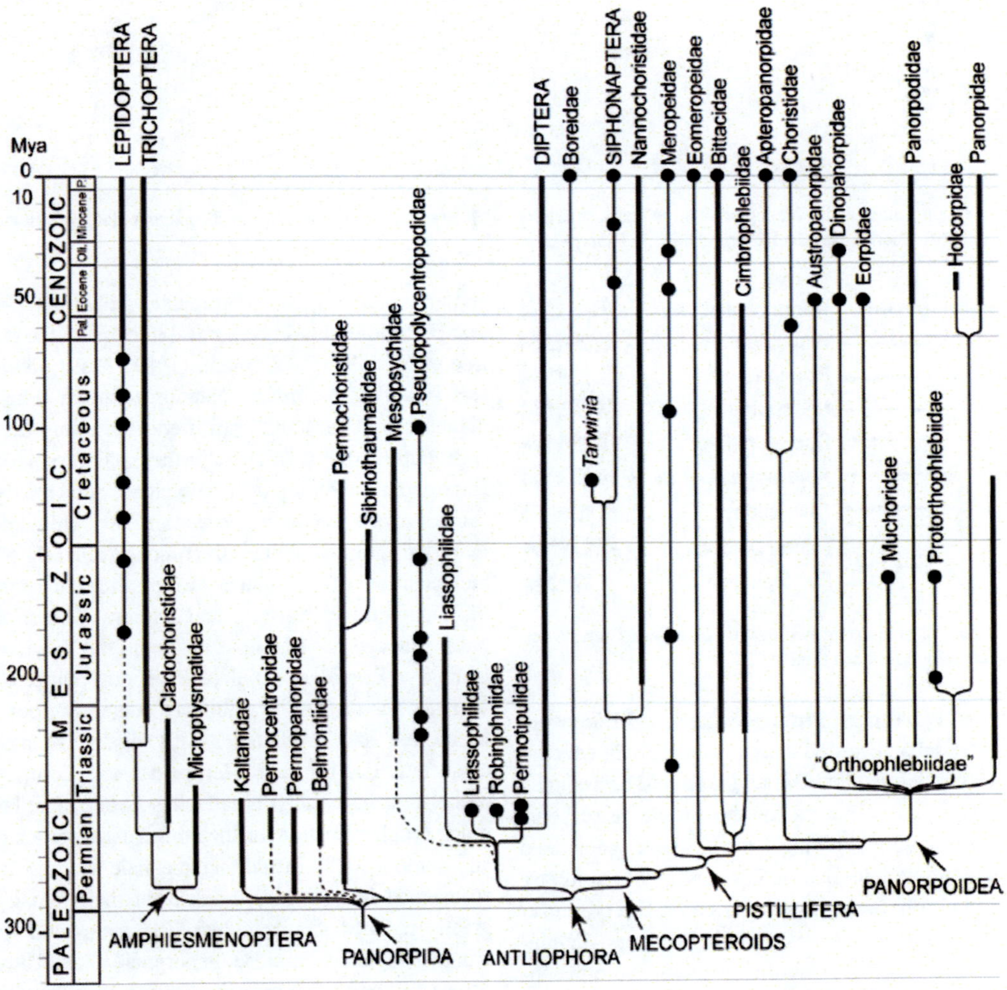

Fig. 3.1 Phylogeny of Panorpida. Thick lines denote diverse occurrences, dots denote occasional occurrences, and dashed lines denote uncertain relationships (revised from Grimaldi and Engel 2005)

J.-S. Wang, B.-Z. Hua, *A Color Atlas of the Chinese Mecoptera*, https://doi.org/10.1007/978-981-16-9558-2_3

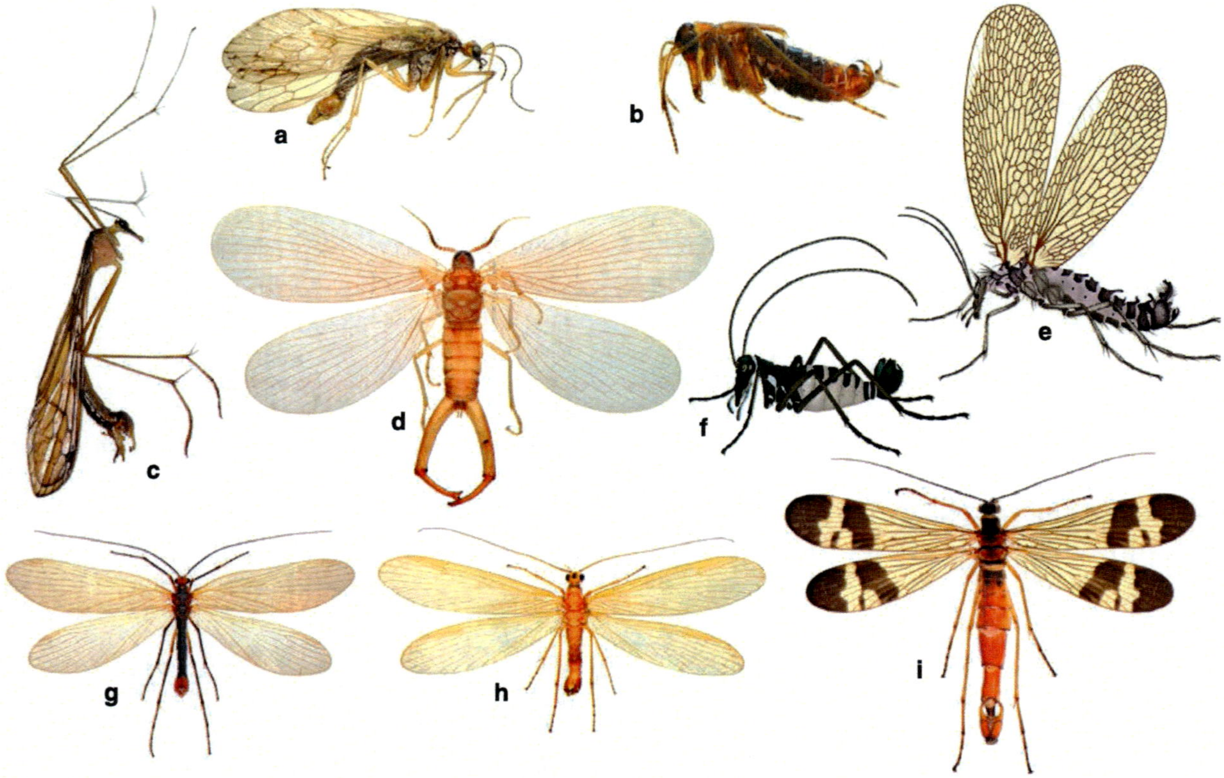

Fig. 3.2 Nine extant families in Mecoptera. (**a**) Nannochoristidae; (**b**) Boreidae; (**c**) Bittacidae; (**d**) Meropeidae; (**e**) Eomeropidae; (**f**) Apteropanorpidae; (**g**) Choristidae; (**h**) Panorpodidae; (**i**) Panorpidae

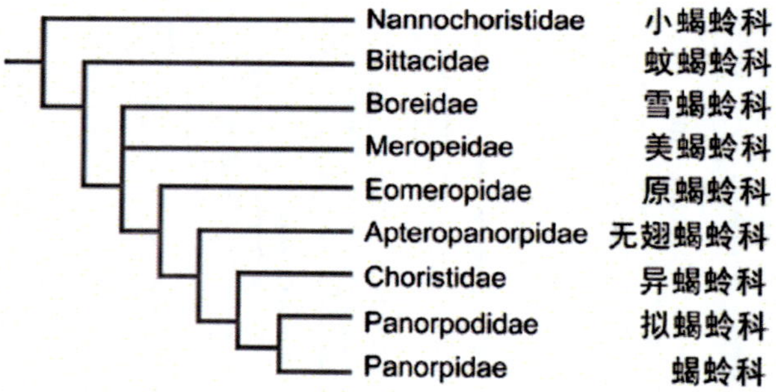

Fig. 3.3 Phylogeny of nine extant families in Mecoptera. Chinese vernacular names are added for each family

(1917) proposed two orders, Protomecoptera and Eumecoptera, which were later regarded as two suborders in Mecoptera. Chou (1963) claimed that Bittacidae should be assigned to the suborder Zygomecoptera for the following reasons: relatively narrower wings; a single tarsal claw; specialized male genitalia; and unable to walk but can only hang themselves in dense shrubs and grasses. Based on the presence/absence and length of wings and genital characters, Kaltenbach (1978) proposed a three-suborder system: Protomecoptera, including Eomeropidae and Meropeidae; Eumecoptera, including Apteropanorpidae, Bittacidae, Choristidae, Nannochoristidae, Panorpodidae, and Panorpidae; and Neomecoptera, including only Boreidae. Mickoleit (1978) noted that Bittacidae should be a basal lineage on the phylogenetic tree of Mecoptera. Willmann (1987, 1989) divided Mecoptera into two suborders: Nannomecoptera (only including Nannochoristidae) and Pistillifera (including the remaining eight families). Pistillifera is further split into two infraorders, Raptipedia (including only Bittacidae) and Opisthogonopora (including the remaining seven families). Hünefeld and Beutel (2005) pointed out that the function and related structures and position of the sperm pumps differed greatly among Siphonaptera, Pistillifera, and Diptera. Therefore, the real basal lineage in Mecoptera should be Nannochoristidae instead of Boreidae, which lack a sperm pump.

After such a long evolutionary course, various families of Mecoptera have exhibited greatly distinctive morphological characters. For instance, Meropeidae and Eomeropidae have reticulate veins on their broad wings, while other families have much simpler venations on their much narrower wings (Dunford and Somma 2008). In addition, the male wings of Boreidae are specialized into clasping hooks, but their female wings are greatly reduced and scale-like. The males and females lack wings in Apteropanorpidae. Panorpodidae and Bittacidae have some short-winged or apterous species. Moreover, these insects show wide distinctions not only in the wings, but also in the rostrum, compound eyes, abdomen, and larval morphology.

4 Geographical Distribution

Members of Mecoptera are widely distributed among all the geographical regions except Antarctica.

4.1 Bittacidae

The species of Bittacidae are widely distributed in all the geographical regions of the world except Antarctica (Fig. 4.1). *Bittacus* is the most widespread genus, while others are mostly constrained in narrower ranges. For instance, *Anabittacus* can only be found in Chile, *Anomalobittacus* in South Africa, *Apterobittacus* in North America, *Issikiella* in South America, *Bicaubittacus* and *Terrobittacus* in south China, and *Harpobittacus* in Australia.

For the distributional and evolutionary explanations, Penny (1975) claimed that the family Bittacidae probably originated in Gondwana and was split into several distinct genera around late Cretaceous (approximately 60 mya). The hangingflies spread across from South to North America, and then migrated into the Eurasian continent. The high diversity at both specific and generic levels in Eurasia, however, indicates that these insects are more likely originated from Laurasia instead of Gondwana. After the ancient supercontinent Pangaea broken apart, the hangingflies spread and formed the distributional pattern as we know today (Ren 1993). Li and Ren (2009) discussed the distribution and evolution of Bittacidae and claimed that Eurasia was likely the area of origin of these insects, and Asia was the differentiat-

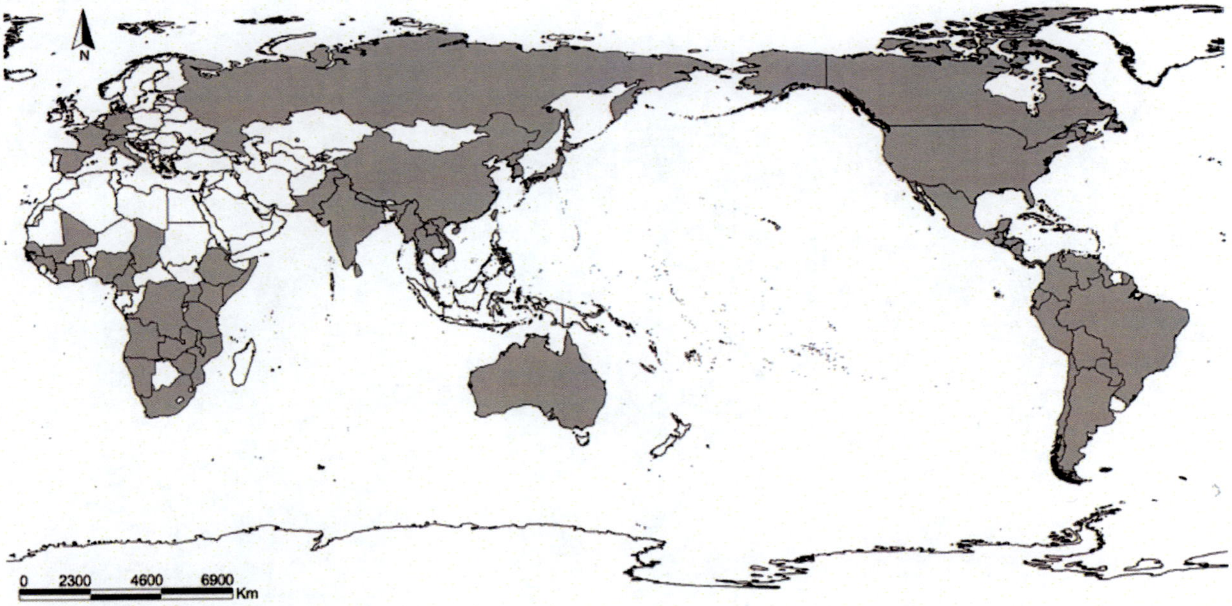

Fig. 4.1 Distribution of Bittacidae in the World

J.-S. Wang, B.-Z. Hua, *A Color Atlas of the Chinese Mecoptera*, https://doi.org/10.1007/978-981-16-9558-2_4

ing and dispersal center where hangingflies started their migration to other regions in Eocene.

China has the maximum species number of hangingflies all over the world. Most of its territories have distribution records of Bittacidae. Only several provincial regions lack records of hangingflies, such as Hebei, Shandong, and Xinjiang (Fig. 4.2).

4.2 Panorpidae

The species of Panorpidae are distributed in the Holarctic and Oriental Regions, including most areas of Europe, Asia, and North America (Fig. 4.3). Byers (1988) presumed that the Panorpidae in North America might be migrated from Asia through Beringia during Eocene to Oligocene. East Asia bears

Fig. 4.2 Distribution of Bittacidae in China

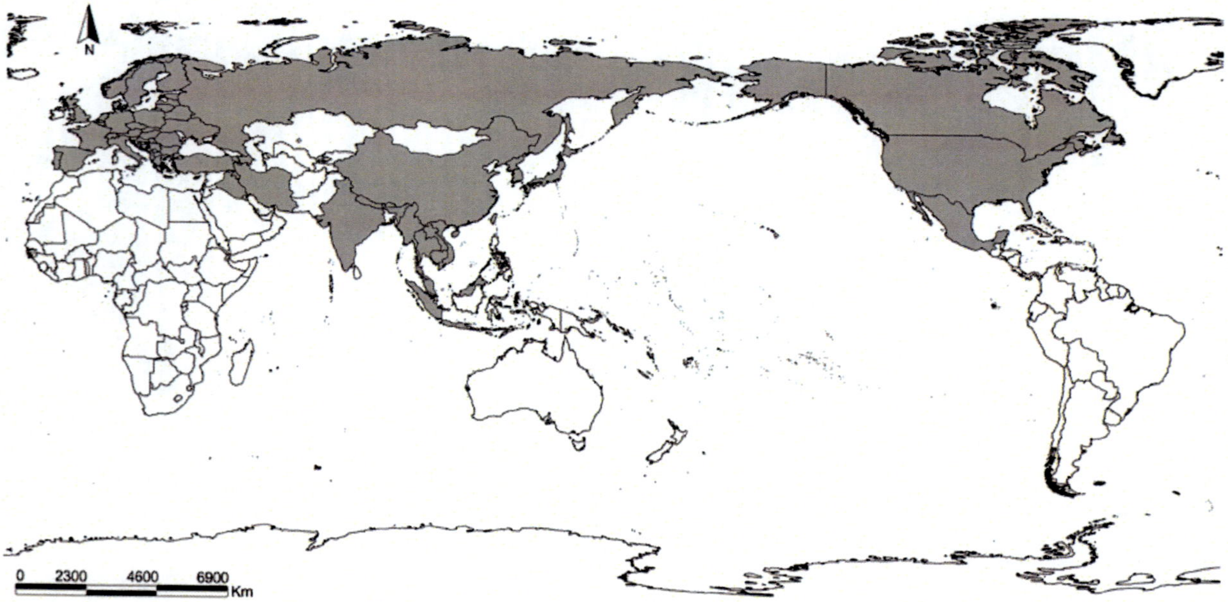

Fig. 4.3 Distribution of Panorpidae in the World

the highest diversity of Panorpidae than other regions, suggesting that it is the center of origin and dispersal of Panorpidae.

China has the maximum number of recorded species of Panorpidae in the world. Most provinces have distribution records of Panorpidae. Only a few provinces lack records of scorpionflies, including Hebei, Shandong, and Xinjiang (Fig. 4.4).

4.3 Panorpodidae

The species of Panorpodidae are distributed in the northern Pacific Rim, including the USA, the Korean peninsula, Japan, and China (Fig. 4.5). Fossil records have been found in Europe and South America, indicating that this family likely had a larger distribution range in Eurasia and North &

Fig. 4.4 Distribution of Panorpidae in China

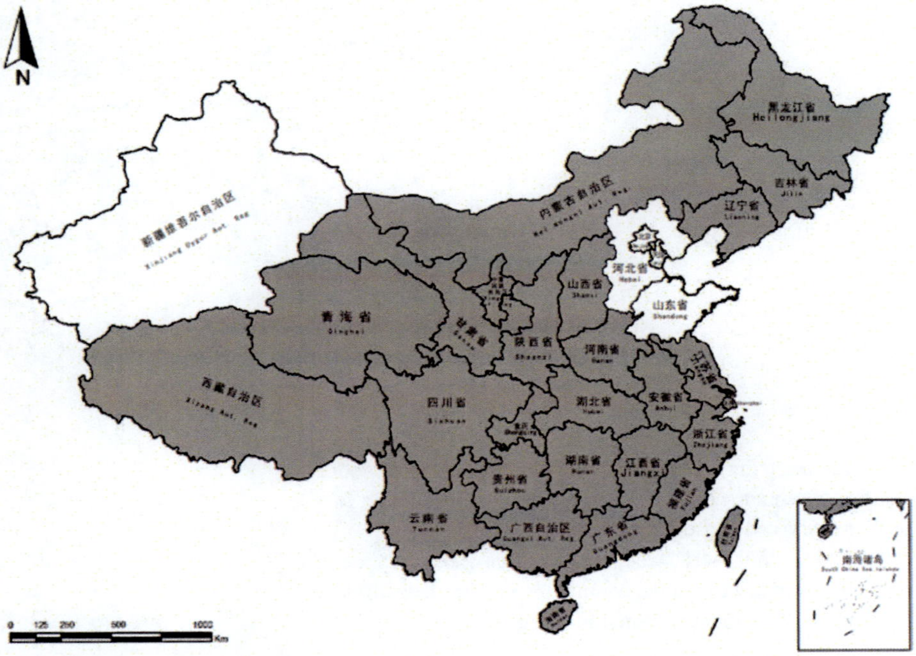

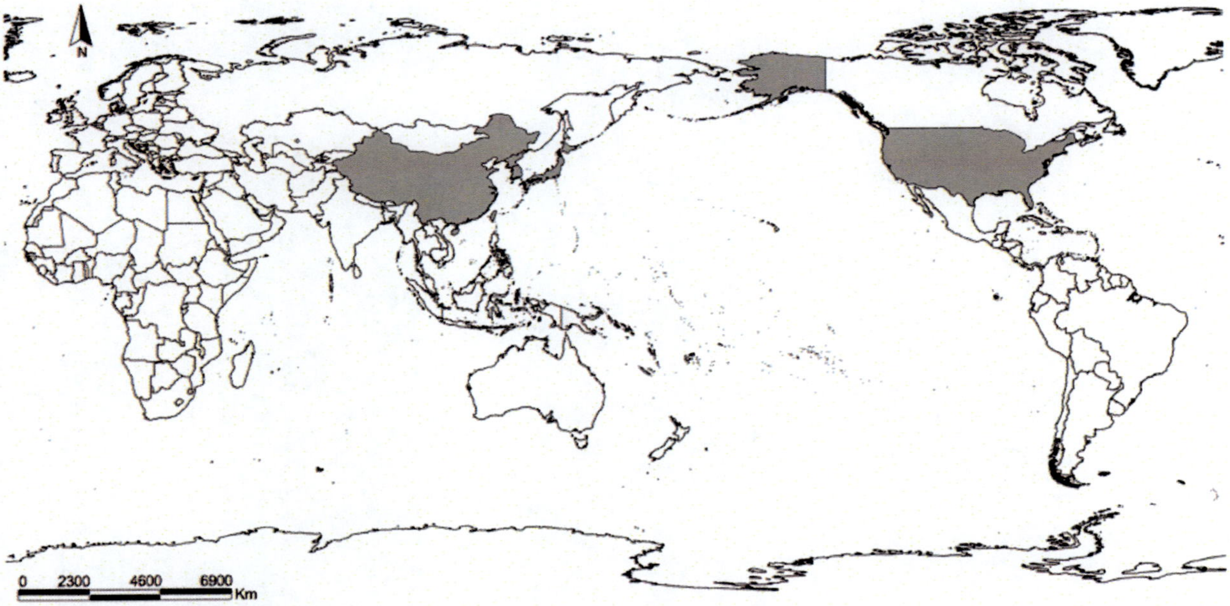

Fig. 4.5 Distribution of Panorpodidae in the World

Fig. 4.6 Distribution of Panorpodidae in China

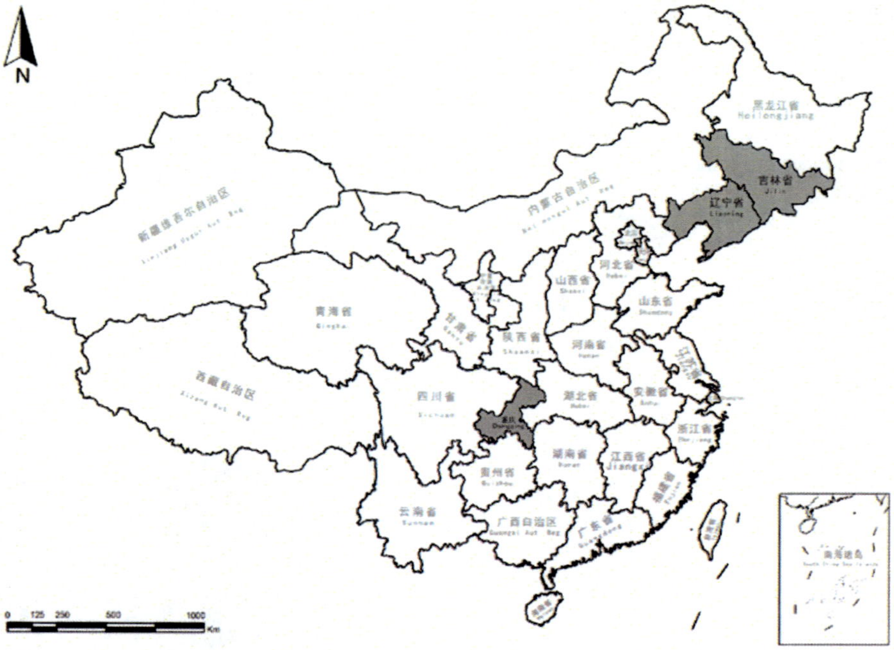

South America during the Eocene (Hu and Hua 2016). After a series of climate changes and mass extinctions, only a few of them survived to the present day.

In China, the short-faced scorpionflies have only been recorded from the Changbai Mountains in Jilin, Kuandian in Liaoning, and Beibei in Chongqing (Fig. 4.6).

Morphology

5

Main Diagnostic Characters of Mecoptera

- Head more or less elongated ventrad into a rostrum, with chewing mouthparts located at the distal end of the rostrum;
- Wings normally narrow with venations close to primitive pattern (Comstock–Needham System); fore- and hindwings usually similar in size, shape, and venations;
- Male genital segments enlarged to some extent, and some of them curved upward resembling the stinger of scorpions;
- Cerci usually short and small; females lacking an ovipositor (Fig. 5.1).

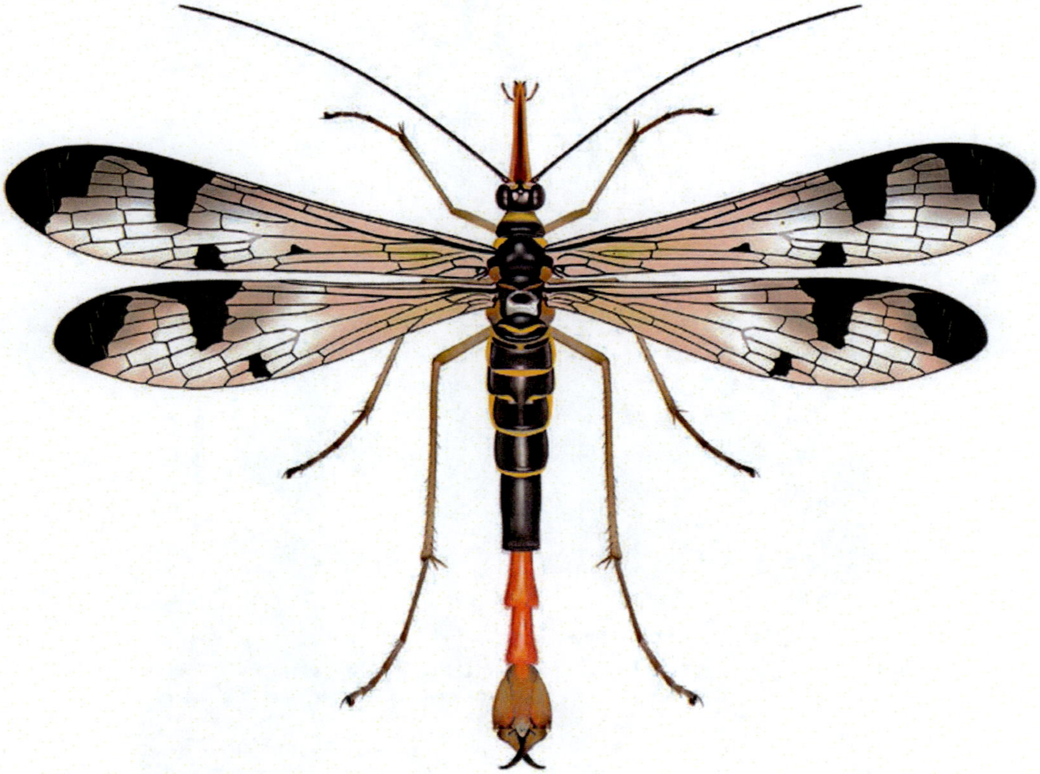

Fig. 5.1 Male adult of Mecoptera. Male adult of *Neopanorpa nielseni*, dorsal view

J.-S. Wang, B.-Z. Hua, *A Color Atlas of the Chinese Mecoptera*, https://doi.org/10.1007/978-981-16-9558-2_5

5.1 Bittacidae (Figs. 5.2, 5.3, 5.4, 5.5, 5.6, 5.7, 5.8, and 5.9)

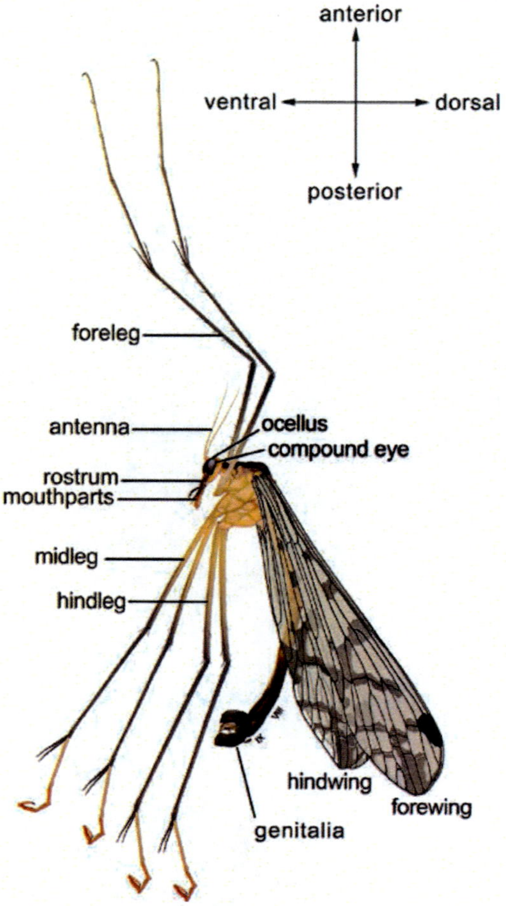

Fig. 5.2 Male adult of Bittacidae. *Bittacus strigatus*, lateral view

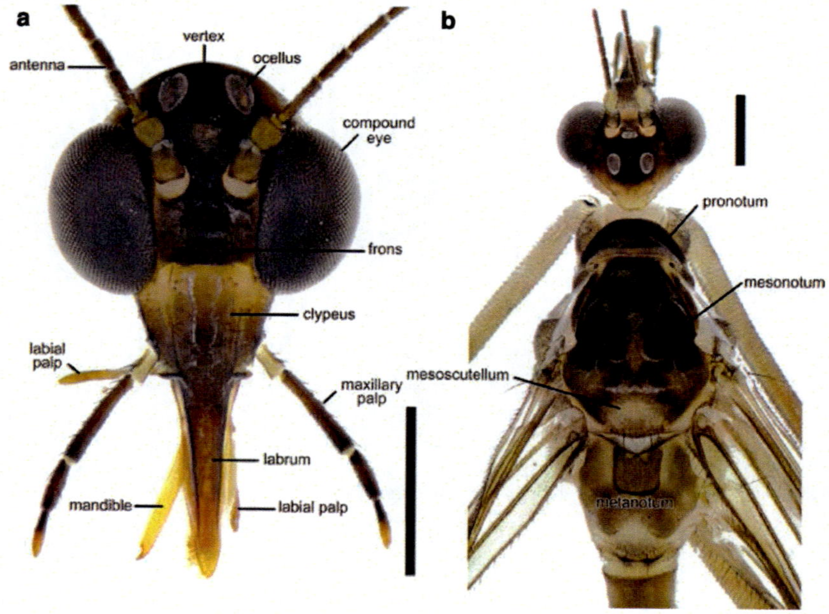

Fig. 5.3 Head and thorax of Bittacidae. (**a**) Head of *Terrobittacus angustus*, frontal view (*Scale bar*: 0.5 mm); (**b**) Thorax of *T. rostratus*, dorsal view (*Scale bar*: 0.5 mm)

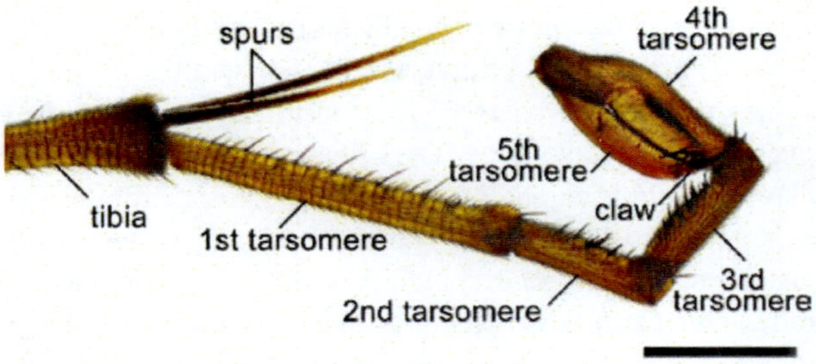

Fig. 5.4 Terminal portion of leg of Bittacidae. Hind leg of *Bittacus sinensis*, lateral view (*Scale bar*: 1.0 mm)

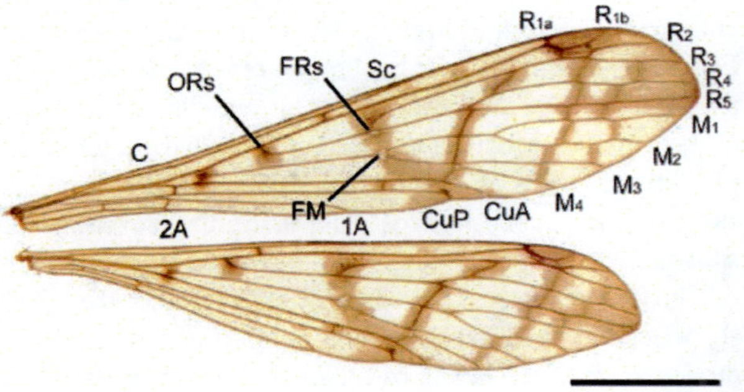

Fig. 5.5 Wings of Bittacidae. Wings of *Bittacus strigatus* (*Scale bar*: 5.0 mm)

Fig. 5.6 Male genitalia of Bittacidae. *Bittacus planus*, lateral view (*Scale bar*: 5.0 mm)

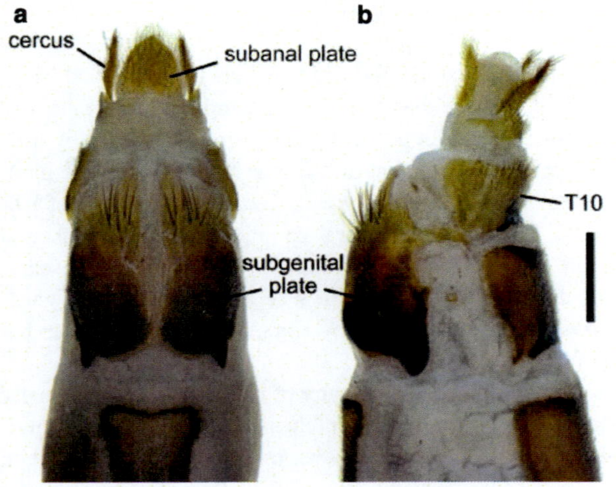

Fig. 5.7 Female genitalia of Bittacidae. *Bicaubittacus burmanus* (*Scale bar*: 0.5 mm). (**a**) ventral view; (**b**) lateral view

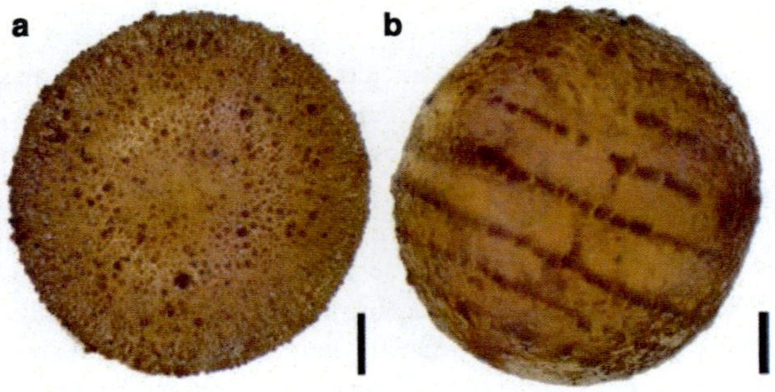

Fig. 5.8 Spherical eggs of Bittacidae. (**a**) *Bittacus planus*; (**b**) *Terrobittacus implicatus* (*Scale bars*: 0.2 mm)

Fig. 5.9 Cuboidal eggs of Bittacidae. *Bittacus cirratus* (*Scale bar*: 0.2 mm)

5.1.1 General Information

Insects in the family Bittacidae are commonly called hangingflies, because that they are very similar to the crane flies in Diptera, but can be differentiated from the latter by their two pairs of wings that are similar in size, while the hindwings of crane flies are greatly reduced into a pair of halteres. Therefore, they are commonly called "mosquito scorpionflies" literally in Chinese. In Japan, they are commonly called "ガガンボモドキ"(gaganbomodoki, gaganbo = crane fly, modoki = something that is similar but different).

The adults have three pairs of raptorial legs, and usually hang themselves using the prehensile tarsi of front and middle legs from the edge of leaves or twigs of plants. They can neither walk nor stand on a surface (Dunford and Somma 2008; Bicha 2018).

5.1.2 Adult

Head. The diagnostic characters in the head include the color, the stripes on ocellar triangle, the size of dorsal ocelli, and the morphology of compound eyes and antennae. The head is usually yellowish, dark, or reddish brown. In some species, the dorsum and rostrum are different in color. The ocellar triangle is generally surrounded by a deep pattern. Some species bear distinct ocellar bristles between the ocelli. The compound eyes are mostly long ellipsoidal. The size, shape of compound eyes and the distance between them are often important traits of identification. The scape and pedicel of antenna are short and thick, while the flagellum is filiform and sometimes ciliated.

Thorax. Strong and slightly humped. The pronotum is narrow and often bears a few black setae along its anterior margin. The meso- and metanotum are variously colored and sometimes are uniformly tawny or dark brown. In some species, the notum is different in color between the anterior and posterior portions. Some species possess a distinct notal stripe on the notum. Pleura are usually yellowish brown or cream-colored.

Hangingflies have three pairs of slender raptorial legs. The legs are usually uniformly yellowish or reddish brown. In some species, the two ends of femur and tibia are darkened. The length of basitarsus and the number of black spines on the two sides of the fourth tarsomere are frequently used in taxonomy. The fifth tarsomere with a single claw can fold back and secure with the inner margin of the enlarged fourth tarsomere. The inner margins of these two tarsomeres bear strong spines that allow them to form a retractable clasp. This structure allows hangingflies to hang themselves or catch preys. The generic characters of hangingflies are mainly concentrated on the wings. When describing the wings, we mainly use the following abbreviations and terms: A, anal vein; Av, anal crossvein; C, costa; CuA, anterior Cubitus; CuP, posterior Cubitus; CuV, cubital crossvein; FM, fork of Media; FM3+4, fork of M3 and M4; FRs, fork of radial sector; h, humeral crossvein; M, Media; nygmata (circular and chitinized spots on wing membrane); OM, origin of Media; ORs, origin of radial sector; Pcv, pterostigmal crossvein; pterostigma; Rs, radial sector; Sc, subcosta; Scv, subcostal crossvein; thyridium (a translucent spot on FM).

The frequently used characters of wings include the color, shape, ending position of Sc, the number of Pcv, coalesce condition of Rs, the length of 1A and its ending position, the coalesce condition between 1A and CuP in hindwings, etc. The shapes are varied greatly among different genera. For instance, the wing base is broad or stalked, and the apex is acute and slightly hook-like or rounded. The bands and spots are also important traits in taxonomy.

Abdomen. The epandrium (tergum IX, T9) in males is modified into a pair of claspers, which are normally called "epandrial lobes." The shape of these lobes are significant characters in taxonomy. The gonopods in Bittacidae are not used as grasping organs, since the gonostylus is greatly reduced. The male aedeagus is located in the cavity formed by joined gonocoxites. In most Chinese species, the apex of aedeagus is greatly elongated into a coiled filament that is called the "penisfilum". In addition, there is often a pair of aedeagal lobes on the two basal sides of aedeagus. The tergum X is reduced to various extents. The T11 and S11, called proctigers, are modified and elongated, and are regarded as upper and lower branches. The upper branch of proctiger often extends beyond the two epandrial lobes, with its apex enlarged, hook-like, or tufted in some species. The lower branch is usually shorter, hook-like, and curving ventrad.

The main diagnostic character in females lies in the morphology of the subgenital plate. On the ventral side of A8 and A9, the subgenital plate are fused from two valves (gonocoxosternites VIII), and sometimes with a membranous area medially. A pair of acute processes each projects from the lateral margin of the subgenital plate distally. Several thick setae are on the apical half. The T10 is usually reduced and extends ventrad, surrounding the base of cerci.

5.1.3 Egg

Mostly spherical. The chorion is smooth in most species *Bittacus*, but is covered with latitude-like ornaments in some *Terrobittacus* species. However, the eggs are nearly cuboidal in a few species, such as *Bittacus cirratus*.

5.1.4 Larva

Eruciform. The head is hypognathous, with a pair of compound eyes and a single median ocellus. The body is furnished with rows of thick protuberances terminated in clavate setae on each somite. The first eight abdominal segments each bear a pair of ventral prolegs. The anal segment is terminated in a protrusible sucker (Fig. 5.10). The larvae mostly live on the ground and feed on dead arthropods or other kinds of organic materials. Once matured, the larvae build a chamber in earth and prepare to pupate.

5.1.5 Pupa

The pupae of hangingflies are decticous and exarate, with their heads and abdomens capable of making limited movements. The pupae resemble adults but with undeveloped rostrum, abdomen, and wings. Their antennae, legs, and wings are loosely attached to the body (Fig. 5.11).

Fig. 5.10 Larvae of Bittacidae. (**a**) *Bittacus planus*, lateral view (*Scale bar*: 1.0 mm); (**b**) *Terrobittacus implicatus*, lateral view (*Scale bar*:1.0 mm)

Fig. 5.11 Pupa of Bittacidae. *Bicaubittacus longiprocessus*. (**a**) Lateral view; (**b**) Ventral view; (**c**) Dorsal view (*Scale bar*: 1.0 mm)

5.2 Panorpidae (Figs. 5.12, 5.13, 5.14, 5.15, 5.16, 5.17, 5.18, 5.19, 5.20, 5.21, 5.22, 5.23, 5.24, 5.25, 5.26, 5.27, 5.28, 5.29, and 5.30)

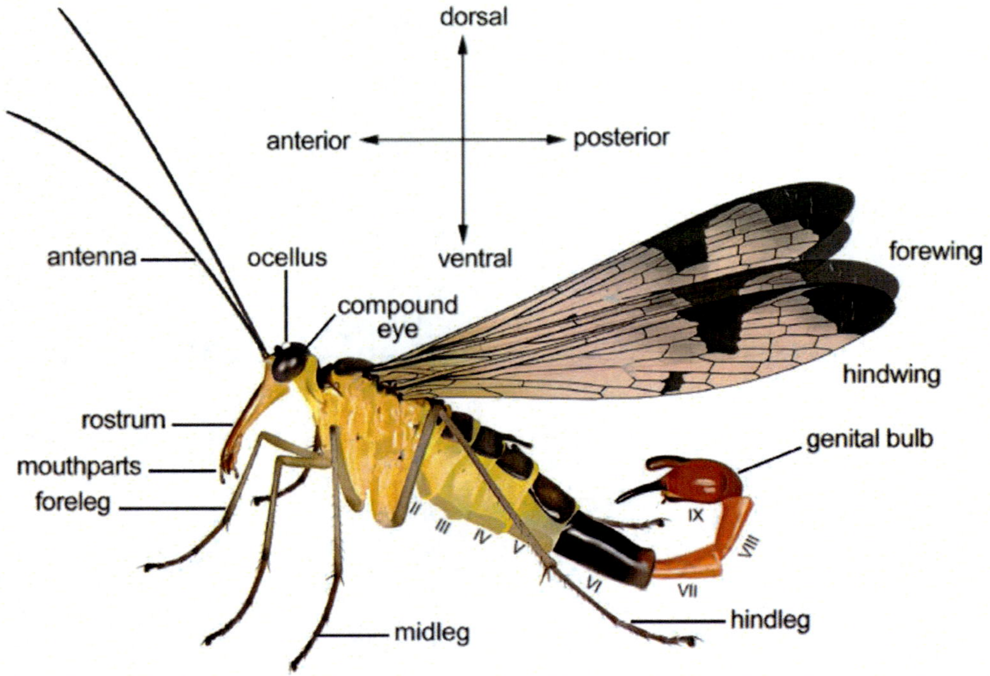

Fig. 5.12 Male adult of Panorpidae. *Neopanorpa nielseni*, lateral view

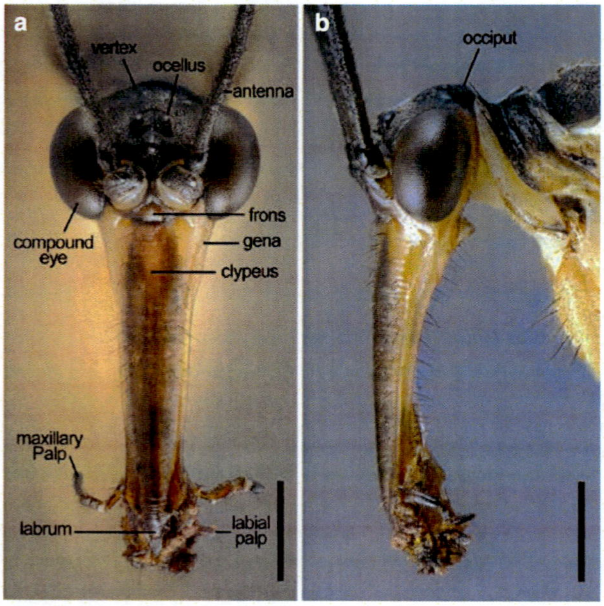

Fig. 5.13 Head of Panorpidae. *Cerapanorpa obtusa*. (**a**) Frontal view; (**b**) Lateral view (*Scale bars*: 1.0 mm)

5.2.1 General Information

The insects of Panorpidae are commonly called scorpion-flies, because the male genitalia are enlarged and recurved backward, resembling the stinger of scorpions. They are also called "tail-lifting insects" literally in Chinese, and " 举尾虫, シリアゲムシ"(shiriagemushi, shiri = tail, age = lifting, mushi = insects) in Japanese.

5.2.2 Adult

Head. Prolonged ventrad into a rostrum, with a greatly elongated clypeus and a small, triangular labrum. The chewing mouthparts are located at the end of the rostrum. In general, the rostrum is relatively slender in Neopanorpinae (*Leptopanorpa* and *Neopanorpa*) than in Panorpinae (other genera). The rostra are mostly yellow, yellowish, or dark brown, with a pair of dark longitudinal stripes frontally in some species. The compound eyes are mostly long ellipsoidal. The vertex is usually yellow, yellowish, dark brown, black, or with various patterns. Three ocelli are located on a protuberated area, called the ocellar tri-

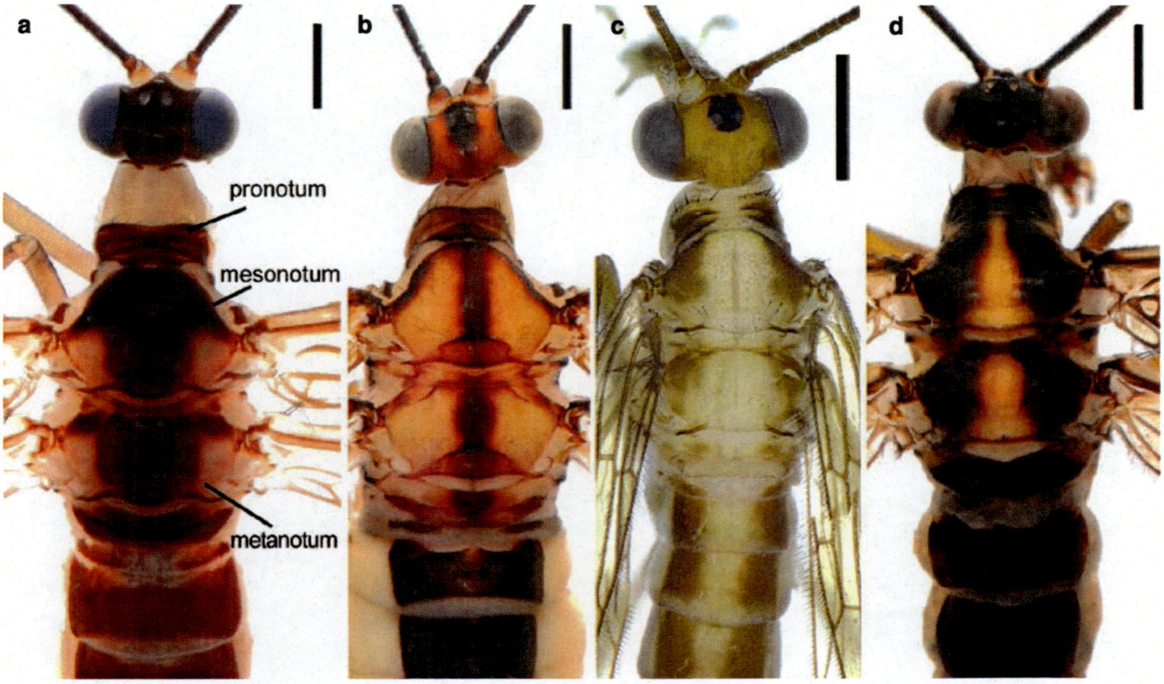

Fig. 5.14 Thoraces of Panorpidae (*Scale bars*: 1.0 mm). (**a**) *Neopanorpa pulchra*; (**b**) *Neopanorpa caveata*; (**c**) *Panorpa kunmingensis*; (**d**) *Panorpa sonani*

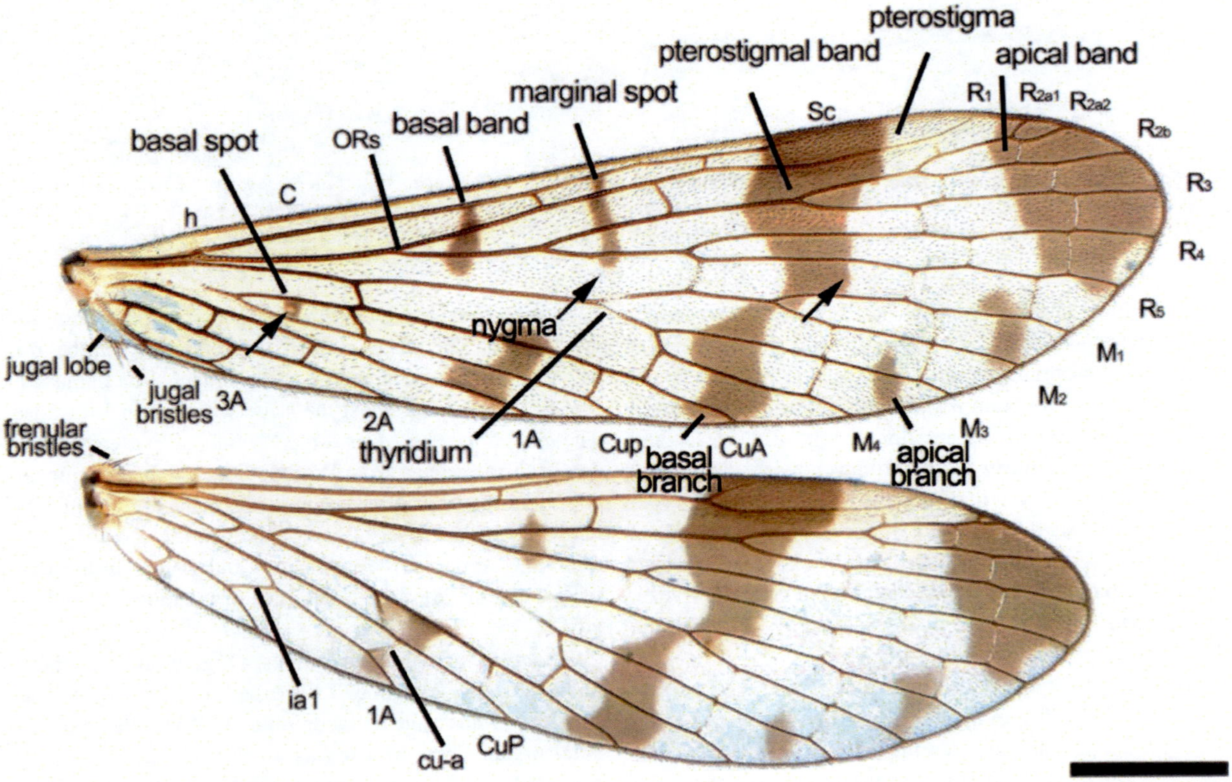

Fig. 5.15 Wings of Panorpidae. Right fore- and hindwing of *Panorpa communis* (*Scale bar*: 2.0 mm)

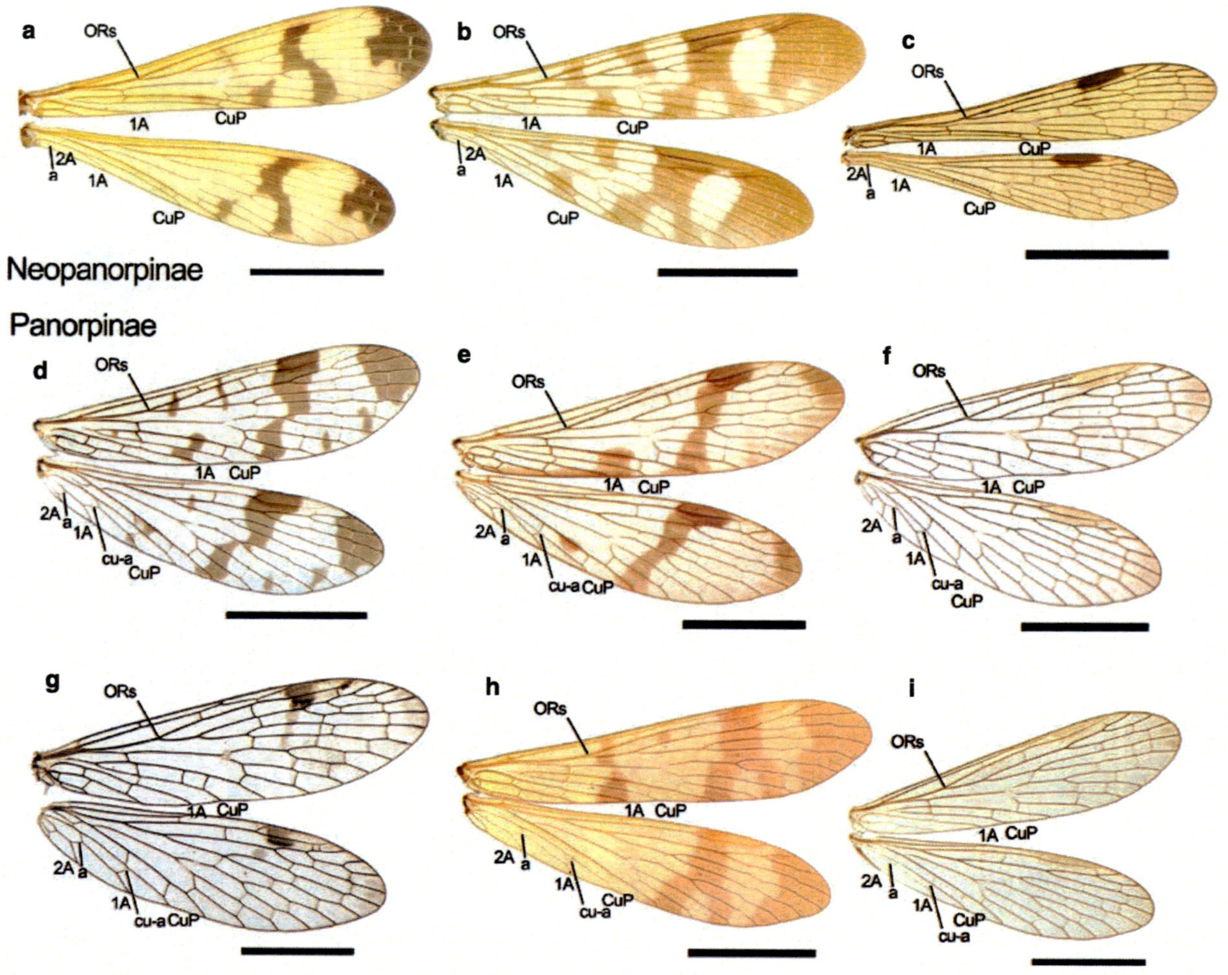

Fig. 5.16 Wings of Panorpidae (*Scale bars*: 5.0 mm). Right wings, dorsal view. **(a–c)** Neopanorpinae; **(d–i)** Panorpinae. **(a)** *Neopanorpa chillcotti*; **(b)** *Neopanorpa pulchra*; **(c)** *Leptopanorpa cingulata*; **(d)** *Panorpa communis*; **(e)** *Furcatopanorpa longihypovalva*; **(f)** *Dicerapanorpa diceras*; **(g)** *Megapanorpa grandis*; **(h)** *Sinopanorpa tincta*; **(i)** *Cerapanorpa obtusa*

angle. Various patterns enclose the ocellar triangle in some species. The antennae have short and thick scape and pedicel, and slender and filiform flagellum.

Thorax. The pronotum is saddle-shaped and often bears a row of thick setae on its anterior margin. The meso- and metanotum are variously colored from yellowish, dark brown to black, with a paler or darkened median stripe in some species. The legs are slender, with a pair of spurs at the distal end of tibia. The pretarsus has a pair of claws, which bear comb-like teeth on their inner margins. In Neopanorpinae, the second apical tooth is considerably enlarged; in Panorpidae, the teeth are uniformly sized.

Two pairs of membranous wings are hyaline or translucent, mostly colorless or tinged with yellow or smoky brown. Varied markings on wings are important traits. In most species, the wing markings are arranged in a series of regular patterns, and are named as follows from the base to the apex: basal spot, basal band, marginal spot, pterostigmal band, and apical band. The posterior half of the pterostigmal band is often split into a basal and an apical branch. The nomenclature of venation is almost the same with that in Bittacidae. The curved base of M_4 is a typical trait of Panorpidae, and can be differentiated from Panorpodidae. In the assignments of subfamilies and genera, significant traits are whether 1A ends beyond the level of ORs or not in the forewings, the number of crossveins between 1A and 2A (a) in the forewings, and the presence or absence of ia1 in the hindwings, etc.

Abdomen. In male scorpionflies, a highly variable process on the posterior margin of T3 is called the notal organ, and a smaller protuberance on the middle of T4 is called the postnotal organ. They are used as a clamp device to hold the wings of the female during copulation. The presence or absence, and the morphology of these organs are very impor-

Fig. 5.17 Terminal portion of hindlegs of Panorpidae (*Scale bars*: 0.5 mm). Right hindleg, ventral view. (**a**) *Leptopanorpa cingulata*; (**b**) *Neopanorpa chillcotti*; (**c**) *N. muelleri*; (**d**) *N. brisi*; (**e**) *N. nielseni*; (**f**) *Dicerapanorpa magna*; (**g**) *Furcatopanorpa longihypovalva*; (**h**) *Sinopanorpa tincta*; (**i**) *Panorpa japonica*; (**j**) *P. reflexa*

Fig. 5.18 Male abdomen of Panorpidae (*Scale bars*: 2.0 mm). Lateral views. (**a–f**) Abdomen; (**g–j**) A6–A11. (**a**) *Neopanorpa chillcotti*; (**b**) *N. choui*; (**c**) *Leptopanorpa linyejiei*; (**d**) *Panorpa duanyu*; (**e**) *P. communis*; (**f**) *Furcatopanorpa longihypovalva*; (**g**) *Dicerapanorpa magna*; (**h**) *Megapanorpa gaokaii*; (**i**) *Sinopanorpa nangongshana*; (**j**) *Cerapanorpa obtusa*

Fig. 5.19 Male adult of Panorpidae. *Leptopanorpa linyejiei*, lateral view (Mount Ciremai, Java Island, Indonesia)

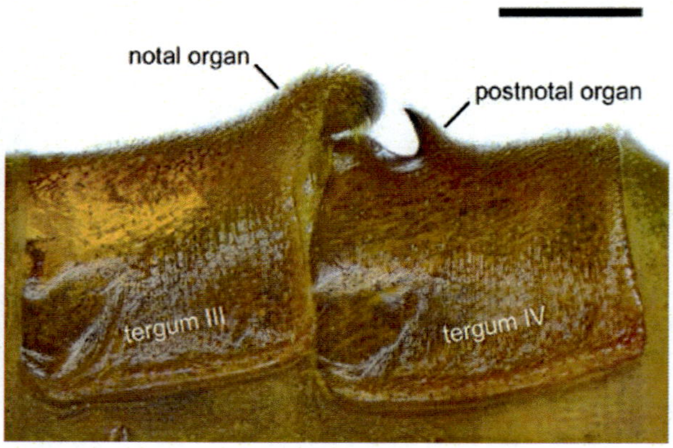

Fig. 5.20 Male T3 and T4 of Panorpidae (*Scale bar*: 0.5 mm). *Panorpa amurensis*, lateral view (showing the notal and postnotal organs)

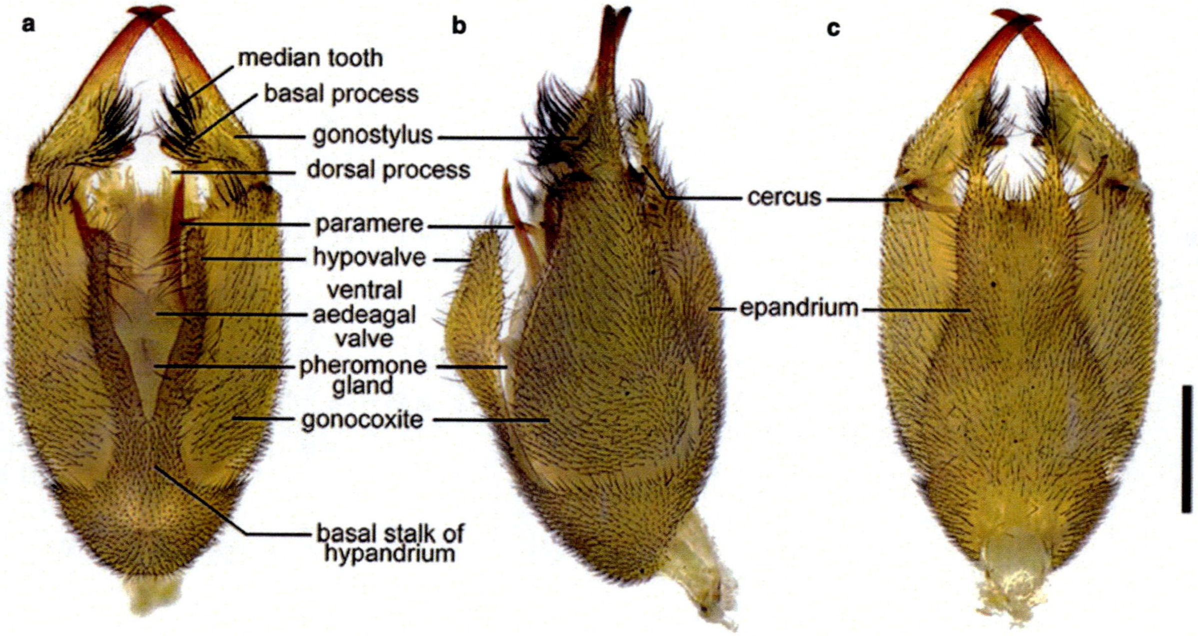

Fig. 5.21 Male genitalia of Panorpidae. Male genital bulb of *Panorpa guttata* (*Scale bar*: 0.5 mm). (**a**) Ventral view; (**b**) Right-lateral view; (c) Dorsal view

tant diagnostic characters not only for species, but also genera and subfamilies.

The terga and sterna of male abdominal segments VI–VII (A6–A8) are fused altogether without membranous pleura. In some species, the A6 has one or two finger-like horns on the dorsal apex, called "anal horns." A7 and A8 are shorter or longer than A6. Some of these two segments are cylindrical, some are constricted basally and inverted cone-shaped, and the others are intensively constricted and stalk-like basally. The abdomen (especially A6–A8) is greatly elongated in the males of *Leptopanorpa*, and a few species of *Neopanorpa* and *Panorpa*.

Male genitalia. The terminal segments are enlarged and called the "genital bulb," which turns backward and resembles the stinger of scorpions. Epandrium (tergum IX, T9) has its terminal margin truncated or emarginate and forming a pair of slender setose lobes laterally. On the ventral surface of the epandrium are sometimes a pair of epandrial lobes

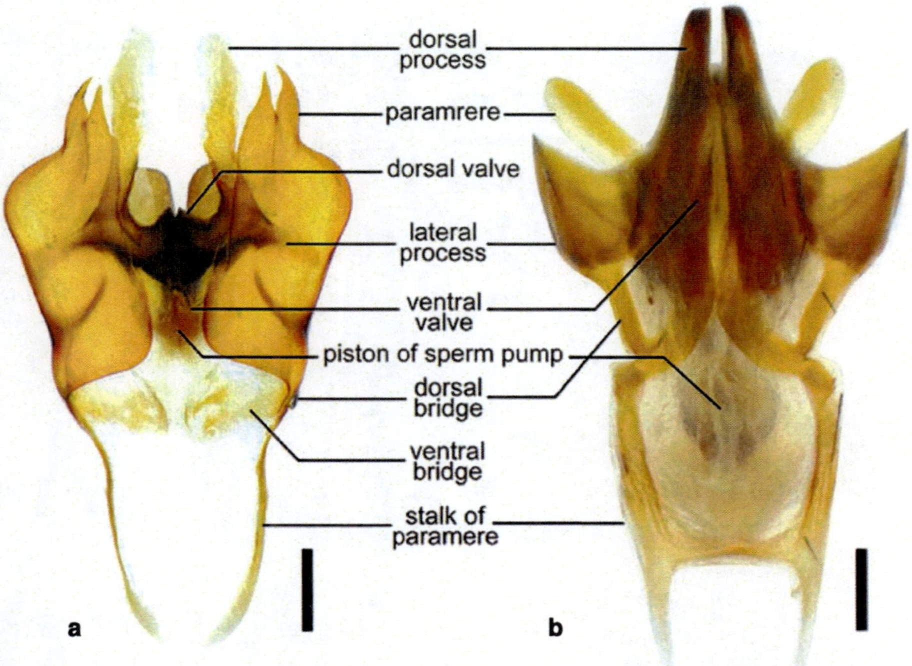

Fig. 5.22 Aedeagal complex of Panorpidae (*Scale bars*: 0.2 mm). Ventral view. (**a**) *Panorpa communis*; (**b**) *Neopanorpa longiprocessa*

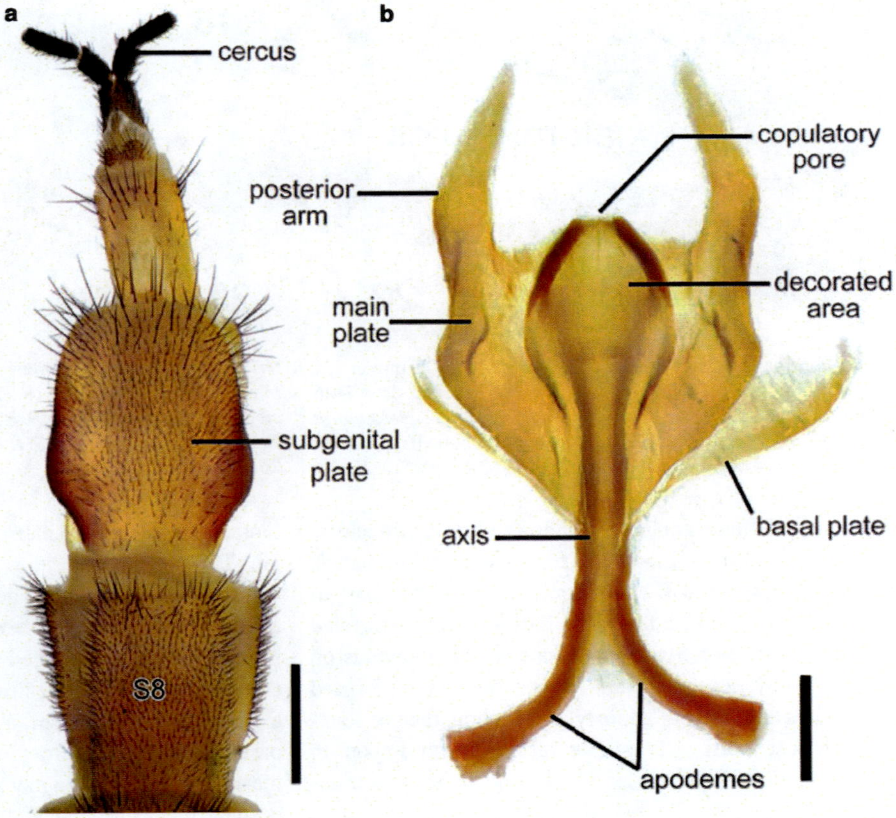

Fig. 5.23 Female genitalia of Panorpidae. *Panorpa communis*, ventral view. (**a**) T8–T11, ventral view (*Scale bar*: 0.5 mm); (**b**) Medigynium, ventral view (*Scale bar*: 0.2 mm)

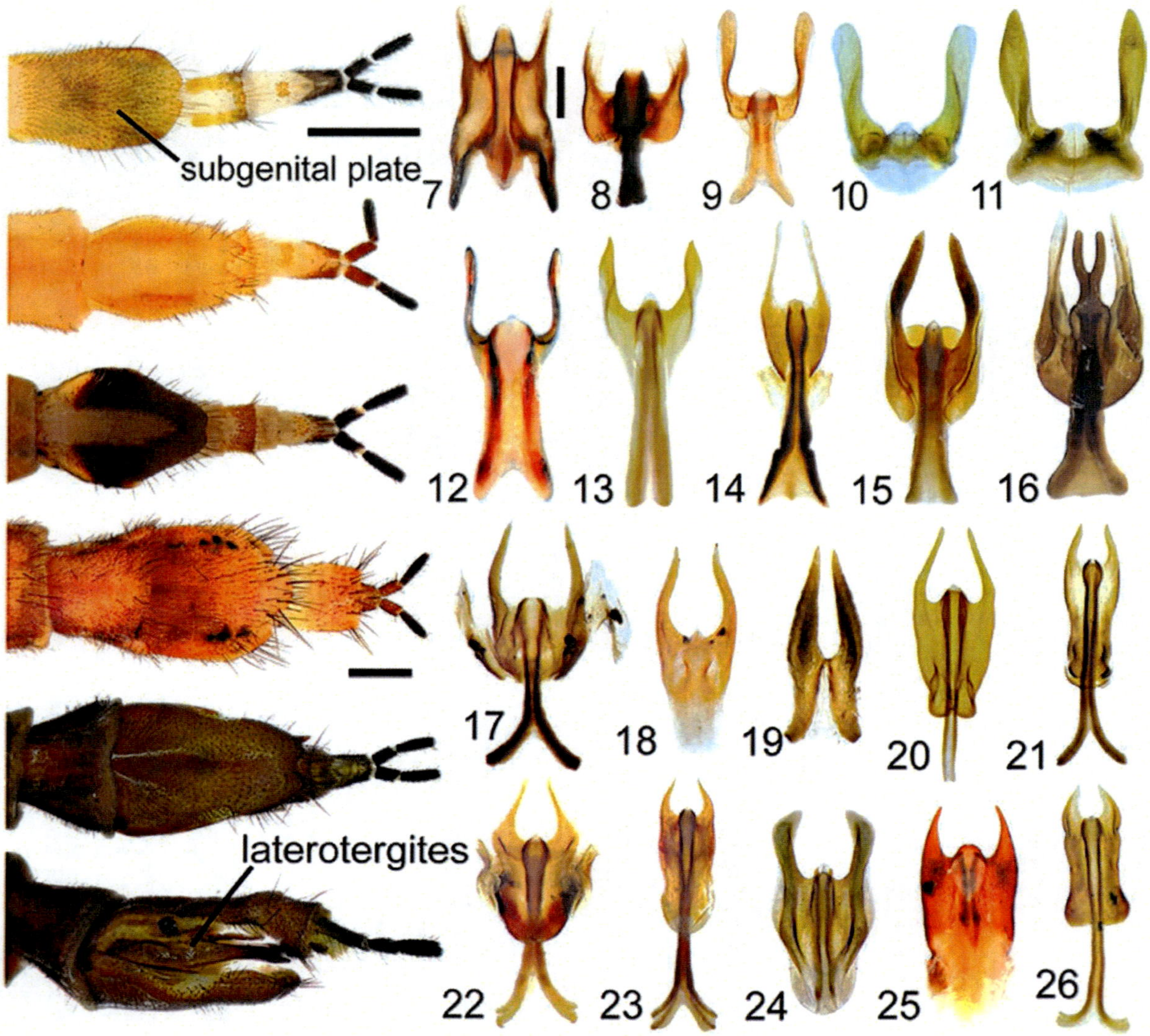

Fig. 5.24 Female genitalia of Panorpidae (*Scale bars*: 0.5 mm in 1, 4; 0.2 mm in 7). 1–6. T9–T11, ventral view except 6 in lateral view; 7–26. Medigynium, ventral view. 1. *Leptopanorpa cingulata*; 2. *Neopanorpa chillcotti*; 3. *P. gressitti*; 4. *Dicerapanorpa magna*; 5, 6, *Megapanorpa grandis*; 7. *L. nematogaster*; 8. *L. peterseni*; 9. *N. nigritis*; 10. *N. sauteri*; 11. *N. nielseni*; 12. *P. amurensis*; 13. *P. waongkehzengi*; 14. *P. kunmingensis*; 15. *P. implicata*; 16. *Furcatopanorpa longihypovalva*; 17. *P. sibirica*; 18. *P. cladocerca*; 19. *P. emeishana*; 20. *P. quadrifasciata*; 21. *P. aurea*; 22. *P. cornigera*; 23. *P. davidi*; 24. *D. magna*; 25. *M. grandis*; 26. *Sinopanorpa tincta*

(structures inhomogeneous to epandrial lobes in Bittacidae, which are entirely modified from epandrium itself). Hypandrium (sternum IX, S9) is often split into a pair of hypovalves distally. On the inner base of each hypovalve there is sometimes a hypandrial process. The gonopods on A9 are greatly developed and formed by a pair of fused gonocoxites and a pair of gonostyli. In general, the gonocoxites each bear a tuft of long setae on their ventral apex. In some species, the inner margin of the gonocoxite has a triangular protuberance. The paired gonocoxites are fused at basal half to form the main body of the genital bulb. Usually, the pincer-like gonostyli each have a subtriangular median tooth, and a variously shaped basal process on the inner margin. The gonostyli serve to grasp the female's genital segment during copulation. The aedeagus complex is modified from two pairs of phallomeres, bilaterally symmetrical, and is set in the cavity formed by the fused gonocoxites. It is composed of a pair of ventral valves, a pair of dorsal valves, a pair of dorsal processes, a pair of lateral processes, dorsobasally, a piston of sperm pump, and ventrally, a pair of parameres. Parameres are a pair of sclerotized structures located besides the aedeagus, and are variously shaped and are significant in taxonomy. Usually, the paramere is connected to the lateral aedeagal process through a dorsal bridge, and to

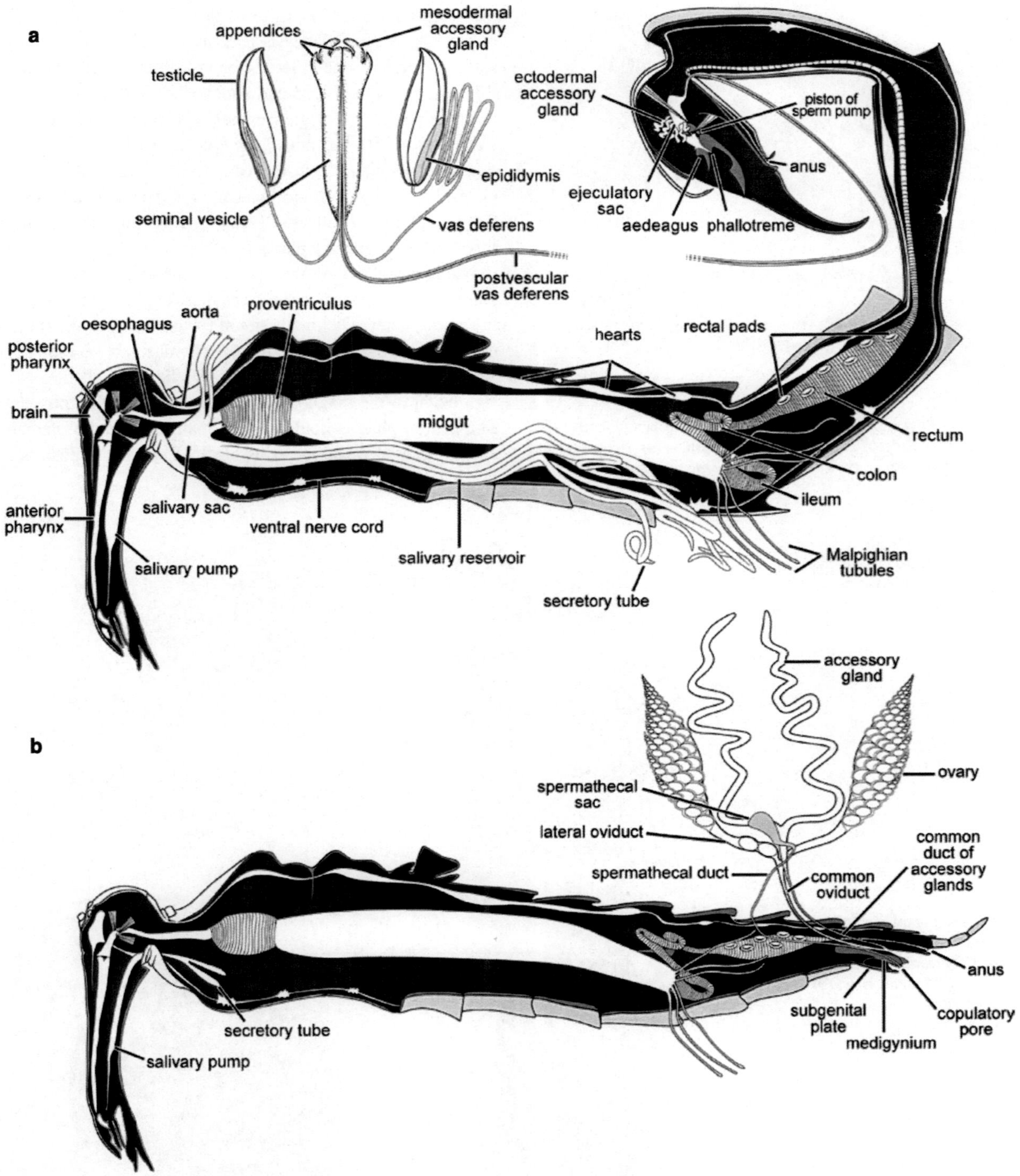

Fig. 5.25 Anatomy of Panorpidae. *Panorpa nuptialis.* **(a)** Male; **(b)** Female

Fig. 5.26 Larvae of Panorpidae (I). Fourth instar larvae, lateral view. Red arrow points to furrows on head capsule. (**a**) Neopanorpinae: *Neopanorpa nielseni*; (**b**) Panorpinae: *Panorpa amurensis*

the ventral aedeagal valves through a ventral bridge. A pair of sex pheromone glands is present ventral to the base of aedeagus, and can be inflated during the courtship. A10 and A11 are greatly reduced and usually covered by the epandrium dorsally. A pair of unsegmented cerci is located on the two sides of T10. The anus is on the apex of A11.

Female genitalia. On the ventral side of A9, there is a plate-like structure ventrally, called the subgenital plate. Subgenital plate is located below epandrium, superficially fused by two separate plates (gonocoxosternites VIII), and rounded, truncated, or V-shaped emargination terminally. A pair of laterotergites is usually reduced, or greatly developed in some species. The medigynium (or genital plate, or internal skeleton) is inside the genital chamber above the subgenital plate. The morphology of the genital plate is significant in the diagnose of species based on the females. The genital plate is made up of a main plate and a median axis. A pair of posterior arms are located on the posterior

Fig. 5.27 Larvae of Panorpidae (II). Fourth instar larva of *Cerapanorpa obtusa*

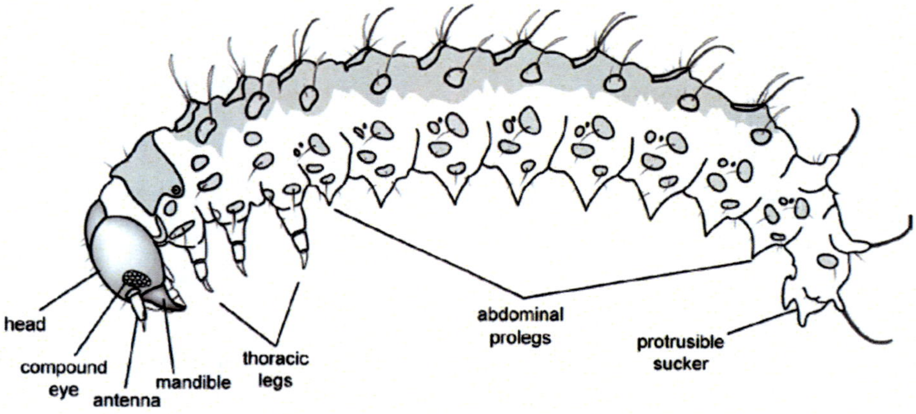

Fig. 5.28 Larval head of Panorpidae. Fourth instar larva of *Megapanorpa grandis*. (**a**) Lateral view; (**b**) Frontal view (*Scale bar*: 0.5 mm)

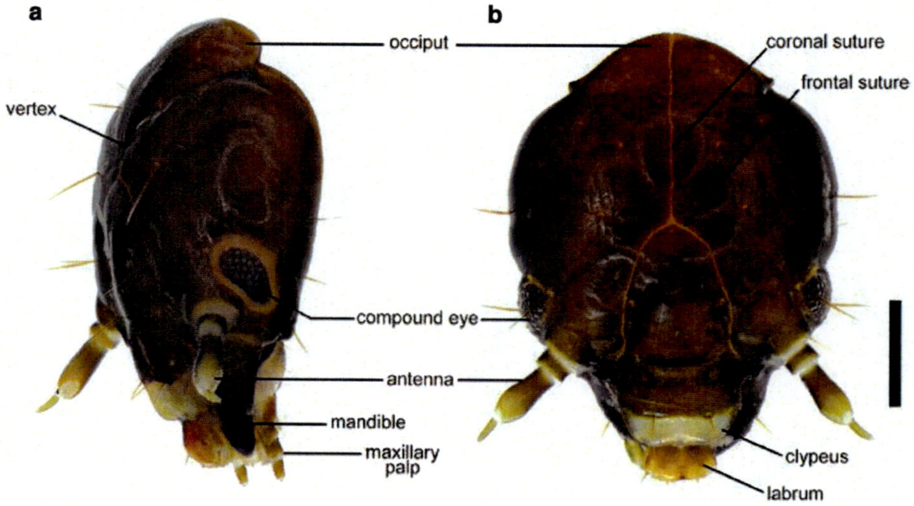

Fig. 5.29 Eggs of Panorpidae. (**a**) A female *Dicerapanorpa magna* labors by inserting her terminal abdomen into the soil; (**b**) An egg cluster of *Neopanorpa nielseni* in the soil; (**c**) An egg of *Cerapanorpa obtusa* (*Scale bar*: 0.1 mm)

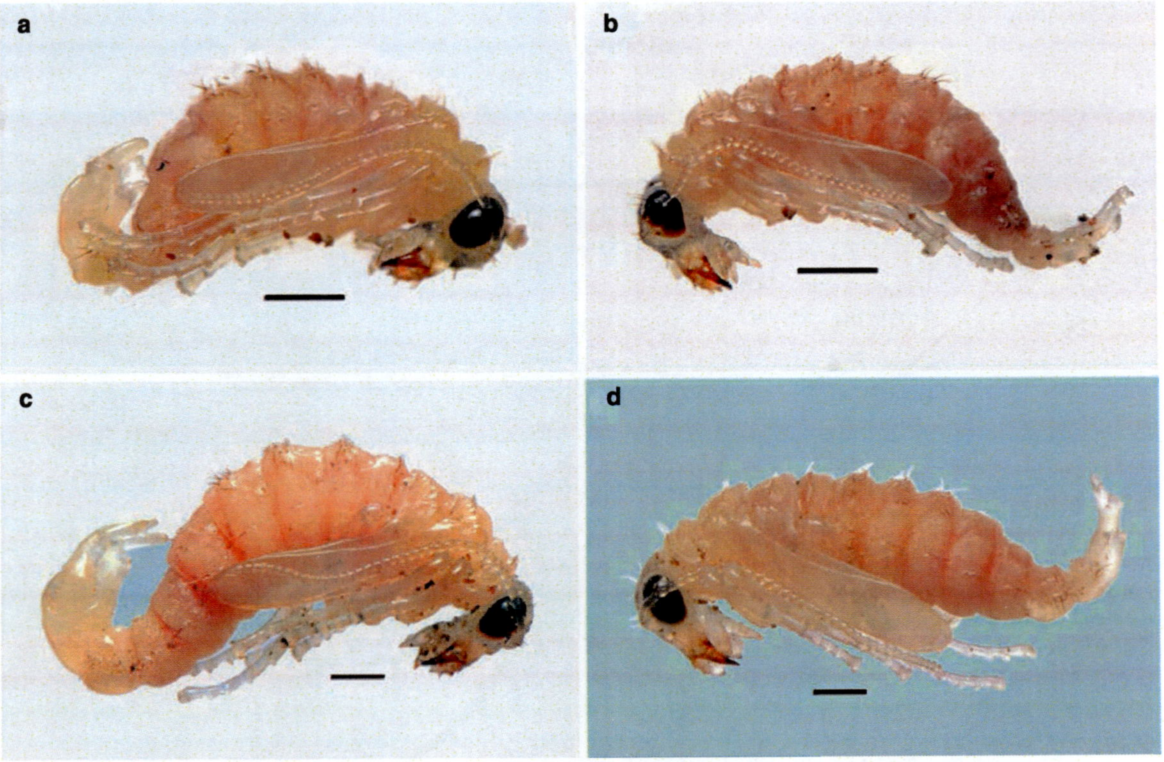

Fig. 5.30 Pupae of Panorpidae. Pupae, lateral view (*Scale bars*: 1.0 mm). (**a, c**) Male; (**b, d**) Female. (**a, b**) *Panorpa kunmingensis*; (**c, d**) *Panorpa amurensis*

portion of the main plate. The copulatory pore lies in the terminal end of the enlarged and decorated area of axis. The anterior half of the axis is split into a pair of parallel or diverged apodemes, with their anterior apexes concealed in or extending beyond the main plate. The spermathecal duct lies in the axis and opening at the copulatory pore. In some species, the main plate is accompanied by one or two pairs of dorsal, lateral, or basal plates.

5.2.3 Egg

The eggs of scorpionflies are usually long ellipsoidal, and furnished with sclerotized, polygonal networks on the surface of the chorion. The eggs are usually laid in shallow layers of soft soil, which is penetrated by the female's terminal abdomen.

5.2.4 Larva

Eruciform, with sparse short setae on the cylindrical trunk. The head bears a pair of prominent compound eyes instead of stemmata on the lateral side. They possess three pairs of thoracic legs and eight pairs of ventral prolegs. The prolegs are in A1–A8, and lack muscles and can only be moved by variations in turgor pressure or indirectly by muscles inserted in the adjacent parts of the sternites. When the larvae are at rest on a flat surface, the prolegs support the ventral surface of the abdominal segments except the last two from contact with the substrate (Hinton 2009). A protrusible sucker is on the last abdominal segment (A11, = telson).

5.2.5 Pupa

Decticous exarate, and resemble fully-grown adults but with much shortened rostrum, wings, and abdomen. The male pupa has enlarged and recurved male genitalia.

5.3 Panorpodidae (Fig. 5.31, 5.32, 5.33, 5.34, 5.35, and 5.36)

5.3.1 General Information

The Panorpodidae are commonly called short-faced scorpionflies due to their short rostra. They are very similar to the true scorpionflies in Panorpidae, but have shorter abdomen. They are also called "scorpionfly-like insects" literally in Chinese, and "シリアゲモドキ" (shiriagemodoki, shiri = tail, age = lifting, modoki = something that is similar but different) in Japanese.

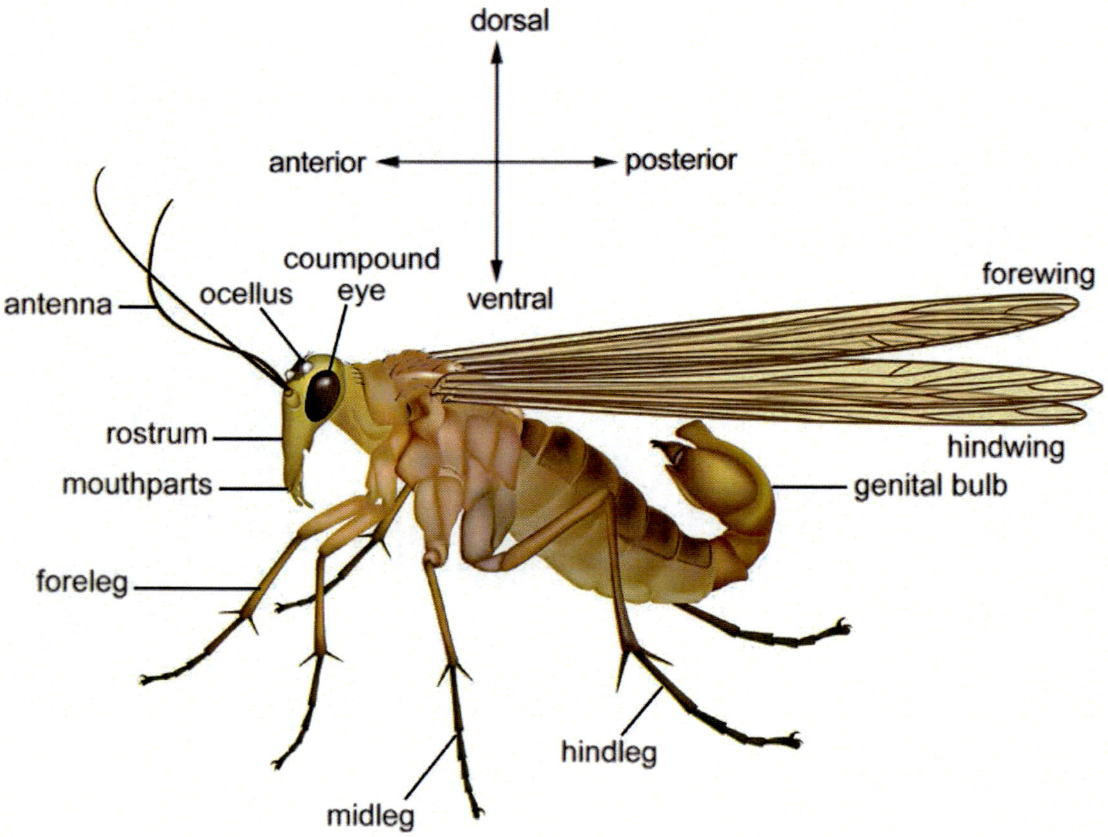

Fig. 5.31 Male adult of Panorpodidae. *Panorpodes komaensis,* lateral view

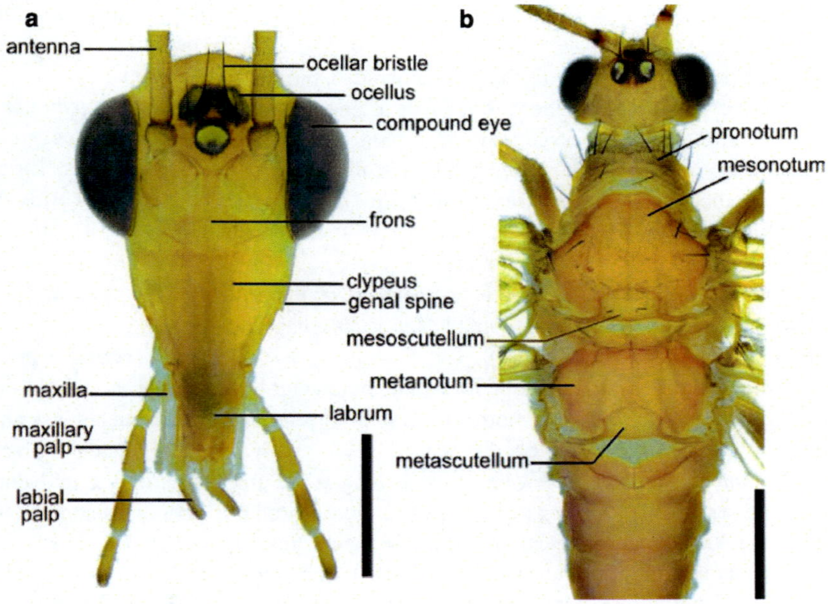

Fig. 5.32 Head and thorax of Panorpodidae. *Panorpodes paradoxa*. (**a**) Head, frontal view (*Scale bar*: 1.0 mm); (**b**) Head and thorax, dorsal view (*Scale bar*: 2.0 mm)

Fig. 5.33 Terminal portion of leg of Panorpodidae. Last two tarsomeres and pretarsus of *Panorpodes komaensis*, showing teeth-lacking claws (*Scale bar*: 0.2 mm)

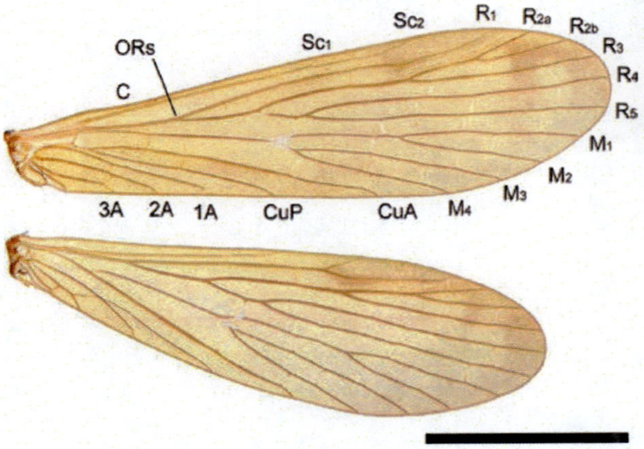

Fig. 5.34 Wings of Panorpodidae. *Panorpodes kuandianensis* (*Scale bar*: 5.0 mm)

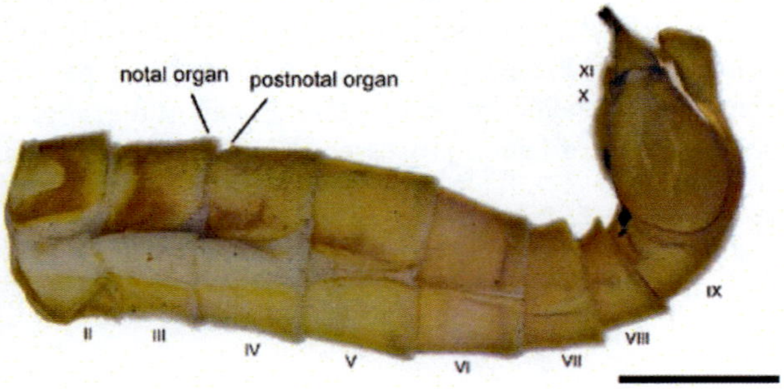

Fig. 5.35 Male abdomen of Panorpodidae. *Panorpodes komaensis*, lateral view (*Scale bar*: 2.0 mm)

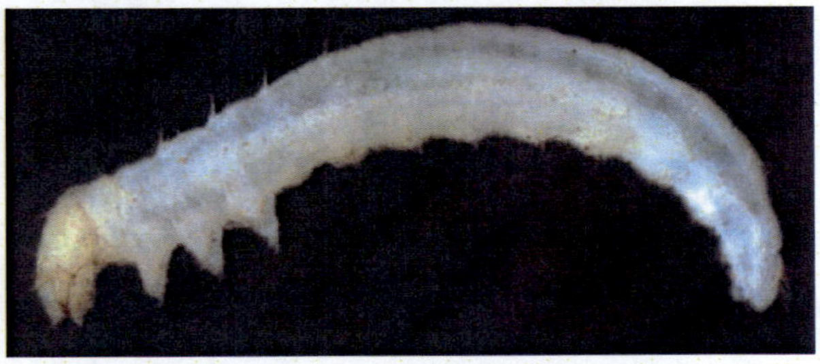

Fig. 5.36 Larva of Panorpodidae. *Panorpodes komaensis*, lateral view (©Lu Jiang)

5.3.2 Adult

Head. The head is similar to that of Panorpidae, but has a rather shorter rostrum. In addition, a pair of genal teeth is distinct and can be observed from the frontal aspect. The labrum has a deep V-shaped emargination distally. Four distinct ocellar bristles are present in the ocellar triangle.

Thorax. In general, three pairs of thick setae are on the anterior margin of pronotum, while two pairs are on the posterior margin. A pair of claws is simple and lacks comb-like teeth on their inner margin.

Wings. The vein M_4 in both fore- and hindwings are straight and not curved as in Panorpidae. In some species, the females' wings are more or less reduced.

Abdomen. A6–A8 of males are nearly as long as the proceeding segments, not elongated as in Panorpidae. In addition, the A7 and A8 are membranous in the pleural region, with separated tergum and sternum, different from those fully-fused cylindrical segments in Panorpidae.

Male genitalia. The genital bulb of Panorpodidae is similar to that of Panorpidae but with shorter stalk and gonostylus. The aedeagus is also composed of a pair of ventral valves, a pair of dorsal valves, a pair of dorsal processes, a pair of lateral processes, dorso-basally, a piston of sperm pump, and ventrally, a pair of parameres. The paramere is not well developed and usually fused with aedeagus.

Female genitalia. The subgenital plate bears a pair of separated sclerites (gonocoxosternites VIII) that are not fused as in Panorpidae. The medigynium is simple and minute, lacking posterior arms and axis.

5.3.3 Larva

The larvae are eruciform with a hypognathous head and sparse short setae on the trunk. They have one pair of stout antennae but lack any distinct visual organs. They have three pairs of stout thoracic legs and eight slightly protruded prolegs on the midventral line. A protrusible sucker is on the terminal segment.

5.3.4 Pupa

The pupal stage remains unknown for the Chinese species in Panorpodidae.

6 Biology

6.1 Bittacidae (Figs. 6.1, 6.2, 6.3, 6.4, 6.5)

Members in the Bittacidae are mostly univoltine insects. They overwinter in diapause as eggs or larvae (Byers and Thornhill 1983). The adults usually emerge from April to August next year.

Hangingflies rely on a high moisture. They are frequently found in the cool forests at the elevations higher than 1000 m. The suitable habitats that hangingflies prefer include dense herbs and shrubs aside a mountain stream, or an open field under the woods. Owing to their specialized prehensile legs, they are unable to walk like other mecopterans, but can only hang themselves on other objects with forelegs and sometimes also midlegs. They are quite strict for the humidity of their habitats. Once left the high-moisture habitats, they might soon be moribund and perish.

Hangingflies are predatory insects. They mainly adopt three predatory strategies: (1) hang themselves on plants

Fig. 6.1 Habitats of Bittacidae (I). (**a**) Forests around a farmyard (Tiantaishan National Forest Park, Shaanxi); (**b**) Dense ground cover beside a mountain stream (Great Snow Mountain, Yongde, Yunnan); (**c**) Grasses and shrubs under a subtropical evergreen broad-leaved forest (Gongshan, Yunnan); (**d**) An alpine meadowland (Cangshan Mountain, Dali, Yunnan)

J.-S. Wang, B.-Z. Hua, *A Color Atlas of the Chinese Mecoptera*, https://doi.org/10.1007/978-981-16-9558-2_6

Fig. 6.2 Habitats of Bittacidae (II). Female adult of *Terrobittacus rostratus* hanging on a leaf in the field (Tangli Mountain, Yongde, Yunnan)

Fig. 6.3 Feeding habits of Bittacidae (I). *Bittacus strigatus* catching a crane fly (Fangxian, Hubei)

Fig. 6.4 Feeding habits of Bittacidae (II). Feeding larvae of *Bittacus planus* with housefly puparia

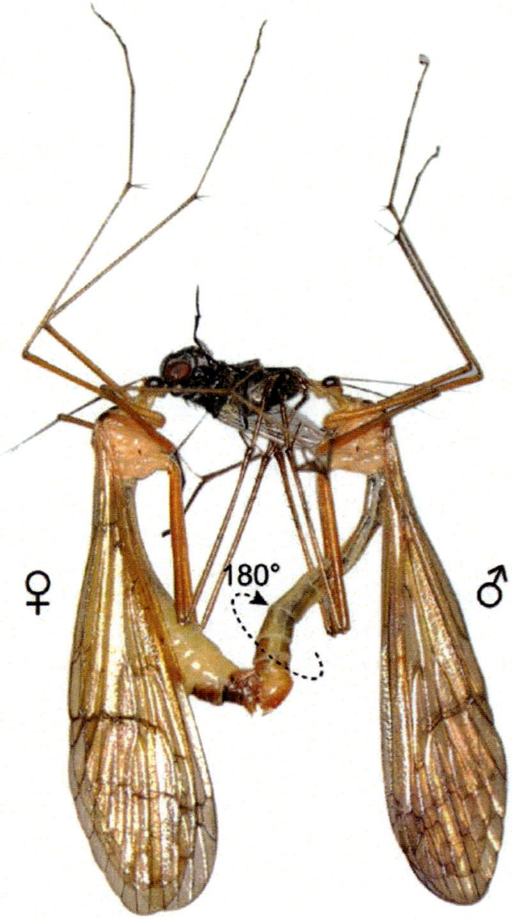

Fig. 6.5 Copulation of Bittacidae. A pair of *Bittacus planus* in a belly-to-belly copulation while eating a prey together. The male's abdomen is twisted about 180° (©Lu Jiang)

using forelegs, and ambush the contacted small insects using their hindlegs; (2) proactively catch preys by sweeping hindlegs while hanging; (3) catch preys in flight (Byers and Thornhill 1983). In some rare scenes, they rob or steal prey from other individuals, even try to attack their own kind. Once a prey is caught, the hangingfly may cut the exoskeleton open using the mandibles, then inserts the rostrum into the victim's body and enjoys a meal. The residue will then be discarded (Tan and Hua 2006). The larvae of hangingflies are saprophagous like those of Panorpidae. They feed on various kinds of organic debris in the soil, especially dead arthropods.

Hangingflies exhibit a special courtship behavior called "nuptial feeding." When a male catches a prey, he first feeds on it briefly. If the prey is proper, he hovers over the vegetation or hangs himself on a leaf while carrying the prey with his hindlegs. Later, he everts his abdominal sex pheromone glands and disperses a "sexual perfume" to lure the females (Thornhill 1976; Kaltenbach 1978).

Hangingflies are unique in Mecoptera for conducting a face-to-face hanging copulation. This is rare even in the whole Insecta. During copulation, the male of *Bittacus planus* twists his abdomen approximately 180° to accommodate the belly-to-belly mating position. In addition, his paired epandrial lobes are used to grasp the female's abdomen, providing an assistance for the genital contact between the mating pair (Gao and Hua 2013). During the mating course, the prey might be fed by the male and the female from both sides, or only fed by a single individual. In *Terrobittacus implicatus*, the male provides a prey as a nuptial gift to the female and twists his abdomen about 180° to form a belly-to-belly hanging mating position. The male's epandrial lobes are used to clamp the female's subgenital plate with the aid of the female abdomen swelling; and inserts his greatly elongated penisfilum into the female's copulatory pore with the cooperation of the upper branch of proctiger, which bears numerous short setae on the ventral groove (Wei et al. 2020). However, the male of *Bittacus mastrillii* provides a salivary mass as a nuptial gift to the female (Iwasaki 1996).

Similar nuptial feeding behavior can be found in many other insects, such as bushcrickets (Orthoptera: Tettigoniidae) (Wedell 1993), dance flies (Diptera: Empididae) (LeBas et al. 2004), ornate moths (Lepidoptera: Erebidae) (Kelly et al. 2012), and scorpionflies (Mecoptera: Panorpidae).

6.2 Panorpidae (Figs. 6.6, 6.7, 6.8, 6.9, 6.10)

The Panorpidae are univoltine or bivoltine insects. They usually overwinter in diapause as grown larvae. In the northern China, the adults begin their emergence in late April. In the southern provinces such as Taiwan, Guangdong, and Hainan, however, scorpionflies can almost be found during the whole year, with different species emerging in different months. For example, *Panorpa peterseana* (Taiwan) emerges in March, most species of *Neopanorpa* in April, *P. parallela* (Yunnan) in late July, *P. jinhuaensis* (Jinhua) in April and October (bivoltine), and *P. insularis* (Hainan) was recorded in November and December.

Relatively, scorpionflies have a wider range of humidity requirements than hangingflies. They prefer cool temperature, and are usually absent in some hot seasons or areas. In the northern China, scorpionflies emerge from early summer to late autumn, and inhabit cool mountains of an elevation over 1000 m. In northeastern China, however, they can live well in plains with high latitudes. In the southern China, scorpionflies have their distributional ranges as low as 100 m in plains or hills, or as high as above 3000 m in mountains. Scorpionflies are usually active on the leaves of grasses and shrubs. They often hide below the leaves under direct sunlight, rains, or strong winds, and are hard to be find. A "feign death" strategy is utilized to run away from danger by falling to the ground or the deep shade of vegetable when they encounter disturbance.

Scorpionflies exhibit two categories of wing postures. Most of them hold wings widely over two sides of abdomen and V-shaped. However, some of them hold wings roof-like, and usually covering the abdomen dorsally. Scorpionflies who hold wings roof-like often inhabit dense and low (20–60 cm) herbaceous vegetation (Byers 1996; Bicha 2006), prefer to be active among the deep shadowed twigs and leaves (Wang and Gong 2021), and tend to be most abundant in deep shaded ravines. Presumably, by holding their wings together over the abdomen, these scorpionflies receive more advantages in avoiding obstacles, and can move with more flexibilities among the dense vegetation (Wang and Gong 2021).

The adults are saprophagous, mainly feeding on dead or dying arthropods, rarely on pollens, nectar, or fruits. They are also recorded to feed on dead vertebrates, even human corpse in a weird case. Therefore, they can be valuable in the forensic research (Lindgren et al. 2015). They often forage by means of smelling. In a shortage of food, a troop of two to a dozen individuals may gather and feed on a single prey. The stronger individuals will attack and try to prey on the smaller, weaker, or crippled ones in some situations. They often tear the prey's exoskeleton apart with their mandibles, and then insert their rostrum inside the prey's body in order to feed. They usually consume the soft tissue and fluids, but discard the tough exoskeletons. A weird but unique behavior of scorpionflies is that they stealthily eat the host's prey as kleptoparasites of web-building spiders (Fig. 6.11). To avoid being attacked by the spider, scorpionflies can regurgitate a brown enteric fluid to dissolve the web that might entangle them (Thornhill 1975) (Figs. 6.12, 6.13, and 6.14).

The courtship and copulation of scorpionflies are complicated and diverse (Mader 2020). The most appealing behavior, nuptial feeding, has attracted research interests from many scholars. Three mating strategies are usually adopted by the male scorpionflies: (1) providing a dead arthropod; (2) providing a salivary mass, which is often attached to leaves of plants for the female to feed; (3) raping without providing any gifts. The notal organ, postnotal organ, anal horn, and gonopods serve to grasp the corresponding body parts of the female, and maintain the mating posture.

The typical mating posture of scorpionflies is in a V-shape. The male of *Neopanorpa longiprocessa* possess an extremely elongated notal organ with which it can hold the female's wings effectively, and provide nothing to his mate (Zhong and Hua 2013b). Similarly, the male of *Panorpa amurensis* also uses his clamp-like notal organ to hold the anterior margin of the female's wings, but provides a dead arthropod as a nuptial gift instead of a salivary mass to earn the mating opportunity (Ma and Hua 2011). The two finger-like anal horns on the A6, and the stalked base of A7 of male *Dicerapanorpa magna* serve to hold the female's terminalia and maintain the mating posture. Meanwhile, he also uses his notal organ to hold the anterior margin of the female's wings, and provides a salivary mass as a nuptial gift (Zhong

Fig. 6.6 Habitats of Panorpidae. (**a**) Alpine meadow, 3000 m (Cangshan Mountain, Yunnan); (**b**) Tropical rainforest, 800 m (Mount Gede-Pangrango, Java, Indonesia); (**c**) Country road with dense ground covers aside, 200 m (Fukuoka, Kyushu, Japan); (**d**) Deep alpine ravine, 2500 m (Gongshan, Yunnan); (**e**) Tropical alpine forest, 2000 m (Tengger Mountains, Java, Indonesia); (**f**) Suburban farmyard, 100 m (Jinhua, Zhejiang); (**g**) Alpine forest, 2000 m (Taibai Mountain, Shaanxi)

Fig. 6.7 Wing postures of Panorpidae. (**a**) *Panorpa bicornuta* (Toyama, Honshu, Japan); (**b**) *P. kunmingensis* (West Hill, Kunming, Yunnan); (**c**) *P. gressitti* (Liuxihe, Conghua, Guangdong); (**d**) *P. arakavae* (Toyama, Honshu, Japan); (**e**) *Furcatopanorpa longihypovalva* (Tiantaishan Forest Park, Baoji, Shaanxi); (**f**) *P. parallela* (Tangli Mountain, Yongde, Yunnan); (**g**) *P. amurensis* (Changbai Mountains, Jilin); (**h**) *P. tetrazonia* (Jinhua, Zhejiang); (**i**) *Leptopanorpa nematogaster* (Mount Gede-Pangrango, Java, Indonesia); (**j**) *Neopanorpa lungtausana* (Nanling Mountains, Guangdong)

Fig. 6.8 Feeding habits of Panorpidae (I). Several individuals of *Panorpa amurensis* are feeding a dead leaf beetle larva (Changbai Mountains, Jilin)

et al. 2015a). Peculiarly, the male of *Furcatopanorpa longihypovalva* lacks a notal organ on A3 and anal horns on A6, and provides liquid salivary secretion through a mouth-to-mouth mode to the female. Therefore, they copulate in an atypical O-shape, with only their mouthparts and genitalia connected to each other (Zhong et al. 2015b).

The larvae live on the surface of soil, in the layer of organic residue, or in the soil, and feed on various organic debris, predominantly, dead insects and other arthropods (Byers and Thornhill 1983; Jiang and Hua 2013, 2015). The epedaphic larvae of *Cerapanorpa*, *Dicerapanorpa*, and *Megapanorpa* are darkish dorsally as camouflage, and possess well-developed locomotory dorsal processes as adaptations likely to avoid potential predators (Jiang et al. 2019; Wang and Hua 2019b); while the euedaphic larvae of *Neopanorpa* are pale on their trunks, with shallow furrows,

Fig. 6.9 Feeding habits of Panorpidae (II). (**a**) *Panorpa pingjiangensis* feeding dead crane fly (Lushan Mountain, Jiangxi); (**b**) *Cerapanorpa obtusa* feeding dead frog (Huoditang, Ningshan, Shaanxi); (**c**) *Leptopanorpa linyejiei* feeding nectar (Mount Ciremai, Java, Indonesia); (**d**) *P. dichotoma* feeding on human finger injury (Kyushu, Japan)

Fig. 6.10 Feeding habits of Panorpidae (III). *Neopanorpa* sp. larva feeding on chopped mealworm larvae

Fig. 6.11 Kleptoparasitism of Panorpidae. A female *Dicerapanorpa tjederi* tries to stealthily feed on a spider's prey and do not getting entangled by the silk (Yulong Snow Mountain, Yunnan)

Fig. 6.12 Courtship and copulation of Panorpidae (I). no, notal organ; sm, salivary mass. (**a**, **b**) *Neopanorpa longistipitata*; (**c**) *Leptopanorpa linyejiei*; (**d**) *P. amurensis*; (**e**) *Dicerapanorpa magna* (©Wen Zhong); (**f**) *Sinopanorpa tincta* (©Wen Zhong); (**g**) *Cerapanorpa obtusa*; (**h**) I. *Neopanorpa longiprocessa* (©Wen Zhong)

reduced antennae, shortened setae, flattened compound eyes on the head capsules, and short dorsal processes on the trunk, characters appear to be adaptations for the underground lifestyle (Jiang et al. 2019). Once matured, the larvae build a chamber in the soil and prepare to pupate.

6.3 Panorpodidae

The adults of Panorpodidae have merely been observed to feed on vegetative materials, such as epidermis and mesophyll of leaves (Byers 1997), or the sap leaked from the injury of twigs (Fig. 6.15). Their larval feeding habits remain a mystery to date (Byers 1997; Jiang et al. 2014).

A similar courtship and mating pattern to Panorpidae was observed in Panorpodidae. The males and females of *Brachypanorpa* assume a V-shaped copulation as in Panorpidae (Byers 1997). The male of *Panorpodes komaensis* offers a salivary mass as nuptial gift to the female and starts a V-shaped copulation, then twists the terminal segments of the female by 180° and changes to an end-to-end position (Tong et al. 2017).

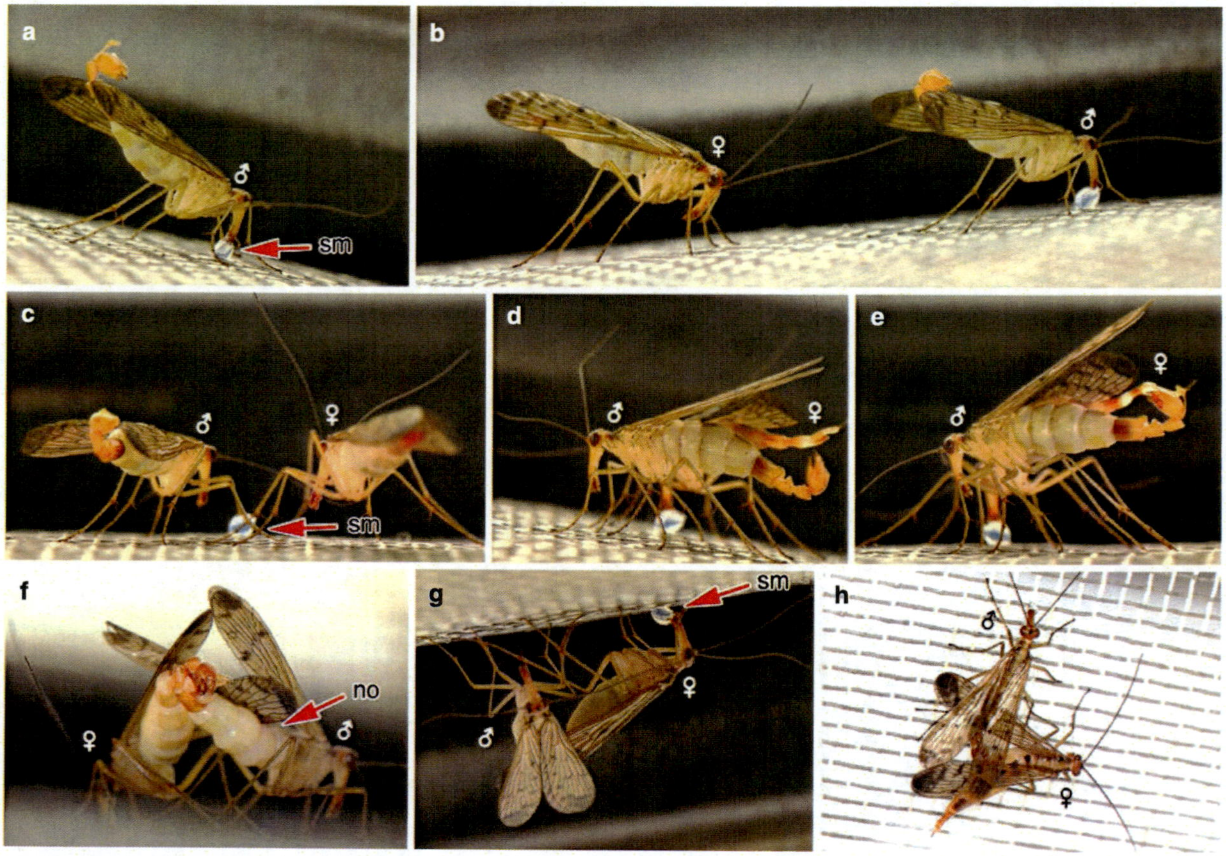

Fig. 6.13 Courtship and copulation of Panorpidae (II). *Panorpa kunmingensis* (©Xin Tong). (**a**) A male raises genital bulb, emitting sexual pheromone, and secretes a salivary mass; (**b**) A female responds and approaches the male; (**c**) The male delivers a salivary mass to the female; (**d, e**) The male tries to control the female's genital segments using his gonostyli while the female is feeding the salivary mass; (**f, g**) Genitalia are coupled between the two sexes, meanwhile, the male tries to clamp the costal margin of the female's wings on one side using his notal and postnotal organs; consequently, a typical V-shaped mating posture is conducted; (**h**) The female's wings are still clamped by the male after her genitalia detached. no. notal organ; sm, salivary mass

Fig. 6.14 Natural enemies of Panorpidae. (**a**) A female scorpionfly being entangled by the spider web; (**b**) A female scorpionfly being caught by a robberfly; (**c**) A female scorpionfly being caught by a parasitic wasp; (**d**) A female scorpionfly being infected by a mite

Fig. 6.14 (continued)

Fig. 6.15 Feeding habits of Panorpodidae. A male *Panorpodes paradoxa* is feeding sap from the injury of a twig (Toyama, Honshu, Japan)

Collecting and Rearing

7

7.1 Collecting Techniques

Scorpionflies, short-faced scorpionflies, and hangingflies can usually be caught by sweeping insect nets on sight (Fig. 7.1). Generally, scorpionflies rest on the upper surface of leaves, and can easily be found from above. However, hangingflies usually hang themselves on the edge of leaves or twigs, and catching them usually requires a lower observing point of view. Some species prefer to be active among the deep shadowed twigs and leaves (Wang and Gong 2021), and the sweeping net can indisputably be impeded by the vegetation. Under this situation, the collector may gently beat the lush vegetation and awaits the hiding scorpionflies or hangingflies to fly out and resettle on the nearby vegetation, and then catch them by the net or hand (Fig. 7.2). A few species such as *Panorpa peterseana* Issiki, 1929 from Taiwan are active in high tree-canopy blossoms, therefore an extended handle is needed for the sweeping net (Bicha et al. 2020). After being captured, the specimens should be immediately immersed with ethanol solution (or other preserving agents), or put in a plastic or mesh cage to prevent them from escaping (Fig. 7.2). The cage should be supplied with leaves, twigs, soft tissue, or other materials to keep the insects to cling onto and stay alive in a high humidity for subsequent processes. A few ventilative holes are needed to avoid asphyxiation. Some fruit pieces or dead insects can also be provided to avoid the starving scorpionflies from killing the weak or moribund ones. Hangingflies, however, should better be solely caged, because they are very easily to be entangled by their slender legs, and committing a severe, mutual slaughter if imprisoned in a small space.

Scorpionflies and hangingflies can also be collected using light (Fig. 7.3) and Malaise traps (Fig. 7.4). Although some authors reported occasional or incidental collections by means of these techniques (Bicha et al. 2020), we found that these are sometimes far more efficient than the conventional net-sweeping, especially in areas with abundant individuals of mecopterans. In one of our collecting sites, a light trap yielded more than 80 individuals in a single night; and in another site, a Malaise trap yielded more than 300 individuals during one week. Hangingflies have been observed to hunt some flies or moths attracted by the light; and scorpionflies have been observed to feed on the dead insects around the light trap. Because scorpionflies and hangingflies are relatively weak fliers, these collecting equipment need to be placed in an area which harbors many individuals, or at least near to the area. Prior to the collection, a general investigation on the individual density is required.

Fig. 7.1 Collection techniques of Mecoptera (I). The first author searches for scorpionflies among dense herbaceous ground cover (Jiajin Mountain, Sichuan, ©Hao Xu)

Due to the saprophytism of many scorpionflies, dead organisms can be used as baits, such as squid pieces (Bicha et al. 2020), rotting fruits, shrimp pieces, or dead insects (Fig. 7.5). In the case of squid pieces, scorpionflies tend to be the first visitor before flies, ants, and hornets. Once the bait is occupied by the ants, scorpionflies refuse to consume it (Bicha et al. 2020). The baits are efficient to draw individuals from thorny or rigid vegetation to allow them to be more easily collected (Bicha et al. 2020). In the pitfall traps for the saprophagous beetles (burying beetles, ground beetles, scarab beetles, etc.), scorpionflies have been occasionally found, soaking in the fluids. When using beating sheet to collect beetles, some scorpionflies fall into the sheet due to their

J.-S. Wang, B.-Z. Hua, *A Color Atlas of the Chinese Mecoptera*, https://doi.org/10.1007/978-981-16-9558-2_7

Fig. 7.2 Collecting techniques of Mecoptera (II). (**a**) Insect net and digital camera; (**b**) Holding a scorpionfly by fingers; (**c, d**) Preserving the specimens in centrifuge tubes using 75–95% ethanol; (**e**) Plastic boxes for temporarily keeping the scorpionflies alive

Fig. 7.3 Collecting techniques of Mecoptera (III). (**a**) Light trap in the form of a detachable tent; (**b**) A male *Bicaubittacus burmanus* attracted by the light; (**c**) A male *Neopanorpa hunanensis* attracted by the light

Fig. 7.4 Collecting techniques of Mecoptera (IV). The collectors are setting a Malaise Trap in the field

Fig. 7.5 Collecting techniques of Mecoptera (V). Several adults of *Panorpa amurensis* are attracted by a dead honeybee. One pair is copulating while the female is feeding

"feign death" strategy. Woodward flail-vac seed stripper, a device for collecting the seeds of wiregrass in the USA, can sweep in and accumulate many individuals of *Panorpa rufa*, which lives in the dense grass and can hardly be collected by the traditional means (Bicha et al. 2020).

Due to the fragility of the antennae and legs, it is better to handle the living scorpionflies with fingers or forceps by gently holding their legs, wing bases, or thorax to avoid struggling as far as possible (Fig. 7.6). After being handled by the fingers, scorpionflies usually regurgitate an enteric fluid, which can give the fingers a brown tinge (Fig. 7.6). Males of some species can attack the collector's fingers using their pincer-like gonostyli, but most species cannot bring a substantial damage. However, in some strong gonostyli-bearing species, such as *Neopanorpa nielseni*, *Panorpa amurensis*, and *Sinopanorpa tincta*, the attacks can cause an acupunctural pain or even a slight bleeding. When the insects are hungry and thirsty, they occasionally lick or try to gnaw the human skin (Fig. 6.9d).

7.2 Specimen Preservation

Scorpionfly specimens keep the best condition in 80–95% ethanol under −20°C for further usage. In lower concentrations of ethanol such as 75%, the body color of the specimens can easily fade away; and in higher concentrations such as 99.9%, some fragile body parts can be easily shrinked or broken due to dehydration, even if handled gently. It is better to hold one specimen in its own tube to avoid squeezing and scratching, which may severely damage the specimens.

However, the specimens can easily dry out if the preserving tubes are not properly sealed or lacking continuous curations. Dry, pinned specimens are relatively more stable in form and can last at least a hundred years in a good facility.

7.3 Rearing

Artificial rearing is a requisite for biological, behavioral, cytological, genetic, morphological, and other kinds of studies. The immature stages of mecopterans are usually hard to be find because they often live underground. Therefore, we usually obtain the immature stages by artificial rearing of the adults. This also ensures that the identity of the species to keep consistent in different life stages.

Once the living adults were caught, they should be put in a plastic or mesh cage. The bottom of the cage should be supplied with wet soil or soft tissue. Scorpionflies and hangingflies generally do not like to stay on the ground; therefore, leaves, twigs, soft tissue, or other materials should be provided to keep the insects to cling onto. Keep the conspecific individuals in a same cage.

Provide the hangingflies with living, flying insects such as houseflies, fruitflies, or crane flies as the food source; while the scorpionflies with freshly cut pieces of insect, such as houseflies, mealworms, or locusts. Raise the ratio between males and females to ensure that every female is pregnant. If the females are mature, feed, and copulated, they will begin to labor in 1–3 days.

Female hangingflies usually labor by randomly drop the eggs one by one to the ground. Once the eggs fall to the soil, they are very hard to be separated. Therefore, soft, white, and wet tissue can be used as the substrate in the cage. Keep the eggs in shallow layers of clean and moist soil, and provide adequate ventilation to avoid mold infestation. The larvae will crack the chorion using a sharp egg burster on its head. The larvae camouflage by adhering soils and other debris on

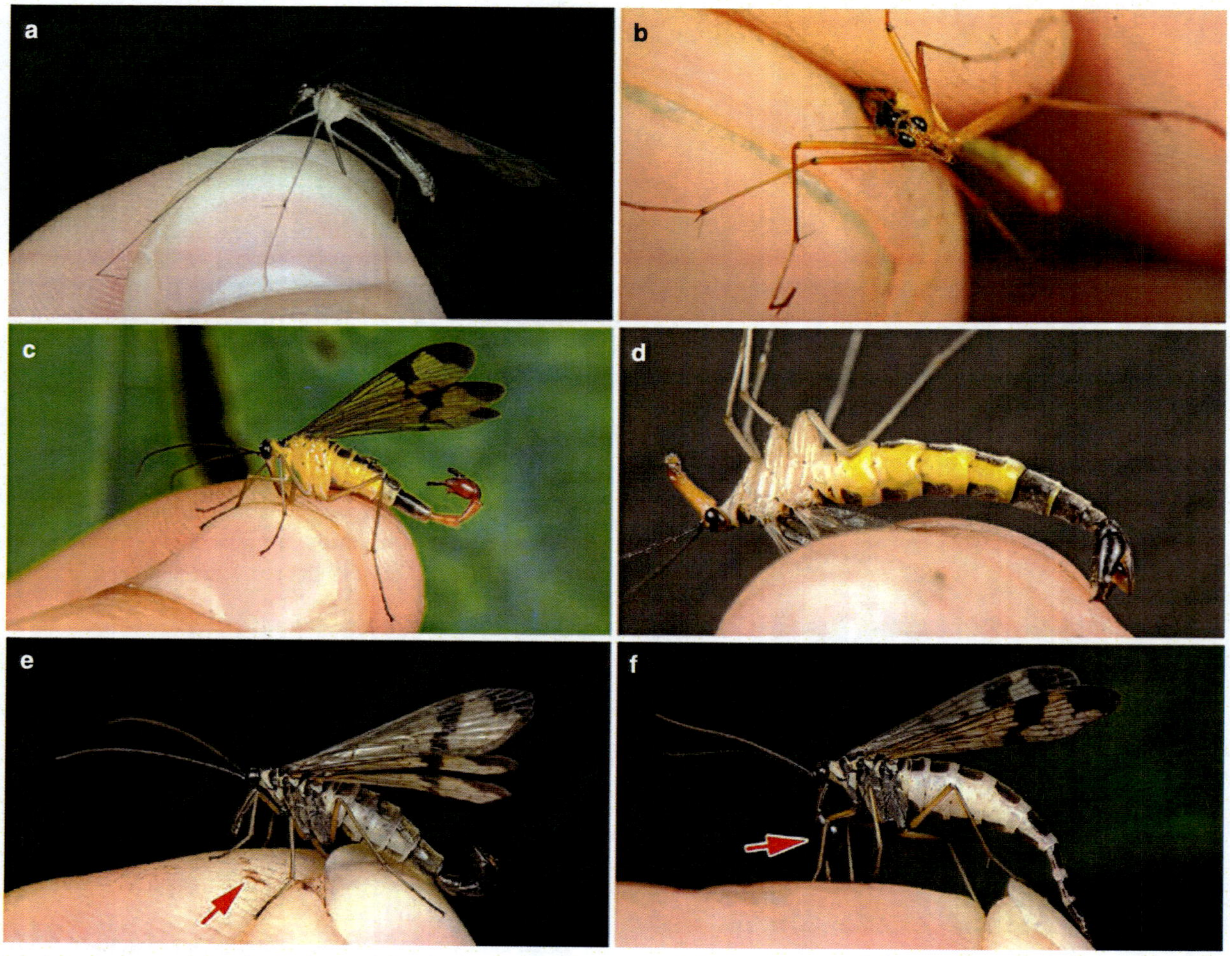

Fig. 7.6 Collecting techniques of Mecoptera (VI). (**a**) Holding legs of a female *Terrobittacus rostratus* by fingers (Tangli Mountain, Yongde, Yunnan); (**b**) Holding thorax of a female *T. xiphicus* by fingers (Tianping Mountain, Sangzhi, Hunan); (**c**) Holding legs of a male *Neopanorpa nielseni* by fingers (Tangli Mountain, Yongde, Yunnan); (**d**) Holding wings of a male *Furcatopanorpa longihypovalva* by fingers, showing the male is attacking the finger using gonostyli (Qinling Mountains, Shaanxi); (**e**) Holding legs of a male *Panorpa sibirica* by fingers (Changbai Mountains, Jilin). Red arrow indicating the brown stain on the finger; (**f**) Holding legs of a female *P. sibirica* by fingers (Changbai Mountains, Jilin). Red arrow indicating a drop of brown enteric fluid regurgitated by the female

their protuberances and setae-bearing backs. Chopped pieces of insects can be provided as the food source, such as housefly puparia, mealworm pupae, or fruitflies. Remove the residual carcasses in time to avoid mold infestation. Usually, hangingflies overwinter as diapausing eggs. The eggs should be put in a relatively dry and cold condition during winter, and they will hatch when the temperature and moisture raise in spring.

Female scorpionflies lay eggs by inserting the terminal abdomen into shallow layers of soft soil. A small cumulation of a few to dozens of eggs are laid on the soil crevices. The eggs usually hatch in 5–20 days. Chopped pieces of insects can be provided as the food source. Scorpionflies usually overwinter as diapausing fourth (final) instar larvae. Provide deep soil (4–5 cm or deeper) for the euedaphic *Neopanorpa* larvae, which spend most of their lifetime under the surface. The larval stage lasts 12 days to several months, depending on the species, the food sources, and the temperatures. When mature, the fourth instar larvae cease feeding and become quiescent, and stay in a soil cell in a prepupal condition. The pupal stage usually lasts 10–21 days.

8 Guide to the Use of This Atlas

8.1 Terminology

The morphological terminology follows Wang and Hua (2021).

The following acronyms are applied in the description part: A1, the first abdominal segment; S1, the first sternum; and T1, the first tergum (and so forth for other segments).

8.2 Specimen Repositories

Specimens are deposited in, interchanged, or loaned from the following institutions:

- **ANIC** Australian National Insect Collections, Canberra, Australia
- **CANU** Capital Normal University, Beijing, China
- **CUFT** Central South University of Forestry and Technology, Changsha, China
- **DALU** Biological Science Museum, Dali University, Dali, China
- **ECAU** Entomological Museum, China Agricultural University, Beijing, China
- **EDKU** Entomological Division (formerly Snow Entomological Museum), Biodiversity Institute, University of Kansas, Lawrence, USA
- **ISWU** Insect Collection of Southwest University, Chongqing, China
- **IZAS** Institute of Zoology, Chinese Academy of Sciences, Beijing, China
- **KYSU** Entomological Laboratory, Kyushu University, Fukuoka, Japan
- **MYNU** Mianyang Normal University, Mianyang, China
- **NAKU** Nankai University, Tianjin, China
- **NMCZ** Department of Entomology, National Museum, Prague, Czech
- **NWAU** Entomological Museum, Northwest A&F University, Yangling, China
- **OMGM** Omogo Mountain Museum, Kumakogen, Japan
- **SCAU** South China Agricultural University, Guangzhou, China
- **SHNU** Insect Collection of Shanghai Normal University, Shanghai, China
- **SYSU** Sun Yat-sen University, Guangzhou, China
- **TJNM** Tianjin Natural History Museum, Tianjing, China

8.3 Maps

The base maps were obtained from SimpleMappr (https://www.simplemappr.net/) and modified with Adobe Illustrator CC and Adobe Photoshop CC.

Provincial/state-level regions are fully colored, and black dots indicate detailed localities. Red indicates that the species is endemic to China, and gray indicates that the species has also been recorded from the other countries.

8.4 Figure Credits

All the figures used in this atlas were taken by the first author except those specified.

J.-S. Wang, B.-Z. Hua, *A Color Atlas of the Chinese Mecoptera*, https://doi.org/10.1007/978-981-16-9558-2_8

Part II

Taxonomy

Family Bittacidae Handlirsch, 1906

9

(Hangingflies) [蚊蝎蛉科]

Hangingflies have three pairs of slender raptorial legs, each bearing a single tarsal claw. They can only hang themselves on twigs or edges of leaves, but are unable to stand or walk on a surface (Fig. 9.1).

More than 200 species have been described worldwide. Over 40 species have been recorded in China (Zhang et al. 2020) and 37 of them are included in this atlas.

Fig. 9.1 Adult of Bittacidae. A female *Bittacus* sp., lateral view (©Bo Lei)

Key to genera of Chinese Bittacidae

1. Gonocoxites with a pair of curved prolongations apically (Fig. 9.2a) .... ***Bicaubittacus*** Tan & Hua
– Gonocoxites lacking such prolongations 2
2. Epandrial lobes shorter than half gonocoxites (Fig. 9.2b) .... ***Terrobittacus*** Tan & Hua
– Epandrial lobes longer than half gonocoxites (Fig. 9.2c) .... ***Bittacus*** Latreille

J.-S. Wang, B.-Z. Hua, *A Color Atlas of the Chinese Mecoptera*, https://doi.org/10.1007/978-981-16-9558-2_9

Fig. 9.2 Male genitalia of Bittacidae (*Scale bars*: 0.5 mm). (**a**) *Bicaubittacus appendiculatus;* (**b**) *Terrobittacus implicatus;* (**c**) *Bittacus planus*

9.1 Genus *Bicaubittacus* Tan & Hua, 2009

(Bicaudate Hangingflies) [双尾蚊蝎蛉属]

This genus can be recognized by the male gonocoxites with a pair of long curved distal prolongations (Fig. 9.3).

Five species are endemic to China, and all of them are included in this atlas.

Fig. 9.3 Male adult of *Bicaubittacus. Bicaubittacus burmanus* in the field (Tangli Mountain, Yongde, Yunnan)

Key to species of *Bicaubittacus* in this atlas (Five species, males)

1. Epandrial lobe with two caudal prongs 2
– Epandrial lobe with one caudal prong 3
2. Ventral caudal prong longer than dorsal .... ***Bicaubittacus mengyangicus*** Tan & Hua
– Ventral caudal prong shorter than dorsal .... ***Bicaubittacus burmanus*** (Tjeder)
3. Epandrial lobe truncated near bifurcated caudal prong and bearing 2–3 acute teeth .... ***Bicaubittacus yangi*** Tan & Hua
– Epandrial lobe evenly tapering towards caudal prong 4
4. Epandrial lobe with finger-like process directed ventrad near base of ventral margin .... ***Bicaubittacus appendiculatus*** (Esben-Petersen)
– Epandrial without such process on ventral margin .... ***Bicaubittacus longiprocessus*** (Huang & Hua)

9.1.1 *Bicaubittacus appendiculatus* (Esben-Petersen, 1927)

(Appendicular Bicaudate Hangingfly) [具肢双尾蚊蝎蛉] (Figs. 9.4, 9.5, and 9.6)

Measurements. Forewing length 17.0–19.0 mm.

Morphology. Vertex yellowish brown. Antennae with yellowish brown scape and pedicel, and blackish brown and ciliated flagellum. Ocellar triangle black and protruded. Rostrum blackish brown. Pronotum unevenly deep brown. Meso- and metanotum with pale, gourd-shaped median stripe. Legs sordidly yellow, with apical end of femur and both ends of tibia darkened. Wing membrane hyaline, slightly tinged with sordid yellow. Apex indistinctly tinged with smoky brown. Male A1–A8 yellowish brown. T8 bearing V-shaped emargination on posterior margin mesally.

Male epandrial lobes peculiarly shaped. Ventral margin protruding ventrad and forming finger-like process subbasally, and bearing row of long setae from middle to end of slender tail distally. Apex of gonocoxites with pair of long and curved gooseneck-shaped prolongations bearing row of long setae along hind margins. Gonostylus boot-shaped with truncated apex. Penisfilum coiled into small loop. Cerci short and clavate. Upper branch of proctiger long and slightly branched at apex in L-shape, while lower branch short and curved ventrad.

Emergence. June to July.

Distribution. Yunnan (Shangri-La, Yunlong).

Fig. 9.4 Male adult of *Bicaubittacus appendiculatus*, dorsal view (Shangri-La, Yunnan) (*Scale bar*: 5.0 mm) [DALU]

Fig. 9.5 Male genitalia of *Bicaubittacus appendiculatus*, lateral view (*Scale bar*: 0.5 mm)

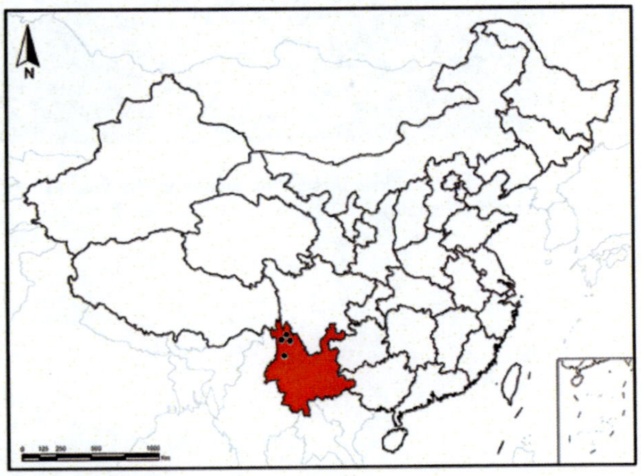

Fig. 9.6 Distribution of *Bicaubittacus appendiculatus*

9.1.2 *Bicaubittacus burmanus* (Tjeder, 1973)

(Myanmar Bicaudate Hangingfly) [缅甸双尾蚊蝎蛉] (Figs. 9.7, 9.8, and 9.9)

Measurements. Forewing length 18.0–19.0 mm.

Morphology. Vertex yellowish brown to reddish brown. Ocellar triangle black. Antennae blackish brown. Nota blackish brown with yellowish brown sides. Legs yellowish brown. Wing membrane hyaline, tinged with smoky yellow, and devoid of markings, except apex indistinctly darkened. Male A1–A8 yellowish brown. T8 with V-shaped emargination on posterior margin mesally.

Male epandrial lobes long rectangular, and forked apically into pair of slender prongs; dorsal margin curled inward and forming three small processes, with middle one more slender than others; ventral margin protruding ventrad at middle. Gonocoxites with pair of long and curved distal prolongations bearing row of long bristles along hind margin, and divergent with each other. Foliate lobe basal to each prolongation. Gonostylus slender and bearing an inner tooth. Aedeagal lobes broad. Penisfilum coiled into small loop. Upper branch of proctiger gooseneck-shaped, and bearing tuft of long setae on enlarged region below its acute apex; lower branch short and curved ventrad.

Emergence. May to June.

Distribution. Yunnan (Yongde); Myanmar (Kambaiti).

Fig. 9.7 Male adult of *Bicaubittacus burmanus* in the field (Tangli Mountain, Yongde, Yunnan)

Fig. 9.8 Male genitalia of *Bicaubittacus burmanus*, lateral view (*Scale bar*: 0.5 mm)

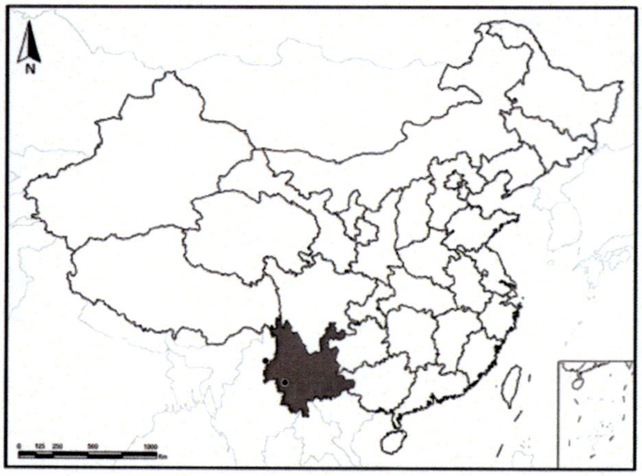

Fig. 9.9 Distribution of *Bicaubittacus burmanus*

9.1.3 *Bicaubittacus longiprocessus* (Huang & Hua, 2005)

(Long-process Bicaudate Hangingfly) [长突双尾蚊蝎蛉] (Figs. 9.10, 9.11, and 9.12)

Measurements. Forewing length 16.5–17.5 mm.

Morphology. Vertex yellowish brown, ocellar triangle darkened, rostrum blackish brown, antennae yellowish brown. Nota yellowish brown. Wing membrane hyaline, tinged with smoky yellow, and devoid of markings. Legs yellowish brown, with apical end of femur and both ends of tibia darkened. Pterostigma yellowish brown. Male A1–A8 yellowish brown. T8 with V-shaped emargination on posterior margin mesally.

Male epandrial lobes with basal half subtrapezoidal, and curled inward at dorsal and ventral margins; apical half extending into tail-like prong curving dorsad bearing row of long bristles on dorsal margin; dorsal margin with large process on inner surface at middle, and ventral margin with smaller process on inner surface at distal 2/3. Gonocoxites with pair of long and curved distal prolongations bearing row of long sparse bristles along hind margin. Aedeagal lobes slender. Penisfilum coiled into small loop. Cerci clavate. Upper branch of proctiger long and greatly curved cephalad at apex, while lower branch short and curved ventrad.

Emergence. April to June.

Distribution. Fujian (Nanjing), Guangdong (Shaoguan, Meizhou), Jiangxi (Jinggangshan).

Fig. 9.10 Male adult of *Bicaubittacus longiprocessus* in the field (Shaoguan, Guangdong)

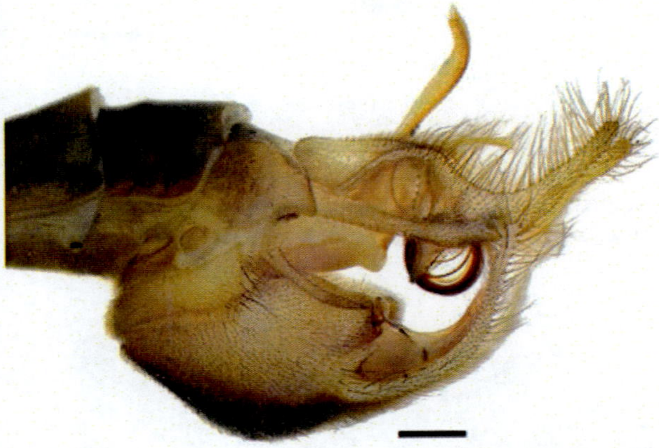

Fig. 9.11 Male genitalia of *Bicaubittacus longiprocessus*, lateral view (*Scale bar*: 0.5 mm)

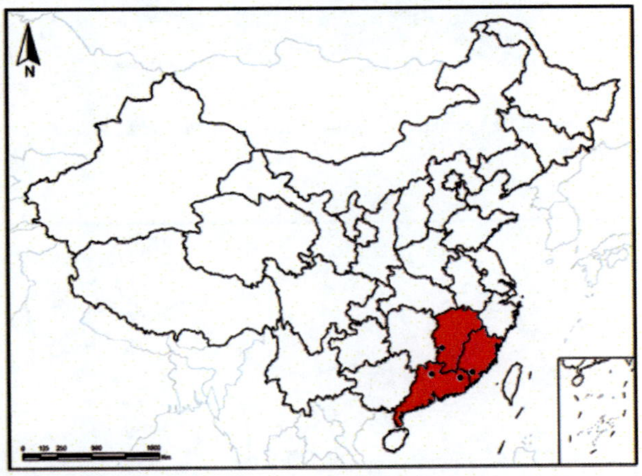

Fig. 9.12 Distribution of *Bicaubittacus longiprocessus*

9.1.4 *Bicaubittacus mengyangicus* Tan & Hua, 2009

(Mengyang Bicaudate Hangingfly) [勐养双尾蚊蝎蛉] (Figs. 9.13, 9.14, and 9.15)

Measurements. Forewing length 16.0–17.5 mm.

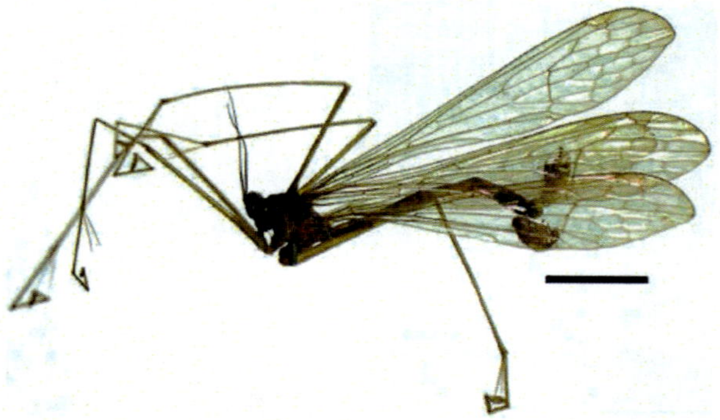

Fig. 9.13 Male adult of *Bicaubittacus mengyangicus*, lateral view (Shaoguan, Guangdong) (*Scale bar*: 5.0 mm) [IZAS]

Fig. 9.14 Male genitalia of *Bicaubittacus mengyangicus*, lateral view (*Scale bar*: 0.5 mm)

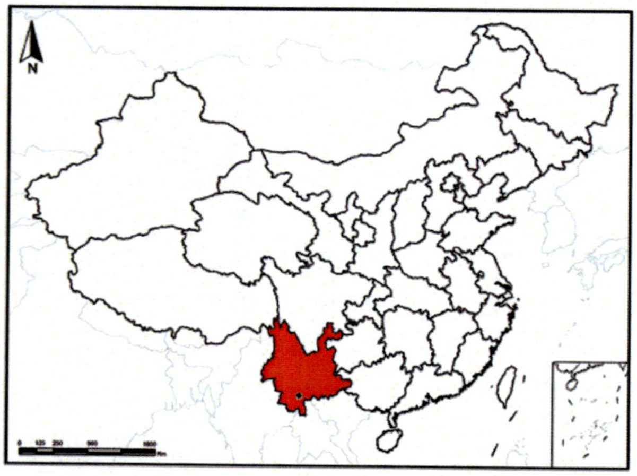

Fig. 9.15 Distribution of *Bicaubittacus mengyangicus*

Morphology. Vertex yellowish brown, ocellar triangle darkened, rostrum blackish brown, antennae yellowish brown. Nota yellowish brown. Wing membrane hyaline, tinged with smoky yellow, and devoid of markings. Legs yellowish brown, with apical end of femur and both ends of tibia darkened. Pterostigma yellowish brown. Male A1–A8 yellowish brown. T8 with V-shaped emargination on posterior margin mesally.

Male epandrial lobes oblong; distal margin forked into short, acute process and long, obtuse prong. Gonocoxites with pair of long and curved distal prolongations bearing row of long sparse bristles along hind margin. Aedeagal lobes slender. Penisfilum coiled into small loop. Cerci clavate. Upper branch of proctiger long and slightly curved cephalad at apex, while lower branch short and curved ventrad.

Emergence. May.

Distribution. Yunnan: Xishuangbanna (Xiaomengyang).

9.1.5 *Bicaubittacus yangi* Tan & Hua, 2009

(Yang's Bicaudate Hangingfly) [杨氏双尾蚊蝎蛉] (Figs. 9.16, 9.17, and 9.18)

Measurements. Forewing length 17.0 mm.

Morphology. Vertex yellowish brown. Ocellar triangle darkened. Antennae blackish brown. Meso- and metanotum dark brown anteriorly and paler near wing bases. Legs yellowish brown, with apical end of femur and both ends of tibia darkened. Wing membrane hyaline, and tinged with smoky yellow. Markings absent. Apex tinged with indistinct light brown. Pterostigma light yellowish brown. Male A1–A8 yellowish brown. T8 with V-shaped emargination on posterior margin mesally.

Male epandrial lobes peculiarly shaped: ventral margin protruding ventrad subbasally and forming finger-like process; truncated apex forming 2–3 short and acute teeth, and tail-like prong slightly bifurcated apically. Apex of gonocox-

Fig. 9.16 Male adult of *Bicaubittacus yangi*, dorsal view (Gunping, Tianlin, Guangxi) (*Scale bar*: 5.0 mm) [ECAU]

ites with pair of long and curved gooseneck-shaped prolongations bearing row of long sparse bristles along hind margin, and divergent with each other. Aedeagal lobes shovel-like. Penisfilum coiled into small loop. Cerci short and clavate. Upper branch of proctiger long and slightly branched at apex in L-shape, while lower branch short and curved ventrad.

Emergence. May.

Distribution. Guangxi (Tianlin).

Fig. 9.17 Male genitalia of *Bicaubittacus yangi*, lateral view (*Scale bar*: 0.5 mm)

Fig. 9.18 Distribution of *Bicaubittacus yangi*

9.2 Genus *Bittacus* Latreille, 1802

(Typical Hangingflies) [蚊蝎蛉属]

Bittacus is the largest genus in Bittacidae. More than 100 species have been described worldwide. Forty-one species have been recorded from China and 26 of them are included in this atlas (Fig. 9.19).

Fig. 9.19 Adult of *Bittacus*. Male adult of *Bittacus strigatus* in the field (Sanzuo'an, Fangxian, Hubei)

Key to species of *Bittacus* in this atlas
(26 species, males)

1. Wings devoid of markings....2
– Wings with distinct markings....3
2. Epandrial lobes rectangular; upper branch of proctiger nearly as long as epandrial lobes and enlarged apically in goose head-shape.... ***Bittacus choui*** Hua & Tan
– Epandrial lobes short quadrilateral; upper branch of proctiger shorter than epandrial lobes and not enlarged apically.... ***Bittacus puripennis*** Cai & Hua
3. Gonocoxites each with distal prolongation....4
– Gonocoxites lacking distal prolongation....5
4. Epandrial lobes with two prolongations along ventral margin; distal prolongations of gonocoxites finger-like.... ***Bittacus hainanicus*** Tan & Hua
– Epandrial lobes lacking prolongations along ventral margin; distal prolongations of gonocoxites conical.... ***Bittacus ussuriensis*** Plutenko
5. Epandrial lobes simple in structure, lacking distal prolongations....6
– Epandrial lobes varied in shape, complicated in structure, branched or with prolongations....15
6. Antennae almost as long as body.... ***Bittacus longantennatus*** Chen, Tan, & Hua
– Antennae shorter than body....7
7. Epandrial lobes thick curved ventrad in distal half....8
– Epandrial lobes flat and not curved ventrad in distal half....9
8. Epandrial lobes greatly elongated, longer than gonocoxites.... ***Bittacus strigatus*** Hua & Chou
– Epandrial lobes less elongated, shorter than gonocoxites.... ***Bittacus shaoguanensis*** Zhang, Du & Hua
9. Epandrial lobes distinctly longer than gonocoxites....10
– Epandrial lobes shorter or nearly as long as gonocoxites....11
10. Epandrial lobes with tuft of long setae apically; upper branch of proctiger slightly exceeding epandrial lobe.... ***Bittacus pieli*** Navás
– Epandrial lobe with short setae apically; upper branch of proctiger distinctly exceeding epandrial lobe for long distance.... ***Bittacus lii*** Zhou
11. Epandrial lobes long rectangular, three times as long as wide; upper branch of proctiger extremely elongated.... ***Bittacus setigerus*** Chen, Tan, & Hua
– Epandrial lobes short and broad; upper branch of proctiger relatively short....12
12. Aedeagal lobes very broad with hook-like acute projection.... ***Bittacus diaoluoshanus*** Chen & Hua
– Aedeagal lobes narrow and simple....13
13. Body yellowish brown; epandrial lobes rounded at apex.... ***Bittacus planus*** Cheng
– Body dark brown to black; epandrial lobes subtrapezoidal and truncated at apex....14
14. Epandrial lobes with distinct longitudinal ridge along midline.... ***Bittacus trapezoideus*** Huang & Hua
– Epandrial lobes lacking ridge along midline.... ***Bittacus sinicus*** Issiki
15. Epandrial lobes trilobed, with median lobe greatly elongated and almost twice as long as others.... ***Bitttacus zhejiangicus*** Tan & Hua
– Epandrial lobes single or bilobed....16
16. Each femur bearing pale, circular marking subapically; upper branch of proctiger extremely elongated, longer than gonocoxites, and sickle-shaped in distal portion.... ***Bittacus malaisei*** Tjeder
– Femora lacking pale markings; upper branch of proctiger less modified....17
17. Epandrial lobes subtrapezoidal in basal portion, with ventro-caudal portion elongated into hook-like process....18
– Epandrial lobes not modified as above....19
18. Hook-like process small, shorter than half-length of basal subtrapezoidal portion; aedeagal lobes simple.... ***Bittacus acutus*** Zhang, Du & Hua
– Hook-like process large, longer than half-length of basal subtrapezoidal portion; aedeagal lobes bifurcated.... ***Bittacus bifurcatus*** Hua & Tan
19. Epandrial lobes subtriangular, with dorso-caudal portion greatly elongated and tapering towards acute apex.... ***Bittacus obscurus*** Huang & Hua
– Epandrial lobes not modified as above....20

20. Epandrial lobes with distal portion greatly elongated and finger-like 22
– Epandrial lobes not modified as above 24
21. Distal portion of epandrial lobes finger-like and relatively straight ..... ***Bittacus monastyrskiyi*** Bicha
– Distal portion of epandrial lobes finger-like and curled dorsad in distal portion 23
22. Finger-like processes of epandrial lobes shorter than basal subtrapezoidal portion ..... ***Bittacus flavidus*** Huang & Hua
– Finger-like processes of epandrial lobes longer than basal subtrapezoidal portion ..... ***Bittacus longilobus*** Zhang, Du & Hua
23. Epandrial lobes with basal portion subtriangular, distal portion elongated and finger-like ..... ***Bittacus cirratus*** Tjeder
– Epandrial lobes not modified as above 26
24. Epandrial lobes with longitudinal groove near ventral margin, and emarginated at end of groove; upper branch of proctiger with pair of finger-like processes on dorsal surface subapically ..... ***Bittacus tienmushana*** Cheng
– Epandrial lobes not modified as above 26
25. Epandrial lobes shortly bifurcated into two acute apexes ..... ***Bittacus sinensis*** Walker
– Epandrial lobes unfurcated, with finger-like process at ventro-caudal portion ..... ***Bittacus gressitti*** Cheng

9.2.1 *Bittacus acutus* Zhang, Du & Hua, 2020

(Acute Hangingfly) [尖角蚊蝎蛉] (Figs. 9.20, 9.21, and 9.22)

Measurements. Forewing length 18.0–19.0 mm.

Morphology. Vertex and rostrum yellowish brown. Ocellar triangle dark brown. Antennae blackish brown, with flagellum filiform and ciliated. Pronotum blackish brown and bearing one stout seta on each side at anterior margin. Meso- and metanotum yellowish brown. Pleura and legs light yellowish brown, with apical end of femur and both ends of tibia darkened. Wing membrane hyaline, yellowish brown. Pterostigma brown. Four brown markings, one each at ORs, FRs, OM, and subapical portion of CuP. Male A1–A8 yellowish brown.

Male epandrial lobes yellowish brown, subtrapezoidal with rounded dorsal margin, and an acute, hook-shaped process at ventro-apical portion. Gonocoxites with V-shaped membranous area posteriorly. Gonostylus short, blunt distally with process on inner side. Penisfilum slender and greatly coiled into loop; aedeagal lobe broad.

Emergence. July.

Distribution. Guangxi (Kitten Mountain).

Fig. 9.21 Male genitalia of *Bittacus acutus*, lateral view (*Scale bars*: 0.5 mm)

Fig. 9.20 Male adult of *Bittacus acutus*, lateral view (Kitten Mountain, Guangxi) (*Scale bar*: 5.0 mm) [NWAU]

Fig. 9.22 Distribution of *Bittacus acutus*

9.2.2 *Bittacus bifurcatus* Hua & Tan, 2008

(Bifurcate Hangingfly) [双叉蚊蝎蛉] (Figs. 9.23, 9.24, and 9.25)

Measurements. Forewing length 16.0 mm.

Morphology. Vertex and rostrum yellowish brown. Ocellar triangle dark brown and bearing two bristles behind middle ocellus. Antennae blackish brown, with flagellum filiform and ciliated. Pronotum blackish brown and bearing one stout seta on each side at anterior margin. Meso- and metanotum dark brown with yellowish brown scutella. Pleura and legs light yellowish brown, with apical end of femur and both ends of tibia darkened. Wing membrane hyaline. Deepened smoky markings along FRs, ORs, and end of CuP in forewings. Pterostigma distinct and brown. Male A1–A8 yellowish brown to blackish brown.

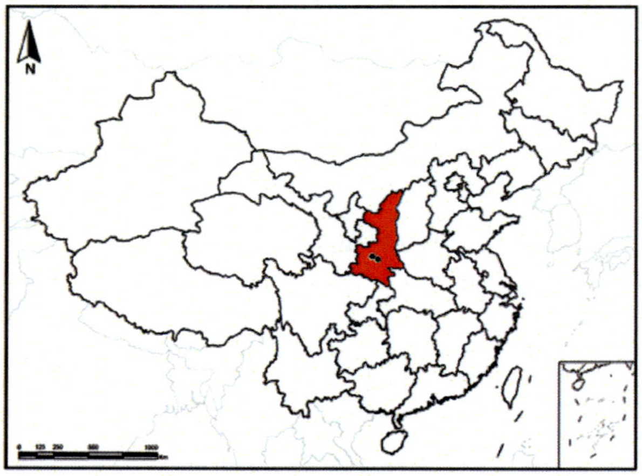

Fig. 9.25 Distribution of *Bittacus bifurcatus*

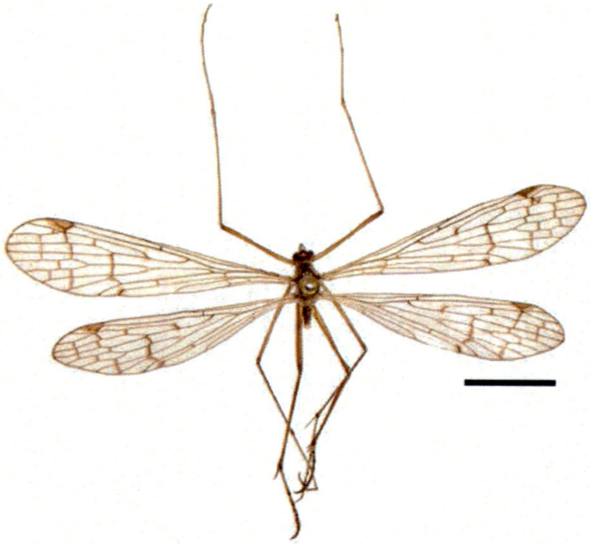

Fig. 9.23 Male adult of *Bittacus bifurcatus* with abdomen removed, dorsal view (Zhuque Forest Park, Huyi, Shaanxi) (*Scale bar*: 5.0 mm) [NWAU]

Fig. 9.24 Male genitalia of *Bittacus bifurcatus*, lateral view (*Scale bar*: 0.5 mm)

Male epandrial lobes yellowish brown; half of dorsal margin protruding dorsad; ventral margin elongated and forming hook-like tail at apex. Gonocoxites with V-shaped membranous area posteriorly. Gonostylus slender and L-shaped, and bearing small process on inner side. Aedeagal lobe bifurcated into broad branch and another slender branch. Penisfilum coiled into small loop. Upper branch of proctiger long and protruding between epandrial lobes, while lower branch short and hook-shaped.

Emergence. July to August.

Distribution. Shaanxi (Huyi, Meixian, Taibai, Zhouzhi).

9.2.3 *Bittacus choui* Hua & Tan, 2007

(Chou's Hangingfly) [周氏蚊蝎蛉] (Figs. 9.26, 9.27, 9.28, and 9.29)

Measurements. Forewing length 15.0–16.0 mm.

Morphology. Vertex and rostrum light yellowish brown. Ocellar triangle black. Dark ocellar stripe extending to the inner margin of compound eyes laterally. Pronotum yellowish brown. Meso- and metanotum unevenly yellowish brown. Pleura and legs light yellowish brown, with apical end of femur and both ends of tibia darkened. Wing membrane hyaline and slightly tinged with light yellow. Markings absent. Male A1–A8 dark brown. T8 with deep V-shaped emargination on middle of posterior margin.

Male epandrial lobes long rectangular, light yellowish brown and translucent, contrast to dark brown gonocoxites. Gonocoxites hemispherical, bearing membranous V-shaped area at apex. Gonostylus short, broad basally, and rounded apically. Aedeagal lobes rounded. Penisfilum coiled into loop. Upper branch of proctiger slender, enlarged gooseheadshaped and bearing tuft of dense setae apically, while lower branch long hook-shaped and curved ventrad.

Emergence. July to August.

Distribution. Chongqing (Nanchuan), Hubei (Shennongjia), Shaanxi (Nanzheng), Sichuan (Emeishan).

Fig. 9.26 Male adult of *Bitttacus choui*, lateral view (Liping National Forest Park, Nanzheng, Shaanxi) (*Scale bar*: 5.0 mm) [NWAU]

Fig. 9.27 Female adult of *Bittacus choui* in the field (Mount Jinfo, Nanchuan, Chongqing)

Fig. 9.28 Male genitalia of *Bittacus choui*, lateral view (*Scale bars*: 0.5 mm)

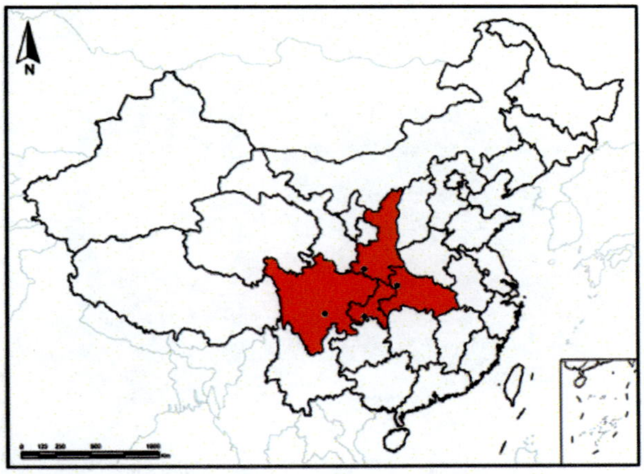

Fig. 9.29 Distribution of *Bittacus choui*

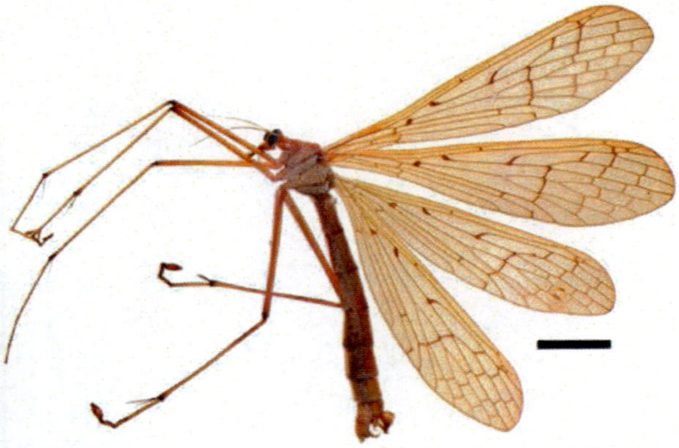

Fig. 9.30 Male adult of *Bittacus cirratus*, lateral view (Tongliao, Inner Mongolia, NWAU) (*Scale bar*: 5.0 mm) [NWAU]

9.2.4 *Bittacus cirratus* Tjeder, 1956

(Cirrated Hangingfly) [环带蚊蝎蛉] (Figs. 9.30, 9.31, and 9.32)

Measurements. Forewing length 24.0–26.0 mm.

Morphology. Vertex and rostrum yellowish brown, with apex of rostrum slightly darkened. Antennae yellowish brown, with flagellum ciliated. Ocellar triangle dark brown. Pronotum yellowish light brown. Meso- and metanotum unevenly yellowish brown with pale scutella. Pleura and legs yellowish brown, with apical end of femur and both ends of tibia darkened. Wing membrane hyaline and slightly tinged with light yellow. Four brown spots on OM, ORs, FRs, and near end of CuP. All crossveins with surrounding blurry stripes. Male A1–A8 yellowish brown. T8 with deep V-shaped emargination on posterior margin.

Male epandrial lobes subtriangular in lateral view, protruding and forming finger-like prong apically. Dense black spines on the inner surface. Gonocoxites yellowish brown, protruding backward apically. Gonostylus short and bearing smooth process on inner side. Aedeagal lobes long and acute apically.

Fig. 9.31 Male genitalia of *Bittacus cirratus*, lateral view (*Scale bar*: 0.5 mm)

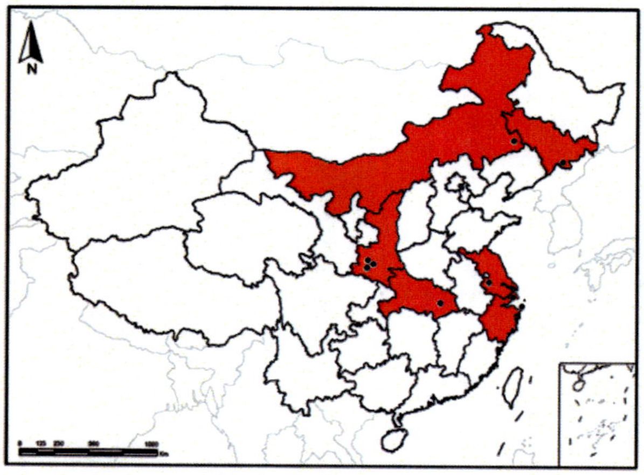

Fig. 9.32 Distribution of *Bittacus cirratus*

Penisfilum coiled into loop. Cerci slender and approximately half-length of gonocoxites. Upper branch of proctiger broad with lateral lobe on each side, and bearing row of yellow setae on each lobe marginally; lower branch shorter and narrower.

Emergence. July to August.

Distribution. Hubei (Wuchang), Inner Mongolia (Tongliao), Jiangsu (Baohua Mountain, Laoshan Mountain), Jilin (Linjiang), Shaanxi (Foping, Meixian, Zhouzhi), Shanghai (Sheshan), Zhejiang (Zhoushan).

9.2.5 *Bittacus diaoluoshanus* Chen & Hua, 2011

(Diaoluoshan Hangingfly) [吊罗山蚊蝎蛉] (Figs. 9.33, 9.34, and 9.35)

Measurements. Forewing length 19.7–20.2 mm.

Morphology. Vertex and rostrum yellowish brown. Ocellar triangle black. Antennae brown and ciliated. Pronotum yellowish brown. Meso- and metanotum dark brown with pale median stripe. Pleura and legs yellowish brown, with apical end of femur and both ends of tibia darkened. Wing

Fig. 9.33 Male adult of *Bittacus diaoluoshanus*, lateral view (Diaoluo Mountain, Hainan) (*Scale bar*: 5.0 mm) [NWAU]

Fig. 9.34 Male genitalia of *Bittacus diaoluoshanus*, lateral view (*Scale bar*: 0.5 mm)

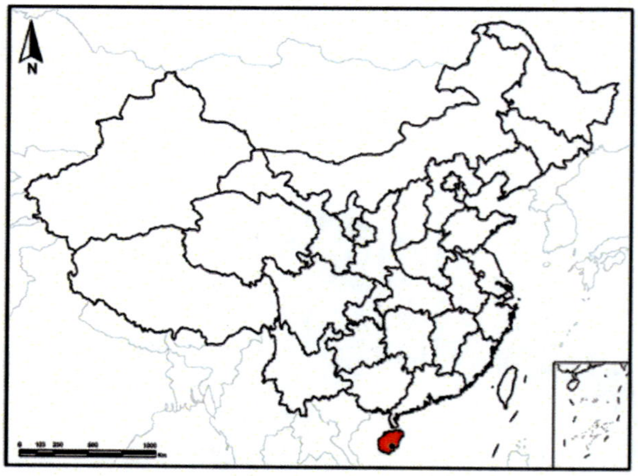

Fig. 9.35 Distribution of *Bittacus diaoluoshanus*

membrane hyaline and slightly tinged with yellow. Several brown spots on OM, ORs, FRs, respectively. Pterostigma brown and distinct. Brown stripe extending from pterostigma to wing apex. All crossveins with surrounding blurry stripes. Male A1–A8 yellowish brown. A8 with deep V-shaped emargination on middle of posterior margin.

Male epandrial lobes subtrapezoidal, and truncated at apex. Gonocoxites with V-shaped membranous area on apex. Gonostylus short and bearing small process on inner side. Aedeagal lobes broad. Penisfilum coiled into loop. Cerci short and clavate. Upper branch of proctiger acute and bearing tuft of setae apically, and stretching out between paired epandrial lobes. Lower branch short and curved ventrad.

Emergence. May.

Distribution. Hainan (Diaoluo Mountain).

Fig. 9.37 Male genitalia of *Bittacus flavidus*, lateral view (*Scale bar*: 0.5 mm)

9.2.6 *Bittacus flavidus* Huang & Hua, 2005

(Yellow Hangingfly) [淡黄蚊蝎蛉] (Figs. 9.36, 9.37, and 9.38)

Measurements. Forewing length 22.0–25.0 mm.

Morphology. Vertex unevenly yellowish brown and darkened laterad. Rostrum yellowish brown. Ocellar triangle dark brown. Antennae brown and ciliated. Pronotum with two long black setae on each side of anterior margin, and one seta on each side of posterior margin. Meso- and metanotum yellowish brown with pale median stripe almost as wide as scutella. Pleura and legs yellowish brown, with apical end of femur and both ends of tibia darkened. Wing membrane hyaline, slightly tinged with yellow. Pterostigma brown and distinct. Several brown spots on OM, ORs, and FRs, respectively. All crossveins with surrounding blurry stripes. Terminal portion of R_5 with large smoky stripe. Male A1–A8 yellowish brown and gradually darkening caudad. T8 with deep V-shaped emargination on posterior margin.

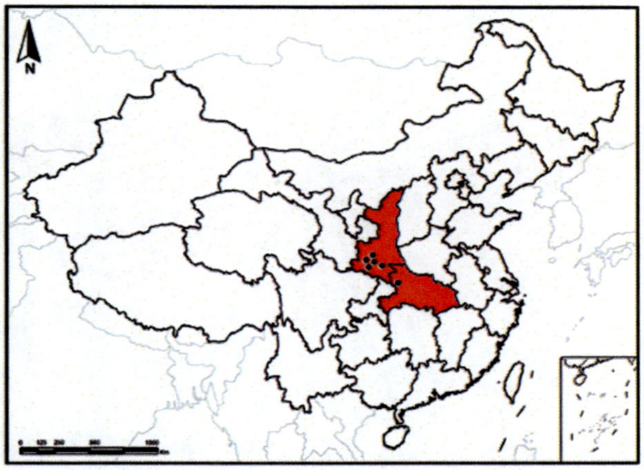

Fig. 9.38 Distribution of *Bittacus flavidus*

Fig. 9.36 Male adult of *Bittacus flavidus*, lateral view (Liping National Forest Park, Nanzheng, Shaanxi, NWAU) (*Scale bar*: 5.0 mm)

Male epandrial lobes subtrapezoidal, forming hook-like process on apex, and bearing longitudinal ridge above ventral margin. Gonocoxites with V-shaped emargination on posterior margin. Aedeagal lobes broad basally, acute on apex. Penisfilum short and coiled into loop. Cerci long and extending beyond gonocoxites apically, with acute apexes. Upper branch of proctiger slender and bearing tuft of setae apically, and stretching out between paired epandrial lobes. Lower branch short and curved ventrad.

Emergence. July to August.

Distribution. Hubei (Shennongjia), Shaanxi (Ningshan, Nanzheng, Taibai, Meixian, Zhouzhi).

9.2.7 *Bittacus gressitti* Cheng, 1957

(Gressitt's Hangingfly) [嘉理斯蚊蝎蛉] (Figs. 9.39, 9.40, and 9.41)

Measurements. Forewing length 23.0–25.0 mm.

Morphology. Vertex unevenly yellowish brown. Rostrum yellowish brown. Ocellar triangle dark brown. Antennae brown and ciliated. Pronotum with two long black setae on each side of anterior margin, and one seta on each side of posterior margin. Meso- and metanotum yellowish brown with indistinct pale median stripe almost as wide as scutella. Pleura and legs yellowish brown, with apical end of femur and both ends of tibia darkened. Wing membrane hyaline, slightly tinged with yellow. Pterostigma brown and distinct. Several brown spots on OM, ORs, and FRs, respectively. All crossveins with surrounding blurry stripes. Male A1–A8 yellowish brown and gradually darkening caudad. T8 with deep V-shaped emargination on posterior margin.

Male epandrial lobes subtriangular, forming finger-like process on apex. Gonocoxites with V-shaped emargination on posterior margin. Aedeagal lobes broad basally and acute apically. Penisfilum short and coiled into loop. Cerci long and slightly clavate. Upper branch of proctiger slender and bearing tuft of setae apically. Lower branch short and curved ventrad.

Emergence. July to August.

Distribution. Guangdong (Chebaling Mountains, Yinna Mountain).

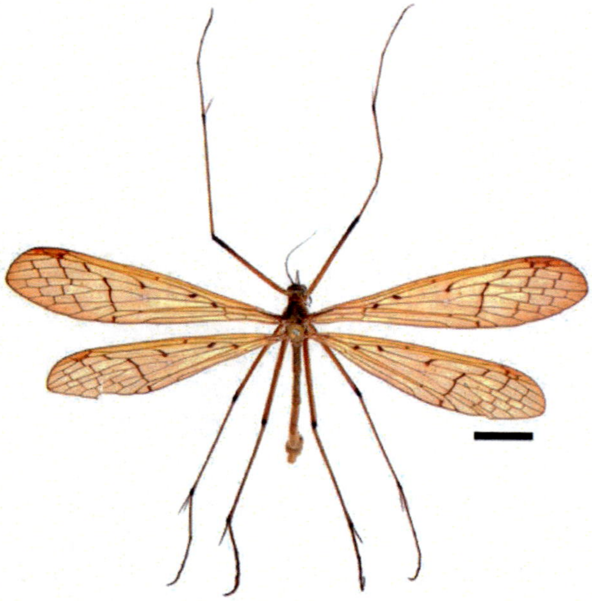

Fig. 9.39 Male adult of *Bittacus gressitti*, dorsal view (Chebaling, Guangdong) (*Scale bar*: 5.0 mm) [DALU]

Fig. 9.40 Male genitalia of *Bittacus gressitti*, lateral view (*Scale bar*: 0.5 mm)

Fig. 9.41 Distribution of *Bittacus gressitti*

9.2.8 *Bittacus hainanicus* Tan & Hua, 2008

(Hainan Hangingfly) [海南蚊蝎蛉] (Figs. 9.42, 9.43, and 9.44)

Measurements. Forewing length 18.0–19.0 mm.

Morphology. Vertex and frons amber. Rostrum yellowish brown. Ocellar triangle black. Antennae brown and ciliated. Pronotum blackish brown, without distinct setae on anterior and posterior margins. Meso- and metanotum yellowish brown with pale median stripe. Pleura unevenly brown. Legs yellowish brown, with apical end of femur and both ends of tibia darkened. Wing membrane hyaline, slightly tinged with yellow. Pterostigma brown and distinct. Several brown spots on OM, ORs, and FRs, etc. Brown stripe extending along R_5. All crossveins with surrounding blurry stripes. Male abdomen yellowish brown. T7 and T8 each with deep V-shaped emargination on posterior margin.

Fig. 9.42 Female adult of *Bittacus hainanicus*, lateral view (Jianfengling, Ledong, Hainan, NWAU) (*Scale bar*: 5.0 mm) [NWAU]

Fig. 9.43 Male genitalia of *Bittacus hainanicus*, lateral view (*Scale bar*: 0.5 mm)

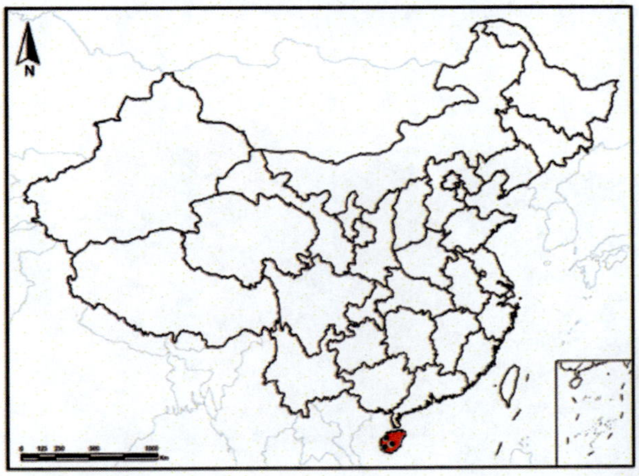

Fig. 9.44 Distribution of *Bittacus hainanicus*

Male epandrial lobes peculiarly shaped: ventral margin curled inward and bearing two small processes mesially; dorsal margin with V-shaped cleft at middle; triangular process next to cleft; apex extending caudad and forming long and truncated prolongation. Gonocoxites each with an acute process apically, which connected to aedeagus by translucent and tongue-shaped plate. Gonostylus subconical, and bearing three long bristles on apex. Aedeagal lobe broad and protruding caudad to form arm-shaped process. Penisfilum coiled into small loop. Cerci short and clavate. Upper branch of proctiger slender and bearing tuft of setae apically. Lower branch of proctiger short and curved ventrad.

Emergence. May.

Distribution. Hainan (Changjiang, Ledong, Wuzhishan).

Remarks. The male gonocoxites of *Bittacus hainanicus* bear finger-like distal prolongations, suggesting a possible homology to those greatly elongated distal prolongations in *Bicaubittacus* spp.

9.2.9 *Bittacus lii* Zhou, 2003

(Li's Hangingfly) [李氏蚊蝎蛉] (Figs. 9.45, 9.46, and 9.47)

Measurements. Forewing length 19.0–21.0 mm.

Fig. 9.45 Male adult of *Bittacus lii*, lateral view (Badagong Mountains, Sangzhi, Hunan, NWAU) (*Scale bar*: 5.0 mm) [NWAU]

Fig. 9.46 Male genitalia of *Bittacus lii*, lateral view (*Scale bar*: 0.5 mm)

Fig. 9.47 Distribution of *Bittacus lii*

Morphology. Vertex yellowish brown. Ocellar triangle black. Black stripe passing through ocellar triangle and extending to compound eyes laterally. Rostrum deep brown basally and paler apically. Antennae brown and ciliated. Pronotum blackish brown, without distinct setae on posterior margin. Meso- and metanotum unevenly yellowish brown with pale yellow median stripe. Pleura yellowish brown. Legs yellowish brown with apical end of femur and both ends of tibia darkened. Wing membrane hyaline, slightly tinged with yellowish brown. Pterostigma brown and distinct. Multiple smoky brown spots on joints of veins. Almost all veins with surrounding blurry stripes. Male abdomen yellowish brown. T8 with deep V-shaped emargination on middle of posterior margin.

Male epandrial lobes broadly fused basally, long subtrapezoidal with rounded apexes, bearing two ridge-like prominences on inner surface, and slightly longer than gonocoxites. Gonocoxites semispherical, with V-shaped membranous area on apex; gonostylus very short. Aedeagal lobes small with rounded apexes. Upper branch of proctiger long and extending between epandrial lobes, with tuft of setae on apex; lower branch short and curved ventrad.

Emergence. July to August.

Distribution. Guangxi (Kitten Mountain), Guizhou (Jiangkou), Hubei (Wufeng), Hunan (Badagong Mountains).

Fig. 9.48 Male adult of *Bittacus longantennatus* in the field (Jiajin Mountain, Sichuan)

Fig. 9.49 Male genitalia of *Bittacus longantennatus*, lateral view (*Scale bar*: 0.5 mm)

9.2.10 *Bittacus longantennatus* Chen, Tan & Hua, 2013

(Long-Antennae Hangingfly) [长角蚊蝎蛉] (Figs. 9.48, 9.49, and 9.50)

Measurements. Forewing length 19.0–21.0 mm.

Morphology. Vertex deep brown. Ocellar triangle dark brown. Black stripe passing through ocellar triangle and extending to compound eyes laterally. Antennae longer than femora. Pronotum blackish brown, without distinct setae on anterior and posterior margins. Meso- and metanotum unevenly blackish brown, paler at scutella and median stripe of metanotum, but devoid of pale median stripe on

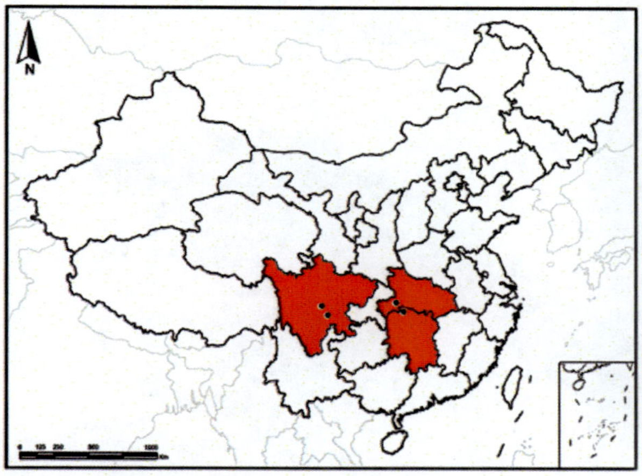

Fig. 9.50 Distribution of *Bittacus longantennatus*

Fig. 9.51 Male adult of *Bittacus longilobus* in the field (Yinna Mountain, Meizhou, Guangdong)

mesonotum. Pleura yellowish brown. Legs yellowish brown with apical end of femur and both ends of tibia darkened. Wing membrane intensively tinged with smoky brown markings. Pterostigma brown and distinct. Male A1–A8 blackish brown. T8 with shallow emargination on posterior margin.

Male epandrial lobes subtrapezoidal in lateral view and slightly shorter than gonocoxites, and bearing dense black spines on the inner surface of each truncated apex. Gonocoxites semispherical, with V-shaped membranous area on apex; gonostylus very short, emarginated apically, and bearing small process on inner surface. Aedeagal lobes short with rounded apexes. Upper branch of proctiger stretching out between epandrial lobes, with tuft of long setae on apex; lower branch short and curved ventrad.

Emergence. July to August.

Distribution. Sichuan (Mount Emei, Jiajin Mountain, Tianquan), Hubei (Badong), Hunan (Shimen).

Fig. 9.52 Male genitalia of *Bittacus longilobus*, lateral view (*Scale bar*: 0.5 mm)

9.2.11 *Bittacus longilobus* Zhang, Du & Hua, 2020

(Long-lobe Hangingfly) [长瓣蚊蝎蛉] (Figs. 9.51, 9.52, and 9.53)

Measurements. Forewing length 19.0–22.0 mm.

Morphology. Vertex deep brown. Ocellar triangle dark brown. Antennae blackish brown, with flagellum filiform and ciliated. Pronotum blackish brown and bearing one stout seta on each side at anterior margin. Meso- and metanotum dark brown with yellowish brown scutella. Pleura and legs light yellowish brown, with apical end of femur and both ends of tibia darkened. Wing membrane hyaline, deep yellowish brown. Pterostigma brown. Six markings, one each at OM, ORs, Scv, FRs, and ends of CuA and CuP; crossveins and apical margin shaded with dark brown. Male A1–A8 yellowish brown.

Male epandrial lobes yellowish brown, subtrapezoidal in basal half; dorso-distal portion slightly elongated into short finger-like process; ventro-distal portion greatly elongated into long finger-like process nearly as long as basal half. Cerci greatly elongated, slender. Gonocoxites with V-shaped membranous area posteriorly. Gonostylus short. Penisfilum slender and greatly coiled into loop; aedeagal lobe short. Upper branch of proctiger greatly elongated and bearing tuft of long setae apically.

Emergence. May.

Distribution. Guangdong (Yinna Mountain).

Fig. 9.53 Distribution of *Bittacus longilobus*

Fig. 9.55 Male genitalia of *Bittacus malaisei*, lateral view (*Scale bar*: 0.5 mm)

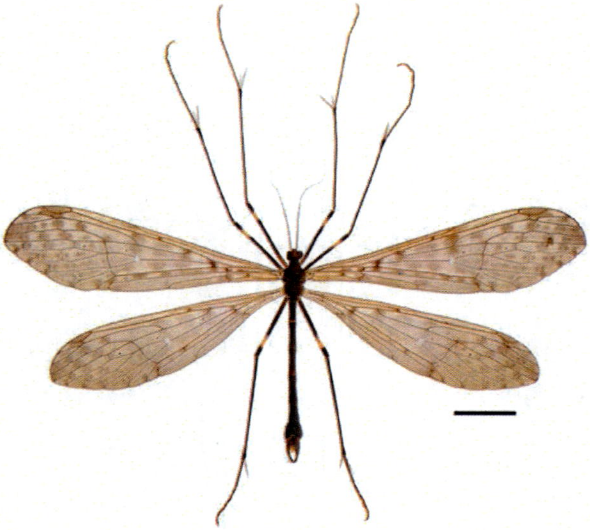

Fig. 9.54 Male adult of *Bittacus malaisei*, dorsal view (Pianma, Yunnan) (*Scale bar*: 5.0 mm) [DALU]

Fig. 9.56 Distribution of *Bittacus malaisei*

9.2.12 *Bittacus malaisei* Tjeder, 1973

(Malaise's Hangingfly) [马来氏蚊蝎蛉] (Figs. 9.54, 9.55, and 9.56)

Measurements. Forewing length 23.0 mm.

Morphology. Vertex yellowish brown. Pronotum dark brown, with two long setae along anterior and posterior margin, respectively. Meso- and metanotum dark brown. Pleura pale brown. Legs mostly dark brown, with pale subapical ring at each femur. Wing membrane subhyaline, tinged with grayish brown; pterostigma brown; markings deep brown and scattered into numerous spots; nygmata circled by dark brown; thyridium whitish and conspicuous. Terga and sterna mostly dark brown and covered with short sparse setae. T8 greatly emarginated in V-shape at posterior margin.

Male epandrial lobes greatly elongated, subtrapezoidal in basal half and finger-like in distal half, with finger-like accessory lobe at ventral side of the inner surface, and bearing numerous long stout setae at apex and along distal half of ventral margin. Gonostylus small and covered with numerous thin setae. Aedeagus greatly elongated and stout basally, with pair of small subtriangular aedeagal lobes, and greatly elongated penisfilum straight basally and greatly coiled into loop distally. Upper branch of proctiger greatly elongated and sclerotized, with distal 1/3 greatly enlarged, tapering towards pointed apex, bent anteriorly, and furnished with numerous thin microtrichia. Lower branch of proctiger approximately 2/3 as long as upper branch, slender, slightly arcuate.

Emergence. June, September.

Distribution. Yunnan (Lushui); Myanmar (Kambaiti).

9.2.13 *Bittacus monastyrskiyi* Bicha, 2007

(Monastyrskiy's Hangingly) [莫氏蚊蝎蛉] (Figs. 9.57, and 9.58)

Measurements. Forewing length 24.8 mm.

Morphology. Vertex deep brown. Ocellar triangle dark brown. Black stripe passing through ocellar triangle and extending to compound eyes laterally. Rostrum yellowish brown. Pronotum unevenly dark brown. Meso- and metanotum unevenly blackish brown without median stripe but only paler at middle. Pleura yellowish brown. Legs yellowish brown with apical end of femur and both ends of tibia darkened. Wing membrane translucent, tinged with yellowish brown, with dense brown markings. Pterostigma deep brown and slender. Apical margin abruptly turning backwards in right angle. Pterostigma deep brown and slender. Darkened stripe extending from pterostigma to end of R_5. Several distinct spots on Av, OM, ORs, and FRs. Almost all crossveins surrounded by smoky and blurred stripes. Male terga brown medially, yellowish brown laterally, and darker on several distal segments. T8 with deep V-shaped emargination on posterior margin.

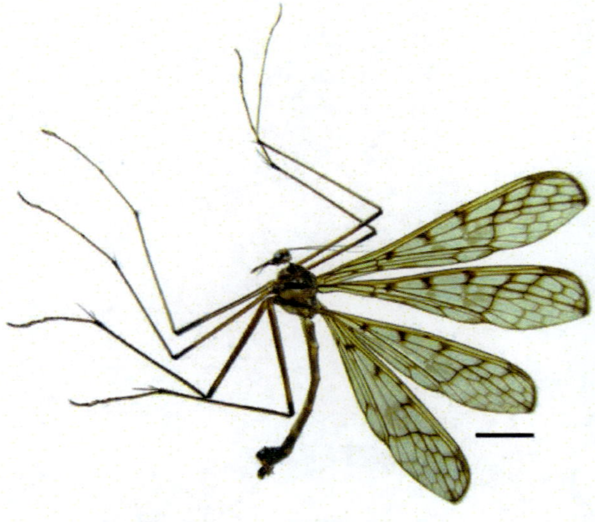

Fig. 9.57 Male adult of *Bittacus monastyrskiyi*, lateral view (Laohutiao, Napo, Guangxi) (*Scale bar*: 5.0 mm) [NWAU]

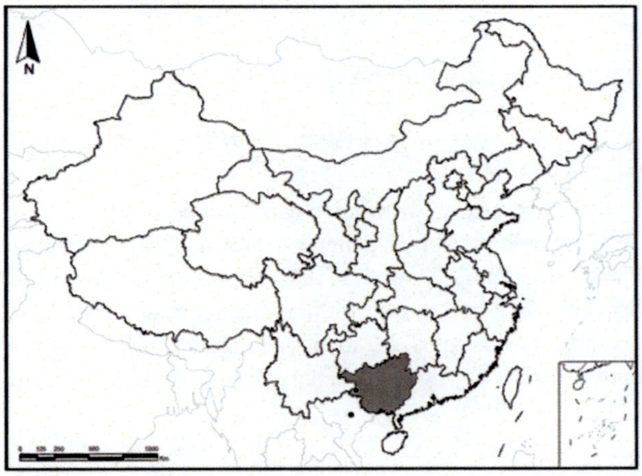

Fig. 9.58 Distribution of *Bittacus monastyrskiyi*

Male epandrial lobes long and splits at dorsal margin medially into two lobes, divided into short dorsal and long ventral branch. Two small acute protuberances under base of the upper branch, and on the inner surface of ventral margin of epandrial lobe, respectively. Gonocoxites with V-shaped membranous area on apex. Gonostylus very small and curved mesad. Aedeagal lobes semicircular. Penisfilum very long and coiled into small loop. Cerci nearly half-length of gonocoxites. Upper branch of proctiger slender and protruding between epandrial lobes, and bearing stout spine apically. Lower branch short and curved ventrad.

Emergence. July.

Distribution. Guangxi (Napo); Vietnam.

9.2.14 *Bittacus obscurus* Huang & Hua, 2005

(Obscure Hangingfly) [暗蚊蝎蛉] (Figs. 9.59, 9.60, and 9.61)

Measurements. Forewing length 19.0–20.0 mm.

Morphology. Vertex and frons yellowish brown. Ocellar triangle black. Black stripe passing through ocellar triangle and extending to compound eyes laterally. Antennae yellowish brown and ciliated. Pronotum unevenly yellowish brown, bearing two long black setae on anterior and posterior margins. Mesonotum and metanotum dark brown with two sides darkened. Pleura and coxae dark brown to light yellow. Apical end of femur and both ends of tibia dark brown. Wing membrane hyaline and tinged with light brown, and darkened at apex. Some blurred smoky stripes along apical crossveins. Male abdomen A1–A6 yellowish brown. A7–A9 dark brown. T8 with deep V-shaped emargination on posterior margin.

Male epandrial lobes axe-shaped in lateral view, with an elongated acute apex dorso-posteriorly, and rounded ventro-posteriorly. Gonocoxites with V-shaped membranous area on apex. Gonostylus extremely reduced. Aedeagal lobes very long and narrow, with small process curving mesad on apex. Penisfilum very long and coiled into loop. Upper branch of proctiger slender and gooseneck-like, with apex curved cephalad and bearing tuft of thin setae. Lower branch very long, curved ventrad basally and straight distally, and tapering towards apex.

Emergence. July to August.

Distribution. Henan (Songxian), Hubei (Shennongjia), Shaanxi (Baoji, Taibai, Zhouzhi, Lan'gao, Meixian, Nanzheng, Ningshan).

Fig. 9.59 Male adult of *Bittacus obscurus* in the field (Tiantaishan National Forest Park, Baoji, Shaanxi, ©Lu Jiang)

Fig. 9.60 Male genitalia of *Bittacus obscurus*, lateral view (*Scale bar*: 0.5 mm)

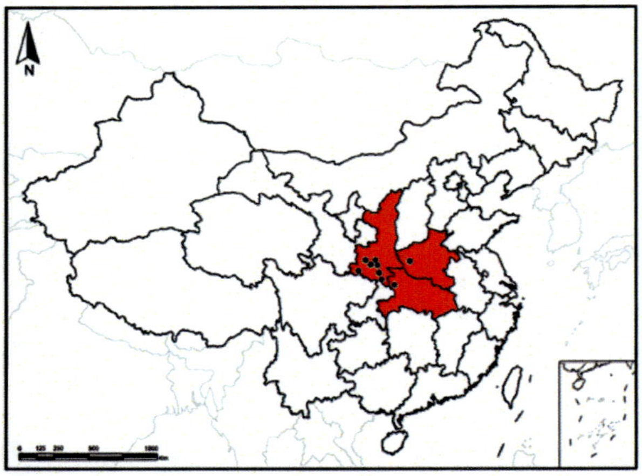

Fig. 9.61 Distribution of *Bittacus obscurus*

Fig. 9.62 Male adult of *Bittacus pieli*, lateral view (Dapan Mountain, Pan'an, Zhejiang, NWAU) (*Scale bar*: 5.0 mm) [NWAU]

9.2.15 *Bittacus pieli* Navás, 1935

(Father Piel's Hangingfly) [璧尔蚊蝎蛉] (Figs. 9.62, 9.63, and 9.64)

Measurements. Forewing length 24.0–26.0 mm.

Morphology. Vertex yellowish brown. Ocellar triangle black. Rostrum brown. Antennae ciliated. Pronotum unevenly dark brown, and without distinct setae on anterior and posterior margins. Mesonota and metanotum yellowish brown with two sides darkened. Pleura and coxae pale

Fig. 9.63 Male genitalia of *Bittacus pieli*, lateral view (*Scale bar*: 0.5 mm)

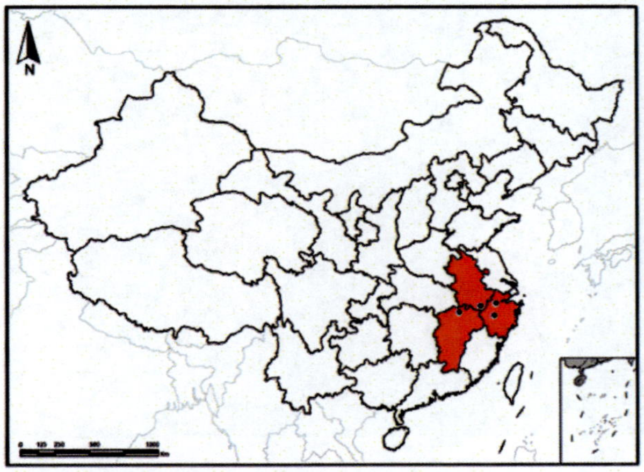

Fig. 9.64 Distribution of *Bittacus pieli*

brown. Legs yellowish brown, with apical end of femur and both ends of tibia dark black. Wing membrane hyaline and tinged with pale yellow. Pterostigma brown and distinct. Several deep brown spots on Av, OM, ORs, and FRs, respectively. Some blurred smoky stripes along apical crossveins. Male abdomen yellowish brown. T4–T6 each with tuft of long setae posteriorly. T7 and T8 each with deep V-shaped emargination on posterior margin.

Male epandrial lobes broadly fused at base, and oblong in lateral view, with dorsal margin curled inward. The rounded apex with tuft of long setae. Gonocoxites rounded and bearing tuft of long setae apically, gonostylus greatly reduced. Aedeagal lobes short and broad. Penisfilum very long and coiled into loop. Cerci short and clavate. Upper branch of proctiger slightly extending beyond epandrial lobes, and bearing tuft of setae apically. Lower branch broad basally and tapering towards acute apex, and curling ventrad.

Emergence. July to August.

Distribution. Anhui (Huangshan), Jiangxi (Lushan), Zhejiang (Lin'an, Pan'an).

9.2.16 *Bittacus planus* Cheng, 1949

(Flat Hangingfly) [扁蚊蝎蛉] (Figs. 9.65, 9.66, and 9.67)

Measurements. Forewing length 19.0–21.0 mm.

Morphology. Vertex yellowish brown, rostrum dark brown. Compound eyes black, ocellar triangle dark brown. Antennae ciliated. Pronotum unevenly yellowish brown, bearing two distinct black setae on anterior margin. Meso- and metanotum dark brown laterally and pale yellow mesally. Pleura and legs yellowish brown. Wing membrane hyaline, tinged with yellowish brown. Pterostigma indistinct. No distinct markings. Male A1–A8 yellowish brown. T7 and T8 with deep V-shaped emargination on posterior margin.

Male epandrial lobes subtrapezoidal in lateral view, shorter than gonocoxites, and bearing dense black spines on apical inner margins. Gonocoxites rounded and bearing V-shaped membranous area posteriorly. Gonostylus greatly reduced. Aedeagal lobes broad and rounded apically. Penisfilum long and coiled into loop. Upper branch of proctiger extending beyond epandrial lobes and bearing tuft of setae on apex. Lower branch short and tapering towards apex.

Emergence. July to August.

Distribution. Beijing (Xishan), Henan (Songxian), Shanxi (Taihang Mountains), Shaanxi (Baoji, Huyi, Ningshan, Taibai).

9.2.17 *Bittacus puripennis* Cai & Hua, 2006

(Clear-winged Hangingfly) [净翅蚊蝎蛉] (Figs. 9.68, 9.69, and 9.70)

Measurements. Forewing length 15.0–16.0 mm.

Morphology. Vertex and rostrum yellowish brown. Ocellar triangle black. Black stripe passing through ocellar triangle and extending to compound eyes laterally. Pronotum unevenly dark brown. Meso- and metanotum dark brown, with pale yellow median stripe. Pleura and legs yellow. Wing membrane hyaline and slightly tinged with yellow. Pterostigma broad and light brown. No distinct markings. Male A1–A6 yellowish brown. T7 and T8 dark brown and each with deep V-shaped emargination on posterior margin.

Male epandrial lobes pale yellow, translucent, subtrapezoidal in lateral view, tapering towards apex, and bearing dense black spines on apical inner margins. Gonocoxites deep brown, semispherical, and bearing V-shaped membranous area posteriorly. Gonostylus greatly reduced, with apex

Fig. 9.65 Adults of *Bittacus planus* in the field (Tiantaishan National Forest Park, Baoji, Shaanxi, ©Lu Jiang) Left, male; right, female.

Fig. 9.66 Male genitalia of *Bittacus planus*, lateral view (*Scale bar*: 0.5 mm)

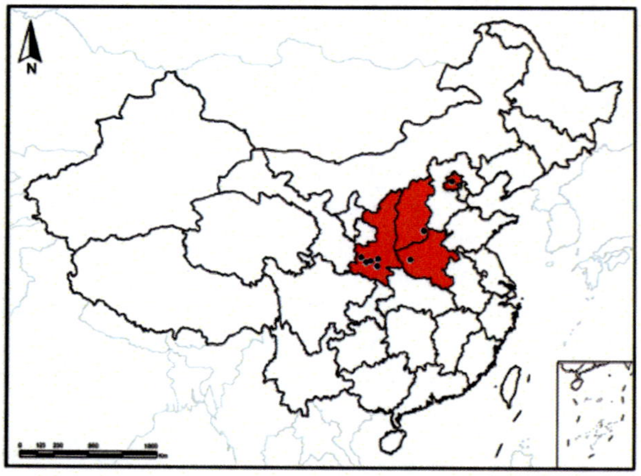

Fig. 9.67 Distribution of *Bittacus planus*

slightly emarginate on middle of posterior margin and V-shaped, and covered by slender setae. Aedeagal lobes broad and rounded apically. Penisfilum short and coiled into loop. T10 saddle-shaped dorsally, and extending laterad at base and enclosing cerci. Cerci extending to 2/3 of epandrial lobes, and rounded at apexes. Upper branch of proctiger extending beyond epandrial lobes and bearing tuft of long setae on apex. Lower branch short and tapering towards apex.

Emergence. July to August.

Distribution. Shaanxi (Nanzheng), Sichuan (Wolong).

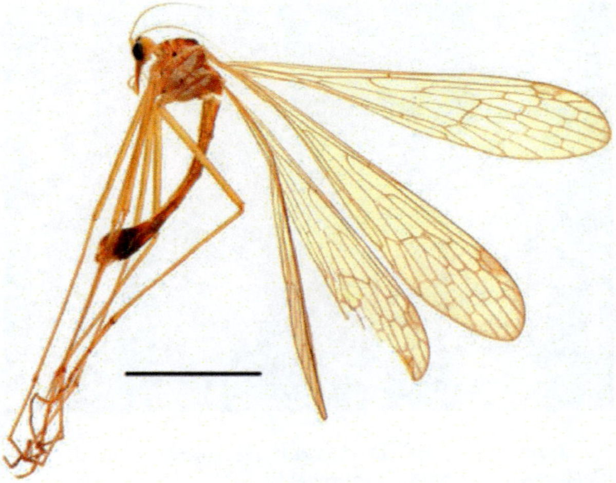

Fig. 9.68 Male adult of *Bittacus puripennis*, lateral view (Liping National Forest Park, Nanzheng, Shaanxi) (*Scale bar*: 5.0 mm) [NWAU]

Fig. 9.69 Male genitalia of *Bittacus puripennis*, lateral view (*Scale bar*: 0.5 mm)

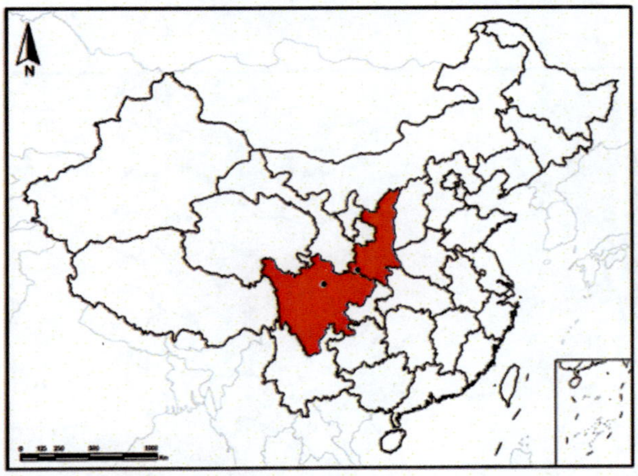

Fig. 9.70 Distribution of *Bittacus puripennis*

9.2.18 *Bittacus setigerus* Chen, Tan & Hua, 2013

(Setae-bearing Hangingfly) [具刺蚊蝎蛉] (Figs. 9.71, 9.72, and 9.73)

Measurements. Forewing length 21.5–22.0 mm.

Morphology. Vertex yellowish brown. Ocellar triangle black. Rostrum dark brown basally and paler apically.

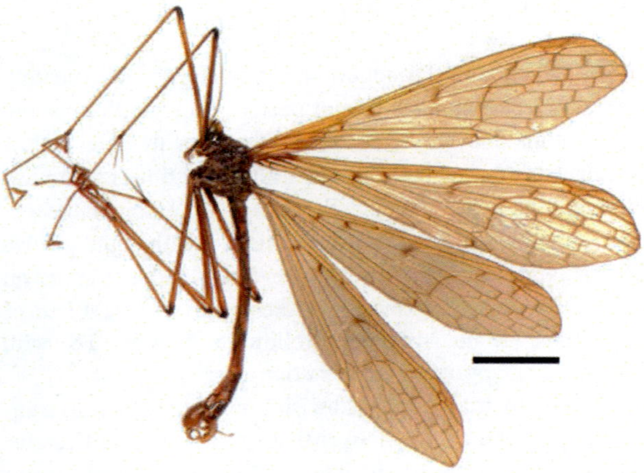

Fig. 9.71 Male adult of *Bittacus setigerus*, lateral view (Fanjing Mountain, Jiangkou, Guizhou) (*Scale bar*: 5.0 mm) [NWAU]

Fig. 9.72 Male genitalia of *Bittacus setigerus*, lateral view (*Scale bar*: 0.5 mm)

Fig. 9.73 Distribution of *Bittacus setigerus*

Fig. 9.74 Male adult of *Bittacus shaoguanensis* in the field (Caojiaowan, Shaoguan, Guangdong)

Antennae ciliated. Pronotum unevenly dark brown, bearing two stout setae on each side of posterior margin. Meso- and metanotum unevenly yellowish brown, with pale yellow median stripe as wide as scutella. Pleura and legs yellowish brown, with apex of femur and both ends of tibia darkened. Wing membrane translucent and tinged with light yellow. Pterostigma broad and brown. Several cloudymarkings on OM, ORs, FRs, CuP and along crossveins in apical half of forewings. Male abdomen yellowish brown. T8 with V-shaped emargination on posterior margin.

Male epandrial lobes yellowish brown, oblong with truncated apex, and slightly longer than gonocoxites, with series of black short setae on the inner surface and ventral margin medially, and row of dark brown long setae near apex. Gonocoxites semispherical, with V-shaped membranous area posteriorly; gonostylus greatly reduced and covered with long setae. Aedeagal lobes truncated apically. Penisfilum long and coiled into loop. Cerci clavate with slightly acute apexes. Upper branch of proctiger slender, with apical portion curved cephalad and bearing tuft of setae on posterior margin of apex. Lower branch short and hook-like.

Emergence. July to August.

Distribution. Guizhou (Jiangkou), Hubei (Shennongjia, Badong).

9.2.19 *Bittacus shaoguanensis* Zhang, Du & Hua, 2020

(Shaoguan Hangingfly) [韶关蚊蝎蛉] (Figs. 9.74, 9.75, and 9.76)

Measurements. Forewing length 24.0–26.0 mm.

Morphology. Vertex dark brown. Ocellar triangle black. Rostrum deep brown. Antennae brown and ciliated. Nota dark brown. Pleura and legs dark yellowish brown, with apex of femur and both ends of tibia darkened. Wing membrane

Fig. 9.75 Male genitalia of *Bittacus shaoguanensis*, lateral view (*Scale bar*: 0.5 mm)

Fig. 9.76 Distribution of *Bittacus shaoguanensis*

hyaline, tinged with dark yellowish brown. Pterostigma dark brown and distinct. Three distinct tawny brown markings, one each at OM, ORs, and Scv; pale brown clouding along apical crossveins, two middle streaks in V-shape; apical margin slightly infuscate. Male A1–A6 yellowish brown. A7 and A8 dark brown.

Male epandrial lobes dark brown, broad, shorter than gonocoxites, with apical halves slightly curved ventrad. Apex of epandrial lobe protruded inward and bearing dense spines. Gonocoxites slightly emarginated apically at middle. Aedeagal lobes broad. Penisfilum slender and coiled into loop. Cerci slender. Upper branch of proctiger slender and gooseneck-like, with apical portion enlarged, curved cephalad, and bearing tuft of setae. Lower branch stout and curved ventrad.

Emergence. April.

Distribution. Guangdong (Shaoguan).

9.2.20 *Bittacus sinensis* Walker, 1853

(Northern Chinese Hangingfly) [华北蚊蝎蛉] (Figs. 9.77, 9.78, and 9.79)

Measurements. Forewing length 22.0–23.0 mm.

Morphology. Vertex and frons amber. Ocellar triangle dark brown. Rostrum deep brown. Antennae thread-like and ciliated. Pronotum yellowish brown, and without distinct setae on anterior and posterior margins. Meso- and metanotum mostly dark brown and with yellowish brown scutella. Pleura and coxae yellowish brown. Apex of femur and both ends of tibia darkened, the rest parts of legs yellowish brown. Wing membrane tinged with light yellow. Apex of both fore- and hindwings rounded. Pterostigma broad and yellowish brown. Several small spots on OM, ORs, FRs, CuP, and along apical crossveins. Male terga and S1–S8 yellowish brown. T8 not emarginate on posterior margin.

Male epandrial lobes subtrapezoidal, and divided posteriorly into short upper branch and long lower branch. Both branches with dense black setae on inner margin apically. Gonocoxites rounded at middle of hind margin and bearing dense setae. Aedeagal lobes broad and long, and acute apically. Penisfilum slender and coiled into small loop. Cerci 2/3 as long as epandrial lobes, and acute apically. Upper branch of proctiger long, extending beyond epandrial lobes, and bearing tuft of setae on apex. Lower branch curved ventrad with acute apex.

Emergence. July to August.

Distribution. Liaoning (Qingyuan, Shenyang), Shaanxi (Huangling); Korea (Gyeonggi-do); Russia (Primorsky Krai).

Fig. 9.77 Female adult of *Bittacus sinensis* in the field (Shenyang Forest Park, Liaoning, ©Lu Jiang)

Fig. 9.78 Male genitalia of *Bittacus sinensis*, lateral view (*Scale bar*: 0.5 mm)

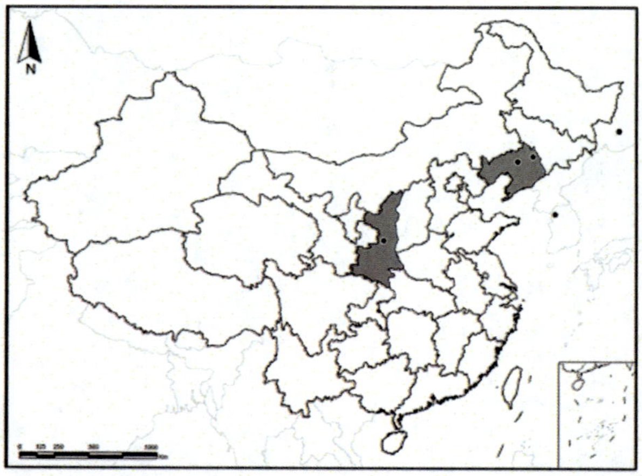

Fig. 9.79 Distribution of *Bittacus sinensis*

Fig. 9.80 Male adult of *Bittacus sinicus*, lateral view (Mount Emei, Emeishan, Sichuan) (*Scale bar*: 5.0 mm) [NWAU]

Fig. 9.81 Female adult of *Bittacus sinicus* in the field (Mount Jinfo, Chongqing)

9.2.21 *Bittacus sinicus* Issiki, 1931

(Western Chinese Hangingfly) [华西蚊蝎蛉] (Figs. 9.80, 9.81, 9.82, and 9.83)

Measurements. Forewing length 17.5–19.3 mm.

Morphology. Vertex brown. Ocellar triangle black. Transverse black band passing through ocellar triangle and extending to compound eyes laterally. Rostrum dark brown. Antennae blackish brown and ciliated. Pronotum yellowish brown. Meso- and metanotum mostly dark brown and with pale yellow median stripe. Episterna of mesothorax and metathorax dark brown, while epimera pale white. Legs unevenly yellowish brown, with apex of femur and both ends of tibia darkened. Wing membrane hyaline and tinged with light yellowish brown. Pterostigma brown and distinct. Dense smoky brown stripes along veins. Male T2–T6 dark brown medially and yellowish brown posteriorly, forming distinct light and dark patterns. T7 and T8 dark brown and slightly emarginate on posterior margin.

Male epandrial lobes light brown and translucent, subtrapezoidal, and with rounded apex. Gonocoxites with hind margin slightly protruded dorsad, and paler at middle. Gonostylus small and bearing distinct process near middle portion. Aedeagal lobes short and narrow, and rounded apically. Penisfilum slender and coiled into small loop. Cerci short and approximately half-length of gonocoxites. Upper branch of proctiger slender, extending beyond epandrial

Fig. 9.82 Male genitalia of *Bittacus sinicus*, lateral view (*Scale bar*: 0.5 mm)

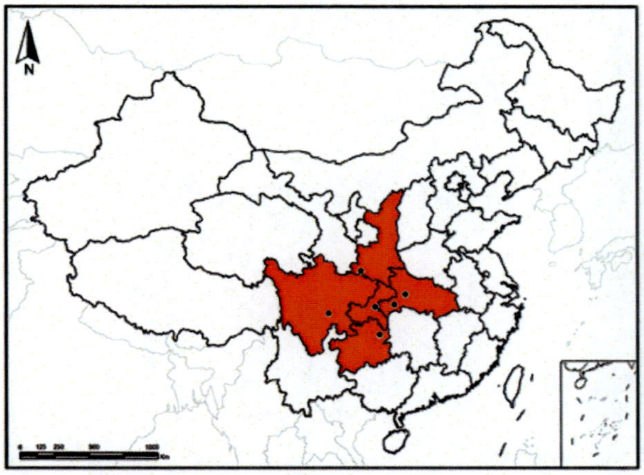

Fig. 9.83 Distribution of *Bittacus sinicus*

lobes, and bearing tuft of setae on apex. Lower branch short and curved ventrad with acute apex.

Emergence. July to August.

Distribution. Chongqing (Shizhu), Hubei (Enshi, Xingshan), Guizhou (Jiangkou), Shaanxi (Nanzheng), Sichuan (Mount Emei).

9.2.22 *Bittacus strigatus* Hua & Chou, 1998

(Strigate Hangingfly) [纹翅蚊蝎蛉] (Figs. 9.84, 9.85, and 9.86)

Measurements. Forewing length 20.0–26.0 mm.

Morphology. Vertex yellowish brown. Ocellar triangle black. Rostrum deep brown. Antennae brown and ciliated. Nota dark brown without pale median stripes. Pleura and legs yellowish brown, with apex of femur and both ends of tibia darkened. Wing membrane hyaline, slightly tinged with light yellow. Pterostigma dark brown and distinct. Dense smoky brown stripes arranged in four transversal lines, with two of them connected posteriorly in V-shape. Male A1–A6 yellowish brown. A7 and A8 dark brown.

Male epandrial lobes dark brown, very broad and longer than gonocoxites, with apical halves curved ventrad. The apex of epandrial lobe protruded inward and bearing dense spines. Gonocoxites slightly emarginated apically at middle. Aedeagal lobes broad. Penisfilum slender and coiled into loop. Cerci slender. Upper branch of proctiger slender and gooseneck-like, with apical portion curved cephalad and bearing tuft of setae. Lower branch stout and curved ventrad.

Emergence. July to August.

Distribution. Henan (Songxian), Hubei (Shennongjia, Fangxian), Shaanxi (Lan'gao, Ningshan, Taibai).

Fig. 9.84 Male adult of *Bittacus strigatus* in the field (Sanzuo'an, Fangxian, Hubei)

Fig. 9.85 Male genitalia of *Bittacus strigatus*, lateral view (*Scale bar*: 0.5 mm)

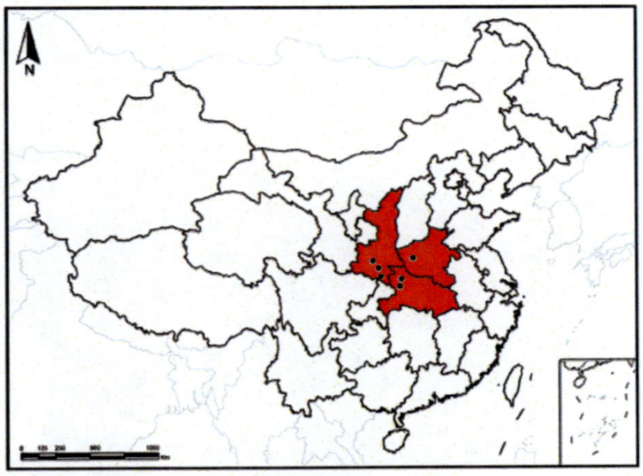

Fig. 9.86 Distribution of *Bittacus strigatus*

9.2.23 *Bittacus tienmushana* Cheng, 1957

(Tianmushan Hangingfly) [天目山蚊蝎蛉] (Figs. 9.87, 9.88, and 9.89)

Measurements. Forewing length 26.0–27.0 mm.

Morphology. Vertex yellowish brown. Ocellar triangle black. Rostrum dark brown. Antennae yellowish brown and ciliated. Two antennal sockets confused mesally and forming an oblong depression. Pronotum unevenly deep brown. Mesonotum mostly dark brown, and paler towards hind portion and metanotum. Pleura and legs yellowish brown, with apex of femur and both ends of tibia darkened. Wing membrane hyaline and slightly tinged with yellow. Pterostigma light brown and indistinct. Several dark brown spots on OM, ORs, FRs, and ending of CuP, respectively. Deepened smoky stripes along crossveins. Male abdomen yellowish brown. T8 with deep V-shaped emargination on posterior margin.

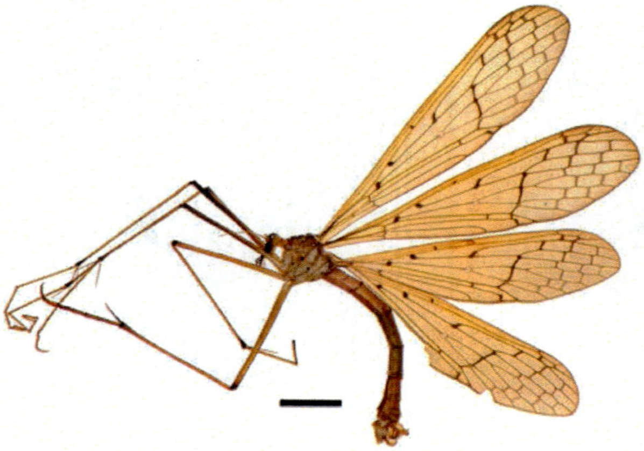

Fig. 9.87 Male adult of *Bittacus tienmushana*, lateral view (West Tianmu Mountain, Lin'an, Zhejiang) (*Scale bar*: 5.0 mm) [NWAU]

Fig. 9.88 Male genitalia of *Bittacus tienmushana*, lateral view (*Scale bar*: 0.5 mm)

Fig. 9.89 Distribution of *Bittacus tienmushana*

Male epandrial lobes subtriangular in lateral view; longitudinal groove slightly above ventral margin; apex slightly emarginate at end of groove. Gonocoxites protruded apically and forming pair of small processes. Gonostylus short and boot-shaped with dense setae. Aedeagal lobes long and broad with slightly acute apexes. Penisfilum slender and coiled into loop. Upper branch of proctiger extending beyond epandrial lobes, bearing pair of curved finger-like processes subapically, and tuft of setae on central apex. Lower branch intensively curved ventrad.

Emergence. May to July.

Distribution. Zhejiang (Longwang Mountain, Tianmu Mountains).

9.2.24 *Bittacus trapezoideus* Huang & Hua, 2005

(Trapezoidal Hangingfly) [四边蚊蝎蛉] (Figs. 9.90, 9.91, and 9.92)

Measurements. Forewing length 17.0–19.0 mm.

Morphology. Vertex brown. Ocellar triangle dark brown. Black transversal band passing through ocellar triangle and extending to compound eyes laterally. Antennae ciliated on apical 2/3. Pronotum unevenly brown to dark brown. Meso- and metanotum dark brown with light median stripe. Pleura and legs yellowish brown, with apex of femur and both ends of tibia darkened. Wing membrane hyaline and slightly tinged with brown. Series of smoky dark brown markings covering almost entire wing surface. Male T1–T6 blackish brown anteriorly and yellowish brown posteriorly. T7 and T8 yellowish brown.

Male epandrial lobes yellowish white, half as long as gonocoxites, and subtrapezoidal in lateral view with truncated apex. Longitudinal groove on middle. Gonocoxites with V-shaped membranous area posteriorly. Gonostylus short and

Fig. 9.90 Adults of *Bittacus trapezoideus* in the field (Tiantaishan National Forest Park, Baoji, Shaanxi) Left, male; right, female

Fig. 9.91 Male genitalia of *Bittacus trapezoideus*, lateral view (*Scale bar*: 0.5 mm)

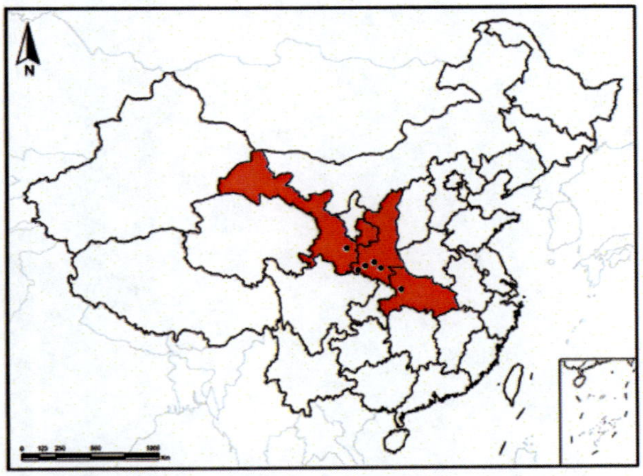

Fig. 9.92 Distribution of *Bittacus trapezoideus*

small. Aedeagal lobes rounded. Penisfilum slender and coiled into loop. Cerci slender but slightly shorter than epandrial lobes. Upper branch of proctiger slender and protruding between epandrial lobes and bearing tuft of setae apically.

Emergence. July to August.

Distribution. Gansu (Tianshui), Hubei (Shennongjia), Shaanxi (Baoji, Meixian, Ningshan, Nanzheng, Taibai, Zhouzhi).

9.2.25 *Bittacus ussuriensis* Plutenko, 1985

(Ussurian Hangingfly) [乌苏里蚊蝎蛉] (Figs. 9.93, 9.94, and 9.95)

Measurements. Forewing length 16.0–19.0 mm.

Morphology. Vertex dark brown. Ocellar triangle black. Rostrum unevenly brown. Antennae thread-like and ciliated. Pronotum unevenly brown. Mesonotum deep brown basally and yellowish brown distally. Metanotum light yellow. Pleura and legs light yellowish brown. Wing membrane hyaline, slightly tinged with yellow. Pterostigma indistinct. Several black spots present. Apical crossveins with surrounding smoky stripes. Male A1–A6 yellowish brown, T7 and T8 deep brown.

Male epandrial lobes yellowish white, and peculiar in shape: dorsal margin with small triangular process subbasally; outer margin ear-like; inner surface with longitudinal ridge. Gonocoxites shorter than epandrial lobes; each bearing conical prolongation. Aedeagal lobes indistinct. Penisfilum coiled into small loop. Upper branch of proctiger long and slightly emarginated subapically, and bearing row of long setae on apex. Lower branch stout and curved ventrad.

Emergence. July to August.

Distribution. Liaoning (Kuandian); Russia (Far East).

Fig. 9.93 Female adult of *Bittacus ussuriensis* in the field (Huabo Mountain, Kuandian, Liaoning, ©Lu Jiang)

Fig. 9.94 Male genitalia of *Bittacus ussuriensis*, lateral view (*Scale bar*: 0.5 mm)

Fig. 9.95 Distribution of *Bittacus ussuriensis*

9.2.26 *Bittacus zhejiangicus* Tan & Hua, 2008

(Zhejiang Hangingfly) [浙江蚊蝎蛉] (Figs. 9.96, 9.97, and 9.98)

Measurements. Forewing length 23.5 mm.

Morphology. Vertex and frons yellowish brown. Rostrum brown. Antennae ciliated. Pronotum unevenly yellowish brown and lacking distinct setae on anterior margin. Meso- and metanotum dark brown and bearing grayish brown median stripe, almost equal width with scutella. Pleura and legs yellowish brown, with apex of femur and both ends of tibia darkened. Wing membrane translucent and slightly tinged with yellow. Pterostigma brown and distinct. Several black spots on ORs, FRs, OM, and ending of CuP, respectively. Crossveins with surrounding smoky stripes. Male abdomen yellowish brown. T8 deeply emarginate in V-shaped on posterior margin.

Fig. 9.96 Male adult of *Bittacus zhejiangicus*, lateral view (Fengyang Mountain, Longquan, Zhejiang) (*Scale bar*: 5.0 mm) [NWAU]

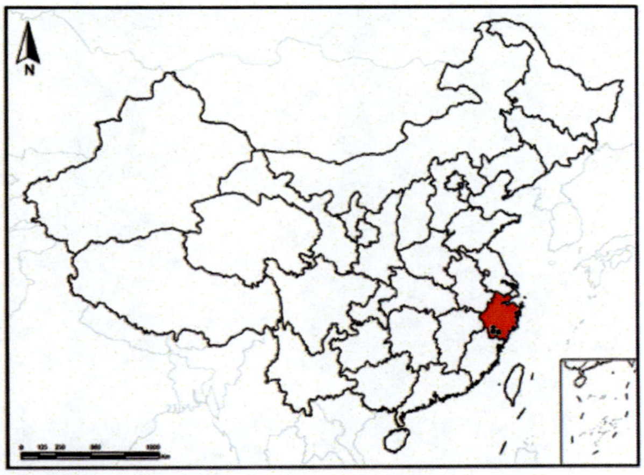

Fig. 9.98 Distribution of *Bittacus zhejiangicus*

Fig. 9.97 Male genitalia of *Bittacus zhejiangicus*, lateral view (*Scale bar*: 0.5 mm)

Male epandrial lobes longer than gonocoxites. Dorsal margin splits and bearing rounded short lobe subbasally; middle portion elongated and terminated in curved finger-like apex; ventral margin with an acute lobe subbasally. All these lobes with dense black spines on inner surfaces. Gonocoxites rounded apically. Gonostylus simple and unbranched. Cerci slender and acute apically. Upper branch of proctiger broad and protruding beyond epandrial lobes, and bearing tuft of long setae apically. Lower branch short and curved ventrad.

Emergence. July.

Distribution. Zhejiang (Longquan, Qingyuan, Taishun).

9.3 Genus *Terrobittacus* Tan & Hua, 2009

(Terro-Hangingflies) [地蚊蝎蛉属] (Figs. 9.99

Terrobittacus is similar to *Bittacus* in appearance, but can be differentiated from the latter by the male epandrial lobes shorter than half the length of gonocoxites.

The genus is endemic to China and comprises six species. All of them are included in this atlas.

Key to species of *Terrobittacus* in this atlas
(Six species, males)

1. Nota with distinct pale median stripe; gonocoxites with 1–3 pairs of long bristles apically; cercus greatly elongated and longer than epandrial lobes .... ***Terrobittacus wui*** (Zhou)
– Nota lacking median stripe; gonocoxites lacking long bristles; cercus shorter than epandrial lobes 2
2. Upper branch of proctiger relatively slender and bearing tuft of setae at apex 3
– Upper branch of proctiger enlarged, greatly curved, and hook-like apically 4
3. Upper branch of proctiger saber-shaped and slightly curved ventrad with acute apex .... ***Terrobittacus xiphicus*** Tan & Hua
– Upper branch of proctiger long stick-shaped and straight, with rounded apex .... ***Terrobittacus echinatus*** (Hua & Huang)
4. Wings devoid of markings except faint darkening at apex .... ***Terrobittacus rostratus*** Du & Hua
– Wings with cloudy spots along crossveins 5
5. Epandrial lobes subtriangular in lateral view; upper branch of proctiger same wide at basal and distal portions .... ***Terrobittacus implicatus*** (Huang & Hua)
– Epandrial lobes clavate in lateral view; upper branch of proctiger narrow at base and enlarged at distal portion .... ***Terrobittacus angustus*** Du & Hua

Fig. 9.99 Male adult of *Terrobittacus*. *Terrobittacus implicatus* in the field (Liping, Shaanxi)

9.3.1 *Terrobittacus angustus* Du & Hua, 2017

(Angular Terro-Hangingfly) [狭瓣地蚊蝎蛉] (Figs. 9.100, 9.101, and 9.102)

Measurements. Forewing length 18.5–19.0 mm.

Morphology. Vertex and frons yellowish brown.

Ocellar triangle black, with two bristles above median ocellus. Rostrum yellowish brown. Pronotum dark brown with two to four stout setae on anterior margin. Mesonotum deep brown laterally, and paler mesally and posteriorly, with yellowish brown scutellum. Metanotum yellowish brown. Pleura and coxae cream-colored or yellowish brown. Legs reddish brown with apex of femur and both ends of tibia darkened. Wing membrane hyaline and tinged with yellowish brown. Pterostigma yellowish brown and distinct. Four smoky spots on ORs, FRs, OM, and ending of CuP, respectively. Crossveins with surrounding smoky stripes. Smoky stripe along distal portion of R_5 and broadened towards apex. Male terga yellowish brown to dark brown. Sterna yellowish brown. T8 slightly emarginate posteriorly.

Male epandrial lobes narrow, clavate in lateral view with rounded apexes; series of black spines on the inner surface of apex; longitudinal ridge along middle. Gonocoxites rounded apically. Gonostylus short. Aedeagal lobes broad basally with acute apexes. Penisfilum slender and coiled into loop. T10 greatly reduced mesally and forming pair of sclerites surrounding lender cerci. Upper branch of proctiger very long and intensively sclerotized, bilobed dorsally and bearing deep groove ventrally, and curved ventrad into hook apically. Lower branch short and curved ventrad.

Emergence. June to August

Distribution. Yunnan (Yongde).

9.3.2 *Terrobittacus echinatus* (Hua & Huang, 2008)

(Pricky Terro-Hangingfly) [具刺地蚊蝎蛉] Figs. 9.103, 9.104, and 9.105

Measurements. Forewing length 16.0–17.0 mm.

Morphology. Vertex yellowish brown. Ocellar triangle black, with two indistinct bristles above median ocellus. Rostrum deep brown. Pronotum dark brown. Meso- and metanotum deep brown laterally, and paler mesally and posteriorly. Pleura and legs yellowish brown, with apex of femur and both ends of tibia darkened. Wing membrane hyaline and tinged with light yellow. Pterostigma brown and distinct. Four smoky dark spots on ORs, FRs, OM, and ending of CuP, respectively. Crossveins with surrounding smoky stripes. Male terga yellowish brown. T8 slightly emarginate in V-shaped posteriorly.

Male epandrial lobes yellowish brown, bearing series of black spines on apical inner region, and longitudinal ridge on middle. Gonocoxites rounded apically. Gonostylus short.

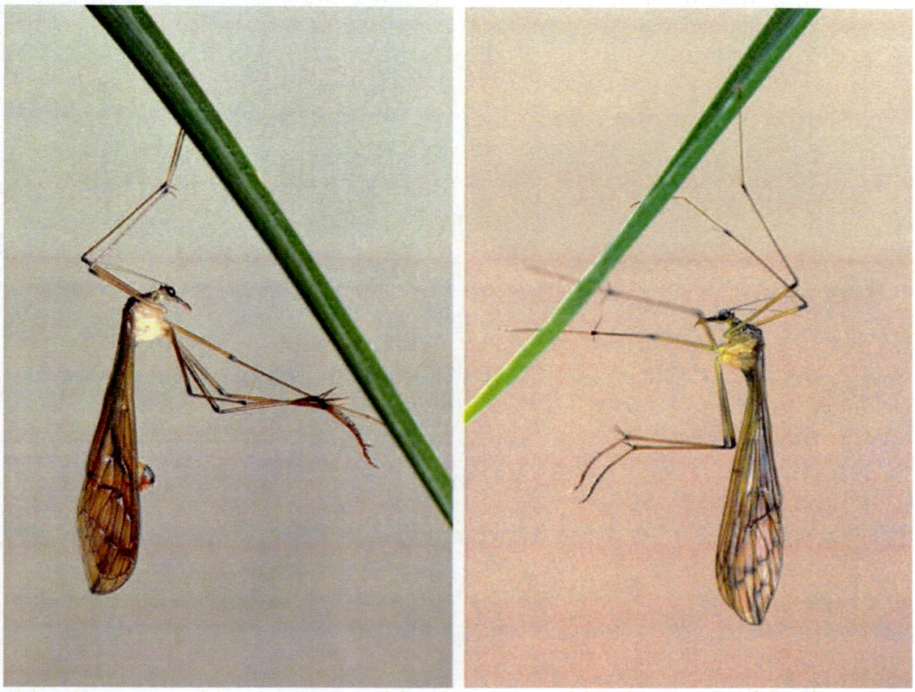

Fig. 9.100 Adults of *Terrobittacus angustus* in the field (Daxue Mountain, Yongde, Yunnan). Left, male; right, female

Fig. 9.101 Male genitalia of *Terrobittacus angustus*, lateral view (*Scale bar*: 0.5 mm)

Aedeagal lobes broad basally and acute apically. Penisfilum slender and coiled into loop. T10 greatly reduced mesally and forming pair of sclerites surrounding slender cerci. Upper branch of proctiger slender, bearing tuft of long setae at acute apex. Lower branch hook-like and curved ventrad.

Emergence. July to August.

Distribution. Fujian (Shaowu), Henan (Songxian), Shaanxi (Chang'an, Foping).

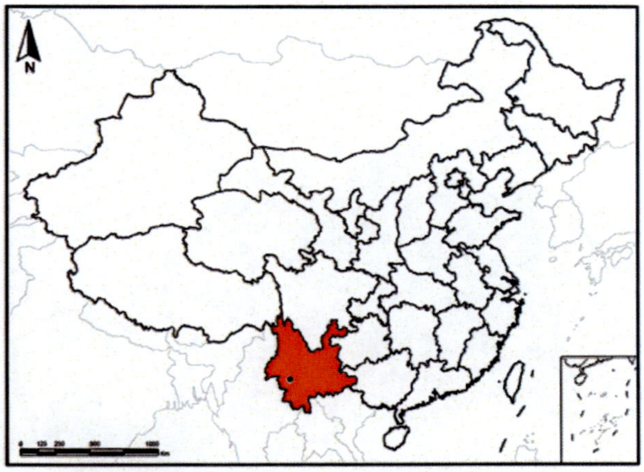

Fig. 9.102 Distribution of *Terrobittacus angustus*

9.3.3 *Terrobittacus implicatus* (Huang & Hua, 2006)

(Entangled Terro-Hangingfly) [缠绕地蚊蝎蛉] (Figs. 9.106, 9.107, and 9.108)

Measurements. Forewing length 15.0–16.0 mm.

Morphology. Vertex yellowish brown. Ocellar triangle black. Rostrum yellowish brown with deepened flanks. Antennae ciliated. Pronotum yellowish brown, bearing two black setae on anterior margin and two to four setae on pos-

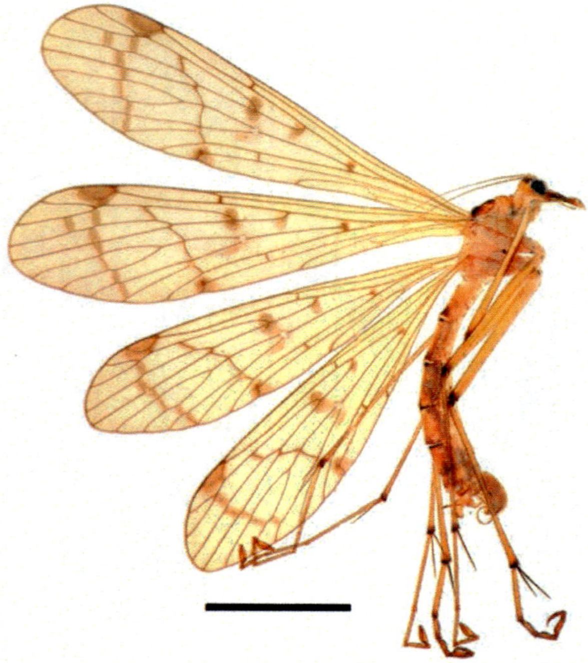

Fig. 9.103 Male adult of *Terrobittacus echinatus*, lateral view (Foping, Shaanxi) (*Scale bar*: 5.0 mm) [NWAU]

Fig. 9.104 Male genitalia of *Terrobittacus echinatus*, lateral view (*Scale bar*: 0.5 mm)

terior margin. Mesonotum deep brown anteriorly and light brown posteriorly. Metanotum light brown. Pleura yellowish brown. Legs yellowish brown with apex of femur and both ends of tibia darkened. Wing membrane hyaline and tinged with light yellow. Pterostigma yellowish brown and distinct. Several smoky dark spots on ORs, FRs, OM, and ending of CuP. Almost all crossveins with surrounding smoky stripes.

Fig. 9.105 Distribution of *Terrobittacus echinatus*

Male terga yellowish brown. T8 slightly emarginate in V-shaped posteriorly.

Male epandrial lobes considerably short, subtriangular in lateral view, and bearing dense black spines on apical inner margin. Gonocoxites with V-shaped membranous area posteriorly. Gonostylus short. T10 greatly reduced mesally and forming pair of sclerites surrounding short cerci. Aedeagal lobes acute apically. Penisfilum slender and coiled into loop. Upper branch of proctiger intensively sclerotized, with deep groove ventrally; row of dense short setae in groove; apex curved caudad and hook-like. Lower branch short and curved ventrad.

Emergence. July to August.

Distribution. Shaanxi (Lan'gao, Nanzhen).

9.3.4 *Terrobittacus rostratus* Du & Hua, 2017

(Rostrate Terro-Hangingfly) [钩瓣地蚊蝎蛉] (Figs. 9.109, 9.110, 9.111, and 9.112)

Measurements. Forewing length 14.0–16.0 mm.

Morphology. Vertex blackish brown. Frons brown. Ocellar triangle black and bearing two bristles above median ocellus. Rostrum light brown basally and blackish brown distally. Pronotum dark brown, bearing two setae on anterior margin. Meso- and metanotum mostly dark brown but paler mesally. Pleura and legs reddish or yellowish brown. Wing membrane hyaline and tinged with light yellow or grayish brown. Pterostigma light brown and indistinct. Markings absent but apical portion tinged with smoky brown. Male terga yellowish brown to dark brown. Sterna yellowish brown. T8 slightly emarginate on posterior margin.

Male epandrial lobes yellowish brown, relatively short, boot-shaped in lateral view with rounded apexes, and bearing

Fig. 9.106 Adults of *Terrobittacus implicatus* in the field (Liping National Forest Park, Nanzheng, Shaanxi). Left, male; right, female

Fig. 9.107 Male genitalia of *Terrobittacus implicatus*, lateral view (*Scale bar*: 0.5 mm)

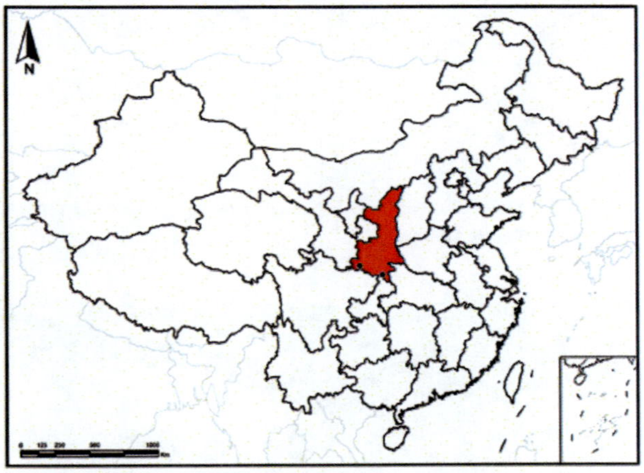

Fig. 9.108 Distribution of *Terrobittacus implicatus*

series of black stout spines on the inner surface apically. Gonocoxites rounded apically. Gonostylus short. Aedeagal lobes acute apically. Penisfilum slender and coiled into loop. T10 greatly reduced mesally and forming pair of sclerites surrounding slender cerci. Upper branch of proctiger greatly sclerotized and with two broad lateral flanks dorsally, deep groove bearing row of short setae ventrally, and curved caudad into beak apically. Lower branch short and curved ventrad.

Emergence. July to August.

Distribution. Yunnan (Yongde).

9.3.5 *Terrobittacus wui* Zhou, 2001 [New Combination]

(Wu's Terro-Hangingfly) [吴氏地蚊蝎蛉] (Figs. 9.113, 9.114, and 9.115)

Original combination: *Bittacus wui* Zhou, 2001

Terrobittacus longisetus Tan & Hua, 2009, **new synonym**

Measurements. Forewing length 20.0–21.0 mm.

Morphology. Vertex yellowish brown. Ocellar triangle black, bearing two bristles above median ocellus.

Fig. 9.109 Male adult of *Terrobittacus rostratus* in the field (Tangli Mountain, Yongde, Yunnan)

Fig. 9.110 Female adult of *Terrobittacus rostratus* in the field (Tangli Mountain, Yongde, Yunnan)

Fig. 9.111 Male genitalia of *Terrobittacus rostratus*, lateral view (*Scale bar*: 0.5 mm)

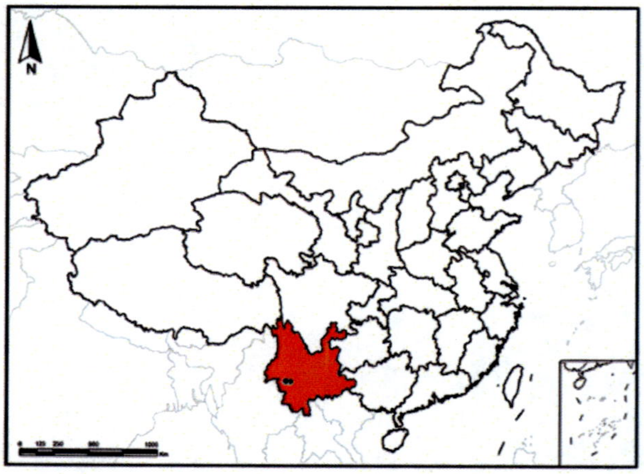

Fig. 9.112 Distribution of *Terrobittacus rostratus*

Rostrum yellowish brown with deepened flanks. Antennae ciliated. Pronotum yellowish brown, bearing four black setae on anterior margin. Meso- and metanotum dark brown with distinct pale median stripe. Legs light yellowish brown. Wing membrane hyaline and tinged with light yellow. Pterostigma light yellow and indistinct. Markings absent except apical crossveins with surrounding smoky stripes. Male terga yellowish brown with each segment darkening caudad. T8 slightly emarginate in V-shape posteriorly.

Male epandrial lobes considerably short, shovel-shaped with rounded apexes, and bearing tuft of long setae on dorsal margins. Series of black and stout spines on the inner surface of each lobe apically. Longitudinal ridge on middle of each lobe. Gonocoxites semispherical, bearing V-shaped membranous area posteriorly, and one to three pairs of rather long setae on membranous area. Gonostylus small. Aedeagal lobes acute apically. Penisfilum slender and coiled into loop. T10 greatly reduced mesally and forming pair of sclerites surrounding cerci. Cerci approximately 2/3 as long as gono-

Fig. 9.113 Male adult of *Terrobittacus wui* in the field (Lushan Mountain, Lushan, Jiangxi)

Fig. 9.114 Male genitalia of *Terrobittacus wui*, lateral view (*Scale bar*: 0.5 mm)

Fig. 9.115 Distribution of *Terrobittacus wui*

coxites and slightly rounded apically. Upper branch of proctiger intensively sclerotized and broad, strongly curved caudad apically and sickle-like. Lower branch short and curved ventrad.

Emergence. July to August.

Distribution. Guizhou (Leigong Mountain), Hunan (Zhangjiajie, Sangzhi), Jiangxi (Lushan), Zhejiang (Lin'an).

Remarks. *T. longisetus* was originally described from the Badagongshan Mountains (Hunan), and *T. wui* was from the Tianmu Mountains (Zhejiang). By comparing specimens from many localities, we confirm that these two are conspecific. Therefore, *T. longisetus* is treated a junior synonym of *T. wui*.

9.3.6 *Terrobittacus xiphicus* Tan & Hua, 2009

(Sword-Bearing Terro-Hangingfly) [刀地蚊蝎蛉] (Figs. 9.116, 9.117, and 9.118)

Measurements. Forewing length 15.0–16.0 mm.

Morphology. Vertex reddish brown. Ocellar triangle black and with transverse black stripe extending laterally to compound eyes. Rostrum yellowish brown. Antennae ciliated. Pronotum unevenly dark brown. Meso- and metanotum yellowish brown mesally and shining black laterally. Metanotum yellowish brown. Pleura and legs yellowish brown. Wing membrane hyaline and tinged with light yellow. Pterostigma black brown and distinct. Several brown spots on ORs, FRs, OM, and ending of CuP. Smoky tinges along crossveins. Male terga yellowish brown and gradually darkened caudally. T8 emarginate in V-shape posteriorly.

Male epandrial lobes shorter than half-length of gonocoxites, bearing dozens of black spines on inner surfaces apically. Acute tooth protrudes on ventral margin subbasally. Gonocoxites semispherical and bearing tuft of sparse setae. Gonostylus short and curved inward. Aedeagal lobes acute apically. Penisfilum considerably short and coiled in approximately three circles. T10 greatly reduced mesally and forming pair of sclerites surrounding slender and acute cerci. Upper branch of proctiger saber-shaped and bearing tuft of setae on apex. Lower branch short and hook-like.

Emergence. June to August.

Distribution. Guizhou (Jiangkou), Hunan (Sangzhi).

Fig. 9.116 Adults of *Terrobittacus xiphicus* in the field (Tianping Mountain, Sangzhi, Hunan). Left, male; right, female

Fig. 9.117 Male genitalia of *Terrobittacus xiphicus*, lateral view (*Scale bar*: 0.5 mm)

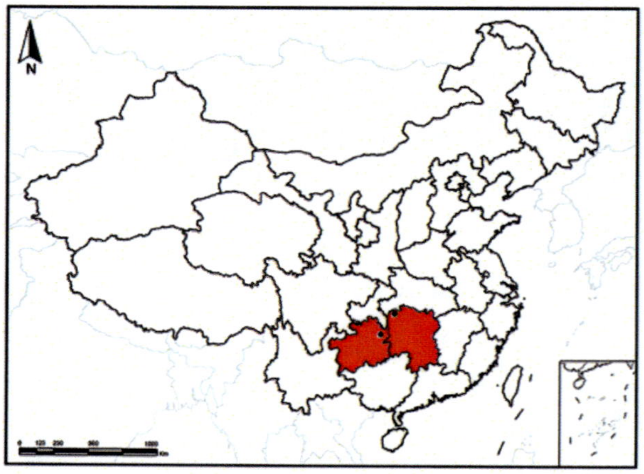

Fig. 9.118 Distribution of *Terrobittacus xiphicus*

10 Family Panorpidae Latreille, 1802

(Scorpionflies) [蝎蛉科]

Panorpidae are commonly called scorpionflies. Their heads are prolonged ventrally into a rostrum, with the chewing mouthparts located at the tip of the latter (Fig. 10.1). Their male genitalia are enlarged and recurved backwards, which superficially resemble the stinger of scorpions.

Panorpidae is the largest family in Mecoptera, with more than 500 extant species described worldwide. Over 200 species have been recorded from China, and 134 of them are included in this atlas.

Currently, members of Panorpidae are categorized into two subfamilies, **Neopanorpinae** Wang & Hua, 2021 and **Panorpinae** Latreille, 1802, respectively. According to Wang and Hua (2021), species in the genera *Dicerapanorpa*, *Neopanorpa*, and *Panorpa* can be further assigned into several species groups by morphological criteria and phylogenetic relationships (Fig. 10.2).

Fig. 10.1 Adult of Panorpidae. *Panorpa* sp., frontal view (©Bo Lei)

J.-S. Wang, B.-Z. Hua, *A Color Atlas of the Chinese Mecoptera*, https://doi.org/10.1007/978-981-16-9558-2_10

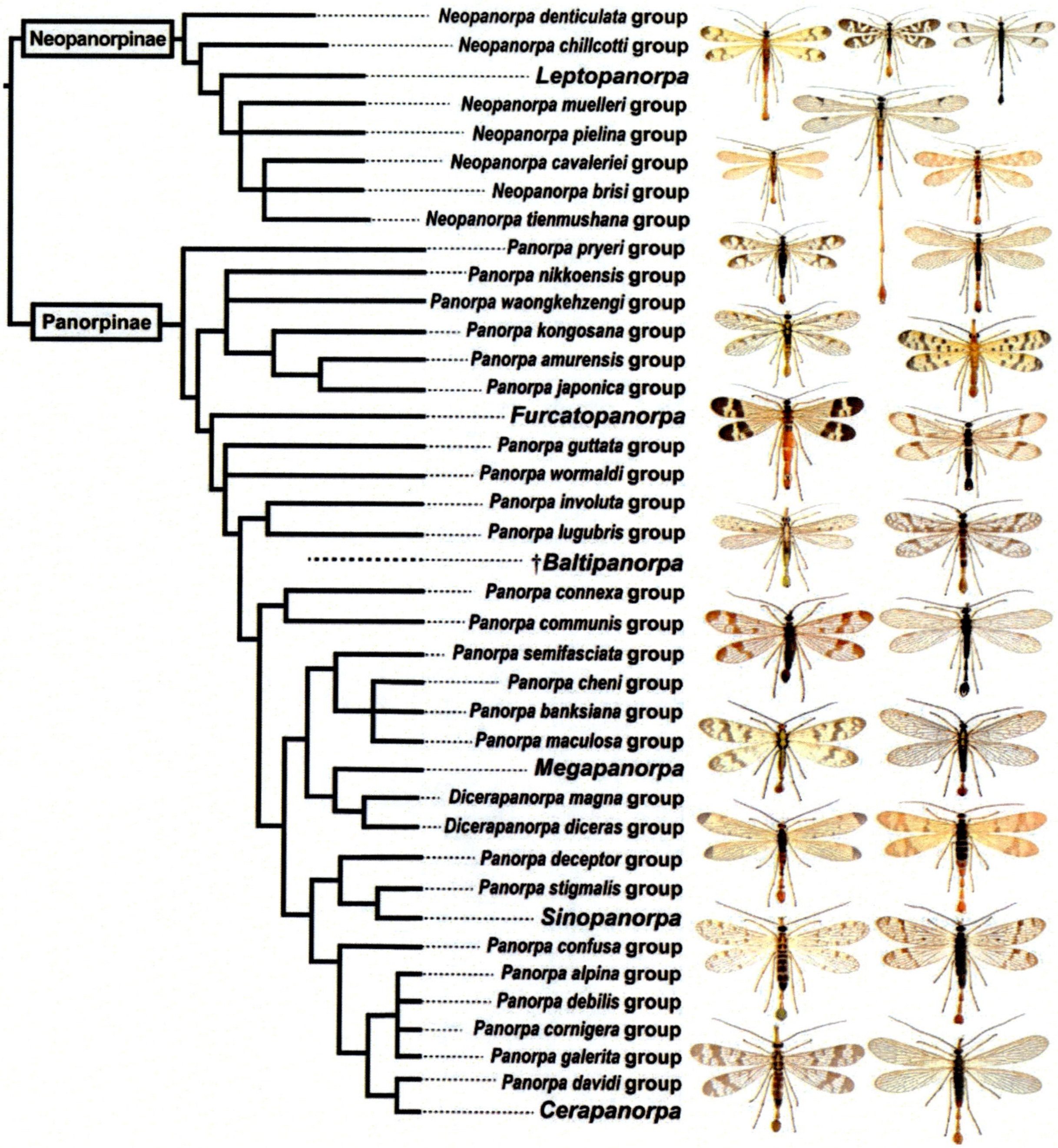

Fig. 10.2 Phylogeny of subfamilies, genera, and species groups in Panorpidae (revised from Wang and Hua 2021)

Key to Subfamilies, Genera, and Species Groups of Panorpidae

(Two subfamilies, nine genera, and 34 species groups; including 2 groups in ***Dicerapanorpa***, 8 in ***Neopanorpa***, and 24 in ***Panorpa***. Modified from Wang and Hua 2021)

1. Forewing with 1A usually ending proximal to ORs, only one anal cross-vein a between 1A and 2A; hindwing with anterior ending of a proximal to fork of CuP + 1A, cu-a absent; pretarsal claws with second preapical tooth distinctly larger than others; male notal organ on T3 greatly developed, more or less exceeding rounded or flat postnotal organ on T4; larvae with shallow furrows on head capsule (**Neopanorpinae** Wang & Hua) 2
– Forewing with 1A usually ending distal to ORs, at least two anal cross-veins a between 1A and 2A; hindwing with anterior ending of a distal to fork of CuP + 1A, cu-a present; pretarsal claws with preapical teeth similar in size; male notal organ on T3 usually short and flat, postnotal organ on T4 small and acute; larvae lacking shallow furrows on head capsule (**Panorpinae** Latreille) 3
2. Wings broad at base, with ratio of forewing widths at ending of M_4 to 1A < 2; male epandrial lobes absent ....***N. denticulata* group** (India)
– Wings relatively narrow at base, with ratio of forewing widths at ending of M_4–1A ≥ 2; male epandrial lobes present 3
3. Male hypovalves with outer margin simple, and unconstricted basally ....***N. chillcotti* group** (Himalayas)
– Male hypovalves with outer margin curled dorsad and constricted basally 4
4. Male abdomen much longer than wings; A7 and A8 extremely elongated, at least four times as long as wide and twice as long as A5; A9 with elongated basal stalk; epandrium with terminal projection; female medigynium with apodemes of axis curved dorsad ....***Leptopanorpa*** MacLachlan (Indonesia)
– Male abdomen usually less elongated; A7 and A8 at most three times as long as wide and twice as long as A5; A9 lacking distinct basal stalk; epandrium lacking terminal projection; female medigynium with apodemes of axis simple (***Neopanorpa*** van der Weele) 5
5. Male hypandrial processes absent; parameres greatly reduced ....***N. muelleri* group** (Indonesia, Malaysia)
– Male hypandrial processes present; parameres well-developed 6
6. Male T3 with small tooth-like projection under notal organ ***N. brisi* group** (southwestern China)
– Male T3 without projection under notal organ 7
7. Meso- and metanotum with dark median stripe narrower than scutellum; epandrial lobes small ....***N. tienmushana* group** (southern China; Japanese Ryukyu Islands)
– Meso- and metanotum with dark median stripe usually broader than scutellum; epandrial lobes greatly developed 8
8. Male notal organ extremely elongated, exceeding to hind border of A6; postnotal organ depressed with series of long setae N. choui group (merged with N. pielina group in this Atlas)
– Male notal organ shorter; postnotal organ greatly developed 9
9. Male T4 with membranous region on anterior portion ....***N. pielina* group** (Oriental Region)
– Male T4 lacking membranous region of anterior portion ....***N. cavaleriei* group** (Oriental Region)
10. Male notal and postnotal organs greatly elongated, exceeding A7; notal organ with four pairs of long setae subapically; postnotal organ slightly longer than notal organ ....†***Baltipanorpa*** (Baltic amber)
– Male notal and postnotal organs usually less elongated; notal organ lacking long setae subapically; postnotal organ shorter than notal organ 11
11. Maxillary palps lacking sclerotized ring basal to third segment; male T7 emarginate at apex; dorsal aedeagal valves surrounded by lateral wall formed by ventral valves and dorsal processes ....***P. pryeri* group** (Japan)
– Maxillary palps with sclerotized ring basal to third segment; male T7 simple; dorsal aedeagal valves simple 12
12. Male T7 usually cylindrical, unconstricted or only slightly constricted at base 13
– Male T7 greatly constricted at base 20
13. Male epandrium not emarginate or indistinctly emarginate; parameres usually crossed subbasally; female medigynium with poorly developed main plate 14
– Male epandrium greatly emarginate apically; parameres not crossed subbasally; female medigynium with well-developed main plate 18

14. Male notal organ extremely developed, exceeding apex of A8; hypandrium bearing long setae on apex of basal stalk .... ***P. nikkoensis* group** (Japan)
– Male notal organ less developed; hypandrium lacking long setae on apex of basal stalk 15
15. Male gonostyli shorter than half-length of gonocoxites .... ***P. waongkehzengi* group** (China)
– Male gonostyli longer than half length of gonocoxites 16
16. Male gonostyli greatly elongated, longer than gonocoxites. .... ***P. kongosana* group** (China, North and South Korea)
– Male gonostyli less elongated, approximately as long as gonocoxites 17
17. Male gonocoxites bearing subtriangular projection directed caudo-mesad on basal third of inner margin; distal half of gonostyli greatly curved dorsad; female medigynium with apodemes greatly elongated, longer than main plate .... ***P. amurensis* group** (northeastern China; North and South Korea; Russian Far East)
– Male gonocoxites lacking projection on basal third of inner margin; distal half of gonostyli simple; female medigynium with apodemes concealed in or slightly extending beyond main plate .... ***P. japonica* group** (Japan; southern China)
18. Male notal and postnotal organs absent; median tooth of gonostylus greatly enlarged, serrate on inner margin; female medigynium with posterior apex of axis greatly elongated and bifurcated .... ***Furcatopanorpa*** (central China)
– Male notal and postnotal organs present; median tooth of gonostylus simple; female medigynium with posterior apex of axis short and simple 19
19. Male lateral processes of aedeagus simple, dorsal processes greatly constricted at base, neck-like; female medigynium lacking dorsal plate .... ***P. guttata* group** (central and southwestern China)
– Male lateral processes of aedeagus greatly elongated, dorsal processes simple; female medigynium bearing broad dorsal plate .... ***P. wormaldi* group** (Japan; southern China)
20. Male gonocoxital concavity narrow and shallow, not exceeding to middle of gonocoxites with subtrapezoidal bottom; gonocoxites lacking fused lobes on dorsal apex; ventral valves of aedeagus projected ventrad, beak-like 21
– Male gonocoxital concavity relatively broad and deep, exceeding middle of gonocoxites with rounded bottom; gonocoxites with pair of fused lobes on dorsal apex; ventral valves of aedeagus simple 22
21. Male abdomen relatively short, not exceeding apex of wings .... ***P. involuta* group** (Mexico)
– Male abdomen greatly elongated, exceeding apex of wings .... ***P. lugubris* group** (Mexico; USA)
22. Male gonocoxites with medial spine 23
– Male gonocoxites lacking medial spine 24
23. Male gonocoxites lacking or with less developed terminal plate; parameres usually simple; female medigynium with decorated area of axis simple .... ***P. connexa* group** (Europe)
– Male gonocoxites bearing well-developed terminal plate; parameres usually bifurcated subapically; female medigynium with decorated area of axis greatly enlarged .... ***P. communis* group** (Europe; northeastern China; North and South Korea; Russian Far East)
24. Male lateral processes of aedeagus reduced; female A9 distinctly wider than A8; anterior margin of medigynium less sclerotized than posterior portion 25
– Male lateral processes well-developed; female A9 approximately as wide as A8; anterior margin of medigynium uniformly sclerotized as posterior portion 31
25. Male aedeagus greatly elongated, with apex greatly exceeding apex of gonocoxites 26
– Male aedeagus less elongated, with apex concealed in aedeagal concavity, approximate to apex of gonocoxites, or only slightly exceeding apex of gonocoxites 29

26. Male genitalia with M-shaped process at joint of two gonocoxites ventrally; female T9 with lateral margin greatly curved ventrad and enclosing subgenital plate .... ***P. semifasciata* group** (southwestern China)
– Male genitalia lacking an M-shaped process at joint of two gonocoxites; female T9 simple 27
27. Male hypovalves extremely narrow, thread-like; parameres bifurcated .... ***P. cheni* group** (southern China)
– Male hypovalves relatively broad and stripe-like; parameres simple 28
28. Male A6 with single anal horn on dorsal apex. .... ***P. banksiana* group** (Canada, USA)
– Male A6 lacking anal horns .... ***P. maculosa* group** (Canada, USA)
29. Male A6 with single anal horn; female with laterotergites of A9 greatly elongated, stick-like, and fused with medigynium on inner margin .... ***Megapanorpa*** Wang & Hua (central and southwestern China)
– Male A6 with pair of anal horns; female with laterotergites short and simple .... (***Dicerapanorpa*** Zhong & Hua) 30
30. Wings with well-developed markings; apical, pterostigmal, basal bands broad and dense .... ***D. magna* group**
– Wings with markings reduced; apical and pterostigmal bands greatly reduced or absent; basal band absent .... ***D. diceras* group**
31. Male A6 beveled at apex 32
– Male A6 truncated at apex 34
32. Male T6 with tuft of stout setae dorsally; base of A7 greatly constricted, stalk-like .... ***Sinopanorpa*** Cai & Hua (central China)
– Male T6 lacking tuft of stout setae dorsally; A7 usually simple 33
33. Male notal organ on T3 greatly elongated, stick-like, exceeding middle of T4; A6–A8 greatly elongated; female axis of medigynium slender .... ***P. stigmalis* group** (central, southern, and southwestern China)
– Male notal organ on T3 less elongated, not reaching middle of T4; A6–A8 less elongated; female axis of medigynium stout .... ***P. deceptor* group** (eastern and southern China)
34. Male dorsal aedeagal processes lacking swollen membranous process dorso-subapically .... ***P. confusa* group** (Canada; USA)
– Male dorsal aedeagal processes with swollen membranous process dorso-subapically 35
35. Male genitalia with aedeagal hamulus present; apical third of dorsal processes simple 36
– Male genitalia lacking aedeagal hamulus; apical third of dorsal processes greatly curved ventrad 39
36. Male A6 lacking anal horns .... ***P. alpina* group** (Europe)
– Male A6 with single anal horn 37
37. Male gonostylus with accessory lobe on ventral surface; aedeagal hamulus simple and rounded .... ***P. galerita* group** (Canada, USA)
– Male gonostylus lacking accessary lobe; aedeagal hamulus bifurcated and slender 38
38. Male aedeagal hamulus with two branches mostly coalesced; dorsal processes bifurcated subapically .... ***P. cornigera* group** (northeastern China; Japan; North and South Korea; Russian Far East)
– Male aedeagal hamulus with two branches divergent; dorsal processes simple .... ***P. debilis* group**
39. Male A6 lacking anal horns; parameres sigmoidally twisted and usually crossed subdistally; female medigynium with ventral plate simple .... ***P. davidi* group** (central, southern, and southwestern China)
– Male A6 with single anal horn; parameres simple; female medigynium with ventral plate split .... ***Cerapanorpa*** Gao, Ma, & Hua (central, southern, and southwestern China)

Subfamily Neopanorpinae Wang & Hua, 2021

10.1 Genus *Neopanorpa* van der Weele, 1909

(Neoscorpionflies, Neopanorpas) [新蝎蛉属] (Fig. 10.3)
Members in the genus *Neopanorpa* have relatively longer rostrums and wings compared with other Chinese genera. More than 170 species have been described worldwide. Over 100 species are recorded from China. Forty-nine species are included in this atlas.

Fig. 10.3 Male adult of ***Neopanorpa***. ***Neopanorpa*** **sp.** in the field (©Bo Lei)

Neopanorpa chillcotti Group

This group of species can be differentiated from their congeners by the lacking of epandrial processes. Approximately 12 known species in this group are distributed in the southern slobe of the Himalayas, covering Bhutan, southwestern China (Tibet), Nepal, and northern India (Assam, Darjeeling, Sikkim). Four species have been recorded from China and all of them are included in this atlas.

Key to Species of Neopanorpa chillcotti Group in This Atlas
(4 species, males)

1. Wings devoid of markings; hypovalves bifurcated terminally .... ***Neopanorpa wuchaoi*** Wang
– Wings with dense markings; hypovalves simple 2
2. Wings strongly tinged with yellow; apex of hypovalves broadened and truncated ....***Neopanorpa chillcotti*** Byers
– Wings not tinged with yellow; apex of hypovalves not modified as above 3
3. Hypovalves extremely slender .... ***Neopanorpa hushengchangi*** Hua & Chou
– Hypovalves broad, and tapering towards acute apex .... ***Neopanorpa liuxingyuei*** Wang

10.1.1 *Neopanorpa chillcotti* Byers, 1971

(Chillcott's Neoscorpionfly) [契尔科提新蝎蛉] (Fig. 10.4)
Measurements. Forewing length 15.0–16.0 mm.

Morphology. Vertex black. Ocellar triangle black. Rostrum yellowish brown to deep brown with two diffuse brown longitudinal stripes frontally. Pronotum dark brown. Meso- and metanotum, pleura and legs yellowish-brown. Wing membrane hyaline and slightly tinged with light yellow. Markings yellowish brown. Forewing apical band semicircular with incomplete hind border, usually with acute elongation innerly; pterostigmal band slender and usually not exceeding R_1 anteriorly; basal branch of pterostigmal band usually separated; marginal spot inconspicuous; basal spot greatly reduced into small spot. Male T1–T5 sordid yellowish brown. Notal organ greatly developed, long clavate, slightly enlarged terminally and extending to middle of T5, and covering rounded postnotal organ. A6 cylindrical, approximately twice as long as A5, and with black margin terminally. A7 and A8 approximately as long as A6 and slightly constricted basally.

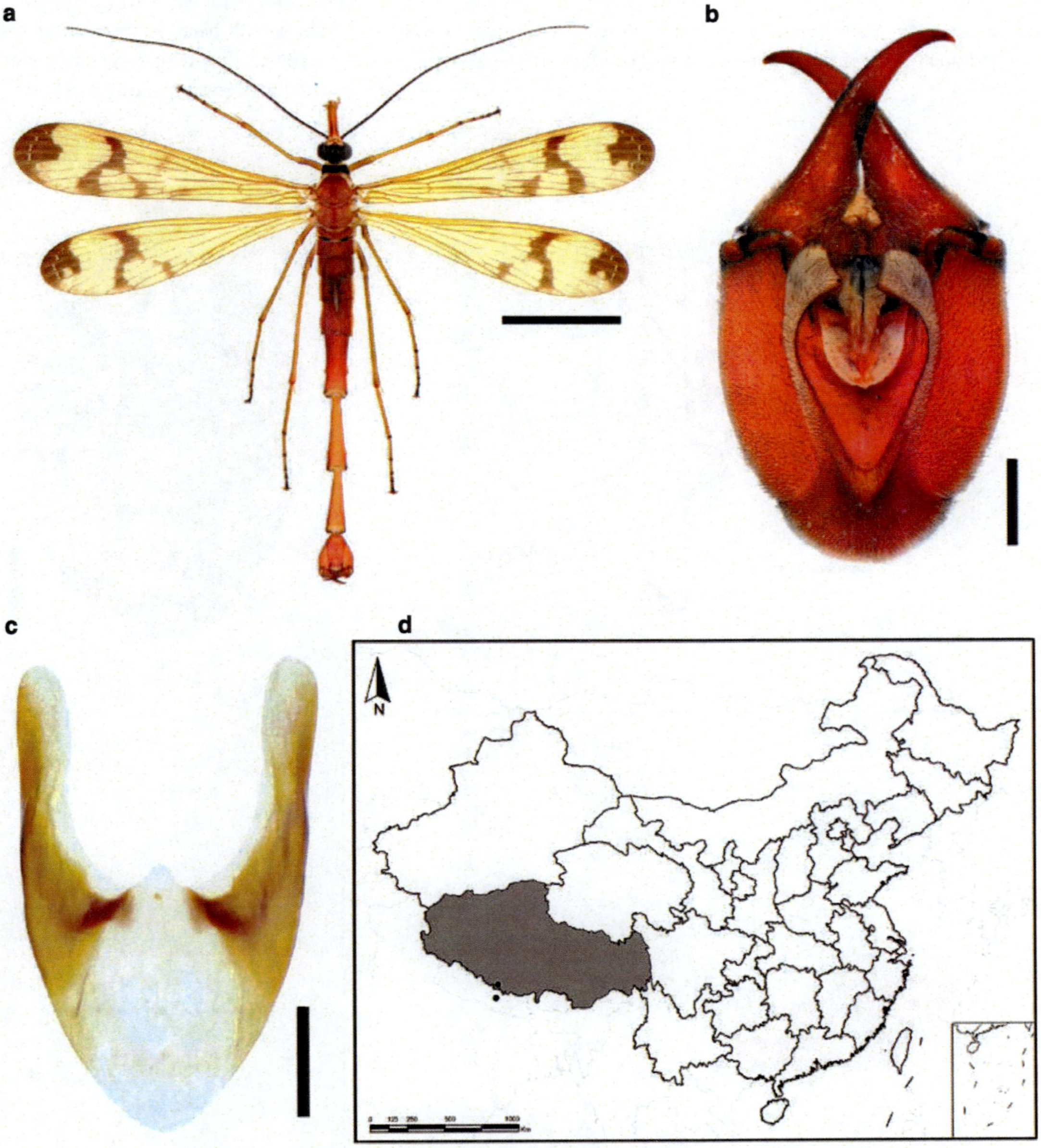

Fig. 10.4 (**a**) Male adult of ***Neopanorpa chillcotti***, dorsal view (Gyirong, Tibet) (*Scale bar*: 5.0 mm). (**b**) Male genital bulb of ***Neopanorpa chillcotti***. (*Scale bar*: 0.5 mm) (**c**) Female medigynium of ***Neopanorpa chillcotti***. (*Scale bar*: 0.2 mm) (**d**) Distribution of ***Neopanorpa chillcotti***, ventral view

Male genital bulb long oval. Epandrium long and broad, and rounded at apex; epandrial lobes broad with truncated apex. Hypandrium with short basal stalk. Hypovalves slender, divergent basally and convergent apically, with numerous long setae on expanded apical portion. Gonostylus slender, bearing obtuse triangular median tooth and large oblong basal process. Paramere thread-like, fused to aedeagus basally and free distally. Aedeagus conical, bearing pair of stout lateral processes, with finger-like dorsal valves greatly elongated caudad.

Female subgenital plate broadly oval, with V-shaped terminal emargination. Mediygnium with broad U-shaped main plate, posterior arms approximately as long as main plate with rounded apex; axis entirely concealed and lacking distinct apodemes.

Emergence. May to August.

Distribution. Tibet (Gyirong); Nepal (Kathmandu).

10.1.2 *Neopanorpa hushengchangi* Hua & Chou, 1999

(Hu's Neoscorpionfly) [胡氏新蝎蛉] (Fig. 10.5)

Measurements. Forewing length 13.5 mm.

Morphology. Vertex black mesally but yellowish brown laterally. Ocellar triangle black. Rostrum yellowish brown with two darkened stripes. Pronotum dark brown. Meso- and metanotum yellowish brown with black median stripe. Pleura and legs light yellowish brown. Wing membrane hyaline and tinged with light yellow, with light brown markings. In forewings, api-

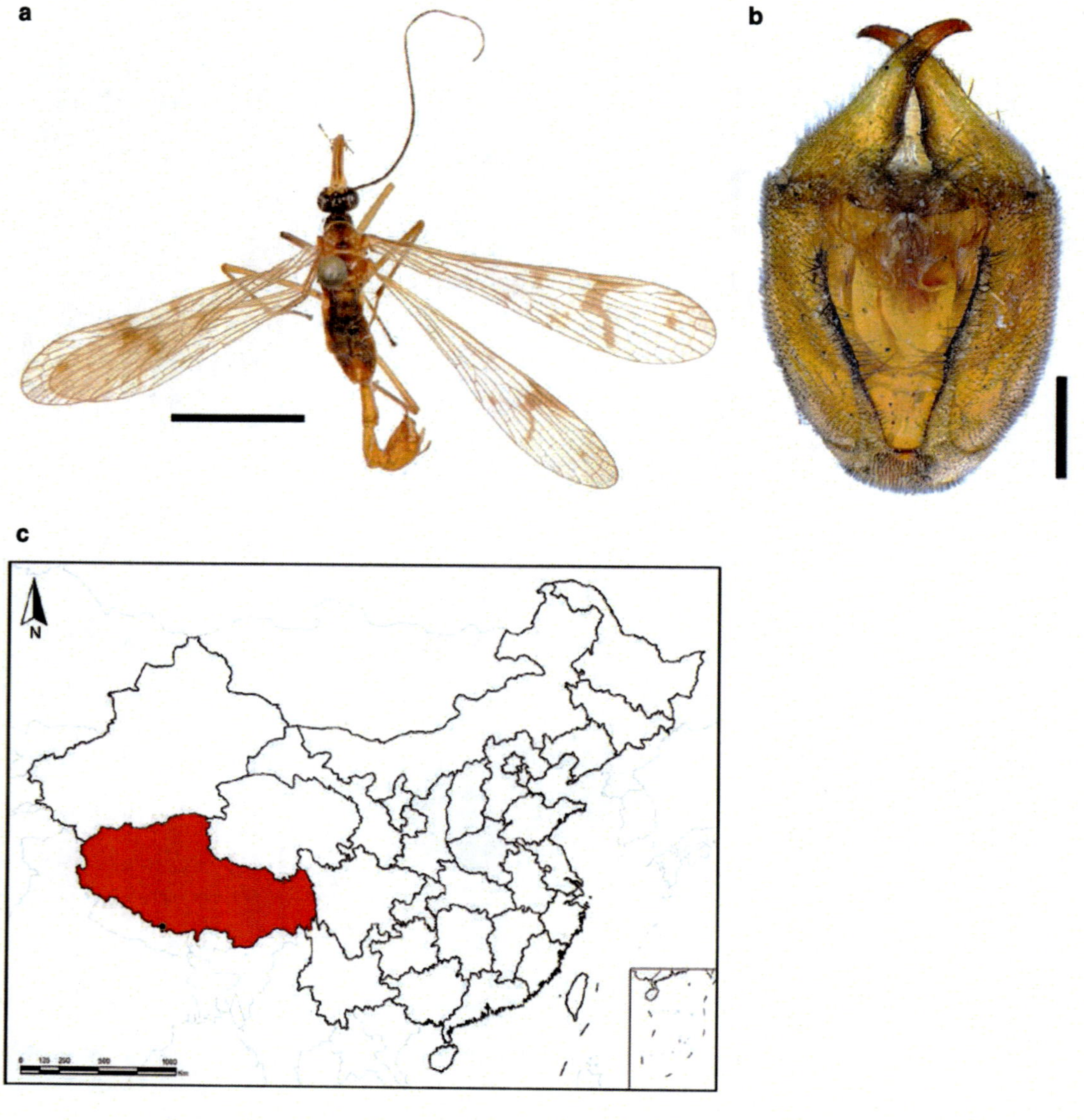

Fig. 10.5 (a) Male adult of ***Neopanorpa hushengchangi***, dorsal view (Zham Port, Nyalam County, Tibet) (*Scale bar*: 5.0 mm) **[NWAU]**. **(b)** Male genital bulb of ***Neopanorpa hushengchangi***, ventral view (*Scale bar*: 0.5 mm). **(c)** Distribution of ***Neopanorpa hushengchangi***

cal band incomplete and represented by small spot between R_3 and M_1; pterostigmal band narrow and with reduced basal branch and stout apical branch; marginal spot conspicuous; basal band only represented by small spot near end of CuP; basal spot absent. Male T1–T5 unevenly yellowish brown. Notal organ long, extending to two-third of T5. A6 yellowish brown. A7 and A8 yellowish brown and constricted basally.

Male genital bulb oval. Epandrium long and broad with truncated apex. Hypandrium with greatly shortened basal stalk. Hypovalves slender and divergent in V-shape. Gonostylus approximately half as long as gonocoxites, bearing obtuse triangular median tooth and large basal process. Aedeagus conical with stout lateral processes.

Female unknown.

Emergence. July to August.
Distribution. Tibet (Nyalam).

10.1.3 *Neopanorpa liuxingyuei* Wang, 2021

(Xing-Yue Liu's Neoscorpionfly) [刘氏新蝎蛉] (Fig. 10.6)
Measurements. Forewing length 15.8–16.0 mm.

Morphology. Vertex, ocellar triangle, and occiput shining black. Meso- and metanotum mostly black with small brown spot on each side of posterior portion. Pleura and legs yellowish brown. Wings narrow basally with rounded apex. Membrane subhyaline and tinged with grayish brown; markings brown; pterostigma dark brown. Forewing apical,

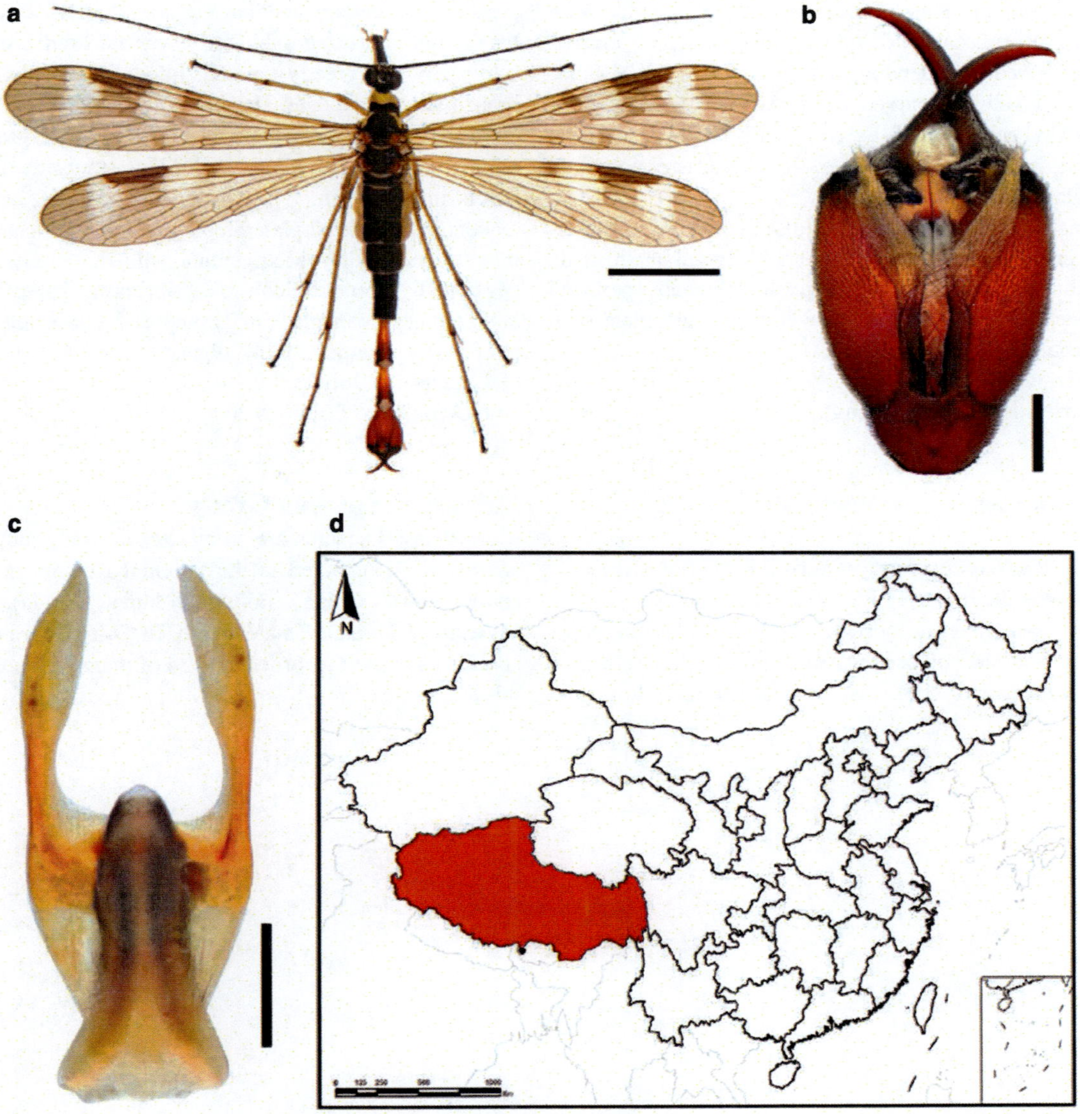

Fig. 10.6 (a) Male adult of ***Neopanorpa liuxingyuei***, dorsal view (Kambu Maqu, Yadong, Tibet, NWAU) (*Scale bar*: 5.0 mm) **[DALU]**. (b) Male genital bulb of ***Neopanorpa liuxingyuei***, ventral view (*Scale bar*: 0.5 mm). (c) Female medigynium of ***Neopanorpa liuxingyuei***, ventral view (*Scale bar*: 0.2 mm). (d) Distribution of ***Neopanorpa liuxingyuei***

pterostigmal bands and marginal spot broad and faded; area between apical and pterostigmal bands whitish and distinctly lighter than remainder of wing. Notal organ triangular in basal half and finger-like in distal half, extending beyond approximate middle of T4; postnotal organ rounded. T6 black and cylindrical. A7 and A8 reddish brown basally and gradually darkening towards apex, greatly constricted at base.

Male genital bulb stout, oval, mostly reddish brown with distal half dark brown. Epandrium broad, slightly constricted in distal 1/4, with pale and narrowed apex, and slightly emarginated terminally; epandrial lobes stout. Hypovalves approximately three times as long as basal stalk of hypandrium, narrow in basal 2/5 and slightly broader in distal 3/5, tapering to subacute apex, bearing sparse long bristles along inner margin, extending slightly beyond base of gonostyli. Gonostyli slightly shorter than gonocoxites, slender, with large basal process and subtriangular median process. Parameres slender, extending slightly beyond apex of gonocoxites, with slightly curved, hook-like apex; ventral aedeagal valves pale and V-shaped distally; dorsal aedeagal processes broad and semi-foliate; lateral processes acute and triangular.

Female subgenital plate broadly oval, with V-shaped terminal emargination. Mediygnium with broad main plate, posterior arms longer than main plate with acute apex, and spatulate in distal half; axis longer than main plate with diverged apodemes.

Emergence. August.

Distribution. Tibet (Yadong).

10.1.4 *Neopanorpa wuchaoi* Wang, 2021

(Chao Wu's Neoscorpionfly) [吴氏新蝎蛉] (Fig. 10.7)

Measurements. Forewing length 12.6 mm.

Morphology. Vertex mostly black, large yellowish brown spot on each side of occiput. Rostrum slender, yellowish brown with pair of brown longitudinal stripes. Meso- and metanotum mostly yellowish brown with narrow black median stripe. Wing membrane hyaline and strongly tinged with yellow; markings entirely reduced; pterostigma brown. T1–T5 yellowish brown. Notal organ on T3 extending approximately to 4/5 of T4, finger-like and clavate terminally; postnotal organ on T4 rounded. A6 yellowish brown, cylindrical. A7 yellowish brown, greatly constricted basally and gradually enlarged towards truncated apex; A8 similar to A7 but shorter.

Male genital bulb yellowish brown, long elliptical. Epandrium broad and tapering towards apex, with deep O-shaped emargination and forming pair of finger-like processes laterally, epandrial lobes small. Hypandrial processes absent. Hypovalves long, bifurcated into pair of finger-like processes terminally. Gonostyli slightly shorter than gonocoxites, with basal process and median tooth greatly elongated; median tooth shortly bifurcated. Parameres bifurcated subbasally; ventral branch enlarged subbasally and greatly tapering into pointed apex; dorsal branch slightly sinuate. Dorsal aedeagal processes greatly elongated and broad, with subtriangular, acute process in middle of inner margin, and 10–12 long stout setae in distal half of inner margin.

Female subgenital plate broadly oval, with V-shaped terminal emargination. Medigynium with main plate greatly developed, subtrapezoidal; posterior arms approximately half as long as main plate; axis completely concealed in main plate with apodemes closely aligned.

Emergence. July.

Distribution. Tibet (Cona).

Neopanorpa pielina Group

More than 60 species are recognized in this group. They are mainly distributed in the Oriental Region, including southwestern China, northern India (Simla), Laos, Myanmar, Thailand, and Vietnam. In China, more than 40 species have been recorded, and 34 of them are included in this atlas.

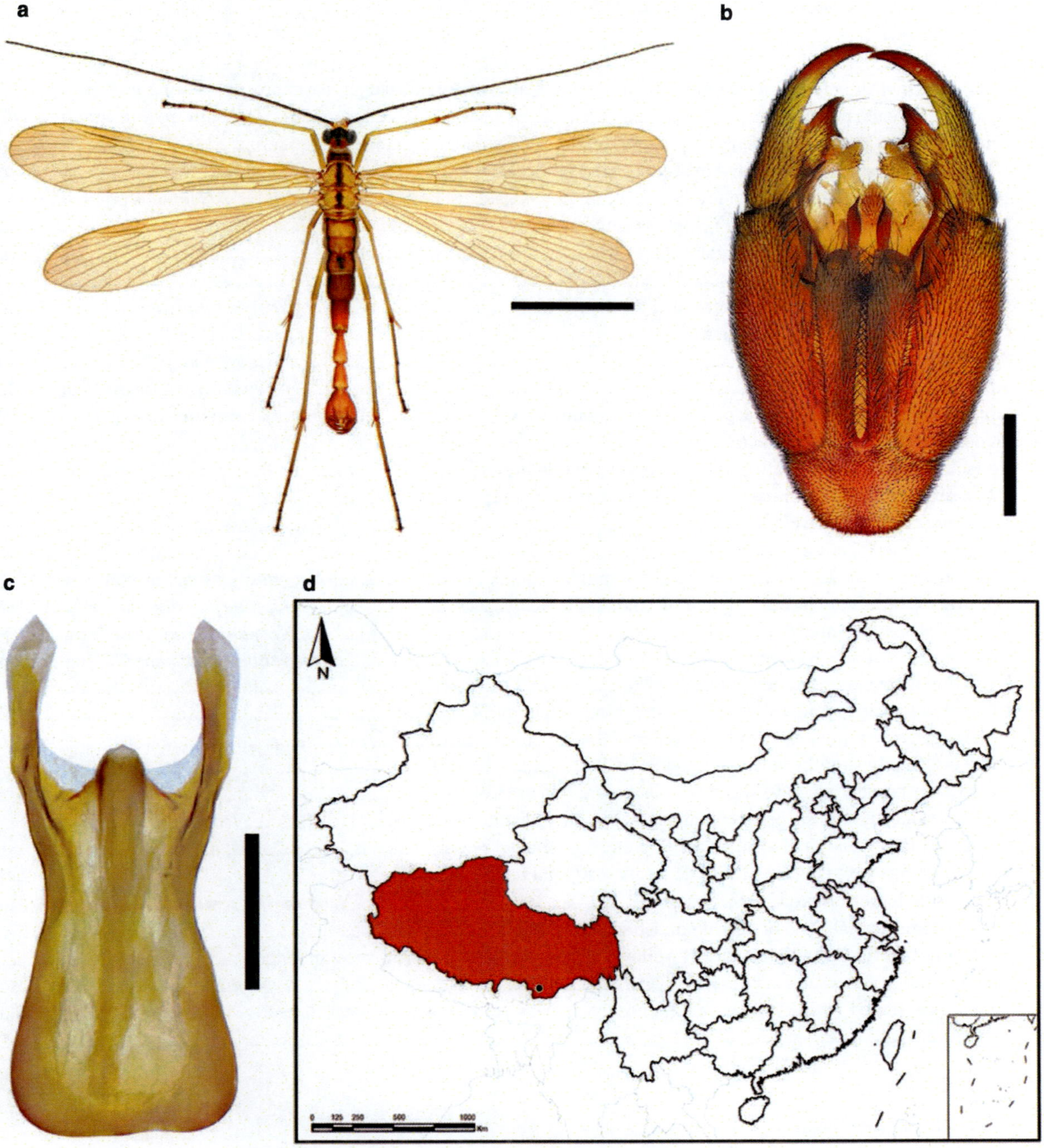

Fig. 10.7 (**a**) Male adult of ***Neopanorpa wuchaoi***, dorsal view (Cona, Tibet, NWAU) (*Scale bar*: 5.0 mm) **[DALU]**. (**b**) Male genital bulb of ***Neopanorpa wuchaoi***, ventral view (*Scale bar*: 0.5 mm). (**c**) Female medigynium of ***Neopanorpa wuchaoi***, ventral view (*Scale bar*: 0.2 mm). (**d**) Distribution of ***Neopanorpa wuchaoi***

Key to Species of Neopanorpa pielina Group in This Atlas
(34 species, males)

1. Wing markings greatly developed and covering more than half area of wings; basal process bearing greatly elongated, slender, and acute process .... ***Neopanorpa harmandi*** (Navás)
– Wing markings less developed; basal process lacking such process 2
2. Notal organ greatly elongated and exceeding hind border of A7; T4 greatly elongated, almost twice as long as T5, and bearing row of dense setae along midline .... ***Neopanorpa choui*** Cheng
– Notal organ not extending to A7; T4 not modified as above 3
3. A4–A6 each bearing postnotal organ 4
– Only A4 bearing postnotal organ 6
4. Gonostylus lacking tuft of long setae .... ***Neopanorpa longistipitata*** Wang & Hua
– Gonostylus bearing tuft of long setae 5
5. Abdomen yellowish brown .... ***Neopanorpa luojishana*** Wang & Hua
– Abdomen black .... ***Neopanorpa setigera*** Wang & Hua
6. Gonostylus greatly elongated and longer than gonocoxites .... ***Neopanorpa moganshanensis*** Zhou & Wu
– Gonostylys shorter than gonocoxites 7
7. T4 lacking membranous frame approximal to postnotal organ 8
– T4 bearing membranous frame approximal to postnotal organ 11
8. Wing markings well-developed 9
– Wing markings greatly reduced 10
9. Hypovalves slender and greatly constricted at middle .... ***Neopanorpa hualizhongi*** Hua & Chou
– Hypovalves broad and indistinctly constricted at middle .... ***Neopanorpa hainanica*** Hua & Chou
10. Hypovalves pointed apically .... ***Neopanorpa minuta*** Chou & Wang
– Hypovalves relatively rounded apically .... ***Neopanorpa puripennis*** Chou & Wang
11. Notal organ not exceeding hind border of T4 12
– Notal organ greatly elongated and exceeding hind border of T4 27
12. Notal organ extending nearly to hind border of T4 13
– Notal organ only slightly exceeding middle of T4 14
13. Hypovalves broad diamond-shaped with lateral margin greatly folded .... ***Neopanorpa pielina*** Navás
– Hypvalves narrow and stripe-like with lateral margin indistinctly folded .... ***Neopanorpa mutabilis*** Cheng
14. Wing markings greatly reduced, no dense pterostigmal band present 15
– Wing markings well-developed with dense pterostigmal band present 21
15. Wings with dense and broad apical band .... ***Neopanorpa semiorbiculata*** Wang & Hua
– Wings lacking apical band or bearing indistinct apical band 16
16. Wings tinged with yellow; nota mostly yellowish brown with very narrow median stripe .... ***Neopanorpa tincta*** Wang & Hua
– Wings not tinged with yellow; nota entirely black, or yellowish brown laterally with broad, black median stripe 17

17. Wings with water mark-like brown markings .... ***Neopanorpa quadristigma*** Wang & Hua
– Wings nearly devoid of markings 18
18. A7 and A8 dark brown to black .... ***Neopanorpa nigritis*** Carpenter
– A7 and A8 entirely or partly light brown 19
19. Hypovalves not exceeding apex of gonocoxites .... ***Neopanorpa malaisei*** Byers
– Hypovalves exceeding apex of gonocoxites 20
20. A7 and A8 entirely light brown.......... ***Neopanorpa pennyi*** Byers
– A7 and A8 light brown at base with apex dark brown .... ***Neopanorpa lifashengi*** Hua & Chou
21. Genital bulb black; dorsal process of aedeagus greatly enlarged and elongated .... ***Neopanorpa magnatitilana*** Wang & Hua
– Genital bulb yellowish brown; dorsal process of aedeagus simple 22
22. Head yellowish brown with black longitudinal stripe at middle .... ***Neopanorpa diancangshanensis*** Wang & Hua
– Head uniformly colored, dark brown or black 23
23. Basal stalk of hypandrium extremely shortened; hypovalves longer than basal stalk .... ***Neopanorpa maai*** Cheng
– Basal stalk of hypandrium long; hypovalves shorter than basal stalk 24
24. Basal process of gonostylus greatly enlarged and bowl-shaped 25
– Basal process of gonostylus not modified as above 26
25. Apical band lacking inner projection .... ***Neopanorpa tienpingshana*** Chou & Wang
– Apical band bearing inner projection .... ***Neopanorpa chaohsiufui*** Wang & Hua
26. Hypovalves shorter than basal stalk of hypandrium .... ***Neopanorpa triangulata*** Wang & Hua
– Hypovalves nearly as long as basal stalk of hypandrium .... ***Neopanorpa leigongshana*** Zhou & Zhou
27. Wings devoid of markings .... ***Neopanorpa zhengyucheni*** Wang
– Wings with well-developed markings 28
28. Additional band between apical and pterostigmal bands absent 29
– Additional band between apical and pterostigmal bands present 30
29. Apical band with inner projection .... ***Neopanorpa xingmini*** Wang & Hua
– Apical band lacking inner projection .... ***Neopanorpa tibetensis*** Hua & Chou
30. Hypandrial process greatly elongated 31
– Hypandrial process simple 32
31. Hypandrial process conical and can be observed from ventral aspect; hypovalves slender .... ***Neopanorpa jigongshanensis*** Hua
– Hypandrial process finger-like and concealed by broad hypovalves .... ***Neopanorpa longiprocessa*** Hua & Chou
32. Hypovalves narrower than base of gonostylus .... ***Neopanorpa retina*** Chou & Li
– Hypovalves broader than base of gonostylus 33
33. Hypovalves bearing subtriangular process on lateral margin near apex .... ***Neopanorpa mangshanensis*** Chou & Wang
– Hypovalves bearing subtriangular process on lateral margin near middle .... ***Neopanorpa ovata*** Cheng

10.1.5 *Neopanorpa chaohsiufui* Wang & Hua, 2017

(Hsiu-Fu Chao's Neoscorpionfly) [修复新蝎蛉]
(Fig. 10.8)
=*Neopanorpa chaoi* Zhou, 2003, preoccupied by *N. chaoi* Cheng, 1957

Measurements. Forewing length 12.0–13.0 mm.

Morphology. Vertex dark brown. Ocellar triangle black. Rostrum yellowish brown. Pronotum black. Meso- and metanotum yellowish brown laterally with black median stripe. Pleura and legs yellowish brown. Wing membrane hyaline with brown markings. In forewings, apical band incomplete posteriorly with slender projection innerly; pterostigmal complete, with broad basal branch and slightly reduced apical branch; marginal spot small; basal band split into two small spots; basal spot absent. Male T1–T5 dark brown. Notal organ short subtriangular. A6 dark brown. A7 and A8 yellowish brown and constricted basally.

Male genital bulb oval. Epandrium long broad, with truncated apex. Hypandrium with broad basal stalk. Hypovalves curled apically and overlapped mesally, forming subcircular window subbasally. Parameres parallel and convergent apically. Aedeagus conical, with stout lateral processes.

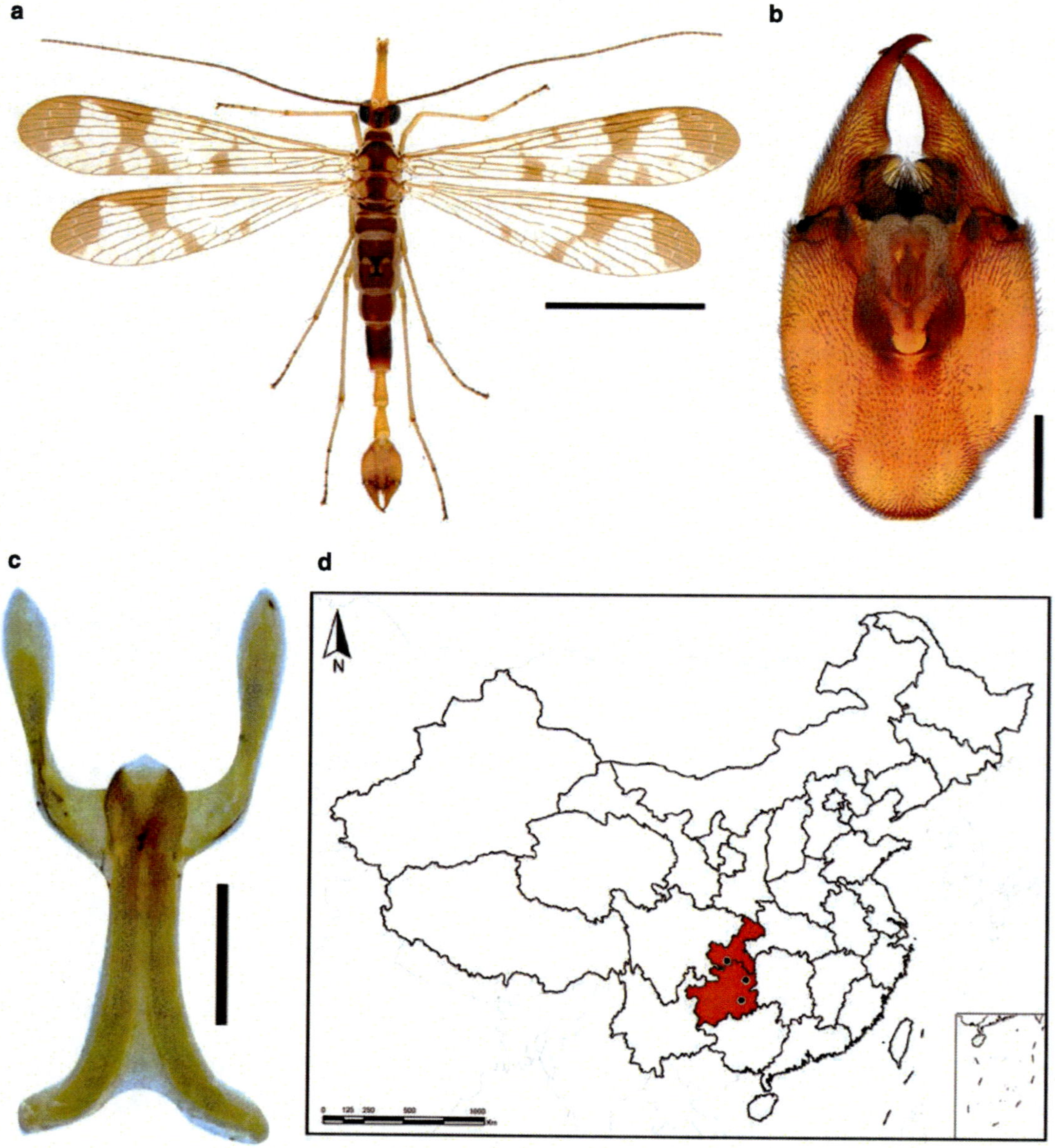

Fig. 10.8 (**a**) Male adult of ***Neopanorpa chaohsiufui***, dorsal view (Mount Jinfo, Nanchuan, Chongqing) (*Scale bar*: 5.0 mm) **[NWAU]**. (**b**) Male genital bulb of ***Neopanorpa chaohsiufui***, ventral view (*Scale bar*: 0.5 mm). (**c**) Female medigynium of ***Neopanorpa chaohsiufui***, ventral view (*Scale bar*: 0.2 mm). (**d**) Distribution of ***Neopanorpa chaohsiufui***

Female subgenital plate broadly oval, with V-shaped terminal emargination. Medigynium slender; axis slightly longer than posterior arms with greatly diverged apodemes.

Emergence. June to August.

Distribution. Chongqing (Mount Jinfo), Guizhou (Fanjing Mountain, Leigong Mountain).

10.1.6 *Neopanorpa choui* Cheng, 1949

(Chou's Neoscorpionfly) [周氏新蝎蛉] (Fig. 10.9)
Measurements. Forewing length 13.5–14.0 mm.

Morphology. Vertex black. Ocellar triangle black. Rostrum yellowish brown. Pronotum dark brown. Meso- and metanotum yellowish brown, with broad black median stripe. Pleura and legs yellowish brown. Wing membrane hyaline, slightly tinged with light yellow, markings brown. In forewings, apical band represented by series of scattered spots; pterostigmal with detached basal and apical branches; basal band split into two spots, or only represented by spot on terminal portion of CuP; marginal spot inconspicuous; basal spot absent. Male T1–T5 yellowish brown. Notal organ extremely elongated and extending beyond apex of A7 and bearing numerous cephalad-curling setae on ventral surface, with basal half arched dorsad and apex curved dorsad. T4 with row of short and cephalad-curling setae on dorso-mesal line, and nearly twice length of T3. A6 cylindrical. A7 and A8 slightly constricted basally.

Male genital bulb broad oval. Epandrium tapering caudad, and slightly emarginate terminally. Hypandrium with short broad basal stalk. Hypovalves narrow at base and acute at apex, broadest at distal 2/3, with sparse and long bristles on inner and outer margins. Gonostylus slightly shorter than gonocoxites, with obtuse triangular median tooth and oblong basal process. Paramere thin basally and foliate distally, with apex split into two acute tips. Aedeagus short and stout, not exceeding apex of gonocoxites, with stout lateral processes.

Female subgenital plate broadly oval, with V-shaped terminal emargination. Medigynium slender; axis slightly shorter than posterior arms with greatly diverged apodemes; posterior arms slender, greatly notched apically.

Emergence. June to July.

Distribution. Sichuan (Dujiangyan, Ya'an).

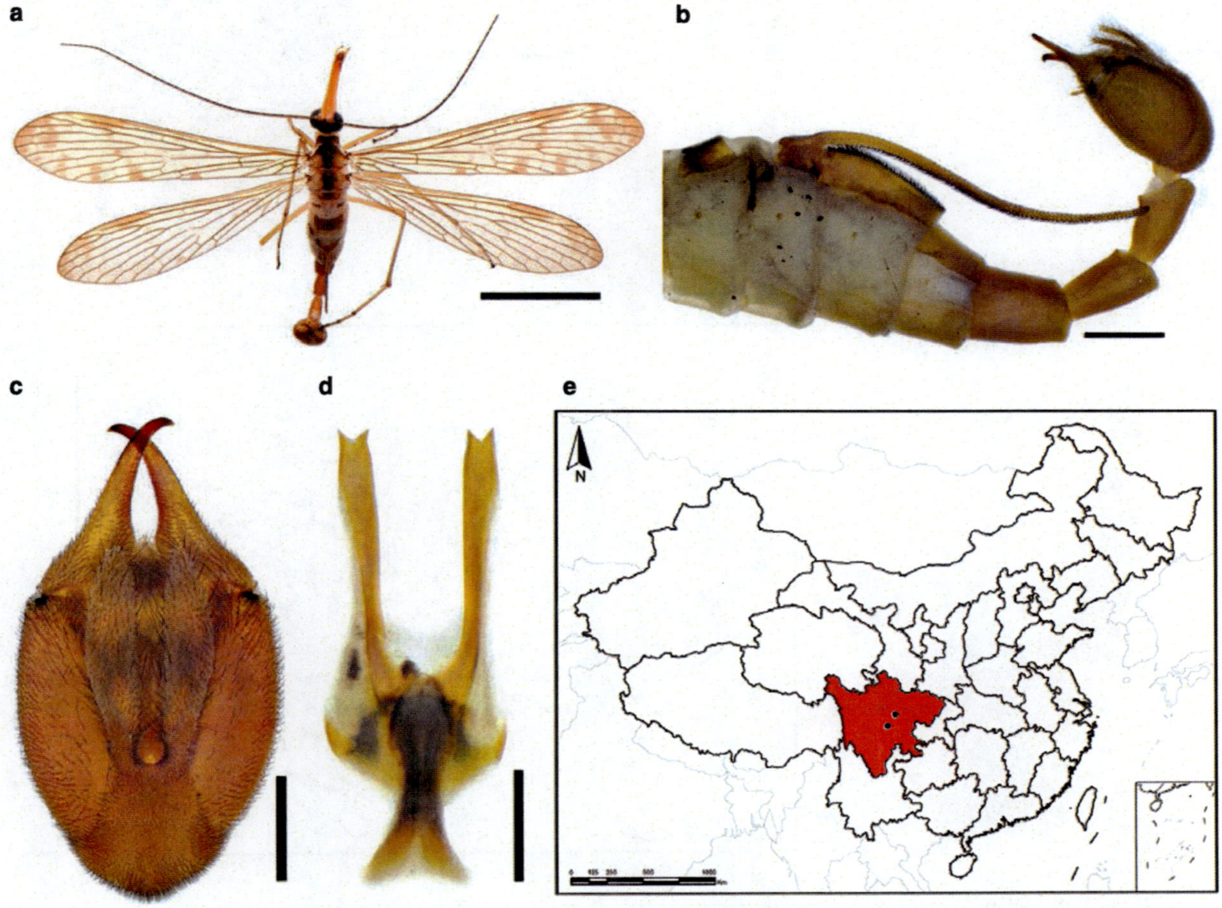

Fig. 10.9 (a) Male adult of ***Neopanorpa choui***, dorsal view (Qingcheng Mountain, Dujiangyan, Sichuan) (*Scale bar*: 5.0 mm) **[NWAU]. (b)** Male abdomen of ***Neopanorpa choui***, lateral view. (c) Male genital bulb of ***Neopanorpa choui***, ventral view. (d) Distribution of ***Neopanorpa choui***

10.1.7 *Neopanorpa diancangshanensis* Wang & Hua, 2018

(Diancangshan Neoscorpionfly) [点苍山新蝎蛉] (Fig. 10.10)
Measurements. Forewing length 12.5–13.8 mm.

Morphology. Vertex yellowish brown with longitudinal dark brown marking extending from ocellar triangle to occiput. Rostrum yellowish brown. Nota yellowish brown with black narrow longitudinal median stripe. Wing membrane hyaline with markings brown. In forewings, apical band incomplete, notched posteriorly with acute inner projection; pterostigmal band broad at base, forked posteriorly; basal band reduced to spot on posterior margin; basal and marginal spots absent. Male T1–T5 black. Notal organ elongated and extending to middle of T4, postnotal organ slightly produced. A6 black, cylindrical. A7 and A8 yellowish brown, slightly constricted basally.

Male genital bulb broad oval. Epandrium tapering caudad, and slightly emarginate terminally. Hypandrium with broad basal stalk. Hypovalves slender, twisted ventro-mesad, and widely divergent distally, bearing small hypandrial processes subbasally. Gonostylus brown with blunt median tooth and cup-like basal process. Ventral valves of aedeagus spatulate, widely divergent; dorsal valves broad and stout, concealed in ventral aspect; dorsal processes greatly expanded into large, subtriangular lobes. Paramere short and spatulate.

Female subgenital plate elliptical with deep V-shaped emargination terminally. Medigynium with axis extremely

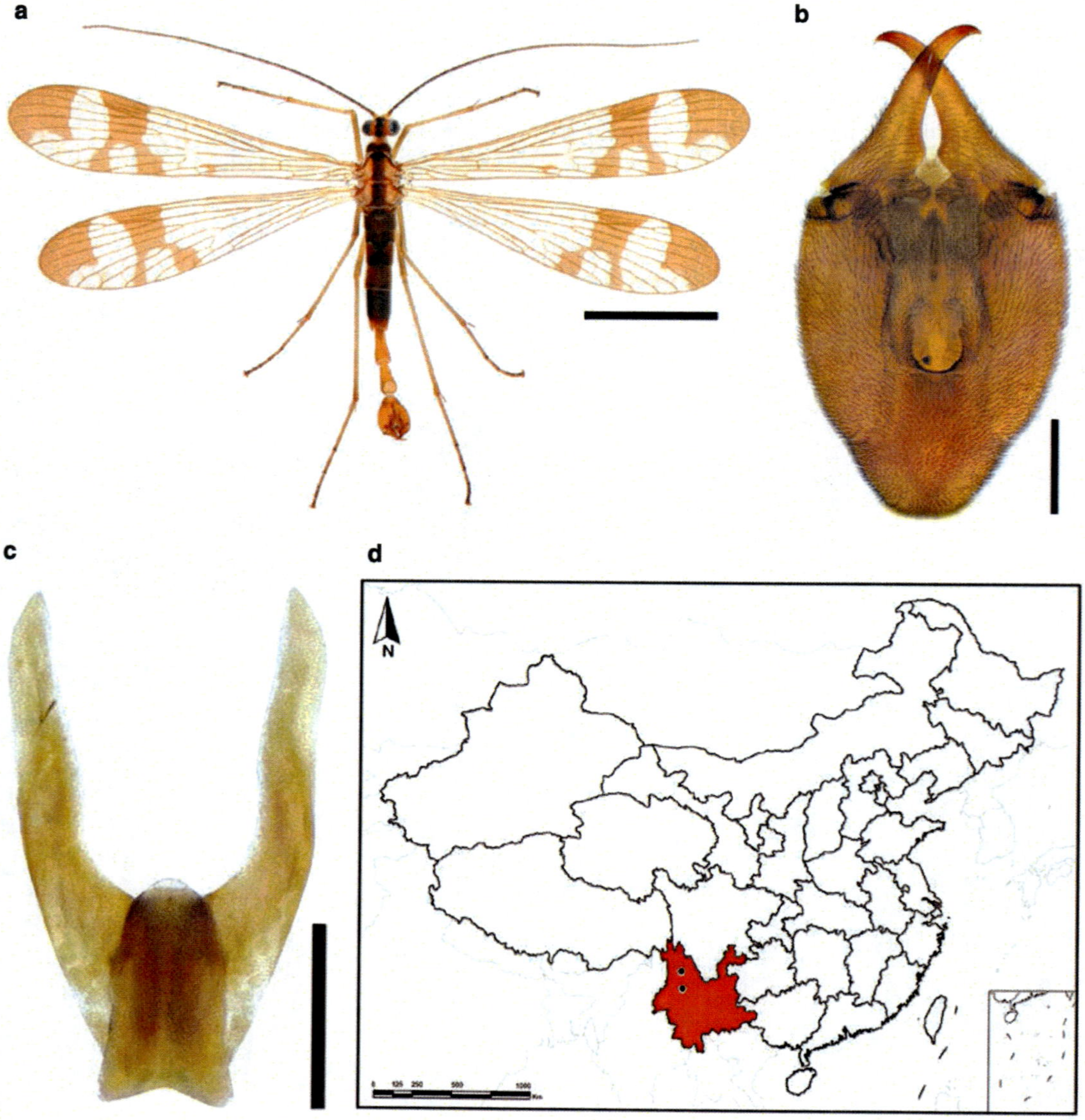

Fig. 10.10 (**a**) Male adult of ***Neopanorpa diancangshanensis***, dorsal view (Cangshan Mountain, Yunnan) (*Scale bar*: 5.0 mm) [NWAU]. (**b**) Male genital bulb of ***Neopanorpa diancangshanensis***, ventral view (*Scale bar*: 0.5 mm). (**c**) Female medigynium of ***Neopanorpa diancangshanensis***, ventral view (*Scale bar*: 0.2 mm). (**d**) Distribution of ***Neopanorpa diancangshanensis***

short; slender posterior arms twice as long as axis; posterior arms acute terminally.

Emergence. June to August.

Distribution. Yunnan (Cangshan Mountain, Tiger Leaping Gorge).

10.1.8 *Neopanorpa hainanica* Hua & Chou, 1998

(Hainan Neoscorpionfly) [海南新蝎蛉] (Fig. 10.11)

Measurements. Forewing length 12.0 mm.

Morphology. Vertex black. Ocellar triangle black. Rostrum yellowish brown. Nota mostly yellowish brown, with black median stripe. Pleura and legs yellowish brown. Wing membrane hyaline with markings brown. In forewings, apical band broad with long hyaline spot at posterior portion; pterostigmal band broad with thin apical branch and broad basal branch and complete, and nearly connected to apical band on anterior margin; basal band split into two spots; marginal spot conspicuous; basal spot absent. Male T1–T5 black. Notal organ triangular, covering rounded postnotal organ. A6 dark brown and long cylindrical. A7 and A8 yellowish brown and constricted basally.

Male genital bulb long oval. Epandrium long broad with truncated apex. Hypandrium with shorter basal stalk. Hypovalves slender and slightly convergent medially, diver-

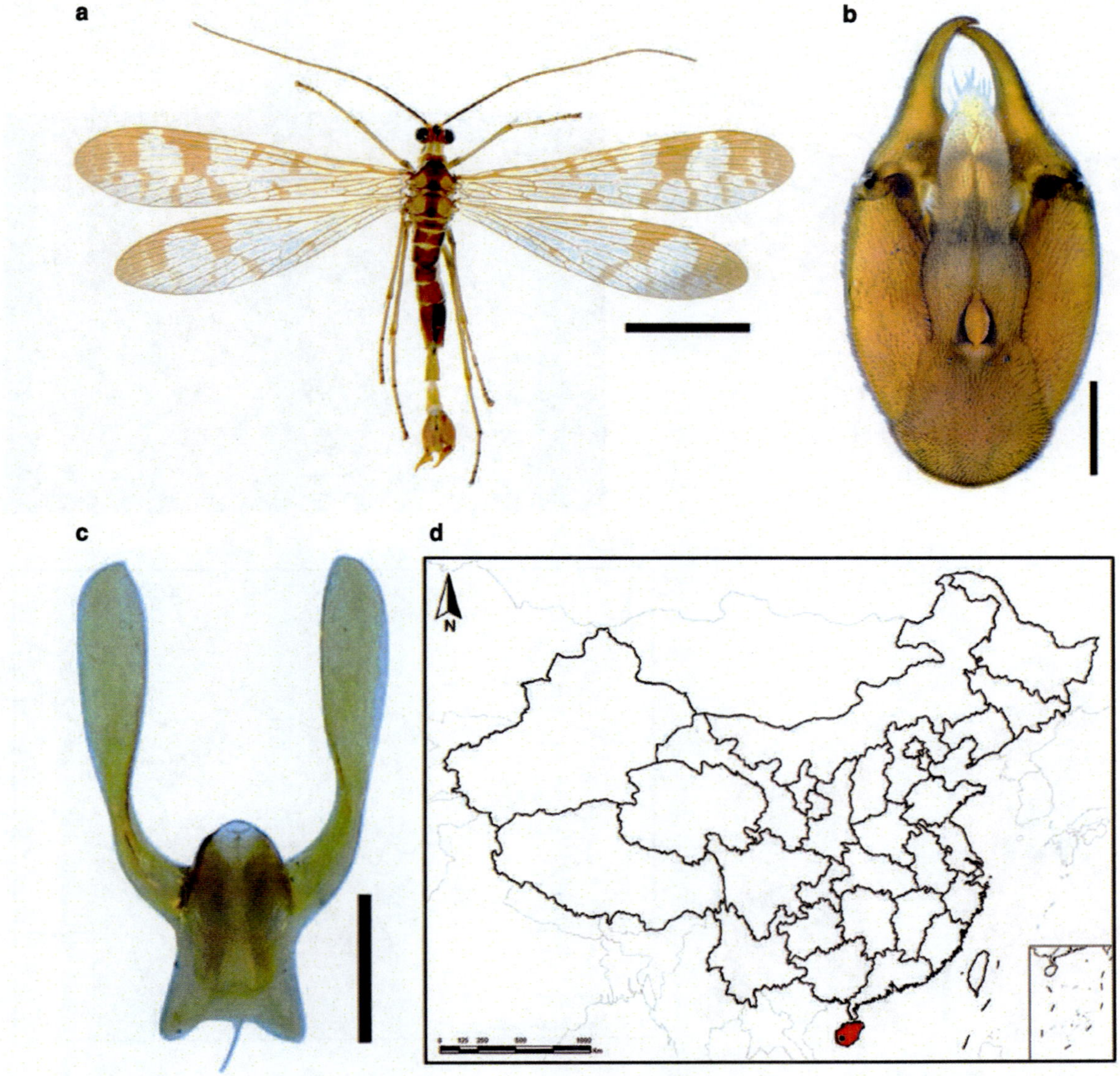

Fig. 10.11 (**a**) Male adult of ***Neopanorpa hainanica***, dorsal view (Jianfengling, Ledong, Hainan) (*Scale bar*: 5.0 mm) **[NWAU]**. (**b**) Male genital bulb of ***Neopanorpa hainanica***, ventral view (*Scale bar*: 0.5 mm). (**c**) Female medigynium of ***Neopanorpa hainanica***, ventral view (*Scale bar*: 0.2 mm). (**d**) Distribution of ***Neopanorpa hainanica***

gent distally, and exceeding apex of gonocoxites. Gonostylus with obtuse triangular median tooth and stout basal process. Aedeagus conical, with pair of stout lateral processes.

Female subgenital plate oval with deep V-shaped emargination terminally. Medigynium with axis short; slender posterior arms three times as long as axis; posterior arms rounded terminally.

Emergence. February to October.

Distribution. Hainan (Ledong).

10.1.9 *Neopanorpa harmandi* Navás, 1908

(Harmand's Neoscorpionfly) **[哈氏新蝎蛉]** (Fig. 10.12)

Measurements. Forewing length 14.0–15.0 mm.

Morphology. Vertex black. Ocellar triangle black. Rostrum yellowish brown. Pronotum dark brown. Meso- and metanotum yellowish brown with distinct black median stripe. Pleura and legs yellowish brown. Wing membrane hyaline with dense brown markings. In forewings, apical band broad; pterostigmal band very broad and connected to apical band anteriorly, with basal branch broad, and apical branch separated as small triangular spot connected to apical band occasionally; basal band very broad and connected to basal branch of pterostigmal band posteriorly; marginal band broad triangular and connected to pterostigmal band posteriorly; basal spot greatly elongated into stripe. Male T1–T5 dark brown. Notal organ slightly elongated and con-

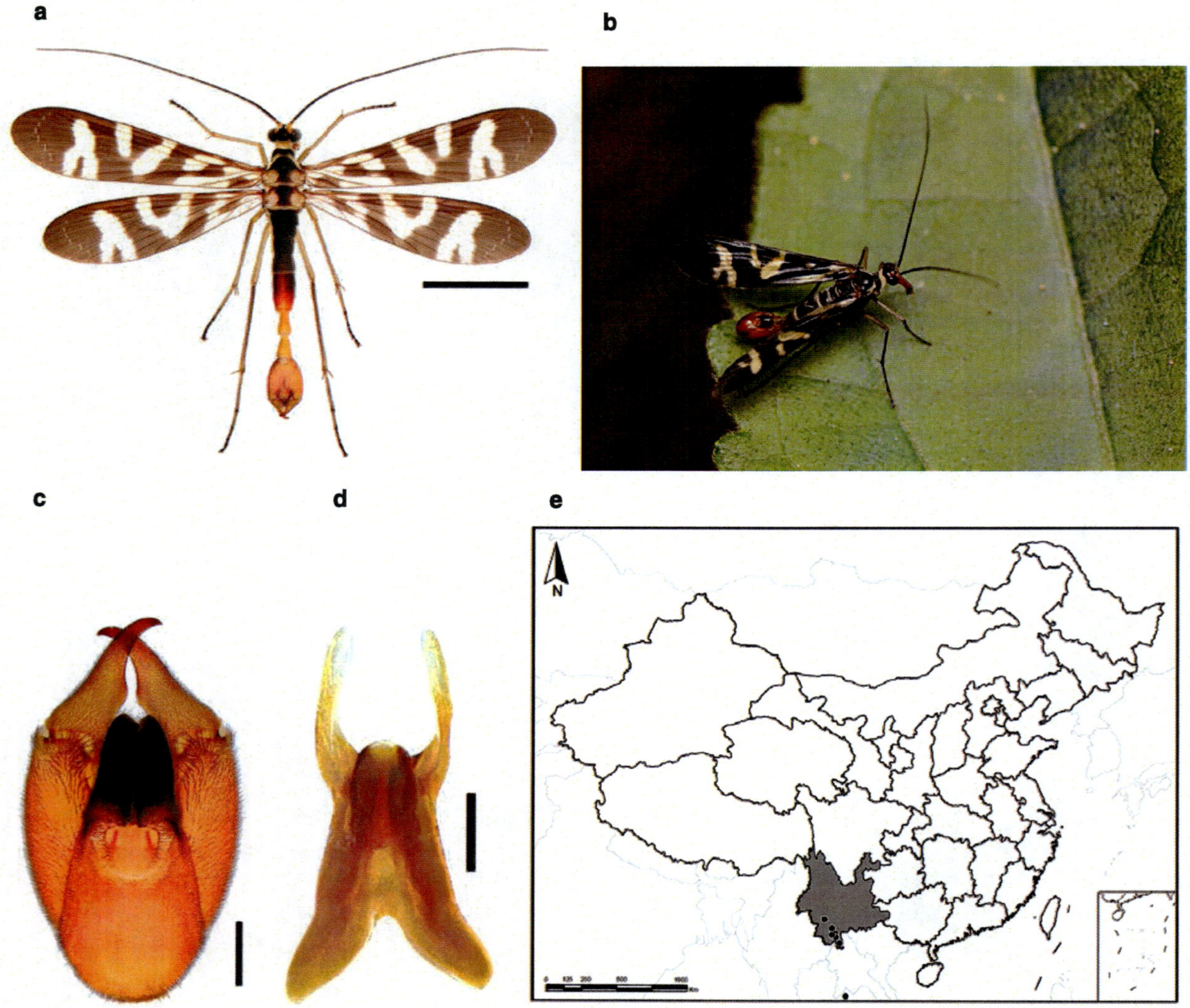

Fig. 10.12 (**a**) Male adult of ***Neopanorpa harmandi*** (*Scale bar*: 5.0 mm). (**b**) Male adult of ***Neopanorpa harmandi*** in the field (Sun River, Pu'er, Yunnan, © Chao Wu). (**c**) Male genital bulb of ***Neopanorpa harmandi*** (*Scale bar*: 0.5 mm) (**d**) Female medigynium of ***Neopanorpa harmandi***, ventral view (*Scale bar*: 0.5 mm). (**e**) Distribution of ***Neopanorpa harmandi***

ical. A6 dark brown and gradually turned to yellowish brown caudad. A7 and A8 yellowish brown and constricted basally.

Male genital bulb oval. Epandrium long broad, with truncated apex. Hypandrium with long broad basal stalk. Hypovalves subtriangular and slightly overlapped mesally, forming circular window subbasally. Gonostylus with obtuse triangular median tooth and bifurcated basal process. Aedeagus conical with pair of stout lateral processes.

Female subgenital plate oval, with V-shaped terminal emargination. Mediygnium with reduced main plate; posterior arms slender; axis almost three times as long as posterior arms with extremely stout and diverged apodemes.

Emergence. May to June.

Distribution. Yunnan (Jingdong, Shuangjiang, Xishuangbanna); Cambodia, Thailand.

10.1.10 *Neopanorpa hualizhongi* Hua & Chou, 1998

(Hua's Neoscorpionfly) [华氏新蝎蛉] (Fig. 10.13)
Measurements. Forewing length 12.0 mm.

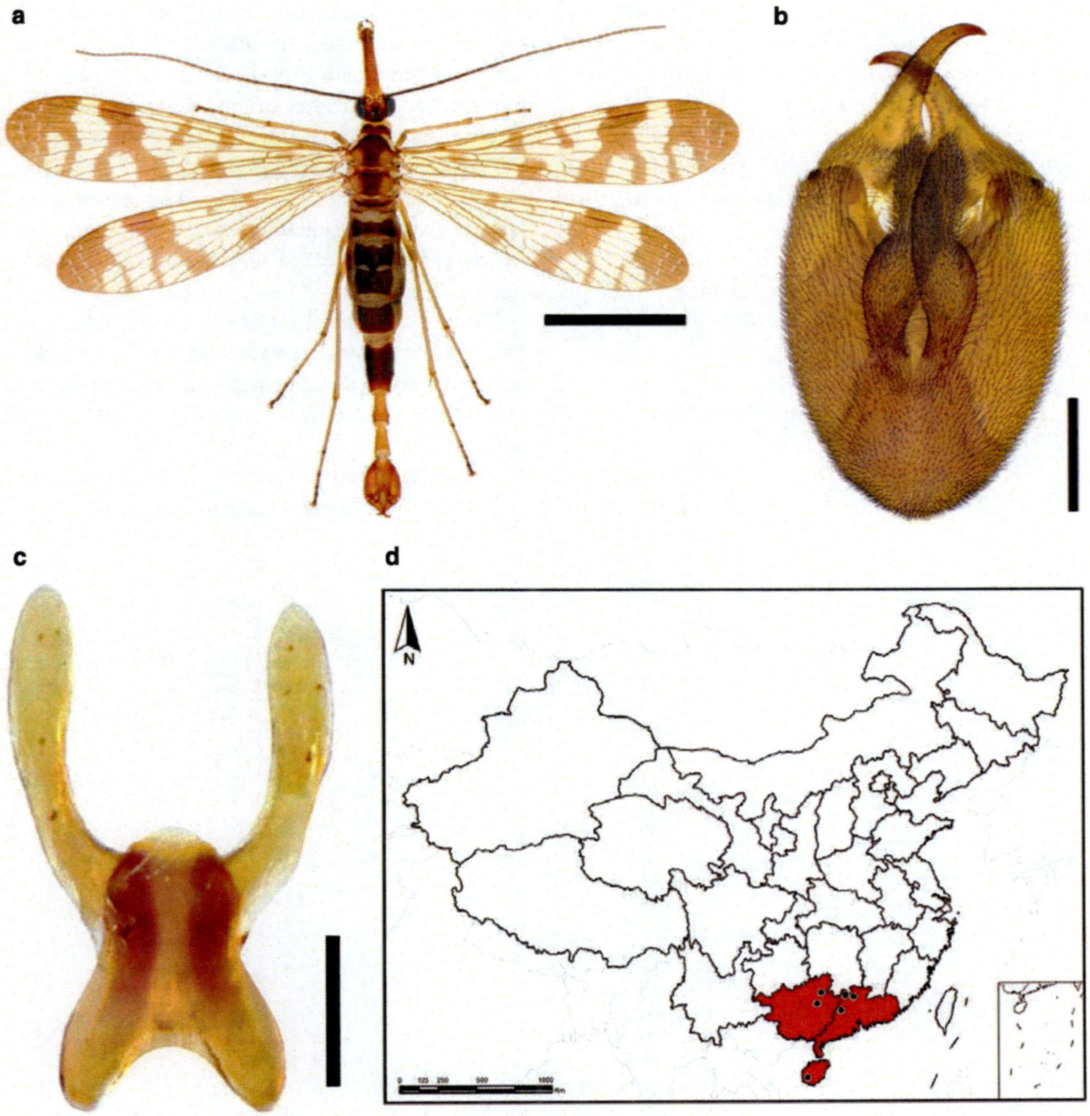

Fig. 10.13 (**a**) Male adult of ***Neopanorpa hualizhongi***, dorsal view (Jinxiu, Guangxi) (*Scale bar*: 5.0 mm) [NWAU]. (**b**) Male genital bulb of ***Neopanorpa hualizhongi***, ventral view (*Scale bar*: 0.5 mm). (**c**) Female medigynium, ventral view (*Scale bar*: 0.2 mm). (**d**) Distribution of ***Neopanorpa hualizhongi***

Morphology. Vertex black. Ocellar triangle black. Rostrum yellowish brown with darkened flanks. Pronotum dark brown. Meso- and metanotum light brown, with black median stripe. Pleura and legs yellowish brown. Wing membrane hyaline with brown markings. In forewings, apical band broad and with hyaline window posteriorly; pterostigmal band broad and with broad apical branch and thin basal branch; additional stripe between apical and pterostigmal bands; marginal spot conspicuous; basal band split into two spots; basal spot absent. Male T1–T5 dark brown. Notal organ short triangular. A6 dark brown. A7 and A8 yellowish brown and slightly constricted basally.

Male genital bulb oval. Epandrium long and broad with truncated apex. Hypandrium with long and broad basal stalk. Hypovalves slender and slightly overlapped mesally, and greatly constricted at middle. Gonostylus with obtuse triangular median tooth and large basal process. Aedeagus conical, with pair of stout lateral processes.

Female subgenital plate oval, with V-shaped terminal emargination. Mediygnium with reduced main plate; posterior arms slender and subacute terminally; axis shorter than posterior arms with stout and diverged apodemes.

Emergence. March to July.

Distribution. Hainan (Ledong), Guangxi (Jinxiu, Yangshuo), Guangdong (Fengkai, Lianzhou, Ruyuan).

10.1.11 *Neopanorpa jigongshanensis* Hua, 1998

(Jigongshan Neoscorpionfly) [鸡公山新蝎蛉] (Fig. 10.14)

Measurements. Forewing length 12.0 mm.

Morphology. Vertex yellowish brown. Dark brown stripe extending from hind border of frons to occiput, and passing through black ocellar triangle. Rostrum yellowish brown with two darkened longitudinal stripes laterally. Pronotum dark brown. Meso- and metanotum yellowish brown, with distinct black median stripe. Pleura and legs yellowish brown. Wing membrane hyaline and tinged with light yellow, with brown markings. In forewings, apical band broad with large hyaline window posteriorly; pterostigmal band complete with well-developed apical and basal branches; marginal spot large; basal band split into two spots; basal spot absent. Male T1–T5 dark brown. Notal organ long, stick-like, extending beyond hind border of T4. Trapezoidal membranous area on anterior portion of T4. A6 dark brown at basal half and gradually becoming yellowish brown towards apex. A7 and A8 yellowish brown and constricted basally.

Male genital bulb oval. Epandrium long and broad, with truncated apex. Hypandrium with broad basal stalk. Hypovalves slender and divergent in V-shape, and with acute triangular tooth on inner margin subbasally. Gonostylus with obtuse triangular median tooth, and large bowl-shaped basal process. Aedeagus laterally flattened with acute apex; ventral valves protruding basad and hook-like; lateral processes stout.

Female subgenital plate oval, with V-shaped terminal emargination. Mediygnium with reduced main plate; posterior arms slender and subacute terminally; axis shorter than posterior arms with stout and diverged apodemes.

Emergence. July.

Distribution. Henan (Tongbai, Xinyang).

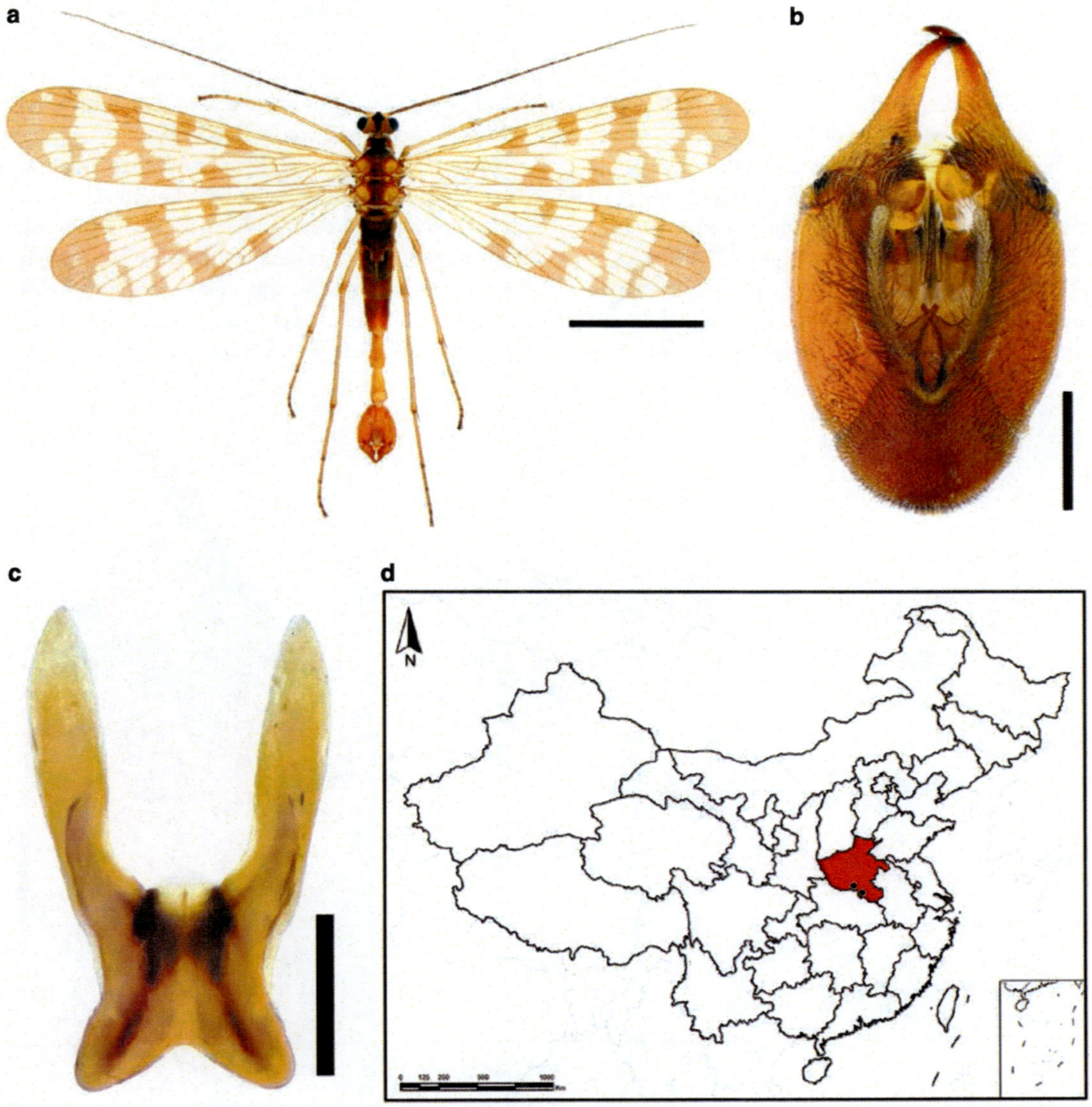

Fig. 10.14 (**a**) Male adult of ***Neopanorpa jigongshanensis***, dorsal view (Tongbai Mountain, Tongbai, Henan) (*Scale bar*: 5.0 mm) [NWAU]. (**b**) Male genital bulb of ***Neopanorpa jigongshanensis***, ventral view (*Scale bar*: 0.5 mm). (**c**) Female medigynium of ***Neopanorpa jigongshanensis***, ventral view (*Scale bar*: 0.2 mm). (**d**) Distribution of ***Neopanorpa jigongshanensis***

10.1.12 *Neopanorpa leigongshana* Zhou & Zhou, 2007

(Leigongshan Neoscorpionfly) [雷公山新蝎蛉] (Fig. 10.15)

Measurements. Forewing length 13.0 mm.

Morphology. Vertex dark brown. Ocellar triangle black. Rostrum yellowish brown. Pronotum black. Meso- and metanotum dark brown, with slightly yellowish brown flanks. Pleura and legs yellowish brown. Wing membrane hyaline with brown markings. In forewings, apical band incomplete posteriorly and ending at termination of R_5; pterostigmal band incomplete, connected to apical band anteriorly and with broad basal branch and separate reduced apical branch; marginal spot small; basal band split into two small spots; basal spot absent. Male T1–T5 dark brown. Notal organ short subtriangular. A6 dark brown. A7 and A8 yellowish brown and constricted basally.

Male genital bulb oval. Epandrium long broad, with truncated apex. Hypandrium with broad basal stalk. Hypovalves curled apically and overlapped mesally, forming subcircular window subbasally. Parameres parallel and convergent apically. Aedeagus conical, with stout lateral processes.

Female subgenital plate oval, with V-shaped terminal emargination. Mediygnium with reduced main plate; poste-

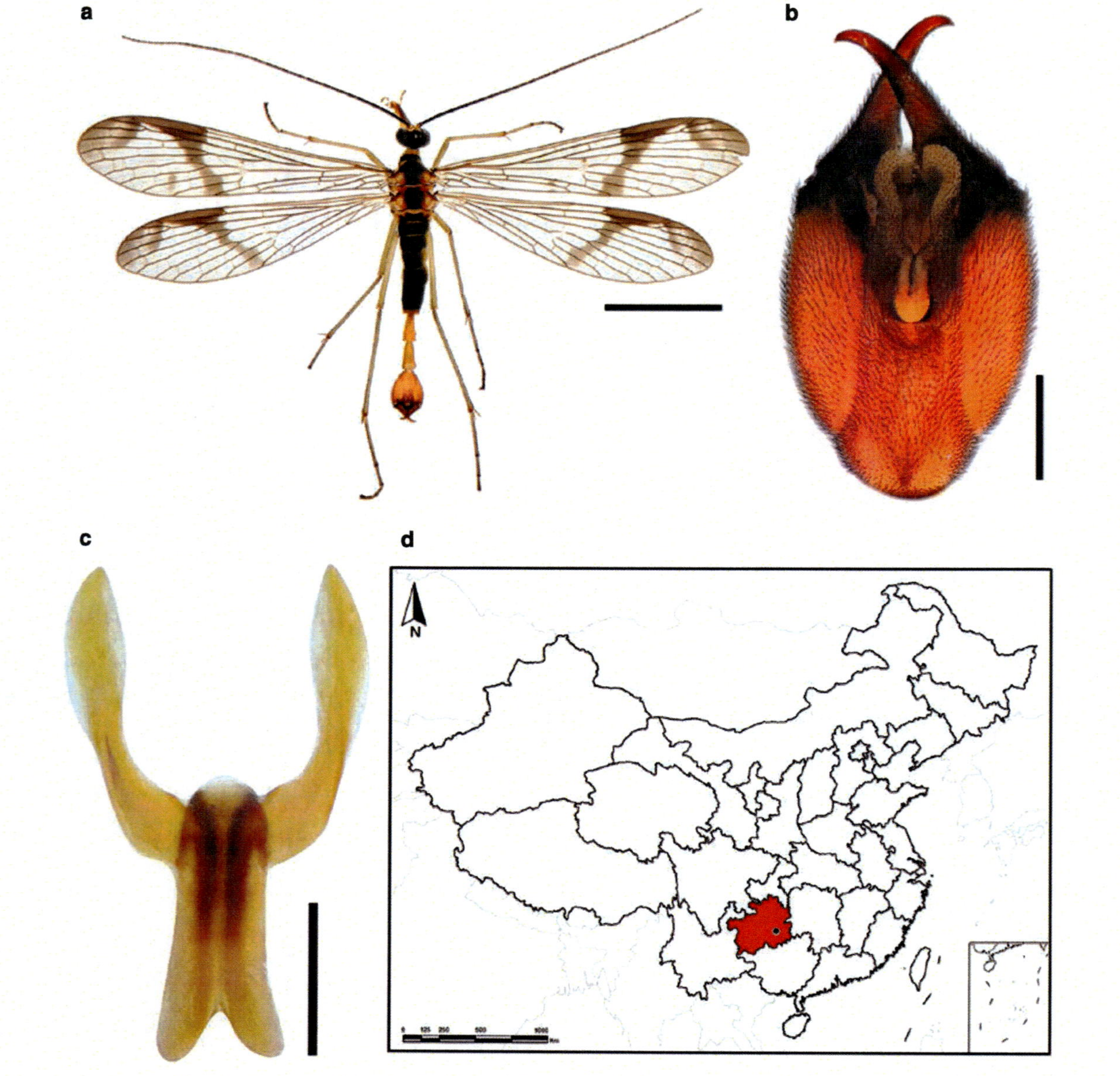

Fig. 10.15 (**a**) Male adult of ***Neopanorpa leigongshana***, dorsal view (Leigong Mountain, Guizhou) (*Scale bar*: 5.0 mm) **[NWAU]**. (**b**) Male genital bulb of ***Neopanorpa leigongshana***, ventral view (*Scale bar*: 0.5 mm). (**c**) Female medigynium of ***Neopanorpa leigongshana***, ventral view (*Scale bar*: 0.2 mm). (**d**) Distribution of ***Neopanorpa leigongshana***

rior arms slender, spatulate in distal half, and subacute terminally; axis slightly shorter than posterior arms with slightly diverged apodemes.

Emergence. June to August.

Distribution. Guizhou (Leigong Mountain).

10.1.13 *Neopanorpa lifashengi* Hua & Chou, 1999

(Li's Neoscorpionfly) [李氏新蝎蛉] (Fig. 10.16)

Measurements. Forewing length 15.0 mm.

Morphology. Vertex dark brown. Ocellar triangle black. Rostrum yellowish brown. Nota mostly dark brown, but latero-posterior corners of metanotum yellowish brown. Pleura and legs yellowish brown. Wing membrane hyaline, devoid of markings or only with reduced and triangular pterostigmal band. Pterostigma brown and distinct. Male T1–T5 dark brown. Notal organ short. A6 dark brown. A7 and A8 yellow in basal half and yellowish brown apical half.

Male genital bulb oval. Epandrium long broad with truncated apex. Hypandrium with broad basal stalk, extending to middle of gonocoxites. Hypovalves yellowish white brown

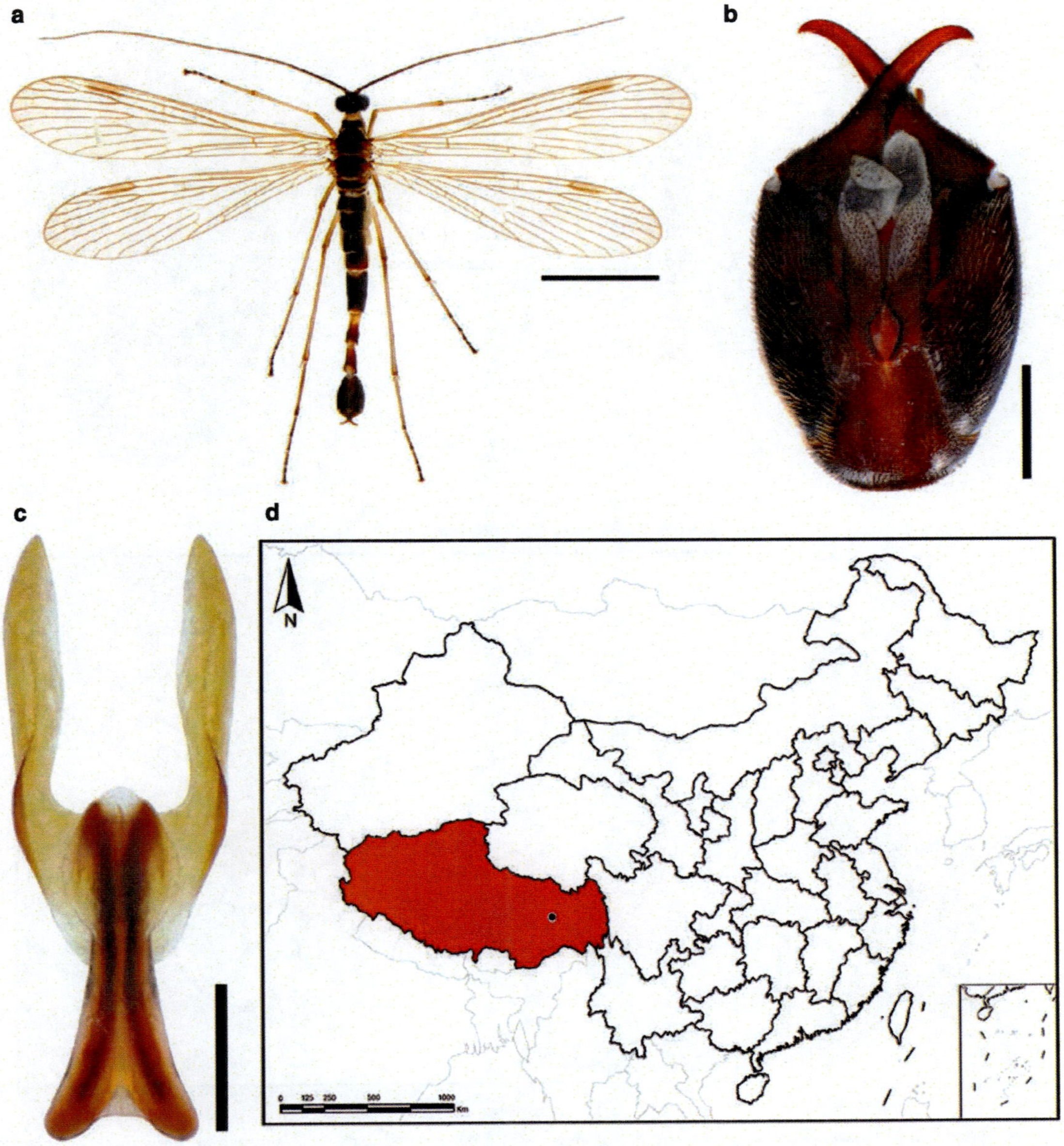

Fig. 10.16 (**a**) Male adult of ***Neopanorpa lifashengi***, dorsal view (Yi'ong, Bomê, Tibet) (*Scale bar*: 5.0 mm) [NWAU]. (**b**) Male genital bulb of ***Neopanorpa lifashengi***, ventral view (*Scale bar*: 0.5 mm). (**c**) Female medigynium of ***Neopanorpa lifashengi*** (*Scale bar*: 0.2 mm). (**d**) Distribution of ***Neopanorpa lifashengi***

and curled ventrad distally. Gonostylus slender with obtuse triangular median tooth and large basal process. Aedeagus conical, with stout lateral processes.

Female subgenital plate oval, with V-shaped terminal emargination. Mediygnium with slightly reduced main plate; posterior arms slender, spatulate in distal half, and subacute terminally; axis slightly shorter than posterior arms with slightly diverged apodemes.

Emergence. April to May.

Distribution. Tibet (Bomê).

10.1.14 *Neopanorpa longiprocessa* Hua & Chou, 1997

(Henan Long-Notal Organ Neoscorpionfly)
[河南新蝎蛉] (Fig. 10.17)

Measurements. Forewing length 13.5–15.0 mm.

Morphology. Vertex deep brown. Ocellar triangle black. Rostrum yellowish brown. Nota dark brown. Pleura creamy yellow. Legs mostly light yellow. Wing membrane hyaline with dark brown markings. In fore-

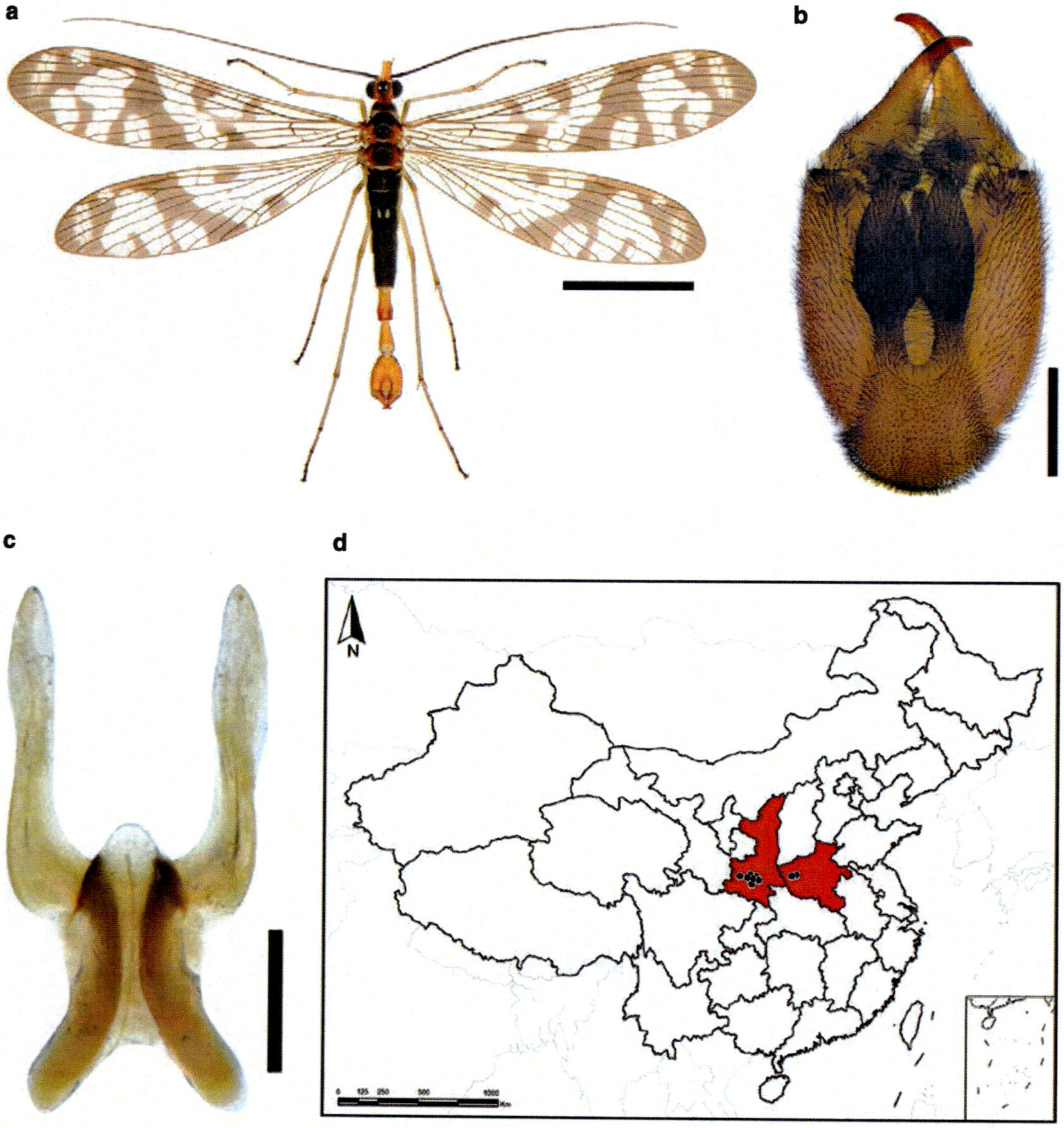

Fig. 10.17 (**a**) Male adult of ***Neopanorpa longiprocessa***, dorsal view (Qinling Mountains, Shaanxi) (*Scale bar*: 5.0 mm) **[NWAU]**. (**b**) Male adult of ***Neopanorpa longiprocessa,*** lateral view (*Scale bar*: 0.5 mm). (**c**) Male genital bulb of ***Neopanorpa longiprocessa***, ventral view (*Scale bar*: 0.2 mm). (**d**) Distribution of ***Neopanorpa longiprocessa***

wings, apical band broad, with two hyaline windows frontally; pterostigmal band with broad basal branch and thin apical branch; additional stripe between apical band and pterostigmal band; marginal spot large; basal band split into two spots; basal spot absent. Male T1–T5 dark brown. Notal organ stout, clavate, and almost extending to hind border of T5. A6 dark brown, cylindrical, and tapering posteriorly. A7 and A8 yellowish brown, slightly constricted basally.

Male genital bulb long oval. Epandrium long and broad, with truncated apex and pair of enlarged processes on ventral surface subapically. Hypandrium with broad basal stalk. Hypovalves broad, flat, and slightly overlapped mesally, forming circular window subbasally. Each hypovalve with finger-like basal process. Gonostylus with obtuse triangular median tooth and large trapezoidal basal process. Parameres fused at basal half with aedeagus and divergent at apical half. Aedeagus conical, with pair of stout lateral processes.

Female subgenital plate oval, with V-shaped terminal emargination. Mediygnium with reduced main plate; posterior arms slender, subacute terminally; axis shorter than posterior arms with greatly diverged apodemes.

Emergence. May to August.

Distribution. Henan (Luanchuan, Songxian), Shaanxi (Baoji, Huyi, Ningshan, Taibai, Zhouzhi).

10.1.15 *Neopanorpa longistipitata* Wang & Hua, 2018

(Yunnan Long-Notal Organ Neoscorpionfly)
[长杆新蝎蛉] (Fig. 10.18)

Measurements. Forewing length 12.0–14.0 mm.

Morphology. Vertex yellow, ocellar triangle covered by dark brown marking extending caudally to occiput. Rostrum unevenly yellowish brown. Pronotum dark brown. Meso- and metanotum yellowish brown, with distinct black median

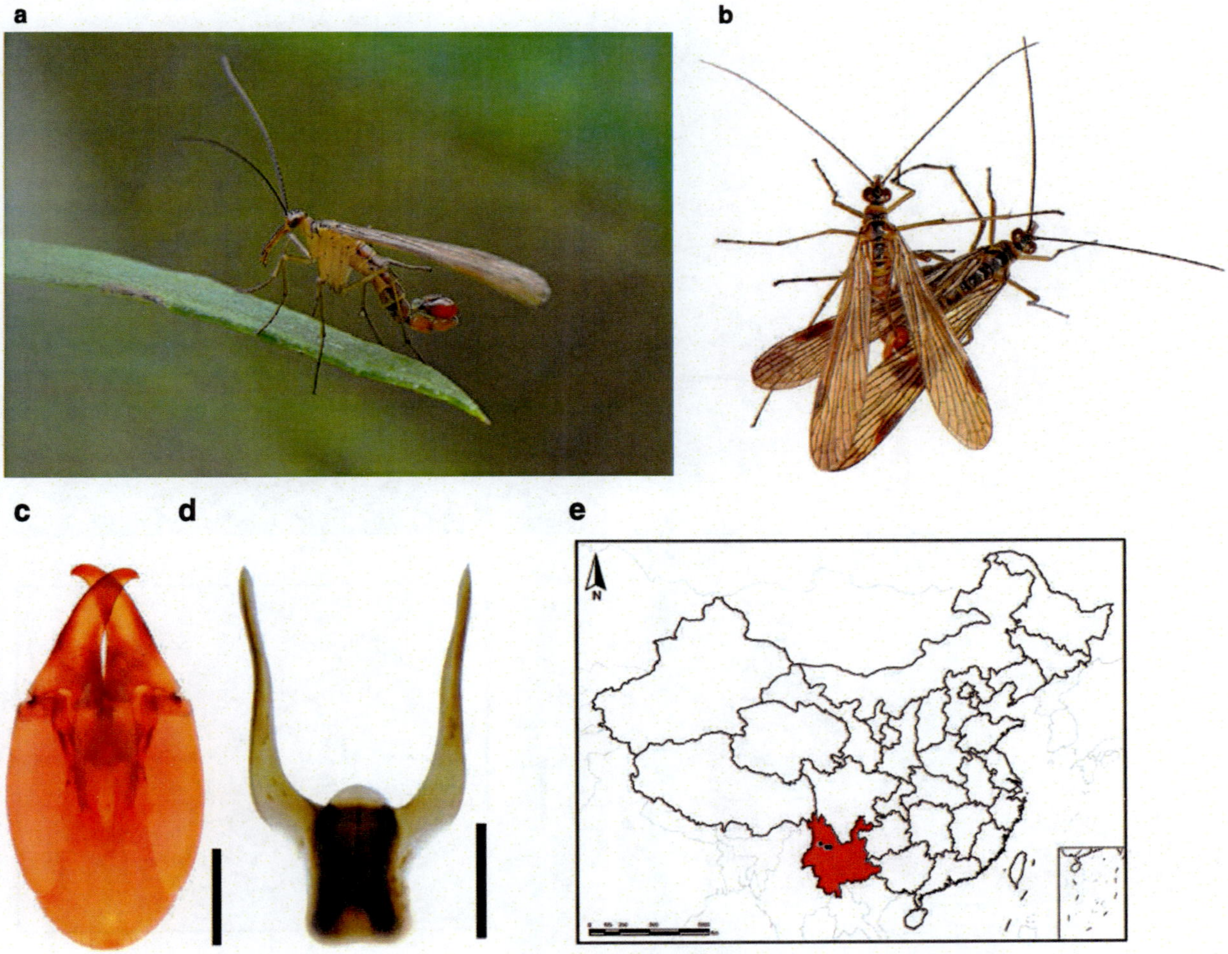

Fig. 10.18 (**a**). Male adult of ***Neopanorpa longistipitata***, lateral view (Cangshan Mountain, Yunnan). (**b**) A mating couple of ***Neopanorpa longistipitata***, dorsal view. (**c**) Male genital bulb of ***Neopanorpa longistipitata***, ventral view (*Scale bar*: 0.5 mm). (**d**) Female medigynium of ***Neopanorpa longistipitata***, ventral view (*Scale bar*: 0.2 mm). (**e**) Distribution of ***Neopanorpa longistipitata***

stripe. Pleura and legs yellowish brown. Wing membrane hyaline and deeply tinged with yellow, lacking distinct markings. In forewings, only greatly reduced apical and pterostigmal bands present. Pterostigma reddish brown and distinct. Male abdomen yellowish to reddish brown. Notal organ greatly elongated and stick-like, extending beyond middle of A6; postnotal organ indistinct. A6 cylindrical and tapering towards apex. A7 and A8 greatly constricted and stalk-shaped basally.

Male genital bulb long elliptical. Epandrium long and broad, with truncated apex. Hypandrium with broad basal stalk. Hypovalves slender. Gonostylus with obtuse triangular median tooth, and large concaved basal process. Aedeagus laterally flattened with acute apex; ventral valves blade-like; lateral processes stout; parameres lender.

Female subgenital plate elliptical with deep V-shaped emargination terminally. Medigynium with axis short and stout; apodemes indistinctly diverged; slender posterior arms three times as long as axis.

Emergence. May to September.

Distribution. Yunnan (Binchuan, Dali, Yunlong).

10.1.16 *Neopanorpa luojishana* Wang & Hua, 2019

(Luojishan Neoscorpionfly) [螺髻山新蝎蛉] (Fig. 10.19)

Measurements. Forewing length 13.5–14.0 mm.

Morphology. Vertex yellow, ocellar triangle covered by dark brown marking extending caudally to occiput. Rostrum unevenly yellowish brown. Pronotum dark brown. Meso- and metanotum yellowish brown, with distinct black median stripe. Pleura and legs yellowish brown. Wing membrane hyaline and slightly tinged with yellow, lacking distinct markings. Pterostigma reddish brown and distinct. Male abdomen yellowish brown. Notal organ greatly elongated and stick-like, extending beyond hind border of T5; postnotal organ indistinct, setose. A6 cylindrical and tapering towards apex. A7 and A8 greatly constricted and stalk-shaped basally.

Male genital bulb long elliptical. Epandrium long and broad, with truncated apex. Hypandrium with broad basal stalk. Hypovalves slender. Gonostylus with obtuse triangular median tooth, large concaved basal process, and bearing numerous long setae in middle of ventral surface. Aedeagus

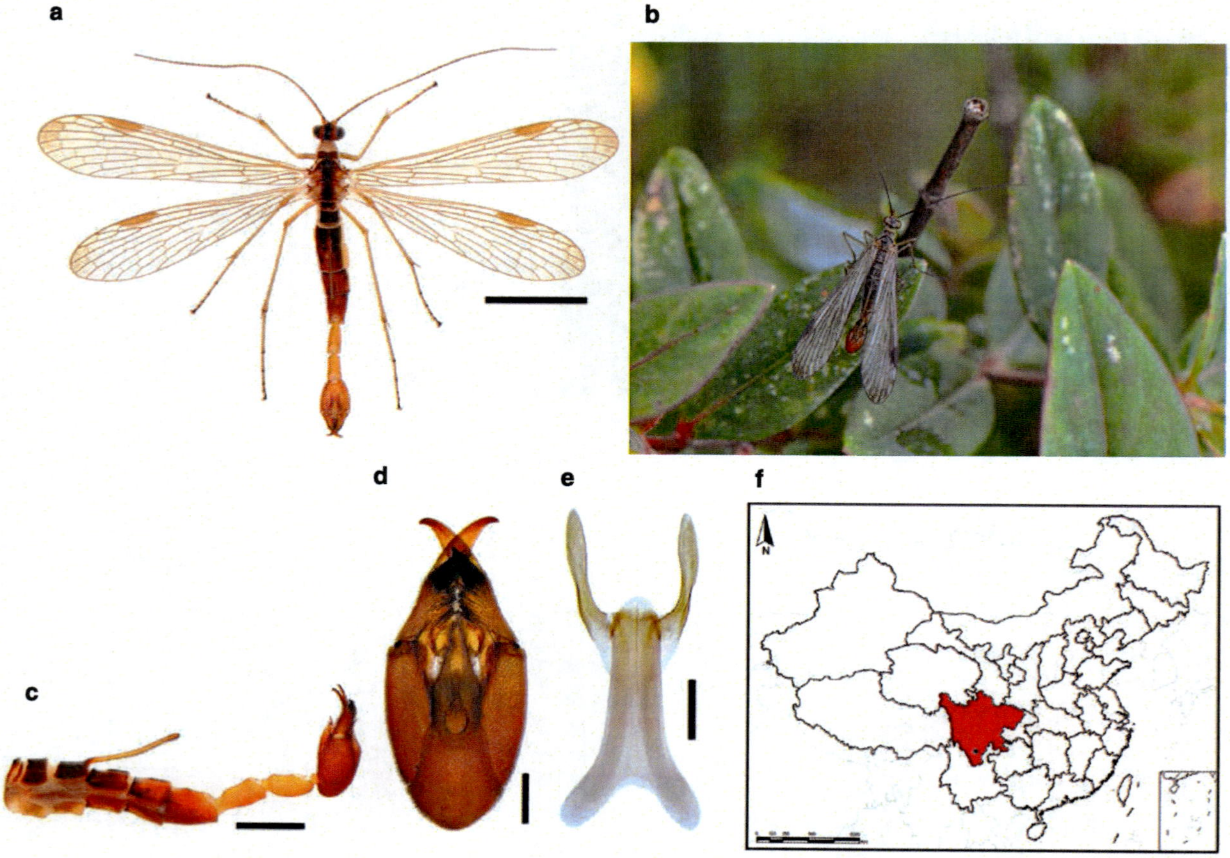

Fig. 10.19 (**a**) Male adult of ***Neopanorpa luojishana***, dorsal view (Luoji Mountain, Sichuan) (*Scale bar*: 5.0 mm) **[NWAU]**. (**b**) Male adult of ***Neopanorpa luojishana*** in the field (Luoji Mountain, Sichuan). (**c**) Male abdomen of ***Neopanorpa luojishana***, lateral view (*Scale bar*: 2.0 mm). (**d**) Male genital bulb of ***Neopanorpa luojishana***, ventral view (*Scale bar*: 0.5 mm). (**e**) Female medigynium of ***Neopanorpa luojishana***, ventral view (*Scale bar*: 0.2 mm). (**f**) Distribution of ***Neopanorpa luojishana***

laterally flattened with acute apex; ventral valves blade-like; lateral processes stout; parameres slightly spatulate.

Female subgenital plate elliptical with deep V-shaped emargination terminally. Medigynium with axis greatly elongated; apodemes greatly diverged basally; slender posterior arms half as long as axis.

Emergence. June.

Distribution. Sichuan (Luoji Mountain).

10.1.17 *Neopanorpa maai* Cheng, 1957

(Maa's Neoscorpionfly) [马氏新蝎蛉] (Fig. 10.20)

Measurements. Forewing length 14.0 mm.

Morphology. Vertex reddish brown. Ocellar triangle enclosed by large black spot. Rostrum reddish brown, with two deep brown longitudinal stripes. Pronotum dark brown. Meso- and metanotum mostly dark brown, with latero-posterior corners reddish brown. Pleura and legs mostly yellowish brown. Wing membrane hyaline, slightly tinged with greyish brown. In forewings, apical band broad, with large hyaline window innerly, and blurred posteriorly; pterostigmal band broad, with separated apical branch; additional spot between apical and pterostigmal bands; basal band scattered and connected to small marginal spot; basal spot absent. Male T1–T5 black. Notal organ on hind border of T3 long clavate, not exceeding hind border of T4; T4 with trapezoidal membranous region on anterior portion mesally. A6 black and tapering towards apex. A7 and succeeding segments reddish brown.

Male genital bulb oval. Epandrium long and broad, with truncated apex. Hypandrium with broad but short basal stalk.

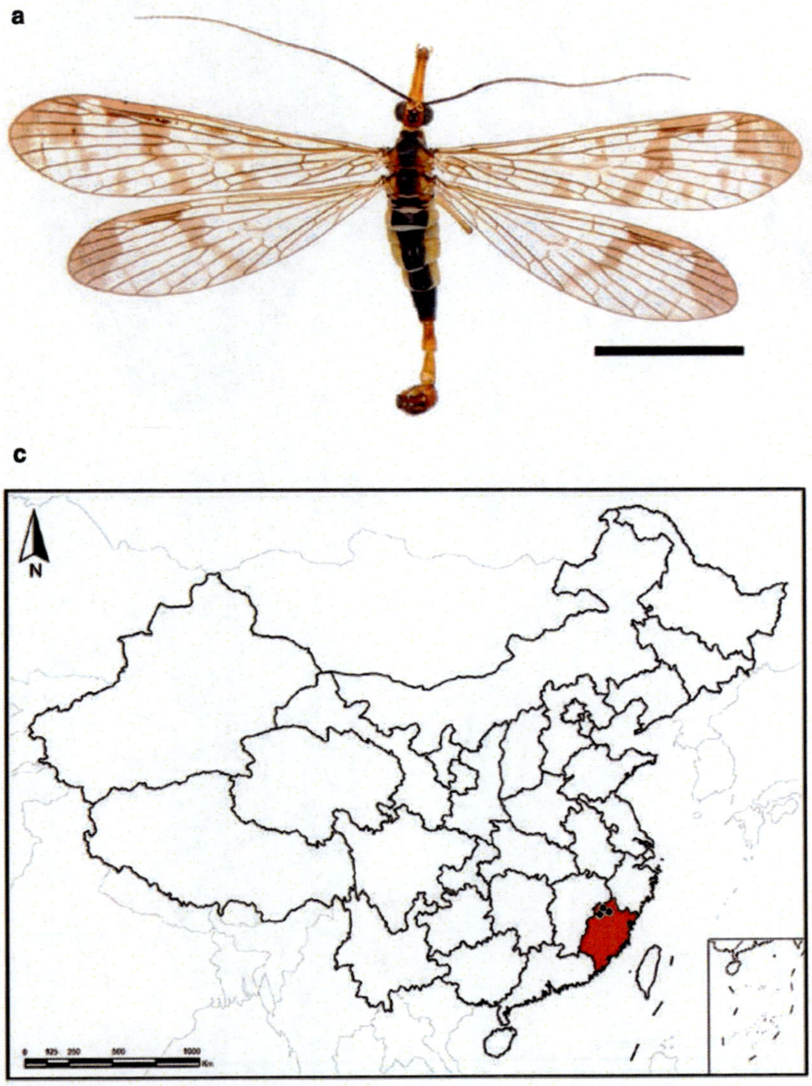

Fig. 10.20 (**a**) Male adult of ***Neopanorpa maai***, dorsal view (Dazhulan, Shaowu, Fujian) (*Scale bar*: 5.0 mm) **[NWAU]**. (**b**) Male genital bulb of ***Neopanorpa maai***, ventral view (*Scale bar*: 0.5 mm). (**c**) Distribution of ***Neopanorpa maai***

Hypovalves slender and stripe-like, exceeding apex of gonocoxites apically. Gonostylus with obtuse triangular median tooth and disc-shaped basal process. Paramere thread-like, not exceeding apex of aedeagus; aedeagus conical, not exceeding apex of gonocoxites; lateral processes of aedeagus stout.

Female subgenital plate oval, with V-shaped terminal emargination. Mediygnium with reduced main plate; posterior arms slender, spatulate in distal half, and subacute terminally; axis slightly shorter than posterior arms with parallel apodemes.

Emergence. April to July.

Distribution. Fujian (Jianyang, Shaowu, Wuyishan).

10.1.18 *Neopanorpa magnatitilana* Wang & Hua, 2018

(Large-Process Neoscorpionfly) [巨突新蝎蛉] (Fig. 10.21)

Measurements. Forewing length 12.5–13.0 mm.

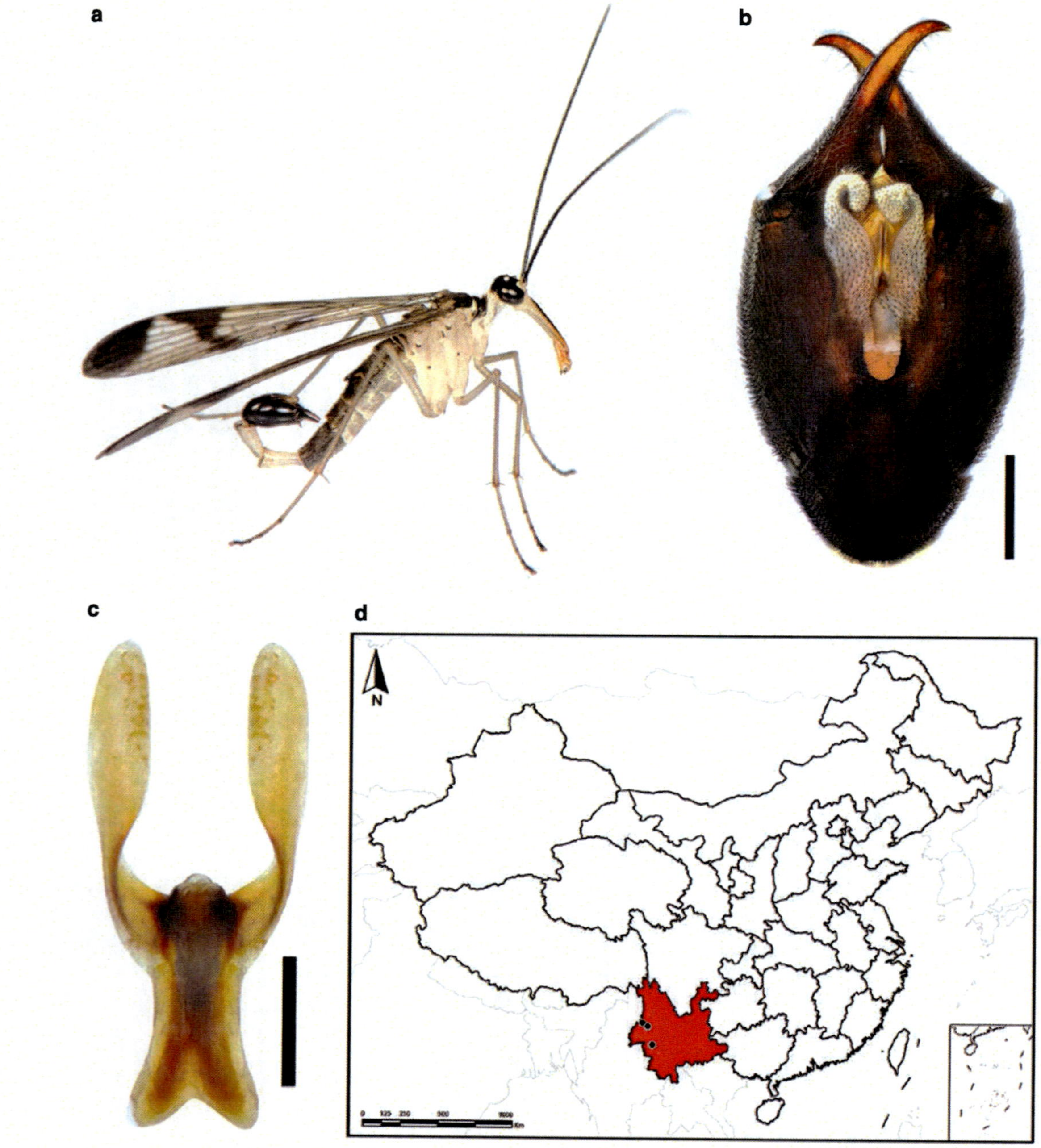

Fig. 10.21 (a) Male adult of ***Neopanorpa magnatitilana***, lateral view (Tangli Mountain, Yongde, Yunnan). (**b**) Male genital bulb of ***Neopanorpa magnatitilana***, ventral view (*Scale bar*: 0.5 mm). (**c**) Female medigynium of ***Neopanorpa magnatitilana***, ventral view (*Scale bar*: 0.2 mm). (**d**) Distribution of ***Neopanorpa magnatitilana***

Morphology. Vertex black. Ocellar triangle black. Rostrum dark brown. Nota yellowish brown with broad, black median stripe. Pleura and legs greyish brown. Wing membrane hyaline, tinged with greyish brown. In forewings, apical band extending to M_1 or M_2; pterostigmal band broad, with apical branch absent and basal branch running obliquely to posterior margin; basal band reduced to small spot. Male T1–T5 black. Notal organ clavate, reaching middle of T4; postnotal organ rounded and setose. A6 black, cylindrical. A7 and A8 light brown and constricted basally.

Male genital bulb long oval, mostly black except whitish hypovalves and apex of epandrium. Epandrium long and broad, slightly emarginate terminally. Hypandrium with broad basal stalk. Hypovalves slender, greatly curving mesad apically. Gonostylus with obtuse triangular median tooth and large cylindrical basal process. Paramers thin. Aedeagus conical, with stout lateral processes; dorsal processes greatly elongated and foliate.

Female subgenital plate elliptical with deep V-shaped emargination terminally. Medigynium with axis greatly elongated and diverged apodemes; posterior arms longer than axis and spatulate.

Emergence. June to August.

Distribution. Yunnan (Gaoligong Mountains, Tangli Mountain, Tengchong, Tongbiguan).

10.1.19 *Neopanorpa malaisei* Byers, 1999

(Malaise's Neoscorpionfly) [马来氏新蝎蛉] (Fig. 10.22)

Measurements. Forewing length 12.0–3.0 mm.

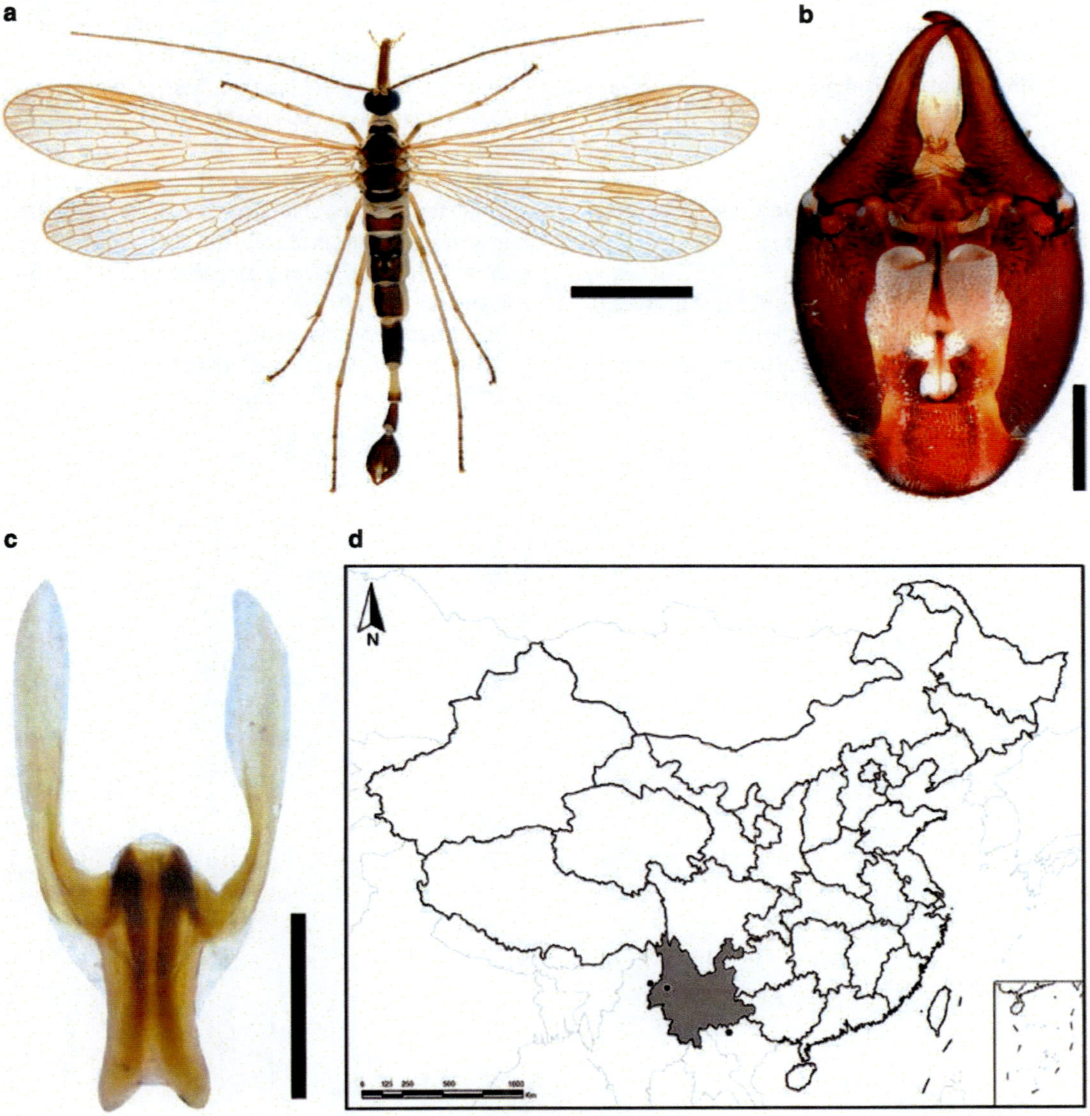

Fig. 10.22 (a) Male adult of ***Neopanorpa malaisei***, dorsal view (Gaoligong Mountains, Yunnan) (*Scale bar*: 5.0 mm) **[NWAU]**. **(b)** Male genital bulb of ***Neopanorpa malaisei***, ventral view (*Scale bar*: 0.5 mm). **(c)** Female medigynium of ***Neopanorpa malaisei***, ventral view (*Scale bar*: 0.2 mm). **(d)** Distribution of ***Neopanorpa malaisei***

Morphology. Vertex black. Ocellar triangle black. Rostrum dark brown. Nota black. Pleura and legs yellowish brown. Wing membrane hyaline and devoid of markings. Pterostigma light brown. Male T1–T5 black. Notal organ long, clavate, reaching middle of T4. A6 black, cylindrical. A7 light brown in basal 4/5 and black in distal 1/5; A8 black.

Male genital bulb oval, mostly dark brown except whitish hypovalves. Epandrium long and broad, greatly emarginate terminally, epandrial processes slender and bifurcated terminally. Hypandrium with short and broad basal stalk. Hypovalves broad, greatly curving mesad apically; hypandrial processes slender and acute. Gonostylus with obtuse triangular median tooth and large cylindrical basal process. Paramere foliate. Aedeagus conical, with stout lateral processes; dorsal processes slightly elongated.

Female subgenital plate elliptical with shallow emargination terminally. Medigynium with axis slightly shorter than spatulate posterior arms.

Emergence. May to August.

Distribution. Yunnan (Gaoligong Mountains); Myanmar (Kachin).

10.1.20 *Neopanorpa mangshanensis* Chou & Wang, 1988

(Mangshan Neoscorpionfly) [莽山新蝎蛉] (Fig. 10.23)

Measurements. Forewing length 13.0 mm.

Morphology. Vertex dark brown. Ocellar triangle black. Rostrum yellowish brown. Pronotum dark brown. Meso- and metanotum mostly dark brown, with latero-posterior corners yellowish brown. Pleura and legs yellowish brown. Wing membrane hyaline with brown markings. In forewings, apical band complete and protruding on inner margin; thin stripe near apical band posteriorly; pterostigmal band with broad basal branch and thin or separated apical branch; basal band split into two small spots; marginal spot small; basal spot inconspicuous. Male T1–T5 dark brown. Notal organ long, stick-like, exceeding hind border of T4 and covering postnotal organ. A6 dark brown, long cylindrical. A7 and A8 yellowish brown, relatively short, and constricted basally.

Male genital bulb oval. Epandrium long and broad, protruding laterad and forming two small processes subapically, and truncated terminally. Hypandrium with broad basal stalk. Hypovalve mostly greyish white, blackish marginally on basal half, and almost oblong in shape. Two hypovalves overlapped mesally, forming diamond-shaped window subbasally. Gonostylus with obtuse triangular median tooth and stout basal process. Parameres broad foliate and divergent apically. Aedeagus conical and with stout lateral processes.

Female subgenital plate oval, with V-shaped terminal emargination. Mediygnium with reduced main plate; posterior arms spatulate in distal half, and subacute terminally; axis at least twice as long as posterior arms with diverged apodemes.

Emergence. May to June.

Distribution. Guangdong (Nanling Mountains), Hunan (Bamian Mountain, Mangshan Mountains).

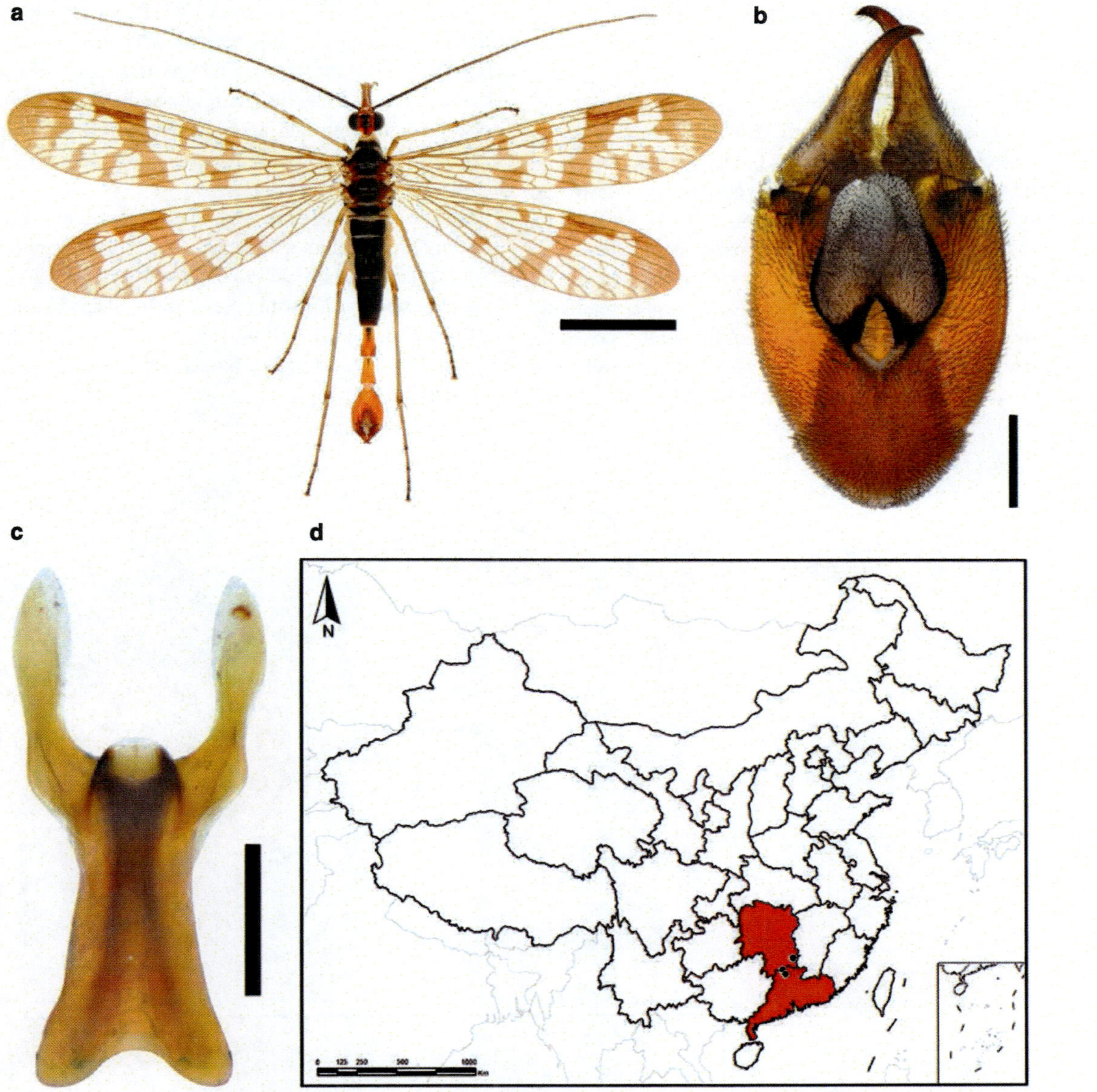

Fig. 10.23 (**a**) Male adult of ***Neopanorpa mangshanensis***, dorsal view (Mangshan Mountains, Yizhang, Hunan) (*Scale bar*: 5.0 mm) [NWAU]. (**b**) Male genital bulb of ***Neopanorpa mangshanensis***, ventral view (*Scale bar*: 0.5 mm). (**c**) Female medigynium of ***Neopanorpa mangshanensis***, ventral view (*Scale bar*: 0.2 mm). (**d**) Distribution of ***Neopanorpa mangshanensis***

10.1.21 *Neopanorpa minuta* Chou & Wang, 1988

(Minute Neoscorpionfly) [小新蝎蛉] (Fig. 10.24)
Measurements. Forewing length 11.0–12.0 mm.

Morphology. Vertex dark brown. Ocellar triangle black. Rostrum yellowish brown. Pronotum dark brown. Meso- and metanotum mostly yellowish brown, with black median stripe. Pleura and legs yellowish brown. Wing membrane hyaline, tinged with yellow, devoid of markings. Pterostigma yellow. Male T1–T5 dark brown. Notal organ relatively short, not exceeding middle of T4 and covering postnotal organ. A6 dark brown, cylindrical. A7 and A8 yellowish brown, relatively short, and constricted basally.

Male genital bulb long oval. Epandrium long and broad, truncated terminally. Hypandrium with broad basal stalk. Hypovalve long and broad, with rounded apex. Gonostylus slender, concaved on outer margin, and bearing obtuse triangular median tooth and acute peach-shaped basal process. Paramere short and stout with acute apex. Aedeagus conical and with stout lateral processes.

Female subgenital plate oval, with V-shaped terminal emargination. Mediygnium with reduced main plate; posterior arms spatulate in distal half, and rounded terminally; axis nearly as long as posterior arms with greatly diverged apodemes.

Emergence. May to June.

Distribution. Hunan (Shuangpai, Yizhang), Jiangxi (Yichun).

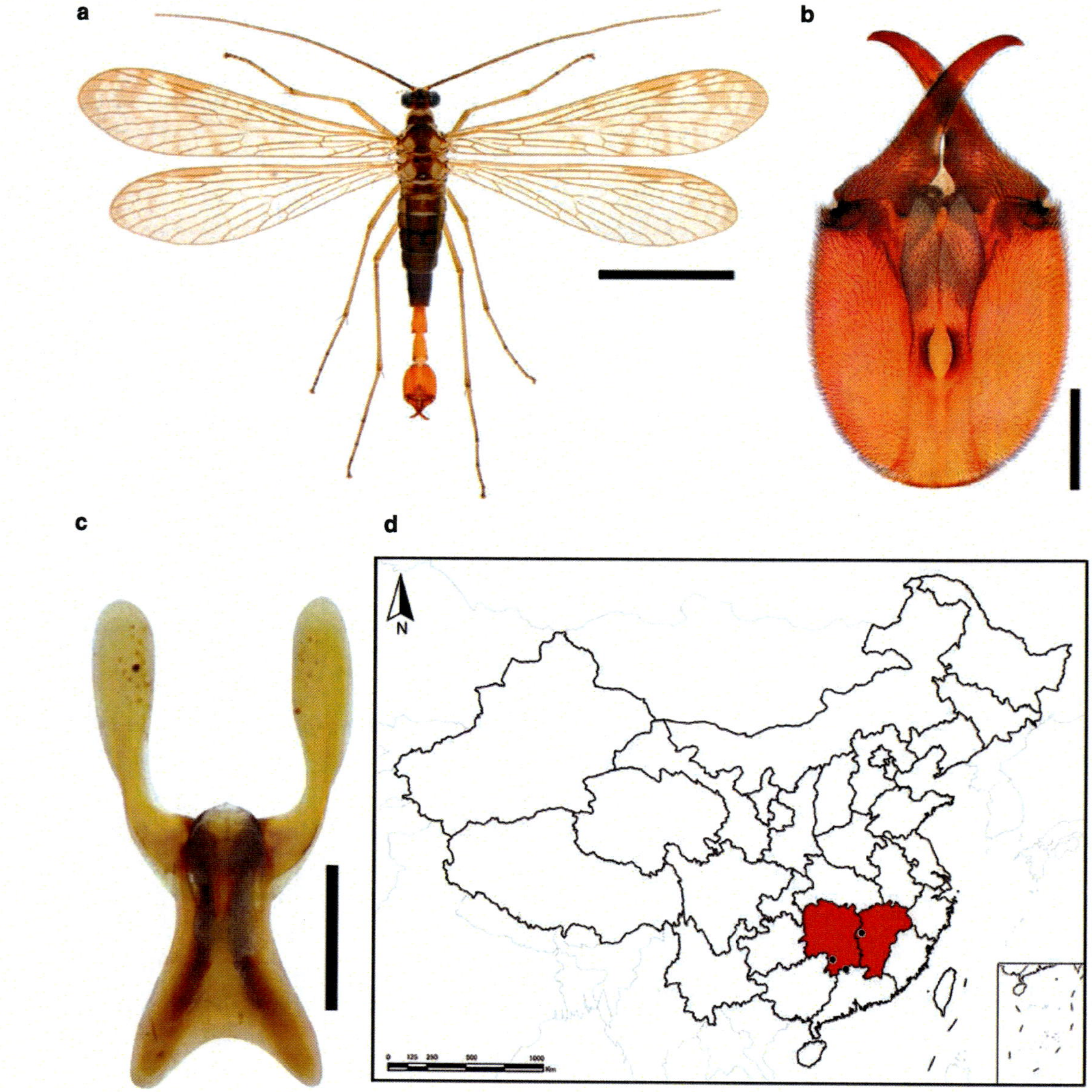

Fig. 10.24 **(a)** Male adult of ***Neopanorpa minuta***, dorsal view (Mangshan Mountains, Yizhang, Hunan) (*Scale bar*: 5.0 mm) **[NWAU]**. **(b)** Male genitalia of ***Neopanorpa minuta***, ventral view (*Scale bar*: 0.5 mm). **(c)** Female medigynium of ***Neopanorpa minuta*** (*Scale bar*: 0.2 mm). **(d)** Distribution of ***Neopanorpa minuta***

10.1.22 *Neopanorpa moganshanensis* Zhou & Wu, 1993

(Moganshan Neoscorpionfly) [莫干山新蝎蛉] (Fig. 10.25)
Measurements. Forewing length 12.5–13.0 mm.

Morphology. Vertex black. Ocellar triangle black. Rostrum deep brown with two dark brown longitudinal stripes laterally. Nota dark brown. Pleura and legs mostly yellowish brown. Wing membrane hyaline, slender, slightly tinged with greyish, and devoid of markings, except darkened tinge on apex. Pterostigma brown. Male T1–T5 black. Notal organ greatly elongated, stick-like, exceeding hind border of A6. A6 black, cylindrical, and tapering toward apex. A7 and A8 dark brown, slightly constricted basally.

Male genital bulb oval. Epandrium long and broad, deeply emarginate in U-shape terminally and forming pair of finger-like processes laterally. Hypandrium with broad basal stalk. Hypovalve short and thin apically, and separated basally. Gonostylus almost as long as gonocoxites, and bearing large triangular basal process on inner margin. Paramere foliate apically. Aedeagus conical and with stout lateral processes.

Female subgenital plate oval with deep V-shaped emargination terminally. Medigynium with rounded main plate; axis greatly reduced with extremely short apodemes; posterior arms slender.

Emergence. May to August.

Distribution. Zhejiang (Fengyang, Mogan, Tianmu, Qingliang Mountains), Fujian (Wuyi Mountains).

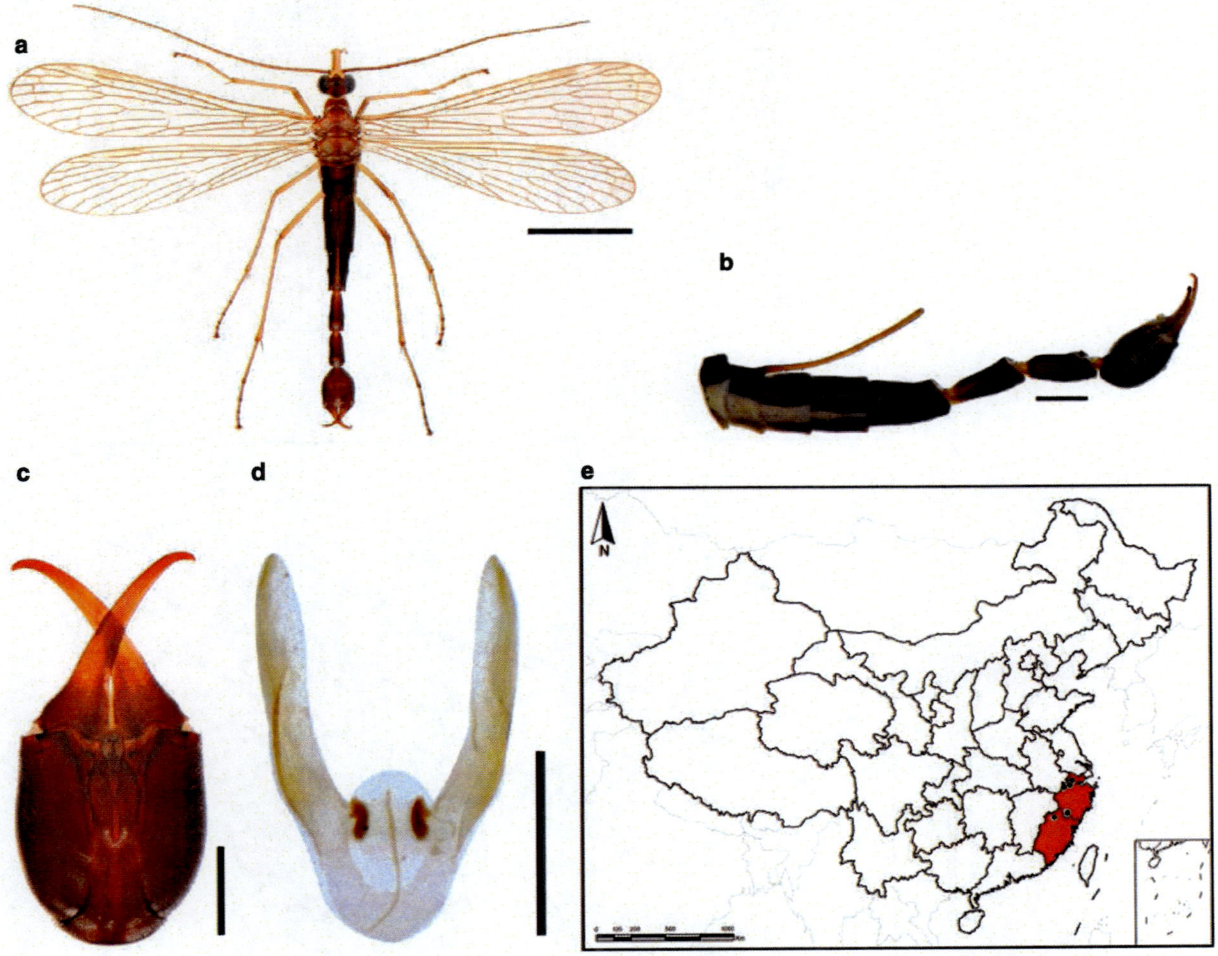

Fig. 10.25 (**a**) Male adult of ***Neopanorpa moganshanensis***, dorsal view (Longquan Mountain, Lishui, Zhejiang) (*Scale bar*: 5.0 mm) [NWAU]. (**b**) Male abdomen of ***Neopanorpa moganshanensis***, lateral view (*Scale bar*: 1.0 mm). (**c**) Male genital bulb of ***Neopanorpa moganshanensis***, ventral view (*Scale bar*: 0.5 mm). (**d**) Female medigynium of ***Neopanorpa moganshanensis***, ventral view (*Scale bar*: 0.2 mm). (**e**) Distribution of ***Neopanorpa moganshanensis***

10.1.23 *Neopanorpa mutabilis* Cheng, 1957

(Mutable Neoscorpionfly) [异新蝎蛉] (Fig. 10.26)
Measurements. Forewing length 12.5–13.0 mm.

Morphology. Vertex black. Ocellar triangle black. Rostrum yellowish brown. Pronotum dark brown. Meso- and metanotum yellowish brown laterally, with broad black median stripe. Pleura and legs mostly yellowish brown. Wing membrane hyaline with brown markings. In forewings, apical band broad, with large hyaline open window posteriorly; pterostigmal band complete with broad basal branch and thin apical branch; additional female oblique stripe between apical and pterostigmal bands; marginal spot slender; basal band split into two irregular spots; basal spot absent. Male T1–T5 dark brown. Notal organ long, stick-like, not exceeding hind border of T4 and covering postnotal organ. A6 dark brown. A7 and A8 yellowish brown, and constricted basally.

Male genital bulb long oval. Epandrium long and broad, and truncated terminally. Hypandrium with broad basal stalk. Hypovalve slender, broad medially and thin basally, with tapering apex extending to apex of gonocoxites. Gonostylus with obtuse triangular median tooth and bowl-shaped basal

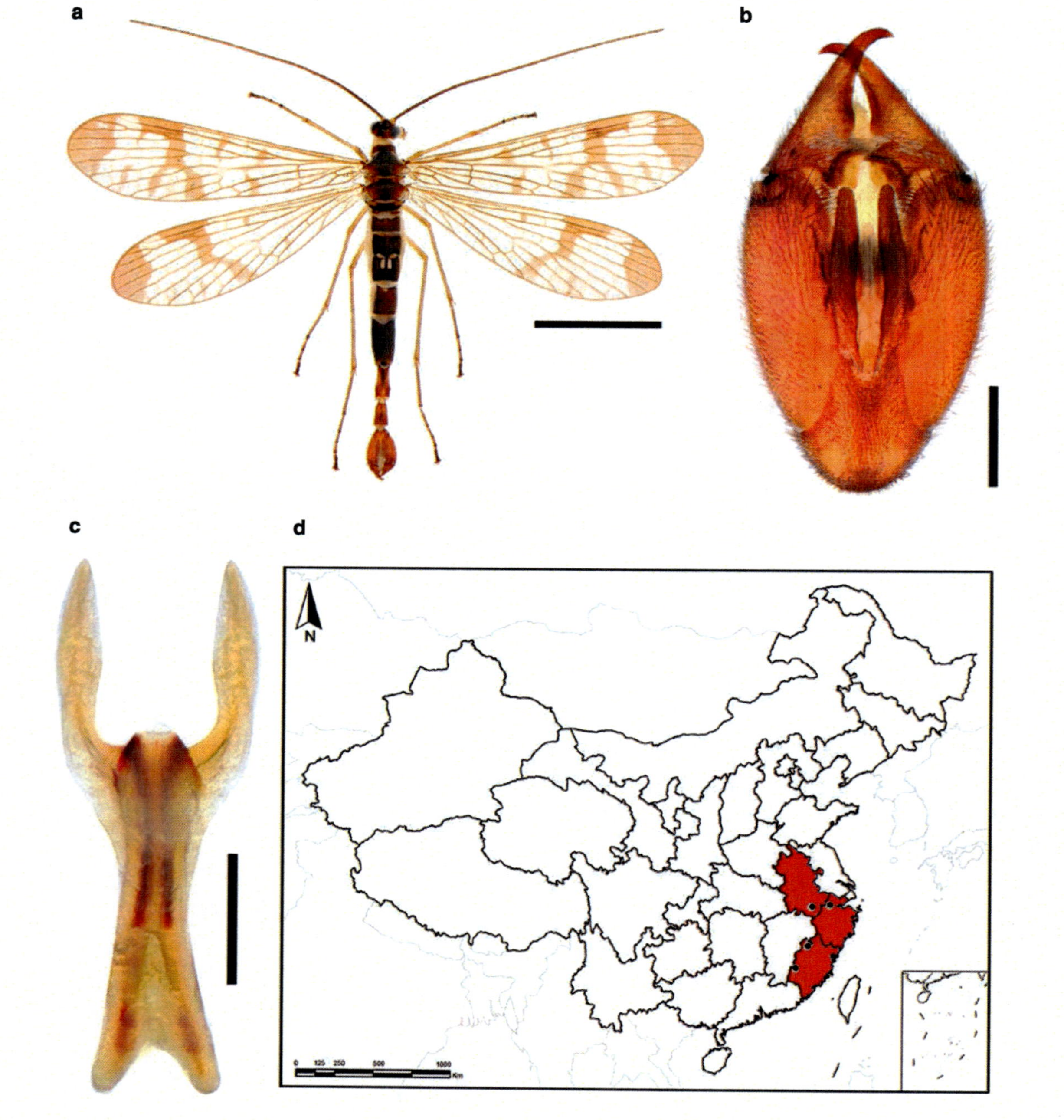

Fig. 10.26 (**a**) Male adult of ***Neopanorpa mutabilis***, dorsal view (West Tianmu Mountain, Lin'an, Zhejiang) (*Scale bar*: 5.0 mm) [NWAU]. (**b**) Male genital bulb of ***Neopanorpa mutabilis***, ventral view (*Scale bar*: 0.5 mm). (**c**) Female medigynium of ***Neopanorpa mutabilis***, ventral view (*Scale bar*: 0.2 mm). (**d**) Distribution of ***Neopanorpa mutabilis***

process on inner margin. Paramere bifurcated, with ventral branch thin and free, and dorsal branch fused with aedeagus. Aedeagus conical, with stout lateral processes.

Female subgenital plate oval with deep V-shaped emargination terminally. Medigynium slender; axis slightly longer than posterior arms with slightly diverged apodemes.

Emergence. April to July.

Distribution. Anhui (Huangshan Mountain), Fujian (Wuyi Mountains), Zhejiang (Tianmu Mountains).

10.1.24 *Neopanorpa nigritis* Carpenter, 1938

(Black Neoscorpionfly) [黑色新蝎蛉] (Fig. 10.27)

Neopanorpa pallivalva Zhou & Zhou, 2006, **new synonym**

Measurements. Forewing length 13.0 mm.

Morphology. Vertex black. Ocellar triangle black. Rostrum dark brown. Nota black. Pleura and legs grayish brown. Wing membrane hyaline, tinged with greyish brown, and devoid of markings. Pterostigma distinct, light reddish brown. Male T1–T5 black. Notal organ long clavate, almost exceeding middle portion of T4. A6 black, cylindrical. A7 and A8 black and constricted basally.

Male genital bulb oval. Epandrium long and broad, slightly emarginate terminally; epandrial lobes large. Hypandrium with relatively short basal stalk. Hypovalves slender, whitish at apical half, and curving mesad apically. Gonostylus with obtuse triangular median tooth and very large cylindrical basal process. Paramere foliate. Aedeagus conical, with stout lateral processes.

Female subgenital plate oval, with V-shaped terminal emargination. Mediygnium with reduced main plate; poste-

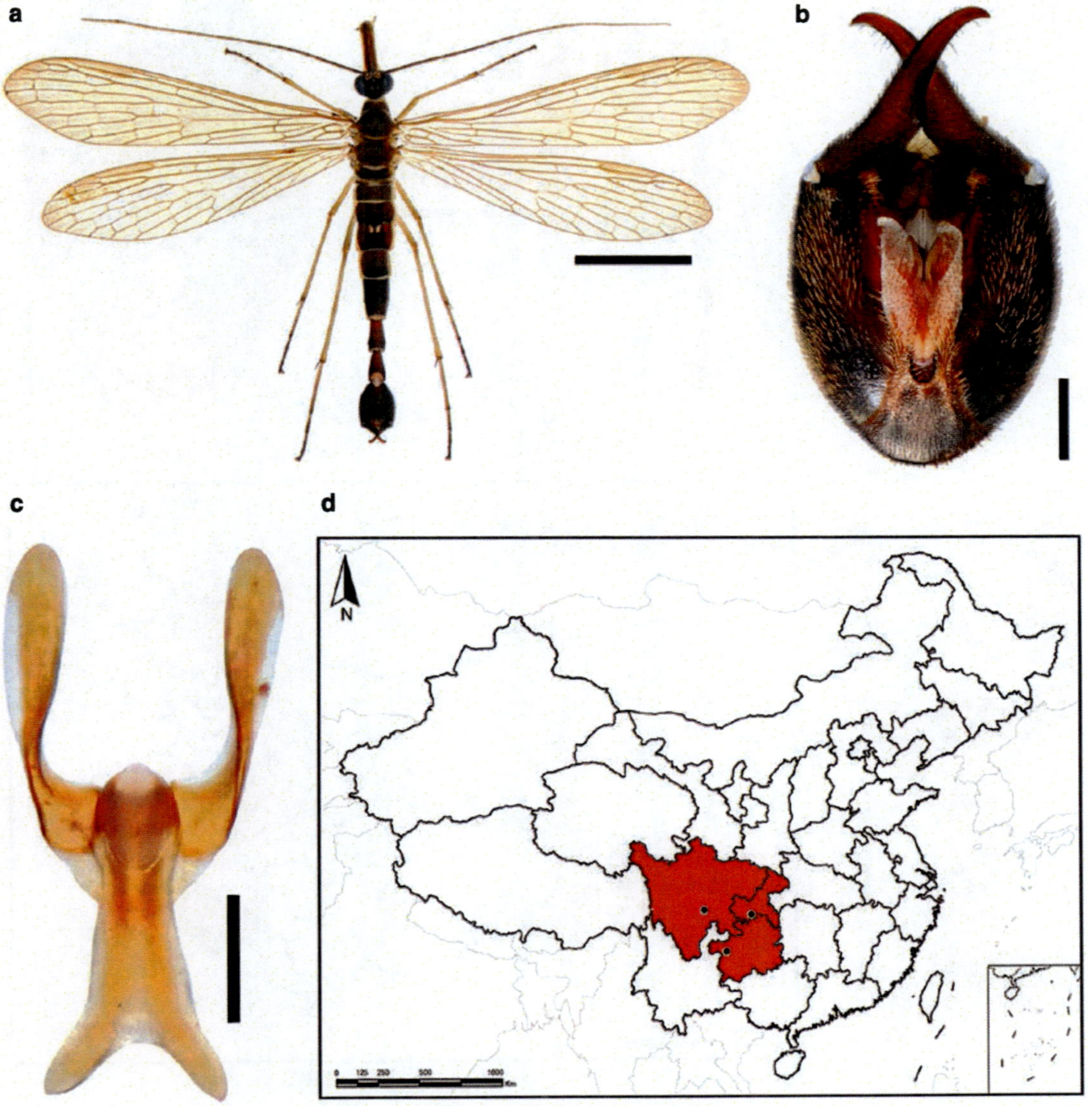

Fig. 10.27 (a) Male adult of ***Neopanorpa nigritis***, dorsal view (Mount Jinfo, Chongqing) (*Scale bar*: 5.0 mm) [NWAU]. (b) Male genital bulb of ***Neopanorpa nigritis***, ventral view (*Scale bar*: 0.5 mm). (c) Female medigynium of ***Neopanorpa nigritis***, ventral view (*Scale bar*: 0.2 mm). (d) Distribution of ***Neopanorpa nigritis***

rior arms spatulate in distal half, and subrounded terminally; axis nearly as long as posterior arms with greatly diverged apodemes.

Emergence. June to July.

Distribution. Chongqing (Jinfo Mountain), Guizhou (Liupanshui), Sichuan (Mount Emei).

Remarks. *Neopanorpa pallivalva* Zhou & Zhou, 2006 shares the identical morphology with *N. nigritis*. Therefore, it is herein treated a junior synonym of the latter.

10.1.25 *Neopanorpa ovata* Cheng, 1957

(Ovate Neoscorpionfly) [卵瓣新蝎蛉] (Fig. 10.28)

Measurements. Forewing length 13.0–14.0 mm.

Morphology. Vertex dark brown. Ocellar triangle black. Rostrum yellowish brown. Pronotum dark brown. Meso- and metanotum mostly dark brown, with latero-posterior corners yellowish brown. Pleura and legs yellowish brown. Wing membrane hyaline with brown markings. In forewings, apical band complete with serrate inner margin; thin stripe obliquely extending from antero-outer portion of pterostigmal band to hind mar-

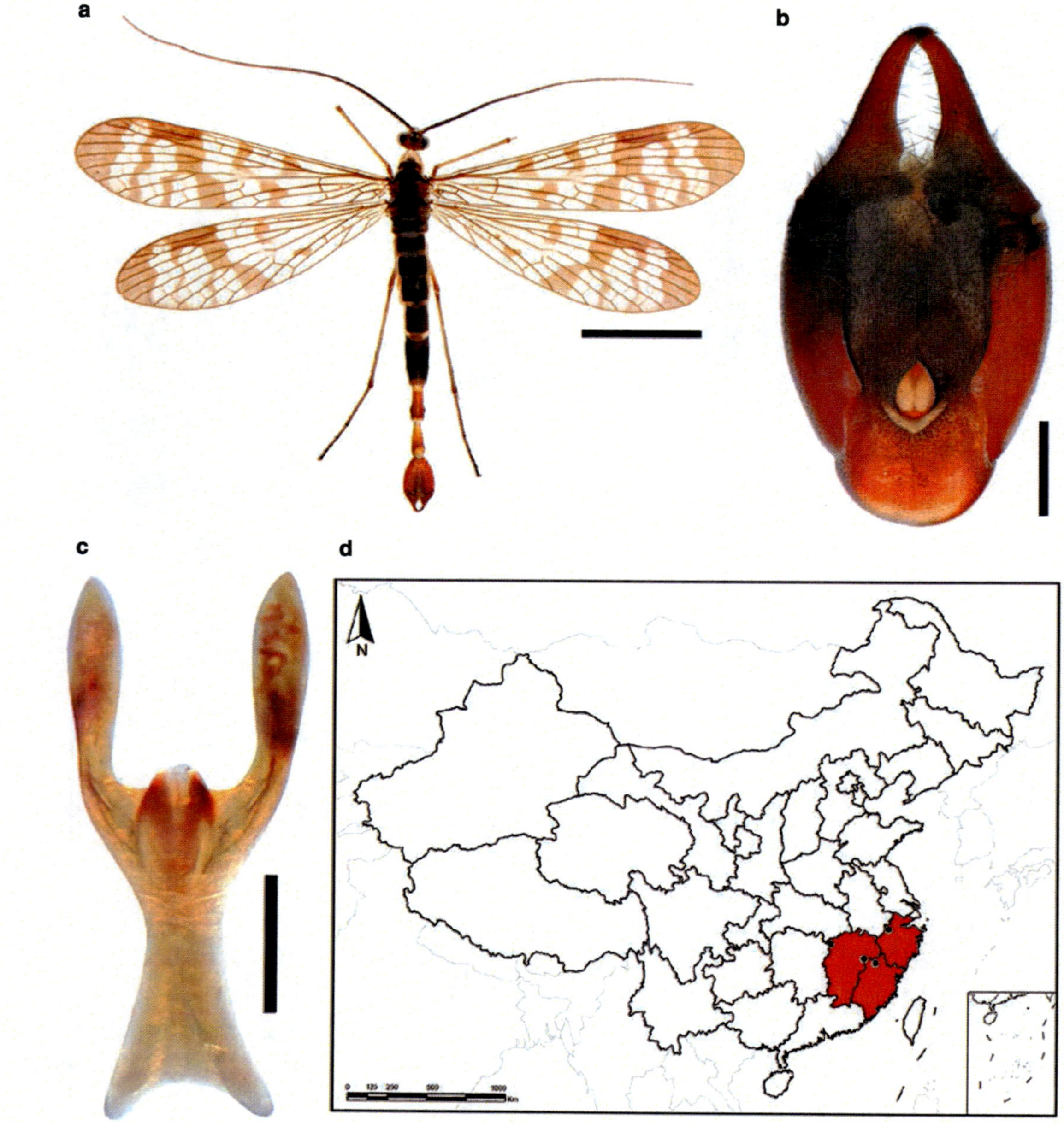

Fig. 10.28 (**a**) Male adult of ***Neopanorpa ovata***, dorsal view (Matou Mountain, Zixi, Jiangxi) (*Scale bar*: 5.0 mm) **[NWAU]**. (**b**) Male genital bulb of ***Neopanorpa ovata***, ventral view (*Scale bar*: 0.5 mm). (**c**) Female medigynium of ***Neopanorpa ovata***, ventral view (*Scale bar*: 0.2 mm). (**d**) Distribution of ***Neopanorpa ovata***

gin; pterostigmal band with broad basal branch and thin apical branch; basal band split into two small spots; marginal spot small; basal spot inconspicuous. Male T1–T5 dark brown. Notal organ long, stick-like, exceeding hind border of T4 and covering rounded postnotal organ. A6 dark brown, long cylindrical. A7 and A8 yellowish brown, relatively short, and constricted basally.

Male genital bulb oval. Epandrium long and broad, protruding laterad and forming two small processes subapically, and truncated terminally; epandrial lobes small. Hypandrium with broad basal stalk. Hypovalve mostly greyish white, blackish marginally on basal half, and almost oblong in shape. Two hypovalves overlapped mesally, forming diamond-shaped window subbasally. Gonostylus with obtuse triangular median tooth and stout basal process. Parameres broad foliate and divergent apically. Aedeagus conical and with stout lateral processes.

Female subgenital plate oval, with V-shaped terminal emargination. Mediygnium with reduced main plate; posterior arms spatulate in distal half, and acute terminally; axis longer than posterior arms with diverged apodemes.

Emergence. May to June.

Distribution. Fujian (Wuyi Mountains), Jiangxi (Zixi), Zhejiang (Qingliang Peak).

10.1.26 *Neopanorpa pennyi* Byers, 1999 [New Record to China]

(Penny's Scorpionfly) [佩尼新蝎蛉] (Fig. 10.29)

Neopanorpa uncata Zhou, 2000, **new synonym**

Measurements. Forewing length 10.5–12.0 mm.

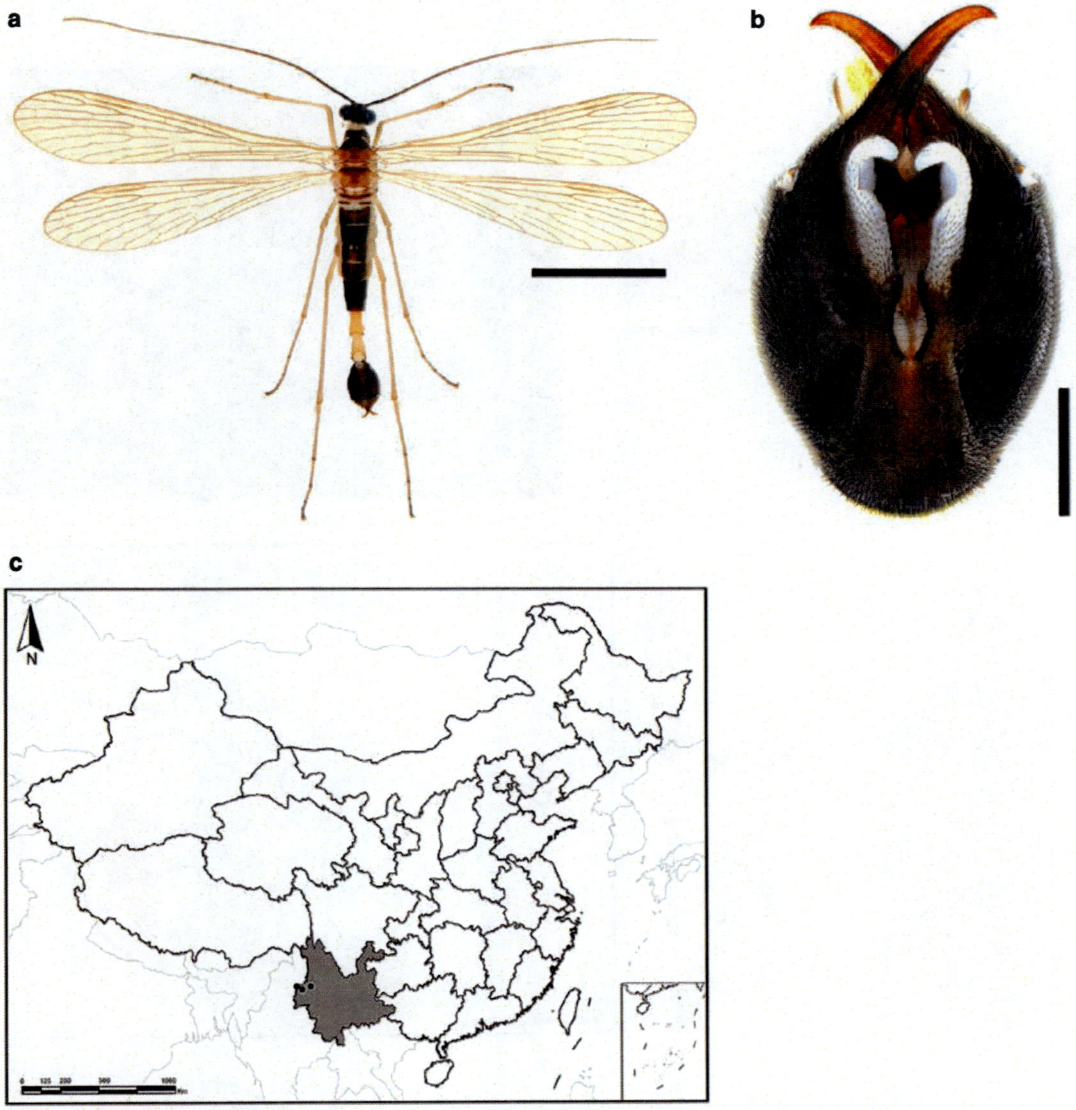

Fig. 10.29 (a) Male adult of ***Neopanorpa pennyi***, dorsal view (Gaoligong Mountains, Baoshan, Yunnan) (*Scale bar*: 5.0 mm) **[NWAU]**. **(b)** Male genital bulb of ***Neopanorpa pennyi***, ventral view (*Scale bar*: 0.5 mm). **(c)** Distribution of ***Neopanorpa pennyi***

Morphology. Vertex black. Ocellar triangle black. Rostrum dark brown. Nota black. Pleura and legs greyish brown. Wing membrane hyaline, tinged with greyish brown, and devoid of markings. Pterostigma distinct, light reddish brown. Male T1–T5 black. Notal organ long clavate, almost extending to middle portion of T4 and slightly enlarged terminally. A6 black, cylindrical. A7 and A8 light yellowish brown.

Male genital bulb oval, black except distal 2/3 of hypovalves white. Epandrium long and broad, slightly emarginate terminally; epandrial lobes greatly developed. Hypandrium with broad and long basal stalk. Hypovalves slender, greatly curving mesad apically. Gonostylus with obtuse triangular median tooth and very large rounded basal process. Paramere slender, curving dorsad, and hook-like. Aedeagus conical, with greatly elongated and acute lateral processes.

Female unknown.

Emergence. June to July.

Distribution. Yunnan (Gaoligong Mountains, Tengchong); Myanmar (Kachin).

Remarks. *Neopanorpa uncata* Zhou, 2000 is morphologically identical to *N. pennyi*. Therefore, it is treated a junior synonym of the latter.

10.1.27 *Neopanorpa pielina* Navás, 1936

(Father Piel's Neoscorpionfly) [璧尔新蝎蛉] (Fig. 10.30)
Measurements. Forewing length 11.5 mm.

Morphology. Vertex yellowish brown. Ocellar triangle black. Rostrum yellowish brown. Pronotum dark brown. Meso- and metanotum yellowish brown with broad black median stripe. Pleura and legs light yellowish brown. Wing membrane hyaline with pterostigma and markings brown. Wing membrane hyaline with brown markings. In forewings,

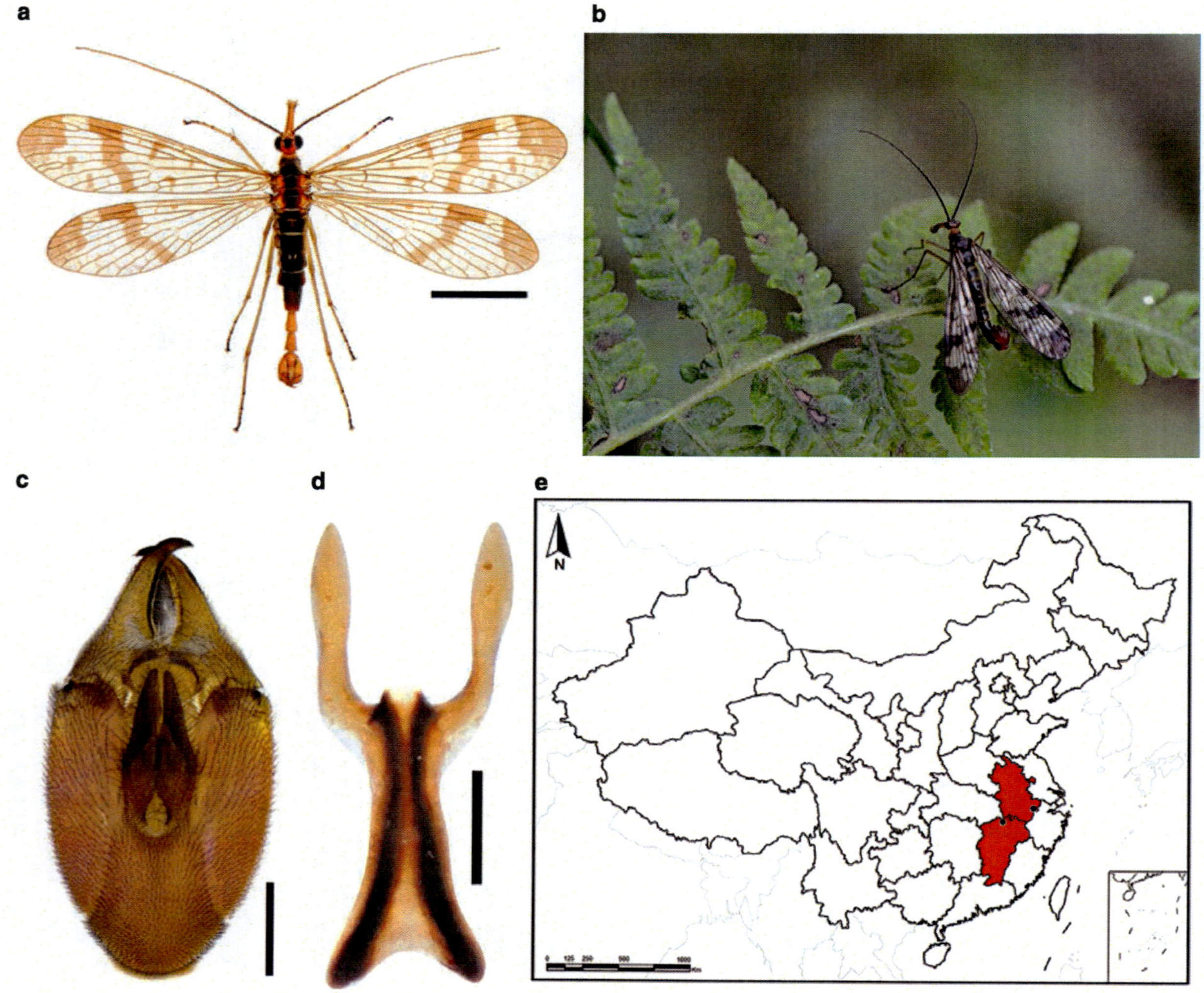

Fig. 10.30 (**a**) Male adult of ***Neopanorpa pielina***, dorsal view (Lushan Mountain, Jiangxi) (*Scale bar*: 5.0 mm) **[NWAU]**. (**b**) Male adult of ***Neopanorpa pielina*** in the field (Lushan Mountain, Jiangxi). (**c**) Male genital bulb of ***Neopanorpa pielina***, ventral view (*Scale bar*: 0.5 mm). (**d**) Female medigynium of ***Neopanorpa pielina***, ventral view (*Scale bar*: 0.2 mm). (**e**) Distribution of ***Neopanorpa pielina***

apical band broad with series of hyaline windows and separated slender stripe posteriorly; pterostigmal band complete, with broad basal branch and thin apical branch; marginal spot small; basal band split into pair of small spots; basal spot absent. Male terga I–V dark brown. Notal organ long, exceeding hind border of T4. A6 dark brown. A7 and A8 yellowish brown, slightly constricted basally.

Male genital bulb long oval. Epandrium long and broad, with truncated apex. Hypandrium with broad basal stalk. Hypovalves thin and long, with lateral margins curled dorsad and apexes slightly acute, forming subcircular window subbasally. Gonostylus with obtuse triangular median tooth and bowl-shaped basal process. Parameres Y-shaped and divergent. Aedeagus conical with stout lateral processes.

Female subgenital plate oval, with V-shaped terminal emargination. Mediygnium with reduced main plate; posterior arms spatulate in distal half, and acute terminally; axis longer than posterior arms with diverged apodemes.

Emergence. April to August.

Distribution. Jiangxi (Lushan Mountain).

10.1.28 *Neopanorpa puripennis* Chou & Wang, 1988

(Pure-Winged Neoscorpionfly) [净翅新蝎蛉] (Fig. 10.31)

Measurements. Forewing length 12.0–13.0 mm.

Morphology. Vertex yellowish brown. Ocellar triangle black. Rostrum yellowish brown. Pronotum yellowish

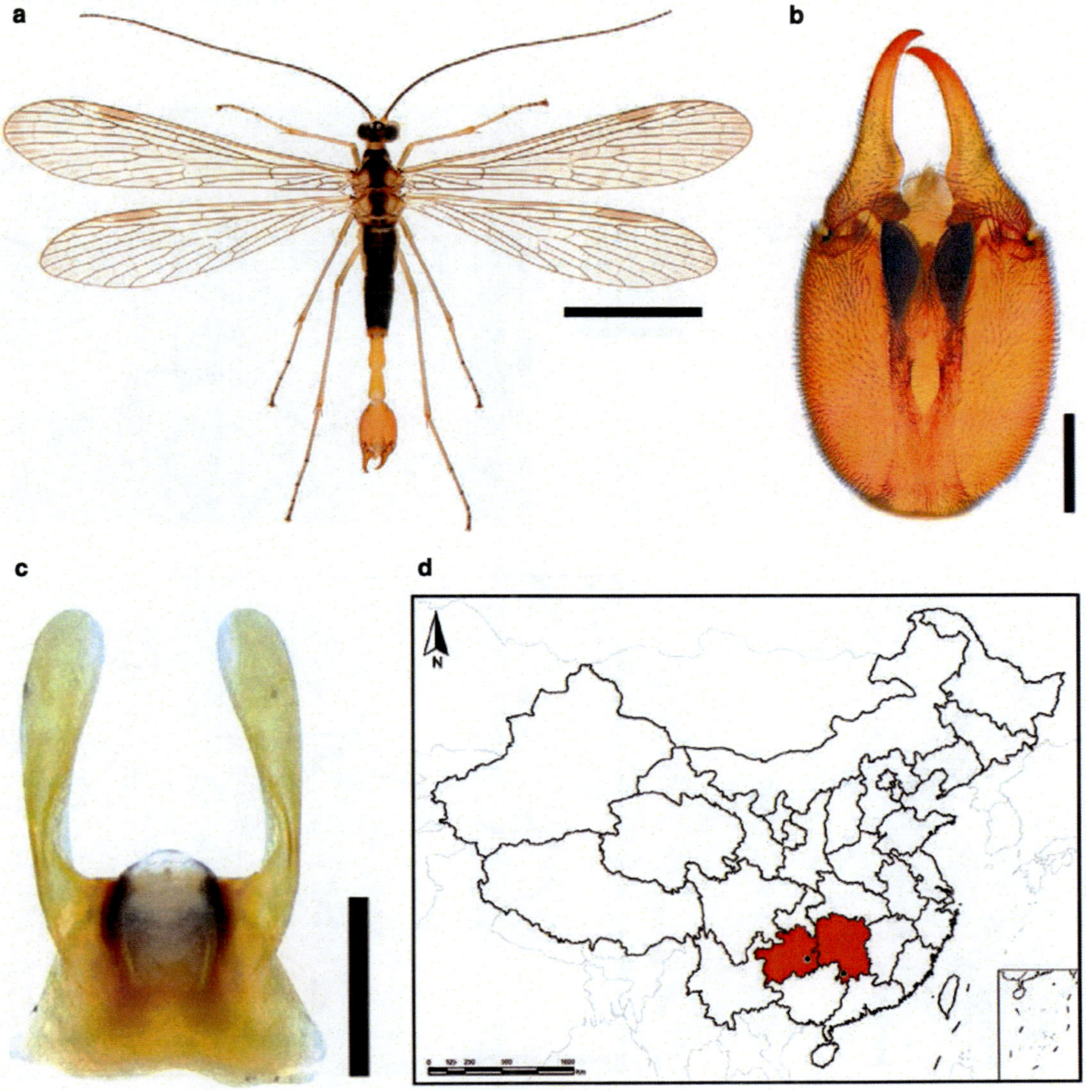

Fig. 10.31 (**a**) Male adult of ***Neopanorpa puripennis***, dorsal view (Leigong Mountain, Guizhou) (*Scale bar*: 5.0 mm) [NWAU]. (**b**) Male genital bulb of ***Neopanorpa puripennis***, ventral view (*Scale bar*: 0.5 mm). (**c**) Female medigynium of ***Neopanorpa puripennis***, ventral view (*Scale bar*: 0.2 mm). (**d**) Distribution of ***Neopanorpa puripennis***

brown. Meso- and metanotum yellowish brown, with broad dark brown median stripe. Pleura and legs light yellowish brown. Wing membrane hyaline, devoid of markings or with greatly reduced, spot-like pterostigmal band. Pterostigma light brown. Male T1–T5 dark brown. Notal organ short and slightly triangular. A6 mostly brown with yellow apex. A7 and A8 yellowish brown, and constricted basally.

Male genital bulb oval. Epandrium long and broad, with truncated apex. Hypandrium with long broad basal stalk. Hypovalves rounded and divergent apically, exceeding apex of gonocoxites. Gonostylus with obtuse triangular median tooth and rounded basal process. Paramere indistinct. Aedeagus with very strong lateral processes; dorsal valves with curled horn-like processes laterally.

Female subgenital plate oval, with V-shaped terminal emargination. Mediygnium with greatly sclerotized main plate; posterior arms slender and spatulate in distal half, and subrounded terminally; axis concealed with indistinct apodemes.

Emergence. July to August.

Distribution. Guizhou (Leigong Mountain), Hunan (Daoxian).

10.1.29 *Neopanorpa quadristigma* Wang & Hua, 2018

(Qradrate-Pterostigma Neoscorpionfly) [方痣新蝎蛉] (Fig. 10.32)

Measurements. Forewing length 10.0–13.6 mm.

Morphology. Vertex black. Ocellar triangle black. Rostrum dark brown. Meso- and metanotum mostly reddish brown with indistinct dark median stripe. Pleura and legs greyish brown. Wing membrane hyaline, tinged with greyish brown, with

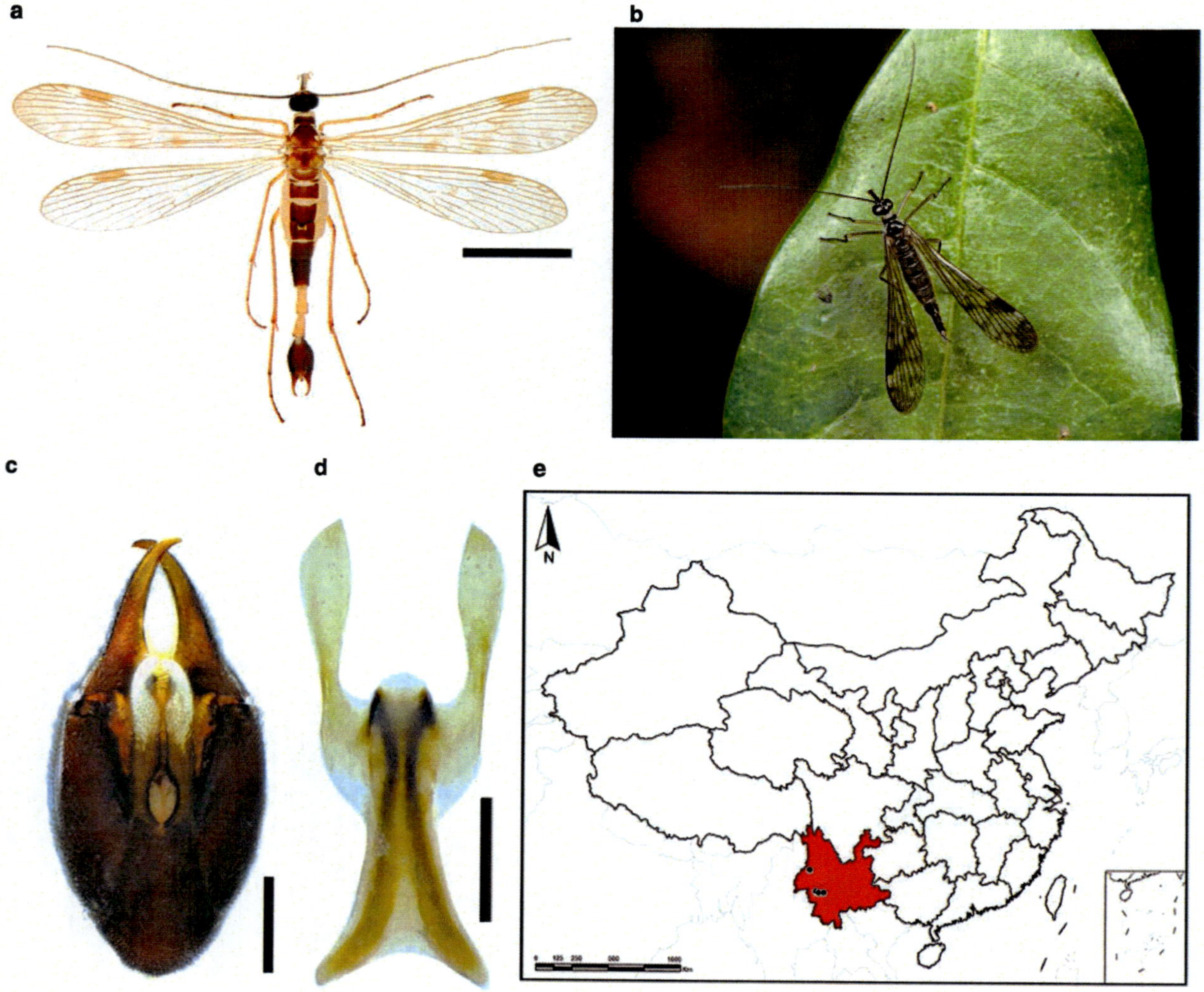

Fig. 10.32 (**a**) Male adult of ***Neopanorpa quadristigma***, dorsal view (Tangli Mountain, Yongde, Yunnan) (*Scale bar*: 5.0 mm) **[NWAU]**. (**b**) Female adult of ***Neopanorpa quadristigma*** in the field (Tangli Mountain, Yongde). (**c**) Male genital bulb of ***Neopanorpa quadristigma***, ventral view (*Scale bar*: 0.5 mm). (**d**) Female medigynium of ***Neopanorpa quadristigma***, ventral view (*Scale bar*: 0.2 mm). (**e**) Distribution of ***Neopanorpa quadristigma***

water mark-like brown markings; pterostigma distinct, dark brown, quadrate. In forewings, apical and pterostigmal bands complete, marginal spot and basal band slightly reduced, basal spot absent. Male T1–T5 black. Notal organ long clavate, almost exceeding middle portion of T4. A6 black, cylindrical. A7 and A8 light yellowish brown and constricted basally.

Male genital bulb oval, black except distal half of hypovalves white. Epandrium long and broad, slightly emarginate terminally; epandrial lobes large. Hypandrium with relatively short basal stalk. Hypovalves slender, whitish at apical half, and curving mesad apically. Gonostylus with obtuse triangular median tooth and very large cylindrical basal process. Paramere foliate. Aedeagus conical, with stout lateral processes.

Female subgenital plate long subrectangular, with shallow V-shaped terminal emargination. Mediygnium with reduced main plate; posterior arms greatly spatulate in distal half, and acute terminally; axis longer than posterior arms with greatly diverged apodemes.

Emergence. May to August.

Distribution. Yunnan (Cangyuan, Gaoligong Mountains, Lincang, Yongde).

10.1.30 *Neopanorpa retina* Chou & Li, 1988

(Retina Neoscorpionfly) [网纹新蝎蛉] (Fig. 10.33)

Measurements. Forewing length 13.0–14.0 mm.

Morphology. Vertex yellowish brown. Ocellar triangle enclosed by large black spot. Rostrum yellowish brown. Pronotum dark brown. Meso- and metanotum mostly dark

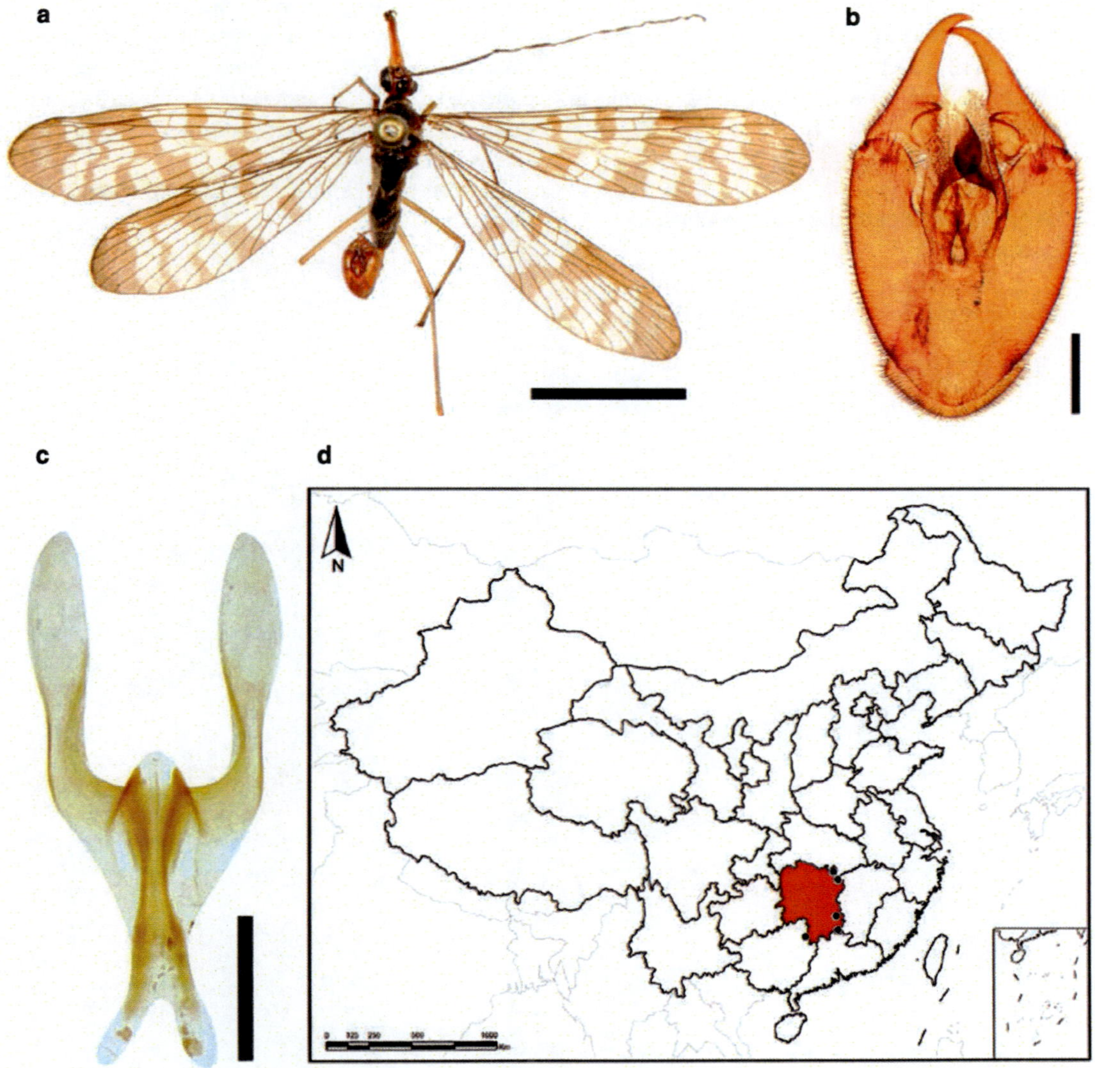

Fig. 10.33 (a) Male adult of ***Neopanorpa retina***, dorsal view (Rucheng, Hunan) (*Scale bar*: 5.0 mm) **[NWAU]**. **(b)** Male genital bulb of ***Neopanorpa retina***, ventral view (*Scale bar*: 0.5 mm). **(c)** Female medigynium of ***Neopanorpa retina***, ventral view (*Scale bar*: 0.2 mm). **(d)** Distribution of ***Neopanorpa retina***

brown, with lateroposterior corners yellowish brown. Pleura and legs yellowish brown. Wing membrane hyaline with brown markings. In forewings, apical band broad with two hyaline windows; pterostigmal band with basal branch and apical branch; base of apical branch thin; additional stripe beyond pterostigmal band and parallel to inner margin of apical band; basal band split medially; marginal spot large; basal spot absent.

Male T1–T5 dark brown. Notal organ exceeding hind border of T4. A6 dark brown and cylindrical. A7 and A8 yellowish brown and constricted basally. Male genital bulb oval. Epandrium long and broad with rounded apex. Hypandrium with long broad basal stalk. Hypovalves rounded, overlapping mesally. Gonostylus with obtuse triangular median tooth and bowl-shaped basal process. Aedeagus short, not exceeding apex of gonocoxites, with stout lateral processes.

Female subgenital plate oval, with V-shaped terminal emargination. Mediygnium with reduced main plate; posterior arms spatulate in distal half, and acute terminally; axis short and thin, with diverged apodemes.

Emergence. April to October.

Distribution. Hunan (Anren, Jiangyong, Pingjiang, Linxiang, Rucheng).

10.1.31 *Neopanorpa semiorbiculata* Wang & Hua, 2018

(Semicircular Neoscorpionfly) [半圆新蝎蛉] (Fig. 10.34)

Measurements. Forewing length 13.0–14.0 mm.

Morphology. Vertex yellowish brown. Ocellar triangle black. Rostrum yellowish brown. Pronotum yellowish brown. Meso- and metanotum yellowish brown laterally, with distinct dark brown median stripe. Pleura and legs light yellowish brown. Wing membrane hyaline, tinged with yellow, bearing only semicircular apical band, or in some individuals greatly reduced, spot-like basal band. Pterostigma light brown. Male T1–T5 dark brown. Notal organ short and slightly triangular. A6 mostly brown with yellow apex. A7 and A8 yellowish brown, and constricted basally.

Male genital bulb oval. Epandrium long and broad, with truncated apex; epandrial lobes greatly developed with truncated apex. Hypandrium with long broad basal stalk. Hypovalves almost half as long as stalk of hypandrium, slightly convergent. Gonostylus with obtuse triangular median tooth and rounded basal process. Paramere slender. Aedeagus conical with stout lateral processes; dorsal processes small, triangular, and acute.

Female subgenital plate oval, with deep V-shaped terminal emargination. Mediygnium with greatly reduced main plate; posterior arms slender and spatulate in distal half; axis slightly shorter than posterior arms; apodemes diverged basally.

Emergence. May to August.

Distribution. Yunnan (Daxue Mountain, Gaoligong Mountains)

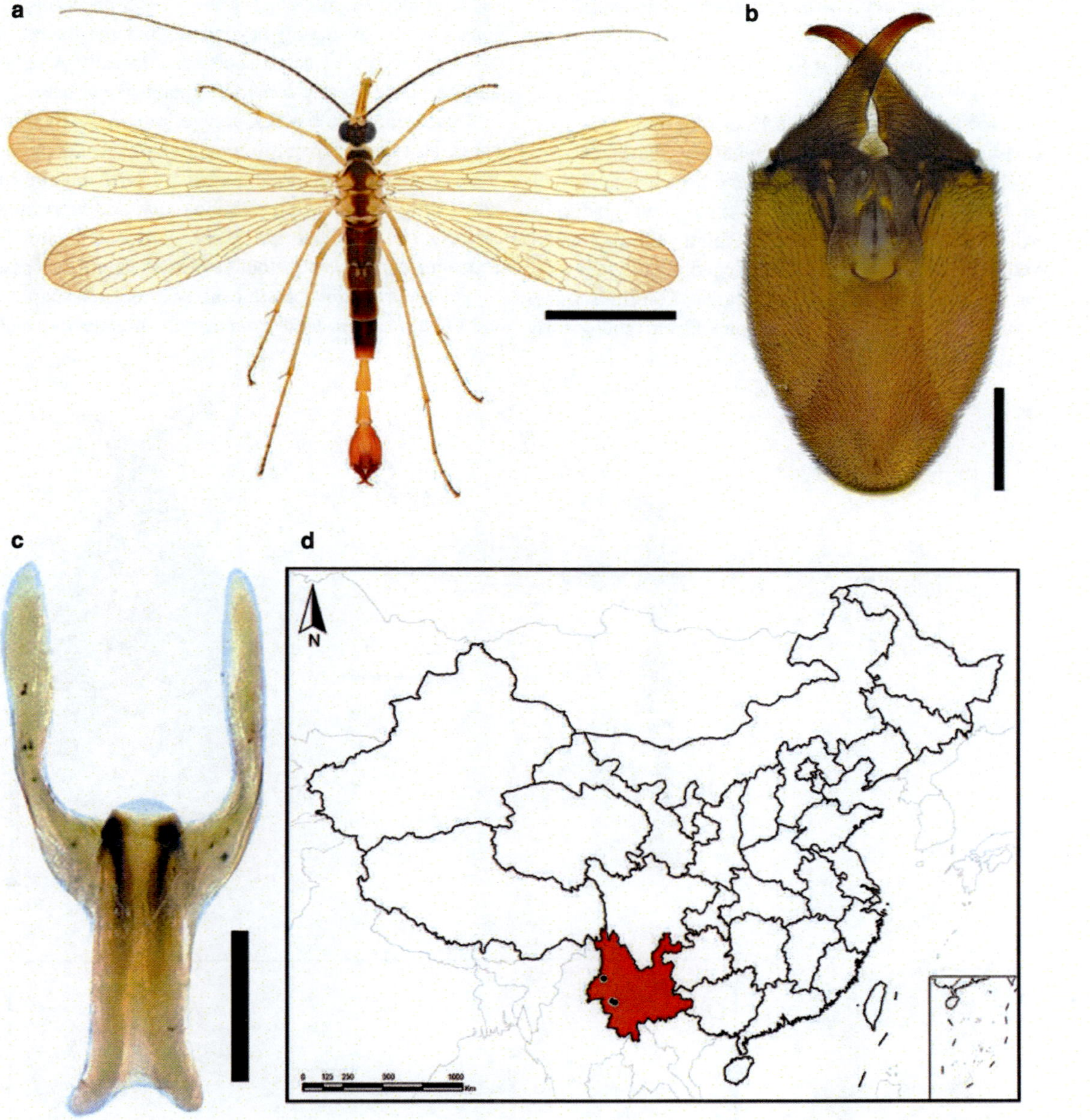

Fig. 10.34 (**a**) Male adult of ***Neopanorpa semiorbiculata***, dorsal view (Daxue Mountain, Yongde, Yunnan) (*Scale bar*: 5.0 mm) **[NWAU]**. (**b**) Male genital bulb of ***Neopanorpa semiorbiculata***, ventral view (*Scale bar*: 0.5 mm). (**c**) Female medigynium of ***Neopanorpa semiorbiculata***, ventral view (*Scale bar*: 0.2 mm). (**d**) Distribution of ***Neopanorpa semiorbiculata***

10.1.32 *Neopanorpa setigera* Wang & Hua, 2018

(Setae-Bearing Neoscorpionfly) [具毛新蝎蛉]
(Fig. 10.35)

Measurements. Forewing length 13.0–14.0 mm.

Morphology. Vertex black, ocellar triangle black; rostrum yellowish brown, with two dark brown longitudinal stripes. Nota dark brown, Meso- and metanotum grading to yellowish brown towards wing base. Pleura yellowish white, legs yellowish brown. Wing slightly tinged with light brown, hyaline with brown distinct pterostigma. Markings indistinct, represented only by vague apical band and greatly reduced pterostigmal band. Male T1–T5 black. Notal organ long clavate, extending beyond A5 but not exceeding A6. T4–T6 with row of dense short setae mesally. A6 black. A7 and A8 black, greatly constricted stalk-like at base.

Male genital bulb black, long elliptical. Epandrium truncated at apex. Hypandrium with long broad basal stalk; hypovalves slender, curled dorsad at lateral margin, forming water drop-shaped window subbasally, and with finger-like process on dorsal surface subbasally. Paramere slender; aedeagus conical, with stout and broad lateral processes.

Female subgenital plate oval, with deep V-shaped terminal emargination. Mediygnium with greatly reduced main

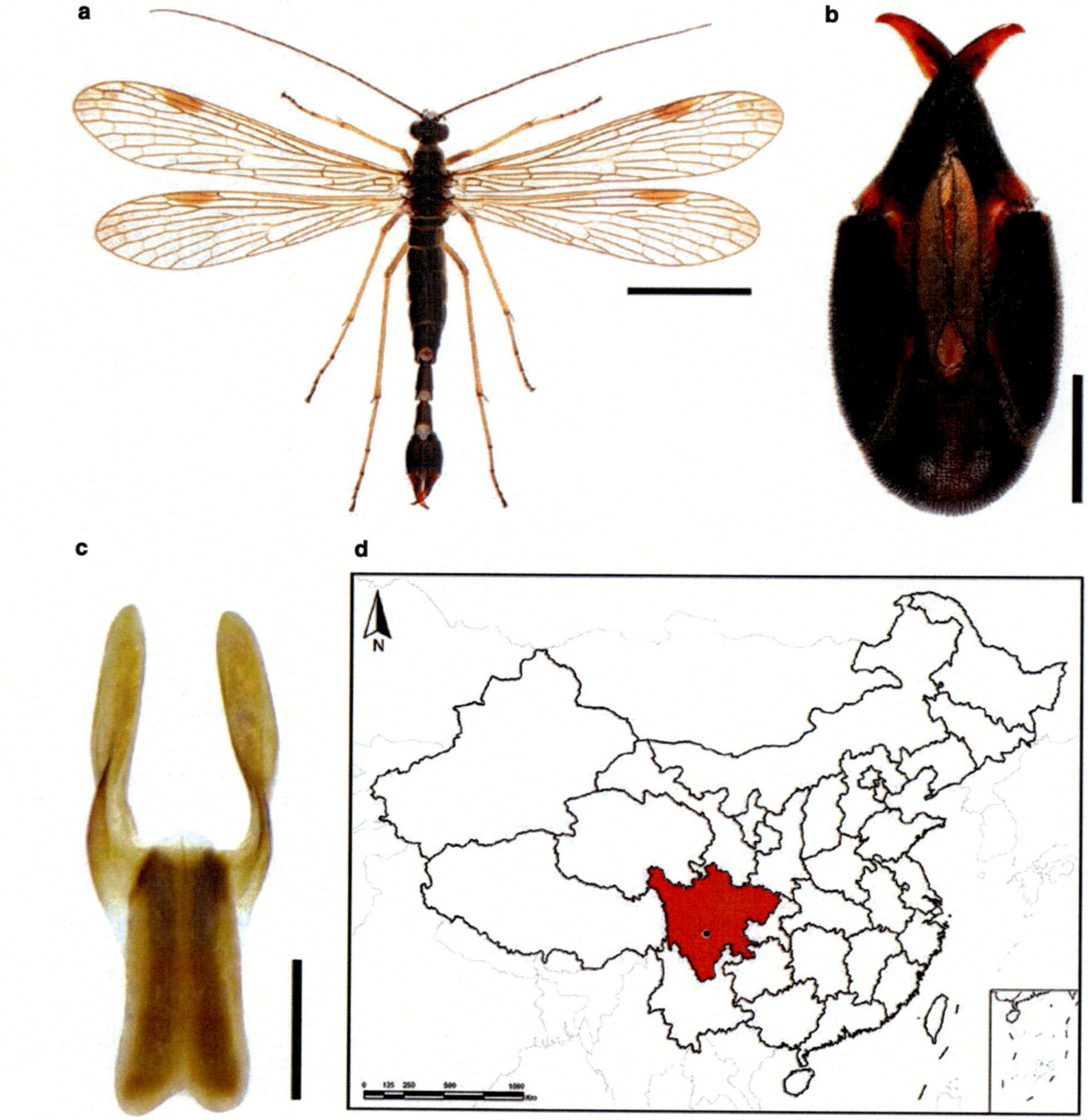

Fig. 10.35 (**a**) Male adult of ***Neopanorpa setigera***, dorsal view (Shimian, Sichuan) (*Scale bar*: 5.0 mm) [NWAU]. (**b**) Male genital bulb of ***Neopanorpa setigera***, ventral view (*Scale bar*: 0.5 mm). (**c**) Female medigynium of ***Neopanorpa setigera***, ventral view (*Scale bar*: 0.2 mm). (**d**) Distribution of ***Neopanorpa setigera***

plate; posterior arms slender and spatulate in distal half; axis nearly as long as posterior arms with stout, nearly parallel apodemes.

Emergence. June to August.

Distribution. Sichuan (Shimian).

10.1.33 *Neopanorpa tibetensis* Hua & Chou, 1999

(Tibetan Neoscorpionfly) [西藏新蝎蛉] (Fig. 10.36)

Measurements. Forewing length 14.5–15.5 mm.

Morphology. Vertex and ocellar triangle black. Rostrum yellowish brown, with two dark brown longitudinal stripes laterally. Pronotum dark brown. Meso- and metanotum mostly dark brown with latero-posterior corners yellowish brown. Pleura and legs yellowish brown. Wing membrane hyaline, slightly tinged with yellow. Markings yellowish brown. In forewings, apical band complete and connected to pterostigmal band anteriorly; pterostigmal band with complete basal and apical branches; basal band incomplete and only represented by small spot-on hind margin; marginal and basal spots absent. Male T1–T5 dark brown and gradually turning into yellowish brown caudad. Notal organ extremely elongated and clavate, extending to middle of T6. A6 darkish on basal half and yellowish brown on apical half, cylindrical. A7 and A8 yellowish brown and constricted basally.

Male genital bulb oval. Epandrium long and broad, with truncated apex. Hypandrium with long and broad basal

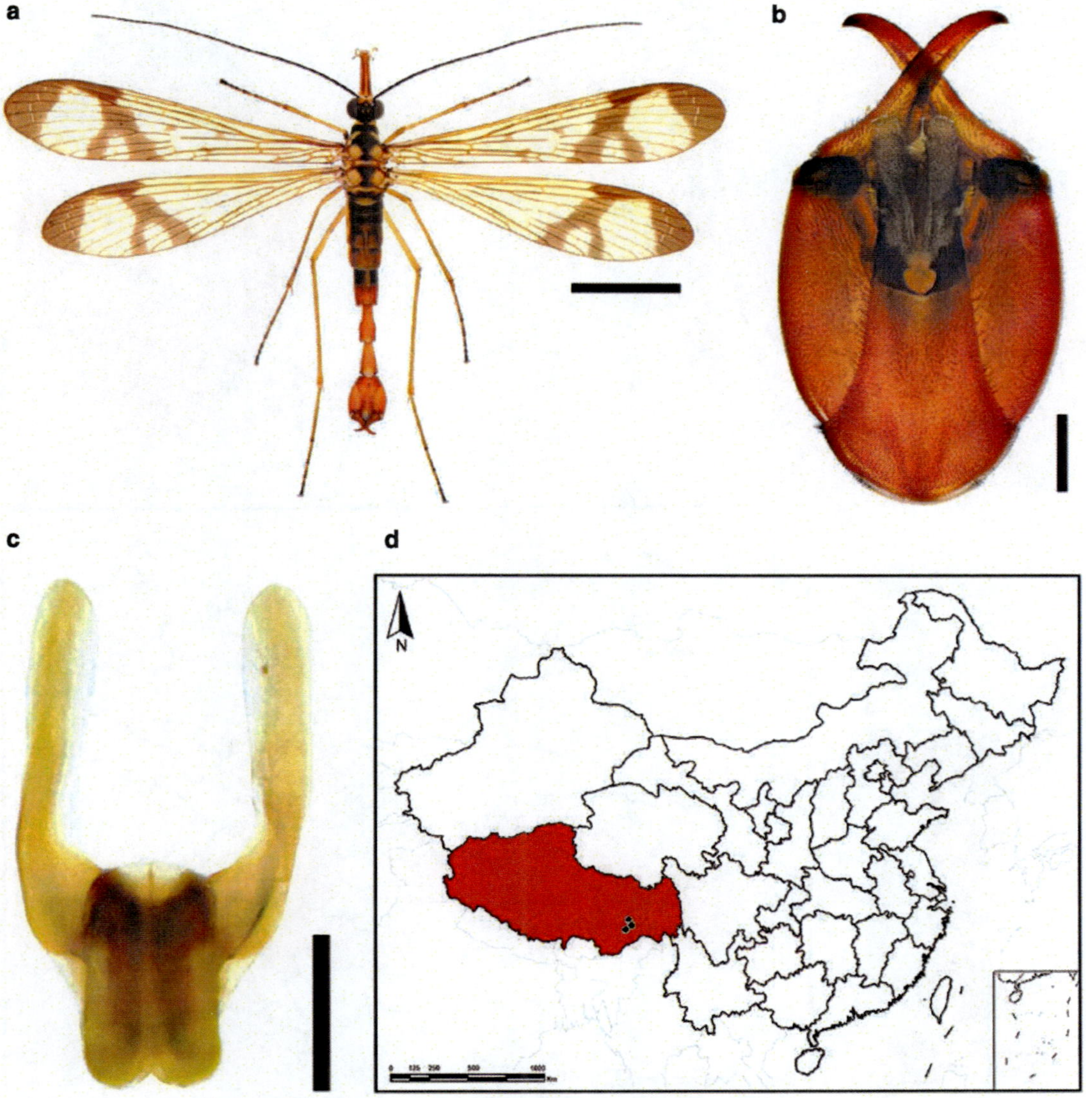

Fig. 10.36 (**a**) Male adult of ***Neopanorpa tibetensis***, dorsal view (Tangmai, Tibet) (*Scale bar*: 5.0 mm) **[NWAU]**. (**b**) Male terminal abdomen of ***Neopanorpa tibetensis***, lateral view (*Scale bar*: 0.5 mm). (**c**) Male genital bulb of ***Neopanorpa tibetensis***, ventral view (*Scale bar*: 0.2 mm). (**d**) Distribution of ***Neopanorpa tibetensis***

stalk. Hypovalves folded ventrad in distal half and forming circular window subbasally. Gonostylus concaved on outer margin, bearing obtuse triangular median tooth, and rounded basal process. Aedeagus conical, with stout lateral processes.

Female subgenital plate oval, with deep V-shaped terminal emargination. Mediygnium with greatly reduced main plate; posterior arms spatulate in distal half and rounded terminally; axis greatly shortened with stout, nearly parallel apodemes.

Emergence. July to September.

Distribution. Tibet (Tangmai, Nyingchi).

10.1.34 *Neopanorpa tienpingshana* Chou & Wang, 1988

(Tianpingshan Neoscorpionfly) [天平山新蝎蛉] (Fig. 10.37)

Measurements. Forewing length 11.5–12.5 mm.

Morphology. Vertex black. Ocellar triangle black. Rostrum yellowish brown. Nota dark brown. Pleura and coxae creamy yellow. Legs mostly light yellowish brown. Wing membrane hyaline with black markings. In forewings, apical band broad; pterostigmal band lacking apical branch; basal band absent or represented only by small spot near hind margin; marginal and

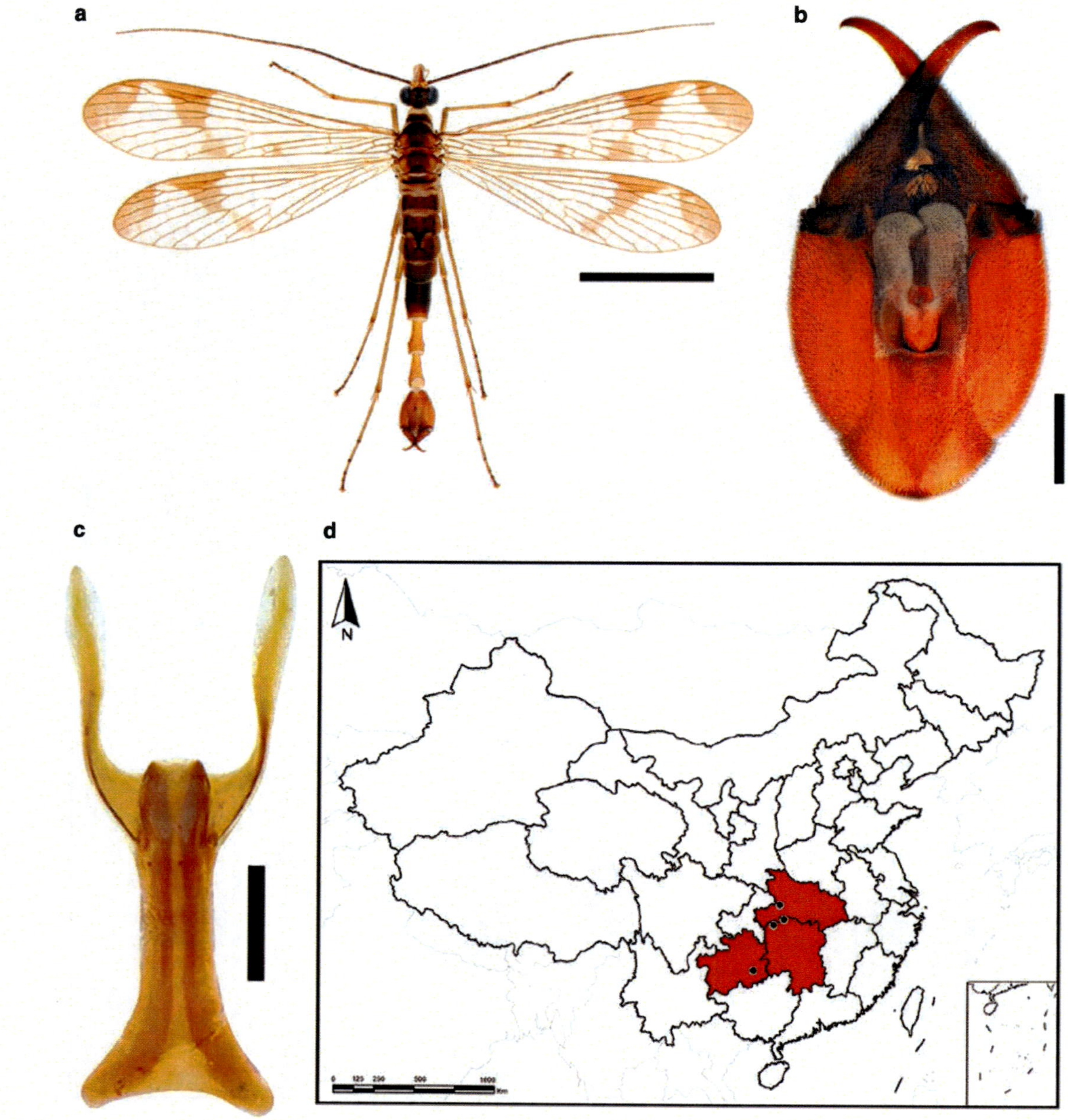

Fig. 10.37 (**a**) Male adult of ***Neopanorpa tienpingshana***, dorsal view (Tianping Mountain, Sangzhi, Hunan) (*Scale bar*: 5.0 mm) **[NWAU]**. (**b**) Male genital bulb of ***Neopanorpa tienpingshana***, ventral view (*Scale bar*: 0.5 mm). (**c**) Female medigynium of ***Neopanorpa tienpingshana***, ventral view (*Scale bar*: 0.2 mm). (**d**) Distribution of ***Neopanorpa tienpingshana***

basal spots absent. Male T1–T5 dark brown. Notal organ moderately long, extending to hind margin of T4, covering rounded postnotal organ. A6 dark brown, cylindrical. A7 and A8 yellowish brown and constricted basally.

Male genital bulb oval. Epandrium long and broad, slightly emarginate at apex. Hypandrium with long and broad basal stalk. Hypovalves twisted apically, forming heart-shaped window subbasally. Gonostylus with obtuse triangular median tooth, and large bowl-shaped basal process with long bristles. Paramere foliate apically. Aedeagus conical and with stout lateral processes.

Female subgenital plate oval, with deep V-shaped terminal emargination. Mediygnium with greatly reduced main plate; posterior arms slender and spatulate in distal half; axis slightly longer than posterior arms with apodemes greatly diverged basally.

Emergence. June to August.

Distribution. Guizhou (Leigong Mountain), Hubei (Badong), Hunan (Sangzhi, Shimen, Yizhang).

10.1.35 *Neopanorpa tincta* Wang & Hua, 2018

(Tinged Neoscorpionfly) [染翅新蝎蛉] (Fig. 10.38)

Measurements. Forewing length 13.0–14.0 mm.

Morphology. Vertex yellowish brown. Ocellar triangle black. Rostrum yellowish brown. Pronotum yellowish brown. Meso- and metanotum yellowish brown laterally, with distinct dark brown median stripe. Pleura and legs light yellowish brown. Wing membrane hyaline, tinged with yellow, bearing only greatly reduced apical and pterostigmal bands. Pterostigma light brown. Male T1–T5 dark brown. Notal organ short and slightly triangular. A6 mostly brown with yellow apex. A7 and A8 yellowish brown, and constricted basally.

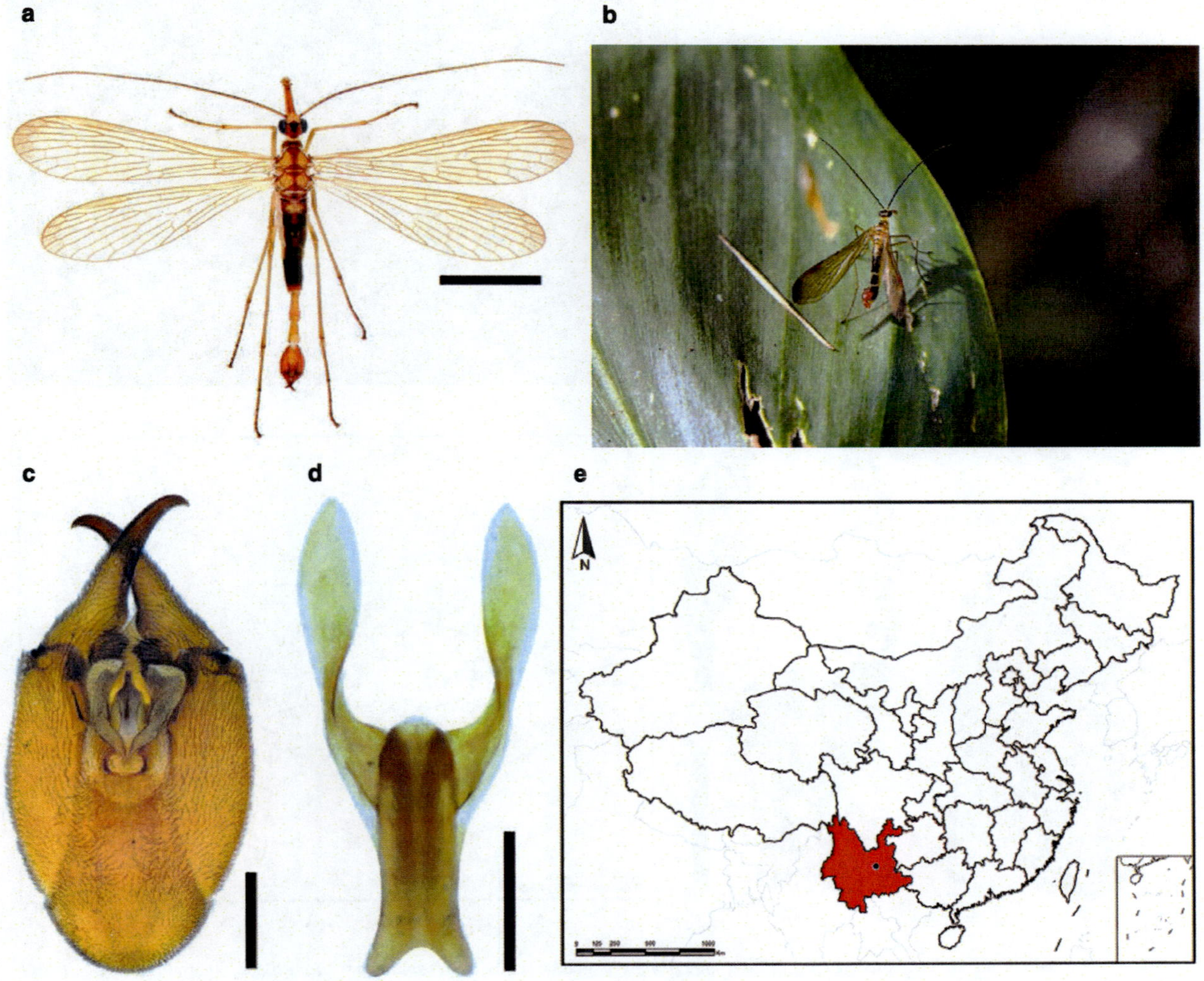

Fig. 10.38 (**a**) Male adult of ***Neopanorpa tincta***, dorsal view (West Hill, Kunming, Yunnan) (*Scale bar*: 5.0 mm) **[NWAU]**. (**b**) Male adult of ***Neopanorpa tincta*** in the field (West Hill, Kunming, Yunnan). (**c**) Male genital bulb of ***Neopanorpa tincta***, ventral view (*Scale bar*: 0.5 mm). (**d**) Female medigynium of ***Neopanorpa tincta***, ventral view (*Scale bar*: 0.2 mm). (**e**) Distribution of ***Neopanorpa tincta***

Male genital bulb oval. Epandrium long and broad, with truncated apex; epandrial lobes greatly developed with truncated apex. Hypandrium with long broad basal stalk. Hypovalves almost half as long as stalk of hypandrium, slightly convergent; hypandrial processes earlobe-shaped. Gonostylus with obtuse triangular median tooth and rounded basal process. Paramere slender. Aedeagus conical with stout lateral processes; dorsal processes small, triangular, and acute.

Female subgenital plate oval, with deep V-shaped terminal emargination. Mediygnium with greatly reduced main plate; posterior arms slender and spatulate in distal half; axis slightly shorter than posterior arms; apodemes greatly diverged basally.

Emergence. May to August.

Distribution. Yunnan (Kunming).

10.1.36 *Neopanorpa triangulata* Wang & Hua, 2018

(Triangular Neoscorpionfly) [三角新蝎蛉] (Fig. 10.39)

Measurements. Forewing length 12.0–14.0 mm.

Morphology. Vertex yellowish brown. Ocellar triangle black. Rostrum yellowish brown. Pronotum yellowish brown. Meso- and metanotum yellowish brown laterally, with distinct dark brown median stripe. Pleura and legs light yellowish brown. Wing membrane hyaline, bearing only semicircular apical band, and oblique, apical branch-lacking pterostigmal band; basal band represented by small spot in some individuals. Pterostigma light brown. Male T1–T5 dark brown. Notal organ short and slightly triangular. A6 mostly brown with yellow apex. A7 and A8 yellowish brown, and constricted basally.

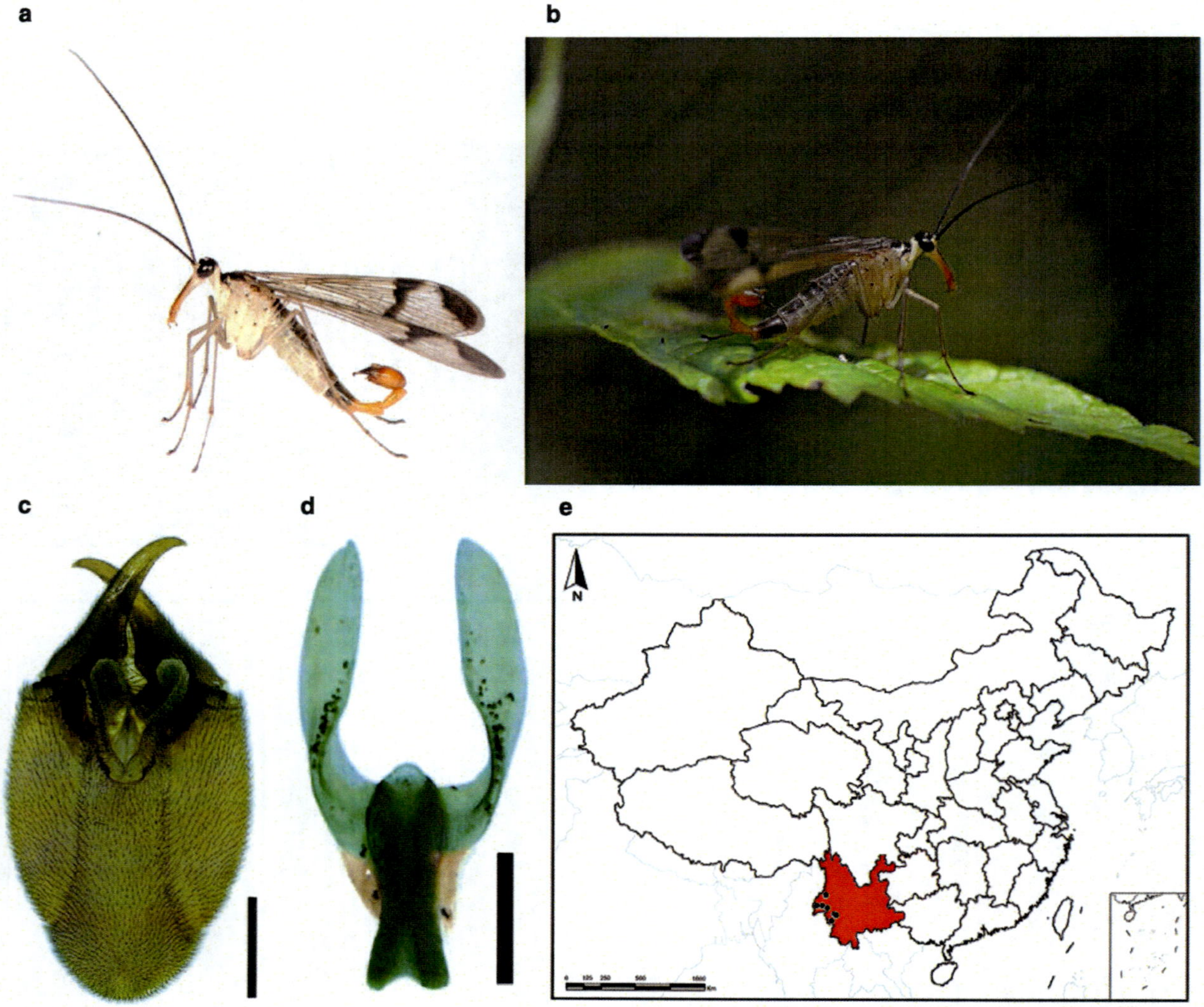

Fig. 10.39 (**a**) Male adult of ***Neopanorpa triangulata***, lateral view (Tangli Mountain, Yongde, Yunnan) (**b**) Male adult of ***Neopanorpa triangulata*** in the field (Tangli Mountain, Yongde, Yunnan). (**c**) Male genital bulb of ***Neopanorpa triangulata***, ventral view (*Scale bar*: 0.5 mm). (**d**) Female medigynium of ***Neopanorpa triangulata***, ventral view (*Scale bar*: 0.2 mm). (**e**) Distribution of ***Neopanorpa triangulata***

Male genital bulb oval, mostly yellowish brown except apex of gonocoxites and basal portion of gonostylus dark. Epandrium long and broad, with truncated apex; epandrial lobes greatly developed with truncated apex. Hypandrium with long broad basal stalk. Hypovalves almost half as long as stalk of hypandrium, slightly convergent; hypandrial lobes earlobe-shaped. Gonostylus with obtuse triangular median tooth and rounded basal process. Paramere slender. Aedeagus conical with stout lateral processes; dorsal processes small, triangular, and acute.

Female subgenital plate oval, with deep V-shaped terminal emargination. Mediygnium with greatly reduced main plate; posterior arms broadly spatulate in distal half; axis shorter than posterior arms; apodemes slightly diverged.

Emergence. May to August.

Distribution. Yunnan (Cangyuan, Gaoligong Mountains, Longling, Ximeng, Yingjiang, Yongde).

10.1.37 *Neopanorpa xingmini* Wang & Hua, 2019

(Xing-Min's Neoscorpionfly) [兴民新蝎蛉] (Fig. 10.40)

Measurements. Forewing length 11.0–11.5 mm.

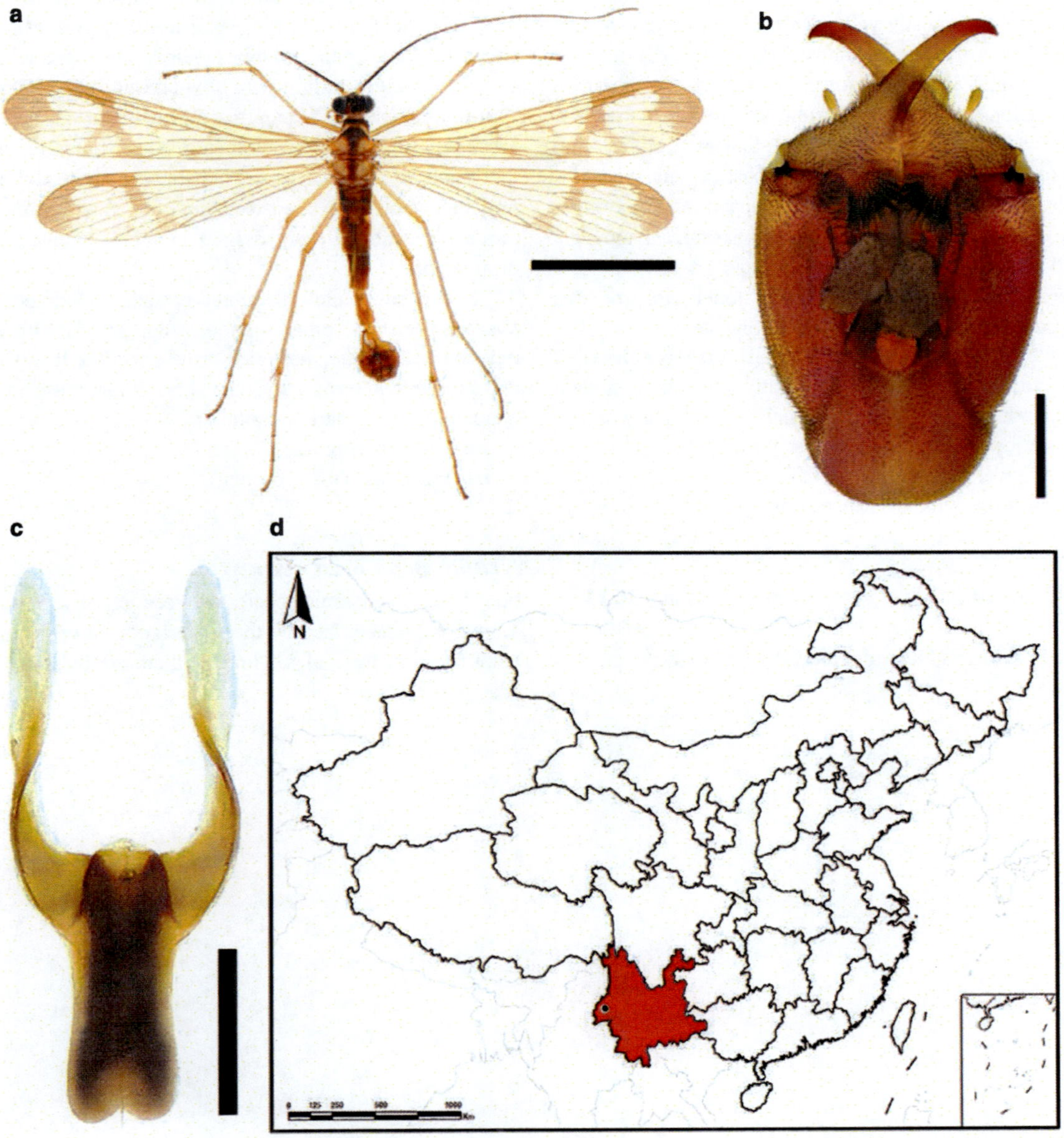

Fig. 10.40 (a) Male adult of ***Neopanorpa xingmini***, dorsal view (Longchuan, Yunnan) (*Scale bar*: 5.0 mm) [NWAU]. (b) Male genital bulb of ***Neopanorpa xingmini***, ventral view (*Scale bar*: 0.5 mm). (c) Female medigynium of ***Neopanorpa xingmini***, ventral view (*Scale bar*: 0.2 mm). (d) Distribution of ***Neopanorpa xingmini***

Morphology. Vertex yellowish brown. Ocellar triangle black. Rostrum yellowish brown. Pronotum yellowish brown. Meso- and metanotum yellowish brown laterally, with distinct dark brown median stripe. Pleura and legs light yellowish brown. Wing membrane hyaline, slightly tinged with yellow, markings brown. Pterostigma light brown. In forewings, apical band with acute inner projection and connected with pterostigmal band along anterior margin; pterostigmal band complete with broad basal branch and slightly reduced apical branch; marginal spot greatly elongated postero-outerward and connected with basal branch of pterostigmal band and forming N-shaped pattern; basal band split into two small spots; basal spot absent. Male T1–T5 dark yellowish brown. Notal organ greatly elongated, stick-like, and extending to middle of A6. A6 mostly yellowish brown with yellow apex. A7 and A8 yellowish brown, and constricted basally.

Male genital bulb stout oval, mostly yellowish brown. Epandrium long and broad, with truncated apex; epandrial lobes greatly developed with acute and hook-like apex. Hypandrium with broad basal stalk. Hypovalves almost as long as stalk of hypandrium, greatly curled ventrad; hypandrial lobes finger-shaped. Gonostylus with indistinct triangular median tooth and rounded basal process; tuft of stout setae on basal process. Paramere slender. Aedeagus conical with stout lateral processes; dorsal processes small, triangular, and subacute.

Female subgenital plate oval, with V-shaped terminal emargination. Mediygnium with greatly reduced main plate; posterior arms spatulate in distal half; axis slightly shorter than posterior arms; apodemes nearly parallel.

Emergence. May.

Distribution. Yunnan (Longchuan).

10.1.38 *Neopanorpa zhengyucheni* Wang, 2021

(Yu-Chen Zheng's Neoscorpionfly) [郑氏新蝎蛉]
(Fig. 10.41)

Measurements. Forewing length 15.0–16.0 mm.

Morphology. Vertex yellowish brown. Ocellar triangle black. Rostrum yellowish brown. Pronotum yellowish brown. Meso- and metanotum yellowish brown laterally, with indistinct dark brown median stripe. Pleura and legs light yellowish brown. Wing membrane hyaline, greatly tinged with yellow, markings sooty brown. Pterostigma light brown. In forewings, only greatly reduced apical and pterostigmal bands present. Male T1–T5 yellowish brown. Notal organ greatly elongated, stick-like, and exceeding hind border of T4; postnotal organ rounded. A6 mostly yellowish brown. A7 and A8 yellowish brown, and constricted basally.

Male genital bulb oval, mostly yellowish brown except apex of gonocoxites dark. Epandrium long and broad, with truncated apex; epandrial lobes greatly developed with truncated apex. Hypandrium with broad basal stalk. Hypovalves shorter than stalk of hypandrium, greatly curled; hypandrial lobes small. Gonostylus with indistinct triangular median tooth and small basal process. Paramere short and acutely projected ventrad at base. Aedeagus conical with stout lateral processes; dorsal processes greatly elongated and drop-shaped.

Female subgenital plate nearly diamond-shaped, with V-shaped terminal emargination. Mediygnium with greatly reduced main plate; posterior arms spatulate in distal half and subrounded terminally; axis shorter than posterior arms; apodemes stout, nearly parallel.

Emergence. August.

Distribution. Tibet (Yadong).

Neopanorpa brisi Group

Four Chinese-endemic species are recognized in this group. They are characterized by the acute tooth on ventral side of notal organ in the males. Three of them are included in this atlas.

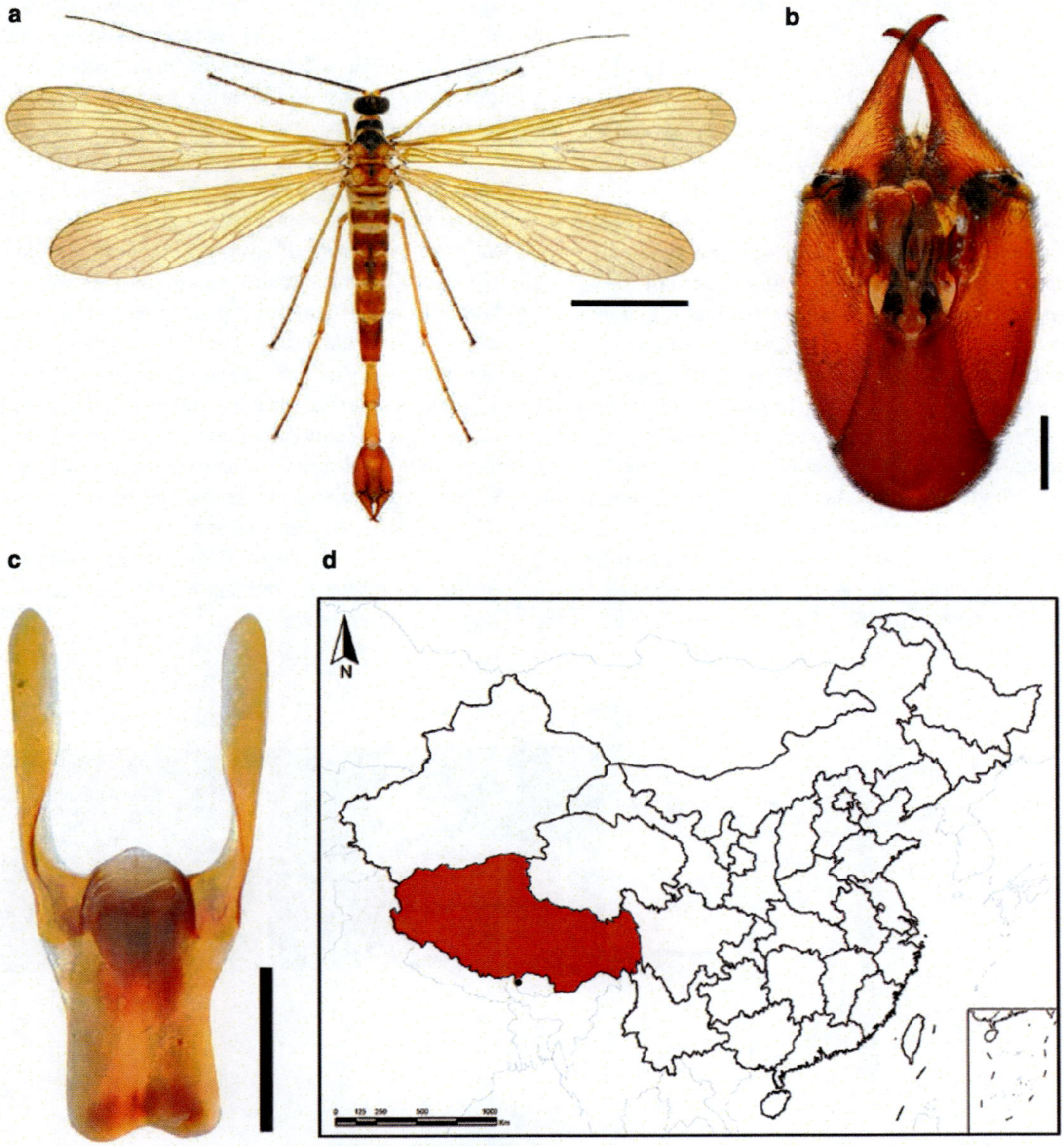

Fig. 10.41 (**a**) Male adult of ***Neopanorpa zhengyucheni***, dorsal view (Kambu Maqu, Yaodong, Shigatse, Tibet) (*Scale bar*: 5.0 mm) [**DALU**]. (**b**) Male genital bulb of ***Neopanorpa zhengyucheni***, ventral view (*Scale bar*: 0.5 mm). (**c**) Female medigynium of ***Neopanorpa zhengyucheni***, ventral view (*Scale bar*: 0.2 mm). (**d**) Distribution of ***Neopanorpa zhengyucheni***

Key to Species of Neopanorpa brisi Group in This Atlas
(Three species, males)

1. T3 lacking small tooth on posterior margin at each side of notal organ; genital bulb broad oval .... ***Neopanorpa fangxianga*** Zhou & Zhou
– T3 bearing small tooth on posterior margin at each side of notal organ; genital bulb long elliptical 2
2. Terminal portion of A6 raised dorsally into broad plate bearing dense setae along hind margin .... ***Neopanorpa pendula*** Qian & Zhou
– Terminal portion of A6 truncated .... ***Neopanorpa brisi*** (Navás)

10.1.39 *Neopanorpa brisi* Navás, 1930

(Bris' Neoscorpionfly) [布里斯新蝎蛉] (Fig. 10.42)
Leptopanorpa brisi Navás, 1930

Neopanorpa banksi Carpenter, 1938, **new synonym**

Neopanorpa dimidiata Navás, 1930, synonymized by Wang & Hua, 2018

Neopanorpa cangshanensis Zhou, 2006, **new synonym**

Neopanorpa pilosa Carpenter, 1945, **new synonym**

Neopanorpa taoi Cheng, 1949, **new synonym**

Neopanorpa validipennis Cheng, 1949, **new synonym**

Neopanorpa varia Cheng, 1949, **new synonym**

Measurements. Forewing length 16.5–17.5 mm.

Morphology. Vertex dark brown. Ocellar triangle black. Rostrum yellowish brown and darkened mesally. Nota black. Pleura and coxae creamy yellow. Legs mostly greenish brown with dark brown tarsi. Wing membrane hyaline, tinged with smoky grey, and devoid of markings or only darkened at apex. Male T1–T5 blackish brown. Notal organ stout and exceeding postnotal organ. Smaller acute tooth on ventral side of notal organ. Beside notal organ two small teeth on posterior margin of T3 laterally. A6 mostly dark brown but reddish brown apically, long, and cylindrical; A7 and A8 reddish brown. A7 slightly constricted at base and bearing pair of rounded teeth on each side of apex.

Male genital bulb long elliptical. Epandrium long and broad, and forming pair of acute processes curved mesad. Hypandrium with long broad basal stalk. Hypovalves long and broad, overlapped mesally, exceeding middle of gonostylus and forming circular window subbasally. Gonostylus shorter than half-length of gonocoxites, and concaved on outer margin subbasally. Paramere blade-like. Aedeagus conical with pair of stout lateral processes.

Female subgenital plate oval with deep V-shaped terminal emargination. Medigynium with broad and greatly sclerotized main plate; axis concealed and lacking distinct apodemes; posterior arms slender and widest at middle.

Emergence. May to August.

Distribution. Guizhou (Chishui, Liupanshui), Sichuan (Garzê, Liangshan), Yunnan (Dali, Kunming, Lijiang, Shangri-La, Yongsheng).

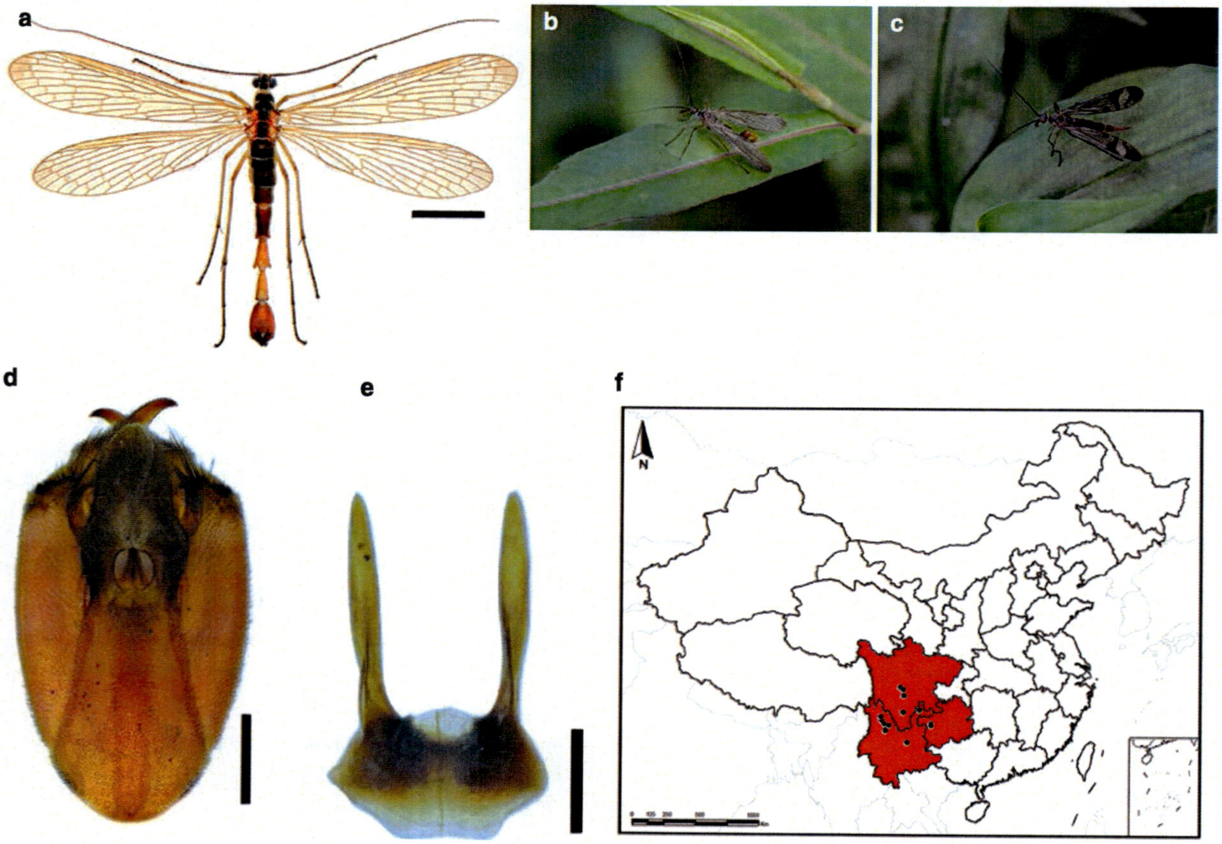

Fig. 10.42 (a) Male adult of ***Neopanorpa brisi***, dorsal view (West Hill, Kunming, Yunnan) (*Scale bar*: 5.0 mm) **[NWAU]**. (b) Male adult of ***Neopanorpa brisi*** in the field (West Hill, Kunming, Yunnan). (c) Female adult of ***Neopanorpa brisi*** in the field (West Hill, Kunming, Yunnan). (d) Male genital bulb of ***Neopanorpa brisi***, ventral view (*Scale bar*: 0.5 mm). (e) Female medigynium of ***Neopanorpa brisi***, ventral view (*Scale bar*: 0.2 mm). (f) Distribution of ***Neopanorpa brisi***

Remarks. This species has wide ranges of geographical distribution and morphological variation and exhibits a distinct sexual dimorphism. Consequently, different individuals can very easily be recognized as several species (Carpenter 1938; Cheng 1949, 1957; Zhou 2006). The detailed information for each synonym is as follows: (1) *N. banksi* was described from Yibin (=Xufu or Suifu, Sichuan) based on a single female; (2) *N. taoi* was described from Lushan Mountain (Sichuan) based on a single male; (3) *N. validipennis* was described from Kangding (=Tachienlu, Sichuan) based on a single male; (4) *N. varia* was described from Haierwa (Sichuan: Shimian) based on a single female; (5) *N. cangshanensis* was described from Chishui (Guizhou) based on a single female. Due to an unknown reason, Zhou (2006) named this species after the Cangshan Mountain (Yunnan: Dali). These populations share the same morphology with *N. brisi*. Therefore, they are treated junior synonyms of *N. brisi*.

10.1.40 *Neopanorpa fangxianga* Zhou & Zhou, 2007

(Fangxiang Neoscorpionfly) [方祥新蝎蛉] (Fig. 10.43)

Measurements. Forewing length 15.0 mm.

Morphology. Vertex black. Ocellar triangle black. Rostrum yellowish brown, with two brown longitudinal stripes laterally. Nota black. Pleura and legs yellowish brown. Wing membrane hyaline and slightly tinged with yellow. Pterostigma yellowish brown. Markings dark brown. In forewings, apical band broad and blurred poste-

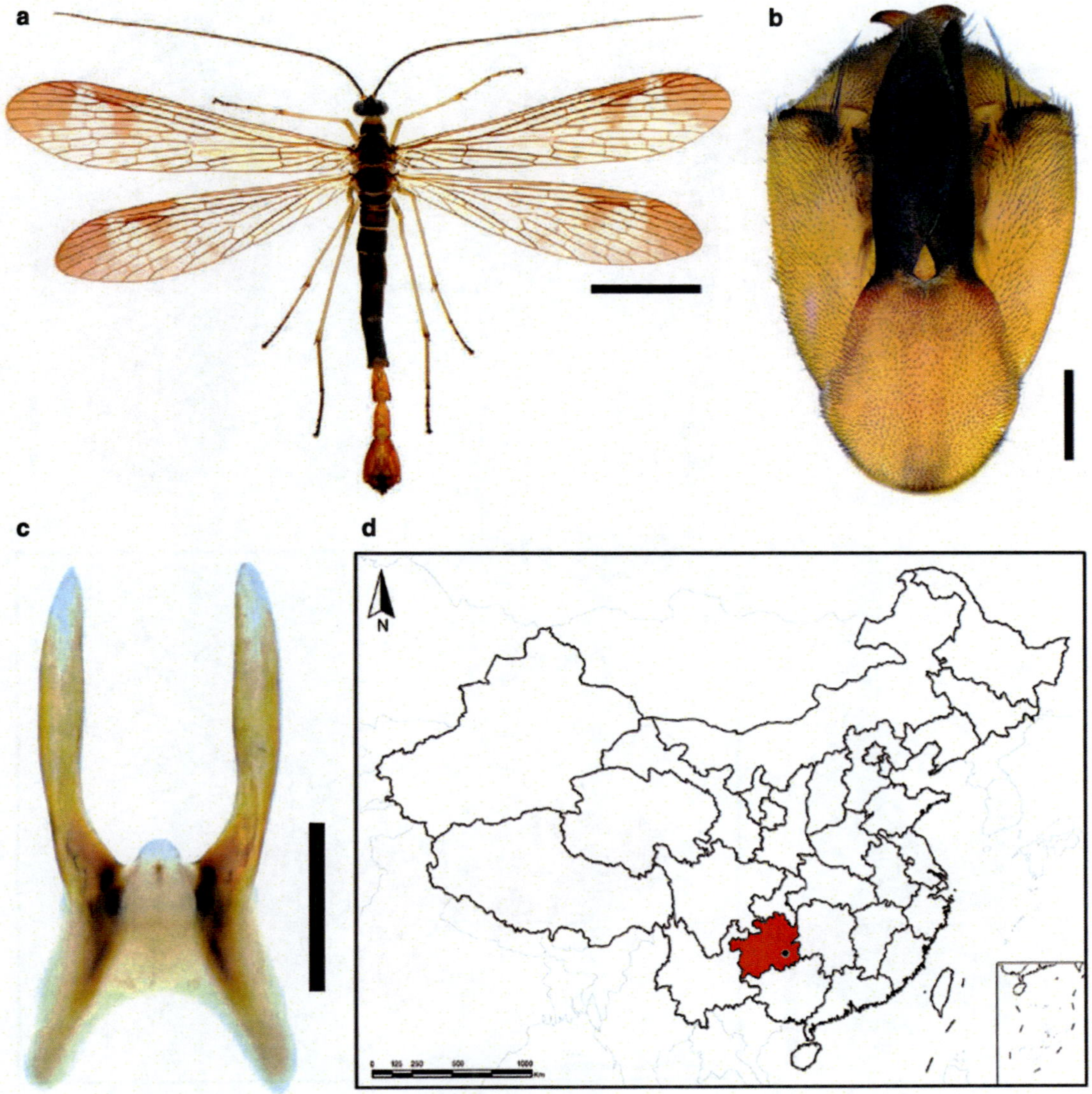

Fig. 10.43 (a) Male adult of ***Neopanorpa fangxianga***, dorsal view (Leigong Mountain, Guizhou) (*Scale bar*: 5.0 mm) **[NWAU]**. (b) Male genital bulb of ***Neopanorpa fangxianga***, ventral view (*Scale bar*: 0.5 mm). (c) Female medigynium of ***Neopanorpa fangxianga***, ventral view (*Scale bar*: 0.2 mm). (d) Distribution of ***Neopanorpa fangxianga***

riorly; pterostigmal band broad, with basal and apical branches reduced and inconspicuous; other markings absent. Male T1–T5 dark brown. Notal organ short clavate, not exceeding middle of T4. Small acute tooth beneath notal organ. A6 black, cylindrical, longer than A5, and bearing semicircular emargination on apex dorsally. A7 and A8 yellowish brown and slightly constricted basally.

Male genital bulb broad oval. Epandrium greatly protruded laterad on latero-apical corners. Hypandrium with broad basal stalk. Hypovalves black, long broad, curling dorsad along outer margins, and overlapping along inner margins, with apexes extending to middle of gonostylus. Gonocoxites with tuft of long bristles on ventral apex. Gonostylus stout and shorter than half-length of gonocoxites, and bearing dense setae on basal half; median tooth inconspicuous; basal process cylindrical. Aedeagus conical with stout lateral processes.

Female subgenital plate oval with deep V-shaped terminal emargination. Medigynium with reduced main plate; axis with distinct and widely divergent apodemes; posterior arms slender and acute terminally.

Emergence. June.

Distribution. Guizhou (Leigong Mountain).

10.1.41 *Neopanorpa pendula* Qian & Zhou, 2001

(Pendulous Neoscorpionfly) [垂齿新蝎蛉] (Fig. 10.44)

Measurements. Forewing length 17.0–17.5 mm.

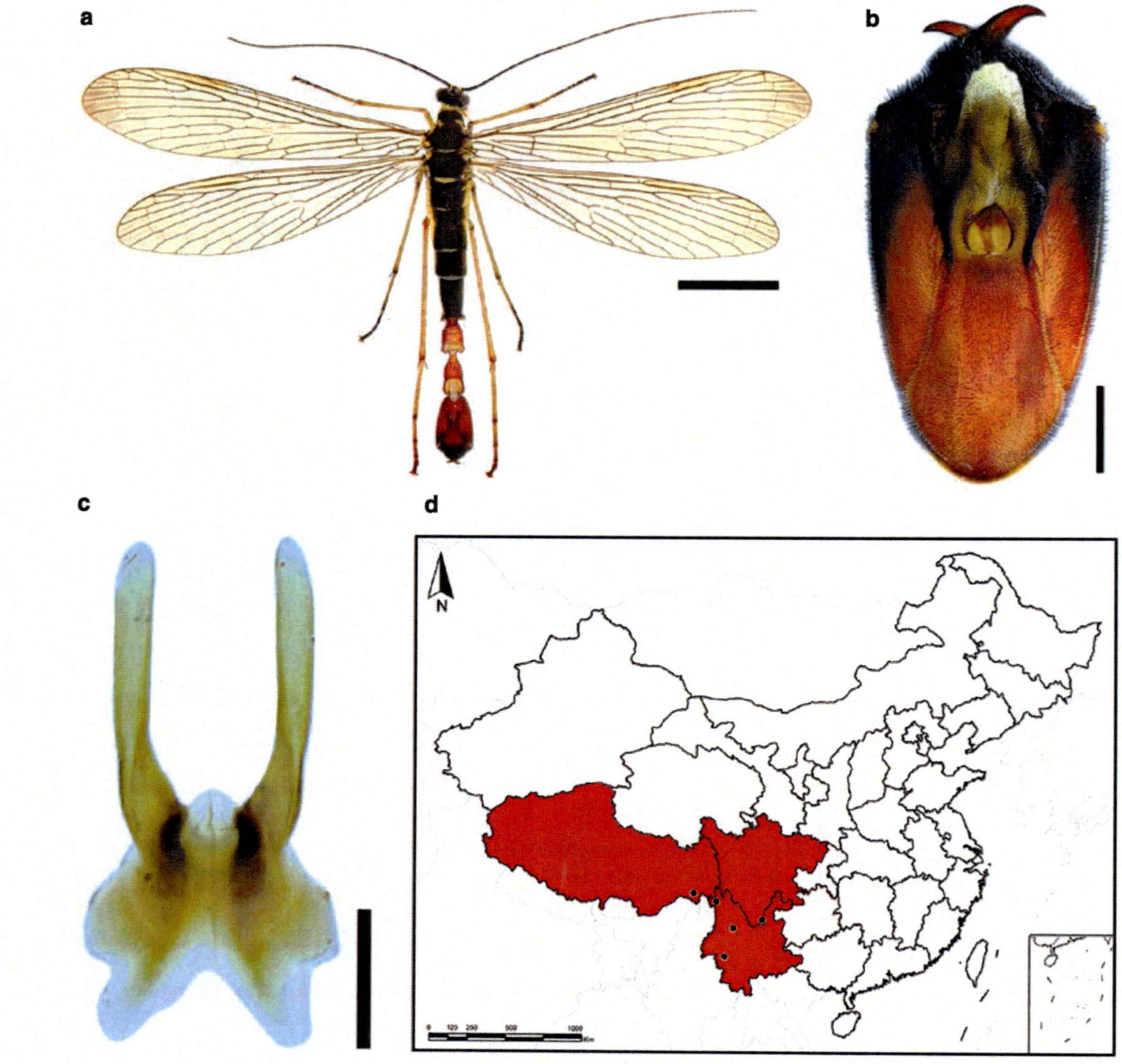

Fig. 10.44 (**a**) Male adult of ***Neopanorpa pendula***, dorsal view (Cangshan Mountain, Dali, Yunnan) (*Scale bar*: 5.0 mm) [**NWAU**]. (**b**) Male genital bulb of ***Neopanorpa pendula***, ventral view (*Scale bar*: 0.5 mm). (**c**) Male genital bulb of ***Neopanorpa pendula***, ventral view (*Scale bar*: 0.2 mm). (**d**) Distribution of ***Neopanorpa pendula***

Morphology. Vertex black. Ocellar triangle black. Rostrum yellowish brown, with pair of dark brown longitudinal stripes laterally. Pronotum black. Meso- and metanotum black, with latero-posterior corners yellowish brown. Pleura and legs light yellowish brown. Wing membrane hyaline, devoid of markings. Pterostigma light brown. Male T1–T5 black. Notal organ long finger-like, exceeding middle of T4, accompanied by pair of small teeth laterally and acute tooth on ventral surface. A6 mostly black and gradually turning into yellowish brown towards apex, and raised dorsally into broad plate bearing dense setae along hind margin. A7 shorter than half-length of A6 and constricted basally. A8 slightly longer than A7.

Male genital bulb long elliptical. Epandrium long broad, with latero-apical corners protruding laterad and acute. Hypandrium with long and broad basal stalk. Hypovalves broad, overlapped mesally, and forming inverted heart-shaped window subbasally. Gonostylus shorter than half-length of gonocoxites, bearing tuft of long curled bristles, inconspicuous median tooth, and stout semicircular basal process. Aedeagus conical with stout lateral processes.

Female subgenital plate oval with deep V-shaped terminal emargination. Medigynium with broad and subtriangular main plate; axis with greatly diverged apodemes; posterior arms slender and subacute terminally.

Emergence. June to July.

Distribution. Sichuan (Liangshan), Tibet (Zayü), Yunnan (Dali, Nujiang, Yongde).

Neopanorpa cavaleriei Group

More than 50 species are recognized in this group. They are mainly distributed in the Oriental Region, including southern and southwestern China, northern India, Laos, Myanmar, Thailand, and Vietnam. In China, approximately 30 species have been recorded, and 18 of them are included in this atlas.

Key to Species of Neopanorpa cavaleriei Group in This Atlas
(18 species, males)

1. Wing markings greatly developed and covering more than half area of wings; all bands and spots intensively fused to each other to form distinctive zigzag pattern .... ***Neopanorpa ocellaris*** (Navás)
– Wing markings less developed and covering less than half area of wings 2
2. Wings devoid of markings .... ***Neopanorpa claripennis*** Carpenter
– Wings with well-developed markings 3
3. Hypovalves extremely reduced and shorter than 1/4 length of basal stalk of hypandrium .... ***Neopanorpa lungtausana*** Cheng
– Hypovalves longer than 1/4 length of basal stalk of hypandrium 4
4. Gonostylus greatly elongated and longer than gonocoxites 5
– Gonostylus shorter than gonocoxites 6
5. Marginal spot elongated posteriorly and forming distinct K-shaped marking with pterostigmal band .... ***Neopanorpa k-maculata*** Cheng
– Marginal spot not elongated posteriorly .... ***Neopanorpa magna*** Issiki
6. Hypovalves tapering into acute apex .... ***Neopanorpa cuspidata*** Byers
– Hypovalves relatively rounded apically 7
7. Hypovalves extremely broad, nearly as wide as base of gonostylus .... ***Neopanorpa spatulata*** Byers
– Hypovalves narrower than base of gonostylus 8
8. Notal organ extremely short, triangular, and not exceeding basal 1/3 of T4 .... ***Neopanorpa sheni*** Hua & Chou
– Notal organ elongated and exceeding basal 1/3 of T4 9
9. Basal process of gonostylus enlarged with truncated apex 10
– Basal process not modified as above 11
10. Notal organ enlarged terminally; gonostylus elongated and nearly as long as gonocoxites .... ***Neopanorpa chelata*** Carpenter
– Notal organ not enlarged terminally; gonostylus shorter than gonocoxites .... ***Neopanorpa gulinensis*** Zhou & Zhou
11. Hypandrium greatly constricted at base of hypovalve .... ***Neopanorpa hei*** Zhou & Fan
– Hypandrium not constricted as above 12
12. Genital bulb long elliptical; gonostylus half as long as gonocoxites .... ***Neopanorpa clara*** Chou & Wang

– Genital bulb oval; gonostylus longer than half length of gonocoxites 13
13. Hypovalves dark brown or black .... ***Neopanorpa siamensis*** Byers
– Hypovalves yellowish brown or grayish brown 14
14. Hypovalves narrow and subacute; paramere broad and foliate .... ***Neopanorpa formosana*** (Navás)
– Hypovalves broad and rounded; paramere narrow 15
15. Meso- and metanotum mostly black, with lateral yellowish brown portion narrower than scutella 16
– Meso- and metanotum with lateral yellowish brown portion wider than scutella 17
16. Wings with apical band notched after R_4; apical branch of pterostigmal band usually reduced .... ***Neopanorpa nielseni*** Byers
– Wings with apical band complete or notched after R_5; apical branch of pterostigmal band well-developed .... ***Neopanorpa hunanensis*** Hua
17. Apical band in forewings with inner projection; epandrial lobes triangular .... ***Neopanorpa cavaleriei*** (Navás)
– Apical band in forewings lacking inner projection; epandrial lobes hook-shaped .... ***Neopanorpa pulchra*** Carpenter

10.1.42 *Neopanorpa cavaleriei* Navás, 1908

(Cavalerie's Neoscorpionfly) [卡氏新蝎蛉] (Fig. 10.45)
Neopanorpa maolanensis Zhou & Bao, 2002, **new synonym**

Measurements. Forewing length 16.0–17.0 mm.

Morphology. Vertex dark brown, ocellar triangle black; rostrum yellowish brown. Pronotum dark brown; Meso- and metanotum dark brown at anterior margin, with broad black median stripe, and yellowish brown at latero-posterior corners. Pleura and legs light yellowish brown. Wing membrane hyaline with brown markings. Apical band with oblique prolongation at middle of inner margin; pterostigmal band with complete apical and basal branches; basal band reduced at middle and forming two large spots; marginal spot slender; basal spot absent. Male T1–T5 black; sterna yellowish white; pleura yellow. Notal organ short clavate, not exceeding hind border of T4; A6 black. A7 and A8 yellowish brown and slightly constricted at base.

Male genital bulb long oval. Epandrium long broad, with truncated apex. Hypandrium with long broad basal stalk. Hypovalves curled along lateral margins and overlapped mesally, forming subcircular window subbasally. Gonostylus slightly shorter than gonocoxites, with obtuse triangular median tooth and rounded basal process bearing curved acute tooth. Paramere almost fused with aedeagus; aedeagus conical with stout lateral processes.

Female subgenital plate oval with deep V-shaped terminal emargination. Medigynium with main plate broad, subtriangular, and greatly sclerotized at lateral portions; axis lacking apodemes; posterior arms slender, slightly spatulate in distal half, and subacute terminally.

Emergence. April to June.

Distribution. Guizhou (Libo).

Remarks. *N. maolanensis* was described from Maolan (Guizhou), where *N. cavaleriei* has also been recorded, and can be differentiated from the latter by the denser wing markings and longer male hypovalves (Zhou and Bao 2002). However, by our examination of a vast number of specimens collected from Maolan, a wide range of variations was found in the wing markings and male hypovalves of *N. cavaleriei*, and the morphology of *N. maolanensis* undoubtfully falls into that range. Therefore, *N. maolanensis* is treated as a junior synonym of *N. cavaleriei*.

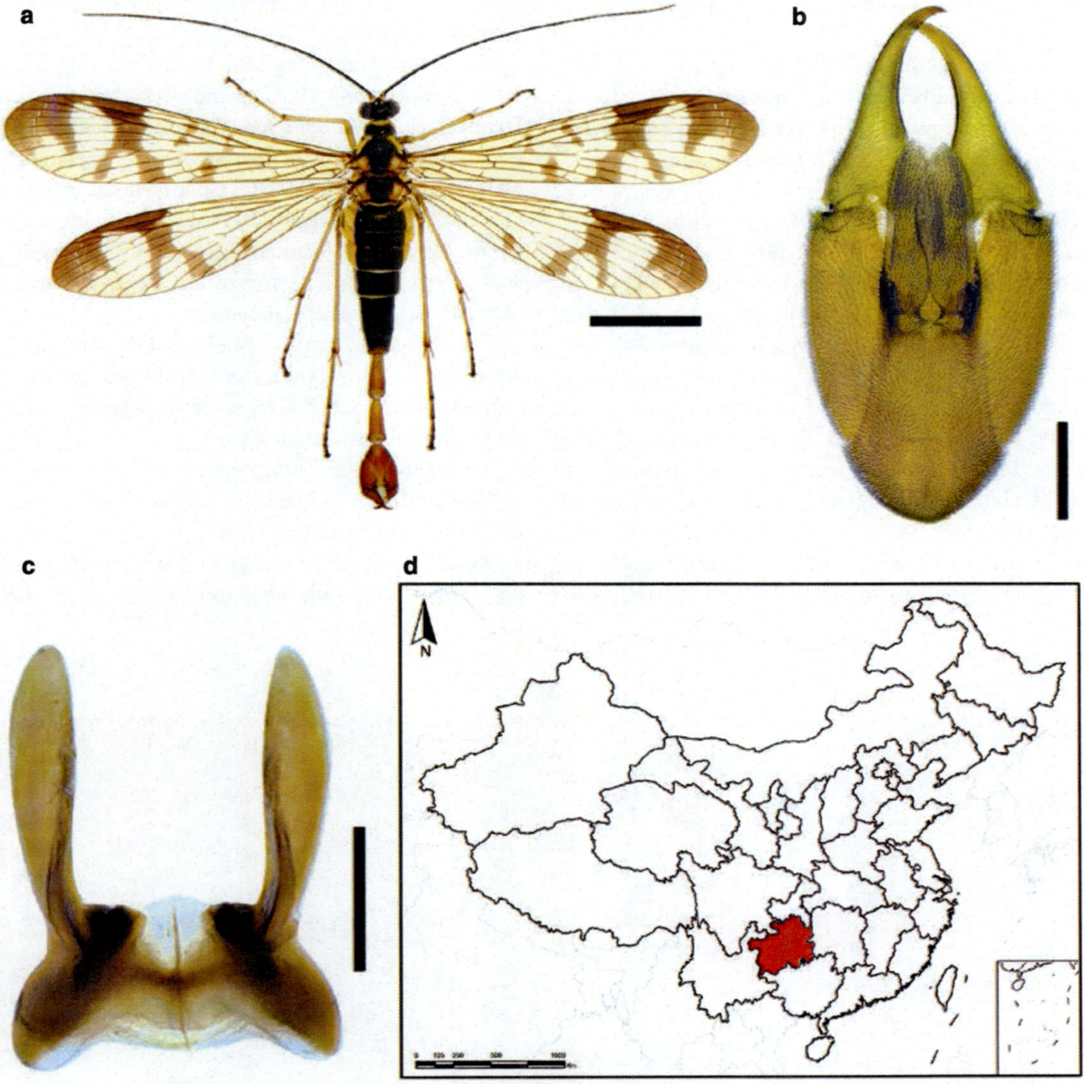

Fig. 10.45 (**a**) Male adult of ***Neopanorpa cavaleriei***, dorsal view (Libo, Guizhou) (*Scale bar*: 5.0 mm) [NWAU]. (**b**) Male genital bulb of ***Neopanorpa cavaleriei***, ventral view (*Scale bar*: 0.5 mm). (**c**) Female medigynium of ***Neopanorpa cavaleriei***, ventral view (*Scale bar*: 0.2 mm). (**d**) Distribution of ***Neopanorpa cavaleriei***

10.1.43 *Neopanorpa chelata* Carpenter, 1938

(Chelate Neoscorpionfly) [爪新蝎蛉] (Fig. 10.46)
Neopanorp parva Carpenter, 1945, **new synonym**
Neopanorpa latipennis Cheng, 1949, **new synonym**
Neopanorpa heii Cheng, 1957, **new synonym**

Measurements. Forewing length 15.0–17.0 mm.

Morphology. Vertex black. Ocellar triangle black. Rostrum yellowish brown to dark brown. Pronotum dark brown. Meso- and metanotum mostly dark brown but yellowish brown laterally. Pleura and legs yellowish brown. Wing membrane hyaline and slightly tinged with light brown with markings dark brown. In forewings, apical band semicircular and broad, and with separate small spot posteriorly; pterostigmal band broad, with broad basal branch, and thin or detached apical branch; marginal spot slender; basal band represented by two irregular spots; basal spot absent. Male T1–T5 dark brown. Notal organ short and clavate, covering rounded postnotal organ. A6 dark brown and long cylindrical. A7 and A8 yellowish brown and long cylindrical, stout, and inconspicuously constricted basally.

Male genital bulb long oval. Epandrium slightly truncated on apical margin. Hypandrium with broad basal stalk. Hypovalves broad and long, slightly exceeding apex of gonocoxites, with outer margins curled dorsad, and inner margins overlapped mesally. Gonostylus nearly as long as gonocoxites, with trapezoidal basal process and obtuse triangular mesal tooth. Paramere fused with aedeagus basally, and curved laterad and foliate apically. Aedeagus conical, with pair of stout lateral processes.

Female subgenital plate oval with deep V-shaped terminal emargination. Medigynium with main plate greatly reduced; posterior arms flat, and diverged in V-shape; axis with median projection and lacking apodemes.

Emergence. May to August.

Distribution. Sichuan (Baoxing, Chongzhou, Dujiangyan, Mount Emei, Ya'an).

Remarks. *N. parva* is only known from the female paratype, which shares the same morphology of *N. chelata* not

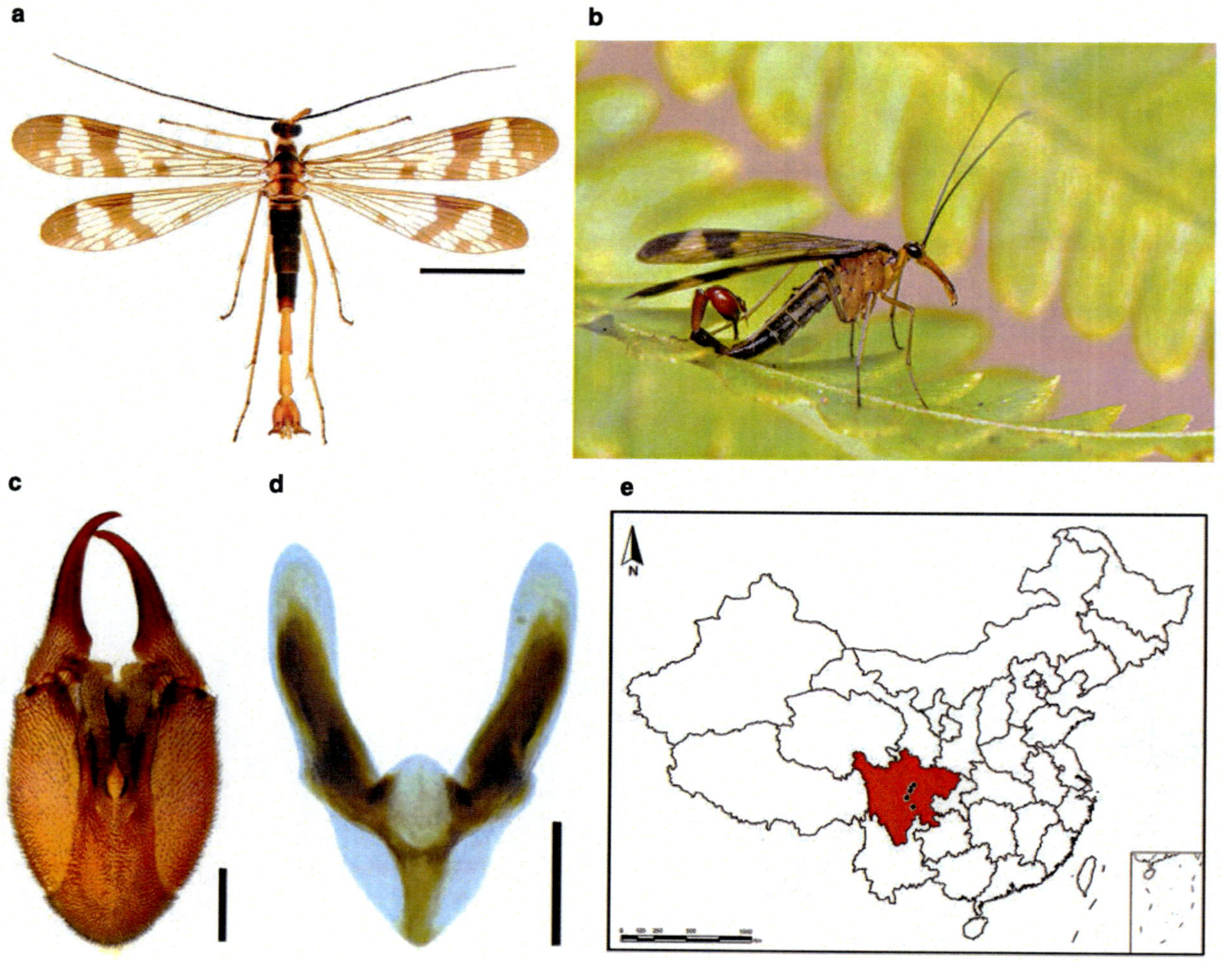

Fig. 10.46 (**a**) Male adult of ***Neopanorpa chelata***, dorsal view (Mount Emei, Sichuan) (*Scale bar*: 5.0 mm) **[NWAU]**. (**b**) Male adult of ***Neopanorpa chelata***, lateral view (Zhougong Mountain, Sichuan). (**c**) Male genital bulb of ***Neopanorpa chelata***, ventral view (*Scale bar*: 0.5 mm). (**d**) Female medigynium of ***Neopanorpa chelata***, ventral view (*Scale bar*: 0.2 mm). (**e**) Distribution of ***Neopanorpa chelata***

only in the body pattern, but also the medigynial morphology. Therefore, it is treated as a junior synonym of *N. chelata*.

N. latipennis is also known only from the female holotype, which bears a Y-shaped pattern between apical and pterostigmal bands in the forewings, and a V-shaped medigynium. These characters are almost identical to some female specimens of *N. chelata*, which bear denser wing markings than the males. Therefore, *N. latipennis* is treated as a junior synonym of *N. chelata*.

N. heii is also known as the female holotype, which can be differentiated from *N. chelata* by the greatly diverged posterior arms in the female medigynium (Cheng 1957). However, after the examination of the topotypical specimens of *N. heii* and *N. chelata* and those from many other localities, we confirm that the medigynium of *N. heii* is identical to that of *N. chelata*, and the greatly diverged posterior arms are deformed from a regular V-shape due to dehydration.

10.1.44 *Neopanorpa clara* Chou & Wang, 1988

(Clear-Marked Neoscorpionfly) [显斑新蝎蛉]
(Fig. 10.47)

Measurements. Forewing length 14.0 mm.

Morphology. Vertex black. Ocellar triangle black. Rostrum yellowish brown. Pronotum black. Meso- and metanotum dark brown. Pleura and legs yellowish brown. Wing membrane hyaline. In forewings, apical band complete; separated spot on hind border near apical band; pterostigmal band complete, with thin apical branch and broad basal branch; marginal spot small; basal band blurred medially and irregular at anterior and posterior ends; basal spot small. Male T1–T5 dark brown. Notal organ long and stick-like, slightly exceeding postnotal organ. A6 mostly black with yellowish brown apex, cylindrical; A7 and A8 yellowish brown and slightly constricted basally.

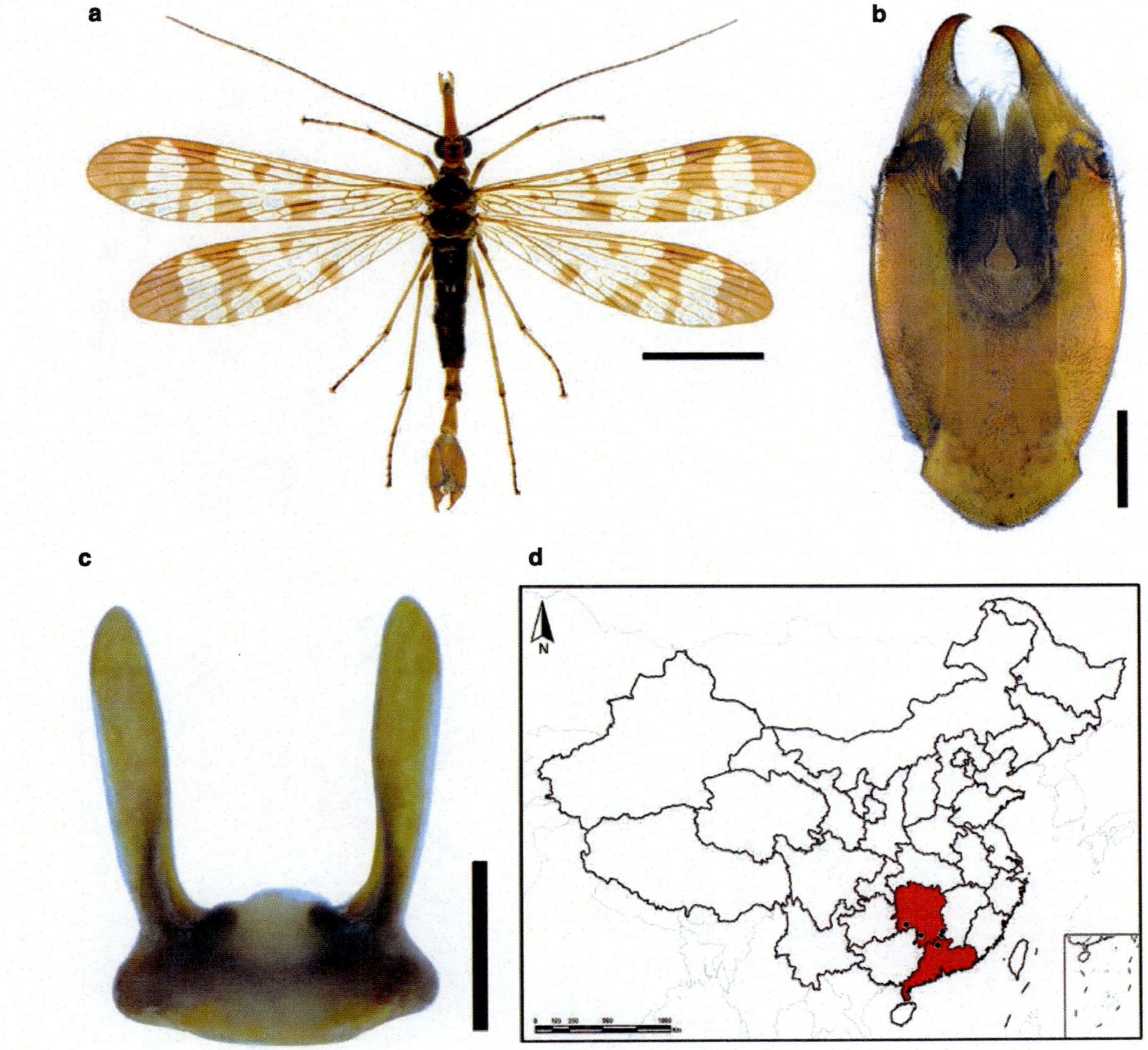

Fig. 10.47 (**a**) Male adult of ***Neopanorpa clara***, dorsal view (Ruyuan, Guangdong) (*Scale bar*: 5.0 mm) [NWAU]. (**b**) Male genital bulb of ***Neopanorpa clara***, ventral view (*Scale bar*: 0.5 mm). (**c**) Female medigynium of ***Neopanorpa clara***, ventral view (*Scale bar*: 0.2 mm). (**d**) Distribution of ***Neopanorpa clara***

Male genital bulb long elliptical. Epandrium long and broad, and slightly emarginate terminally. Hypandrium with long and broad basal stalk. Hypovalves long and overlapped mesally. Gonostylus nearly third as long as gonocoxites, with obtuse triangular median tooth and cylindrical basal process. Paramere foliate. Aedeagus conical, slightly exceeding apex of gonocoxites, with stout lateral processes.

Female subgenital plate oval with deep V-shaped terminal emargination. Medigynium with main plate broad, subtriangular, and greatly sclerotized at lateral portions; axis lacking apodemes; posterior arms slender, slightly spatulate in distal half, and subacute terminally.

Emergence. April to July.

Distribution. Hunan (Bamian Mountain, Rucheng, Suining, Shuangpai).

10.1.45 *Neopanorpa claripennis* Carpenter, 1938

(Clear-Winged Neoscorpionfly) [明翅新蝎蛉] (Fig. 10.48)
Neopanorpa lui Chou & Ran in Chou et al., 1981, **new synonym**

Measurements. Forewing length 13.0 mm.

Morphology. Vertex black. Ocellar triangle black. Rostrum unevenly yellowish brown. Pronotum dark brown. Meso- and metanotum yellowish brown, bearing broad deep brown median stripe. Pleura and legs yellowish brown. Wing membrane hyaline and devoid of markings, except indistinct tinge on apex. Pterostigma inconspicuous. Male T1–T5 black. Notal organ short and covering postnotal organ. A6 black, cylindrical. A7 and A8 yellowish brown and slightly constricted basally.

Male genital bulb oval. Epandrium long and broad, and truncated terminally. Hypandrium with broad basal stalk,

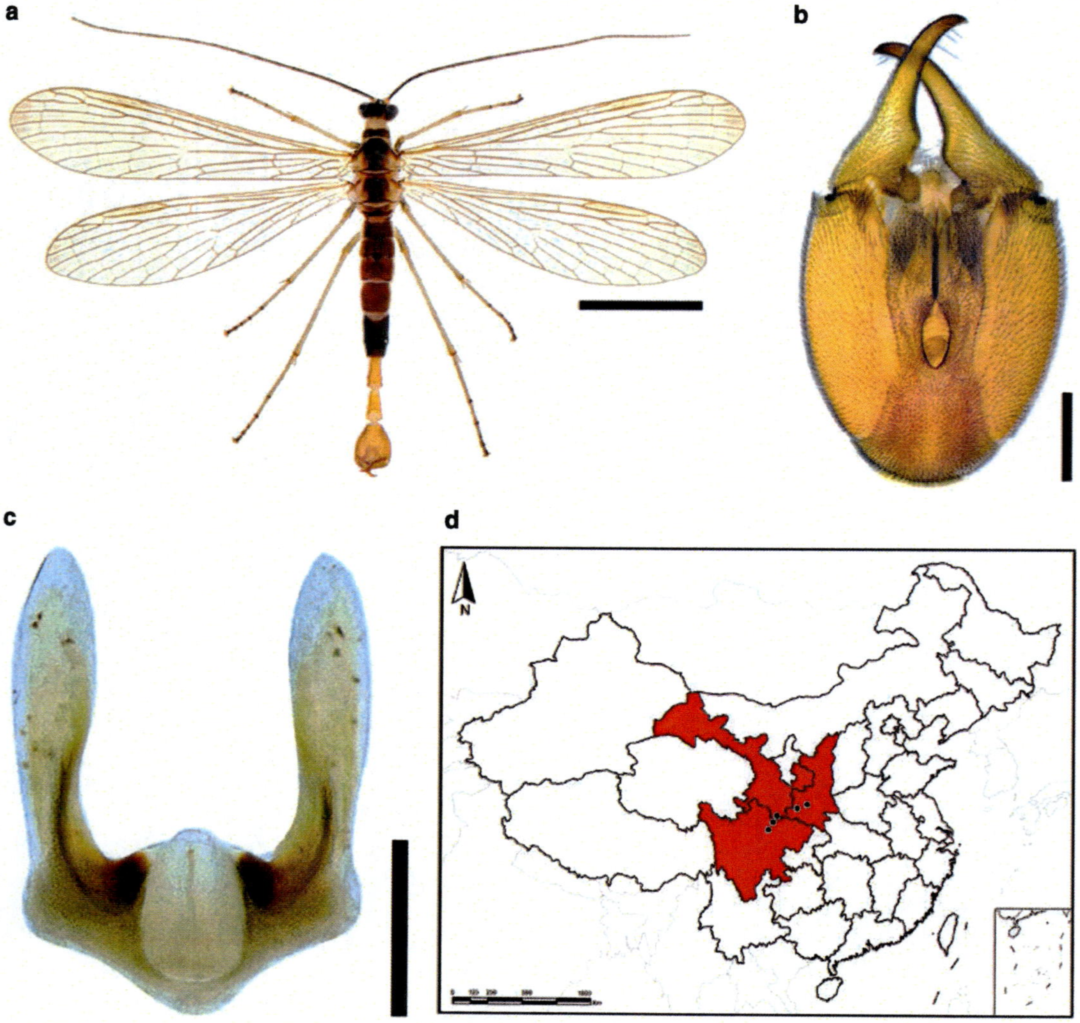

Fig. 10.48 (**a**) Male adult of ***Neopanorpa claripennis***, dorsal (Laohegou, Pingwu, Sichuan) (*Scale bar*: 5.0 mm) [NWAU]. (**b**) Male genital bulb of ***Neopanorpa claripennis***, ventral view (*Scale bar*: 0.5 mm). (**c**) Female mediygnium of ***Neopanorpa claripennis***, ventral view (*Scale bar*: 0.2 mm). (**d**) Distribution of ***Neopanorpa claripennis***

with hypovalves not exceeding apex of gonocoxites. Hypovalves overlapped mesally and forming oval window subbasally. Gonostylus slender and bearing obtuse triangular median tooth and stout basal process. Paramere slender. Aedeagus conical and slightly exceeding apex of gonocoxites, with stout lateral processes.

Female subgenital plate oval with deep V-shaped terminal emargination. Medigynium with main plate reduced, and slightly projected laterad near base of posteiror arms; axis lacking apodemes; posterior arms slender, slightly spatulate in distal half, and subacute terminally.

Emergence. July.

Distribution. Sichuan (Bailuding, Pingwu).

Remarks. *N. lui* was described from western Qinling Mountains (Chou et al. 1981) and conforms *N. claripennis* in the morphology. Therefore, it is treated as a junior synonym of *N. claripennis*.

10.1.46 *Neopanorpa cuspidata* Byers, 1965

(Tipped Neoscorpionfly) [尖齿新蝎蛉] (Fig. 10.49)

Measurements. Forewing length 11.0–13.0 mm.

Morphology. Vertex black. Ocellar triangle black. Rostrum yellowish brown. Pronotum dark brown. Meso- and metanotum dark brown but yellowish brown laterally.

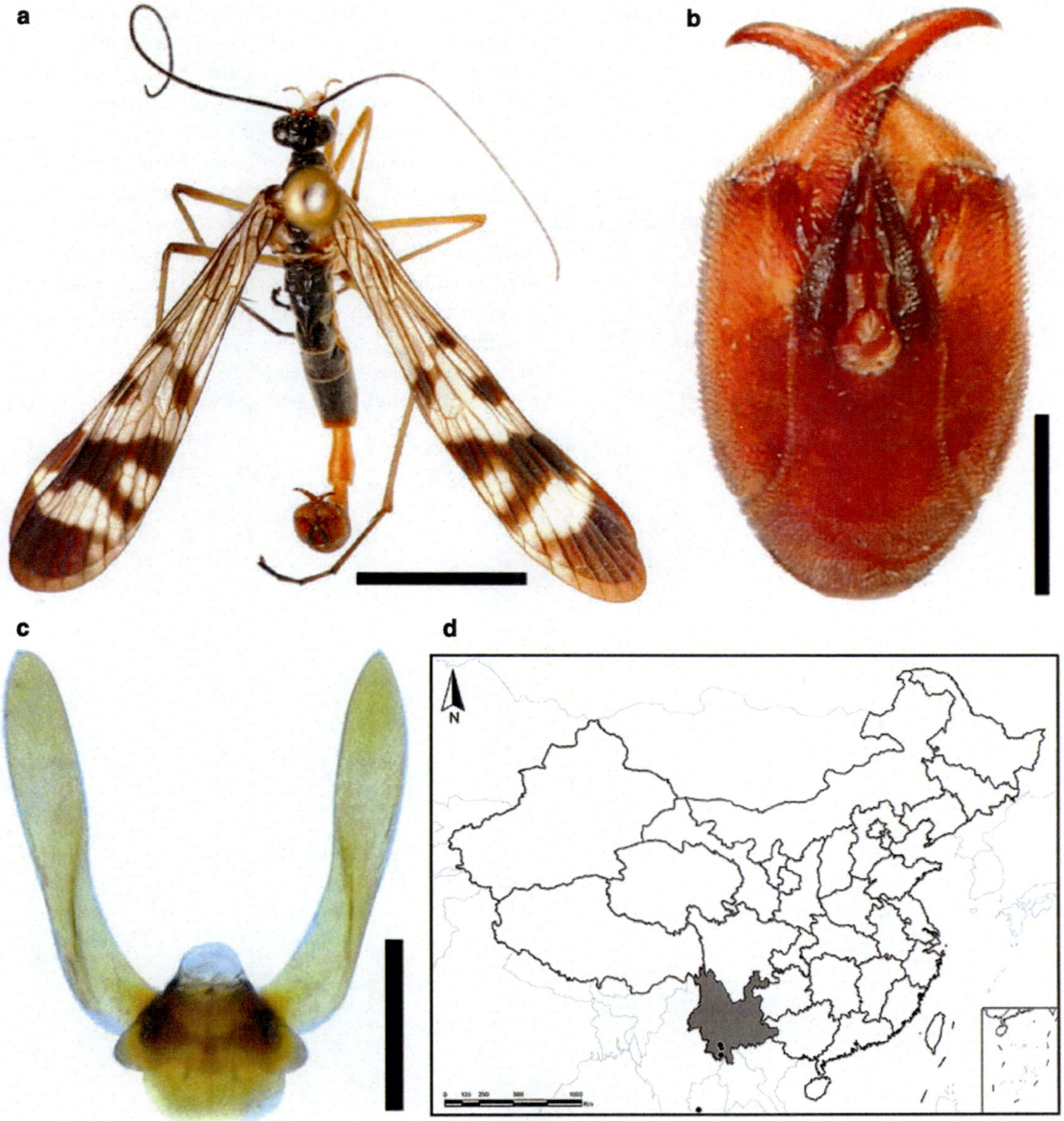

Fig. 10.49 (**a**) Male adult of ***Neopanorpa cuspidata***, dorsal view (Nabanhe, Xishuangbanna, Yunnan) (*Scale bar*: 5.0 mm) **[IZAS]**. (**b**) Male genital bulb of ***Neopanorpa cuspidata***, ventral view (*Scale bar*: 0.5 mm). (**c**) Female medigynium of ***Neopanorpa cuspidata***, ventral view (*Scale bar*: 0.2 mm). (**d**) Distribution of ***Neopanorpa cuspidata***

Pleura and legs yellowish brown. Wing membrane hyaline with markings dark brown. In forewings, apical band broad with small hyaline window posteriorly; pterostigmal band complete; marginal spot broad; basal band split medially and forming two spots; basal spots inconspicuous. Male T1–T5 dark brown. Notal organ well-developed acute triangular, covering rounded postnotal organ. A6 dark brown and long cylindrical. A7 and A8 yellowish brown and slightly constricted basally.

Male genital bulb long oval. Epandrium long and broad, and truncated terminally. Hypandrium with broad basal stalk. Hypovalves exceeding apex of gonocoxites, acute apically, and forming oval window at inner margin subbasally. Gonostylus slender, bearing obtuse triangular median tooth and cup-like basal process. Paramere fused with aedeagus at basal half, and free at distal half. Aedeagus conical, with stout lateral processes.

Female subgenital plate oval with deep V-shaped terminal emargination. Medigynium with main plate subtriangular; axis lacking apodemes; posterior arms slender, slightly spatulate in distal half, and subacute terminally.

Emergence. May to August.

Distribution. Yunnan (Naban River, Menglong, Mengwang); Thailand.

10.1.47 *Neopanorpa formosana* Navás, 1911

(Taiwan Neoscorpionfly) [台湾新蝎蛉] (Fig. 10.50)

Measurements. Forewing length 14.0–15.0 mm.

Morphology. Vertex dark brown. Ocellar triangle black. Rostrum yellowish brown. Nota mostly dark brown except latero-posterior corners of Meso- and metanotum slightly paler. Pleura and legs yellowish brown. Wing membrane hyaline with markings brown. In forewings, apical band broad with small separate spot near hind margin, and with large hyaline spot innerly; pterostigmal band complete, with thin apical branch and broad basal branch, and connected to apical band on anterior margin; basal band split into two small spots; marginal spot conspicuous; basal spot very small and blurred. Male T1–T5 dark brown. Notal organ long, stick-like, extending to middle of T4 and covering rounded postnotal organ. A6 dark brown and cylindrical. A7 and A8 yellowish brown and constricted basally.

Male genital bulb long elliptical. Epandrium long and broad with truncated apex. Hypandrium with long broad basal stalk. Hypovalves black and long, with inner margin overlapped mesally and forming subcircular window subbasally. Gonostylus shorter than half-length of gonocoxites, with obtuse triangular median tooth and stout basal process. Aedeagus conical with stout lateral processes.

Female subgenital plate oval with deep V-shaped terminal emargination. Medigynium with main plate broad, subtriangular, and greatly sclerotized; axis lacking apodemes; posterior arms slender, slightly spatulate in distal half, and acute terminally.

Emergence. March to May.

Distribution. Taiwan (Nantou, New Taipei, Pingtung).

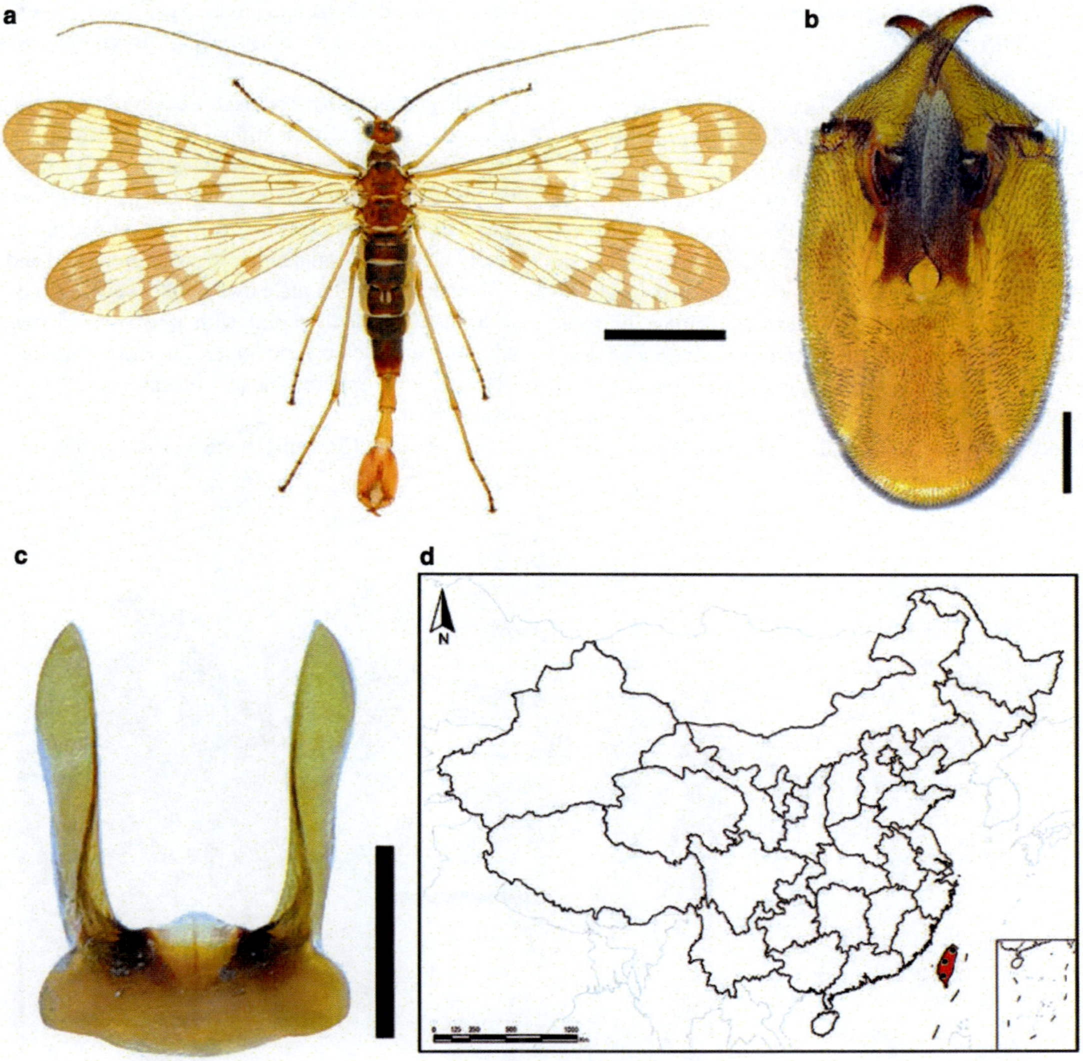

Fig. 10.50 (**a**) Male adult of ***Neopanorpa formosana***, dorsal view (Sanxia, Xinbei, Taiwan) (*Scale bar*: 5.0 mm) [NWAU]. (**b**) Male genital bulb of ***Neopanorpa formosana***, ventral view (*Scale bar*: 0.5 mm). (**c**) Female medigynium of ***Neopanorpa formosana***, ventral view (*Scale bar*: 0.2 mm). (**d**) Distribution of ***Neopanorpa formosana***

10.1.48 *Neopanorpa gulinensis* Zhou & Zhou, 2005

(Gulin Neoscorpionfly) [古蔺新蝎蛉] (Fig. 10.51)

Neopanorpa linjiangensis Zhou, 2005, **new synonym**

Measurements. Forewing length 12.5–13.2 mm.

Morphology. Vertex dark brown. Ocellar triangle black. Rostrum yellowish brown. Pronotum black. Meso- and metanotum yellowish brown laterally with black median stripe. Pleura and legs yellowish brown. Wing membrane hyaline with dark brown markings. In forewings, apical band incomplete with open window posteriorly; pterostigmal band complete, with broad basal branch and slightly reduced apical branch; marginal spot small; basal band split into two small spots; basal spot absent. Male T1–T5 dark brown. Notal organ short finger-like. A6 dark brown. A7 and A8 yellowish brown and constricted basally.

Male genital bulb broad oval. Epandrium long broad, with truncated apex. Hypandrium with broad basal stalk. Hypovalves slightly constricted basally, curled apically, forming long drop-shaped window subbasally. Parameres short and indistinct. Aedeagus conical, with ventral and dorsal valves greatly elongated, and lateral processes elongated and acute.

Female subgenital plate oval with deep V-shaped terminal emargination. Medigynium with undeveloped main plate; axis short and lacking apodemes; posterior arms slender and subacute, with earlobe-shaped processes at outer margin of basal portion.

Distribution. Guizhou (Dashahe, Xishui), Sichuan (Gulin).

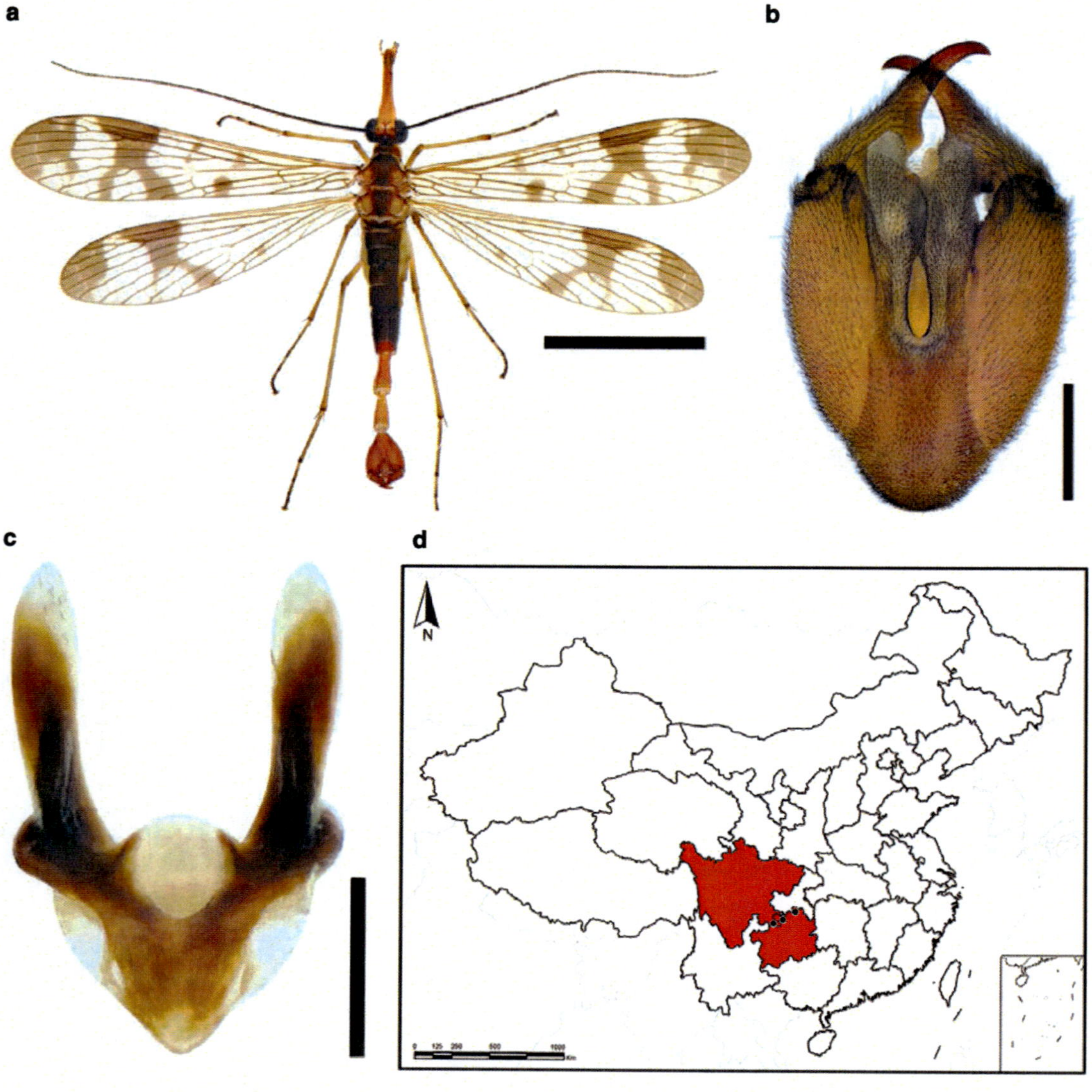

Fig. 10.51 (**a**) Male adult of ***Neopanorpa gulinensis***, dorsal view (Xishui, Guizhou) (*Scale bar*: 5.0 mm) **[NWAU]**. (**b**) Male genital bulb of ***Neopanorpa gulinensis***, ventral view (*Scale bar*: 0.5 mm). (**c**) Female medigynium of ***Neopanorpa gulinensis***, ventral view (*Scale bar*: 0.2 mm). (**d**) Distribution of ***Neopanorpa gulinensis***

Remarks. Although *N. linjiangensis* was described for the specimens collected from Linjiang (Guizhou), it shares the identical morphology to *N. gulinensis* from Gulin (Zhou 2005, fig. 8 v.s. Zhou and Zhou 2005, fig. 4). The former was published in July and the latter in June. Therefore, according to the principle of priority, *N. linjiangensis* is treated as a junior synonym of *N. gulinensis*.

10.1.49 *Neopanorpa hei* Zhou & Fan, 1998

(He's Neoscorpionfly) [何氏新蝎蛉] (Fig. 10.52)
Measurements. Forewing length 11.0–12.0 mm.

Morphology. Vertex dark brown. Ocellar triangle black. Rostrum yellowish brown. Pronotum black. Meso- and metanotum yellowish brown laterally with broad black median stripe. Pleura and legs yellowish brown. Wing membrane hyaline with dark brown markings. In forewings, apical band complete with closed hyaline window innnerly; pterostigmal band complete, with broad basal branch and slightly reduced apical branch; marginal spot distinct; basal band split into two small spots; basal spot indistinct. Male T1–T5 dark brown. Notal organ short finger-like, extending to middle of T4 and slightly enlarged terminally; postnotal organ rounded. A6 dark brown. A7 and A8 yellowish brown and constricted basally.

Male genital bulb broad oval. Epandrium long broad, with rounded apex; epandrial lobes large with truncated apex. Hypandrium with broad basal stalk. Hypovalves greatly constricted basally, overlapping along inner margin, and forming long drop-shaped window subbasally. Parameres slender.

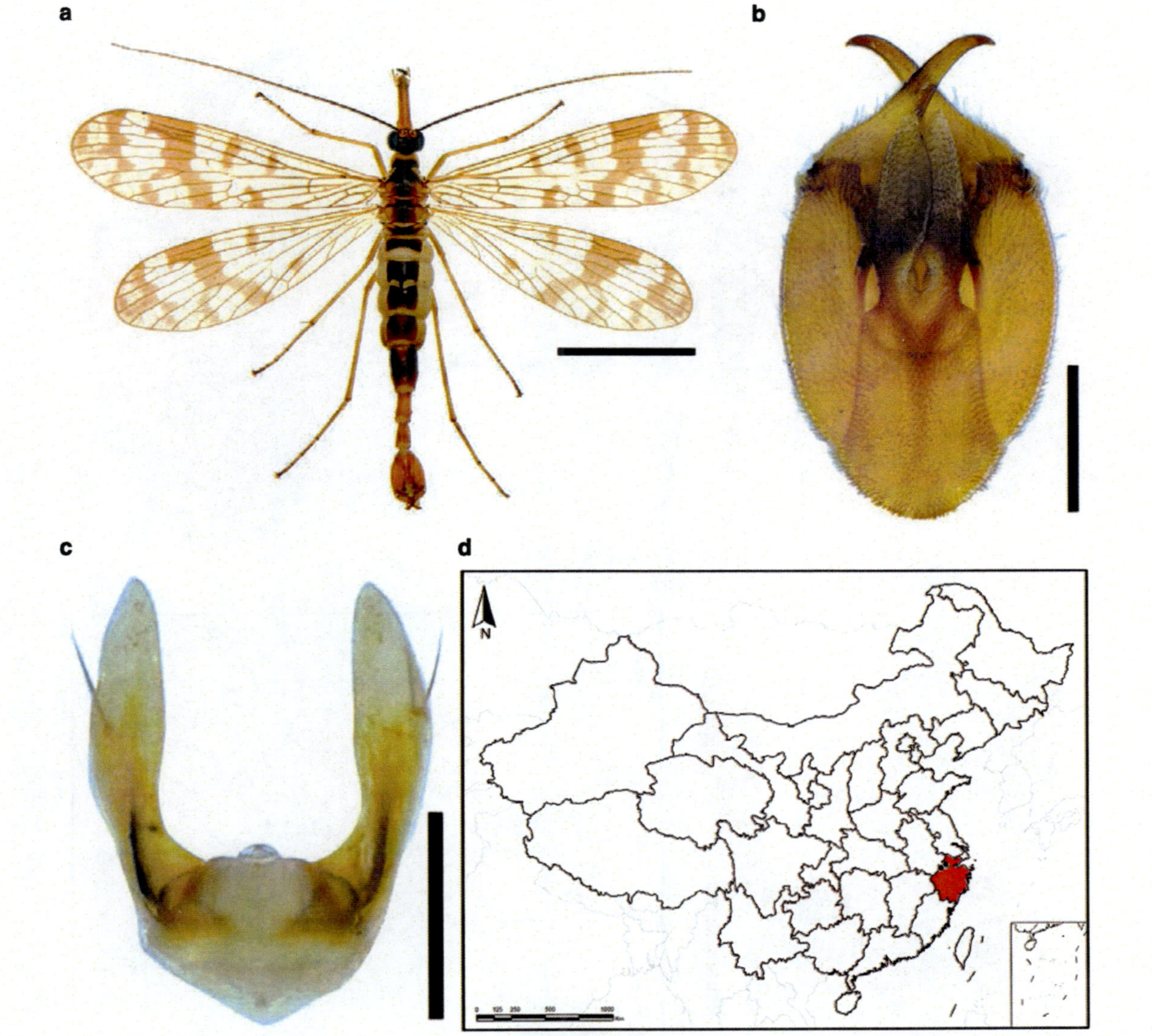

Fig. 10.52 (**a**) Male adult of ***Neopanorpa hei***, dorsal view (West Tianmu Mountain, Zhejiang) (*Scale bar*: 5.0 mm) [**NWAU**]. (**b**) Male genital bulb of ***Neopanorpa hei***, ventral view (*Scale bar*: 0.5 mm). (**c**) Female medigynium of ***Neopanorpa hei***, ventral view (*Scale bar*: 0.2 mm). (**d**) Distribution of ***Neopanorpa hei***

Aedeagus conical, with lateral processes elongated and acute.

Female subgenital plate oval with deep U-shaped terminal emargination. Medigynium with undeveloped main plate; axis short and lacking apodemes; posterior arms long, spatulate in distal half, subacute, and bearing long stout setae on middle of outer margin.

Emergence. May to July.

Distribution. Zhejiang (Longwang Mountain, Tianmu Mountains, Qingliang Peak).

10.1.50 *Neopanorpa hunanensis* Hua, 2002

(Hunan Neoscorpionfly) [湖南新蝎蛉] (Fig. 10.53)
Neopanorpa anchoroides Zhou, 2003, **new synonym**

Measurements. Forewing length 17.0–19.0 mm.

Morphology. Vertex black mesally but yellowish brown laterally. Ocellar triangle black. Rostrum yellowish brown with darkened flanks. Pronotum dark brown. Meso- and metanotum dark brown with paler flanks. Pleura and legs yellowish brown. Wing membrane hyaline with dark brown markings. In forewings, apical band complete with notched inner margin; pterostigmal band broad and connected to apical band on anterior margin; marginal spot small; basal band complete and broad posteriorly; basal spot inconspicuous. Male T1–T5 dark brown. Notal organ short and triangular. A6 dark brown. A7 and A8 yellowish brown and slightly constricted basally.

Male genital bulb long oval. Epandrium long and broad with truncated apex. Hypandrium with long broad basal stalk. Hypovalves long and broad, curled dorsad along outer

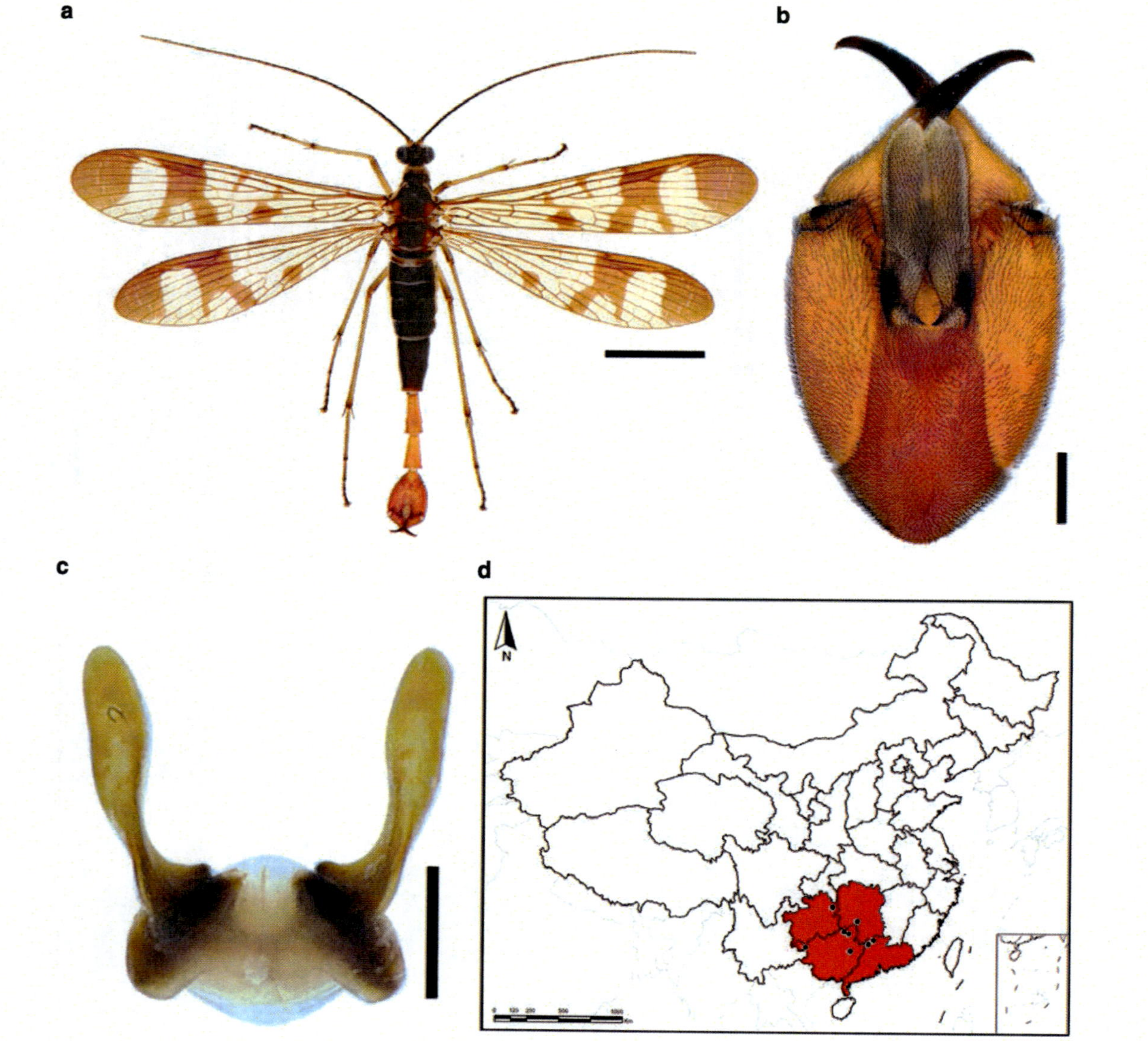

Fig. 10.53 (a) Male adult of ***Neopanorpa hunanensis***, dorsal view (Kitten Mountain, Guangxi) (*Scale bar*: 5.0 mm) **[NWAU]**. **(b)** Male genital bulb of ***Neopanorpa hunanensis***, ventral view (*Scale bar*: 0.5 mm). **(c)** Female medigynium of ***Neopanorpa hunanensis***, ventral view (*Scale bar*: 0.2 mm). **(d)** Distribution of ***Neopanorpa hunanensis***

and inner margins, overlapped mesally, and forming subcircular window subbasally. Paramere short, thin, and divergent apically. Aedeagus conical, slightly exceeding apex of gonocoxites and bearing pair of stout lateral processes.

Female subgenital plate oval with deep V-shaped terminal emargination. Medigynium with main plate broad, subtriangular, and greatly sclerotized at lateral portions; axis lacking apodemes; posterior arms slender, slightly spatulate in distal half, and rounded terminally.

Emergence. May to August.

Distribution. Guangdong (Lianzhou), Guangxi (Jinxiu, Kitten Mountain, Longsheng, Tianlin), Guizhou (Fanjing Mountain), Hunan (Dong'an, Mangshan Mountains).

Remarks. *N. anchoroides* was described from Fanjing Mountain. It shares the same morphology with *N. hunanensis*. Zhou (2003) stated that *N. anchoroides* is peculiar for the anchor-shaped structure in the male genitalia, however, this structure is the piston of the sperm pump (Wang and Hua 2021) and cannot provide enough information in the species recognitions.

10.1.51 *Neopanorpa k-maculata* Cheng, 1952

(K-marked Neoscorpionfly) [K纹新蝎蛉] (Fig. 10.54)

Measurements. Forewing length 18.5–20.0 mm.

Morphology. Vertex yellowish brown, with black longitudinal stripe extending to occiput. Ocellar triangle black. Rostrum yellowish brown. Pronotum dark brown. Mesonotum mostly yellowish brown, with dark brown anterior margin and black median stripe. Metanotum with two dark brown spots on anterior margin and black median stripe. Metanotum with two dark brown spots on anterior margin and black median stripe. Wing membrane hyaline with brown markings. In forewings,

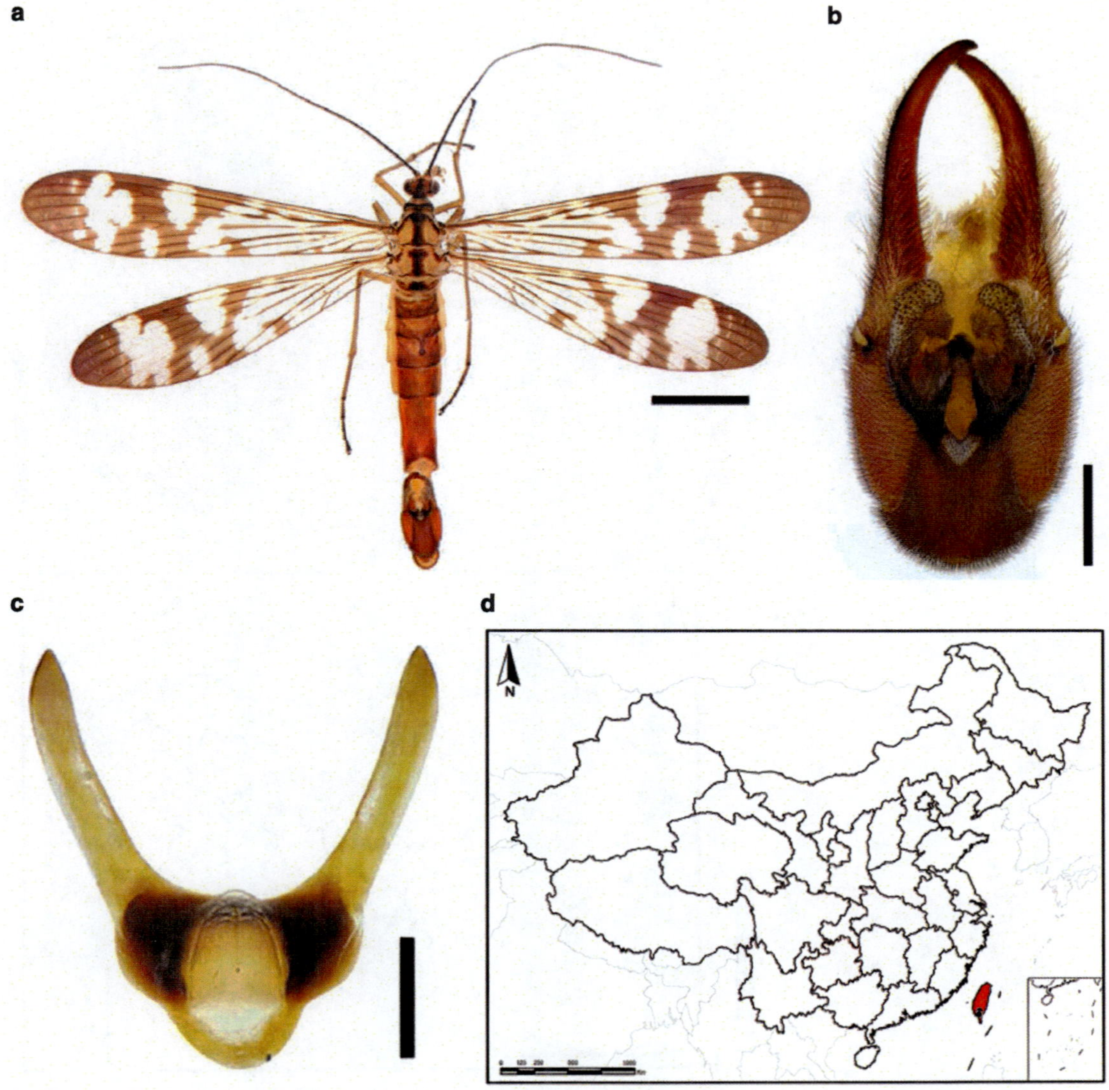

Fig. 10.54 (**a**) Male adult of ***Neopanorpa k-maculata***, dorsal view (Tahan Mountain, Pingtung, Taiwan) (*Scale bar*: 5.0 mm) [**NWAU**]. (**b**) Male genital bulb of ***Neopanorpa k-maculata***, ventral view (*Scale bar*: 0.5 mm). (**c**) Female medigynium of ***Neopanorpa k-maculata***, ventral view (*Scale bar*: 0.2 mm). (**d**) Distribution of ***Neopanorpa k-maculata***

apical band broad with interrupted inner margin, and 2–3 small hyaline spots innerly; pterostigmal band broad with stout basal and apical branches; marginal spot broad, extending postero-apicad to connect basal branch of pterostigmal band, thereby forming distinct K-shaped marking with pterostigmal band; series of interrupted markings near ORs and along hind margin. Male T1–T5 unevenly and sordidly yellowish. Notal organ developed and finger-like, extending slightly beyond middle of T4. A6–A8 distinctly elongated.

Male genital bulb long oval. Epandrium emarginate terminally in V-shape. Hypandrium with broad basal stalk. Hypovalves broad, exceeding apex of gonocoxites, and bearing acute tooth on inner margin subbasally, with distal portions curved mesally and numerous long bristles along inner margin. Gonostylus longer than gonocoxites, bearing triangular median tooth and trapezoidal basal process. Aedeagus conical, with acute parameres.

Female subgenital plate oval with deep V-shaped terminal emargination. Medigynium with main plate reduced and projected semicircularly at base; posterior arms slender and acute terminally; axis lacking apodemes.

Emergence. April to May.

Distribution. Taiwan (Kaohsiung, Pingtung).

10.1.52 *Neopanorpa lungtausana* Cheng, 1957

(Lungtausan Neoscorpionfly) [龙斗山新蝎蛉]
(Fig. 10.55)

Syn. *Neopanorpa brevivalvae* Chou & Wang, 1988

Measurements. Forewing length 12.0–13.5 mm.

Morphology. Vertex black. Ocellar triangle black. Rostrum yellowish brown. Pronotum dark brown. Meso- and metanotum mostly yellowish brown with black median stripe.

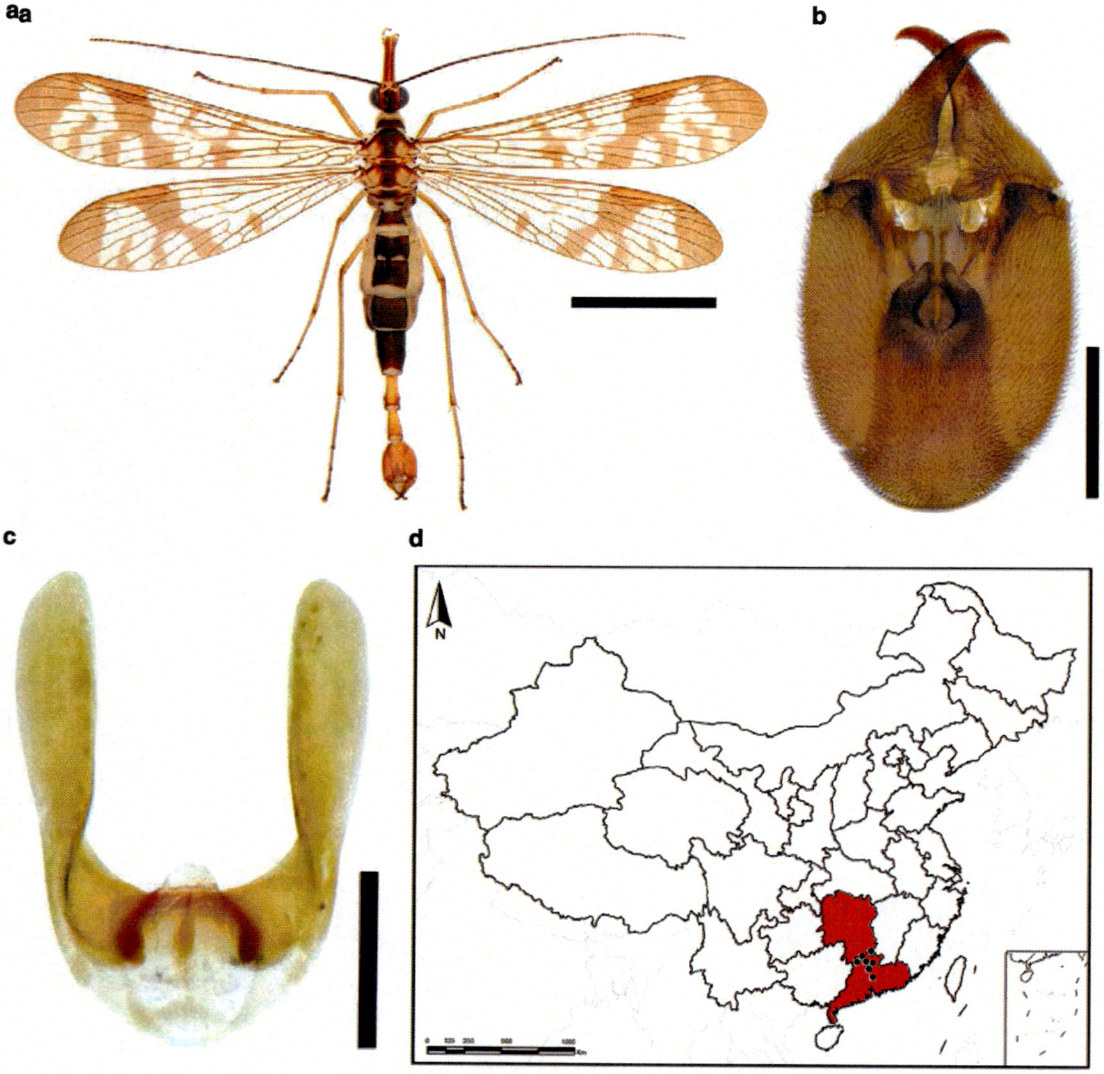

Fig. 10.55 (a) Male adult of ***Neopanorpa lungtausana***, dorsal view (Longtou Mountain, Shaoguan, Guangdong) (*Scale bar*: 5.0 mm) [NWAU]. (b) Male genital bulb of ***Neopanorpa lungtausana***, ventral view (*Scale bar*: 0.5 mm). (c) Female medigynium of ***Neopanorpa lungtausana***, ventral view (*Scale bar*: 0.2 mm). (d) Distribution of ***Neopanorpa lungtausana***

Pleura and legs yellowish brown. Wing membrane hyaline with brown markings. In forewings, apical band complete with large hyaline window innerly and separated stripe aside hind portion; pterostigmal band complete or occasionally with separate apical branch; marginal spots small; basal band split into two spots; basal spot absent. Male T1–T5 dark brown. Notal organ slender and clavate, extending to middle of T4; postnotal organ stout and strongly raised.

Male genital bulb oval. Epandrium long and broad, and truncated terminally. Hypandrium with long broad basal stalk. Hypovalves dark brown and extremely short, curved mesad and forming circular window innerly. Gonostylus with obtuse triangular median tooth and stout basal process. Parameres largely fused with aedeagus and close to each other but abruptly divergent apically. Aedeagus with ventral valves greyish white, membranous, and forming pair of hook-like processes laterally; dorsal valves relatively short and with stout and hook-like lateral processes.

Female subgenital plate oval with deep V-shaped terminal emargination. Medigynium with main plate semicircular; posterior arms broadly spatulate and subacute terminally; axis lacking apodemes.

Emergence. May to June.

Distribution. Hunan (Mountains), Guangdong (Lianzhou, Ruyuan, Shaoguan).

10.1.53 *Neopanorpa magna* Issiki, 1927

(Great Neoscorpionfly) [大新蝎蛉] (Fig. 10.56)

Measurements. Forewing length 20.0–21.0 mm.

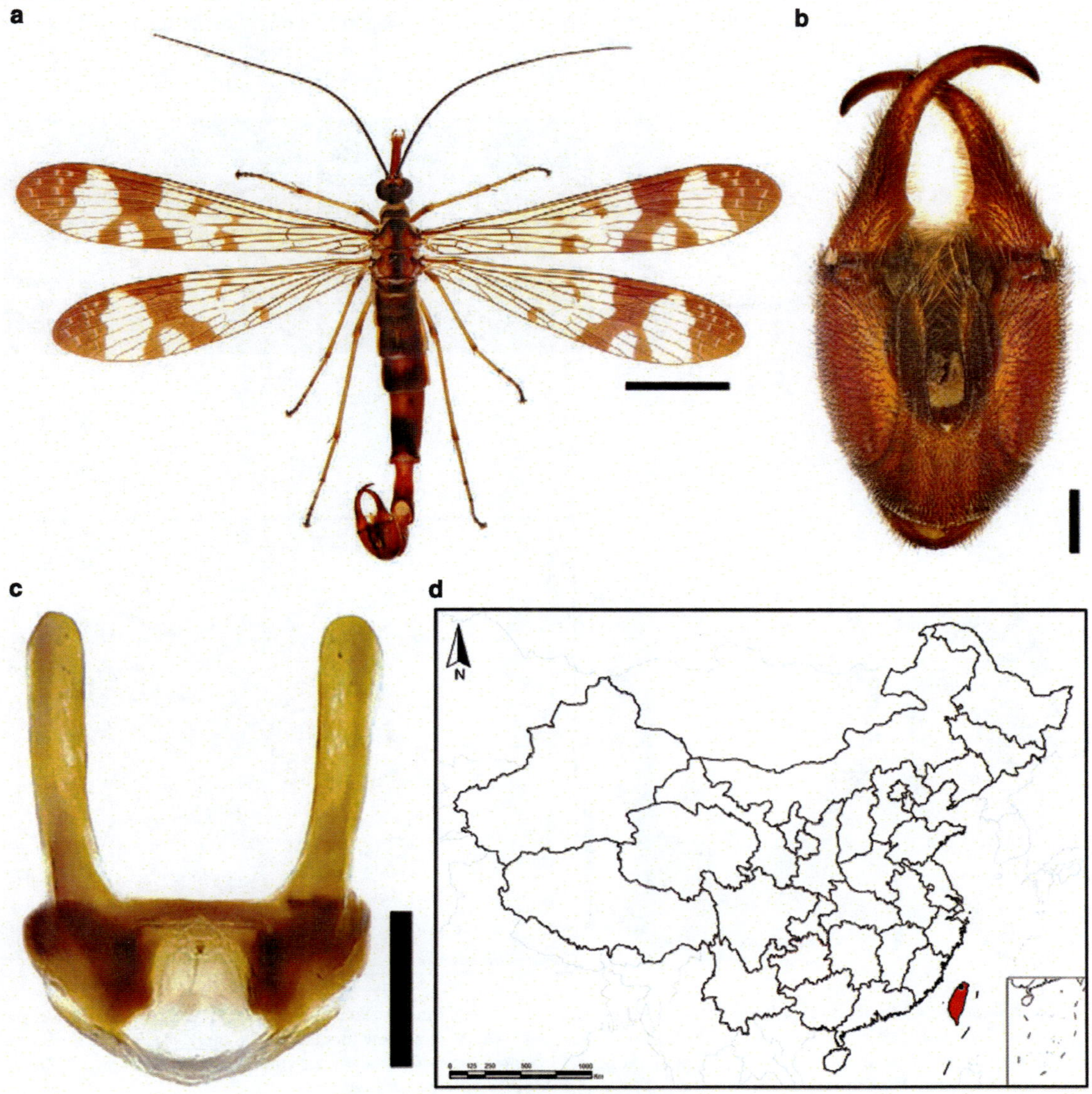

Fig. 10.56 (**a**) Male adult of ***Neopanorpa magna***, dorsal view (Lala Mountain, Taoyuan, Taiwan) (*Scale bar*: 5.0 mm) **[NWAU]**. (**b**) Male genital bulb of ***Neopanorpa magna***, ventral view (*Scale bar*: 0.5 mm). (**c**) Female medigynium of ***Neopanorpa magna***, ventral view (*Scale bar*: 0.2 mm). (**d**) Distribution of ***Neopanorpa magna***

Morphology. Vertex black. Ocellar triangle black. Rostrum reddish brown. Pronotum black. Meso- and metanotum mostly black, with latero-posterior corners reddish brown. Pleura and legs mostly reddish brown. Wing membrane hyaline, slender, with black markings. In forewings, apical band complete; pterostigmal band with basal and apical branches; marginal spot and basal band reduced; basal spot absent. Male T1–T5 dark brown. Notal organ clavate, extending to middle of T4 and covering postnotal organ. A6 dark brown, long cylindrical. A7 and A8 intensively constricted and stalk-like basally.

Male genital bulb oval. Epandrium long and broad, slightly emarginate terminally. Hypandrium with short basal stalk. Hypovalve with projection on inner margin medially. Gonostylus longer than gonocoxites, greatly curved mesad, with triangular median tooth and stout basal process.

Female subgenital plate oval with deep V-shaped terminal emargination. Medigynium with main plate broad, greatly sclerotized laterally, and slightly projected laterally; posterior arms slender and rounded terminally; axis lacking apodemes.

Emergence. July to September.

Distribution. Taiwan (Taoyuan).

10.1.54 *Neopanorpa nielseni* Byers, 1965

(Nielsen's Neoscorpionfly) [尼尔森新蝎蛉] (Fig. 10.57)

Measurements. Forewing length 16.5–17.0 mm.

Morphology. Vertex dark brown. Ocellar triangle dark brown. Rostrum dark brown, with two darkened longitudinal stripes laterally. Pronotum dark brown. Meso- and metanotum mostly dark brown, with latero-posterior corners yellowish brown. Pleura shining yellow. Legs mostly yellowish brown. Wing membrane hyaline with dark brown markings. In forewings, apical band broad and with right angle on inner margin, pterostigmal band with basal branch and apical

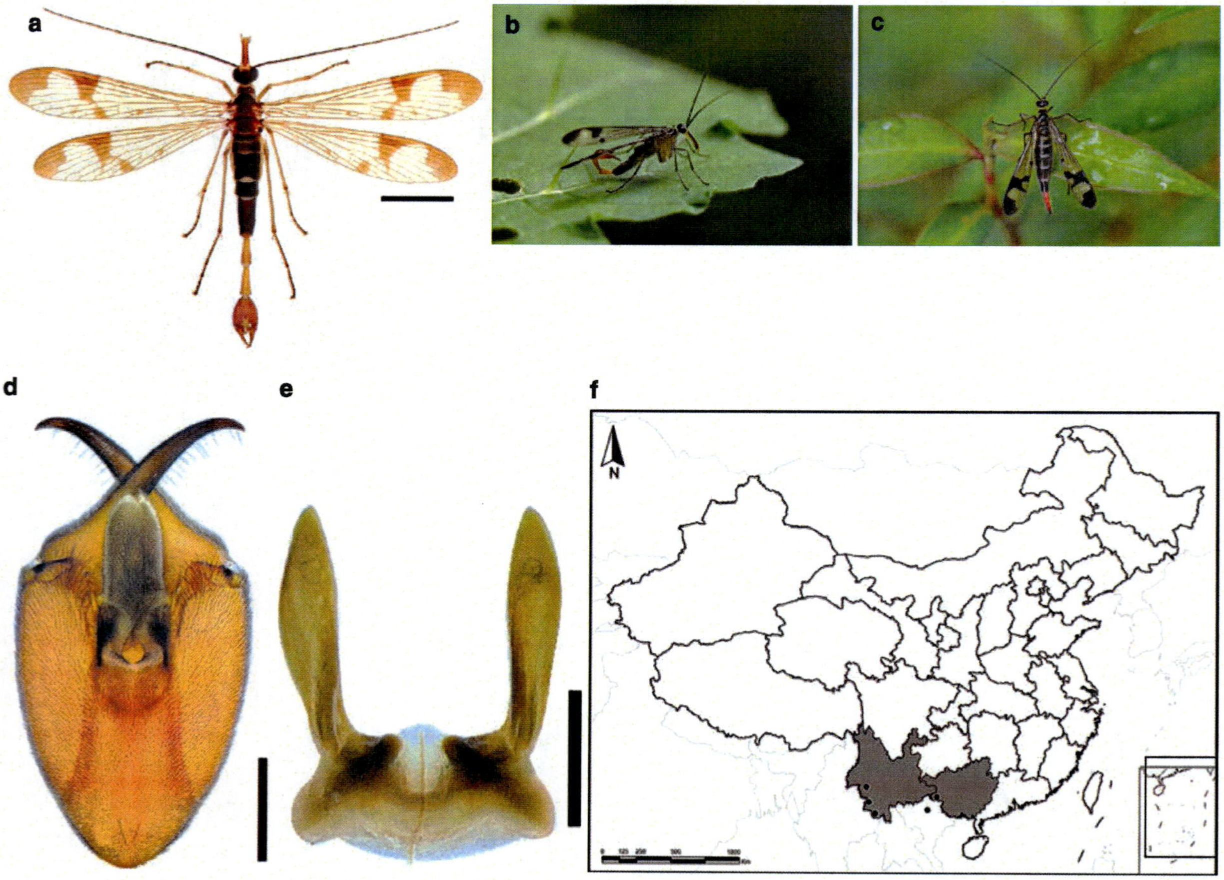

Fig. 10.57 (a) Male adult of ***Neopanorpa nielseni,*** dorsal view (Tangli Mountain, Yongde, Yunnan) (*Scale bar*: 5.0 mm) [NWAU]. (b) Male adult of ***Neopanorpa nielseni*** in the field (Tangli Mountain, Yongde, Yunnan). (c) Female adult of ***Neopanorpa nielseni*** in the field (Tangli Mountain, Yongde, Yunnan). (d) Male genital bulb of ***Neopanorpa nielseni***, ventral view (*Scale bar*: 0.5 mm). (e) Female medigynium of ***Neopanorpa nielseni***, ventral view (*Scale bar*: 0.2 mm). (f) Distribution of ***Neopanorpa nielseni***

branch; apical branch variously reduced in males; basal band reduced into irregular spot near end of CuP; marginal and basal spots absent. Male T1–T5 dark brown; sterna yellowish white; pleura yellow. Notal organ long, clavate, not exceeding hind border of T4. A6 dark brown. A7 and A8 reddish brown and constricted basally.

Male genital bulb long oval. Epandrium long broad, with truncated apex. Hypandrium with long and broad basal stalk. Hypovalves curled along lateral margins and overlapped mesally, forming subcircular window subbasally. Gonostylus slender, bearing obtuse triangular median tooth, and large basal process extending basad and forming acute tooth. Paramere almost fused with aedeagus. Aedeagus conical with stout lateral processes.

Female subgenital plate oval with deep V-shaped terminal emargination. Medigynium with main plate broad, subtriangular, and greatly sclerotized at lateral portions; axis lacking apodemes; posterior arms slender, slightly spatulate in distal half, and subacute terminally.

Emergence. April to August.

Distribution. Guangxi (Napo), Yunnan (Yongde, Xishuangbanna); Vietnam.

10.1.55 *Neopanorpa ocellaris* Navás, 1908

(Ocellar Neoscorpionfly) [之纹新蝎蛉] (Fig. 10.58)

Measurements. Forewing length 13.5–15.0 mm.

Morphology. Vertex black. Rostrum dark reddish brown, with two dark stripes along clypeus. Nota black. Membranous area yellow. Pleura and coxae dark chestnut to black. Pleura dark brown. Legs mostly yellowish brown. Wing membrane hyaline, with well-developed dark brown markings; veins dark brown except whitish cross-veins in apical half;

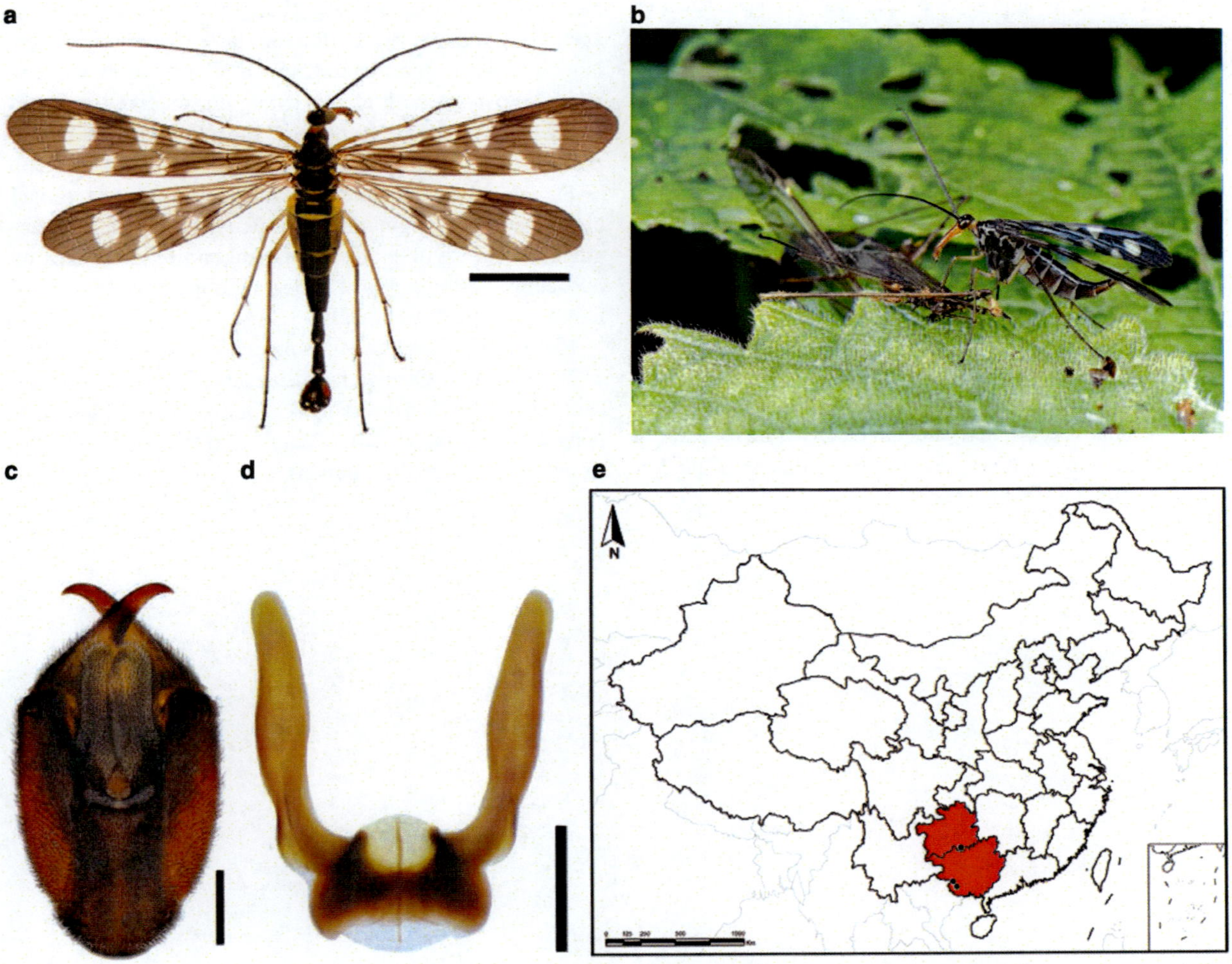

Fig. 10.58 (**a**) Male adult of ***Neopanorpa ocellaris***, dorsal view (Libo, Guizhou) (*Scale bar*: 5.0 mm) **[NWAU]**. (**b**) Female adult of ***Neopanorpa ocellaris*** in the field (Huanjiang, Guangxi, ©Wei-Liang Xie). (**c**) Male genital bulb of ***Neopanorpa ocellaris***, ventral view (*Scale bar*: 0.5 mm). (**d**) Female medigynium of ***Neopanorpa ocellaris***, ventral view (*Scale bar*: 0.2 mm). (**e**) Distribution of ***Neopanorpa ocellaris***

pterostigma brown. In forewings, all bands and spots intensively fused to each other to form distinctive zigzag pattern, leaving several dentate hyaline spots along anterior and posterior margins, and an elliptical hyaline spot between pterostigmal and apical bands.

Male genital bulb dark reddish brown. Epandrium narrowing toward truncated apex, with small epandrial lobes. Hypandrium with broad basal stalk, and split into pair of hypovalves at middle; hypovalves greyish brown, distinctly curled on lateral margin, and slightly overlapping on inner margin; hypandrial processes broad and rounded. Gonostylus shorter than gonocoxites, stout, with median tooth blunt triangular, and basal process subtrapezoidal; small triangular projection at base of basal process. Paramere blade-like, with membranous notch near base. Ventral aedeagal valves broad, laterally flattened with rounded apex; dorsal processes short and rounded; lateral processes short and blunt.

Female subgenital plate oval with deep V-shaped terminal emargination. Medigynium with long, twisted, and narrowly spatulate posterior arms; main plate subtriangular and projected latero-basally; axis lacking apodemes.

Emergence. April to August.

Distribution. Guangxi (Huanjiang, Longzhou, Ningming), Guizhou (Libo).

Remarks. This species was originally described from Sikkim (Navás 1908). However, all the specimens we have seen were collected from Guangxi and Guizhou (southwestern China). *N. ocellaris* was published in the same article with *N. cavaleriei* from Guizhou. In fact, *N. ocellaris* cohabitate with *N. cavaleriei* in Libo, Guizhou (Wang and Hua 2019b). It is possible that Navás erroneously reported this species from Sikkim. Therefore, the record of this species in Sikkim has been tentatively disregarded in this Atlas, rendering *N. ocellaris* a Chinese-endemic.

10.1.56 *Neopanorpa pulchra* Carpenter, 1945

(Fair Neoscorpionfly) [丽新蝎蛉] (Fig. 10.59)

Measurements. Forewing length 14.0–16.0 mm.

Morphology. Vertex dark brown. Ocellar triangle black. Rostrum yellowish brown. Pronotum dark brown. Meso- and metanotum yellowish brown, with broad black median stripe. Pleura and legs light yellowish brown. Wing membrane hyaline with brown markings. In forewings, apical band broad with straight inner margin, and connected to pterostigmal band anteriorly; pterostigmal band complete, with broad basal branch and thin apical branch; marginal spot slender; basal band split into two spots; basal spot absent. Male T1–T5 dark brown. Notal organ stout subtriangular. A6 dark brown, long cylindrical. A7 and A8 yellowish brown and constricted basally.

Male genital bulb long elliptical. Epandrium long and broad with truncated apex. Hypandrium with long broad basal stalk. Hypovalves long, distinctly extending beyond apex of gonocoxites, curled on lateral margins, overlapped mesally, and forming subcircular window subbasally. Gonostylus with obtuse triangular median tooth and large basal process. Aedeagus short with stout lateral processes.

Female subgenital plate oval with deep V-shaped terminal emargination. Medigynium with main plate broad, subtriangular, and greatly sclerotized at lateral portions; axis lacking apodemes; posterior arms slender, slightly spatulate in distal half, and subacute terminally.

Emergence. April to September.

Distribution. Guangdong (Lianzhou, Meizhou, Nanling Mountains, Ruyuan), Guangxi (Jinxiu, Kitten Mountain, Ningming, Longzhou), Hunan (Daoxian), Hainan (Jianfengling, Wuzhi Mountain).

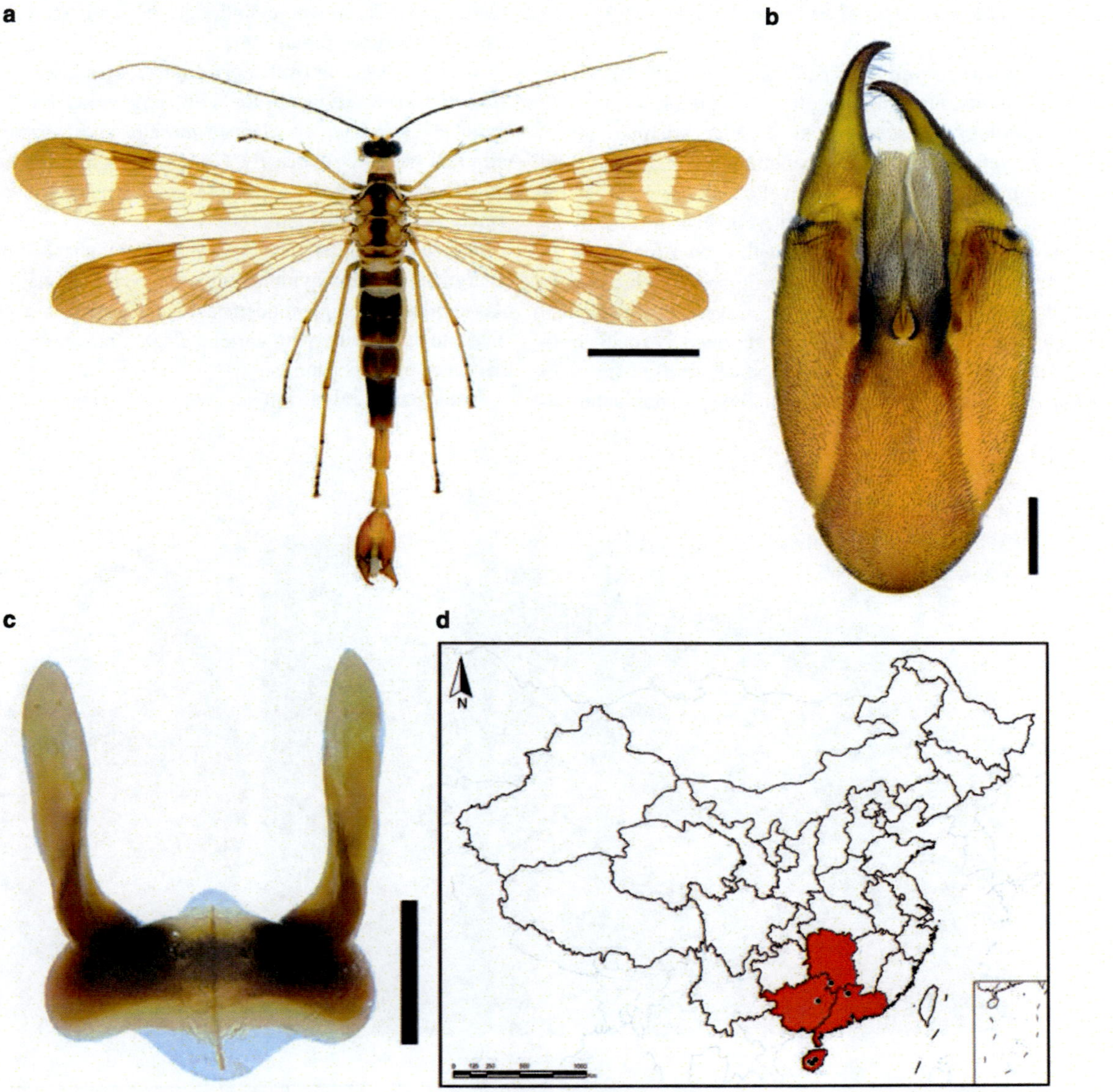

Fig. 10.59 (**a**) Male adult of ***Neopanorpa pulchra***, dorsal view (Jianfengling, Hainan, NWAU) (*Scale bar*: 0.5 mm) **[NWAU]**. (**b**) Male genital bulb of ***Neopanorpa pulchra***, ventral view (*Scale bar*: 0.5 mm). (**c**) Female medigynium of ***Neopanorpa pulchra***, ventral view (*Scale bar*: 0.2 mm). (**d**) Distribution of ***Neopanorpa pulchra***

10.1.57 *Neopanorpa sheni* Hua & Chou, 1997

(Shen's Neoscorpionfly) [申氏新蝎蛉] (Fig. 10.60)
Measurements. Forewing length 11.0–12.0 mm.

Morphology. Vertex black. Ocellar triangle black. Rostrum yellowish brown. Pronotum black. Meso- and metanotum blackish brown, and yellowish brown at latero-posterior corners. Pleura and legs yellowish brown. Wing membrane colorless and hyaline with brown markings. In forewings, apical band broad; pterostigmal band broad with well-developed basal branch and separated thin apical branch; marginal spot very small, represented by small circular spot; basal band split into two small spots; basal spot absent. Male T1–T5 dark brown. Notal organ short subtriangular. A6 dark brown, cylindrical; A7 and A8 yellowish brown, constricted basally.

Male genital bulb oval. Epandrium long and broad, with truncated apex. Hypandrium with long broad basal stalk. Hypovalves slender, overlapped mesally and forming long elliptical window subbasally. Gonostylus with obtuse triangular tooth and large basal process on inner margin. Aedeagus conical, with stout lateral processes.

Female subgenital plate oval with deep V-shaped terminal emargination. Medigynium with undeveloped main plate; axis short and lacking apodemes; posterior arms slender and subacute terminally, with earlobe-shaped processes at outer margin of basal portion.

Emergence. July to August.

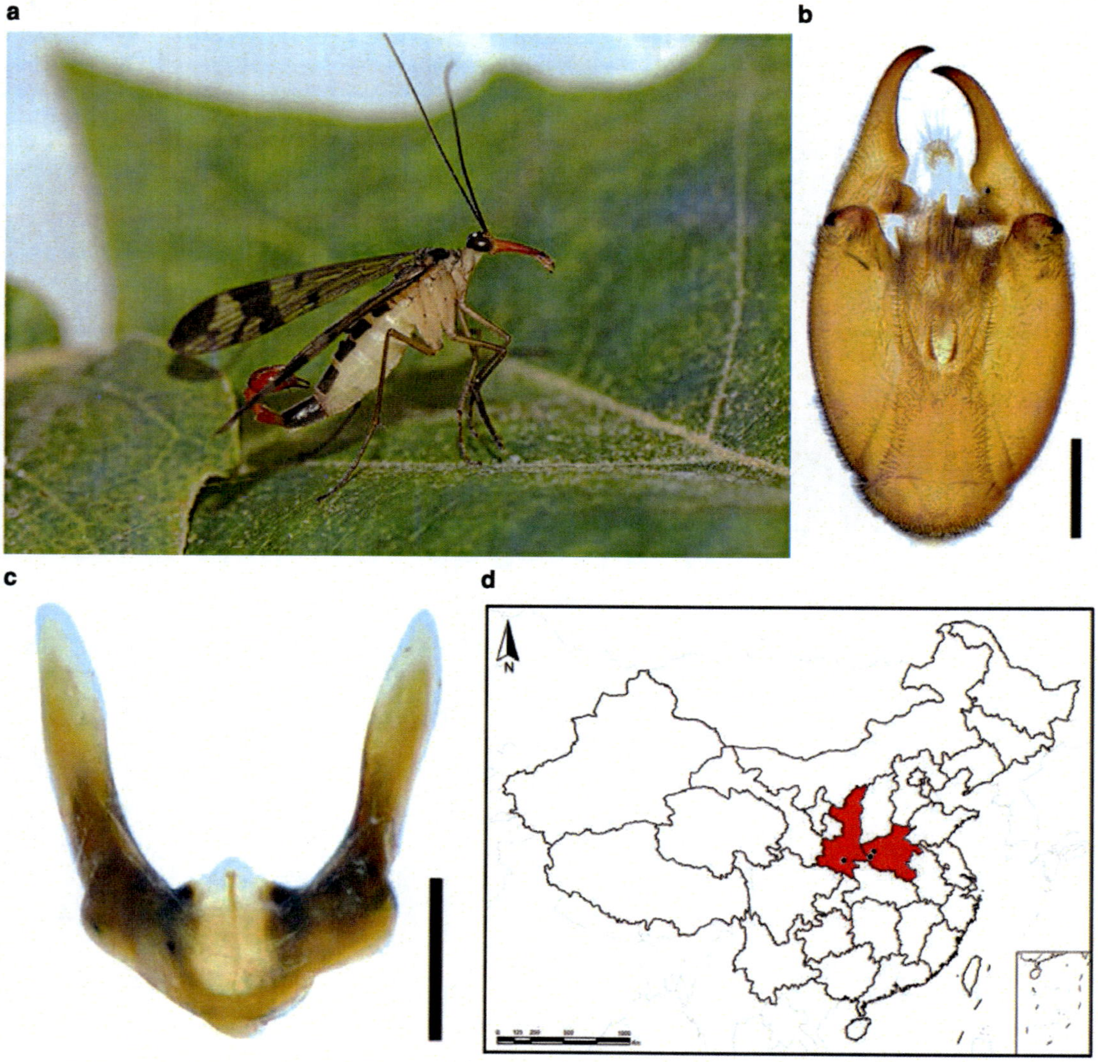

Fig. 10.60 (a) Male adult of ***Neopanorpa sheni*** in the field (Baiyun Mountain, Songxian, Henan, ©Kai Gao). (b) Male genital bulb of ***Neopanorpa sheni***, ventral view (*Scale bar*: 0.5 mm). (c) Female medigynium of ***Neopanorpa sheni***, ventral view (*Scale bar*: 0.2 mm). (d) Distribution of ***Neopanorpa sheni***

Distribution. Henan (Songxian, Luanchuan), Shaanxi (Ningshan).

10.1.58 *Neopanorpa siamensis* Byers, 1965

(Siamese Neoscorpionfly) [泰新蝎蛉] (Fig. 10.61)
Measurements. Forewing length 11.5–14.5 mm.

Morphology. Vertex black mesally but yellowish brown laterally. Ocellar triangle black. Rostrum yellowish brown. Pronotum dark brown. Meso- and metanotum yellowish brown with black median stripe. Pleura and legs yellowish brown. Wing membrane hyaline with dark brown markings. In forewings, apical band complete with small hyaline window posteriorly; pterostigmal band with broad basal branch and reduced apical branch; marginal spot large; basal band split into two large spots; basal spot elongated. Male T1–T5 dark brown. Notal organ short and triangular. A6 dark brown with yellow apex. A7 and A8 yellowish brown and slightly constricted basally.

Male genital bulb long oval. Epandrium long and broad with truncated apex. Hypandrium with long broad basal stalk. Hypovalves long and broad, curled dorsad along outer and inner margins, overlapped mesally, and forming subcircular window subbasally. Paramere long and blade-like. Aedeagus conical, slightly exceeding apex of gonocoxites and bearing pair of stout lateral processes.

Female subgenital plate long oval with deep V-shaped terminal emargination. Medigynium with main plate subtrian-

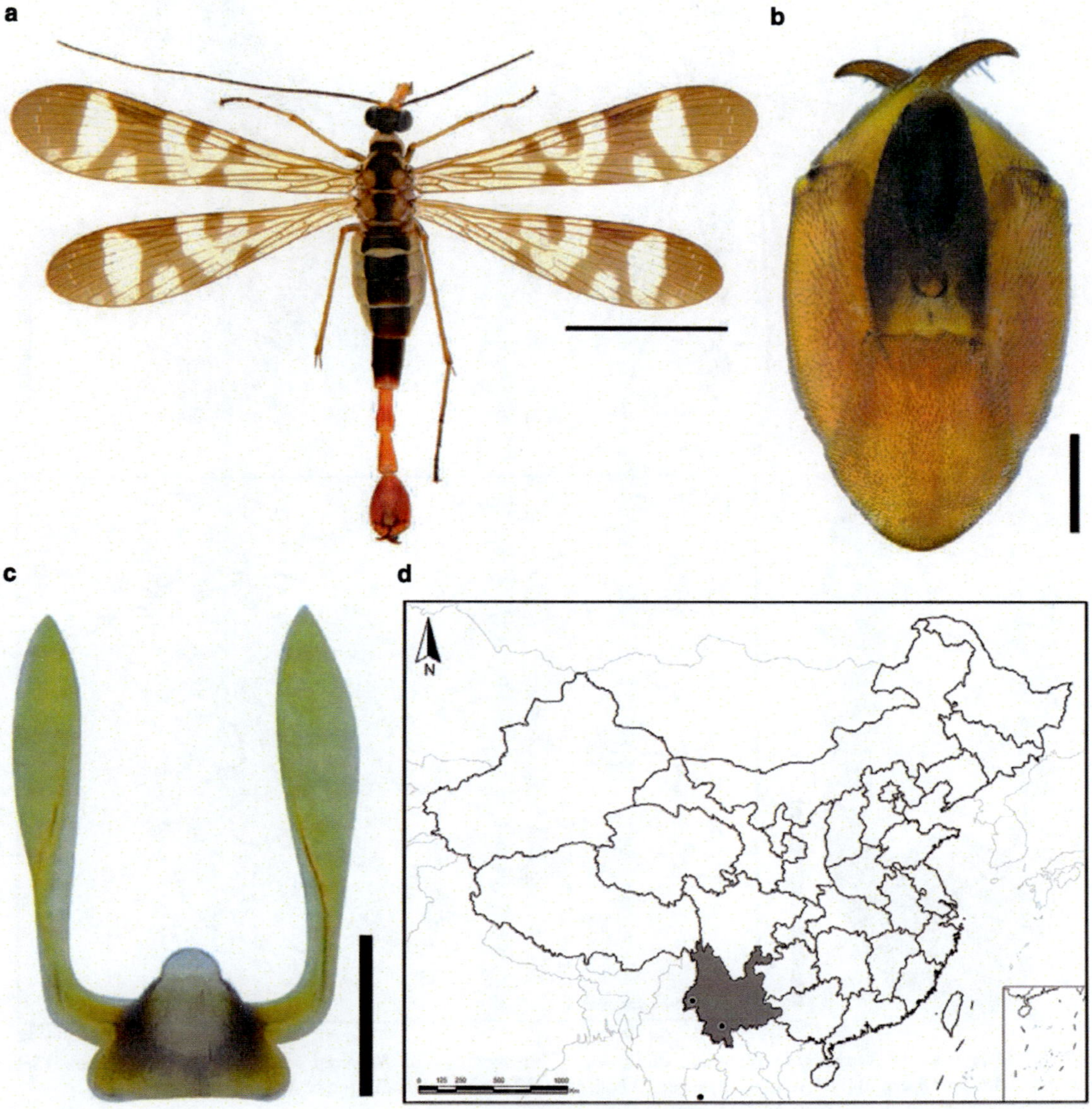

Fig. 10.61 (**a**) Male adult of ***Neopanorpa siamensis***, dorsal view (Dadugang, Pu'er, Yunnna) (*Scale bar*: 5.0 mm) **[NWAU]**. (**b**) Male genital bulb of ***Neopanorpa siamensis***, ventral view (*Scale bar*: 0.5 mm). (**c**) Female medigynium of ***Neopanorpa siamensis***, ventral view (*Scale bar*: 0.2 mm). (**d**) Distribution of ***Neopanorpa siamensis***

gular; axis lacking apodemes; posterior arms long, greatly spatulate in distal half, and acute terminally.

Emergence. May to November.

Distribution. Yunnan (Pu'er, Ruili); Thailand (Chiang Mai).

10.1.59 *Neopanorpa spatulata* Byers, 1965

(Spatulate Neoscorpionfly) [阔瓣新蝎蛉] (Fig. 10.62)

Measurements. Forewing length 13.0–15.0 mm.

Morphology. Vertex black mesally but yellowish brown laterally. Ocellar triangle black. Rostrum yellowish brown. Pronotum dark brown. Meso- and metanotum yellowish brown with black median stripe. Pleura and legs yellowish brown. Wing membrane hyaline with dark brown markings. In forewings, apical band broad and slightly interrupted posteriorly; pterostigmal band with broad basal branch and detached and reduced apical branch; marginal spot small and elongated; basal band split into two oblique small stripes; basal spot small. Male T1–T5 dark brown. Notal organ short and triangular. A6 dark brown with yellow

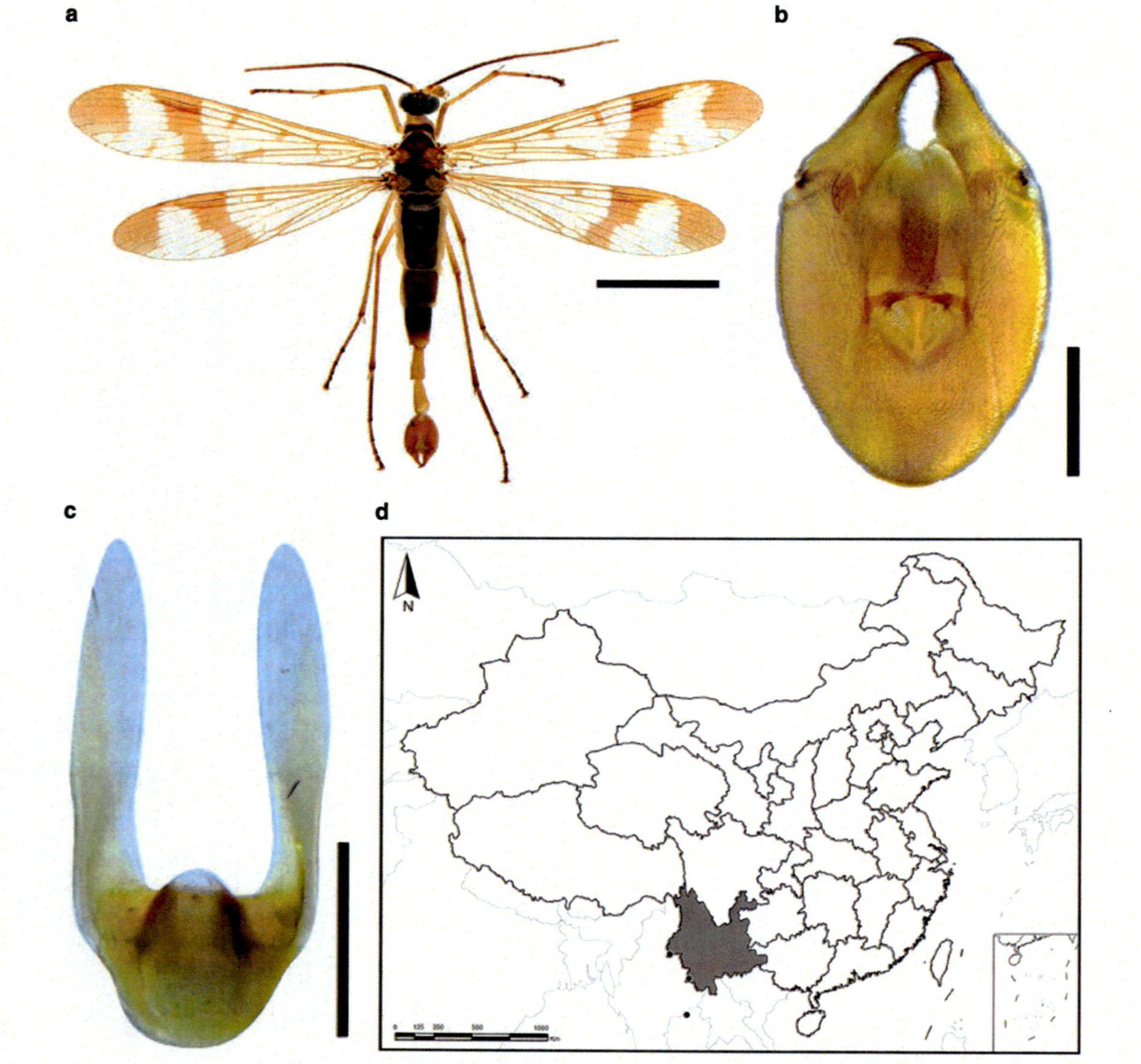

Fig. 10.62 (a) Male adult of ***Neopanorpa spatulata***, dorsal view (Ruili, Yunnan) (*Scale bar*: 5.0 mm) [NWAU]. (b) Male genital bulb of ***Neopanorpa spatulata***, ventral view (*Scale bar*: 0.5 mm). (c) Female medigynium of ***Neopanorpa spatulata***, ventral view (*Scale bar*: 0.2 mm). (d) Distribution of ***Neopanorpa spatulata***

apex. A7 and A8 yellowish brown and slightly constricted basally.

Male genital bulb broad oval. Epandrium long and broad with truncated apex. Hypandrium with long broad basal stalk. Hypovalves broad, curled dorsad along outer and inner margins, overlapped mesally, and forming diamond-shaped window subbasally. Paramere long and blade-like. Aedeagus conical, with stout lateral processes.

Female subgenital plate long oval with deep V-shaped terminal emargination. Medigynium with main plate small; axis lacking apodemes; posterior arms long, greatly spatulate in distal half, and acute terminally.

Emergence. May to November.

Distribution. Yunnan (Menglian, Ruili); Thailand (Chiang Mai).

Neopanorpa tienmushana Group

Approximately eleven species are recognized in this group. They are recorded from the Oriental Region, including southwestern China and Japan (Ryukyu Islands). In China, 10 species have been recorded, and 6 of them are included in this atlas.

Key to Species of Neopanorpa tienmushana Group in This Atlas
(6 species, males)

1. Notal organ extremely elongated, stick-like, and exceeding A6 .... ***Neopanorpa sauteri*** (Esben-Petersen)
– Notal organ shorter, not exceeding T5 2
2. Hypovalves broadening towards beveled apex; inner margin of hypovalves not overlapped .... ***Neopanorpa translucida*** Cheng
– Hypovalves not modified as above; inner margin of hypovalves overlapped 3
3. Wing markings water mark-like; bands and spots greatly entangled and net-like .... ***Neopanorpa caveata***
– Wing markings dense; bands and spots less entangled 4
4. Hypovalves narrow; inner window between two hypovalves bottle-shaped .... ***Neopanorpa gradana*** Cheng
– Hypovalves broad; inner window between two hypovalves subcircular 5
5. Hypovalves subacute terminally .... ***Neopanorpa jiulongensis*** Zhou
– Hypovalves rounded terminally .... ***Neopanorpa tienmushana*** Cheng

10.1.60 *Neopanorpa caveata* Cheng, 1957

(Caveated Neoscorpionfly) [网翅新蝎蛉] (Fig. 10.63)
Measurements. Forewing length 15.0–16.0 mm.

Morphology. Vertex yellowish brown. Ocellar triangle black. Longitudinal black stripe passing through ocellar triangle. Rostrum reddish brown. Pronotum dark brown. Meso- and metanotum yellowish brown, with black longitudinal median stripe. Pleura and legs yellowish brown. Wing membrane hyaline and tinged with light yellow. Markings yellowish brown and reticulate: apical band broad, with slender stripe connected to pterostigmal band, extending inward and connected to apical branch of pterostigmal band; pterostigmal band complete; marginal spot extending outward to connect with basal branch of pterostigmal band, and inward to connect with complete basal band; basal spot absent. Male T1–T5 black. Notal organ long, stick-like, and not exceeding posterior margin of T4. A6 mostly dark brown with yellowish brown apex.

Male genital bulb broad oval. Epandrium long, broad, and truncated apically. Hypandrium with long and broad basal stalk. Hypovalves especially broad, overlapped mesally, exceeding apex of gonocoxites and forming heart-shaped window subbasally. Gonostylus shorter than half-length of

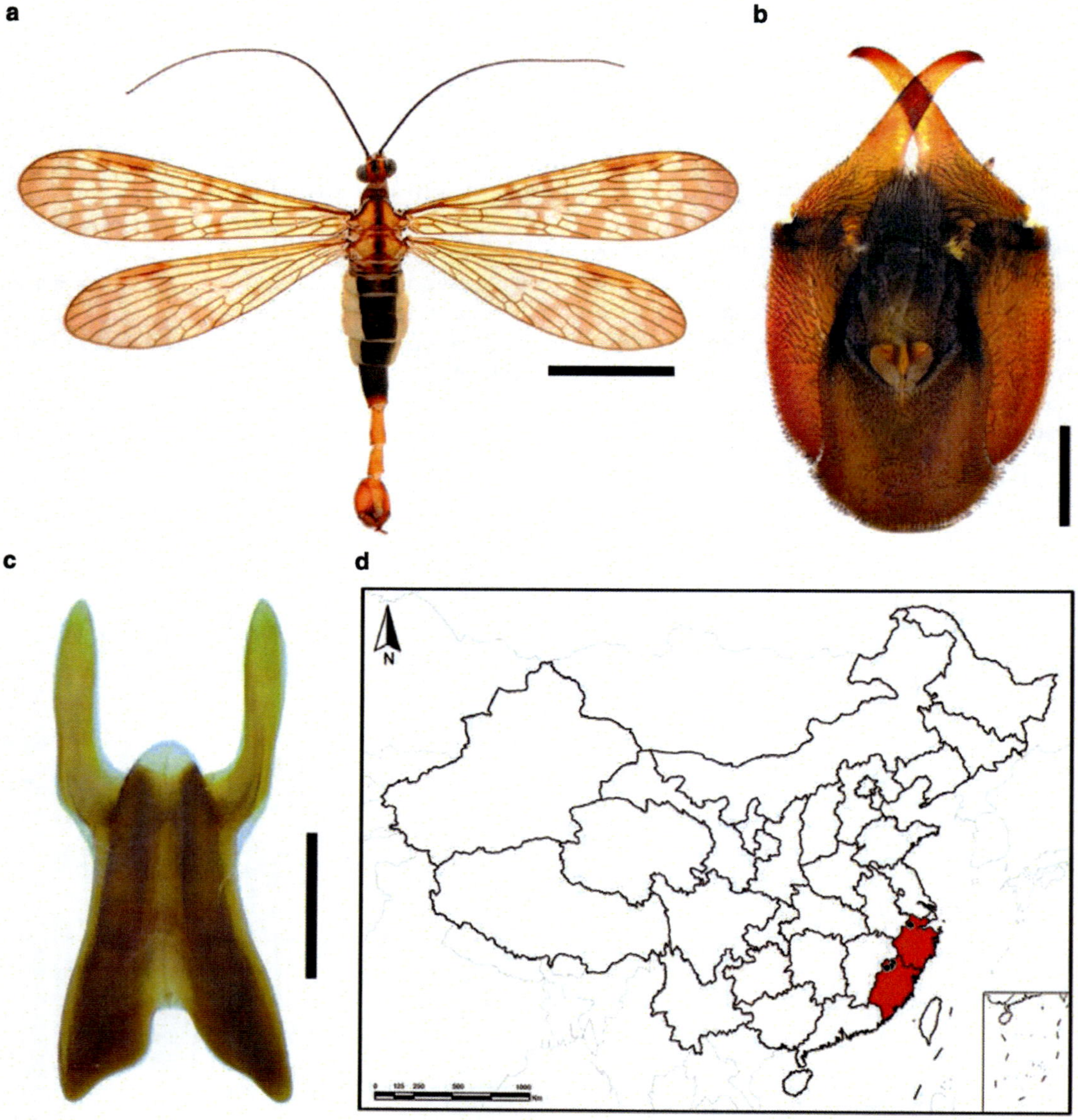

Fig. 10.63 (**a**) Male adult of ***Neopanorpa caveata***, dorsal view (Wuyi Mountains, Fujian) (*Scale bar*: 5.0 mm) [**NWAU**]. (**b**) Male genital bulb of ***Neopanorpa caveata***, ventral view (*Scale bar*: 0.5 mm). (**c**) Female medigynium of ***Neopanorpa caveata***, ventral view (*Scale bar*: 0.2 mm). (**d**) Distribution of ***Neopanorpa caveata***

gonocoxites, slightly concaved on outer margin, and bearing obtuse triangular median tooth and stout basal process. Paramere slender and acute apically. Aedeagus conical with pair of stout lateral processes.

Female subgenital plate oval, with V-shaped terminal emargination. Mediygnium with reduced main plate; posterior arms slender; axis almost two times as long as posteiror arms with extremely stout and diverged apodemes.

Emergence. May to September.

Distribution. Fujian (Shaowu, Jianning, Wuyi Mountains) and Zhejiang (Longwang Mountain).

10.1.61 *Neopanorpa gradana* Cheng, 1952

(Narrow-Valved Neoscorpionfly) [窄瓣新蝎蛉] (Fig. 10.64)

Measurements. Forewing length 12.5–14.0 mm.

Morphology. Vertex yellowish brown. Ocellar triangle black. Longitudinal black stripe passing through ocellar triangle. Rostrum reddish brown. Pronotum dark brown. Meso- and metanotum yellowish brown, with black longitudinal median stripe. Pleura and legs yellowish brown. Wing membrane hyaline and tinged with light yellow. In forewings, markings yellowish brown and reticulate: apical band broad,

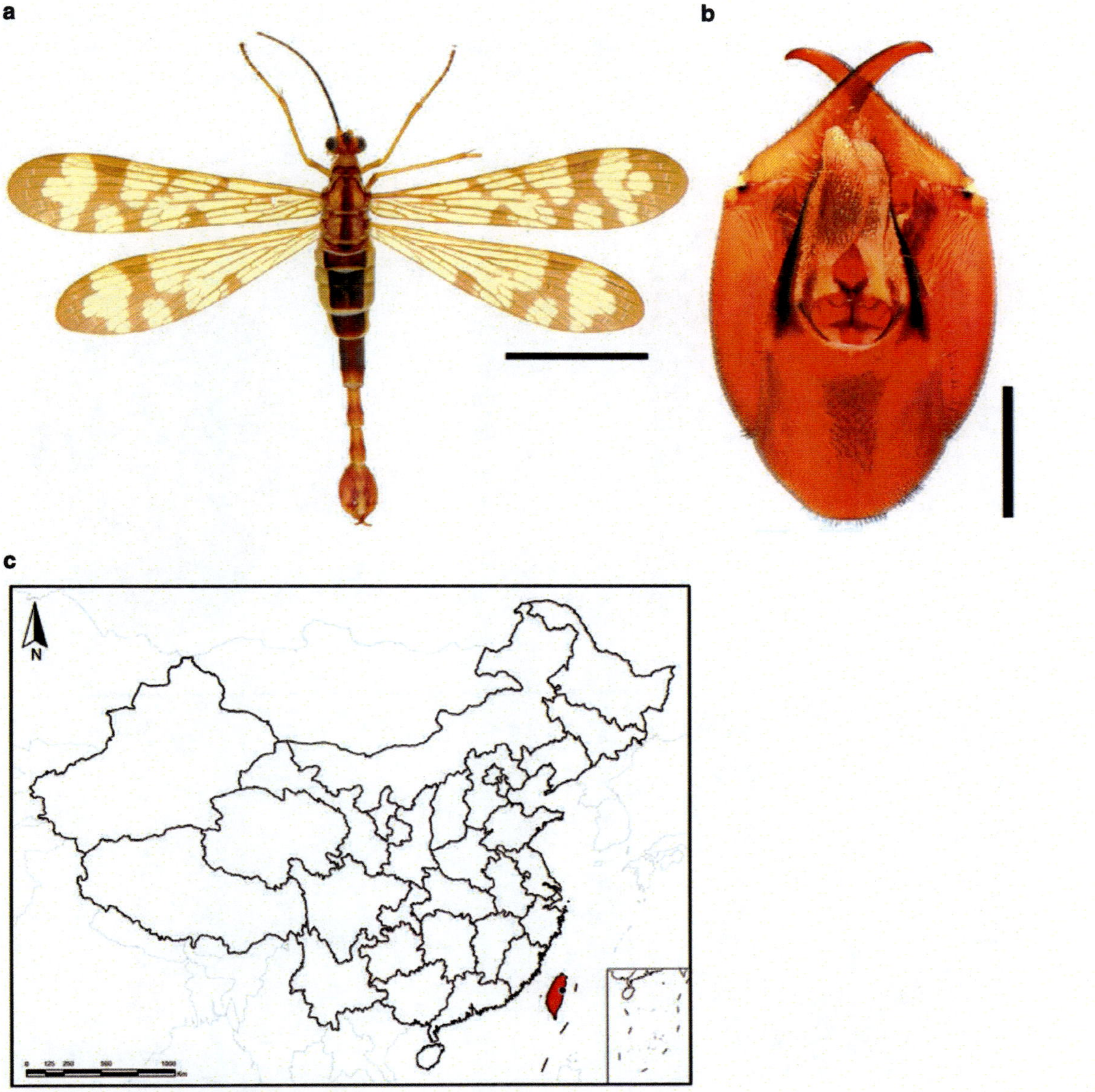

Fig. 10.64 (a) Male adult of ***Neopanorpa gradana***, dorsal view (Hualien, Taiwan) (*Scale bar*: 5.0 mm) **[NWAU]**. **(b)** Male genital bulb of ***Neopanorpa gradana***, ventral view (*Scale bar*: 0.5 mm). **(c)** Distribution of ***Neopanorpa gradana***

with circular thin stripe enclosing small hyaline window on inner margin; pterostigmal band complete; marginal spot extending outward to connect with basal branch of pterostigmal band, and inward to connect with split basal band; basal spot absent. Male T1–T5 black. Notal organ long, stick-like, extending to posterior margin of T4. A6 mostly dark brown with yellowish brown apex. A7 and A8 yellowish brown and constricted basally.

Male genital bulb oval. Epandrium long, broad, and truncated apically; epandrial lobes extremely small. Hypandrium with long and broad basal stalk. Hypovalves narrow, overlapped mesally, exceeding apex of gonocoxites and forming elliptical window subbasally. Gonostylus bearing obtuse triangular median tooth and stout basal process. Paramere stout and finger-like, with stalk greatly developed and broadly expanded. Aedeagus conical with pair of stout lateral processes.

Female unknown.

Emergence. March, October.

Distribution. Taiwan (Hualien).

10.1.62 *Neopanorpa jiulongensis* Zhou, 1993

(Jiulong Neoscorpionfly) [九龙新蝎蛉] (Fig. 10.65)

Measurements. Forewing length 12.5–14.0 mm.

Morphology. Vertex yellowish brown. Ocellar triangle black. Longitudinal black stripe passing through ocellar triangle. Rostrum reddish brown. Pronotum dark brown. Meso- and metanotum yellowish brown, with black longitudinal median stripe. Pleura and legs yellowish brown. Wing membrane hyaline and tinged with light yellow. In forewings, markings yellowish brown and reticulate: apical

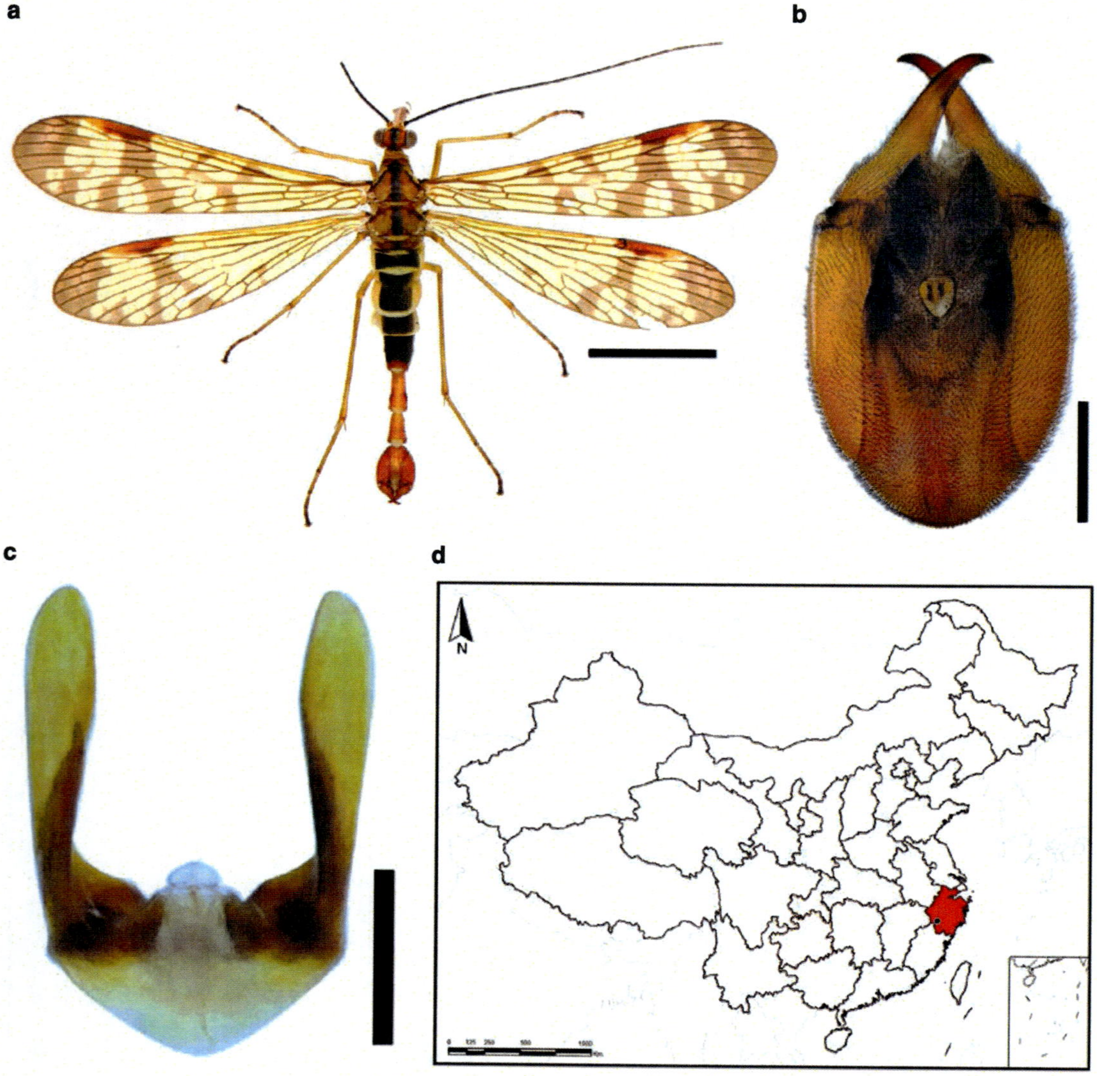

Fig. 10.65 (a) Male adult of ***Neopanorpa jiulongensis***, dorsal view (Jiulong Mountain, Suichang, Zhejiang) (*Scale bar*: 5.0 mm) **[NWAU]**. (b) Male genital bulb of ***Neopanorpa jiulongensis***, ventral view (*Scale bar*: 0.5 mm). (c) Female medigynium of ***Neopanorpa jiulongensis***, ventral view (*Scale bar*: 0.2 mm). (d) Distribution of ***Neopanorpa jiulongensis***

band broad, with two or three small hyaline windows; pterostigmal band complete with broad basal branch and reduced apical branch; marginal spot elongated; basal band split into two spots; basal spot absent. Male T1–T5 black. Notal organ long, stick-like, and not exceeding posterior margin of T4. A6 mostly dark brown with yellowish brown apex.

Male genital bulb broad oval. Epandrium long, broad, and truncated apically. Hypandrium with long and broad basal stalk. Hypovalves especially broad, overlapped mesally, acute distally, exceeding apex of gonocoxites, and forming subcircular window subbasally. Gonostylus with obtuse triangular median tooth and stout basal process. Paramere extremely short and projecting ventrad. Aedeagus conical with pair of stout lateral processes; dorsal processes greatly elongated and acute.

Female subgenital plate oval, with V-shaped terminal emargination. Mediygnium with reduced main plate; posterior arms long, spatulate in distal half and rounded terminally; axis lacking apodemes.

Emergence. May to September.

Distribution. Zhejiang (Jiulong Mountain).

10.1.63 *Neopanorpa sauteri* Esben-Petersen, 1912

(Sauter's Neoscorpionfly) [绍德新蝎蛉] (Fig. 10.66)

Measurements. Forewing length 12.0–14.0 mm.

Morphology. Vertex yellowish brown. Ocellar triangle enclosed by large black spot. Rostrum yellowish brown. Pronotum yellowish brown. Meso- and metanotum mostly yellowish brown, with narrow black median stripe. Pleura and legs yellowish brown. Wing membrane hyaline, tinged with yellow, with dark brown markings. In forewings, apical band irregular along inner margin and with 1–2 large hyaline windows; pterostigmal band complete; marginal spot extending outwards and connected to basal branch of pterostigmal band; basal band narrow at anterior half and broad at posterior half; basal spot split into two small spots. Male T1–T5 dark brown and gradually turning into yellowish brown caudad. Notal organ extremely elongated, clavate, almost extending to hind border of A6. A6 yellowish brown and cylindrical. A7 and A8 yellowish brown and slightly constricted basally.

Male genital bulb oval. Epandrium long and broad, with deep U-shaped terminal emargination. Hypandrium with

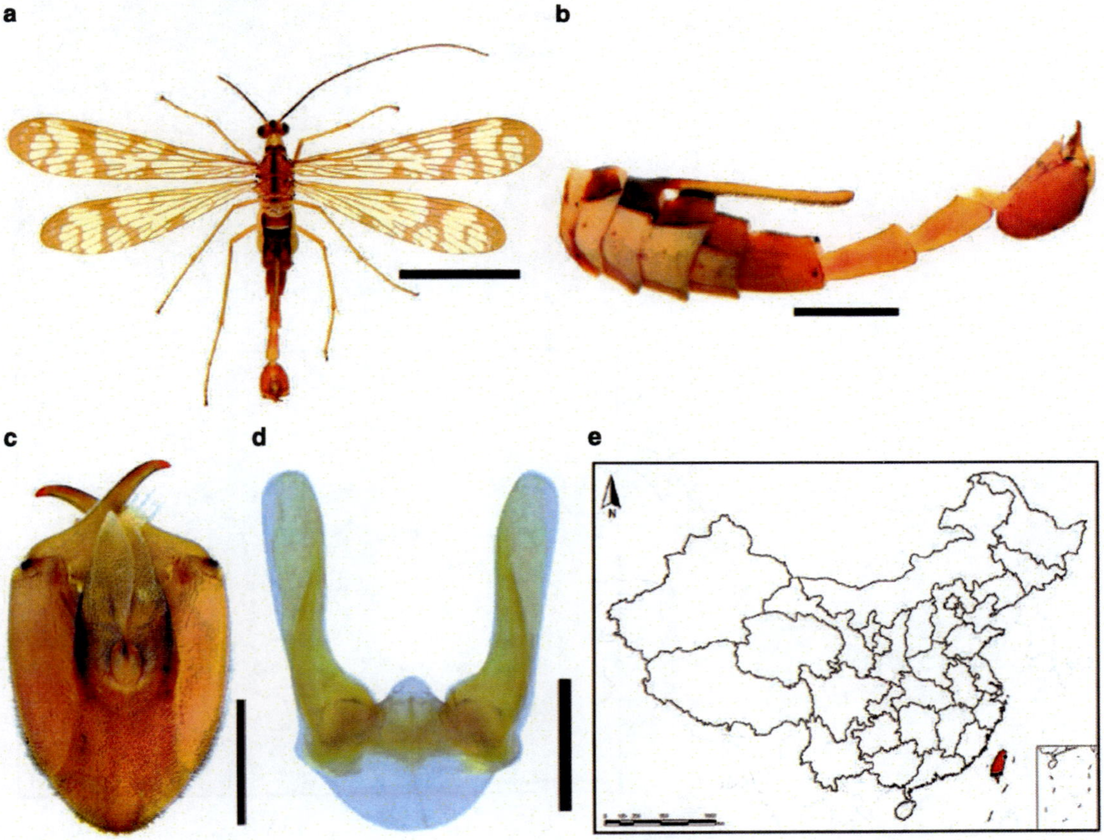

Fig. 10.66 (a) Male adult of ***Neopanorpa sauteri***, dorsal view (Tahan Mountain, Pingtung, Taiwan) (*Scale bar*: 5.0 mm) **[ECAU]**. **(b)** Male abdomen of ***Neopanorpa sauteri***, lateral view (*Scale bar*: 2.0 mm). (c) Male genital bulb of ***Neopanorpa sauteri***, ventral view (*Scale bar*: 0.5 mm). **(d)** Female medigynium of ***Neopanorpa sauteri***, ventral view (*Scale bar*: 0.2 mm). **(e)** Distribution of ***Neopanorpa sauteri***

long broad basal stalk. Hypovalves almost subtriangular, overlapped mesally, and forming large subcircular window subbasally. Gonostylus with obtuse triangular median tooth and large basal process. Aedeagus conical, with stout lateral processes.

Female subgenital plate oval, with V-shaped terminal emargination. Mediygnium with reduced main plate; posterior arms long, spatulate in distal half and rounded terminally; axis lacking apodemes.

Emergence. June to July.

Distribution. Taiwan (Yushan Mountain, Chiayi, Pingtung, Yilan).

10.1.64 *Neopanorpa tienmushana* Cheng, 1957

(Tianmushan Neoscorpionfly) [天目山新蝎蛉]
(Fig. 10.67)

Measurements. Forewing length 13.0–14.0 mm.

Morphology. Vertex yellowish brown. Ocellar triangle black. Rostrum yellowish brown. Pronotum yellowish brown. Meso- and metanotum yellowish brown, with distinct black median stripe; mesonotum black on anterior and latero-anterior margin. Pleura and legs yellowish brown. Wing membrane translucent, tinged with yellow. Markings yellowish brown. In forewings, apical band broad, with large hyaline windows

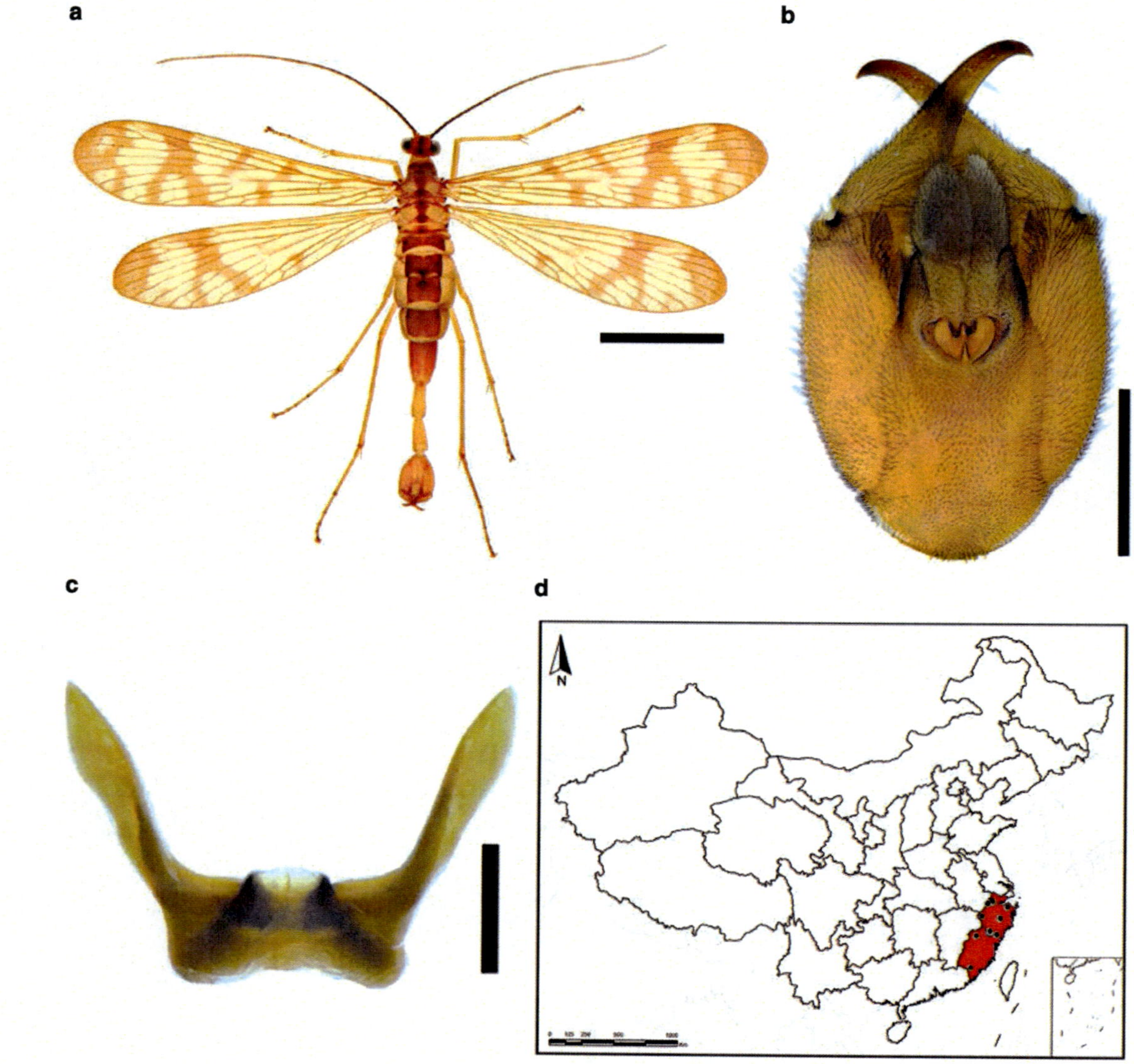

Fig. 10.67 (**a**) Male adult of ***Neopanorpa tienmushana***, dorsal view (West Tianmu Mountain, Lin'an, Zhejiang) (*Scale bar*: 5.0 mm) [NWAU]. (**b**) Male genital bulb of ***Neopanorpa tienmushana***, ventral view (*Scale bar*: 0.5 mm). (**c**) Female medigynium of ***Neopanorpa tiemnushana***, ventral view (*Scale bar*: 0.2 mm). (**d**) Distribution of ***Neopanorpa tienmushana***

innerly and separated stripe near posterior corner; pterostigmal band complete, with broad basal branch and thin apical branch; marginal spot slender and connected to basal band posteriorly; basal spot small. Male T1–T5 yellowish brown. Notal organ exceeding middle of T4. A6 yellowish brown, cylindrical. A7 and A8 yellowish brown and constricted basally.

Male genital bulb oval. Epandrium long and broad, and emarginated in shallow V-shaped on apex. Hypandrium with long and broad basal stalk. Hypovalves very broad and overlapped mesally, forming heart-shaped window subbasally. Paramere short, Y-shaped. Aedeagus conical, with stout lateral processes.

Female subgenital plate oval, with V-shaped terminal emargination. Mediygnium with main plate reduced, projected at latero-basal portions; posterior arms long, spatulate in distal half and rounded terminally; axis lacking apodemes.

Emergence. June to August.

Distribution. Zhejiang (Tianmu Mountains), Fujian (Nanjing).

10.1.65 *Neopanorpa translucida* Cheng, 1957

(Transparent Neoscorpionfly) [透斑新蝎蛉] (Fig. 10.68)

Measurements. Forewing length 14.5–15.0 mm.

Morphology. Vertex yellowish brown. Ocellar triangle enclosed by large black spot. Rostrum yellowish brown.

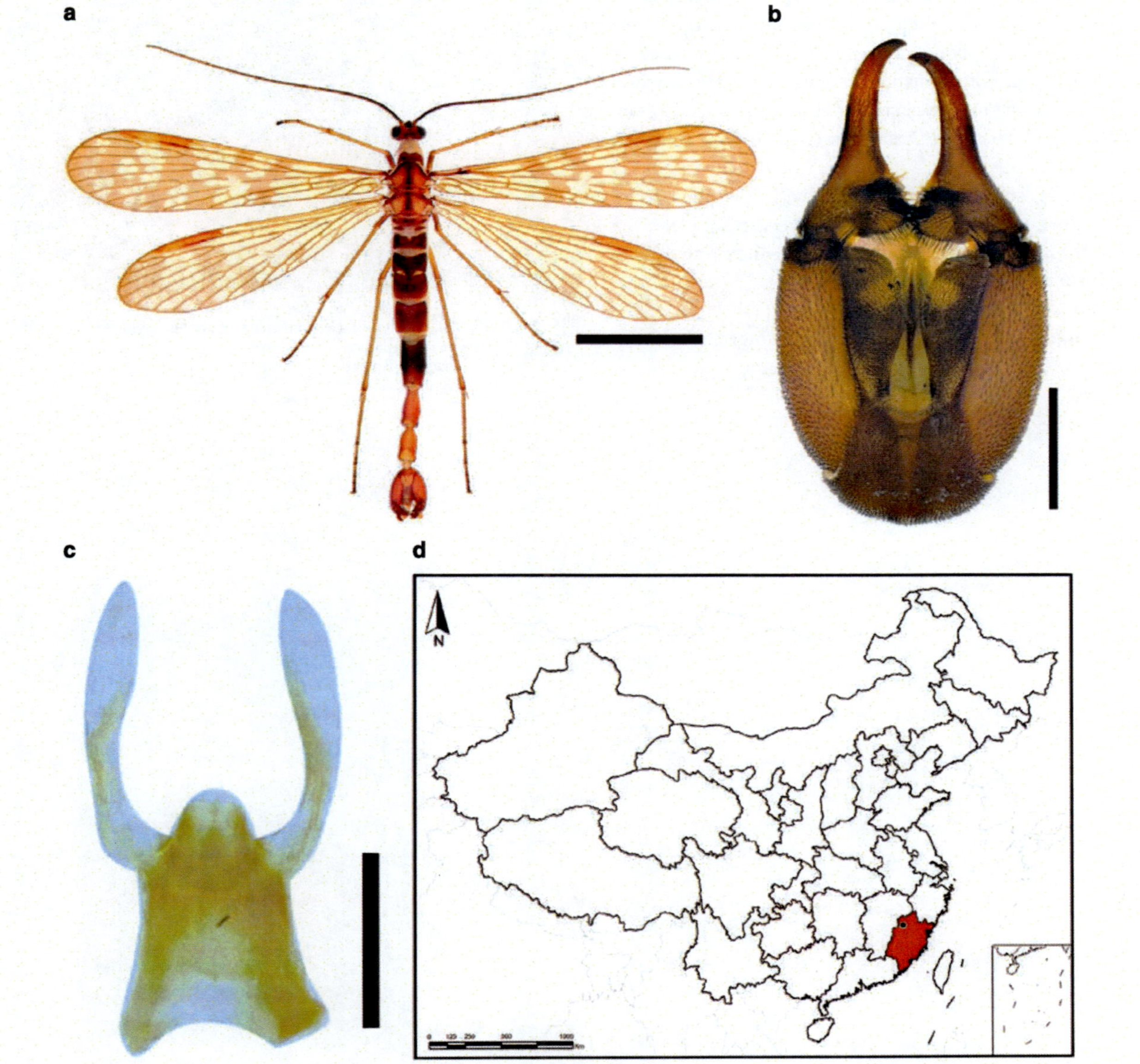

Fig. 10.68 (**a**) Male adult of ***Neopanorpa translucida,*** dorsal view (Dazhulan, Shaowu, Fujian) (*Scale bar*: 5.0 mm) [NWAU]. (**b**) Male genital bulb of ***Neopanorpa translucida,*** ventral view (*Scale bar*: 0.5 mm). (**c**) Female medigynium of ***Neopanorpa translucida***, ventral view (*Scale bar*: 0.2 mm). (**d**) Distribution of ***Neopanorpa translucida***

Pronotum dark brown. Meso- and metanotum mostly yellowish brown, with black anterior margins and distinct black median stripe. Pleura and legs yellowish brown. Wing membrane translucent, tinged with yellow. Markings faint yellowish brown. Pterostigma distinct, reddish. In forewings, apical band irregular along inner margin, with hyaline window posteriorly; pterostigmal band complete, with apical branch constricted basally; marginal spot connected to pterostigmal band apically, and connected to basal band basally; basal spot absent. Male T1–T5 unevenly dark brown. Notal organ slender and stick-like, not exceeding hind border of T4. A6 mostly dark brown and gradually turned into reddish brown towards apex. A7 and A8 reddish brown and constricted basally.

Male genital bulb oval. Epandrium long and broad, with truncated apex. Hypandrium with long and broad basal stalk. Hypovalves broad, curled dorsad on lateral margin, beveled apically, and separated basally. Gonostylus concaved on outer margin, with indistinct median tooth and large disc-shaped basal process bearing dense long setae. Paramere slender and Y-shaped. Aedeagus conical, with stout lateral processes.

Female subgenital plate oval, with V-shaped terminal emargination. Mediygnium with reduced main plate; posterior arms slender; axis slightly shorter than posteiror arms with stout and diverged apodemes.

Emergence. May to August.

Distribution. Fujian (Changting, Guangze, Shaowu).

10.2 Genus *Cerapanorpa* Gao, Ma & Hua, 2016

(Single-horned Scorpionflies, Cerapanorpas) [单角蝎蛉属] (Fig. 10.69)

Cerapanorpa is endemic to China. Its main diagnostic character lies in a distinct finger-like anal horn on the posterior end of T6 in males.

Twenty-two species have been described, and 20 of them are included in this atlas.

Fig. 10.69 Male adult of ***Cerapanorpa***. ***Cerapanorpa obtusa*** in the field

Key to Species of Cerapanorpa in This Atlas
(20 species, males)

1. T5 with additional horn-like projection in hind margin .... ***Cerapanorpa bicornifera*** (Chou & Wang)
– T5 lacking projection in hind margin 2
2. Wings with well-developed markings; apical, pterostigmal and basal bands, marginal and basal spots dense and broad; paramere L-shaped .... ***Cerapanorpa reni*** (Chou)
– Wings with markings variously reduced; paramere not modified as above 3
3. Anal horn short, not longer than 1/5 of A6 .... ***Cerapanorpa brevicornis*** (Hua & Li)
– Anal horn longer than 1/5 of A6 4
4. Parameres bifurcated 5
– Parameres unfurcated 6
5. Parameres with L-shaped lateral branch .... ***Cerapanorpa byersi*** (Hua & Huang)
– Parameres with straight dorsal branch .... ***Cerapanorpa minshana*** Gao, Li & Hua
6. Parameres enlarged basally; basal portion at least three times as wide as distal portion 7
– Parameres same width in basal and distal portion 8
7. Parameres exceeding apex of gonocoxites .... ***Cerapanorpa baimaensis*** Gao & Hua
– Parameres not exceeding apex of gonocoxites .... ***Cerapanorpa centralis*** (Tjeder)
8. Parameres greatly elongated and exceeding basal process of gonostylus 9
– Parameres less elongated and not exceeding basal process of gonostylus 10
9. Parameres nearly extending to apex of gonostylus .... ***Cerapanorpa protrudens*** Gao, Ma & Hua
– Parameres shorter, only slightly exceeding middle of gonostylus .... ***Cerapanorpa emarginata*** (Cheng)
10. Parameres bearing short spines innerly; spines shorter than width of paramere 11
– Parameres bearing long spines innerly; spines longer than width of paramere or as long as latter 12
11. Parameres relatively straight or slightly sinuate, exceeding apex of gonocoxites .... ***Cerapanorpa dubia*** (Chou & Wang)
– Parameres mostly straight with only apex curved inner ward, and not exceeding apex of gonocoxites .... ***Cerapanorpa liupanshana*** Gao, Ma & Hua
12. Parameres greatly sinuate, S- or elbow-shaped 13
– Parameres mostly straight or only curved inner ward at apex 18
13. Parameres with distal half greatly curved and bow-shaped 14
– Paramere not modified as above 15
14. Parameres stout, with inner spines longer than width of paramere .... ***Cerapanorpa sinuata*** Gao, Ma & Hua
– Paramers slender, with inner spines nearly as long as width of paramere .... ***Cerapanorpa taizishana*** Gao & Hua
15. Hypovalves long, extending to basal portion of gonostylus; parameres slightly curved near middle .... ***Cerapanorpa yanggashana*** Gao & Hua
– Hypovalves not exceeding apex of gonocoxites; parameres greatly curved in distal portion 16
16. Parameres with inner spines in distal half shorter than width of paramere; dorsal aedeagal processes greatly projected ventrally .... ***Cerapanorpa nanwutaina*** (Chou)
– Parameres with inner spines longer; dorsal aedeagal processes less projected ventrally 17
17. Hypovalves broad; dorsal aedeagal processes finger-like .... ***Cerapanorpa obtusa*** (Cheng)
– Hypovalves narrow; dorsal aedeagal processes enlarged .... ***Cerapanorpa xuebaodinga*** Gao & Hua
18. Parameres greatly curved inner ward and hook-like at apex .... ***Cerapanorpa qinlingensis*** Gao, Li & Hua
– Parameres not modified as above 19
19. Wings only with faint apical and pterostigmal bands; parameres with spines along inner margin in distal half .... ***Cerapanorpa funiushana*** (Hua & Chou)
– Wings with prominent apical and pterostigmal bands; parameres with spines along inner margin distal to basal stalk .... ***Cerapanorpa wangwushana*** (Hua, Huang & Shen)

10.2.1 *Cerapanorpa baimaensis* Gao & Hua 2019

(Baima Single-Horned Scorpionfly) [白马单角蝎蛉] (Fig. 10.70)

Measurements. Forewing length 12.0–13.0 mm.

Morphology. Vertex black. Rostrum yellowish brown. Nota black. Pleura and legs yellowish brown. Wing membrane hyaline and devoid of markings. Pterostigma light yellow. Male T1–T5 dark brown. Notal organ flat and covering acute postnotal organ. A6 dark brown and bearing single reddish brown anal horn on dorsal apex. A7 and A8 reddish brown and constricted basally.

Male genital bulb oval. Epandrium emarginate apically and forming pair of broad finger-like processes laterally. Hypovalves stripe-like, nearly 2/3 length of gonocoxites, and bearing long bristles along inner margin. Gonocoxites with several long bristles on ventral apex. Gonostylus shorter than gonocoxites and bearing obtuse triangular median tooth and large disc-like basal process. Paramere long, stick-like with acute apex, and bearing numerous spines along inner margin. Aedeagus with ventral valves membranous and short, and dorsal processes finger-like and divergent.

Female subgenital plate elliptical, shallowly emarginate terminally. Medigynium with main plate flat and broad; two pairs of basal plates; posterior arms short and acute; axis lon-

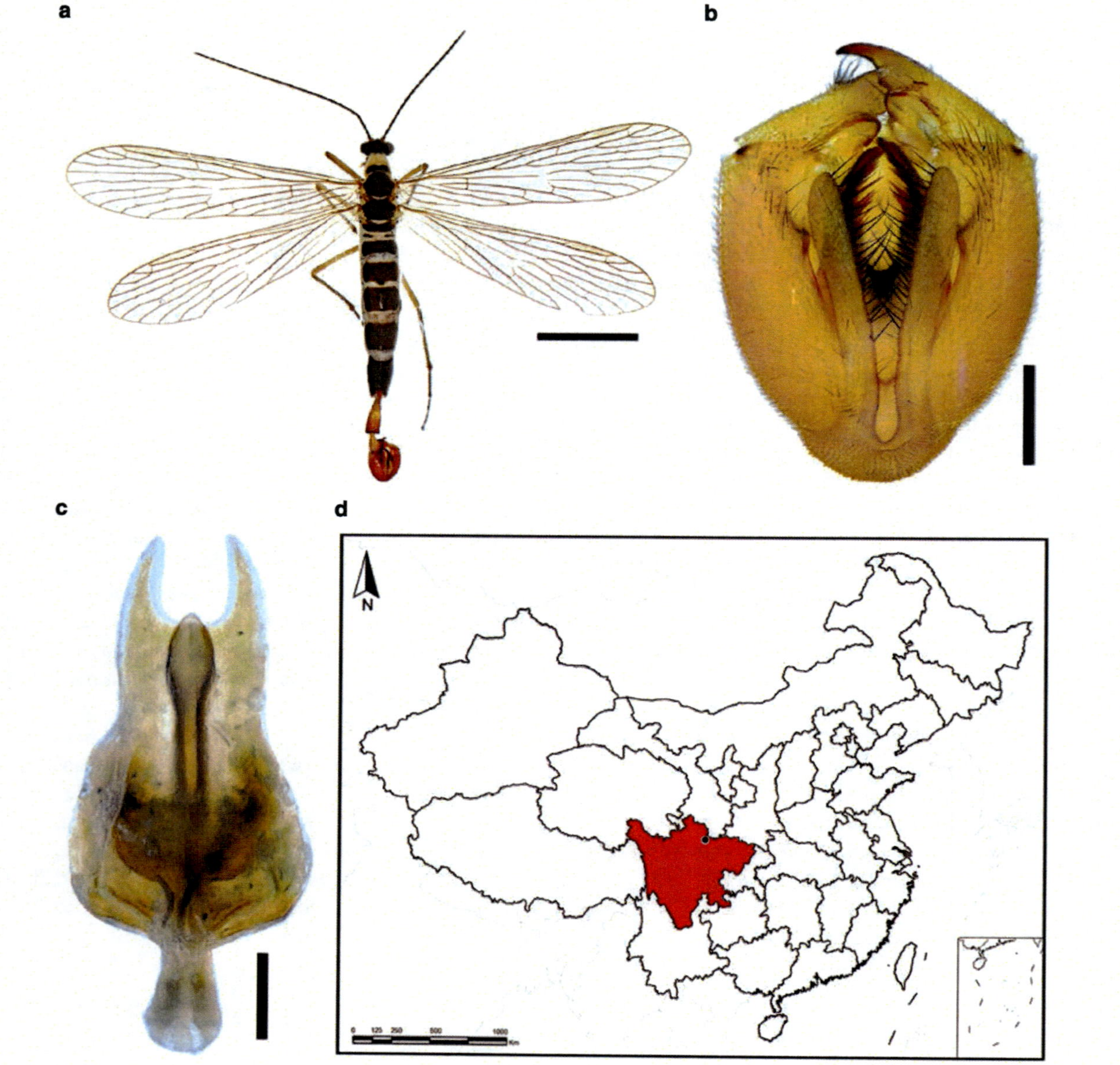

Fig. 10.70 (**a**) Male adult of ***Cerapanorpa baimaensis***, dorsal view (Baima, Pingwu, Sichuan, ©Kai Gao) (*Scale bar*: 5.0 mm) **[NWAU]**. (**b**) Male genital bulb of ***Cerapanorpa baimaensis***, ventral view (*Scale bar*: 0.5 mm). (**c**) Female medigynium of ***Cerapanorpa baimaensis***, ventral view (*Scale bar*: 0.2 mm). (**d**) Distribution of ***Cerapanorpa baimaensis***

ger than main plate with slightly diverged apodemes; apex of apodeme branched.

Emergence. July.

Distribution. Sichuan (Pingwu).

10.2.2 *Cerapanorpa bicornifera* Chou & Wang, 1981

(Bicornous Single-Horned Scorpionfly) [二角单角蝎蛉] (Fig. 10.71)

Measurements. Forewing length 11.0 mm.

Morphology. Vertex black. Rostrum yellowish brown. Nota black. Pleura and legs yellowish brown. Wing membrane hyaline, and devoid of markings except darkened apex. Pterostigma light yellow. Male T1–T5 dark brown. Notal organ flat and covering acute postnotal organ. T5 with small and acute horn-like process on posterior margin. A6 dark brown and bearing single reddish brown anal horn on dorsal apex. A7 and A8 reddish brown and constricted basally.

Male genital bulb oval. Epandrium emarginate apically and forming pair of broad finger-like processes laterally. Hypovalves stripe-like, nearly 2/3 length of gonocoxites, and bearing long bristles along inner margin. Gonocoxites with several long bristles on ventral apex. Gonostylus shorter than gonocoxites and bearing obtuse triangular median tooth and large disc-like basal process. Paramere long, stick-like with acute apex, and bearing numerous spines on apical half.

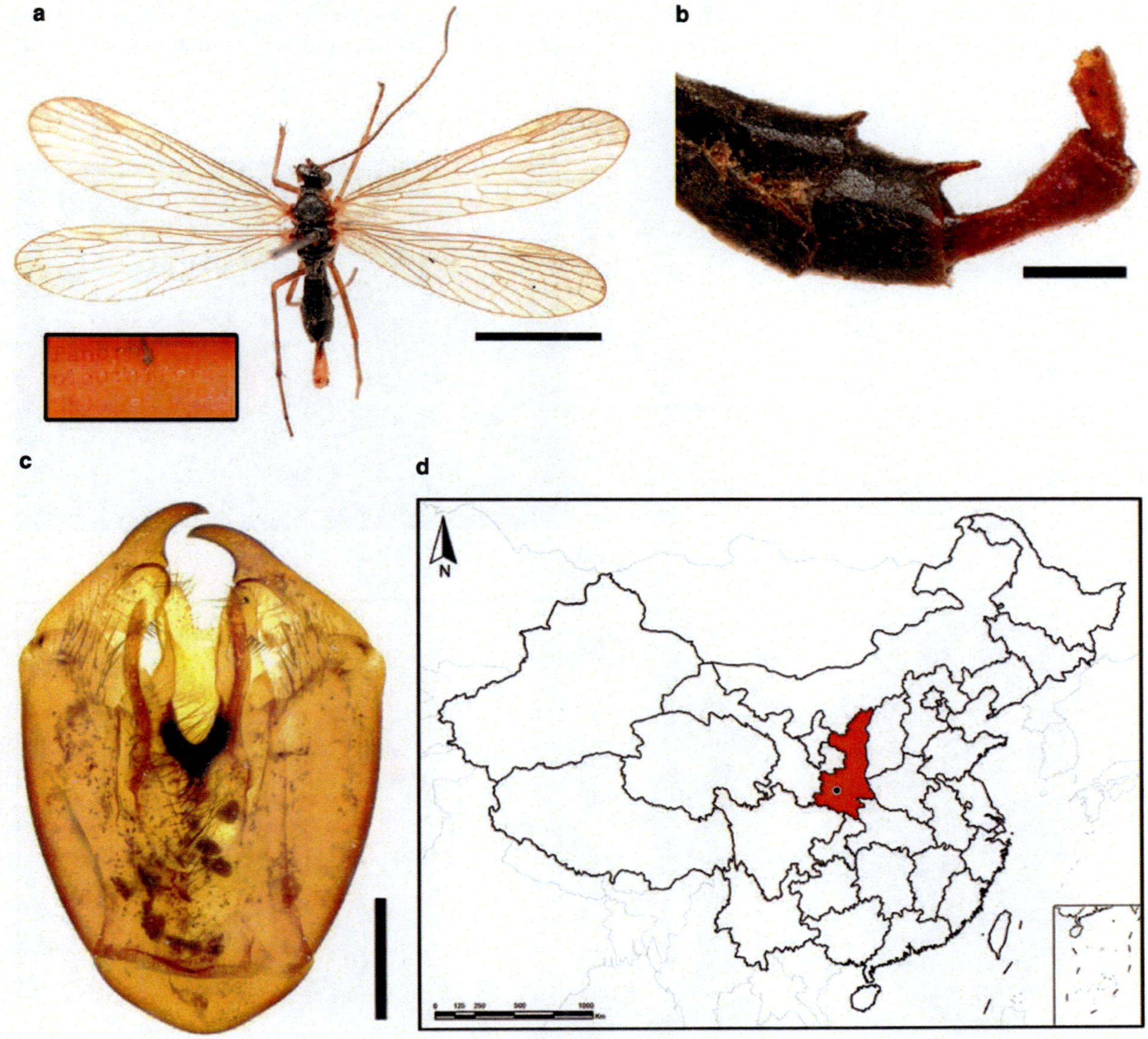

Fig. 10.71 (**a**) Male adult of ***Cerapanorpa bicornifera***, dorsal view (With genitalia removed, Qinling Mountains, Shaanxi) (*Scale bar*: 0.5 mm) **[NWAU]**. (**b**) Male teminal abdomen of ***Cerapanorpa bicornifera***, lateral view (*Scale bar*: 1.0 mm). (**c**) Male genital bulb of ***Cerapanorpa bicornifera***, ventral view (*Scale bar*: 0.5 mm). (**d**) Distribution of ***Cerapanorpa bicornifera***

Aedeagus with ventral valves membranous and short, and dorsal processes finger-like and divergent.

Female unknown.

Emergence. July.

Distribution. Shaanxi (Qinling Mountains).

Remarks: This species has not been rediscovered since its publication in 1981 based on a male holotype. We suppose that the holotype might be a variant of another *Cerapanorpa* species.

10.2.3 *Cerapanorpa brevicornis* Hua & Li, 2007

(Short-Horn Single-Horned Scorpionfly) [短角单角蝎蛉] (Fig. 10.72)

Measurements. Forewing length 13.5–14.0 mm.

Morphology. Vertex dark brown. Ocellar triangle black. Rostrum reddish brown and pale mesally. Nota black. Pleura and legs yellowish brown. Wing membrane hyaline, and devoid of markings except darkened apex. Pterostigma yellow. Male T1–T5 dark brown. Notal organ flat and covering acute postnotal organ. A6 dark brown and bearing single yellowish brown anal horn on dorsal apex. A7 and A8 yellowish brown and constricted basally.

Male genital bulb broad oval. Epandrium emarginate apically and forming pair of finger-like processes laterally. Hypovalves stripe-like, exceeding 2/3 length of gonocoxites, and bearing row of regularly arranged, short bristles along inner margin. Gonocoxites with several long bristles on ventral apex. Gonostylus shorter than gonocoxites and bearing obtuse triangular median tooth and large disc-like basal process. Paramere slender, curved mesad at distal portion, and bearing numerous short spines on apical inner margin. Aedeagus with ventral valves membranous and short, and dorsal processes

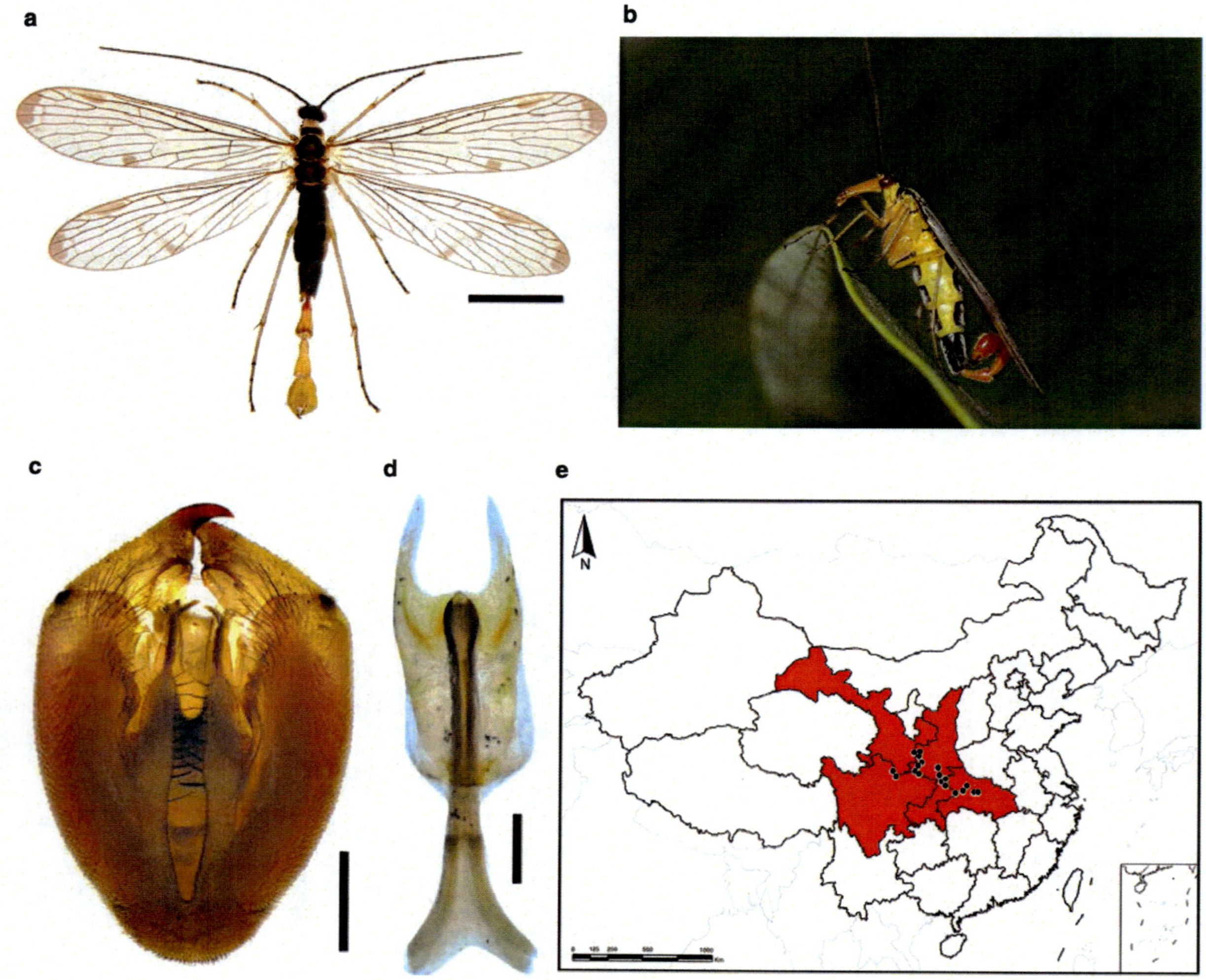

Fig. 10.72 (**a**) Male adult of ***Cerapanorpa brevicornis***, dorsal view (Tiantaishan Forest Park, Baoji, Shaanxi) (*Scale bar*: 5.0 mm). (**b**) Male adult of ***Cerapanorpa brevicornis*** in the field (Tongtian River, Taibai, Shaanxi, ©Kai Gao). (**c**) Male genital bulb of ***Cerapanorpa brevicornis***, ventral view. (**d**) Distribution of ***Cerapanorpa brevicornis***

finger-like apically. Inflated membranous process on dorsal surface of dorsal valve subapically. Lateral processes conspicuous.

Female subgenital plate elliptical, shallowly emarginate terminally. Medigynium with main plate flat and broad; two pairs of basal plates; posterior arms short and acute; axis two times longer than main plate with greatly diverged apodemes; apex of apodemes shortly branched.

Emergence. May to August.

Distribution. Hubei (Shennongjia), Sichuan (Pingwu), Shaanxi (Baoji, Foping, Nanzheng, Ningshan).

10.2.4 *Cerapanorpa byersi* Hua & Huang, 2007

(Byers's Single-Horned Scorpionfly) [拜尔斯单角蝎蛉] (Fig. 10.73)

Measurements. Forewing length 12.5–13.0 mm.

Morphology. Relatively small. Vertex black. Rostrum dark brown with yellowish brown flanks. Nota dark brown. Pleura and legs cream-white or pale brown. Wing membrane hyaline, and devoid of markings except greatly reduced apical band. Veins dark brown. Male T1–T5 dark brown. Notal organ flat and covering acute postnotal organ. A6 dark brown

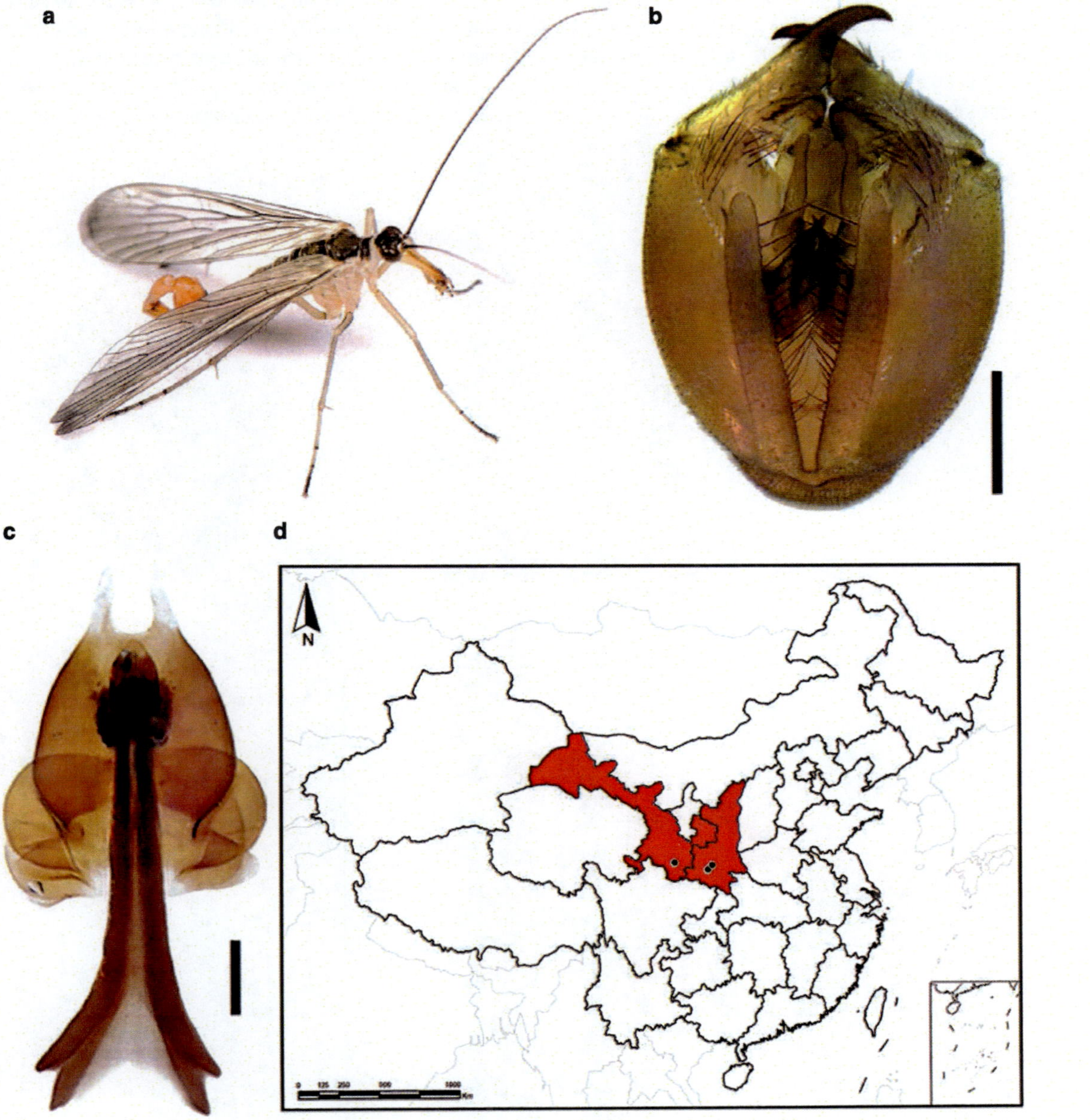

Fig. 10.73 (**a**) Male adult of ***Cerapanorpa byersi*** (Honggou Forest Farm, Taibai, Shaanxi). (**b**) Male genital bulb of ***Cerapanorpa byersi***, ventral view (*Scale bar*: 0.5 mm). (**c**) Female medigynium of ***Cerapanorpa byersi***, ventral view (*Scale bar*: 0.2 mm). (**d**) Distribution of ***Cerapanorpa byersi***

and bearing yellowish brown anal horn on dorsal apex. A7 and A8 yellowish brown and constricted at base.

Male genital bulb globular. Epandrium emarginate terminally in broad U-shape and forming pair of finger-like processes laterally. Hypovalves stripe-like and bearing row of long bristles along inner margin. Gonostylus shorter than gonocoxites and bearing obtuse triangular median tooth and large disc-like basal process. Paramere short and stout, not exceeding apex of gonocoxites, swollen trapezoidally at middle and bearing several comb-like spines along inner margin, and acute L-shaped short branch laterally; dorsal processes of aedeagus extending to apex of gonocoxites with finger-like apexes.

Female subgenital plate elliptical, shallowly emarginate terminally. Medigynium with main plate short and broad; two pairs of basal plates; posterior arms short and acute; axis two times longer than main plate with slightly diverged apodemes; apex of apodemes distinctly branched.

Emergence. June to July.

Distribution. Gansu (Dangchang), Shaanxi (Meixian, Taibai).

10.2.5 *Cerapanorpa centralis* Tjeder, 1936

(Central Single-Horned Scorpionfly) [中部单角蝎蛉] (Fig. 10.74)

Measurements. Forewing length 13.0 mm.

Morphology. Vertex black. Rostrum yellowish brown. Nota black. Pleura and legs yellowish brown.

Wing membrane hyaline, and devoid of markings. Pterostigma light brown. Male T1–T5 dark brown. Notal organ flat and covering acute postnotal organ. A6 dark brown and bearing yellowish brown anal horn on dorsal apex. A7 and A8 yellowish brown and constricted basally.

Male genital bulb oval. Epandrium emarginate terminally in broad U-shape and forming pair of finger-like pro-

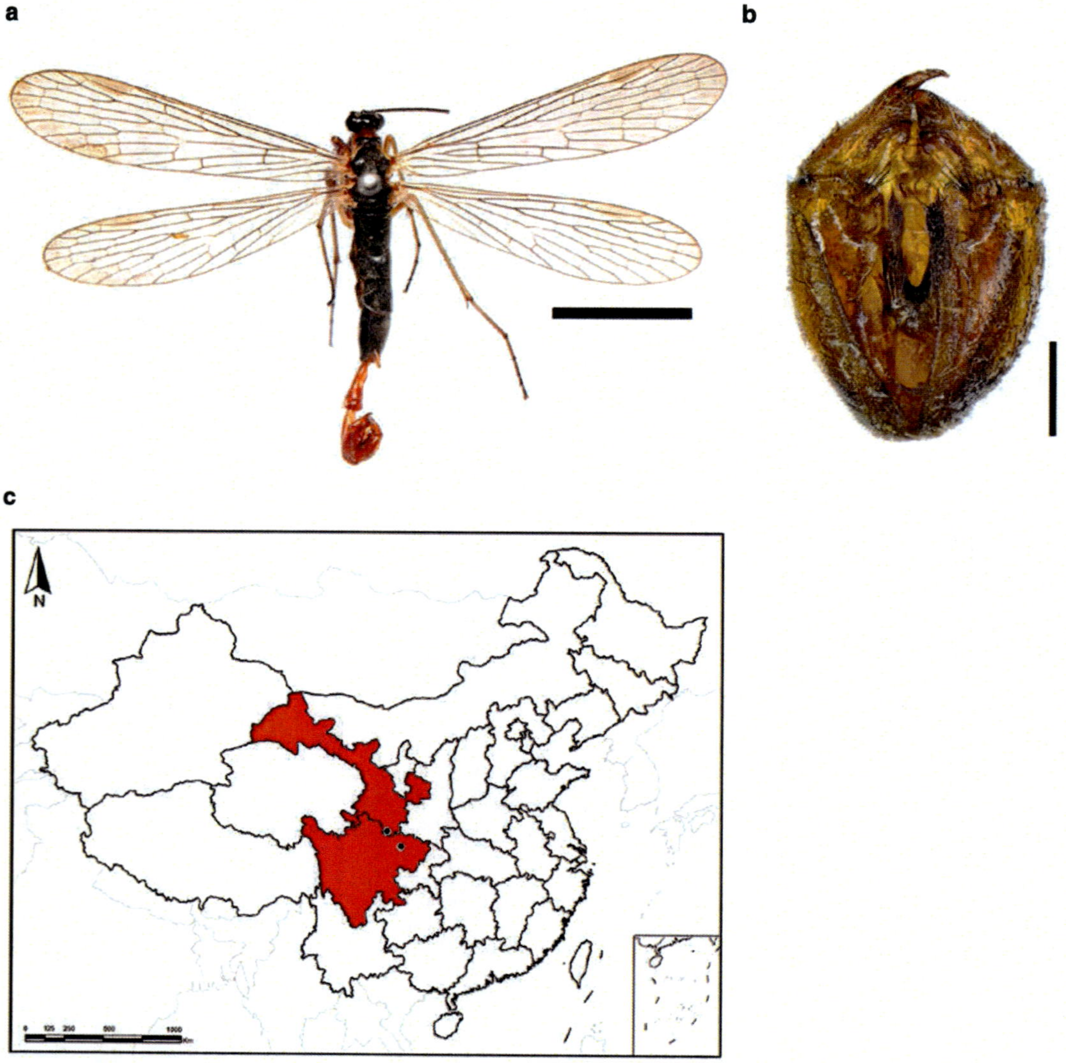

Fig. 10.74 (**a**) Male adult of ***Cerapanorpa centralis***, dorsal view (Wenxian, Gansu) (*Scale bar*: 5.0 mm) **[NWAU]**. (**b**) Male genital bulb of ***Cerapanorpa centralis***, ventral view. (**c**) Distribution of ***Cerapanorpa centralis***

cesses laterally. Hypovalves stripe-like and bearing row of long bristles along inner margin. Gonocoxites with several long setae on ventral apex. Gonostylus shorter than gonocoxites and bearing obtuse triangular median tooth and large disc-like basal process. Paramere stout at basal half and slender at distal half, not exceeding apex of gonocoxites, and bearing several comb-like spines along inner margin. Ventral aedeagal valves membranous and short; dorsal processes exceeding the base of gonocoxites with slightly swollen apexes.

Female subgenital plate elliptical, shallowly emarginate terminally. Medigynium with main plate flat and broad; two pairs of basal plates; posterior arms short and acute; axis longer than main plate with slightly diverged apodemes; apex of apodemes distinctly branched.

Emergence. May to July.

Distribution. Sichuan (Langzhong) and Gansu (Wenxian).

10.2.6 *Cerapanorpa dubia* (Chou & Wang, 1981)

(Dubious Single-Horned Scorpionfly) [拟华山蝎蛉] (Fig. 10.75)

Measurements. Forewing length 13.5–15.0 mm.

Morphology. Vertex black. Rostrum light yellow, yellowish brown, or dark brown. Nota black. Pleura and legs light

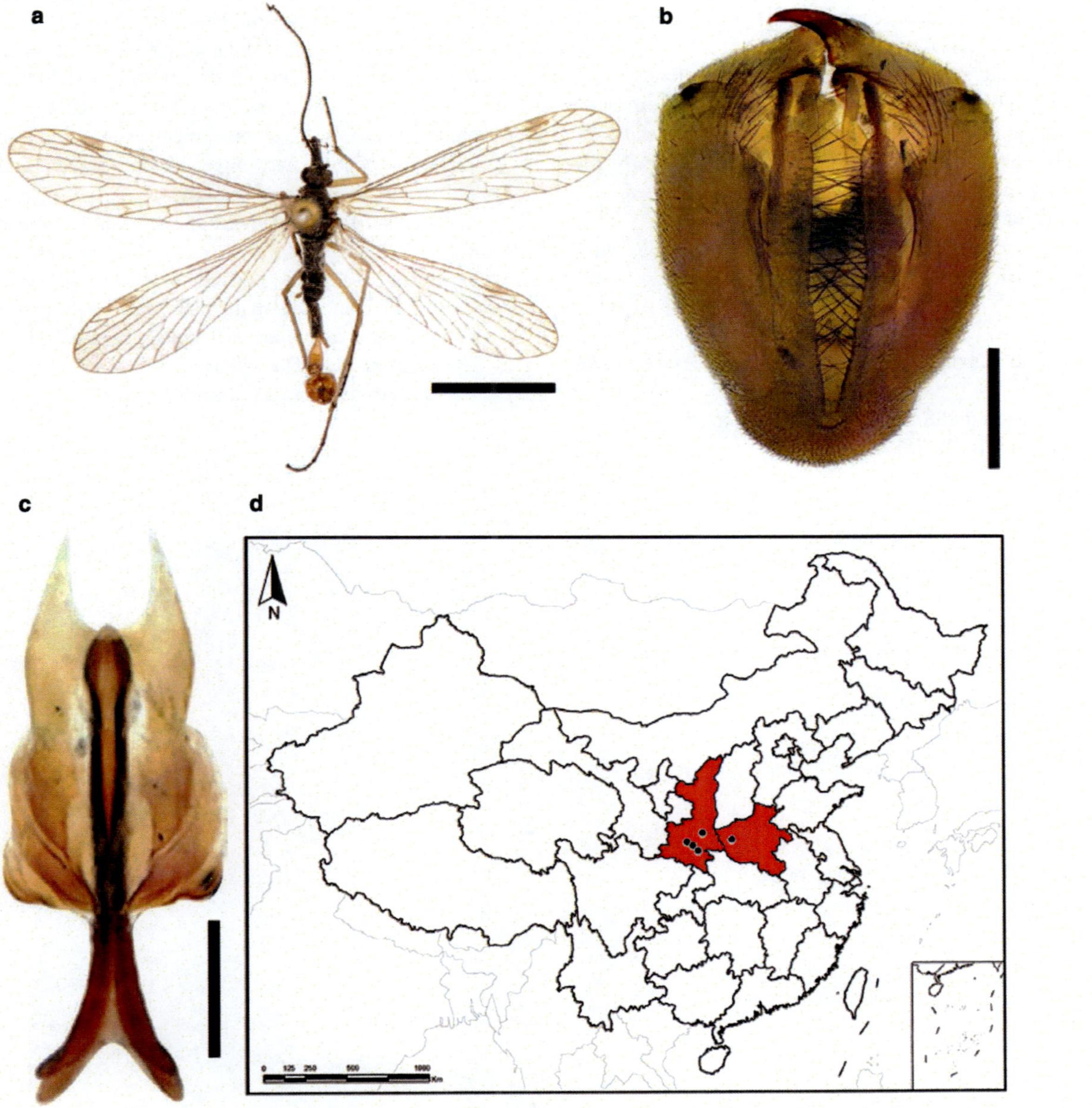

Fig. 10.75 (a) Male adult of ***Cerapanorpa dubia,*** dorsal view (Taibai Mountain, Shaanxi) (*Scale bar*: 5.0 mm) **[NWAU]**. **(b)** Male genital bulb of ***Cerapanorpa dubia***, ventral view (*Scale bar*: 0.5 mm). **(c)** Female medigynium of ***Cerapanorpa dubia***, ventral view (*Scale bar*: 0.2 mm). **(d)** Distribution of ***Cerapanorpa dubia***

yellow. Wing membrane hyaline, and devoid of markings, or darkened apically. Pterostigma light brown and conspicuous. Male T1–T5 dark brown. Notal organ flat and covering acute postnotal organ. A6 dark brown and bearing yellowish brown anal horn on dorsal apex. A7 and A8 yellowish brown and constricted at base.

Male genital bulb oval. Epandrium emarginate terminally in broad U-shape and forming pair of finger-like processes laterally. Hypovalves stripe-like and bearing row of long bristles along inner margin. Gonocoxites with several long setae on ventral apex. Gonostylus shorter than gonocoxites and bearing obtuse triangular median tooth and large disc-like basal process. Paramere slender, straight, or sinuate, and bearing row of short spines on inner margin. Ventral aedeagal valves membranous and short; dorsal processes slender and exceeding base of gonostylus with membranous swollen processes dorsally.

Female subgenital plate elliptical, shallowly emarginate terminally. Medigynium with main plate short and broad; two pairs of basal plates; posterior arms short and acute; axis two times longer than main plate with slightly diverged apodemes; apex of apodemes distinctly branched.

Emergence. June to August.

Distribution. Shaanxi (Meixian, Zhouzhi, Ningshan, Huashan Mountain), Henan (Luanchuan).

10.2.7 *Cerapanorpa emarginata* Cheng, 1949

(Emarginated Single-Horned Scorpionfly) [华山单角蝎 蛉] (Fig. 10.76)

Measurements. Forewing length 13.5–14.0 mm.

Morphology. Vertex black. Rostrum yellowish brown. Nota dark brown. Pleura and legs yellowish brown. Wing membrane hyaline, devoid of markings or only darkened at apex. Pterostigma light yellow and inconspicuous. Male T1–T5 dark brown. Notal organ flat and covering acute postnotal organ. A6 dark brown and bearing yellowish brown anal horn on apex. A7 and A8 yellowish brown and constricted at base.

Male genital bulb broad oval. Epandrium emarginate terminally in broad U-shape and forming pair of finger-like processes laterally. Hypovalves stripe-like and bearing row of sparsely and regularly arranged long bristles along inner margin. Gonocoxites with several long setae on ventral apex. Gonostylus shorter than gonocoxites and bearing obtuse triangular median tooth and large disc-like basal process. Paramere nearly parallel, and not exceeding middle of gonocoxites apically, and bearing row of short and dense comb-like spines on inner margin. Ventral aedeagal valves membranous and short; dorsal processes long and extending to apex of gonocoxites, finger-like at apexes, and with slightly swollen processes dorsally.

Female subgenital plate elliptical, shallowly emarginate terminally. Medigynium with main plate short and broad; two pairs of basal plates; posterior arms short and acute; axis two times longer than main plate with slightly diverged apodemes; apex of apodemes distinctly branched.

Emergence. June to August.

Distribution. Shaanxi (Taibai Mountain).

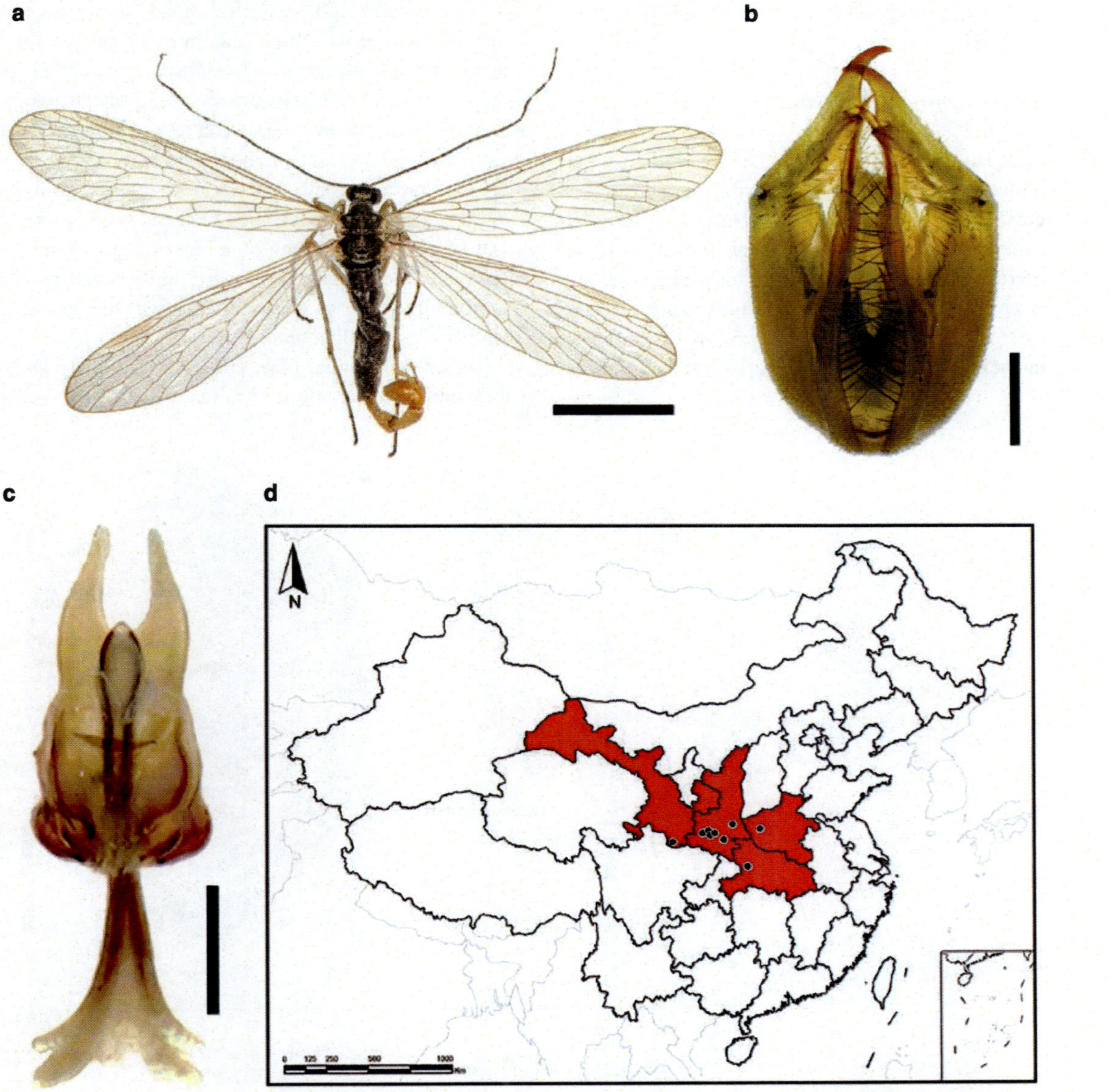

Fig. 10.76 (**a**) Male adult of ***Cerapanorpa emarginata***, dorsal view (Taibai Mountain, Shaanxi) (*Scale bar*: 5.0 mm) [NWAU]. (**b**) Male genital bulb of ***Cerapanorpa emarginata***, ventral view (*Scale bar*: 0.5 mm). (**c**) Female medigynium of ***Cerapanorpa emarginata***, ventral view (*Scale bar*: 0.2 mm). (**d**) Distribution of ***Cerapanorpa emarginata***

10.2.8 *Cerapanorpa funiushana* Hua & Chou, 1997

(Funiushan Single-Horned Scorpionfly)
[伏牛山单角蝎蛉] (Fig. 10.77)
Measurements. Forewing length 13.5–14.0 mm.

Morphology. Vertex black. Rostrum yellowish brown. Nota dark brown. Pleura and legs yellowish brown. Wing membrane hyaline, devoid of markings or only darkened at apex. Pterostigma light yellow and inconspicuous. Male T1–T5 dark brown. Notal organ flat and covering acute postnotal organ. A6 dark brown and bearing yellowish brown anal horn on apex. A7 and A8 yellowish brown and constricted at base.

Male genital bulb broad oval. Epandrium emarginate terminally in broad U-shape and forming pair of finger-like processes laterally. Hypovalves stripe-like and bearing row of sparsely and regularly arranged long bristles along inner margin. Gonocoxites with several long setae on ventral apex. Gonostylus shorter than gonocoxites and bearing obtuse triangular median tooth and large disc-like basal process. Paramere nearly parallel, stick-like, and bearing row of short and dense comb-like spines on inner margin. Ventral aedeagal valves membranous and short; dorsal processes long and extending to apex of gonocoxites, finger-like at apexes, and with slightly swollen processes dorsally.

Female subgenital plate elliptical, shallowly emarginate terminally. Medigynium with main plate short and broad;

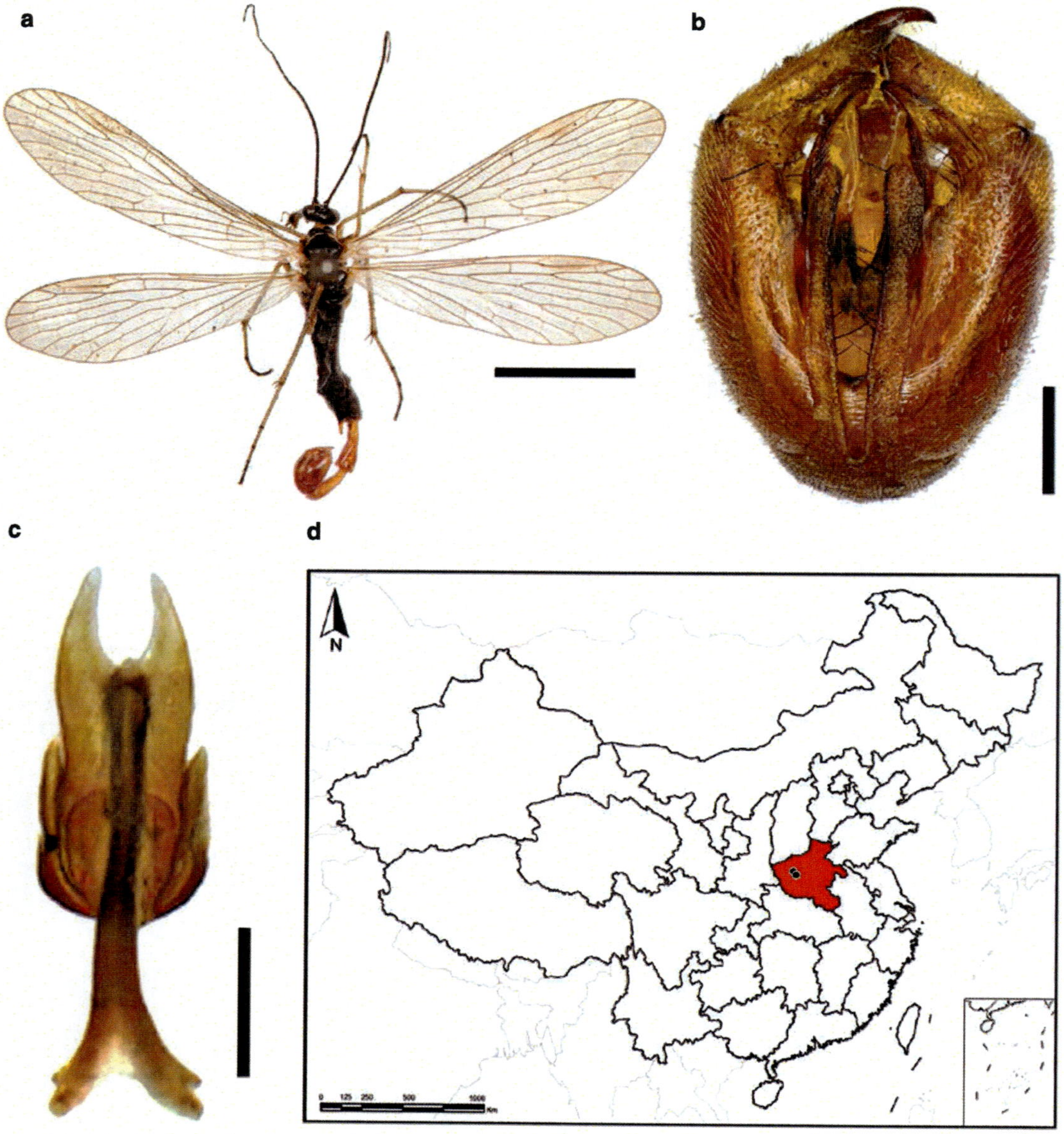

Fig. 10.77 (**a**) Male adult of ***Cerapanorpa funiushana***, dorsal view (Baiyun Mountain, Songxian, Henan) (*Scale bar*: 5.0 mm) **[NWAU]**. (**b**) Male genital bulb of ***Cerapanorpa funiushana***, ventral view (*Scale bar*: 0.5 mm). (**c**) Female medigynium of ***Cerapanorpa funiushana***, ventral view (*Scale bar*: 0.2 mm). (**d**) Distribution of ***Cerapanorpa funiushana***

two pairs of basal plates; posterior arms short and acute; axis two times longer than main plate with slightly diverged apodemes; apex of apodemes distinctly branched.

Emergence. May to August.

Distribution. Henan (Funiu Mountain, Luanchuan).

10.2.9 *Cerapanorpa liupanshana* Gao, Ma & Hua, 2016

(Liupanshan Single-Horned Scorpionfly)
[六盘山单角蝎蛉] (Fig. 10.78)

Measurements. Forewing length 12.0–13.0 mm.

Morphology. Vertex black. Rostrum yellowish brown. Nota black. Pleura and legs light yellow. Wing membrane hyaline, devoid of markings or slightly darkened at apex. Pterostigma inconspicuous. Male T1–T5 dark brown. Notal organ flat and covering acute postnotal organ. A6 dark brown and bearing yellowish brown anal horn on apex. A7 and A8 yellowish brown and constricted at base.

Male genital bulb broad oval. Epandrium emarginate terminally in broad U-shaped and forming pair of finger-like processes laterally. Hypovalves stripe-like and bearing row of sparsely and irregularly arranged long bristles along inner

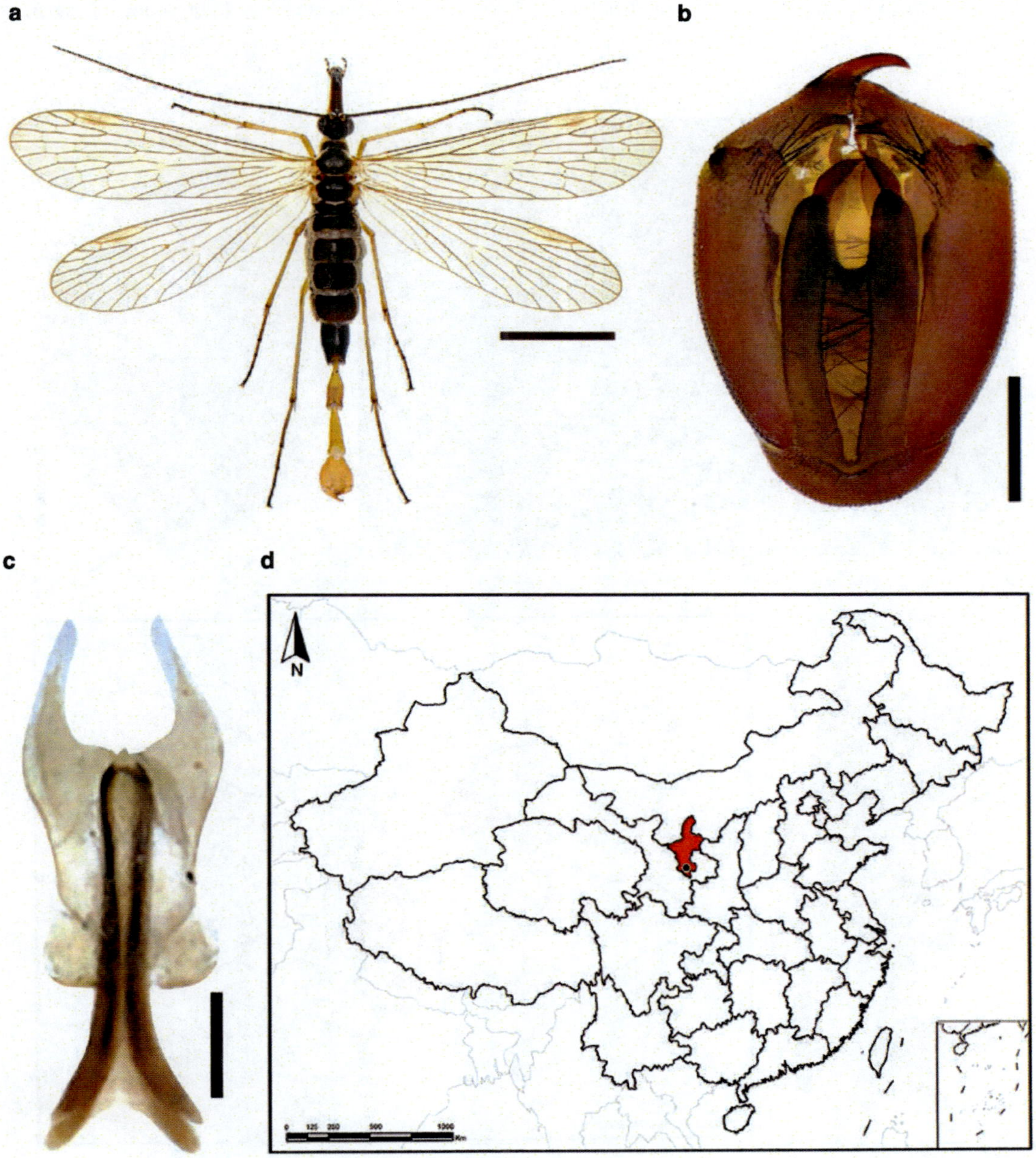

Fig. 10.78 (**a**) Male adult of ***Cerapanorpa liupanshana***, dorsal view (Liupan Mountain, Ningxia) (*Scale bar*: 0.5 mm) **[NWAU]**. (**b**) Male genital bulb of ***Cerapanorpa liupanshana***, ventral view (*Scale bar*: 0.5 mm). (**c**) Female medigynium of ***Cerapanorpa liupanshana***, ventral view (*Scale bar*: 0.2 mm). (**d**) Distribution of ***Cerapanorpa liupanshana***

margin, with apex exceeding inner processes of gonocoxites. Gonocoxites with several long setae on ventral apex. Gonostylus shorter than gonocoxites and bearing obtuse triangular median tooth and large disc-like basal process. Paramere short and not exceeding apex of gonocoxites, with basal portion straight and distal portion curving mesad, and bearing row of short spines along inner margin, spines on inner margin shorter than width of paramere. Aedeagus with ventral valves short and membranous; dorsal processes elongated and finger-like apically.

Female subgenital plate elliptical, shallowly emarginate terminally. Medigynium with main plate broad, greatly incurved on outer margin, and broadest near base of posterior arms; two pairs of basal plates; posterior arms short and acute; axis slightly shorter than two times length of main plate with diverged apodemes; apex of apodemes distinctly branched.

Emergence. June to August.

Distribution. Ningxia (Liupan Mountain).

10.2.10 *Cerapanorpa minshana* Gao, Li & Hua 2020

(Minshan Single-Horned Scorpionfly) [岷山单角蝎蛉] (Fig. 10.79)

Measurements. Forewing length 12.0–13.0 mm.

Morphology. Vertex black. Rostrum yellowish brown. Nota black. Pleura and legs light yellow. Wing membrane

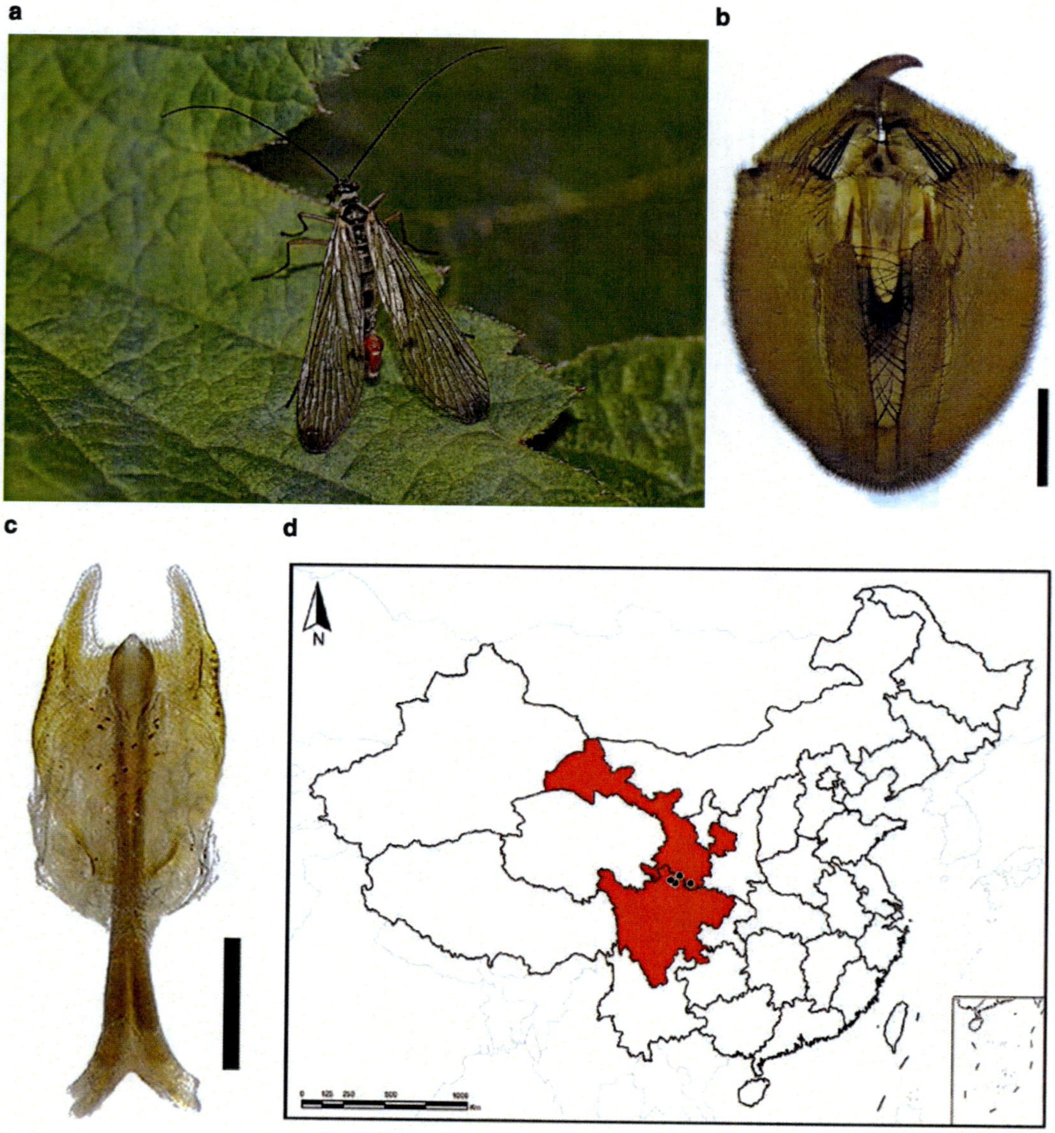

Fig. 10.79 (**a**) Male adult of ***Cerapanorpa minshana*** in the field (Jiuzhaigou, Sichuan, ©Kai Gao). (**b**) Male genital bulb of ***Cerapanorpa minshana***, ventral view (*Scale bar*: 0.5 mm). (**c**) Female medigynium of ***Cerapanorpa minshana***, ventral view (*Scale bar*: 0.2 mm). (**d**) Distribution of ***Cerapanorpa minshana***

hyaline, devoid of markings or slightly darkened at apex. Pterostigma inconspicuous. Male T1–T5 dark brown. Notal organ flat and covering acute postnotal organ. A6 dark brown and bearing yellowish brown anal horn on apex. A7 and A8 yellowish brown and constricted at base.

Male genital bulb broad oval. Epandrium emarginate terminally in broad U-shaped and forming pair of finger-like processes laterally. Hypovalves stripe-like and bearing row of sparsely and irregularly arranged long bristles along inner margin, with apex exceeding inner processes of gonocoxites. Gonocoxites with several long setae on ventral apex. Gonostylus shorter than gonocoxites and bearing obtuse triangular median tooth and large disc-like basal process. Paramere short and thin, not exceeding apex of gonocoxites, bifurcated; ventral branch bearing row of short spines along inner margin; spines longer than width of paramere; dorsal branch shorter and glabrous. Aedeagus with ventral valves short and membranous; dorsal processes elongated and finger-like apically.

Female subgenital plate elliptical, shallowly emarginate terminally. Medigynium with main plate roundly broad; two pairs of basal plates; posterior arms short and acute; axis slightly shorter than two times length of main plate with greatly diverged apodemes; apex of apodemes distinctly branched.

Emergence. June to August.

Distribution. Gansu (Wenxian), Sichuan (Minshan Mountains).

10.2.11 *Cerapanorpa nanwutaina* Chou, 1981

(Nanwutai Single-Horned Scorpionfly) [南五台蝎蛉]
(Fig. 10.80)

Measurements. Forewing length 13.5–14.0 mm.

Morphology. Vertex black. Rostrum dark brown with yellowish flanks. Nota black. Pleura and legs light yellowish

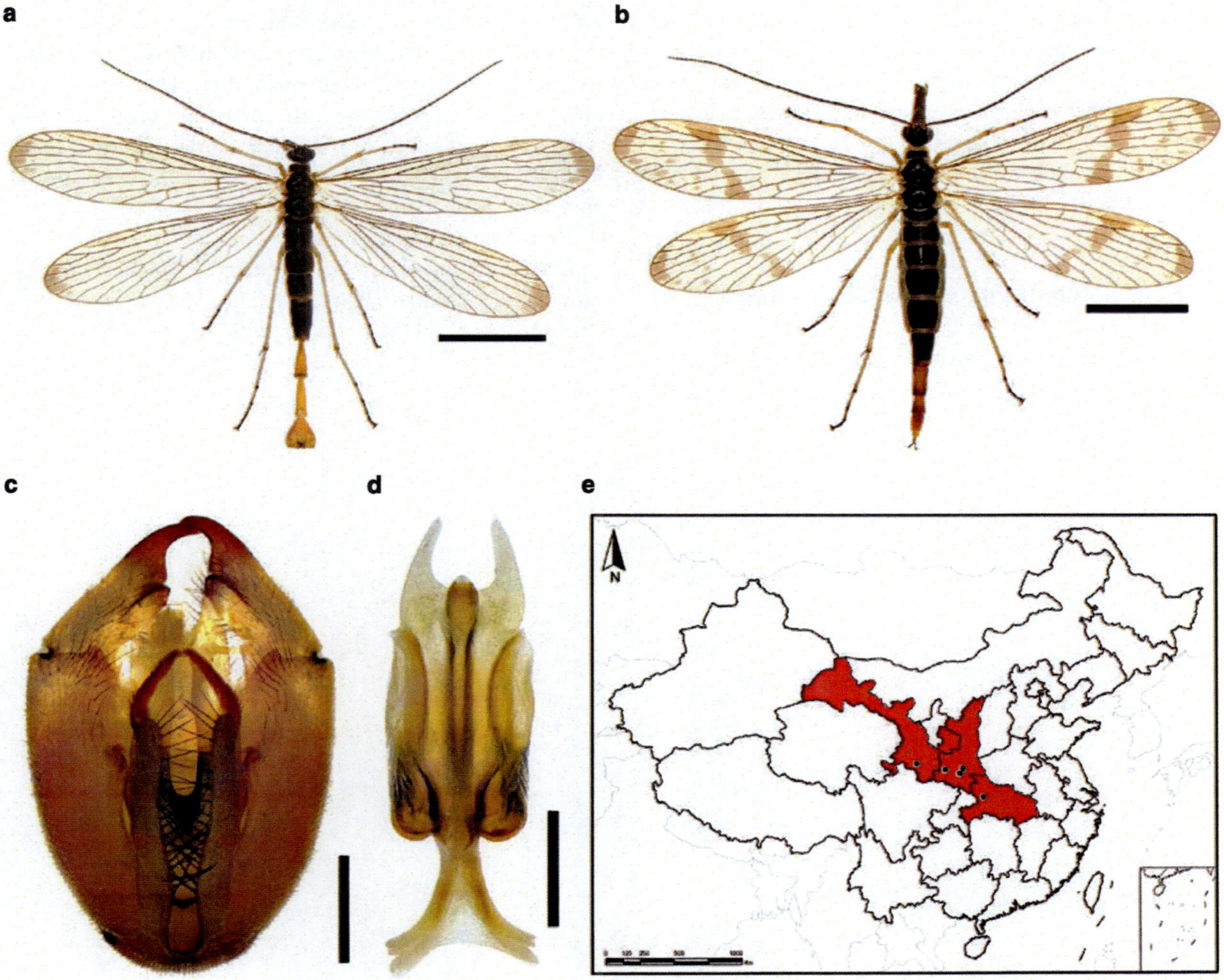

Fig. 10.80 (**a**) Male adult of ***Cerapanorpa nanwutaina***, dorsal view (Tiantaishan National Forest Park, Baoji, Shaanxi) (*Scale bar*: 5.0 mm) [NWAU]. (**b**) Female adult of ***Cerapanorpa nanwutaina***, dorsal view (Tiantaishan National Forest Park, Baoji, Shaanxi) (*Scale bar*: 5.0 mm) [NWAU]. (**c**) Male genital bulb of ***Cerapanorpa nanwutaina***, ventral view (*Scale bar*: 0.5 mm). (**d**) Female medigynium of ***Cerapanorpa nanwutaina*** (*Scale bar*: 0.2 mm). (**e**) Distribution of ***Cerapanorpa nanwutaina***

brown. Wing membrane hyaline without markings or only with apical and pterostigmal bands. Females usually with more markings than males. Male T1–T5dark brown. Notal organ flat and covering acute postnotal organ on T4. A6 dark brown and bearing yellowish brown anal horn on dorsal apex. A7 and A8 yellowish brown and constricted at base.

Male genital bulb oval. Epandrium emarginate terminally in broad U-shape and forming pair of finger-like processes laterally. Hypovalves stripe-like and bearing row of sparsely and regularly arranged long bristles along inner margin, with apex exceeding inner processes of gonocoxites. Gonocoxites with sparse long setae on ventral apex. Gonostylus shorter than gonocoxites and bearing obtuse triangular median tooth and large disc-like basal process. Paramere slender and extending to apex of gonocoxites, with basal portion straight and distal portion geniculate, and bearing row of short spines along inner margin. Aedeagus with ventral valves short and membranous; dorsal processes elongated and broaden towards apexes in lateral view.

Female subgenital plate elliptical, shallowly emarginate terminally. Medigynium with main plate oblong; two pairs of basal plates; posterior arms short and acute; axis slightly shorter than two times length of main plate with greatly diverged apodemes; apex of apodemes distinctly branched.

Emergence. June to August.

Distribution. Gansu (Dangchang), Hubei (Shennongjia), Shaanxi (Baoji, Chang'an, Ningshan).

10.2.12 *Cerapanorpa obtusa* Cheng, 1949

(Obtuse Single-Horned Scorpionfly) [太白单角蝎蛉] (Fig. 10.81)

Measurements. Forewing length 13.5–14.5 mm.

Morphology. Vertex black. Rostrum yellowish brown. Nota black. Pleura and legs light yellowish brown. Wing membrane hyaline with markings highly variable, from absent to only apical and pterostigmal bands present. Females usually with denser markings than males. Male T1–T5 dark brown. Notal organ flat and covering acute postnotal organ on T4. A6 dark brown and bearing yellowish brown anal horn on dorsal apex. A7and A8 yellowish brown and constricted at base.

Male genital bulb broad oval. Epandrium emarginate terminally in broad U-shape and forming pair of finger-like processes laterally. Hypovalves stripe-like and bearing row of sparsely and irregularly arranged long bristles along inner margin, with apexes exceeding inner processes of gonocoxites. Gonocoxites with sparse long setae on ventral apex. Gonostylus shorter than gonocoxites and bearing obtuse triangular median tooth and large disc-like basal process. Paramere extending to apex of gonocoxites, with basal portion straight and distal portion geniculate, and bearing row of long, comb-like spines along inner margin. Aedeagus with ventral valves short and membranous; dorsal processes long and finger-like apically.

Female subgenital plate elliptical, shallowly emarginate terminally. Medigynium with main plate broad, lateral margins greatly incurved; two pairs of basal plates; posterior arms short and acute; axis slightly shorter than two times length of main plate with greatly diverged apodemes; apex of apodemes distinctly branched.

Emergence. June to August.

Distribution. Gansu (Dangchang), Ningxia (Liupan Mountain), Shaanxi (Baoji, Meixian, Ningshan, Taibai, Zhouzhi).

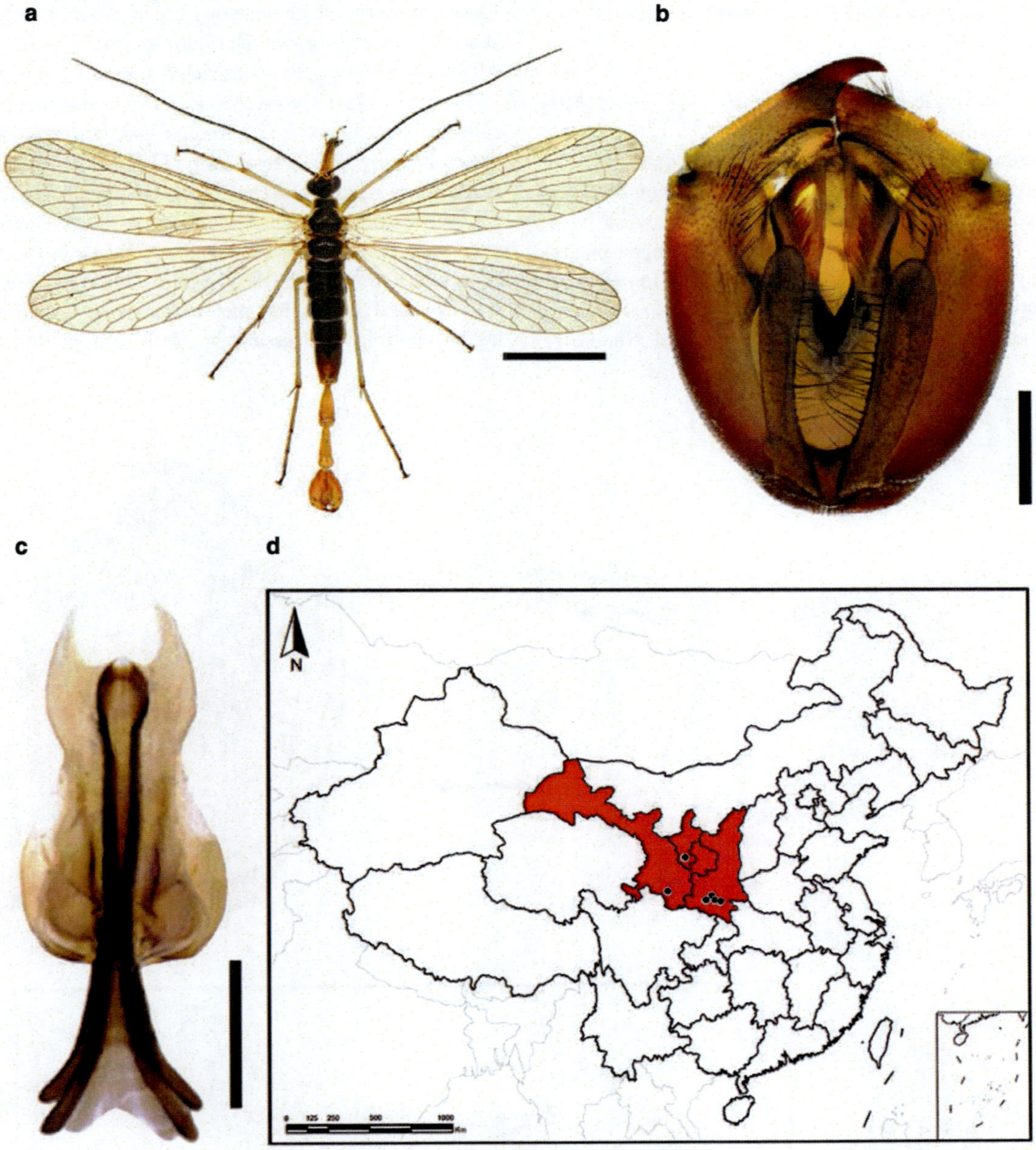

Fig. 10.81 (**a**) Male adult of ***Cerapanorpa obtusa***, lateral view (Tiantaishan National Forest Park, Baoji, Shaanxi) (*Scale bar*: 5.0 mm) [NWAU]. (**b**) Male genital bulb of ***Cerapanorpa obtusa***, ventral view (*Scale bar*: 0.5 mm). (**c**) Female medigynium of ***Cerapanorpa obtusa***, ventral view (*Scale bar*: 0.2 mm). (**d**) Distribution of ***Cerapanorpa obtusa***

10.2.13 *Cerapanorpa protrudens* Gao, Ma & Hua, 2016

(Protruded Single-Horned Scorpionfly) [巨突单角蝎蛉] (Fig. 10.82)

Measurements. Forewing length 12.5–14.2 mm.

Morphology. Vertex dark brown. Rostrum yellowish brown. Nota black. Pleura and legs light yellowish brown. Wing membrane hyaline. Pterostigma inconspicuous. Markings highly variable, from totally absent to only apical and pterostigmal bands present. Females usually with denser markings than males. Male T1–T5 dark brown. Notal organ flat and covering acute postnotal organ. A6 dark brown and bearing yellowish brown anal horn on dorsal apex. A7 and A8 yellowish brown and constricted at base.

Male genital bulb broad oval. Epandrium emarginate terminally in broad U-shape and forming pair of finger-like processes laterally. Hypovalves stripe-like and bearing row of sparsely and regularly arranged long bristles along inner margin, with apexes approximate apex of gonocoxites. Gonocoxites with sparse long setae on ventral apex. Gonostylus shorter than gonocoxites, bearing obtuse triangular median tooth and large disc-like basal process. Paramere extremely elongated, extending beyond middle of gonostylus, with basal portion slightly

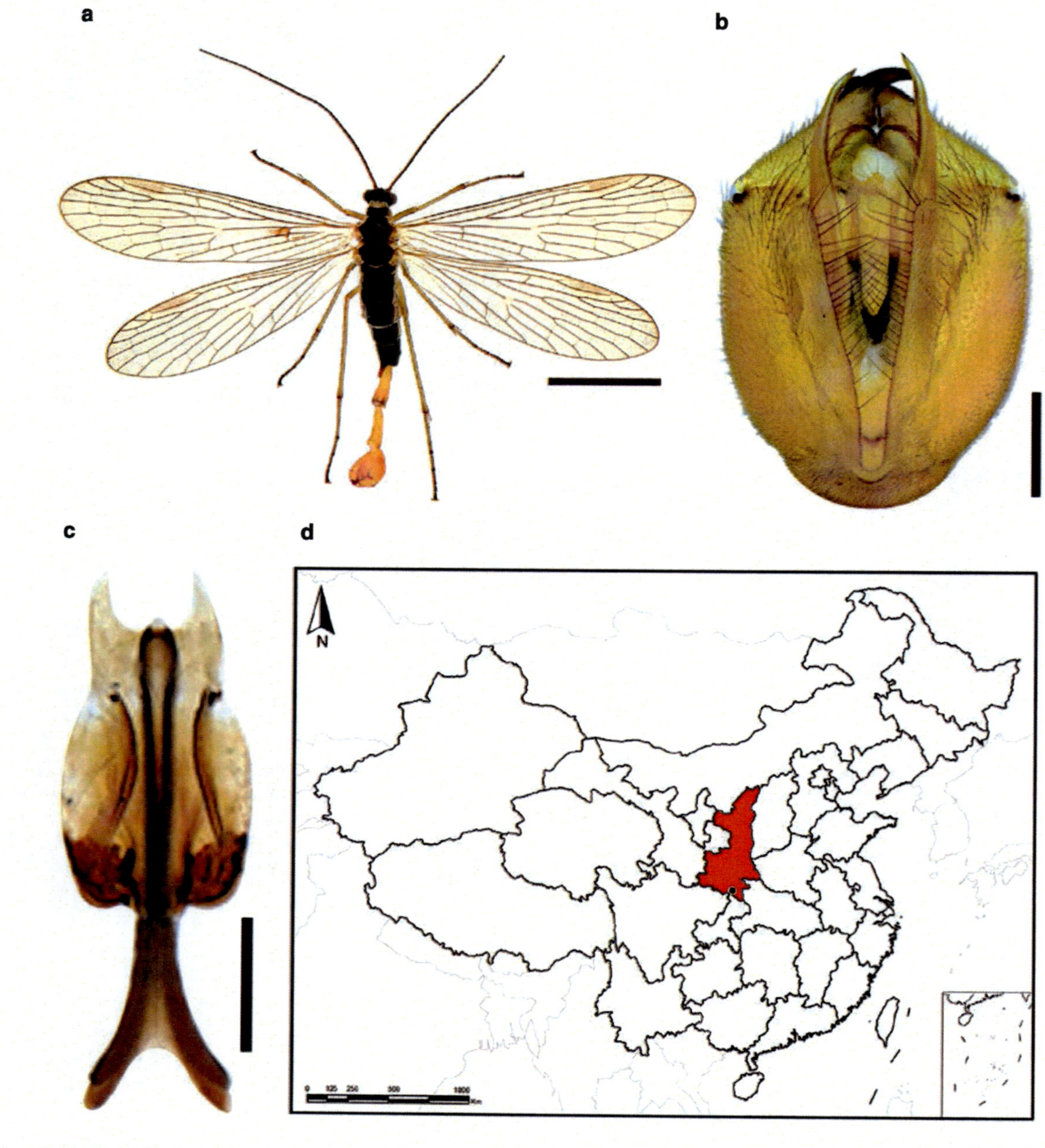

Fig. 10.82 (**a**) Male adult of ***Cerapanorpa protrudens***, dorsal view (Daba Mountains, Ankang, Shaanxi) (*Scale bar*: 5.0 mm) **[NWAU]**. (**b**) Male genital bulb of ***Cerapanorpa protrudens***, ventral view (*Scale bar*: 0.5 mm). (**c**) Female medigynium of ***Cerapanorpa protrudens***, ventral view (*Scale bar*: 0.2 mm). (**d**) Distribution of ***Cerapanorpa protrudens***

geniculate and distal portion curved mesad, and bearing row of short, comb-like spines along inner margin. Aedeagus with ventral valves short and membranous; dorsal processes long and finger-like apically, slightly curved mesad.

Female subgenital plate elliptical, shallowly emarginate terminally. Medigynium with main plate broad, lateral margins slightly incurved; two pairs of basal plates; posterior arms short and acute; axis slightly shorter than two times length of main plate with greatly diverged apodemes; apex of apodemes distinctly branched.

Emergence. July.

Distribution. Shaanxi (Ankang).

10.2.14 *Cerapanorpa qinlingensis* Gao, Li & Hua, 2020

(Qinling Single-Horned Scorpionfly) [秦岭单角蝎蛉] (Fig. 10.83)

Measurements. Forewing length 13.5–14.5 mm.

Morphology. Vertex dark brown. Rostrum yellowish brown. Nota black. Pleura and legs light yellowish brown. Wing membrane hyaline. Pterostigma inconspicuous. Markings highly variable, from totally absent to only apical and pterostigmal bands present. Females usually with denser markings than males. Male T1–T5 dark brown. Notal organ

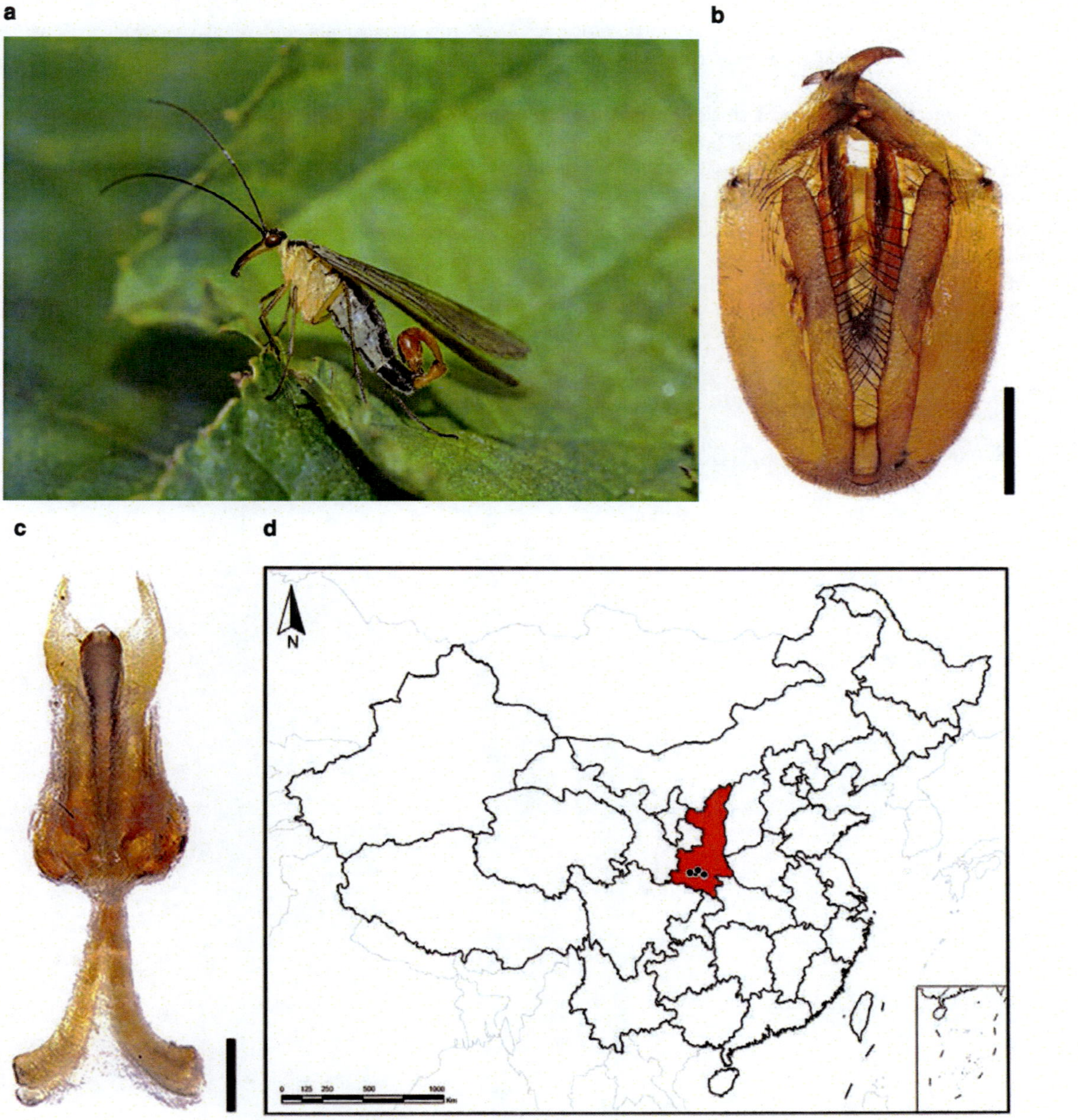

Fig. 10.83 (**a**) Male adult of ***Cerapanorpa qinlingensis*** in the field (Houzhenzi, Zhouzhi, Shaanxi, ©Kai Gao). (**b**) Male genital bulb of ***Cerapanorpa qinlingensis***, ventral view (*Scale bar*: 0.5 mm). (**c**) Female medigynium of ***Cerapanorpa qinlingensis***, ventral view (*Scale bar*: 0.2 mm). (**d**) Distribution of ***Cerapanorpa qinlingensis***

flat and covering acute postnotal organ. A6 dark brown and bearing yellowish brown anal horn on dorsal apex. A7 and A8 yellowish brown and constricted at base.

Male genital bulb broad oval. Epandrium emarginate terminally in broad U-shape and forming pair of finger-like processes laterally. Hypovalves stripe-like and bearing row of sparsely and regularly arranged long bristles along inner margin, with apexes approximate apex of gonocoxites. Gonocoxites with sparse long setae on ventral apex. Gonostylus shorter than gonocoxites, bearing obtuse triangular median tooth and extremely large disc-like basal process. Paramere extending to basal process of gonostylus, bearing row of very short, comb-like spines along inner margin, with apex greatly incurved and hook-like. Aedeagus with ventral valves short and membranous; dorsal processes long and finger-like apically, slightly curved mesad, and greatly enlarged apically.

Female subgenital plate elliptical, shallowly emarginate terminally. Medigynium with main plate broad, lateral margins greatly incurved; two pairs of basal plates; posterior arms short and acute; axis two times as long as main plate with greatly diverged apodemes; apex of apodemes distinctly branched.

Emergence. July.

Distribution. Shaanxi (Foping, Ningshan, Yangxian, Zhouzhi).

10.2.15 *Cerapanorpa reni* (Chou, 1981)

(Ren's Single-Horned Scorpionfly) [任氏单角蝎蛉]
(Fig. 10.84)
Measurements. Forewing length 15.0–16.0 mm.

Morphology. Large Cerapanorpa species. Vertex black. Rostrum yellowish brown or reddish brown. Nota black. Pleura and coxae dark brown. Legs mostly red-dish brown. Wing membrane hyaline. Pterostigma red-dish brown. Markings dark brown. Apical band broad, and connected to small spot near hind portion; pterostigmal band with complete basal and apical branches; marginal spot large; basal band broad; basal spot very small. Male T1–T5 dark brown. Notal organ flat and covering acute postnotal organ. A6 dark brown and bearing reddish-brown anal horn on dorsal apex. A7 and A8 reddish brown and constricted at base.

Male genital bulb broad oval. Epandrium emarginate terminally in broad U-shape and forming pair of finger-like processes laterally. Hypovalves slender stripe-like, and bearing row of sparsely and regularly arranged long bristles along inner margin, with apexes exceeding to inner processes of gonocoxites. Gonocoxites with sparse long setae on ventral apex. Gonostylus shorter than gonocoxites, bearing obtuse triangular median tooth and large disc-like basal process. Paramere L-shaped, with distal portion straight, and bearing row of short spines along inner margin. Aedeagus with ventral valves short and membranous; dorsal processes long and finger-like apically.

Female subgenital plate elliptical, shallowly emarginate terminally. Medigynium with main plate broad, lateral margins slightly incurved; two pairs of basal plates; posterior arms long and stout, subacute terminally; axis two times as long as main plate with greatly diverged apodemes; apex of apodemes distinctly branched.

Emergence. June to July.

Distribution. Shaanxi (Huyi, Ningshan, Taibai, Zhouzhi).

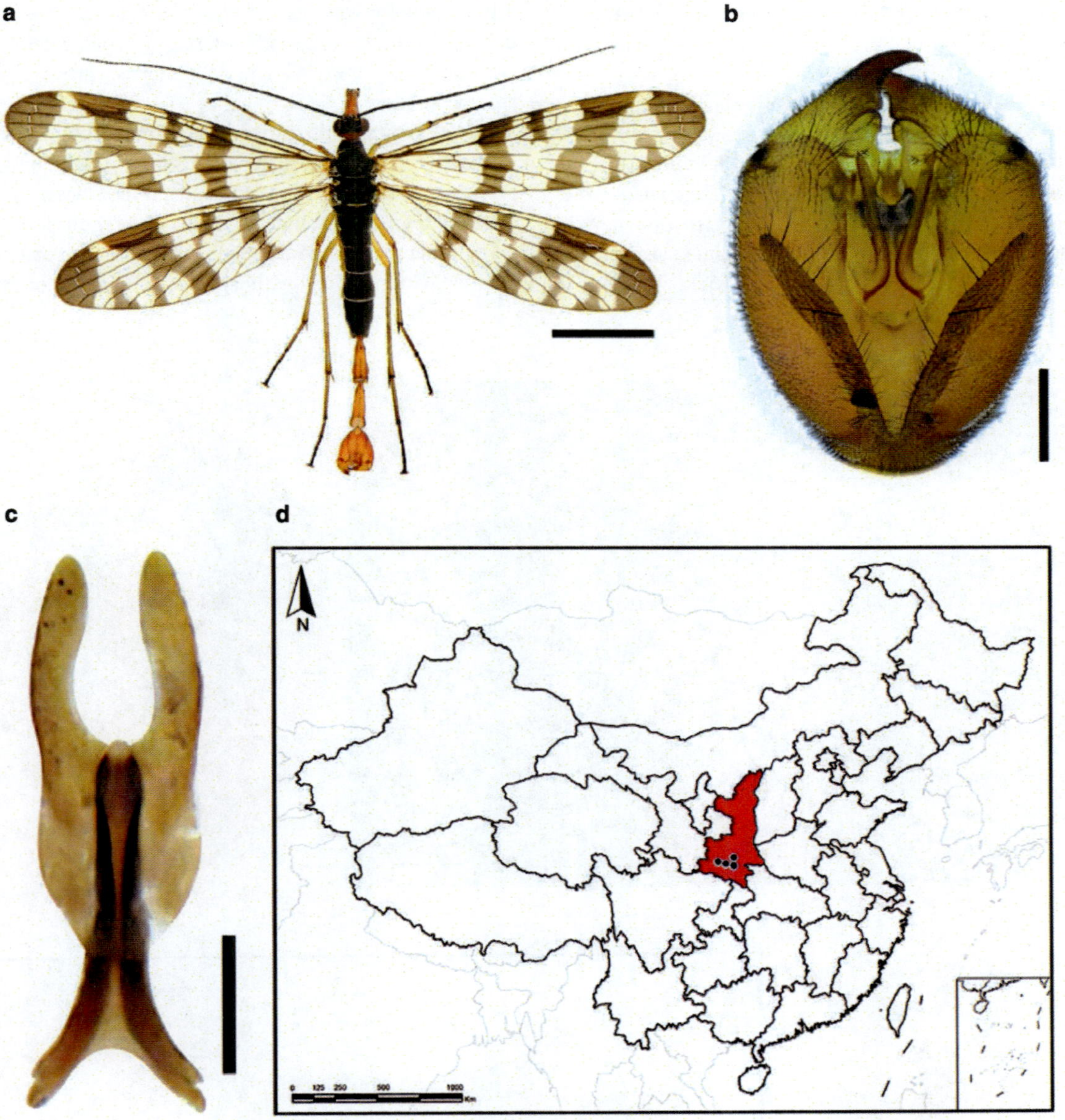

Fig. 10.84 (**a**) Male adult of ***Cerapanorpa reni***, dorsal view (Qinling Mountains, Shaanxi) (*Scale bar*: 5.0 mm) [NWAU]. (**b**) Male genital bulb of ***Cerapanorpa reni***, ventral view (*Scale bar*: 0.5 mm). (**c**) Female medigynium of ***Cerapanorpa reni***, ventral view (*Scale bar*: 0.2 mm). (**d**) Distribution of ***Cerapanorpa reni***

10.2.16 *Cerapanorpa sinuata* Gao, Ma & Hua, 2016

(Sinuated Single-Horned Scorpionfly) [弯曲单角蝎蛉] (Fig. 10.85)

Measurements. Forewing length 12.0–13.0 mm.

Morphology. Vertex black. Rostrum yellowish brown. Nota black. Pleura and legs light yellowish brown. Wing membrane hyaline and devoid of markings, except apex darkened. Pterostigma inconspicuous. Male T1–T5 dark brown. Notal organ flat and covering acute postnotal organ on T4. A6 dark brown and bearing yellowish brown anal horn on dorsal apex. A7 and A8 yellowish brown and constricted at base.

Male genital bulb broad oval. Epandrium emarginate terminally in broad U-shape and forming pair of finger-like processes laterally. Hypovalves stripe-like, bearing row of sparsely regularly arranged long bristles regularly arranged along inner margin, with apexes extending to apex of gonocoxites. Gonocoxites with sparse long setae on ventral

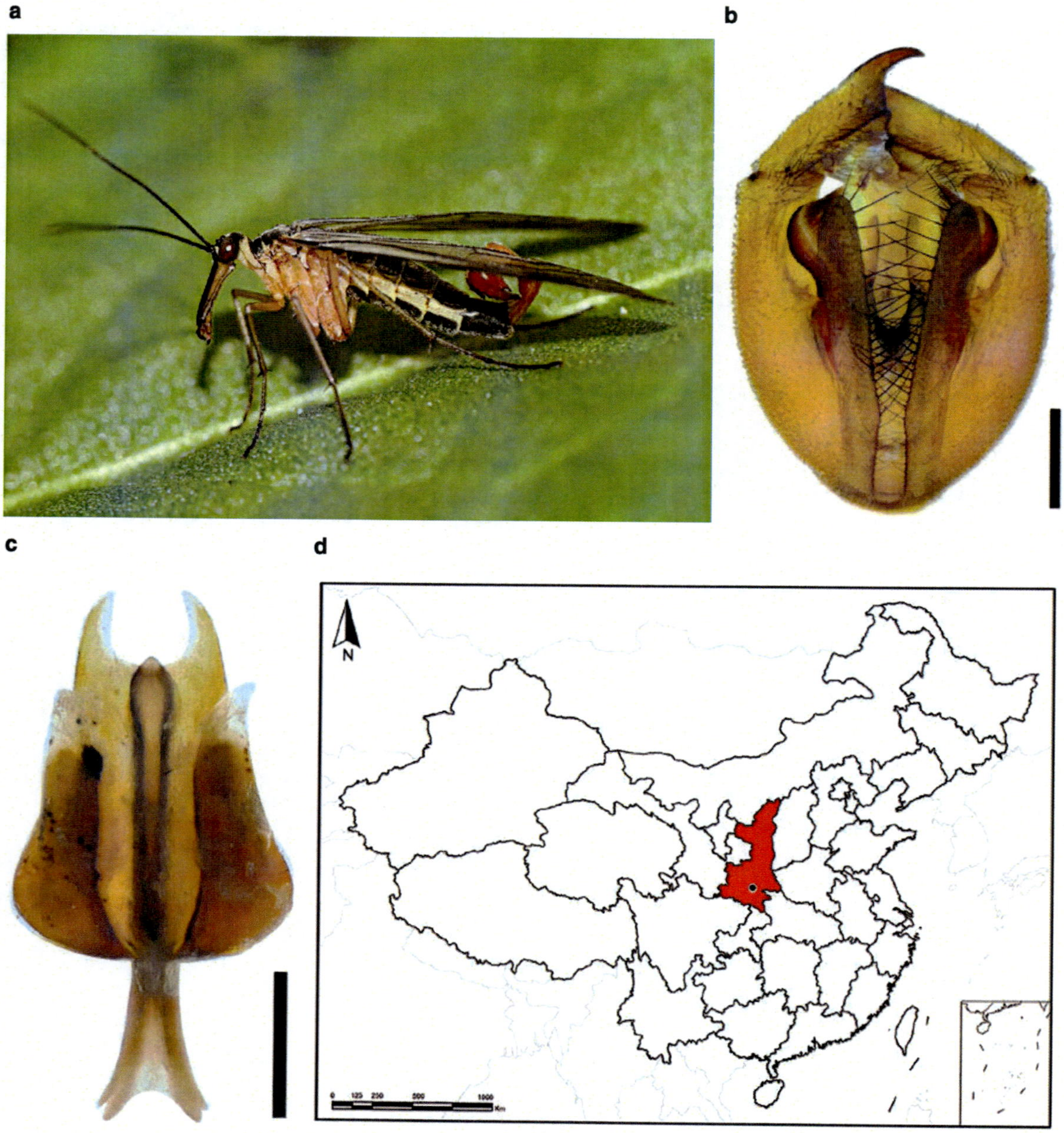

Fig. 10.85 (a) Male adult of ***Cerapanorpa sinuata*** in the field (Pingheliang, Ningshan, Shaanxi, ©Kai Gao). (b) Male genital bulb of ***Cerapanorpa sinuata***, ventral view (*Scale bar*: 0.5 mm). (c) Female medigynium of ***Cerapanorpa sinuata***, ventral view (*Scale bar*: 0.2 mm). (d) Distribution of ***Cerapanorpa sinuata***

apex. Gonostylus shorter than gonocoxites, bearing obtuse triangular median tooth and large disc-like basal process. Paramere greatly sinuate, and bearing dense comb-like spines along inner margin in distal half. Aedeagus with ventral valves short and membranous; dorsal processes long and truncated apically and bearing swollen membranous process on dorsal margin subapically.

Female subgenital plate elliptical, shallowly emarginate terminally. Medigynium with main plate broad, lateral margins greatly incurved; two pairs of very broad basal plates; posterior arms short and acute; axis two times as long as main plate with greatly diverged apodemes; apex of apodemes distinctly branched.

Emergence. June to July.

Distribution. Shaanxi (Ningshan).

10.2.17 *Cerapanorpa taizishana* Gao & Hua, 2019

(Taizishan Single-Horned Scorpionfly) [太子山单角蝎 蛉] (Fig. 10.86)

Measurements. Forewing length 11.5–12.5 mm.

Morphology. Vertex dark brown. Rostrum yellowish brown. Nota black. Pleura and legs light yellowish brown. Wing membrane hyaline. Pterostigma inconspicuous. Markings highly variable, from absent to only apical and pterostigmal bands present. Females usually with denser markings than males. Male T1–T5 dark brown. Notal organ flat and covering acute postnotal organ. A6 dark brown and bearing yellowish brown anal horn on dorsal apex. A7 and A8 yellowish brown and constricted at base.

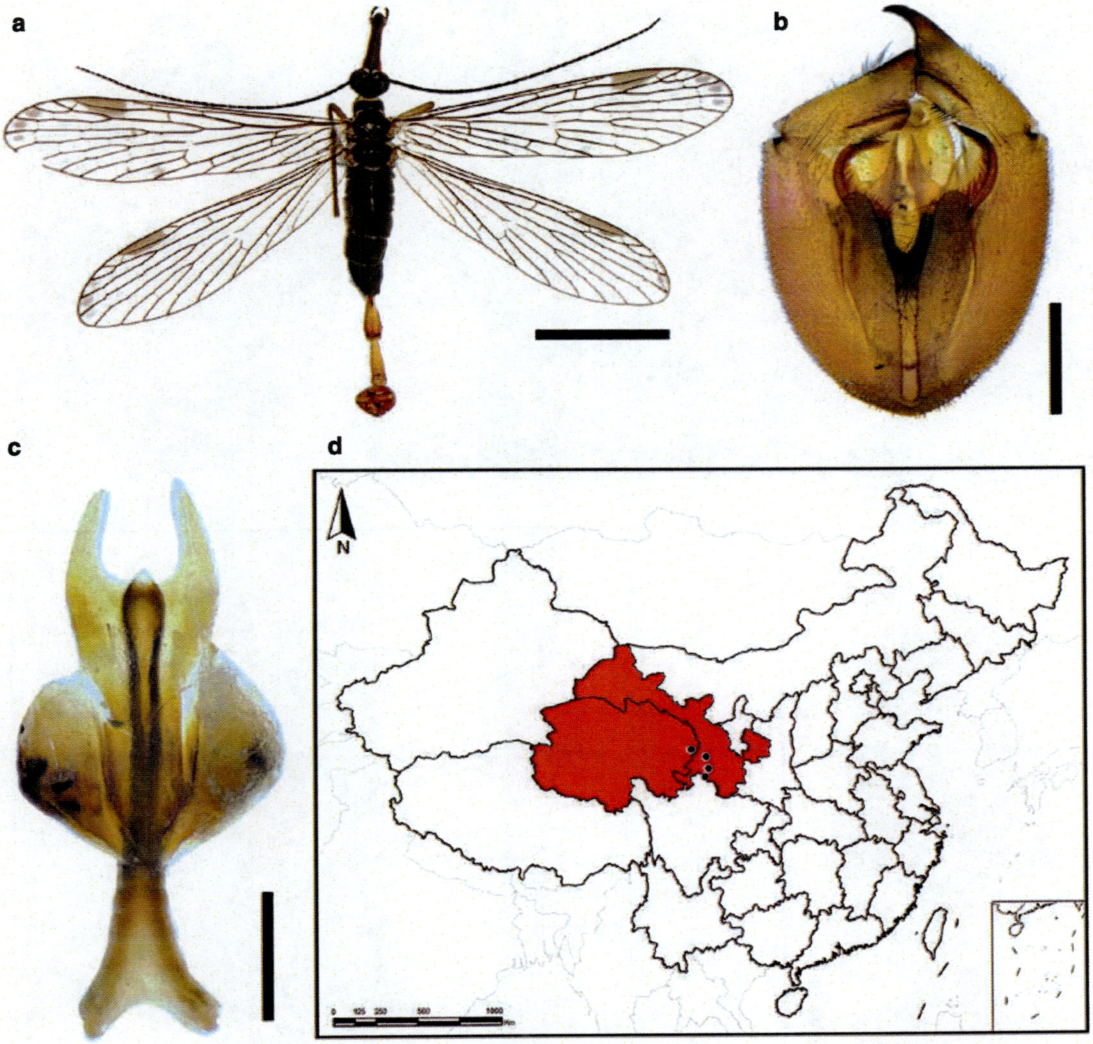

Fig. 10.86 (**a**) Male adult of ***Cerapanorpa taizishana***, dorsal view (Taizi Mountain, Gansu) (*Scale bar*: 5.0 mm) [**NWAU**]. (**b**) Male genital bulb of ***Cerapanorpa taizishana***, ventral view (*Scale bar*: 0.5 mm). (**c**) Female medigynium of ***Cerapanorpa taizishana***, ventral view (*Scale bar*: 0.2 mm). (**d**) Distribution of ***Cerapanorpa taizishana***

Male genital bulb broad oval. Epandrium emarginate terminally in broad U-shape and forming pair of finger-like processes laterally. Hypovalves stripe-like and bearing row of sparsely and regularly arranged long bristles along inner margin, with apexes approximate apex of gonocoxites. Gonocoxites with sparse long setae on ventral apex. Gonostylus shorter than gonocoxites, bearing obtuse triangular median tooth and extremely large disc-like basal process. Paramere slender, greatly incurved and hook-shape. Aedeagus with ventral valves short and membranous; dorsal processes long and finger-like apically, slightly curved mesad, and greatly enlarged apically.

Female subgenital plate elliptical, shallowly emarginate terminally. Medigynium with main plate broad, lateral margins greatly incurved; two pairs of very broad basal plates; posterior arms short and acute; axis slightly shorter than two times length of main plate with greatly diverged apodemes; apex of apodemes distinctly branched.

Emergence. July.

Distribution. Gansu (Hezheng, Lintan, Têwo), Qinghai (Xunhua).

10.2.18 *Cerapanorpa wangwushana* Huang, Hua & Shen, 2004

(Wangwushan Single-Horned Scorpionfly) [王屋山单角 蝎蛉] (Fig. 10.87)

Measurements. Forewing length 13.0–14.0 mm.

Morphology. Vertex black. Rostrum yellowish brown. Nota dark brown. Pleura and legs light yellowish brown. Wing

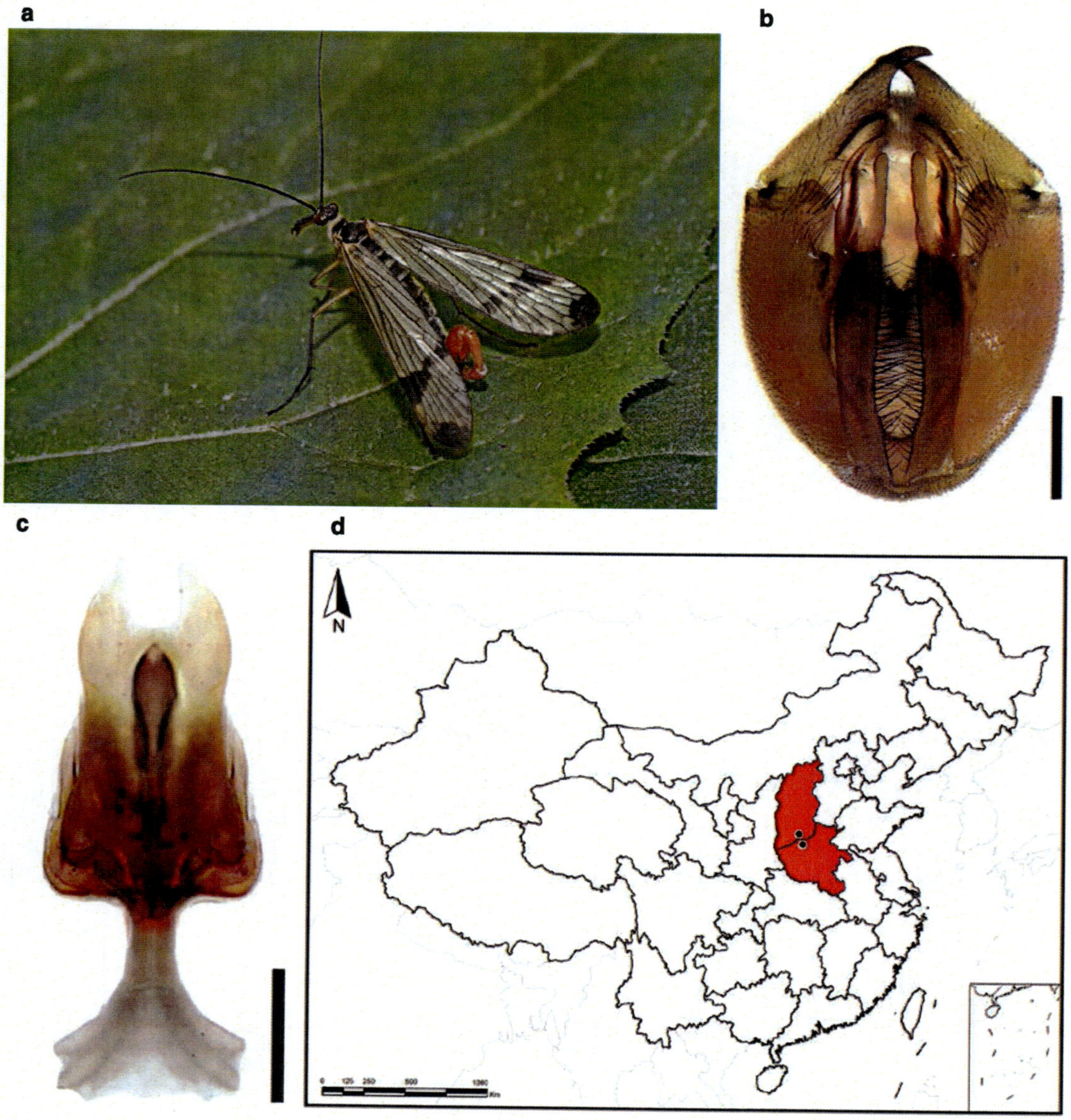

Fig. 10.87 (a) Male adult of ***Cerapanorpa wangwushana*** in the field (Lishan Mountain, Shanxi, ©Kai Gao). (b) Male genital bulb of ***Cerapanorpa wangwushana***, ventral view (*Scale bar*: 0.5 mm). (c) Female medigynium of ***Cerapanorpa wangwushana***, ventral view (*Scale bar*: 0.2 mm). (d) Distribution of ***Cerapanorpa wangwushana***

membrane hyaline with distinct apical and pterostigmal bands. Pterostigmal band only represented by basal branch. Pterostigma light brown and inconspicuous. Male T1–T5 dark brown. Notal organ flat and covering acute postnotal organ. A6 dark brown and bearing yellowish brown anal horn on dorsal apex. A7 and A8 yellowish brown and constricted at base.

Male genital bulb broad oval. Epandrium emarginate terminally in broad U-shape and forming pair of finger-like processes laterally. Hypovalves stripe-like, bearing row of sparsely and regularly arranged long bristles along inner margin, with apexes extending to apex of gonocoxites. Gonocoxites with sparse long setae on ventral apex. Gonostylus shorter than gonocoxites, bearing obtuse triangular median tooth and large disc-like basal process. Paramere slender and wave-shaped, and bearing row of comb-like spines along inner margin. Aedeagus with ventral valves short and membranous; dorsal processes long and finger-like apically.

Female subgenital plate elliptical, shallowly emarginate terminally. Medigynium with main plate broad, lateral margins greatly incurved; two pairs of very broad basal plates; posterior arms short and acute; axis slightly shorter than two times length of main plate with greatly diverged apodemes; apex of apodemes greatly branched.

Emergence. June to July.

Distribution. Henan (Wangwu Mountain), Shanxi (Lishan Mountain).

10.2.19 *Cerapanorpa xuebaodinga* Gao & Hua, 2019

(Xuebaoding Single-Horned Scorpionfly) [雪宝顶单角蝎 蛉] (Fig. 10.88)

Measurements. Forewing length 13.5–14.5 mm.

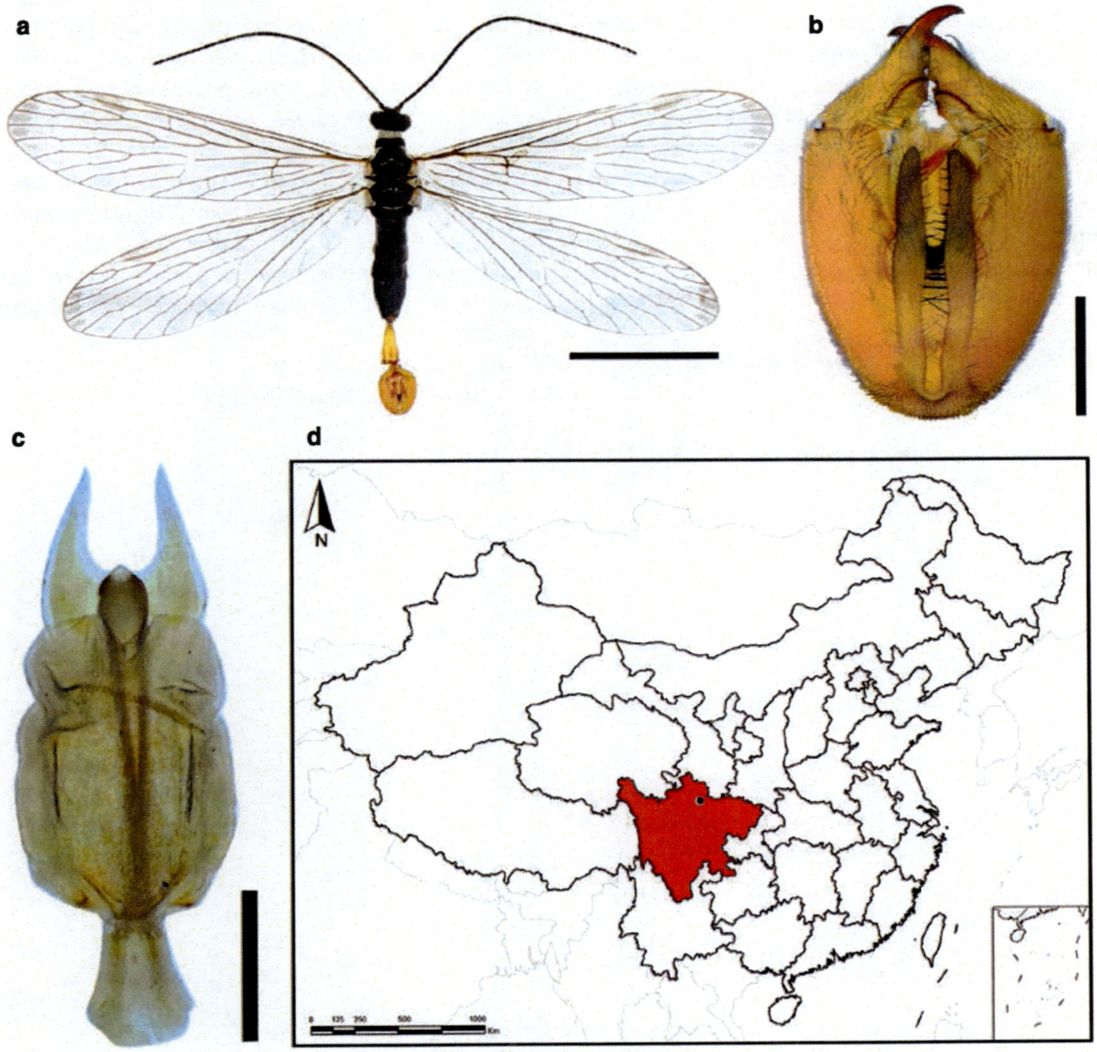

Fig. 10.88 (a) Male adult of ***Cerapanorpa xuebaodinga***, dorsal view (Xuebaoding, Songpan, Sichuan, ©Kai Gao) (*Scale bar*: 5.0 mm) [NWAU]. (b) Male genital bulb of ***Cerapanorpa xuebaodinga***, ventral view (*Scale bar*: 0.5 mm). (c) Female medigynium of ***Cerapanorpa xuebaodinga***, ventral view (*Scale bar*: 0.2 mm). (d) Distribution of ***Cerapanorpa xuebaodinga***

Morphology. Vertex dark brown. Rostrum yellowish brown. Nota black. Pleura and legs light yellowish brown. Wing membrane hyaline. Pterostigma inconspicuous. Markings highly variable, from absent to only apical and pterostigmal bands present. Females usually with denser markings than males. Male T1–T5 dark brown. Notal organ flat and covering acute postnotal organ. A6 dark brown and bearing yellowish brown anal horn on dorsal apex. A7 and A8 yellowish brown and constricted at base.

Male genital bulb broad oval. Epandrium emarginate terminally in broad U-shape and forming pair of finger-like processes laterally. Hypovalves stripe-like and bearing row of sparsely and regularly arranged long bristles along inner margin, with apexes approximate apex of gonocoxites. Gonocoxites with sparse long setae on ventral apex. Gonostylus shorter than gonocoxites, bearing obtuse triangular median tooth and extremely large disc-like basal process. Paramere slender, bearing row of very short, comb-like spines along inner margin, with apical portion greatly bending inward. Aedeagus with ventral valves short and membranous; dorsal processes greatly enlarged apically.

Female subgenital plate elliptical, shallowly emarginate terminally. Medigynium with main plate broad; two pairs of basal plates; posterior arms short and acute; axis slightly longer than main plate with slightly diverged apodemes; apex of apodemes distinctly branched.

Emergence. July.

Distribution. Sichuan (Songpan).

10.2.20 *Cerapanorpa yanggashana* Gao & Hua, 2019

(Yanggashan Single-Horned Scorpionfly) [阳尕山单角蝎 蛉] (Fig. 10.89)

Measurements. Forewing length 13.5–14.5 mm.

Morphology. Vertex dark brown. Rostrum yellowish brown. Nota black. Pleura and legs light yellowish brown. Wing membrane hyaline. Pterostigma inconspicuous. Markings highly variable, from absent to only apical and pterostigmal bands present. Females usually with denser markings than males. Male T1–T5 dark brown. Notal organ flat and covering acute postnotal organ. A6 dark brown and bearing yellowish brown anal horn on dorsal apex. A7 and A8 yellowish brown and constricted at base.

Male genital bulb broad oval. Epandrium emarginate terminally in broad U-shape and forming pair of finger-like processes laterally. Hypovalves stripe-like and bearing row of sparsely and regularly arranged long bristles along inner margin, with apexes approximate apex of gonocoxites. Gonocoxites with sparse long setae on ventral apex. Gonostylus shorter than gonocoxites, bearing obtuse triangular median tooth and extremely large disc-like basal process. Paramere short, bearing row of very short, comb-like spines along inner margin. Aedeagus with ventral valves short and membranous; dorsal processes long and finger-like apically, slightly curved mesad, and greatly enlarged apically.

Female subgenital plate elliptical, shallowly emarginate terminally. Medigynium with main plate narrow, lateral margins greatly incurved; two pairs of basal plates; posterior arms short and acute; axis two slightly longer than main plate with slightly diverged apodemes; apex of apodemes distinctly branched.

Emergence. June.

Distribution. Gansu (Wenxian).

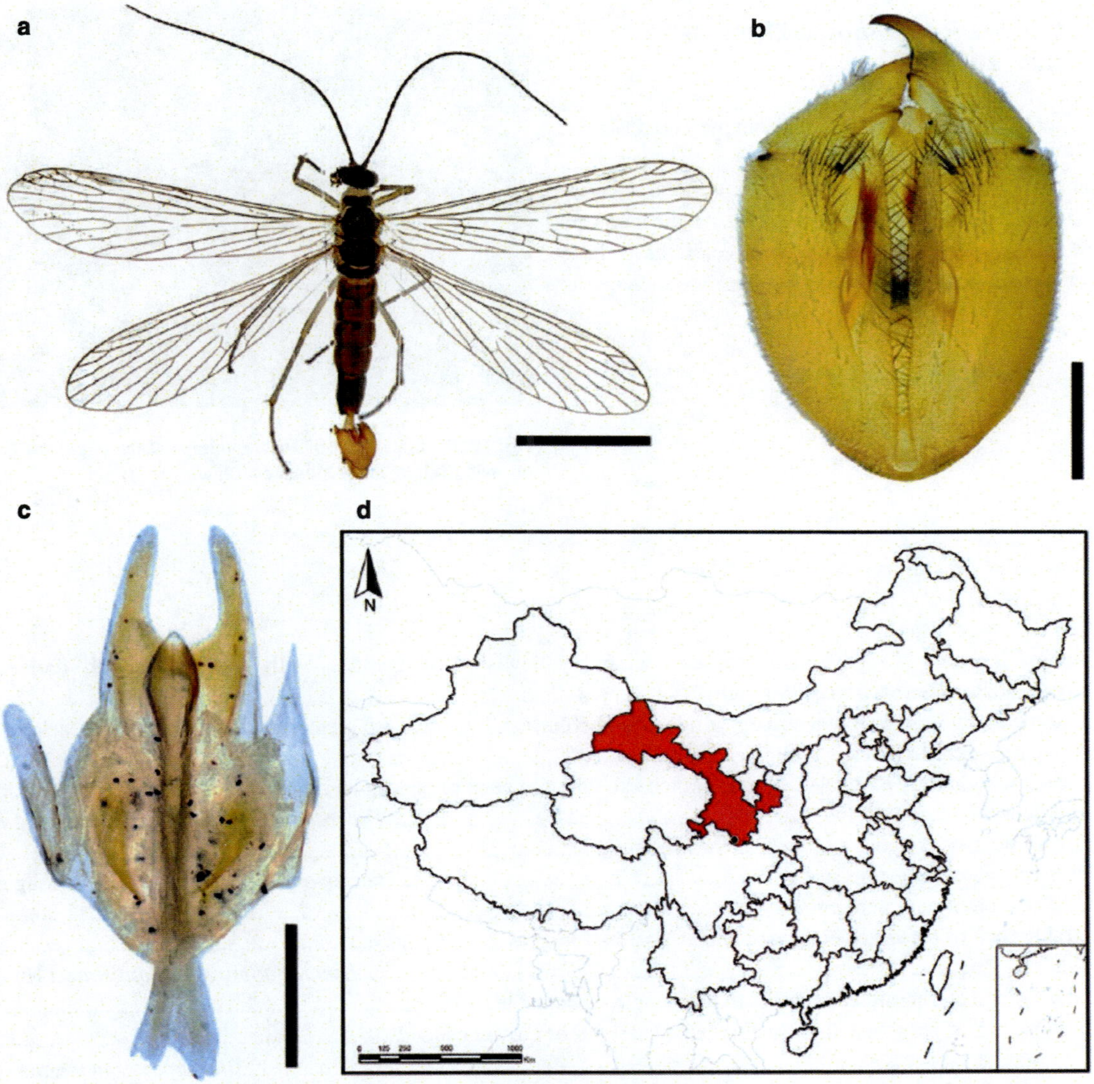

Fig. 10.89 (**a**) Male adult of ***Cerapanorpa yanggashana***, dorsal view (Yanggashan, Wenxian, Ganus, ©Kai Gao) (*Scale bar*: 5.0 mm) [NWAU]. (**b**) Male genital bulb of ***Cerapanorpa yanggashana***, ventral view (*Scale bar*: 0.5 mm). (**c**) Female medigynium of ***Cerapanorpa yanggashana***, ventral view (*Scale bar*: 0.2 mm). (**d**) Distribution of ***Cerapanorpa yanggashana***

10.3 Genus *Dicerapanorpa* Zhong & Hua, 2013

(Two-horned Scorpionflies, Dicerapanorpas) [双角蝎蛉属] (Fig. 10.90)
The main diagnostic character of *Dicerapanorpa* is a pair of finger-like anal horns on the apex of male T6. All the members are endemic to China. Twenty-four species have been described (Wang 2022), and twenty of them are included in this atlas.

Fig. 10.90 Male adult of ***Dicerapanorpa***. ***Dicerapanorpa tjederi*** in the field (Yulong Snow Mountain, Lijiang)

Key to Species of Dicerapanorpa in This Atlas
(20 species, males)

1. Rostrum lacking black frontal stripes; wings yellowish, subhyaline, with well-developed dark brown markings (***Dicerapanorpa magna*** group) 2
– Rostrum with two black frontal stripes; wings mostly colorless, hyaline, with greatly reduced markings or lacking markings (***Dicerapanorpa diceras*** group) 6
2. Paramere with greatly elongated ventral branch extending to apex of aedeagus; middle branch elongated and extending to middle tooth of gonostylus .... ***Dicerapanorpa. baiyunshana*** Zhong & Hua
– Paramere with ventral and middle branches shorter 3
3. Middle branch of paramere straight .... ***Dicerapanorpa shennongensis*** Zhong & Hua
– Middle branch of paramere curved 4
4. Ventral branch of paramere elongated and at least half long as dorsal branch .... ***Dicerapanorpa hualongshana*** Hu & Hua
– Ventral branch of paramere shorter than half length of dorsal branch 5
5. Ventral branch of paramere slightly elongated, straight, or curved caudad .... ***Dicerapanorpa magna*** (Chou)
– Ventral branch of paramere very short and hook-like .... ***Dicerapanorpa minshana*** Hu & Hua
6. Head with black spots on vertex 7
– Head lacking black spots on vertex 11

7. Terga mostly black with indistinct yellow middle stripe; genital bulb black or dark brown .... ***Dicerapanorpa triclada*** (Qian & Zhou)
– Terga black laterally with distinct yellow middle stripe; genital bulb yellowish brown 8
8. Middle branch of paramere bifurcated at distal 1/3 .... ***Dicerapanorpa tjederi*** (Carpenter)
– Middle branch of paramere simple 9
9. Ventral branch of paramere greatly elongated, extending beyond basal process of gonostylus .... ***Dicerapanorpa macula*** Hu, Wang & Hua
– Ventral branch of paramere shorter and not exceeding basal process of gonostylus 10
10. Abdomen with yellow middle stripe wider than half width of T1–T5 .... ***Dicerapanorpa deqenensis*** Hu, Wang & Hua
– Abdomen with yellow middle stripe narrower than half width of T1–T5 .... ***Dicerapanorpa zhongdianensis*** Hu, Wang & Hua
11. Paramere bifurcated .... ***Dicerapanorpa bifurcata*** Hu & Hua
– Paramere trifurcated 12
12. Hypovalve with apical portion nearly as wide as base 13
– Hypovalve greatly expanded apically 15
13. Hypovalve greatly elongated, extending far beyond basal process of gonostylus .... ***Dicerapanorpa zhengkuni*** Hu & Hua
– Hypovalve shorter, not exceeding basal process of gonostylus 14
14. Paramere with middle branch greatly elongated, extending far beyond middle tooth of gonostylus .... ***Dicerapanorpa yijunae*** Hu & Hua
– Paramere with middle branch shorter, not exceeding middle tooth of gonostylus .... ***Dicerapanorpa kimminsi*** (Carpenter)
15. Hypovalves truncated apically .... ***Dicerapanorpa stotzneri*** (Esben-Petersen)
– Hypovalves rounded apically 16
16. Genital bulb broad spherical; paramere with ventral branch shorter than half length of middle branch .... ***Dicerapanorpa tanae*** Hu, Wang & Hua
– Genital bulb long elliptical; paramere with ventral branch at least half long as middle branch 17
17. Paramere with ventral branch half long as middle branch .... ***Dicerapanorpa tenuis*** Hu, Wang & Hua
– Paramere with ventral branch longer than half length of middle branch 18
18. Paramere with three branches closely aligned basally .... ***Dicerapanorpa diceras*** (MacLachlan)
– Paramere with three branches widely separated basally 19
19. Paramere with ventral branch greatly elongated, only slightly shorter than middle branch, and curved laterad in distal portion .... ***Dicerapanorpa luojishana*** Hu & Hua
– Paramere with ventral branch less elongated, nearly 2/3 as long as middle branch, and relatively straight .... ***Dicerapanorpa lativalva*** Hu & Hua

Dicerapanorpa magna Group

10.3.1 *Dicerapanorpa baiyunshana* Zhong & Hua, 2013

(Baiyunshan Two-Horned Scorpionfly) [白云山双角蝎蛉] (Fig. 10.91)

Measurements. Forewing length 16.5–17.5 mm.

Morphology. Vertex yellow. Ocellar triangle black. Rostrum yellowish brown. Pronotum yellowish brown. Nota dark brown laterally with broad yellow median stripe. Pleura and legs light yellowish brown. Wing membrane hyaline and slightly tinged with yellow, with markings brown and blurred marginally. Apical band broad and with large hyaline window; pterostigmal band with broad basal branch and thin apical branch; basal band complete; marginal spot triangular; and basal spot inconspicuous. Male T1–T5 mostly yellow with dark brown flanks. Notal organ flat and covering acute postnotal organ. A6 with pair of finger-like anal horns on apical margin. A7 greatly stalked at base.

Male genital bulb broad oval. Epandrium emarginate terminally in broad U-shape, with pair of finger-like processes laterally. Hypovalves stripe-like and bearing row of sparsely and irregularly arranged long bristles along inner margin, with apexes not reaching apex of gonocoxites and slightly curved mesad. Gonostylus shorter than gonocoxites and bearing acute median tooth and large, bifurcated basal process. Paramere trifurcated: ventral branch long and straight; median branch of paramere not crossed mesally or slightly in

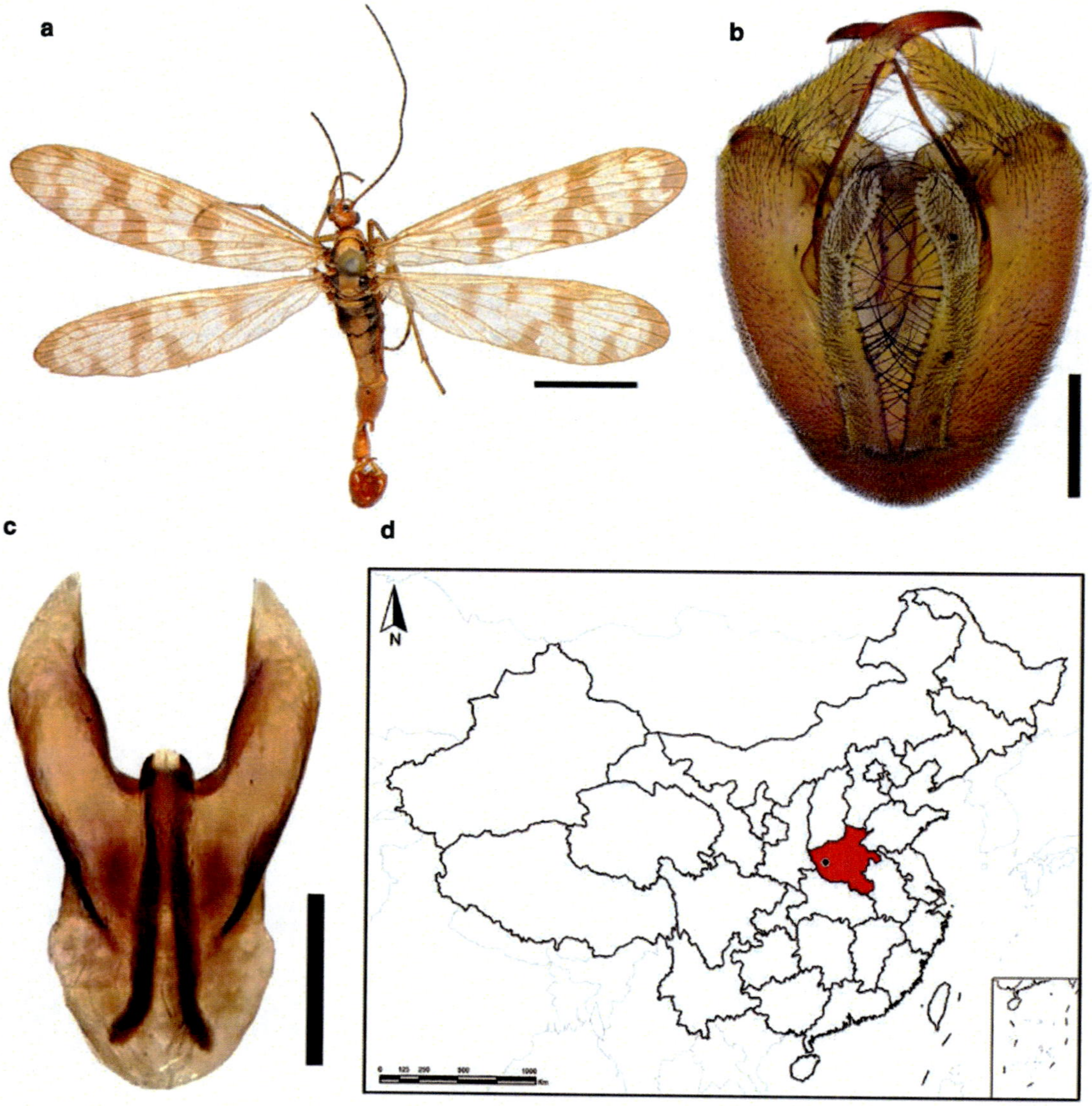

Fig. 10.91 (**a**) Male adult of ***Dicerapanorpa baiyunshana***, dorsal view (Baiyun Mountain, Songxian, Henan) (*Scale bar*: 5.0 mm) [NWAU]. (**b**) Male genital bulb of ***Dicerapanorpa baiyunshana***, ventral view (*Scale bar*: 0.5 mm). (**c**) Female medigynium of ***Dicerapanorpa baiyunshana***, ventral view (*Scale bar*: 0.2 mm). (**d**) Distribution of ***Dicerapanorpa baiyunshana***

contact apically and extending to median tooth of gonostylus; dorsal branch only extending to apex of gonocoxites. Aedeagus short, and not exceeding apex of gonocoxites.

Female subgenital plate broad oval. Medigynium with rounded main plate; posterior arms stout and acute terminally; axis concealed in main plate with slightly diverged apodemes.

Emergence. June to August.

Distribution. Henan (Baiyun Mountain).

10.3.2 *Dicerapanorpa hualongshana* Hu & Hua, 2019

(Hualongshan Two-Horned Scorpionfly) [化龙山双角蝎蛉] (Fig. 10.92)

Measurements. Forewing length 16.5–17.5 mm.

Morphology. Vertex yellow. Ocellar triangle black. Rostrum yellowish brown. Pronotum yellowish brown. Nota dark brown laterally with broad yellow median stripe. Pleura and legs light yellowish brown. Wing membrane hyaline and slightly tinged with yellow, with markings brown and blurred marginally. Apical band broad and with large hyaline window; pterostigmal band with broad basal branch and greatly reduced apical branch; other markings reduced. Male T1–T5 mostly yellow with dark brown flanks. Notal organ flat and covering acute postnotal organ. A6 with pair of finger-like anal horns on apical margin. A7 greatly stalked at base.

Male genital bulb broad oval. Epandrium emarginate terminally in broad U-shape, with pair of finger-like processes laterally. Hypovalves stripe-like and bearing row of sparsely and irregularly arranged long bristles along inner margin, with apexes not reaching apex of gonocoxites and slightly curved mesad. Gonostylus shorter than gonocoxites and bearing acute

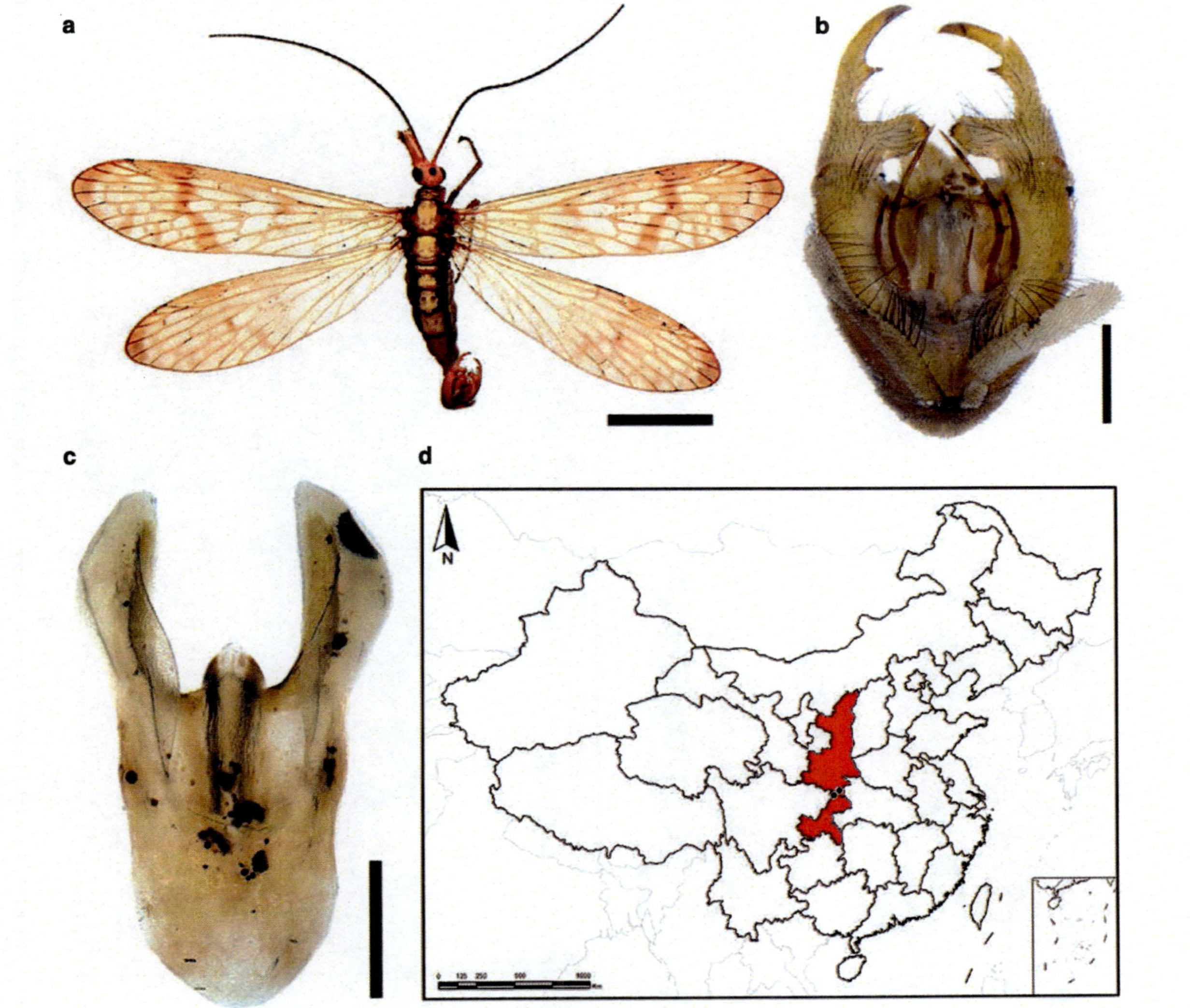

Fig. 10.92 (**a**) Male adult of ***Dicerapanorpa hualongshana***, dorsal view (Hualong Mountain, Shaanxi) (*Scale bar*: 5.0 mm) **[NWAU]**. (**b**) Male genital bulb of ***Dicerapanorpa hualongshana***, ventral view (*Scale bar*: 0.5 mm). (**c**) Female medigynium of ***Dicerapanorpa hualongshana***, ventral view (*Scale bar*: 0.2 mm). (**d**) Distribution of ***Dicerapanorpa hualongshana***

median tooth and large, bifurcated basal process. Paramere trifurcated: ventral branch long and straight, at least half as long as dorsal branch; median branch of paramere not crossed mesally or slightly in contact apically and extending to median tooth of gonostylus; dorsal branch only extending to apex of gonocoxites. Aedeagus short, and not exceeding apex of gonocoxites.

Female subgenital plate broad oval. Medigynium with rounded main plate; posterior arms stout and subacute terminally, greatly projected on outer margin; axis concealed in main plate with short apodemes.

Emergence. June to August.

Distribution. Shaanxi (Hualong Mountain).

10.3.3 *Dicerapanorpa magna* Chou, 1981

(Great Two-Horned Scorpionfly) [大双角蝎蛉]
(Fig. 10.93)

Measurements. Forewing length 16.0–17.5 mm.

Morphology. Vertex mostly yellow. Ocellar triangle enclosed by large black spot. Rostrum yellowish brown. Pronotum yellowish brown. Meso- and metanotum dark brown laterally and bearing broad yellow median stripe. Pleura and legs yellowish brown. Wing membrane tinged with yellow, with markings brown and blurred marginally. In forewings, apical band broad, with large hyaline window innerly, and separated slender spot posteriorly; pterostigmal band with broad basal branch and thin apical branch; basal band complete; marginal and basal spots small. Male T1–T5 mostly yellowish white with dark brown flanks. Notal organ flat and covering acute postnotal organ. A6 yellowish brown, cylindrical with pair of finger-like anal horns on dorsal apex. A7 greatly stalked at base.

Male genital bulb broad oval. Epandrium emarginates terminally in broad U-shape and forms pair of finger-like processes laterally. Hypovalves stripe-like and bearing row of sparsely and irregularly arranged long bristles along inner margin, with rounded apexes extending to apex of gonocox-

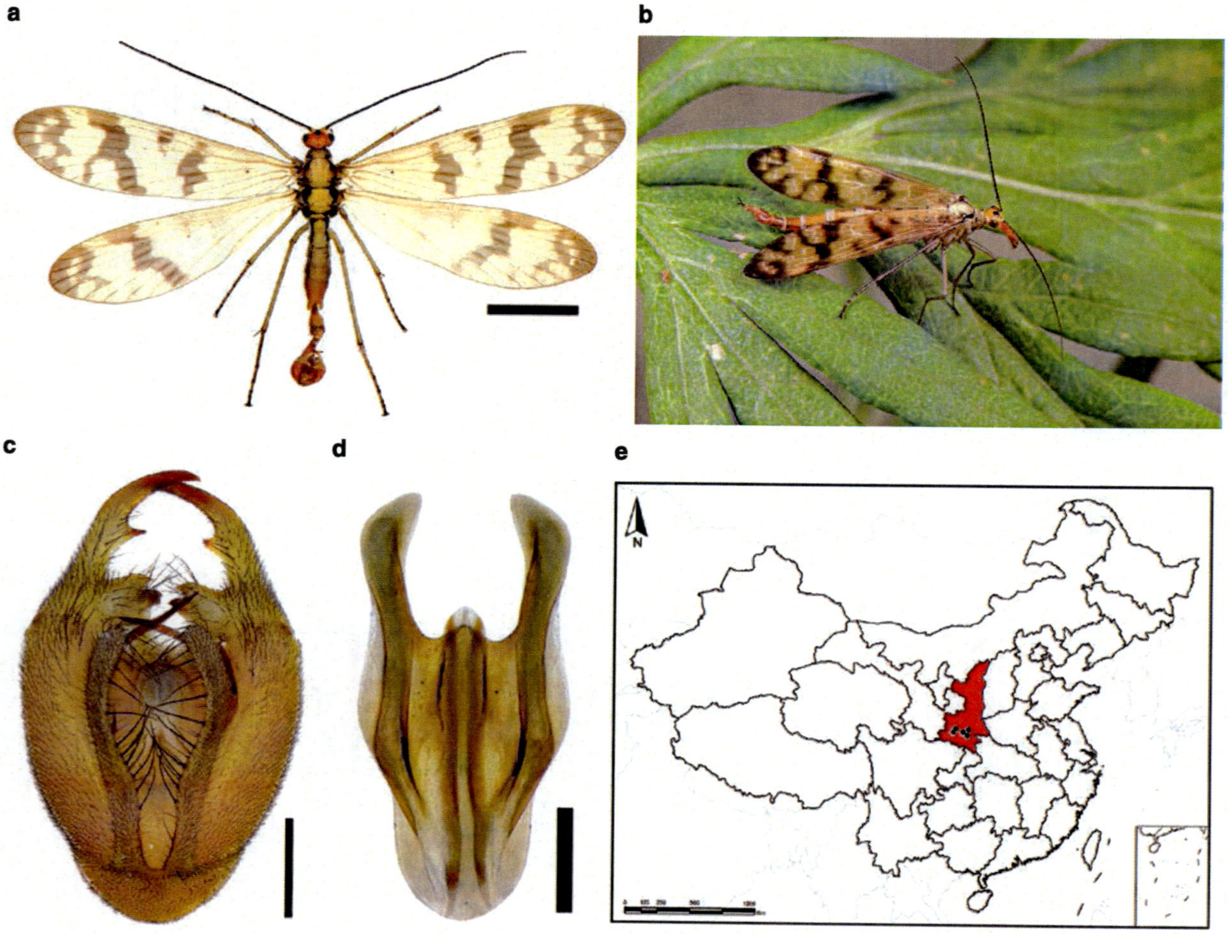

Fig. 10.93 (**a**) Male adult of ***Dicerapanorpa magna***, dorso view (Zhuque Forest Park, Huyi, Shaanxi) (*Scale bar*: 5.0 mm) **[NWAU]**. (**b**) Female adult of ***Dicerapanorpa magna*** in the field (Tiantaishan Forest Park, Baoji, Shaanxi). (**c**) Male genital bulb of ***Dicerapanorpa magna***, ventral view (*Scale bar*: 0.5 mm). (**d**) Female medigynium of ***Dicerapanorpa magna***, ventral view (*Scale bar*: 0.2 mm). (**e**) Distribution of ***Dicerapanorpa magna***

ites. Gonostylus shorter than gonocoxites and bearing acute median tooth and large, bifurcated basal process. Paramere trifurcated: ventral branch short and hook-like; median branch long slender, curving mesad and crossed mesaly; dorsal branch slightly shorter than median branch. Aedeagus short and not exceeding apex of gonocoxites, with long setae on dorsal processes.

Female subgenital plate broad oval. Medigynium with rounded main plate slightly sinuate along lateral margin; posterior arms stout and subacute terminally, greatly curving inward terminally; axis concealed in the main plate with parallel apodemes.

Emergence. May to August.

Distribution. Shaanxi (Baoji, Huyi, Meixian, Ningshan, Taibai, Zhouzhi).

10.3.4 *Dicerapanorpa minshana* Hu & Hua, 2019

(Minshan Two-Horned Scorpionfly) [岷山双角蝎蛉] (Fig. 10.94)

Measurements. Forewing length 16.0–17.5 mm.

Morphology. Vertex mostly yellow. Ocellar triangle enclosed by large black spot. Rostrum yellowish brown. Pronotum yellowish brown. Meso- and metanotum dark brown laterally and bearing broad yellow median stripe. Pleura and legs yellowish brown. Wing membrane tinged with yellow, with markings brown and blurred marginally. In forewings, apical band broad, with large hyaline window innerly, and separated slender spot posteriorly; pterostigmal band with

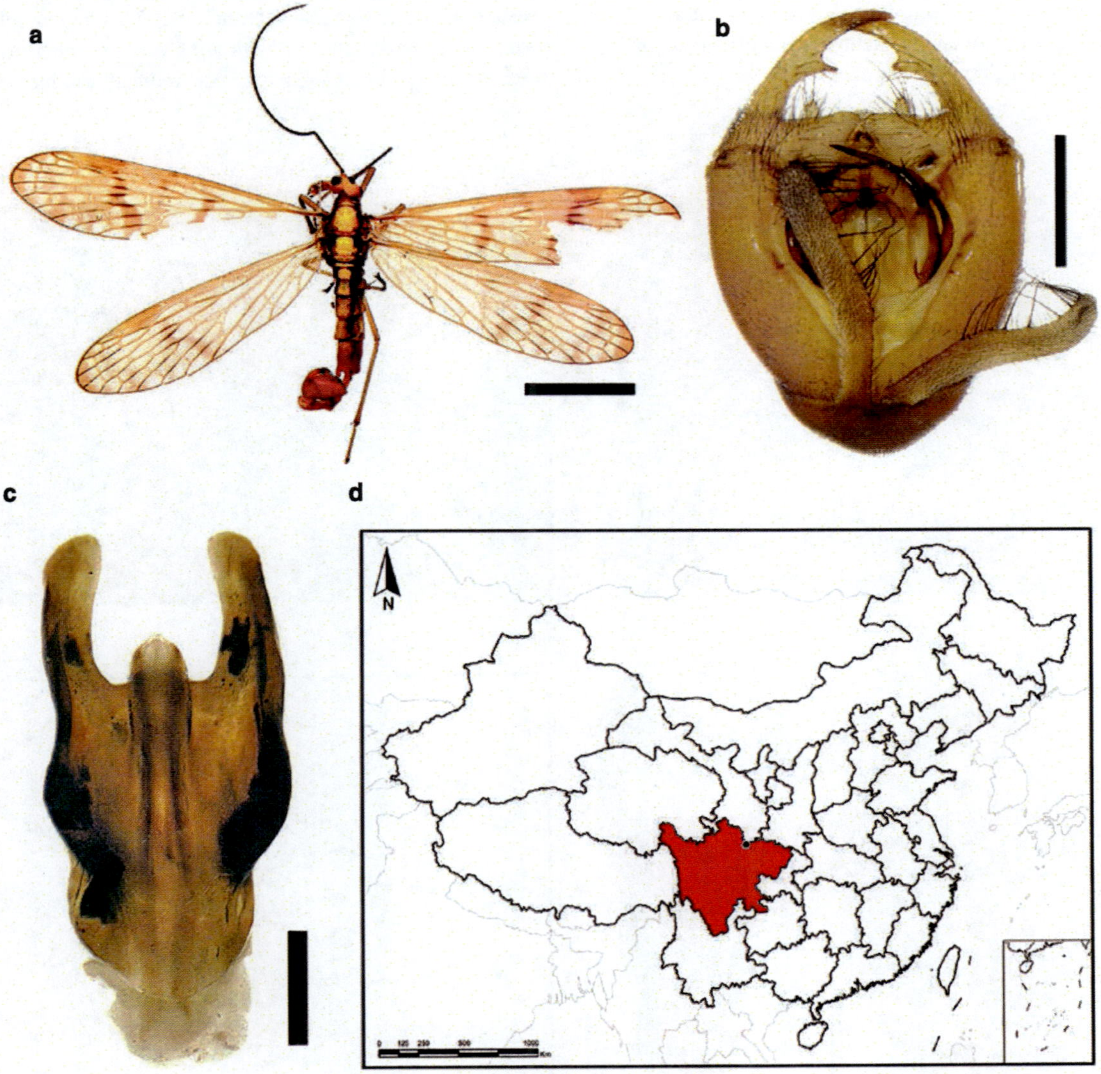

Fig. 10.94 (**a**) Male adult of ***Dicerapanorpa minshana***, dorsal view (Laohegou, Pingwu, Sichuan) (*Scale bar*: 5.0 mm) **[NWAU]**. (**b**) Male genital bulb of ***Dicerapanorpa minshana***, ventral view (*Scale bar*: 0.5 mm). (**c**) Female medigynium of ***Dicerapanorpa minshana***, ventral view (*Scale bar*: 0.2 mm). (**d**) Distribution of ***Dicerapanorpa minshana***

broad basal branch and reduced apical branch; other markings variously reduced. Male T1–T5 mostly yellowish white with dark brown flanks. Notal organ flat and covering acute postnotal organ. A6 yellowish brown, cylindrical with pair of finger-like anal horns on dorsal apex. A7 greatly stalked at base.

Male genital bulb broad oval. Epandrium emarginates terminally in broad U-shape and forms pair of finger-like processes laterally. Hypovalves stripe-like and bearing row of sparsely and irregularly arranged long bristles along inner margin, with rounded apexes extending to apex of gonocoxites. Gonostylus shorter than gonocoxites and bearing acute median tooth and large, bifurcated basal process. Paramere trifurcated: ventral branch extremely short and hook-like; median branch long slender, curving mesad and crossed mesaly; dorsal branch slightly shorter than median branch. Aedeagus short and not exceeding apex of gonocoxites, with long setae on dorsal processes.

Female subgenital plate broad oval. Medigynium with rounded main plate slightly sinuate along lateral margin; posterior arms stout and subacute terminally; axis concealed in main plate with parallel apodemes.

Emergence. May to August.

Distribution. Sichuan (Pingwu, Songpan).

10.3.5 *Dicerapanorpa shennongensis* Zhong & Hua, 2013

(Shennong Two-Horned Scorpionfly) [神农双角蝎蛉] (Fig. 10.95)

Measurements. Forewing length 17.0–17.5 mm.

Morphology. Vertex mostly yellow. Ocellar triangle black. Rostrum yellowish brown. Pronotum yellowish brown. Meso- and metanotum dark brown laterally and bearing broad light-yellow stripe mesally. Pleura and legs yellowish brown. Wing membrane hyaline and slightly tinged with pale yellow, with markings brown and blurred marginally. Apical band broad, with large hyaline window spot; pterostigmal band with broad basal branch and thin apical branch; basal band incomplete or reduced, and disconnected with wing margins anteriorly and posteriorly; marginal and basal spots reduced and inconspicu-

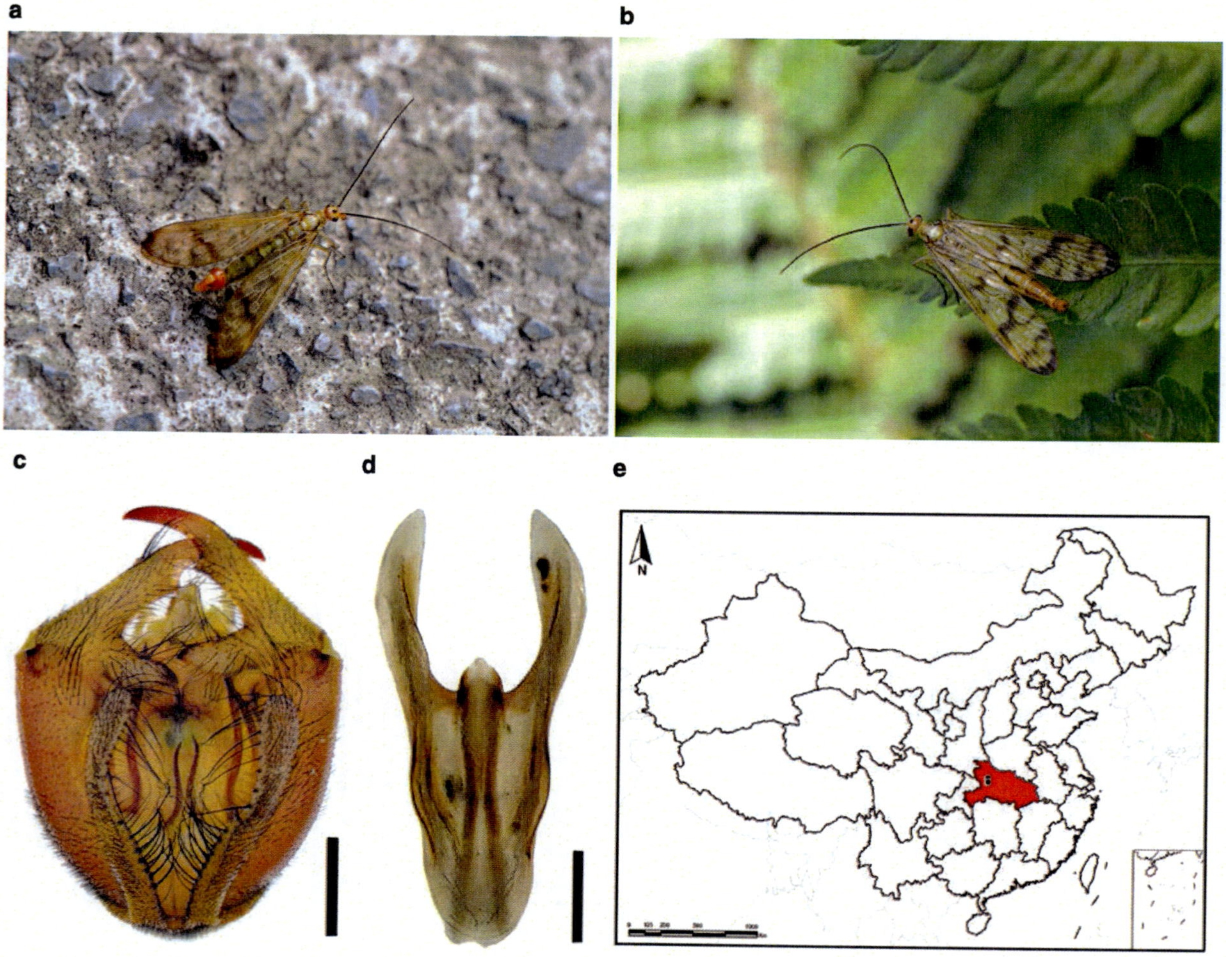

Fig. 10.95 (**a**) Male adult of ***Dicerapanorpa shennongensis*** in the field (Shennongjia, Hubei). (**b**) Female adult of ***Dicerapanorpa shennongensis*** in the field (Shennongjia, Hubei). (**c**) Male genital bulb of ***Dicerapanorpa shennongensis***, ventral view (*Scale bar*: 0.5 mm). (**d**) Female medigynium of ***Dicerapanorpa shennongensis***, ventral view (*Scale bar*: 0.2 mm). (**e**) Distribution of ***Dicerapanorpa shennongensis***

ous. Male T1–T5 mostly pale yellow with dark brown flanks. Notal organ flat and covering acute postnotal organ. A6 yellowish brown with pair of finger-like anal horns on dorsal apex. A7 greatly stalked at base.

Male genital bulb broad oval. Epandrium emarginates terminally in broad U-shape and forms pair of finger-like processes laterally. Hypovalves stripe-like and bearing row of sparsely and irregularly arranged long bristles along inner margin, with rounded apexes extending to apex of gonocoxites. Gonostylus shorter than gonocoxites and bearing acute median tooth and large, bifurcated basal process. Paramere trifurcated: ventral branch short and curved basally; median branch long and relatively straight; and dorsal branch slightly shorter and curved. Aedeagus short and not exceeding apex of gonocoxites.

Female subgenital plate broad oval. Medigynium narrow, with the main plate slightly broadening toward apex; posterior arms slender and subacute terminally; axis concealed in main plate with slightly diverged apodemes.

Emergence. June to August.

Distribution. Hubei (Fangxian, Shennongjia).

Dicerapanorpa diceras Group

10.3.6 *Dicerapanorpa bifurcata* Hu & Hua, 2020

(Bifurcated Two-Horned Scorpionfly) [双叉双角蝎蛉] (Fig. 10.96)

Measurements. Forewing length 15.0–16.5 mm.

Morphology. Vertex yellow. Ocellar triangle enclosed by large black spot. Rostrum yellow, with pair of conspicuous dark

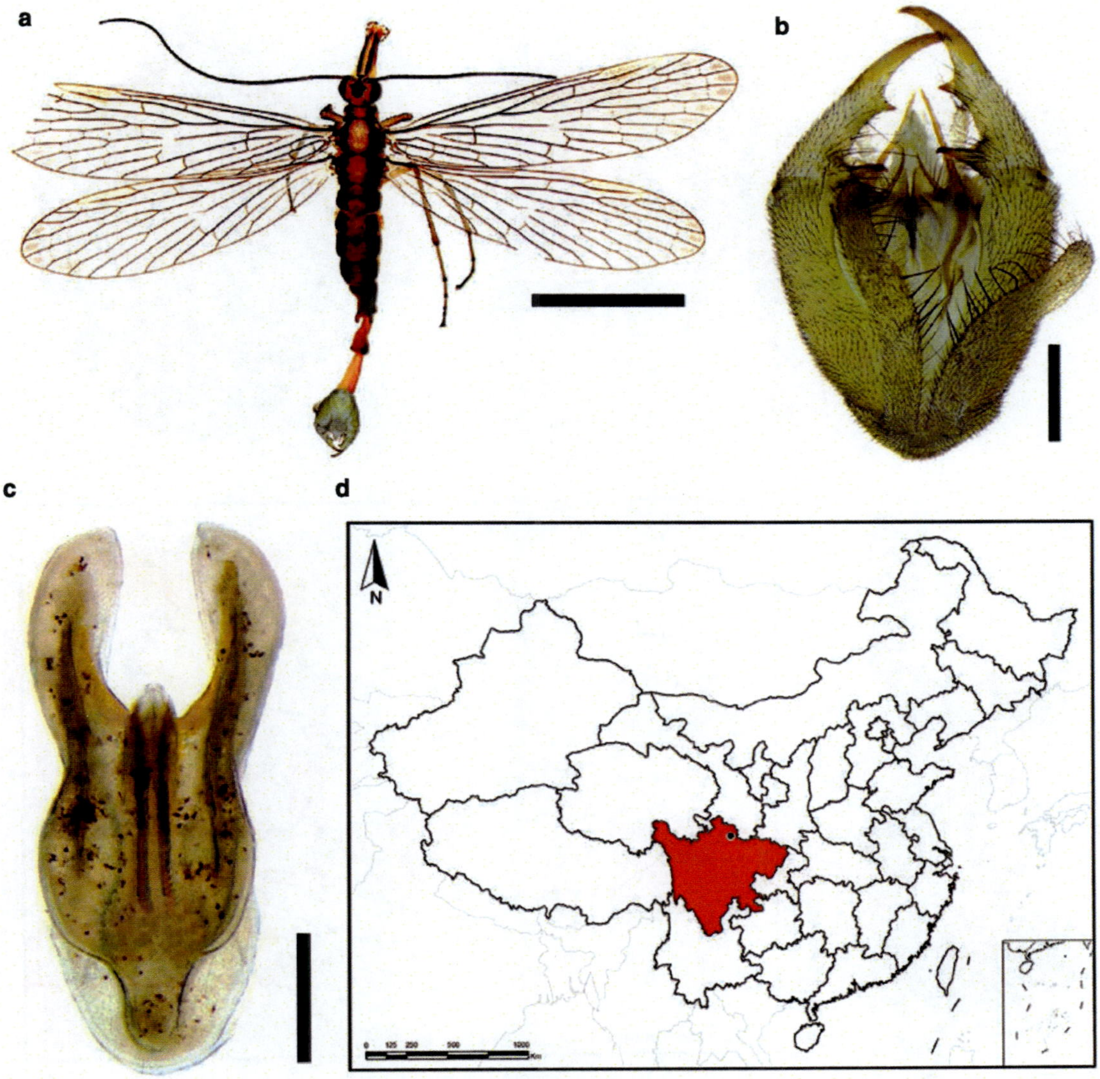

Fig. 10.96 (**a**) Male adult of ***Dicerapanorpa bifurcata***, dorsal view (Wanglang Nature Reserve, Pingwu, Sichuan) (*Scale bar*: 5.0 mm) [NWAU]. (**b**) Male genital bulb of ***Dicerapanorpa bifurcata***, ventral view (*Scale bar*: 0.5 mm). (**c**) Female medigynium of ***Dicerapanorpa bifurcata***, ventral view (*Scale bar*: 0.2 mm). (**d**) Distribution of ***Dicerapanorpa bifurcata***

brown stripes frontally. Pronotum yellow. Meso- and metanotum dark brown laterally. Pleura and legs yellow. Wing membrane hyaline, and devoid of markings, except apex slightly darkened. Pterostigma light yellowish brown. Male T1–T5 mostly yellow with dark brown flanks. Notal organ flat and covering acute postnotal organ. A6 yellowish brown, bearing pair of finger-like anal horns on dorsal apex. A7 greatly stalked at base.

Male genital bulb broad oval. Epandrium emarginate terminally in broad U-shape to form pair of finger-like processes laterally. Hypovalves stripe-like and bearing row of sparsely and irregularly arranged long bristles along inner margin, with expanded rounded apexes extending to apex of gonocoxites. Gonostylus shorter than gonocoxites and bearing acute median tooth and large, bifurcated basal process. Paramere bifurcated: ventral branch absent; median branch extending to median tooth of gonostylus; dorsal branch shorter. Aedeagus short, not exceeding apex of gonocoxites, with membranous ventral valves, and bearing several long setae on dorsal processes.

Female subgenital plate broad with slightly projected apex. Medigynium with rounded main plate; posterior arms broad, subacute terminally, and greatly curving inward terminally; axis concealed in main plate with parallel apodemes.

Emergence. June.

Distribution. Sichuan (Pingwu).

10.3.7 *Dicerapanorpa diceras* MacLachlan, 1894

(Two-Horned Scorpionfly) [双角蝎蛉] (Fig. 10.97)

Measurements. Forewing length 14.5–16.5 mm.

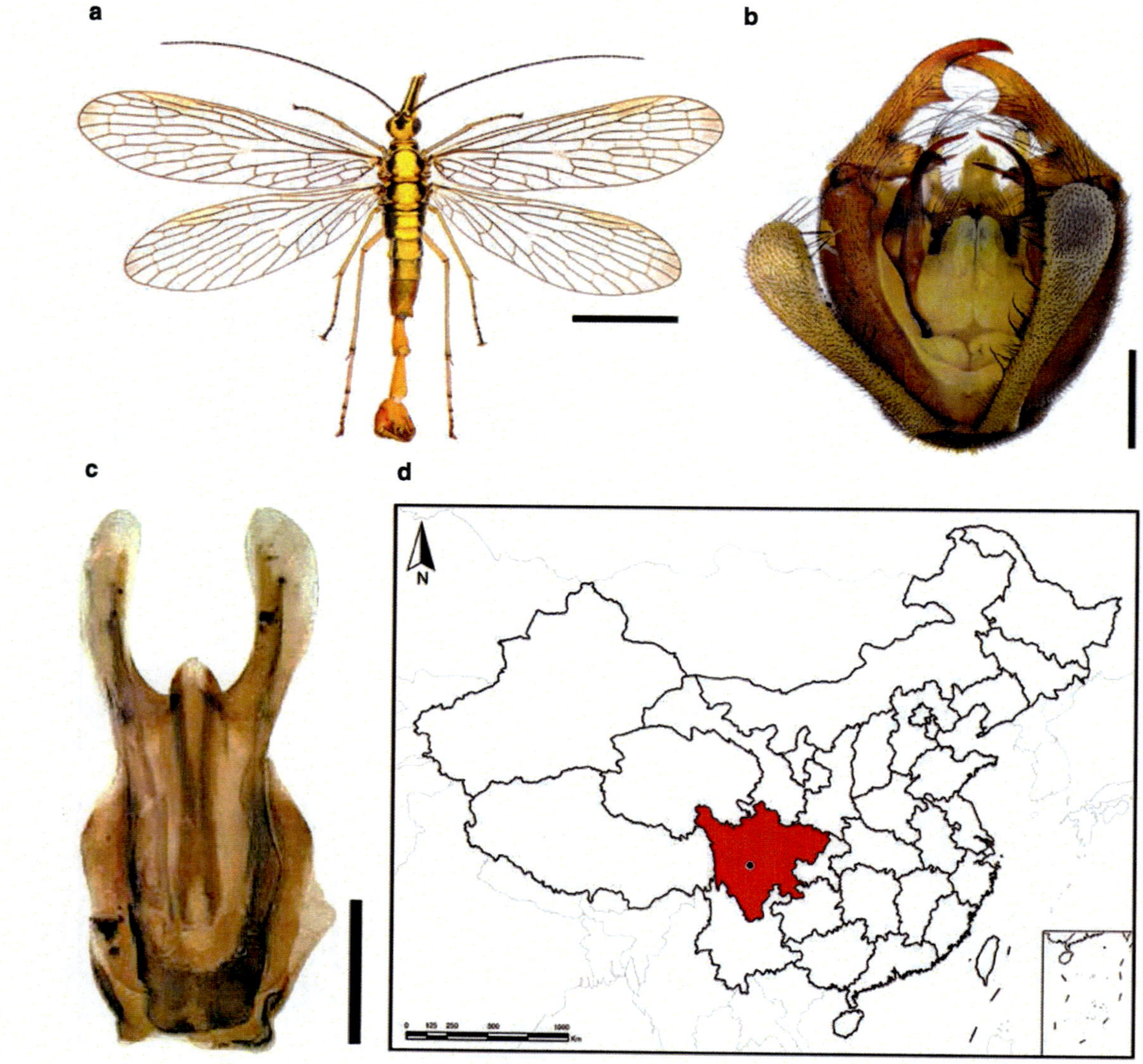

Fig. 10.97 (a) Male adult of ***Dicerapanorpa diceras***, dorsal view (Kangding, Sichuan) (*Scale bar*: 5.0 mm) **[NWAU]**. **(b)** Male genital bulb of ***Dicerapanorpa diceras***, ventral view (*Scale bar*: 0.5 mm). **(c)** Female medigynium of ***Dicerapanorpa***, ventral view (*Scale bar*: 0.2 mm). **(d)** Distribution of ***Dicerapanorpa diceras***

Morphology. Vertex yellow. Ocellar triangle enclosed by large black spot. Rostrum yellow, with pair of conspicuous dark brown stripes frontally. Pronotum yellow. Meso- and metanotum dark brown laterally. Pleura and legs yellow. Wing membrane hyaline, and devoid of markings, except apex slightly darkened. Pterostigma light yellowish brown. Male T1–T5 mostly yellow with dark brown flanks. Notal organ flat and covering acute post-notal organ. A6 yellowish brown, bearing pair of finger-like anal horns on dorsal apex. A7 greatly stalked at base.

Male genital bulb broad oval. Epandrium emarginates terminally in broad U-shape to form pair of finger-like processes laterally. Hypo-valves stripe-like and bearing row of sparsely and irregularly arranged long bristles along inner margin, with expanded rounded apexes extending to apex of gonocoxites. Gonostylus shorter than gonocoxites and bearing acute median tooth and large, bifurcated basal process. Paramere trifurcated: ventral branch straight; median and dorsal branches curved mesad apically. Aedeagus short, not exceeding apex of gonocoxites, with membranous ventral valves, and bearing several long setae on dorsal processes.

Female subgenital plate broad with slightly projected apex. Medigynium long, with oblong main plate; posterior arms broad, subacute terminally, and greatly curving inward terminally; axis concealed in main plate with parallel apodemes.

Emergence. May to June.

Distribution. Sichuan (Kangding).

10.3.8 *Dicerapanorpa deqenensis* Hu, Wang & Hua, 2019

(Dêqên Two-Horned Scorpionfly) [德钦双角蝎蛉]
(Fig. 10.98)

Measurements. Forewing length 14.0–16.0 mm.

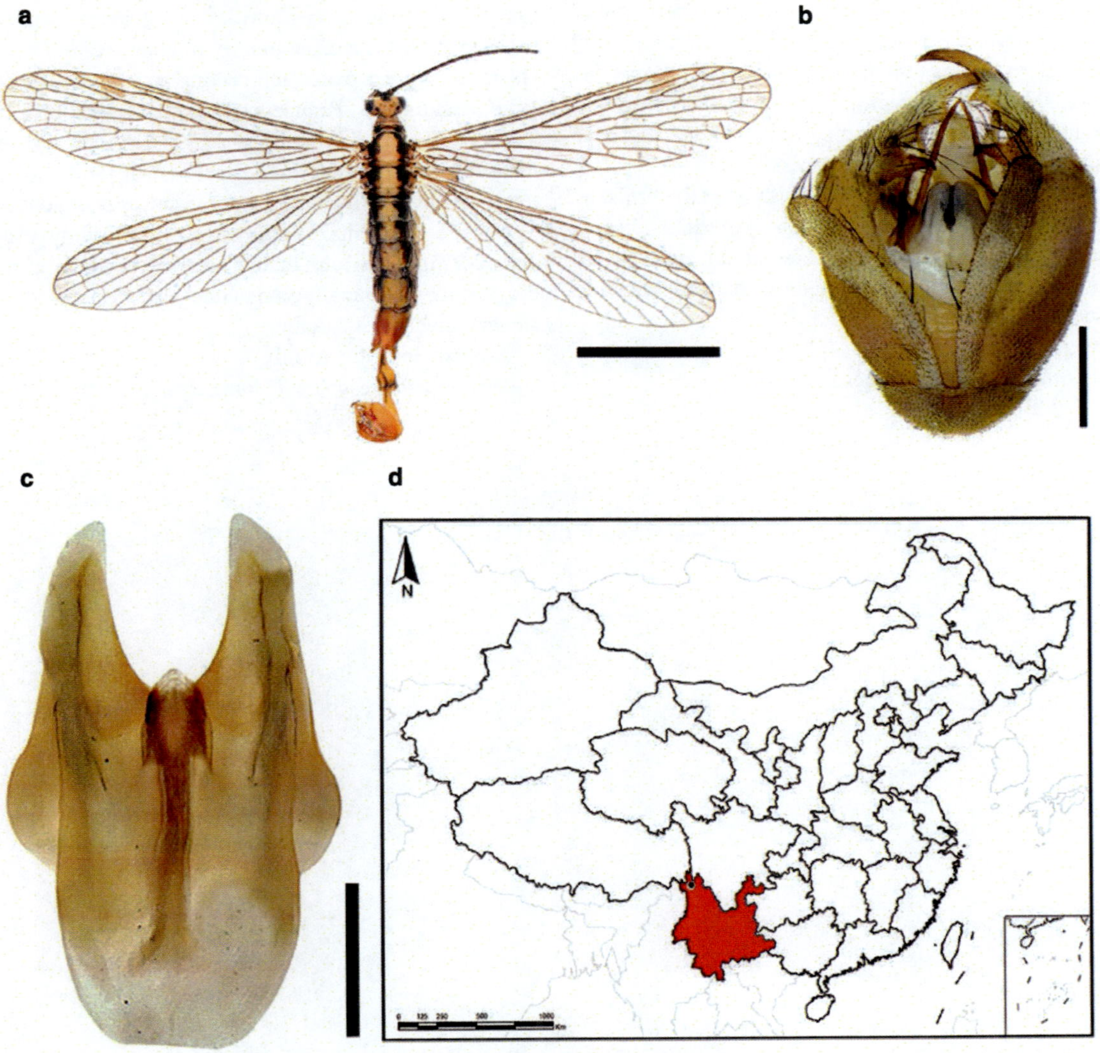

Fig. 10.98 (**a**) Male adult of ***Dicerapanorpa deqenensis***, dorsal view (Baima Snow Mountain, Dêqên, Yunnan) (*Scale bar*: 5.0 mm) **[NWAU]**. (**b**) Male genital bulb of ***Dicerapanorpa deqenensis***, ventral view (*Scale bar*: 0.5 mm). (**c**) Female medigynium of ***Dicerapanorpa deqenensis***, ventral view (*Scale bar*: 0.2 mm). (**d**) Distribution of ***Dicerapanorpa deqenensis***

Morphology. Vertex yellow, with large black spot on each side, and three smaller, blurred spots encircling each large one. Ocellar triangle enclosed by large black spot. Rostrum yellow, with pair of conspicuous dark brown stripes frontally. Pronotum dark brown. Meso- and metanotum mostly yellow middly and dark brown laterally. Pleura and legs yellow. Wing membrane hyaline, and devoid of markings, except greatly reduced pterostigmal band. Pterostigma light yellowish brown. Male T1–T5 mostly yellow with dark brown flanks. Notal organ flat and covering acute post-notal organ. A6 yellowish brown, bearing pair of finger-like anal horns on dorsal apex. A7 greatly stalked at base.

Male genital bulb broad oval. Epandrium emarginate terminally in broad U-shape to form pair of finger-like processes laterally. Hypo-valves stripe-like and bearing row of sparsely and irregularly arranged long bristles along inner margin, with expanded rounded apexes extending to apex of gonocoxites. Gonostylus shorter than gonocoxites and bearing acute median tooth and large, bifurcated basal process. Paramere trifurcated: ventral and dorsal branches straight, reaching basal process of gonostylus; median branch much longer and extending to median tooth of gonostylus. Aedeagus short, not exceeding apex of gonocoxites, with membranous and greatly protruding ventral valves, and bearing several long setae on dorsal processes.

Female subgenital plate broad with slightly projected apex. Medigynium with main plate oblong; earlobe-like process on each side; posterior arms stout, broad, subacute terminally; axis concealed in main plate with slender, parallel apodemes.

Emergence. May to June.

Distribution. Yunnan (Dêqên).

10.3.9 *Dicerapanorpa kimminsi* Carpenter, 1948

(Kimmins's Two-Horned Scorpionfly) [基明斯双角蝎蛉] (Fig. 10.99)

Measurements. Forewing length 15.0–17.0 mm.

Morphology. Vertex mostly yellow. Ocellar triangle enclosed by large black spot. Rostrum yellow and with pair of conspicuous dark brown stripes frontally. Pronotum yellowish brown. Meso- and metanotum dark brown laterally and bearing broad yellow median stripe. Pleura and legs yellowish brown. Wing membrane hyaline, and devoid of markings, except apex slightly darkened. Pterostigma light yellowish brown. Male T1–T5 mostly yellow with dark brown flanks. Notal organ flat and covering acute postnotal organ. A6 yellowish brown and bearing pair of finger-like anal horns on dorsal apex. A7 greatly stalked at base.

Male genital bulb broad oval. Epandrium emarginates terminally in broad U-shape to form pair of finger-like processes laterally. Hypovalves stripe-like and with row of sparsely and irregularly arranged long bristles along inner margin, with rounded apexes extending to apex of gonocoxites. Gonostylus shorter than gonocoxites and bearing acute median tooth and large basal process. Paramere trifurcated: ventral branch short and hook-like; median branch slender and straight, extending to median tooth of gonostylus; dorsal branch short and curving mesad. Aedeagus short and not exceeding apex of gonocoxites.

Female subgenital plate broad with slightly projected apex. Medigynium with main plate oval; posterior arms slender, rounded terminally; axis concealed in main plate with slender, diverged apodemes.

Emergence. May to July.

Distribution. Sichuan (Mount Emei).

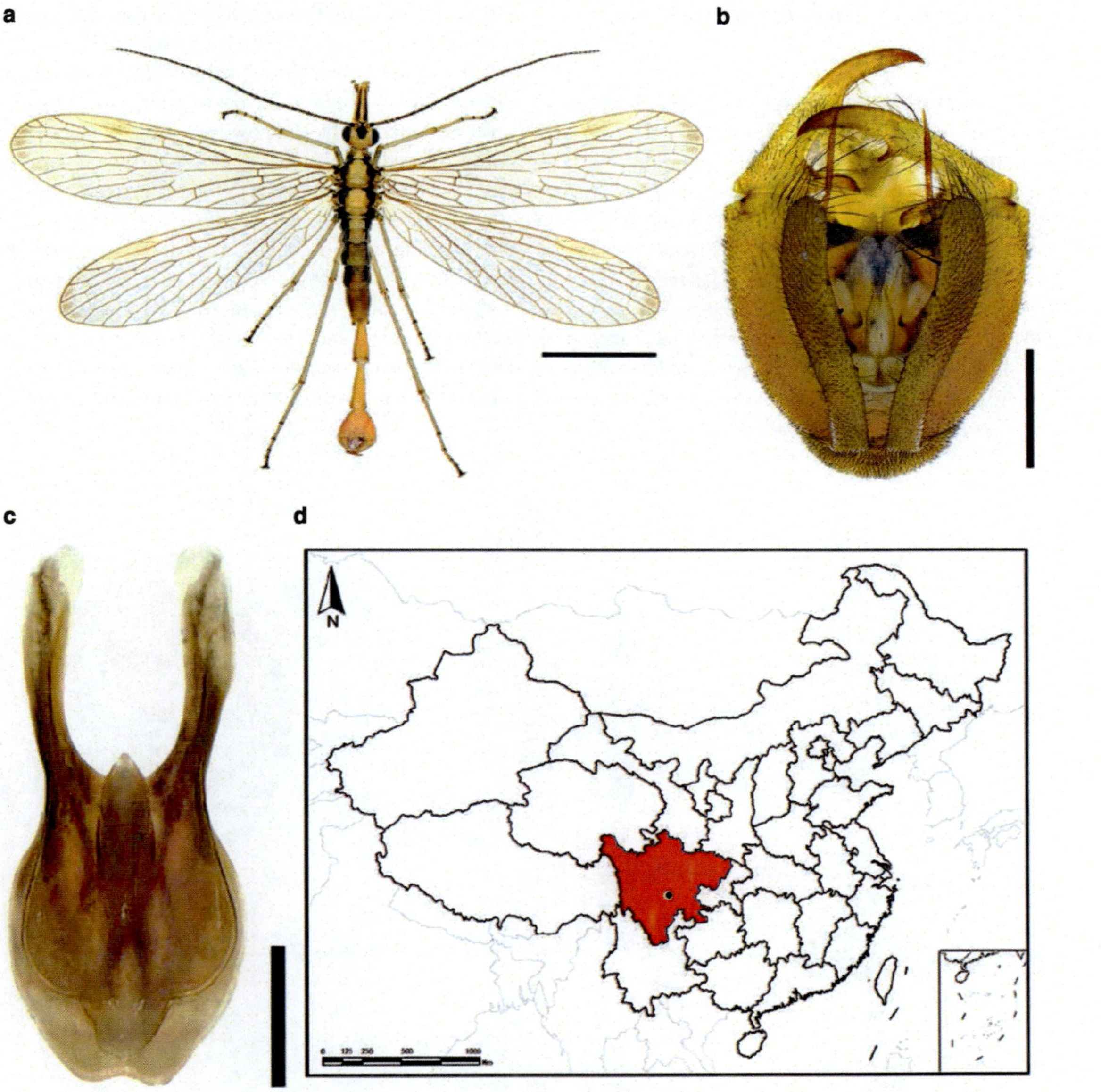

Fig. 10.99 (**a**) Male adult of ***Dicerapanorpa kimminsi***, dorsal view (Mount Emei, Sichuan) (*Scale bar*: 5.0 mm) [NWAU]. (**b**) Male genital bulb of ***Dicerapanorpa kimminsi***, ventral view (*Scale bar*: 0.5 mm). (**c**) Female medigynium of ***Dicerapanorpa kimminsi***, ventral view (*Scale bar*: 0.2 mm). (**d**) Distribution of ***Dicerapanorpa kimminsi***

10.3.10 *Dicerapanorpa lativalva* Hu & Hua, 2019

(Broad-Valve Two-Horned Scorpionfly) [宽瓣双角蝎蛉] (Fig. 10.100)

Measurements. Forewing length 14.0–15.0 mm.

Morphology. Vertex mostly yellow. Ocellar triangle enclosed by large black spot. Rostrum yellow and with pair of conspicuous dark brown stripes frontally. Pronotum yellowish brown. Meso- and metanotum dark brown laterally and bearing broad yellow median stripe. Pleura and legs yellowish brown. Wing membrane hyaline, and devoid of markings. Pterostigma light yellowish brown. Male T1–T5 mostly yellow with dark brown flanks. Notal organ flat and covering acute postnotal organ. A6 yellowish brown and bearing pair of finger-like anal horns on dorsal apex. A7 greatly stalked at base.

Male genital bulb broad oval. Epandrium emarginates terminally in broad U-shape to form pair of finger-like processes laterally. Hypovalves broad, stripe-like and with row of sparsely and irregularly arranged long bristles along inner margin, with rounded apexes extending to apex of gonocoxites. Gonostylus shorter than gonocoxites and bearing acute median tooth and large basal process. Paramere trifurcated: ventral branch elongated and straight; median branch longest and bending inward in distal portion; dorsal branch shorter than median branch and slightly curved. Aedeagus short and not exceeding apex of gonocoxites; dorsal processes broad.

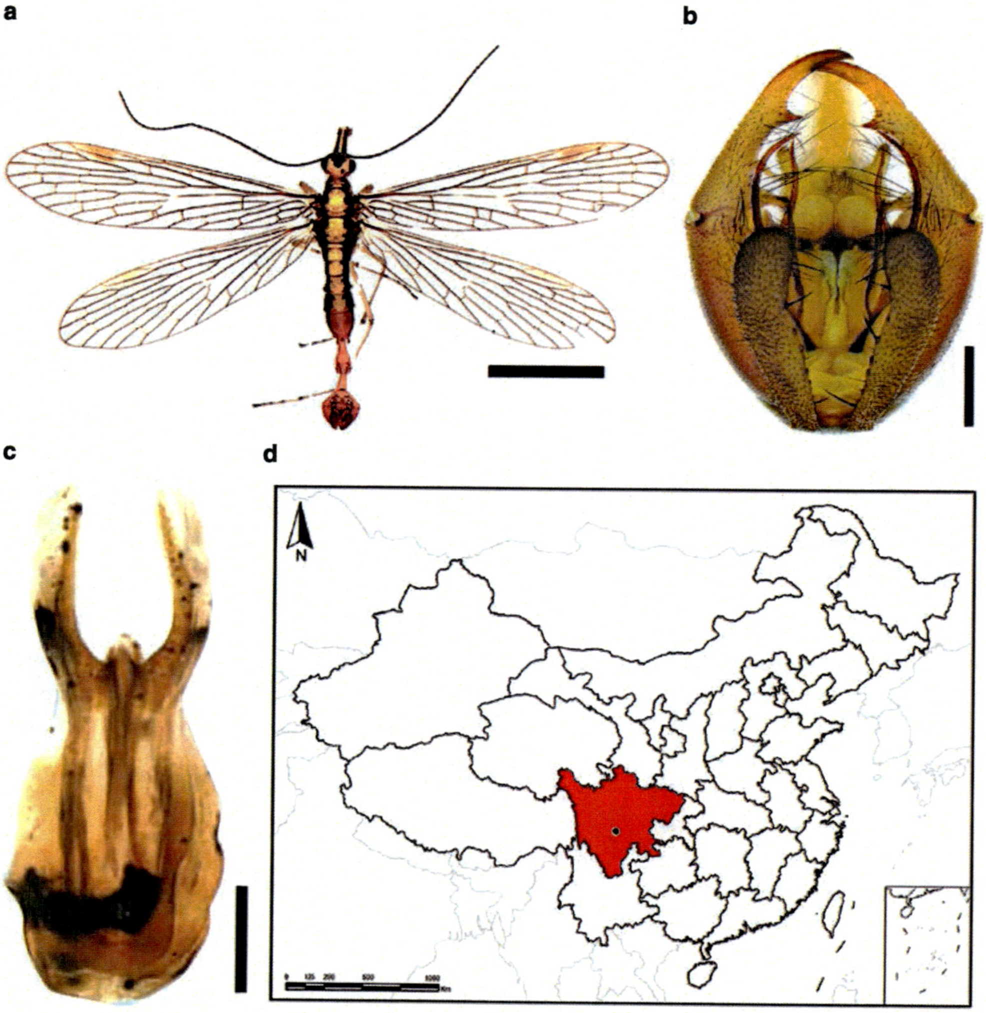

Fig. 10.100 (**a**) Male adult of ***Dicerapanorpa lativalva***, dorsal view (Menghuocheng, Shimian, Sichuan) (*Scale bar*: 5.0 mm) [**NWAU**]. (**b**) Male genital bulb of ***Dicerapanorpa lativalva***, ventral view (*Scale bar*: 0.5 mm). (**c**) Female medigynium of ***Dicerapanorpa lativalva***, ventral view (*Scale bar*: 0.2 mm). (**d**) Distribution of ***Dicerapanorpa lativalva***

Female subgenital plate broad with slightly projected apex. Medigynium with main plate oval; posterior arms slender, rounded terminally; axis concealed in main plate with slender, diverged apodemes.

Emergence. May to July.

Distribution. Sichuan (Shimian).

10.3.11 *Dicerapanorpa luojishana* Hu & Hua, 2019

(Luojishan Two-Horned Scorpionfly) [螺髻山双角蝎蛉] (Fig. 10.101)

Measurements. Forewing length 14.0–15.0 mm.

Morphology. Vertex mostly yellow. Ocellar triangle enclosed by large black spot. Rostrum yellow and with pair of conspicuous dark brown stripes frontally. Pronotum yellowish brown. Meso- and metanotum dark brown laterally and bearing broad yellow median stripe. Pleura and legs yellowish brown. Wing membrane hyaline, and devoid of markings, except darkened apex. Pterostigma light yellowish brown. Male T1–T5 mostly yellow with dark brown flanks. Notal organ flat and covering acute postnotal organ. A6 yellowish brown and bearing pair of finger-like anal horns on dorsal apex. A7 greatly stalked at base.

Male genital bulb broad oval. Epandrium emarginates terminally in broad U-shape to form pair of finger-like processes laterally. Hypovalves stripe-like and with row of sparsely and irregularly arranged long bristles along inner margin, with rounded apexes slightly exceeding apex of gonocoxites. Gonostylus shorter than gonocoxites and bearing acute median tooth and large basal process. Paramere trifurcated: ventral branch elongated and straight; median branch longest and bending inward in distal portion; dorsal branch shorter than median branch and slightly curved.

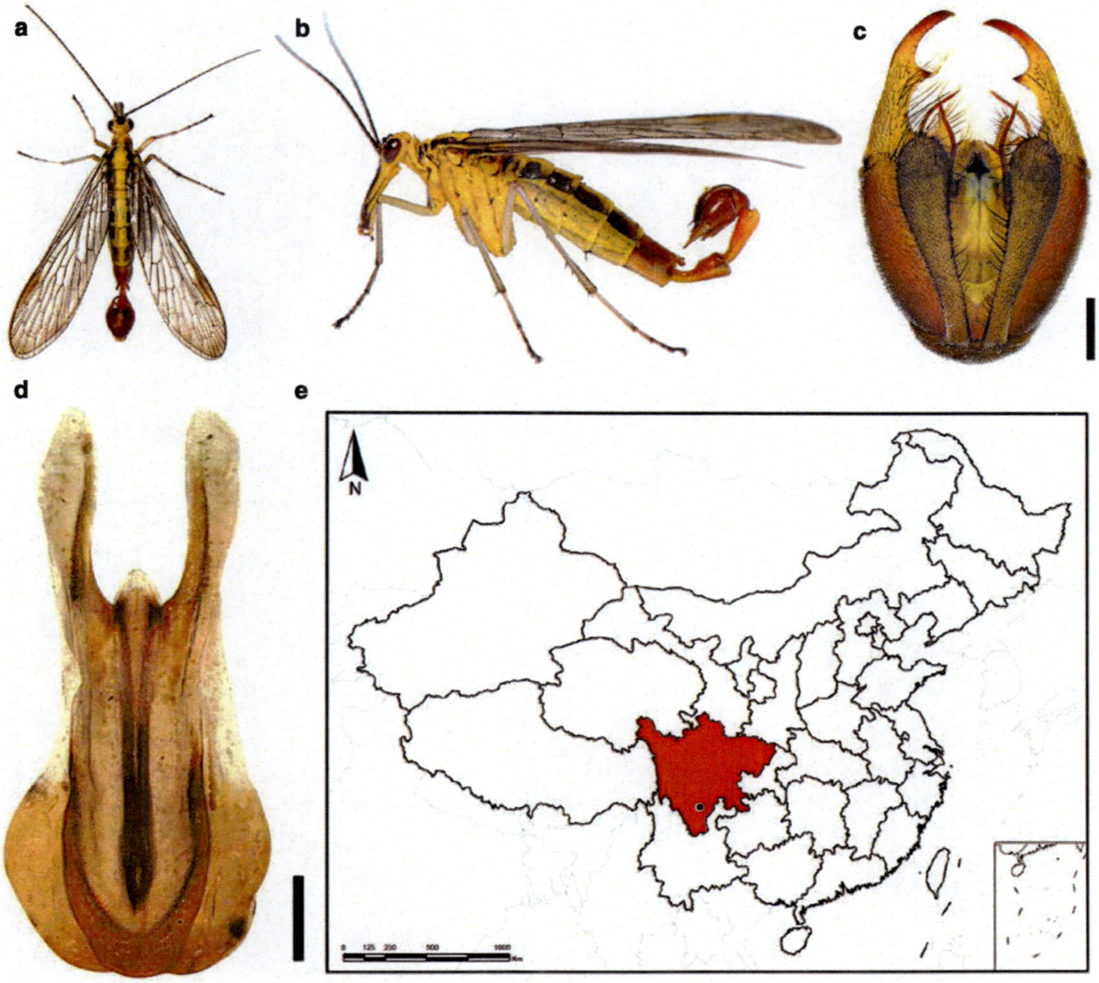

Fig. 10.101 (**a**) Male adult of ***Dicerapanorpa luojishana***, dorsal view (Luoji Mountain, Sichuan). (**b**) Male adult of ***Dicerapanorpa luojishana***, lateral view (Luoji Mountain, Sichuan). (**c**) Male genital bulb of ***Dicerapanorpa luojishana***, ventral view (*Scale bar*: 0.5 mm). (**d**) Female medigynium of ***Dicerapanorpa luojishana***, ventral view (*Scale bar*: 0.2 mm). (**e**) Distribution of ***Dicerapanorpa luojishana***

Aedeagus short and not exceeding apex of gonocoxites; dorsal processes broad.

Female subgenital plate broad with slightly projected apex. Medigynium with main plate pear-shaped; posterior arms slender, rounded terminally; axis concealed in main plate with slender, parallel apodemes.

Emergence. May to July.

Distribution. Sichuan (Luoji Mountain).

10.3.12 *Dicerapanorpa macula* Hu, Wang & Hua, 2019

(Maculate Two-Horned Scorpionfly) [斑头双角蝎蛉] (Fig. 10.102)

Measurements. Forewing length 14.0–15.0 mm.

Morphology. Vertex yellow, with large black spot on each side, and three smaller, blurred spots encircling each large one. Ocellar triangle enclosed by large black spot. Rostrum yellow, with pair of conspicuous dark brown stripes frontally. Pronotum dark brown. Meso- and metanotum mostly yellow middly and dark brown laterally. Pleura and legs yellow. Wing membrane hyaline, and devoid of markings, except greatly reduced pterostigmal band. Pterostigma light yellowish brown. Male T1–T5 mostly yellow with dark brown flanks. Notal organ flat and covering acute post-notal organ. A6 yellowish brown, bearing pair of finger-like anal horns on dorsal apex. A7 greatly stalked at base.

Male genital bulb broad oval. Epandrium emarginates terminally in broad U-shape to form pair of finger-like processes laterally. Hypo-valves stripe-like and bearing row of sparsely and irregularly arranged long bristles along inner margin, with expanded rounded apexes extending to apex of

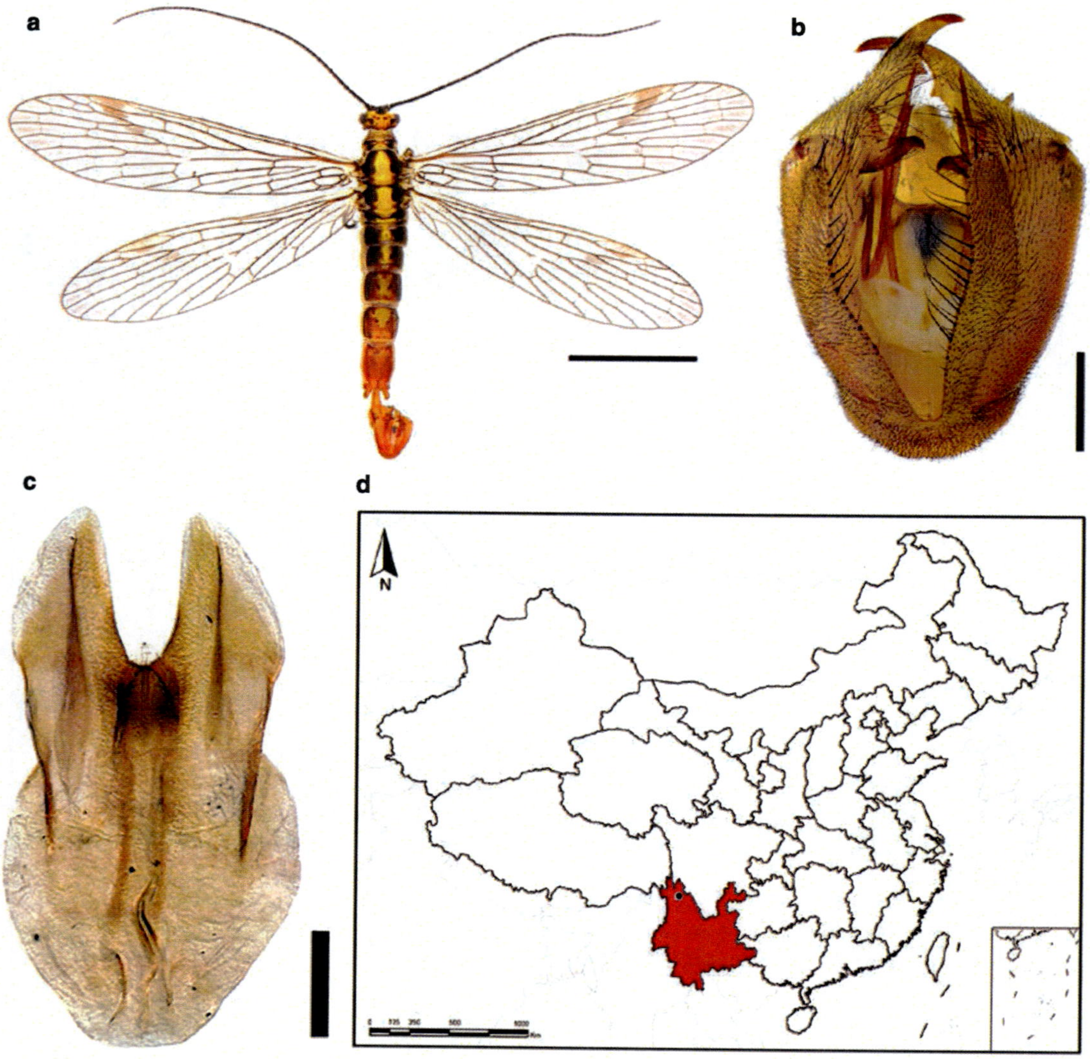

Fig. 10.102 (**a**) Male adult of ***Dicerapanorpa macula***, dorsal view (Xiaozhongdian, Shangri-La, Yunnan) (*Scale bar*: 5.0 mm) **[NWAU]**. (**b**) Male genital bulb of ***Dicerapanorpa macula***, ventral view (*Scale bar*: 0.5 mm). (**c**) Female medigynium of ***Dicerapanorpa macula***, ventral view (*Scale bar*: 0.2 mm). (**d**) Distribution of ***Dicerapanorpa macula***

gonocoxites. Gonostylus shorter than gonocoxites and bearing acute median tooth and large, bifurcated basal process. Paramere trifurcated: all branches straight and elongated, median branch longest and extending to median tooth of gonostylus. Aedeagus short, not exceeding apex of gonocoxites, with membranous and greatly protruding ventral valves, and bearing several long setae on dorsal processes.

Female subgenital plate broad with slightly projected apex. Medigynium with main plate rounded; posterior arms stout, broad, subacute terminally; axis concealed in main plate with slender, parallel apodemes.

Emergence. May to June.

Distribution. Yunnan (Shangri-La).

10.3.13 *Dicerapanorpa stotzneri* Esben-Petersen, 1934

(Stotzner's Two-Horned Scorpionfly) [施托氏双角蝎蛉] (Fig. 10.103)

Measurements. Forewing length 17.0–17.5 mm.

Morphology. Vertex mostly yellow. Ocellar triangle enclosed by large black spot. Rostrum yellowish brown, with two black stripes frontally. Pronotum yellowish brown. Meso- and metanotum dark brown laterally and bearing broad light-yellow stripe mesally. Pleura and legs yellowish brown. Wing membrane hyaline nearly colorless and devoid of markings except slight tinge on apex. Male T1–T5 mostly yellow with dark brown flanks. Notal organ flat and covering acute postnotal organ. A6 yellowish brown, bearing pair of finger-like anal horns on dorsal apex. A7 greatly stalked at base.

Male genital bulb broad oval. Epandrium emarginates terminally in broad U-shape and forms pair of finger-like processes laterally. Hypovalves stripe-like, bearing row of sparsely and irregularly arranged long bristles along inner margin, with expanded and truncated apexes extending to apex of gonocoxites. Gonostylus shorter than gonocoxites, bearing acute median tooth and large basal process. Paramere trifurcated: ventral branch

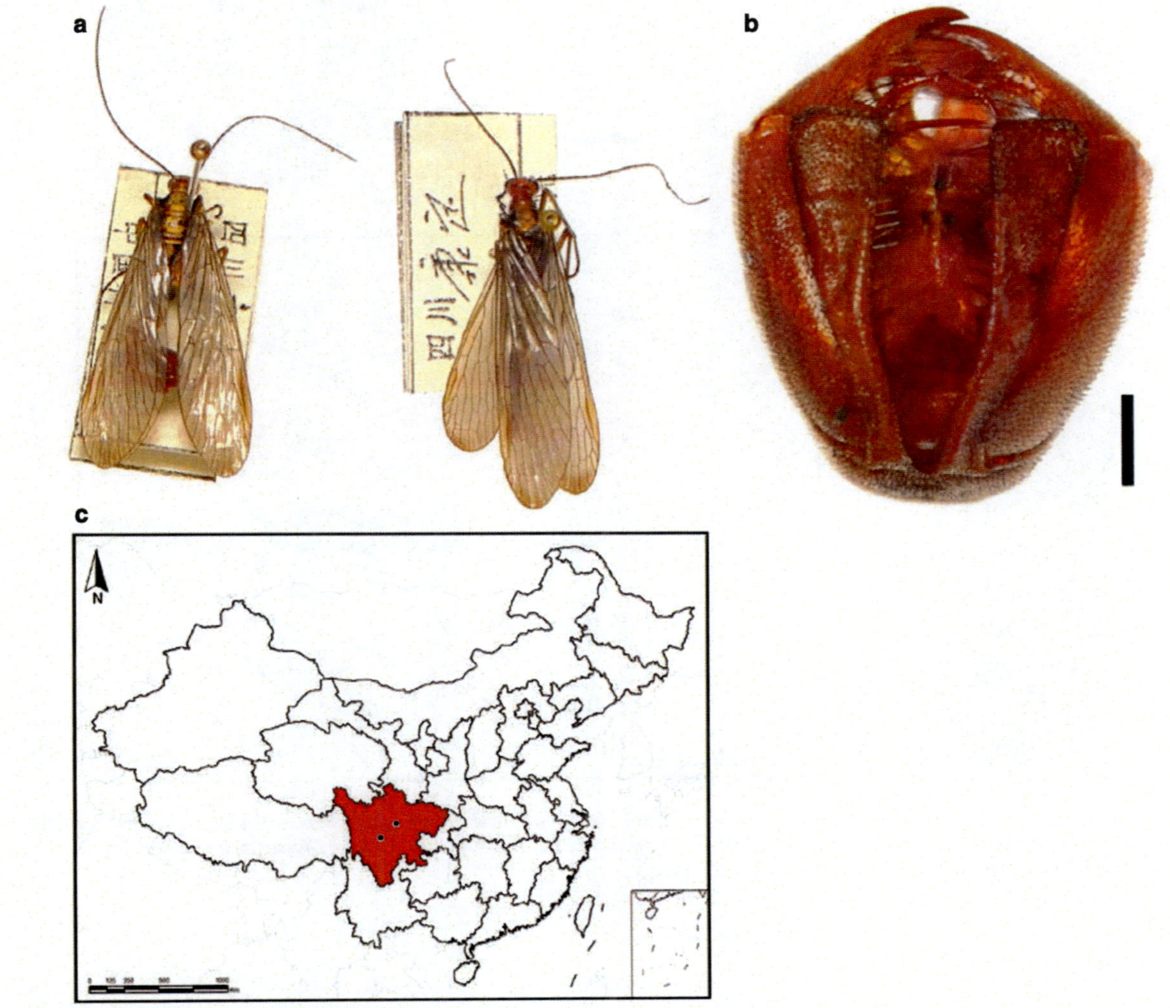

Fig. 10.103 (**a**) Adults of ***Dicerapanorpa stotzneri***, dorsal view (Kangding, Sichuan) **[IZAS]**. Left. Male; Right. Female. (**b**) Male genital bulb of ***Dicerapanorpa stotzneri***, ventral view. (**c**) Distribution of ***Dicerapanorpa stotzneri***

almost extending to apex of aedeagus; median and dorsal branches exceeding apex of gonocoxites and curved mesad apically. Aedeagus short and not exceeding apex of gonocoxites.

Female subgenital plate broad with slightly projected apex. Medigynium with main plate pear-shaped; posterior arms slender, rounded terminally; axis concealed in main plate with slender, diverged apodemes.

Emergence. June to August.

Distribution. Sichuan (Dujiangyan, Kangding).

10.3.14 *Dicerapanorpa tanae* Hu, Wang & Hua, 2019

(Tan's Two-Horned Scorpionfly) [谭氏双角蝎蛉]
(Fig. 10.104)

Measurements. Forewing length 14.0–15.0 mm.

Morphology. Vertex mostly yellow. Ocellar triangle enclosed by large black spot. Rostrum yellow and with pair of conspicuous dark brown stripes frontally. Pronotum yellowish brown. Meso- and metanotum dark brown laterally and bearing

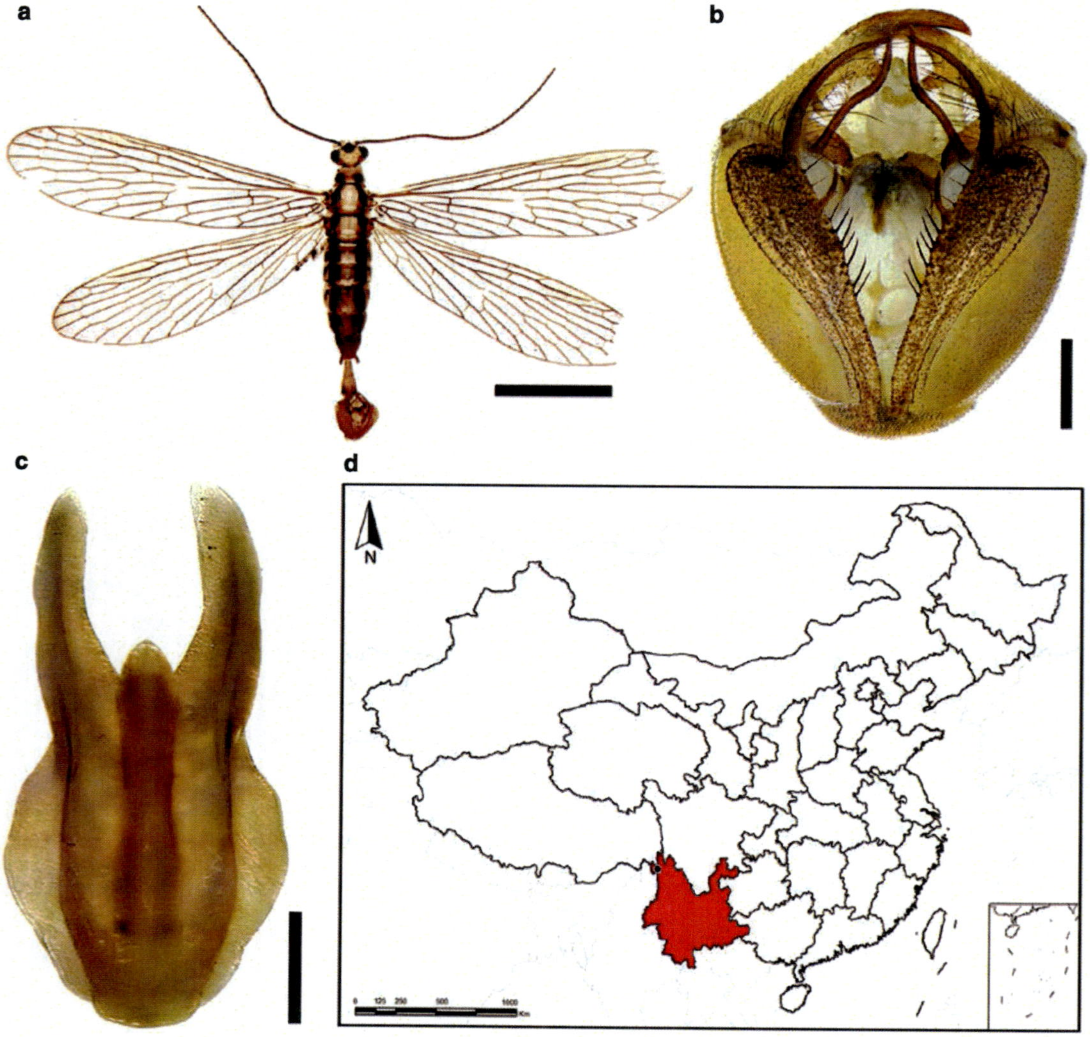

Fig. 10.104 (a) Male adult of *Neopanorpa tanae*, dorsal view (Heiwadi, Gongshan, Yunnan) (*Scale bar*: 5.0 mm) [NWAU]. (b) Male genital bulb of ***Dicerapanorpa tanae***, ventral view (*Scale bar*: 0.5 mm). (c) Female medigynium of ***Dicerapanorpa tanae***, ventral view (*Scale bar*: 0.2 mm). (d) Distribution of ***Dicerapanorpa tanae***

broad yellow median stripe. Pleura and legs yellowish brown. Wing membrane hyaline, and devoid of markings. Pterostigma light yellowish brown. Male T1–T5 mostly yellow with dark brown flanks. Notal organ flat and covering acute postnotal organ. A6 yellowish brown and bearing pair of finger-like anal horns on dorsal apex. A7 greatly stalked at base.

Male genital bulb spherical. Epandrium emarginate terminally in broad U-shape to form pair of finger-like processes laterally. Hypovalves stripe-like and with row of sparsely and irregularly arranged long bristles along inner margin, with rounded apexes slightly exceeding apex of gonocoxites. Gonostylus shorter than gonocoxites and bearing acute median tooth and large basal process. Paramere trifurcated: ventral branch elongated and straight; median branch longest and circularly curved; dorsal branch obliquely extending inward and greatly bending dorsad in distal portion. Aedeagus short and not exceeding apex of gonocoxites; dorsal processes short.

Female subgenital plate broad with slightly projected apex. Medigynium with main plate rounded; posterior arms slender, subacute terminally; axis concealed in main plate with slender, parallel apodemes.

Emergence. May to July.

Distribution. Yunnan (Gongshan).

10.3.15 *Dicerapanorpa tenuis* Hu, Wang & Hua, 2019

(Slender Two-Horned Scorpionfly) [纤细双角蝎蛉]
(Fig. 10.105)

Measurements. Forewing length 14.0–16.0 mm.

Morphology. Vertex mostly yellow. Ocellar triangle enclosed by large black spot. Rostrum yellow and with pair of conspicuous dark brown stripes frontally. Pronotum yellowish brown. Meso- and metanotum dark brown laterally and bearing broad yellow median stripe. Pleura and legs yellowish brown. Wing membrane hyaline, and devoid of markings, except greatly reduced apical band. Pterostigma light yellowish brown. Male T1–T5 mostly

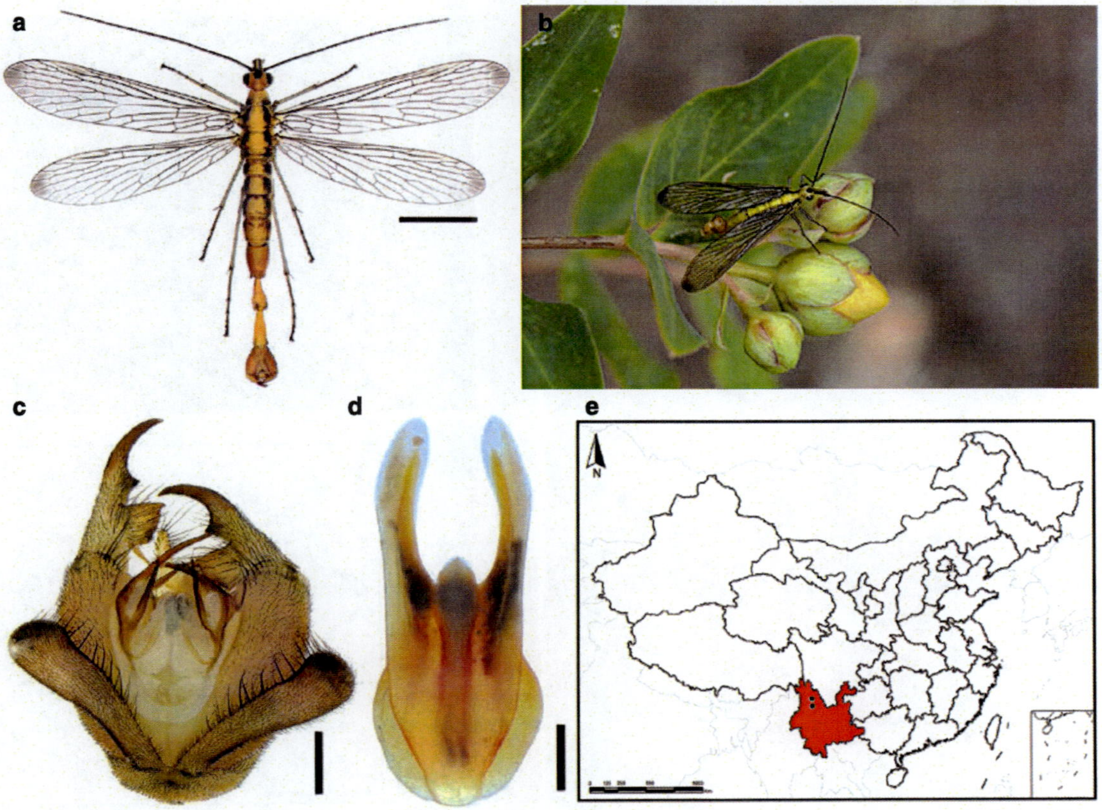

Fig. 10.105 (**a**) Male adult of ***Dicerapanorpa tenuis***, dorsal view (Ma'er Mountain, Heqing, Yunnan) (*Scale bar*: 5.0 mm) **[DALU]**. (**b**) Male adult of ***Dicerapanorpa tenuis*** in the field (Ma'er Mountain, Heqing, Yunnan). (**c**) Male genital bulb of ***Dicerapanorpa tenuis***, ventral view (*Scale bar*: 0.5 mm). (**d**) Female medigynium of ***Dicerapanorpa tenuis***, ventral view (*Scale bar*: 0.2 mm). (**e**) Distribution of ***Dicerapanorpa tenuis***

yellow with dark brown flanks. Notal organ flat and covering acute postnotal organ. A6 yellowish brown and bearing pair of finger-like anal horns on dorsal apex. A7 greatly stalked at base.

Male genital bulb broad oval. Epandrium emarginates terminally in broad U-shape to form pair of finger-like processes laterally. Hypovalves stripe-like and with row of sparsely and irregularly arranged long bristles along inner margin, with rounded apexes slightly exceeding apex of gonocoxites. Gonostylus shorter than gonocoxites and bearing acute median tooth and large basal process. Paramere trifurcated: ventral branch elongated and straight; median branch longest and bending inward; dorsal branch shorter than median branch and slightly curved. Aedeagus short and not exceeding apex of gonocoxites; dorsal processes broad.

Female subgenital plate broad with slightly projected apex. Medigynium with main plate rounded; posterior arms slender, rounded terminally; axis concealed in main plate with slender, parallel apodemes.

Emergence. May to July.

Distribution. Yunnan (Heqing, Lijiang).

10.3.16 *Dicerapanorpa tjederi* Carpenter, 1938

(Tjeder's Two-Horned Scorpionfly) [谢德氏双角蝎蛉] (Fig. 10.106)

Measurements. Forewing length 15.0–16.0 mm.

Morphology. Vertex mostly yellow. Ocellar triangle enclosed by large black spot. Rostrum yellowish brown, with two dark brown stripes frontally. Pronotum yellowish brown. Meso- and metanotum dark brown laterally and bearing broad light-yellow stripe mesally. Pleura and legs yellowish brown. Wing membrane hyaline and colorless, devoid of instinct markings. Male T1–T5 mostly pale yellow with dark brown flanks. Notal organ flat and covering

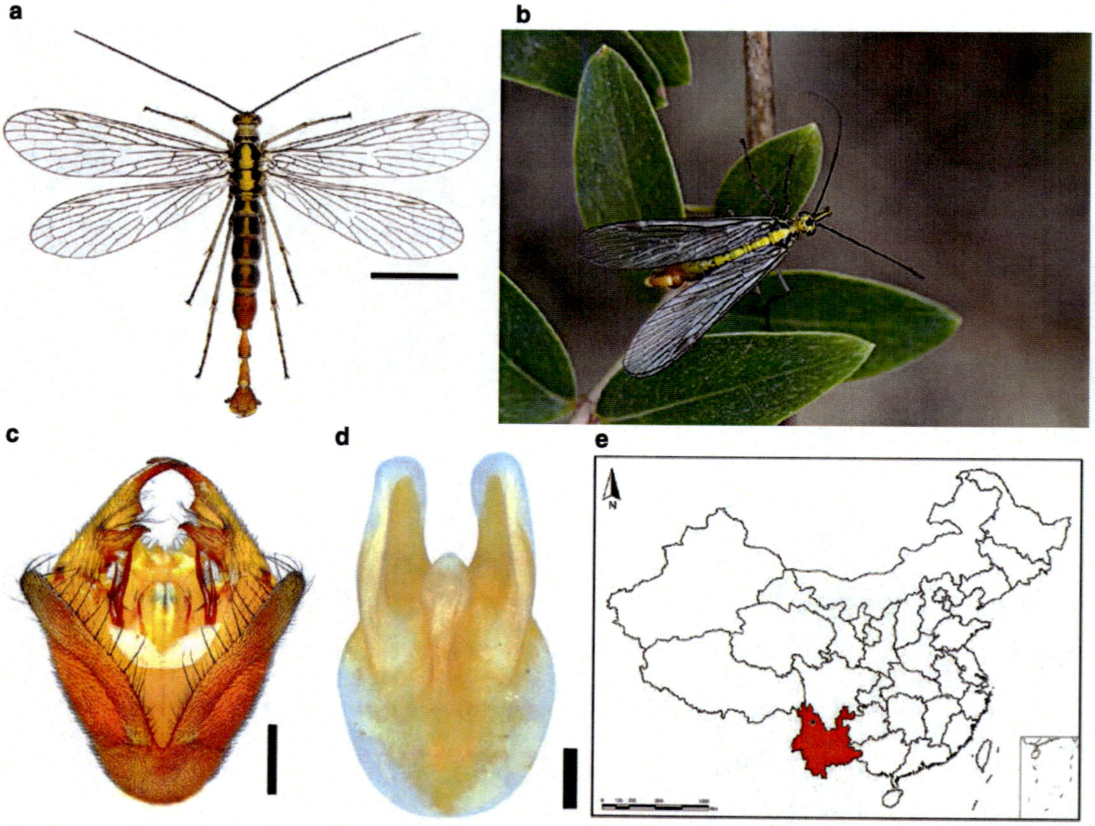

Fig. 10.106 (a) Male adult of ***Dicerapanorpa tjederi***, dorsal view (Yulong Snow Mountain, Lijiang, Yunnan) (*Scale bar*: 5.0 mm) **[DALU]**. (b) Male adult of ***Dicerapanorpa tjederi*** in the field (Yulong Snow Mountain, Lijiang, Yunnan). (c) Male genital bulb of ***Dicerapanorpa tjederi***, ventral view (*Scale bar*: 0.5 mm). (d) Female medigynium of ***Dicerapanorpa tjederi***, ventral view (*Scale bar*: 0.2 mm). (e) Distribution of ***Dicerapanorpa tjederi***

acute postnotal organ. A6 yellowish brown and bearing pair of finger-like anal horns on dorsal apex. A7 greatly stalked at base.

Male genital bulb broad oval. Epandrium emarginates terminally in shallow U-shape and forms pair of finger-like processes laterally. Hypovalves stripe-like and bearing row of sparsely and irregularly arranged long bristles along inner margin, with rounded apexes extending to apex of gonocoxites. Gonostylus shorter than gonocoxites and bearing acute median tooth and large basal process. Paramere trifurcated: ventral branch straight and exceeding apex of aedeagus; median and dorsal branches long and straight, extending to median teeth of gonostylus; median branch usually furcated in distal portion. Aedeagus short and not exceeding apex of gonocoxites.

Female subgenital plate broad with slightly projected apex. Medigynium with main plate rounded; posterior arms stout, rounded terminally; axis concealed in main plate with slender, parallel apodemes.

Emergence. May to June.

Distribution. Yunnan (Lijiang).

10.3.17 *Dicerapanorpa triclada* Qian & Zhou, 2001

(Flower-Meadow Two-Horned Scorpionfly) [花甸双角蝎 蛉] (Fig. 10.107)

Measurements. Forewing length 14.0–15.8 mm.

Morphology. Vertex mostly yellow with a few black spots. Ocellar triangle enclosed by large black spot. Rostrum yellowish brown, with two dark brown stripes frontally. Pronotum yellowish brown. Meso- and metanotum dark brown laterally and bearing broad light-yellow stripe mesally. Pleura and legs dark brown. Wing membrane hyaline and colorless, with distinct, subtriangular pterostigmal band. Male T1–T5 mostly pale yellow with dark brown flanks. Notal organ flat and covering acute postnotal organ. A6–A9 dark brown to black. A6 bearing pair of short finger-like anal horns on dorsal apex. A7 greatly stalked at base.

Male genital bulb broad oval. Epandrium emarginates terminally in a shallow U-shape and forms pair of finger-like

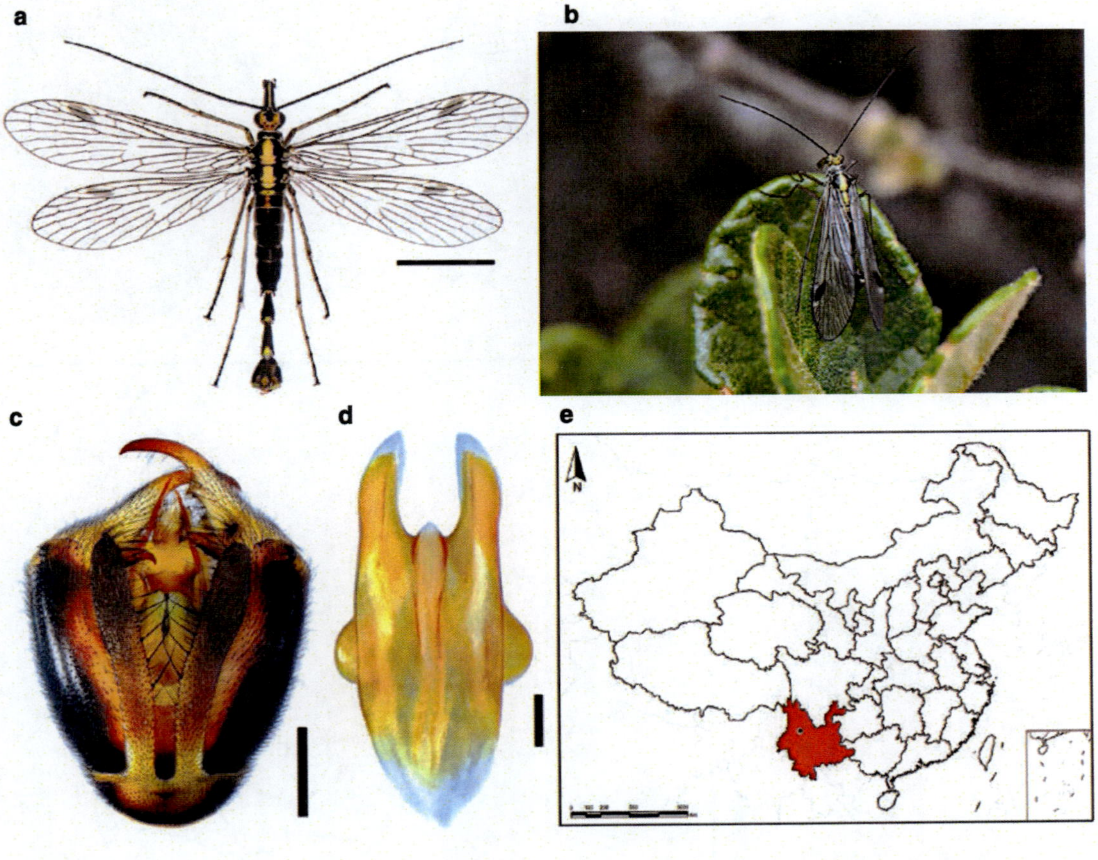

Fig. 10.107 (a) Male adult of ***Dicerapanorpa triclada***, dorsal view (Cangshan Mountain, Dali, Yunnan) (*Scale bar*: 5.0 mm) **[DALU]**. (b) Male adult of *Dicerapanorpa triclada* in the field (Cangshan Mountain, Dali, Yunnan). (c) Male genital bulb of ***Dicerapanorpa triclada***, ventral view (*Scale bar*: 0.5 mm). (d) Female medigynium of ***Dicerapanorpa triclada***, ventral view (*Scale bar*: 0.2 mm). (e) Distribution of ***Dicerapanorpa triclada***

processes laterally. Hypovalves stripe-like and bearing row of sparsely and irregularly arranged long bristles along inner margin, with rounded apexes extending to apex of gonocoxites. Gonostylus shorter than gonocoxites and bearing acute median tooth and large basal process. Paramere trifurcated: ventral branch straight and exceeding apex of aedeagus; median and dorsal branches long and straight, extending to median teeth of gonostylus. Aedeagus short and not exceeding apex of gonocoxites.

Female subgenital plate broad with slightly projected apex. Medigynium with main plate oblong; earlobe-like process on each side; posterior arms stout, broad, subacute terminally; axis concealed in main plate with slender, parallel apodemes.

Emergence. May to June.

Distribution. Yunnan (Cangshan Mountain).

Remarks. Originally, this species bears the Chinese vernacular name "三枝蝎蛉" (literally, trifurcated scorpionfly) (Qian and Zhou 2001). However, most species in the genus *Dicerapanorpa* bear trifurcated parameres in the male genitalia. Therefore, we rename it using the type locality, Huadianba in Cangshan Mountain, Dali.

10.3.18 *Dicerapanorpa yijunae* Hu & Hua, 2019

(Yijun's Two-Horned Scorpionfly) [义君双角蝎蛉] (Fig. 10.108)

Measurements. Forewing length 15.0–17.0 mm.

Morphology. Vertex mostly yellow. Ocellar triangle enclosed by large black spot. Rostrum yellow and with pair of conspicuous dark brown stripes frontally. Pronotum yellowish brown. Meso- and metanotum dark brown laterally and bearing broad yellow median stripe. Pleura and legs yellowish brown. Wing membrane hyaline, and devoid of markings, except apex slightly darkened. Pterostigma light

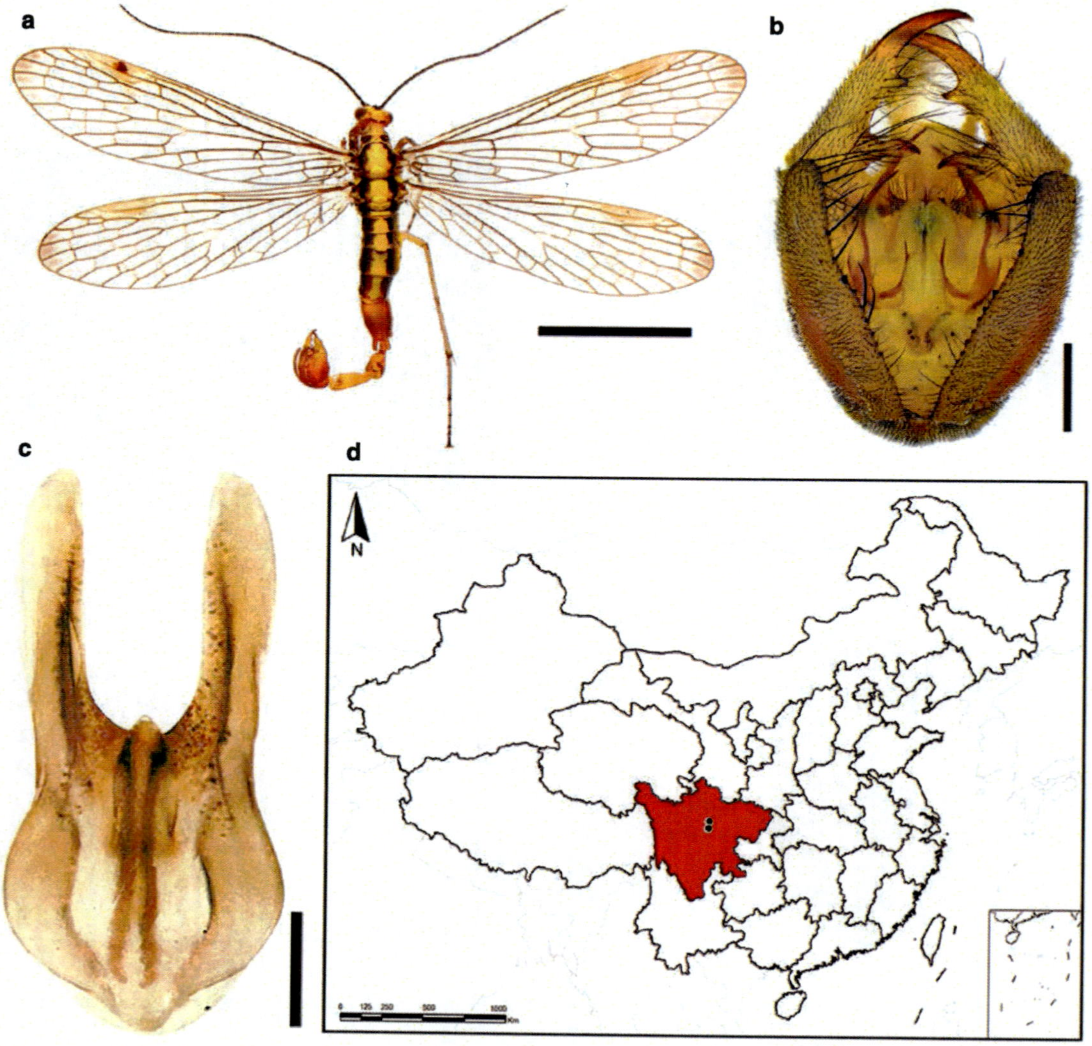

Fig. 10.108 (**a**) Male adult of ***Dicerapanorpa yijunae***, dorsal view (Anzihe, Chongzhou, Sichuan) (*Scale bar*: 5.0 mm) **[NWAU]**. (**b**) Male genital bulb of ***Dicerapanorpa yijunae***, ventral view (*Scale bar*: 0.5 mm). (**c**) Female medigynium of ***Dicerapanorpa yijunae***, ventral view (*Scale bar*: 0.2 mm). (**d**) Distribution of ***Dicerapanorpa yijunae***

yellowish brown. Male T1–T5 mostly yellow with dark brown flanks. Notal organ flat and covering acute postnotal organ. A6 yellowish brown and bearing pair of finger-like anal horns on dorsal apex. A7 greatly stalked at base.

Male genital bulb broad oval. Epandrium emarginate terminally in broad U-shape to form pair of finger-like processes laterally. Hypovalves stripe-like and with row of sparsely and irregularly arranged long bristles along inner margin, with rounded apexes extending to apex of gonocoxites. Gonostylus shorter than gonocoxites and bearing acute median tooth and large basal process. Paramere trifurcated: ventral branch short and hook-like; median branch longest, slender, and straight; dorsal branch shorter and curving mesad. Aedeagus short and not exceeding apex of gonocoxites; dorsal processes bearing numerous long setae.

Female subgenital plate broad with slightly projected apex. Medigynium with main plate short, rounded; posterior arms slender, longer than main plate, rounded terminally; axis concealed in main plate with slender, slightly diverged apodemes.

Emergence. May to July.

Distribution. Sichuan (Chongzhou, Wolong).

10.3.19 *Dicerapanorpa zhengkuni* Hu & Hua, 2020

(Zhengkun's Two-Horned Scorpionfly) [政坤双角蝎蛉] (Fig. 10.109)

Measurements. Forewing length 14.0–15.0 mm.

Morphology. Vertex mostly yellow. Ocellar triangle enclosed by large black spot. Rostrum yellow and with pair of conspicuous dark brown stripes frontally. Pronotum yellowish brown. Meso- and metanotum dark brown laterally

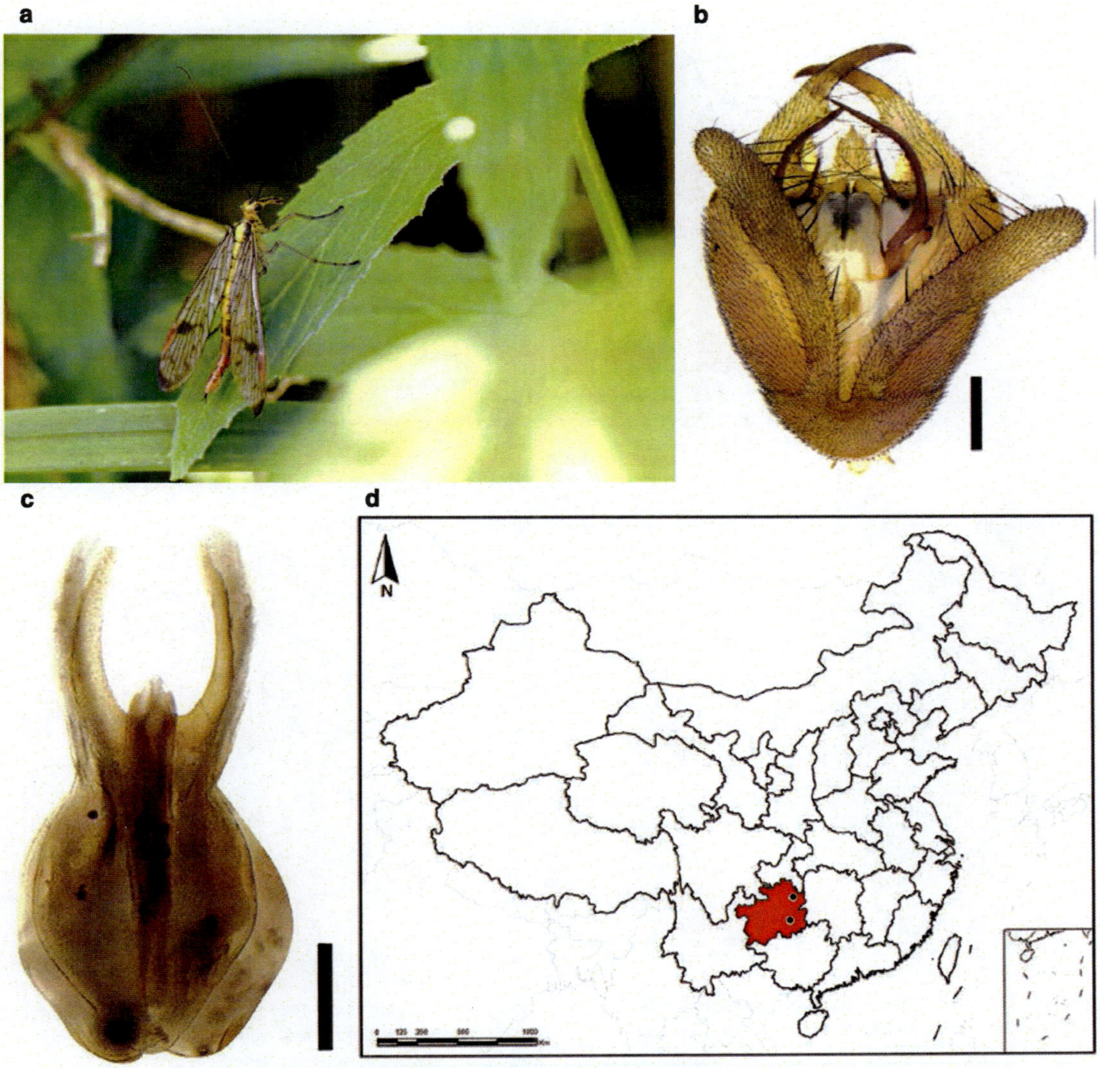

Fig. 10.109 (a) Female adult of ***Dicerapanorpa zhengkuni*** in the field (Leigong Mountain, Guizhou, ©Zhen-Hua Liu). (b) Male genital bulb of ***Dicerapanorpa zhengkuni***, ventral view (*Scale bar*: 0.5 mm). (c) Female medigynium of ***Dicerapanorpa zhengkuni***, ventral view (*Scale bar*: 0.2 mm). (d) Distribution of ***Dicerapanorpa zhengkuni***

and bearing broad yellow median stripe. Pleura and legs yellowish brown. Wing membrane hyaline, and devoid of markings except for dark, greatly reduced apical and pterostigmal band. Pterostigma light yellowish brown. Male T1–T5 mostly yellow with dark brown flanks. Notal organ flat and covering acute postnotal organ. A6 yellowish brown and bearing pair of finger-like anal horns on dorsal apex. A7 greatly stalked at base.

Male genital bulb broad oval. Epandrium emarginates terminally in broad U-shape to form pair of finger-like processes laterally. Hypovalves broad, stripe-like, greatly elongated, extending nearly to median tooth of gonostylus, and with row of sparsely and irregularly arranged long bristles along inner margin. Gonostylus shorter than gonocoxites and bearing acute median tooth and large basal process. Paramere trifurcated: ventral branch extremely short and hook-like; median branch longest and bending inward circularly; dorsal branch shorter than median branch and slightly curved dorsad. Aedeagus short and not exceeding apex of gonocoxites; dorsal processes broad.

Female subgenital plate broad with slightly projected apex. Medigynium with main plate oval; posterior arms slender, slightly shorter than main plate, rounded terminally; axis concealed in main plate with slender, parallel apodemes.

Emergence. May to July.

Distribution. Guizhou (Fanjing Mountain, Leigong Mountain).

10.3.20 *Dicerapanorpa zhongdianensis* Hu, Wang & Hua, 2019

(Zhongdian Two-Horned Scorpionfly) [中甸双角蝎蛉] (Fig. 10.110)

Measurements. Forewing length 13.0–14.0 mm.

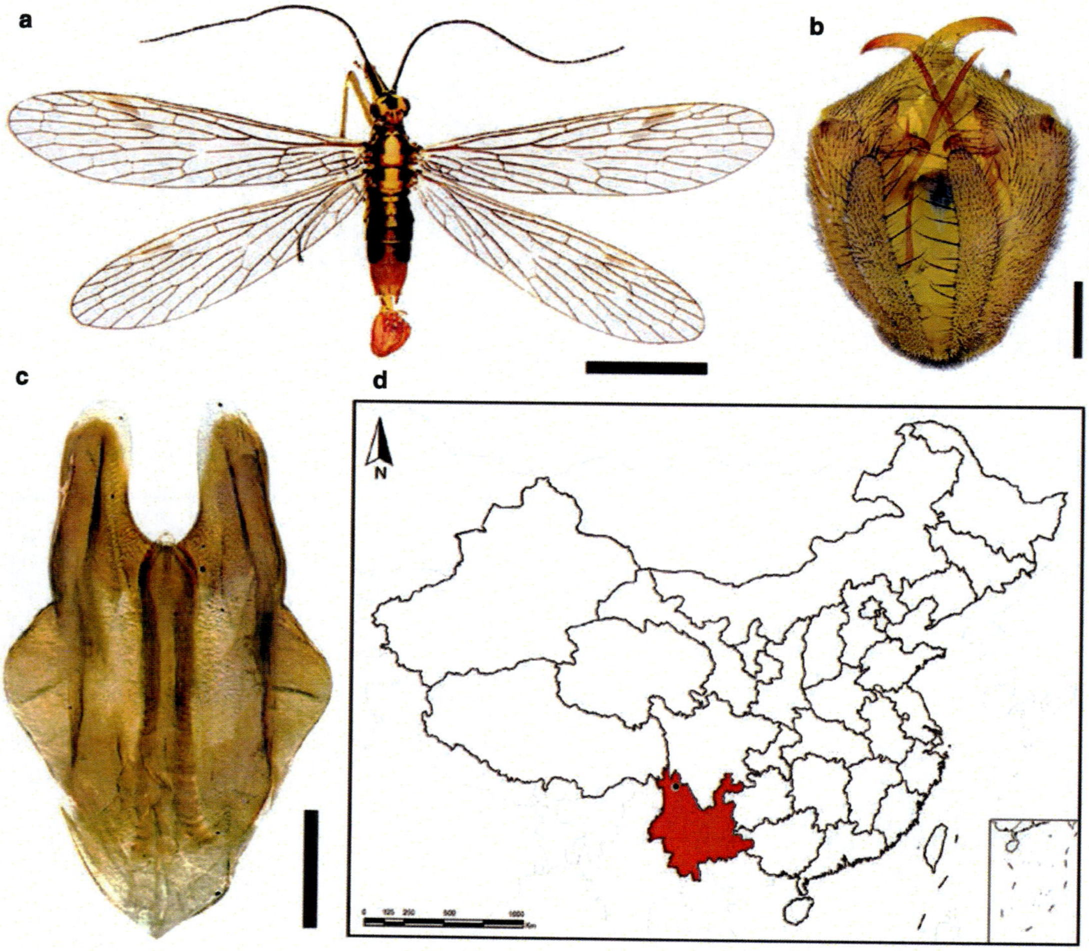

Fig. 10.110 (**a**) Male adult of ***Dicerapanorpa zhongdianensis***, dorsal view (Xiaozhongdian, Shangri-La, Yunnan) (*Scale bar*: 5.0 mm) [NWAU]. (**b**) Male genital bulb of ***Dicerapanorpa zhongdianensis***, ventral view (*Scale bar*: 0.5 mm). (**c**) Female medigynium of ***Dicerapanorpa zhongdianensis***, ventral view (*Scale bar*: 0.2 mm). (**d**) Distribution of ***Dicerapanorpa zhongdianensis***

Morphology. Vertex yellow, with large black spot on each side, and three smaller, blurred spots encircling each large one. Ocellar triangle enclosed by large black spot. Rostrum yellow, with pair of conspicuous dark brown stripes frontally. Pronotum dark brown. Meso- and metanotum mostly yellow middly and dark brown laterally. Pleura and legs yellow. Wing membrane hyaline, and devoid of markings, except greatly reduced pterostigmal band. Pterostigma light yellowish brown. Male T1–T5 mostly yellow with dark brown flanks. Notal organ flat and covering acute post-notal organ. A6 yellowish brown, bearing pair of finger-like anal horns on dorsal apex. A7 greatly stalked at base.

Male genital bulb broad oval. Epandrium emarginates terminally in broad U-shape to form pair of finger-like processes laterally. Hypo-valves stripe-like and bearing row of sparsely and irregularly arranged long bristles along inner margin, with expanded rounded apexes extending to apex of gonocoxites. Gonostylus shorter than gonocoxites and bearing acute median tooth and large, bifurcated basal process. Paramere trifurcated: ventral branch straight, reaching basal process of gonostylus; median branch much longer and extending far beyond median tooth of gonostylus; dorsal branch not exceeding median tooth of gonostylus. Aedeagus short, not exceeding apex of gonocoxites, with membranous and greatly protruding ventral valves, and bearing several long setae on dorsal processes.

Female subgenital plate broad with slightly projected apex. Medigynium with main plate oblong; subtriangular process on each side; posterior arms stout, broad, rounded terminally; axis concealed in main plate with slender, parallel apodemes.

Emergence. May to June.

Distribution. Yunnan (Shangri-La).

10.4 Genus *Furcatopanorpa* Ma & Hua, 2011

In this genus, the notal and postnotal organs are completely lost in the male abdomen, and A6–A8 are not elongated as most scorpionflies. The females have the axis of genital plate extremely elongated posteriorly, and bifurcated terminally. The wings are held roof-like at repose. Their relatively short abdomens are often covered by the wings. Therefore, their genders are hard to determine from dorsal aspect.

A monotypic genus (Fig. 10.111).

Fig. 10.111 Female adult of ***Furcatopanorpa longihypovalva***, dorsal view (Tiantaishan Forest Park, Baoji, Shaanxi)

10.4.1 *Furcatopanorpa longihypovalva* Hua & Cai, 2009

(Long-hypovalved Furcatoscorpionfly) [长瓣叉蝎蛉] (Fig. 10.112)

Measurements. Forewing length 15.5–18.5 mm.

Morphology. Vertex black. Ocellar triangle black. Rostrum yellowish brown. Nota black. Pleura and legs light greyish brown. Wing membrane hyaline with brown pterostigma. Wings held roof-like over body at repose. Markings brown. In forewings, apical band narrow; pterostigmal band broad, lacking apical branch; marginal and basal spots absent; basal band represented by irregular stripe extending form end of CuP to wing base along hind margin. Male terga and sterna I–V black. T3 without notal organ. T4 without postnotal organ. A6 shorter than A5, cylindrical. A7 and A8 shorter than A6 and not constricted basally.

Male genital bulb long elliptical. Epandrium slightly emarginate on apical margin. Hypovalves narrow and very long, extending beyond apex of gonocoxites. Gonostylus shorter than half-length of gonocoxites, with large basal process and mesal tooth surrounded by serrate margin. Paramere slender, greatly curled, or relatively straight, bearing numerous microtrichia. Aedeagus with ventral valves sclerotized, pouch-like; dorsal valves finger-like and divergent; dorsal processes bilobed: ventral lobe greatly enlarged and enclos-

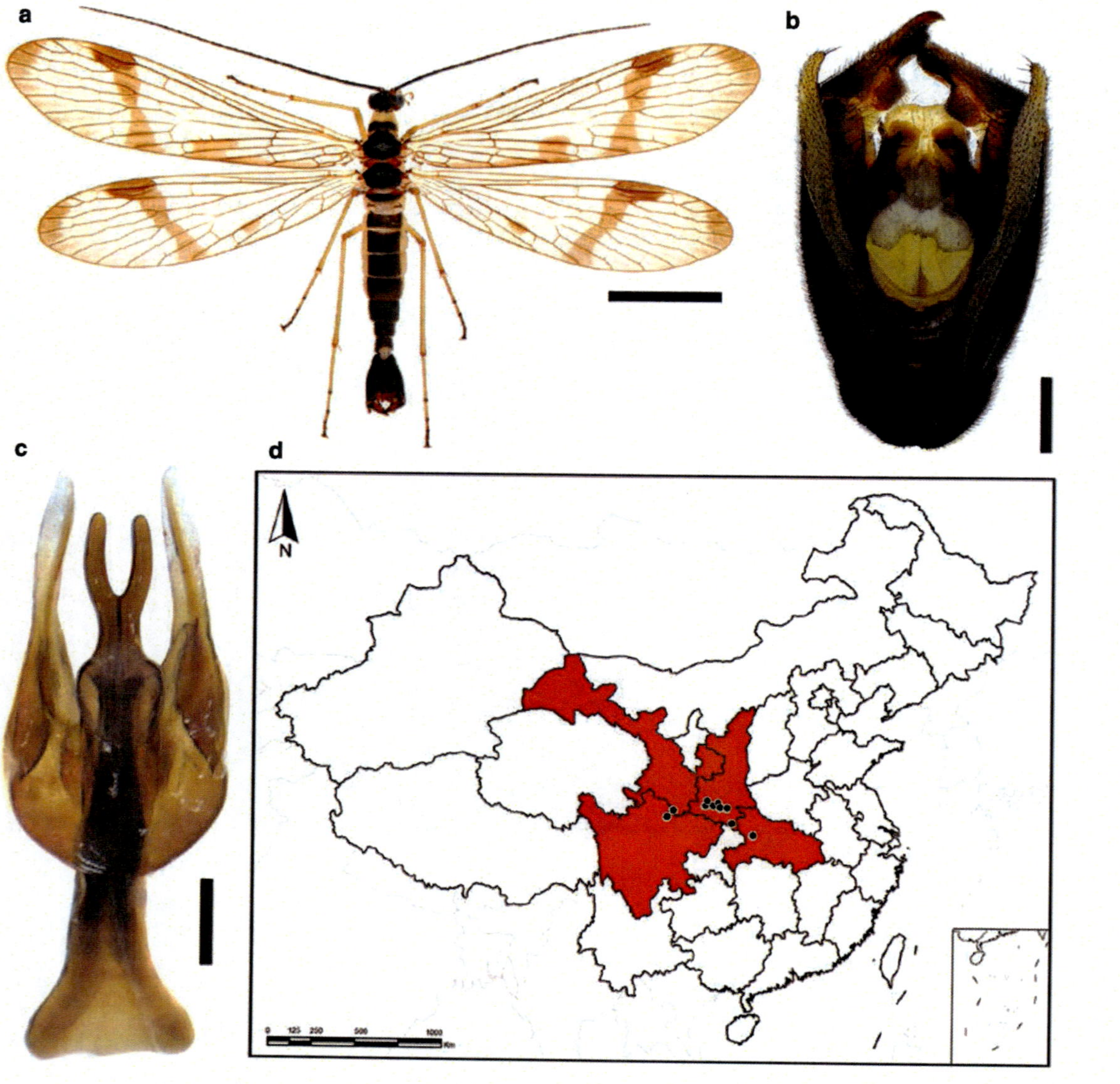

Fig. 10.112 (**a**) Male adult of ***Furcatopanorpa longihypovalva***, dorsal view (Tiantaishan National Forest Park, Baoji, Shaanxi) (*Scale bar*: 5.0 mm) **[NWAU]**. (**b**) Male genital bulb of ***Furcatopanorpa longihypovalva***, ventral view (*Scale bar*: 0.5 mm). (**c**) Female medigynium of ***Furcatopanorpa longihypovalva***, ventral view (*Scale bar*: 0.2 mm). (**d**) Distribution of ***Furcatopanorpa longihypovalva***

ing distal portion of other parts of aedeagus; dorsal lobe extending caudo-dorsad.

Female subgenital plate oval with V-shaped terminal emargination. Medigynium narrow, with main plate greatly folded ventrad on lateral margins; posterior arms slender and acute; axis greatly elongated not only basally but also distally; distal apex of axis greatly bifurcated; apodemes slender and slightly diverged.

Emergence. May to August.

Distribution. Gansu (Wenxian), Hubei (Shennongjia), Shaanxi (Baoji, Liuba, Meixian, Ningshan, Taibai, Zhouzhi), Sichuan (Pingwu).

10.5 Genus *Megapanorpa* Wang & Hua, 2019

(Megascorpionflies, megapanorpas)

Members of *Megapanorpa* can be recognized by the extraordinarily large body size (forewing length usually exceeding 17 mm and width exceeding 3.5 mm), the males bearing a single anal horn, the females bearing a pair of greatly developed laterotergites in A9, and the female medigynium bearing a concealed axis (Fig. 10.113).

Fig. 10.113 Habitus of *Megapanorpa*. *Megapanorpa jiangorum* in the field (Mamadi, Shimian, Sichuan, ©Lu Jiang)

Phylogenetically, this genus is most closely related to *Dicerapanorpa* (Wang and Hua 2021).

Megapanorpa anfracta (Ju & Zhou, 2003) **[New Combination]**

Panorpa anfracta Ju & Zhou, 2003 was published from Jiulong Mountain (Zhejiang: Jiulong). Its remote and isolated distribution seems weird, since all the other congeners were recorded from Shaanxi, Sichuan, Yunnan (Wang and Hua 2019a) and Guizhou (unpublished data). This species bears a single anal horn in male A6 and shares the general morphology with other species of *Megapanorpa*. Therefore, it is transferred into *Megapanorpa* as a new combination.

Six known species are endemic to China. Five of them are included in this atlas.

Key to Species of Megapanorpa in This Atlas
(Five species, males)

1. Wing markings well-developed; anal horn short and acute; parameres bearing small hook-like projection near the middle of outer margin .... ***Megapanorpa wanghongjiani*** Wang & Hua
– Wing markings greatly reduced; anal horn long and finger-like; parameres lacking such projection 2
2. Nota with yellow median stripe 3
– Nota entirely black 4
3. Distal portion of A7 two times as wide as basal stalk; paramere with dorsal branch .... ***Megapanorpa grandis*** Wang & Hua
– Distal portion of A7 narrower; paramere lacking dorsal branch .... ***Megapanorpa absens*** Wang & Hua
4. Distal portion of dorsal branch of paramere greatly curved inner ward .... ***Megapanorpa gaokaii*** Wang & Hua
– Distal portion of dorsal branch slightly curved inner ward .... ***Megapanorpa jiangorum*** Wang & Hua

10.5.1 *Megapanorpa absens* Wang & Hua, 2019

(Absence Megascorpionfly) [缺枝广翅蝎蛉] (Fig. 10.114)
Measurements. Forewing length 15.0–19.0 mm.

Morphology. Vertex and ocellar triangle black. Rostrum mostly black with yellowish portion genae. Pronotum black. Meso- and metanotum black, with yellow median stripe extending from anterior border of mesoscutum to metascutellum, and disconnected at mesopostnotum, forming two bell-shaped patterns. Wing membrane hyaline; markings usually sexually dimorphic: in males, only apex slightly darkened; in females, greatly reduced apical and pterostigmal bands present; pterostigma light brown. T1–T5 black. Notal organ stout, slightly protruded, and covering acute postnotal organ. A6 mostly black, with only anal horn and ventro-apical border yellowish brown. Anal horn conical basally and finger-like apically. A7 and A8 yellowish brown. A7 greatly constricted and stalk-like in basal 1/3, with apex beveled; A8 slightly longer than A7, with apex trumpet-shaped.

Male genital bulb spherical, mostly reddish brown except hypovalves blackish brown. Epandrium greatly emarginated in O-shape and forming pair of finger-like processes laterally. Hypandrium with extremely short basal stalk. Hypovalves stripe-like, almost extending to apex of gonocoxites, broad in basal 3/5 and narrower in distal portion, and bearing several long, stout bristles along inner margin. Gonostylus shorter than gonocoxite, with blunt median tooth and short basal process. Paramere with dorsal bridge strongly developed, two parts connected in U-shape basally; ventral branch slender and glabrous, extending to middle of gonostylus, with tapering apex curved dorsomesally; dorsal branch absent. Aedeagal valves fused mesally and slightly divergent apically, with indistinct small tooth on each dorsal processes dorsolaterally.

Female subgenital plate subtriangular, with shallow V-shaped emargination terminally. Laterotergites slender, exceeding apex of subgenital plate, and bifurcated terminally. Medigynium with short and slightly convergent posterior arms; apodemes of axis slightly divergent; main plate with basal portion membranous and fused with laterotergites.

Emergence. April to July.

Distribution. Sichuan (Chongzhou, Tianquan, Wenchuan, Yingjing).

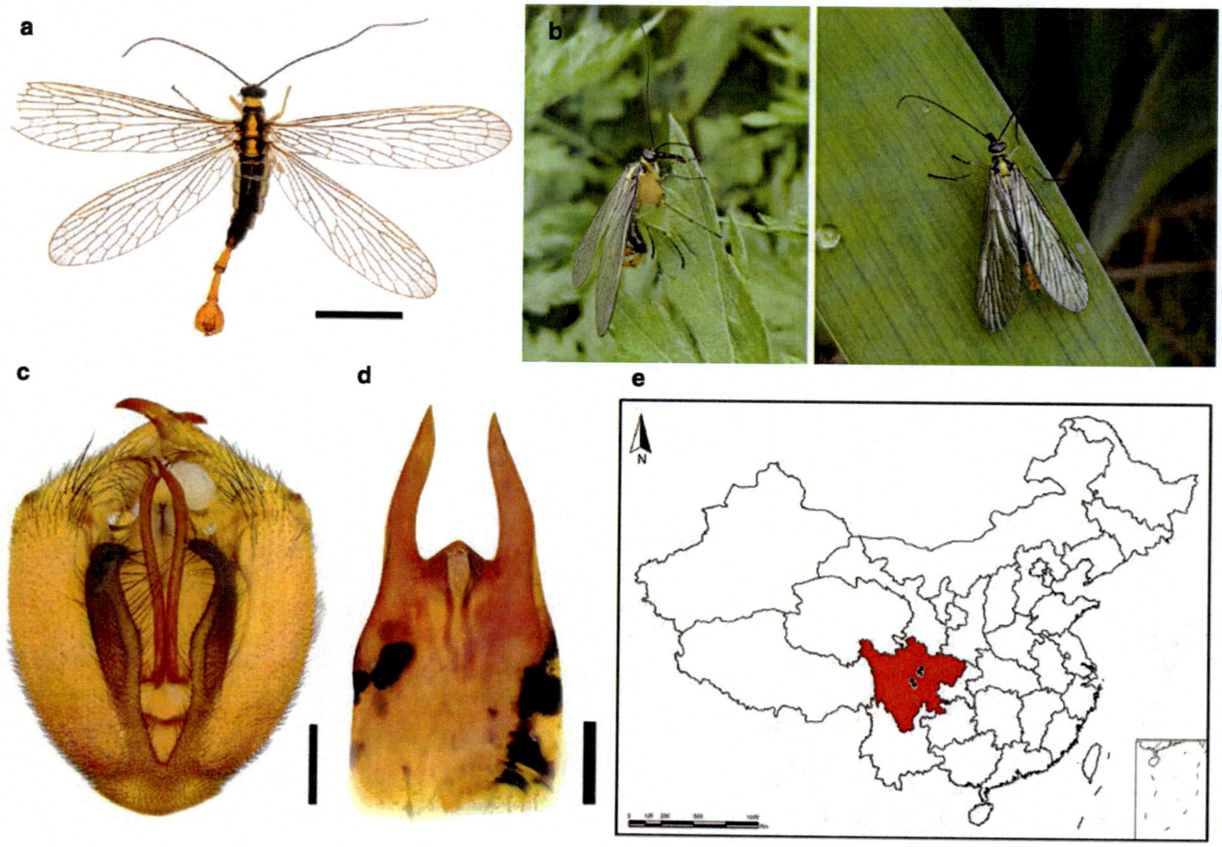

Fig. 10.114 (a) Male adult of ***Megapanorpa absens***, dorsal view (Jiguan Mountain, Chongzhou, Sichuan) (*Scale bar*: 5.0 mm) **[NWAU]**. **(b)** Adults of ***Megapanorpa absens*** in the field (Anzihe, Chongzhou, Sichuan, ©Yi-Jun Chai). Left. Male; Right. Female. **(c)** Male genital bulb of ***Megapanorpa absens***, ventral view (*Scale bar*: 0.5 mm). **(d)** Female medigynium of ***Megapanorpa absens***, ventral view (*Scale bar*: 0.2 mm). **(e)** Distribution of ***Megapanorpa absens***

10.5.2 *Megapanorpa gaokaii* Wang & Hua, 2019

(Gao's Megascorpionfly) [高氏广翅蝎蛉] (Fig. 10.115)
Measurements. Forewing length 16.0–19.0 mm.

Morphology. Vertex and ocellar triangle black. Rostrum mostly black with yellowish portion genae. Nota black. Wing membrane hyaline; markings usually sexually dimorphic: in males, only oblique, greatly reduced pterostignal band present; in females, pterostigmal band reduced into oblique stripe, and basal band represented by small spot; pterostigma light brown. T1–T5 black. Notal organ stout, slightly protruded and covering acute postnotal organ. A6 mostly black, with only anal horn and ventro-apical border yellowish brown. Anal horn conical basally and finger-like apically. A7 and A8 yellowish brown. A7 greatly constricted and stalk-like in basal 1/3, with apex beveled; A8 slightly longer than A7, with apex trumpet-shaped.

Male genital bulb spherical, mostly reddish brown except hypovalves blackish brown. Epandrium greatly emarginated in O-shape and forming pair of finger-like processes laterally. Hypandrium with extremely short basal stalk. Hypovalves stripe-like, almost extending to apex of gonocoxites, broadest and slightly projected in middle of inner margin, and bearing several long, stout bristles along inner margin. Gonostylus shorter than gonocoxite, with blunt median tooth and short basal process. Paramere bifurcated: ventral branch long, straight, and glabrous; dorsal branch bearing numerous microtrichia along inner margin, and greatly curved inward in

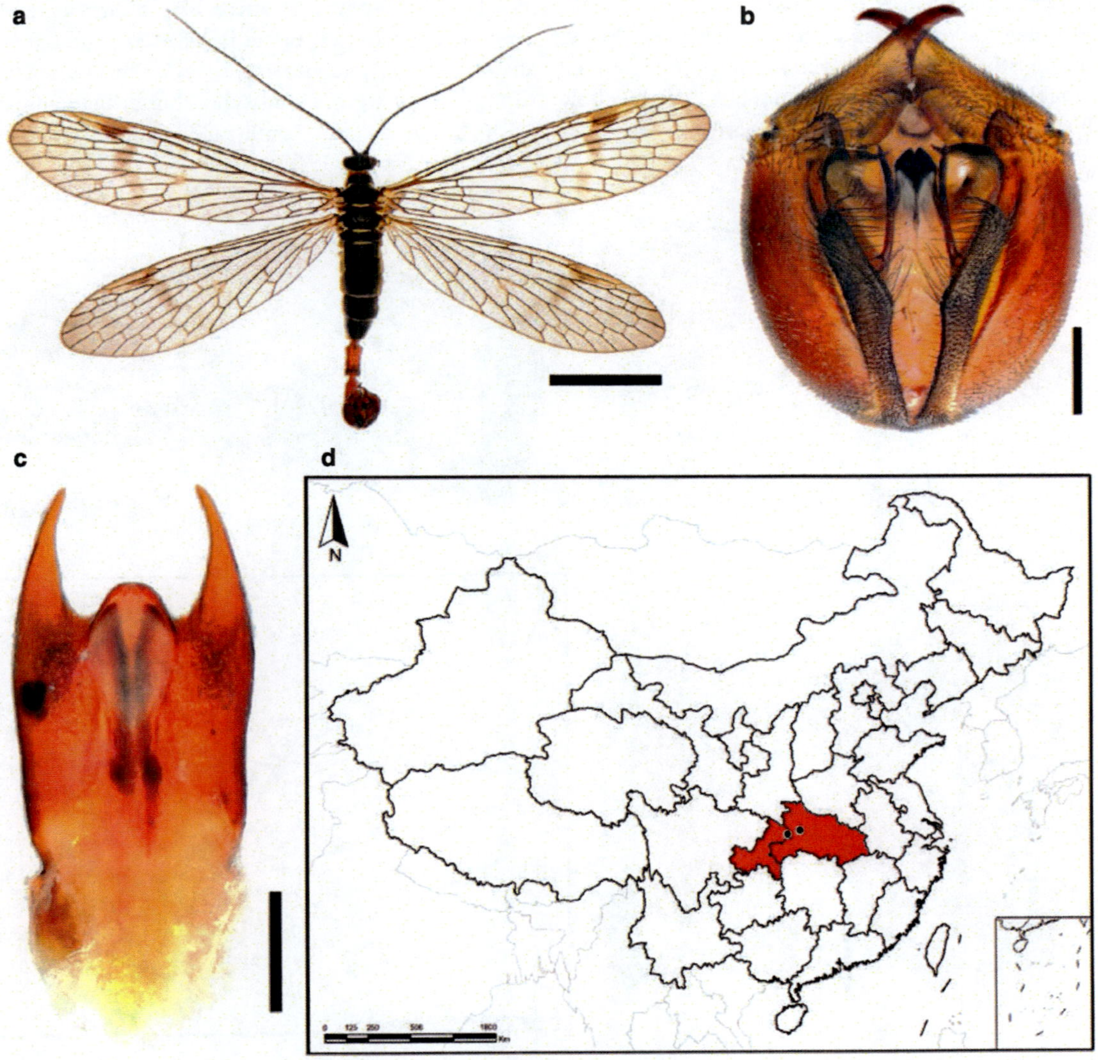

Fig. 10.115 (a) Male adult of ***Megapanorpa gaokaii***, dorsal view (Dajiuhu, Shennongjia, Hubei) (*Scale bar*: 5.0 mm) [NWAU]. (b) Male genital bulb of ***Megapanorpa gaokaii***, ventral view (*Scale bar*: 0.5 mm). (c) Female medigynium of ***Megapanorpa gaokaii***, ventral view (*Scale bar*: 0.2 mm). (d) Distribution of ***Megapanorpa gaokaii***

distal portion. Aedeagal valves fused mesally and slightly divergent apically; dorsal processes greatly protruding caudad and acute.

Female subgenital plate subtriangular, with shallow V-shaped emargination terminally. Laterotergites broad and blade-like, not exceeding apex of subgenital plate. Medigynium with short and slightly convergent posterior arms; apodemes of axis slightly divergent; main plate with basal portion membranous and fused with laterotergites.

Emergence. May to July.

Distribution. Chongqing (Wushan), Hubei (Shennongjia).

10.5.3 *Megapanorpa grandis* Wang & Hua, 2019

(Great Megascorpionfly) [大广翅蝎蛉] (Fig. 10.116)

Measurements. Forewing length 17.5–20.0 mm.

Morphology. Vertex and ocellar triangle black. Rostrum mostly black with yellowish portion genae. Pronotum black. Meso- and metanotum black, with yellow median stripe extending from anterior border of mesoscutum to metascutellum, and disconnected at mesopostnotum, forming two bell-shaped patterns. Wing membrane hyaline; markings usually sexually dimorphic: in males, only apex slightly darkened; in females, greatly reduced apical and pterostigmal bands present; pterostigma light brown. T1–T5 black. Notal organ stout, strongly protruded with acute dorsal apex, and covering acute postnotal organ. A6 mostly black, with only anal horn and ventro-apical border yellowish brown. Anal horn conical basally and finger-like apically. A7 and A8 yellowish brown. A7 greatly constricted and stalk-like in basal 1/3, with apex beveled; A8 slightly longer than A7, with apex trumpet-shaped.

Male genital bulb spherical, mostly reddish brown except hypovalves blackish brown. Epandrium greatly emarginated in O-shape and forming pair of finger-like processes laterally. Hypandrium with extremely short basal stalk. Hypovalves stripe-like, almost extending to apex of gonocoxites, broad in basal 3/5 and narrower in distal portion, and bearing several long, stout bristles along inner margin. Gonostylus shorter than gonocoxite, with blunt median tooth and short basal process.

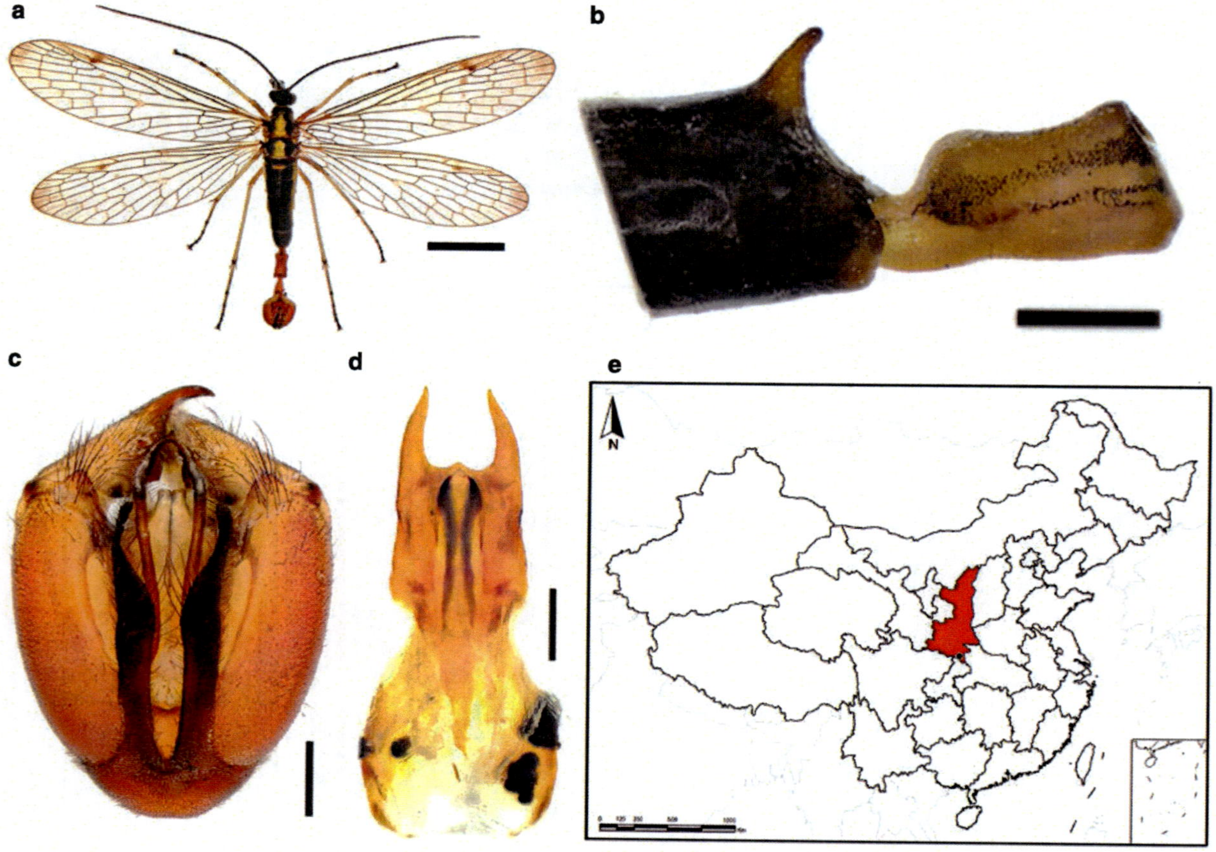

Fig. 10.116 (a) Male adult of ***Megapanorpa grandis***, dorsal view (Hualong Mountain, Shaanxi) (*Scale bar*: 5.0 mm) **[NWAU]. (b)** Male A6 and A7 of ***Megapanorpa grandis***, lateral view (*Scale bar*: 1.0 mm). **(c)** Male genital bulb of ***Megapanorpa grandis***, ventral view (*Scale bar*: 0.5 mm). **(d)** Female medigynium of ***Megapanorpa grandis***, ventral view (*Scale bar*: 0.2 mm). **(e)** Distribution of ***Megapanorpa grandis***

Paramere bifurcated with dorsal bridge strongly developed, two parts connected in U-shape basally; ventral branch slender and glabrous, extending to middle of gonostylus, with tapering apex curved dorsomesally; dorsal branch slender, slightly longer than aedeagus, curved circularly, and greatly vestigial basally. Aedeagal valves fused mesally and slightly divergent apically, with indistinct small tooth on each dorsal processes dorsolaterally.

Female subgenital plate subtriangular, with shallow V-shaped emargination terminally. Laterotergites slender, exceeding apex of subgenital plate, and bifurcated terminally. Medigynium with short and slightly convergent posterior arms; apodemes of axis slightly divergent; main plate with basal portion membranous and fused with laterotergites.

Emergence. June to July.

Distribution. Shaanxi (Hualong Mountain).

10.5.4 *Megapanorpa jiangorum* Wang & Hua, 2019

(Jiang' s Megascorpionfly) [姜氏广翅蝎蛉] (Fig. 10.117)

Measurements. Forewing length 17.0–19.0 mm.

Morphology. Vertex and ocellar triangle black. Rostrum mostly black with yellowish portion genae. Nota black. Wing membrane hyaline; males devoid of markings; in females, pterostigmal band reduced into oblique, short stripe. Pterostigma light brown. T1–T5 black. Notal organ stout, slightly protruded, and covering acute postnotal organ. A6

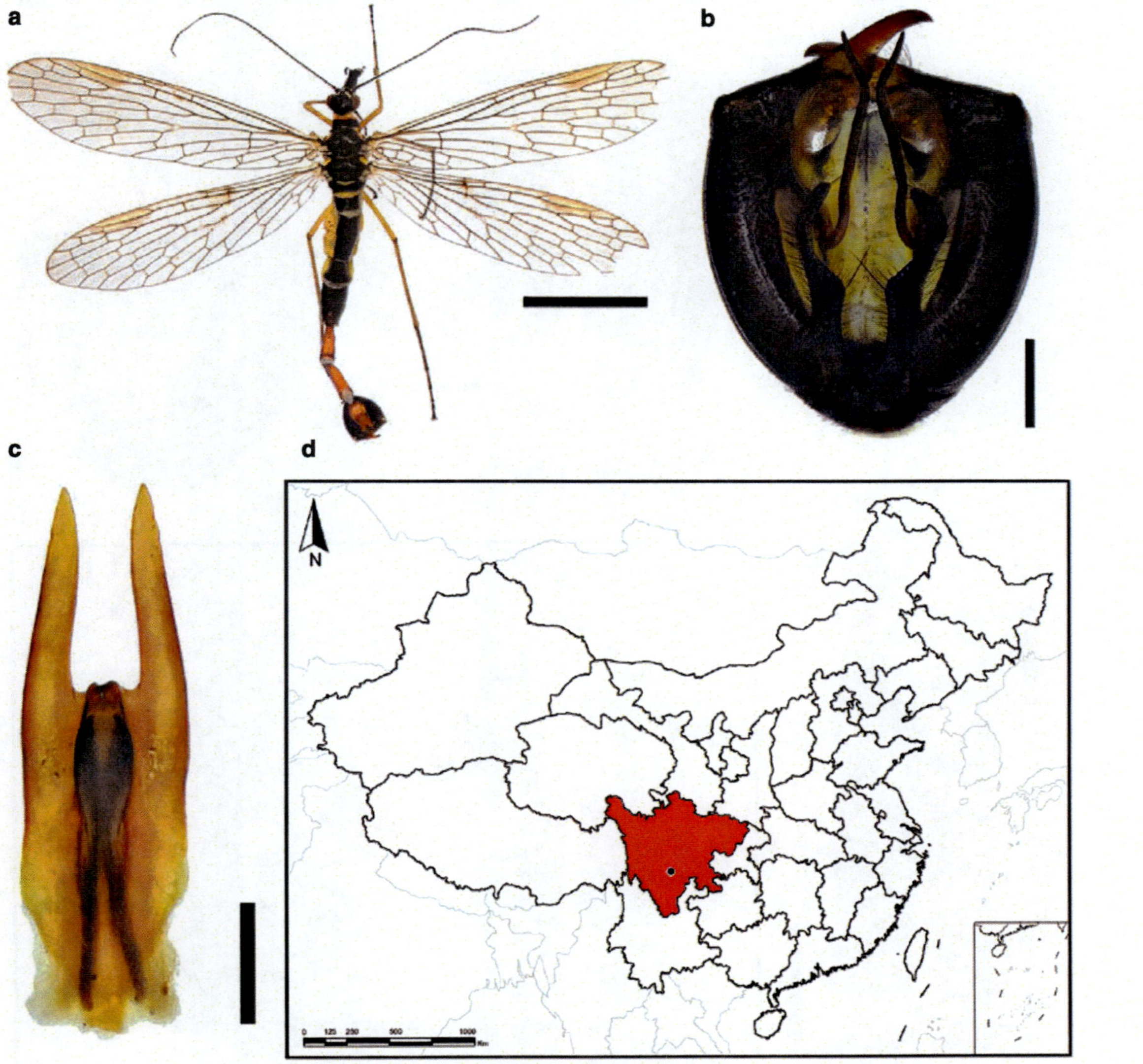

Fig. 10.117 (**a**) Male adult of ***Megapanorpa jiangorum***, dorsal view (Mamadi, Shimian, Sichuan) (*Scale bar*: 5.0 mm) **[NWAU]**. (**b**) Male genital bulb of ***Megapanorpa jiangorum***, ventral view. (**c**) Female medigynium of ***Megapanorpa jiangorum***, ventral view. (**d**) Distribution of ***Megapanorpa jiangorum***

mostly black, with only anal horn and ventro-apical border yellowish brown. Anal horn short, finger-like and acute terminally. A7 and A8 yellowish brown. A7 greatly constricted and stalk-like in basal 1/3, with apex beveled; A8 slightly longer than A7, with apex trumpet-shaped.

Male genital bulb spherical, mostly black. Epandrium greatly emarginater in O-shape and forming pair of finger-like processes laterally. Hypandrium with extremely short basal stalk. Hypovalves stripe-like, almost extending to apex of gonocoxites, broadest and slightly projected in middle of inner margin, and bearing several long, stout bristles along inner margin. Gonostylus shorter than gonocoxite, with blunt median tooth and short basal process. Paramere bifurcated: ventral branch long, glabrous, and bending laterad in distal portion, extending slightly beyond basal process of gonostylus; dorsal branch curving inward, and bearing numerous long spines along inner margin. Aedeagal valves stout.

Female subgenital plate subtriangular, with shallow V-shaped emargination terminally. Laterotergites slender, exceeding apex of subgenital plate, bifurcated terminally. Medigynium narrow with slender and slightly convergent posterior arms; apodemes of axis slightly divergent; main plate with basal portion membranous and fused with laterotergites.

Emergence. May to July.

Distribution. Sichuan (Mount Gongga, Shimian).

10.5.5 *Megapanorpa wanghongjiani* Wang & Hua, 2019

(Wang's Megascorpionfly) [王氏广翅蝎蛉] (Fig. 10.118)

Measurements. Forewing length 17.5–19.0 mm.

Morphology. Vertex and ocellar triangle black. Rostrum mostly black with yellowish portion genae. Nota black. Wing

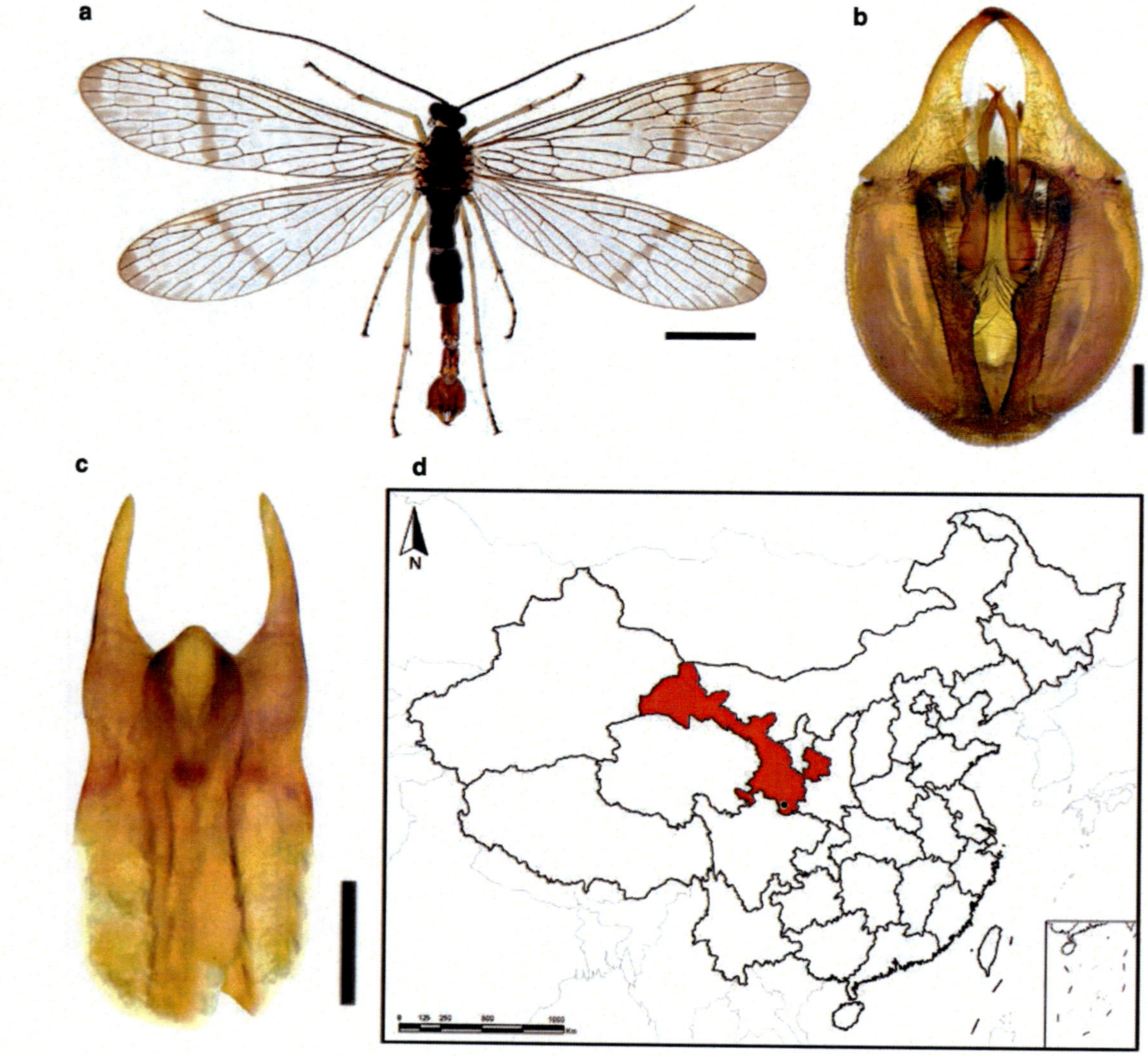

Fig. 10.118 (**a**) Male adult of *Megapanorpa wanghongjiani*, dorsal view (Wenxian, Ganu) (*Scale bar*: 5.0 mm) **[NWAU]**. (**b**) Male genital bulb of ***Megapanorpa wanghongjiani***, ventral view (*Scale bar*: 0.5 mm). (**c**) Female medigynium of ***Megapanorpa wanghongjiani***, ventral view (*Scale bar*: 0.2 mm). (**d**) Distribution of ***Megapanorpa wanghongjiani***

membrane hyaline; in forewings, apical band greatly reduced; pterostigmal band represented by oblique narrow stripe; other markings absent in males; females with reduced basal band. Pterostigma light brown. T1–T5 black. Notal organ stout, slightly protruded, and covering acute postnotal organ. A6 mostly black, with only anal horn and ventro-apical border yellowish brown. Anal horn extremely short, acute terminally. A7 and A8 yellowish brown. A7 greatly constricted and stalk-like in basal 1/4, with apex beveled; A8 slightly longer than A7, with apex trumpet-shaped.

Male genital bulb spherical, reddish brown. Epandrium greatly emarginater in O-shape and forming pair of finger-like processes laterally. Hypandrium with extremely short basal stalk. Hypovalves stripe-like, almost extending to apex of gonocoxites, broadest and greatly projected in middle of inner margin, and bearing several long, stout bristles along inner margin. Gonostylus shorter than gonocoxite, with blunt median tooth and short basal process, and greatly incurved on outer margin. Paramere unfurcated, extending far beyond median tooth of gonostylus, mostly glabrous, and bearing small hook-like projection near middle of outer margin. Aedeagal valves stout, conical.

Female subgenital plate subtriangular, with shallow V-shaped emargination terminally. Laterotergites slender, not exceeding apex of subgenital plate, strongly beveled terminally. Medigynium broad with slender and slightly convergent posterior arms; apodemes of axis slightly divergent; main plate with basal portion membranous and fused with laterotergites.

Emergence. July to August.

Distribution. Gansu (Wenxian).

10.6 Genus *Panorpa* Linnaeus, 1758

Panorpa is the most species genus in Panorpidae, with more than 260 species described worldwide. Over 100 species have been recorded from China. Eighty-nine species are included in this atlas (Fig. 10.119).

Fig. 10.119 Male adult of ***Panorpa sibirica***, dorso-lateral view

Panorpa kongosana Group

This group of species can be recognized by the relatively small body size, the greatly elongated male abdomen, the greatly elongated male gonostylus, and the basally crossed parameres in the male genitalia. Two known species are known from northeastern China, North and South Korea, and all of them are included in this atlas.

Key to Species of the Panorpa kongosana Group in This Atlas
(Two species, males)

1. Wing markings speckled; genital bulb yellow; hypovalves narrow and divergent distally .... ***Panorpa kongosana*** Okamoto
– Wing markings dense and broad; genital bulb dark brown; hypovalves broad and foliate .... ***Panorpa okamotona*** Issiki

10.6.1 *Panorpa kongosana* Okamoto, 1925

(Kongosan Scorpionfly) [金刚山蝎蛉] (Fig. 10.120)
Panorpa macrostyla Hua, 1998, **new synonym**.

Measurements. Forewing length 10.0–12.0 mm.

Morphology. Vertex and ocellar triangle black. Rostrum yellowish brown. Nota dark brown with broad yellow median stripe. Pleura and legs creamy yellow. Wing membrane translucent, tinged with light yellow. Markings dark brown. In forewings, apical, pterostigmal and basal bands greatly reduced into series of scattered spots with high variations. Wings held roof-like over abdomen at repose. Male T1–T5 yellowish brown. Notal organ flat and covering acute postnotal organ. A6 long cylindrical. A7 and A8 long cylindrical and not constricted basally.

Male genital bulb oval, yellowish brown. Epandrium long, broad and rounded at apex. Hypandrium Y-shaped and with long narrow basal stalk. Hypovalves adjoining mesally and slightly divergent apically. Gonostylus longer than gonocoxites, sickle-shaped and bearing inconspicuous median tooth and stout basal process. Paramere bifurcated, with ventral branches crossed near base of aedeagus. Aedeagus slightly exceeding apex of gonocoxites apically.

Female subgenital plate oval with V-shaped terminal emargination. Medigynium with undeveloped main plate; posterior arms slender; axis slightly longer than posterior arms with parallel apodemes.

Emergence. July to August.

Distribution. Jilin (Changbai Mountains, Linjiang), Liaoning (Kuandian, Huanren); North & South Korea.

Remarks. *Panorpa macrostyla* Hua was described from Changbai Mountains, Northeast China (Hua 1998), approximately 380 km north from the type locality of *P. kongosana*, Kumgang Mountain, North Korea (Okamoto 1925). The morphology of *P. macrostyla* properly matches that of *P. kongosana*, especially in the reduced wing markings (Okamoto 1925: pl. IV, figs 3 & 4; pl. V, figs 3 & 4), the apically diverged male hypovalves (Issiki 1933: fig. 7E), and the long sickle-shaped male gonostylus (Issiki 1933: fig. 10H).

Fig. 10.120 (**a**) Male adult of ***Panorpa kongosana*** in the field (Changbai Mountains, Jilin). (**b**) Male adult of ***Panorpa kongosana*** in the field (Changbai Mountains, Jilin). (**c**) Male genital bulb of ***Panorpa kongosana***, ventral view (*Scale bar*: 0.5 mm). (**d**) Female medigynium of ***Panorpa kongosana***, ventral view (*Scale bar*: 0.2 mm). (**e**) Distribution of ***Panorpa kongosana***

Therefore, it is treated as a junior synonym of *P. kongosana* in this study.

10.6.2 *Panorpa okamotona* Issiki, 1927

(Okamoto's Scorpionfly) [冈本蝎蛉] (Fig. 10.121)

Measurements. Forewing length 11.0–12.0 mm.

Morphology. Vertex and ocellar triangle black. Rostrum yellowish brown. Nota dark brown with broad yellow median stripe. Pleura and legs creamy yellow. Wing membrane translucent, tinged with light yellow. Markings dark brown. In forewings, apical band narrow with large inner notch posteriorly, forming acute projection near R_3–R_5; pterostigmal band narrow with basal branch broad and apical branch not reaching hind border; basal band broad and usually not reaching wing margins; marginal and basal spots absent. Wings held roof-like over abdomen at repose. Male T1–T5 dark yellowish brown. Notal organ flat and covering acute postnotal organ. A6 long cylindrical. A7 and A8 long cylindrical and not constricted basally.

Male genital bulb oval, dark yellowish brown. Epandrium long, broad and rounded at apex. Hypandrium Y-shaped and with long narrow basal stalk. Hypovalves broad with rounded apexes, closely aligned along inner margin. Gonostylus longer than gonocoxites, sickle-shaped and bearing inconspicuous median tooth and stout basal process. Paramere

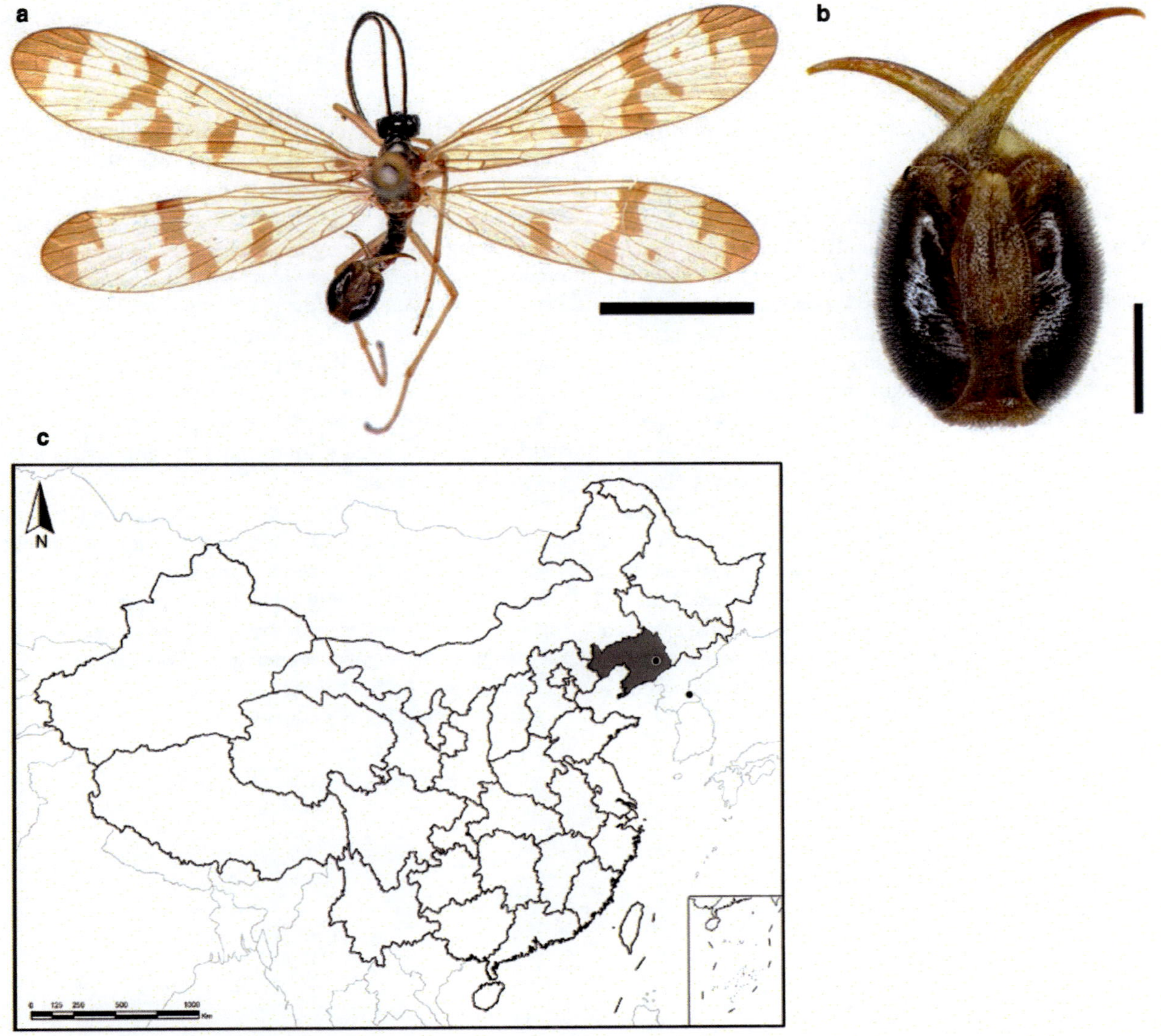

Fig. 10.121 (a) Male adult of ***Panorpa okamotona***, dorsal view (Benxi, Liaoning) (*Scale bar*: 5.0 mm) **[NWAU]**. **(b)** Male genital bulb of ***Panorpa okamotona***, ventral view (*Scale bar*: 0.5 mm). **(c)** Distribution of ***Panorpa okamotona***

bifurcated, with ventral branches crossed near base of aedeagus. Aedeagus slightly exceeding apex of gonocoxites apically.

Female unknown.

Emergence. June to August.

Distribution. Liaoning (Benxi); North & South Korea.

Panorpa waongkehzengi Group

This group of species can be recognized by the relatively small body size, the short male abdomen, and the extremely short male gonostylus. Five species are known from central and eastern China. Three of them are included in this atlas.

Key to Species of the Panorpa kongosana Group in This Atlas
(Three species, males)

1. Parameres elongated and crossed basal to ventral aedeagal valves .... ***Panorpa jinhuaensis*** Wang, Gao & Hua
– Parameres short and not crossed 2
2. Inner margin of hypovalves bearing obtuse projection .... ***Panorpa waongkehzengi*** Navás
– Inner margin of hypovalves lacking projection .... ***Panorpa menqiuleii*** Wang, Gao & Hua

10.6.3 *Panorpa jinhuaensis* Wang, Gao & Hua, 2019

(Jinhua Scorpionfly) [金华蝎蛉] (Fig. 10.122)

Measurements. Forewing length 10.0–11.8 mm.

Morphology. Vertex yellow. Ocellar triangle black. Rostrum mostly yellow with apex darkish. Thorax and abdomen varying from yellow, orange yellow to reddish yellow. Wing membrane densely tinged with yellow. Markings black, with all bands and spots greatly developed. Pterostigma reddish yellow. In forewings, apical band broad, usually with 1–3 hyaline windows enclosing crossveins; separated spot at ending of M_2; apical branch of pterostigmal band variable: intact or detached with pterostigmal band and greatly elongated anteriorly; basal branch of pterostigmal band intact and slightly broader than apical branch; marginal spot C-shaped; basal band split into two large spots; additional transverse short band between basal band and basal spot, and extending from ending of 2A to CuA; basal spot shifted posteriorly along anal margin; additional small spot anterior to 3A. Notal organ slightly projected dorsad and covering acute postnotal organ. A6–A8 short and cylindrical.

Male genital bulb oval, yellow. Epandrium long and broad, evenly tapering toward rounded apex. Hypandrium with short broad stalk and pair of longer hypovalves; each hypovalve tapering toward apex, and bearing long stout setae along inner margin. Gonostylus with rounded median tooth and large bowl-shaped basal process. Paramere slender, extending slightly beyond apex of gonocoxites with spiral and acute apical portion; two parameres crossed basal to ventral aedeagal valves. Dorsal aedeagal processes greatly elongated posteriorly with slightly enlarged and beveled apex, and bearing row of short setae along basal third of inner margin; lateral processes short and stout.

Female subgenital plate oval, slightly tapering toward shallowly emarginate apex. Medigynium with moderately developed main plate; posterior arms slightly shorter than main plate and twisted in distal half; axis longer than posterior arms with diverged apodemes; posterior apex of axis subtriangular and slightly extending beyond main plate.

Emergence. April, October.

Distribution. Zhejiang (Jinhua).

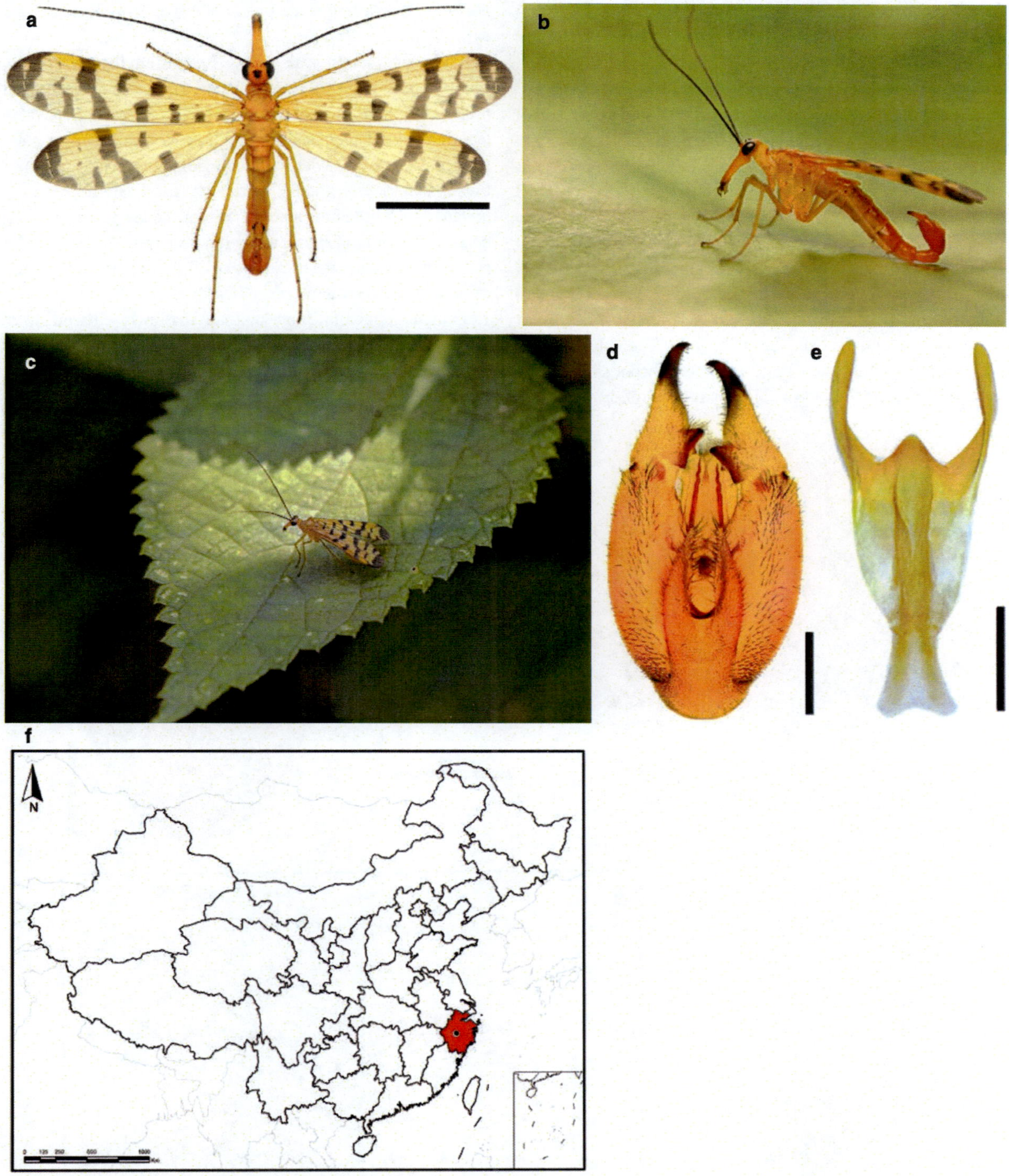

Fig. 10.122 (**a**) Male adult of ***Panorpa jinhuaensis***, dorsal view (Jinhua, Zhejiang) (*Scale bar*: 5.0 mm). (**b**) Male adult of ***Panorpa jinhuaensis*** in the field (Jinhua, Zhejiang). (**c**) Male adult of ***Panorpa jinhuaensis*** in the field (Jinhua, Zhejiang). (**d**) Male genital bulb of ***Panorpa jinhuaensis***, ventral view (*Scale bar*: 0.5 mm). (**e**) Female medigynium of ***Panorpa jinhuaensis***, ventral view (*Scale bar*: 0.2 mm). (**f**) Distribution of ***Panorpa jinhuaensis***

10.6.4 *Panorpa menqiuleii* Wang, Gao & Hua, 2019

(Qiu-Lei Men's Scorpionfly) [门氏蝎蛉] (Fig. 10.123)
Measurements. Forewing length 10.0–12.0 mm.

Morphology. Vertex yellow. Ocellar triangle black, dark transverse pattern on occiput. Rostrum mostly yellow with apex darkish. Thorax and abdomen yellow in males, females similar except terga black. Wing membrane densely tinged with yellow. Markings black, with all bands and spots greatly developed. Pterostigma reddish yellow. In forewings, apical band broad; pterostigmal band with apical branch detached and greatly elongated anteriorly, and connected with apical band along costal margin; basal branch bent inward; marginal spot extending from Sc to beyond R4+5; basal band complete or split into two large spots; basal spot shifted posteriorly along anal margin. Notal organ slightly projected dorsad and covering acute postnotal organ. A6–A8 short and cylindrical.

Male genital bulb oval, yellow. Epandrium long and broad, evenly tapering toward rounded apex. Hypandrium with short broad stalk and pair of longer hypovalves; each hypovalve tapering toward apex, and bearing long stout setae along inner margin. Gonostylus with rounded median tooth and large subtrapezoidal basal process. Paramere short, slightly curved inward, not exceeding apex of ventral aedeagal valves; and bearing row of long spines along inner margin. Aedeagus with dorsal processes constricted neck-like basally and greatly enlarged apically; lateral processes short and stout.

Female subgenital plate oblong, shallowly emarginate at apex. Medigynium with undeveloped main plate; posterior arms slightly shorter than main plate and twisted in distal half; axis half as long as than posterior arms with parallel apodemes.

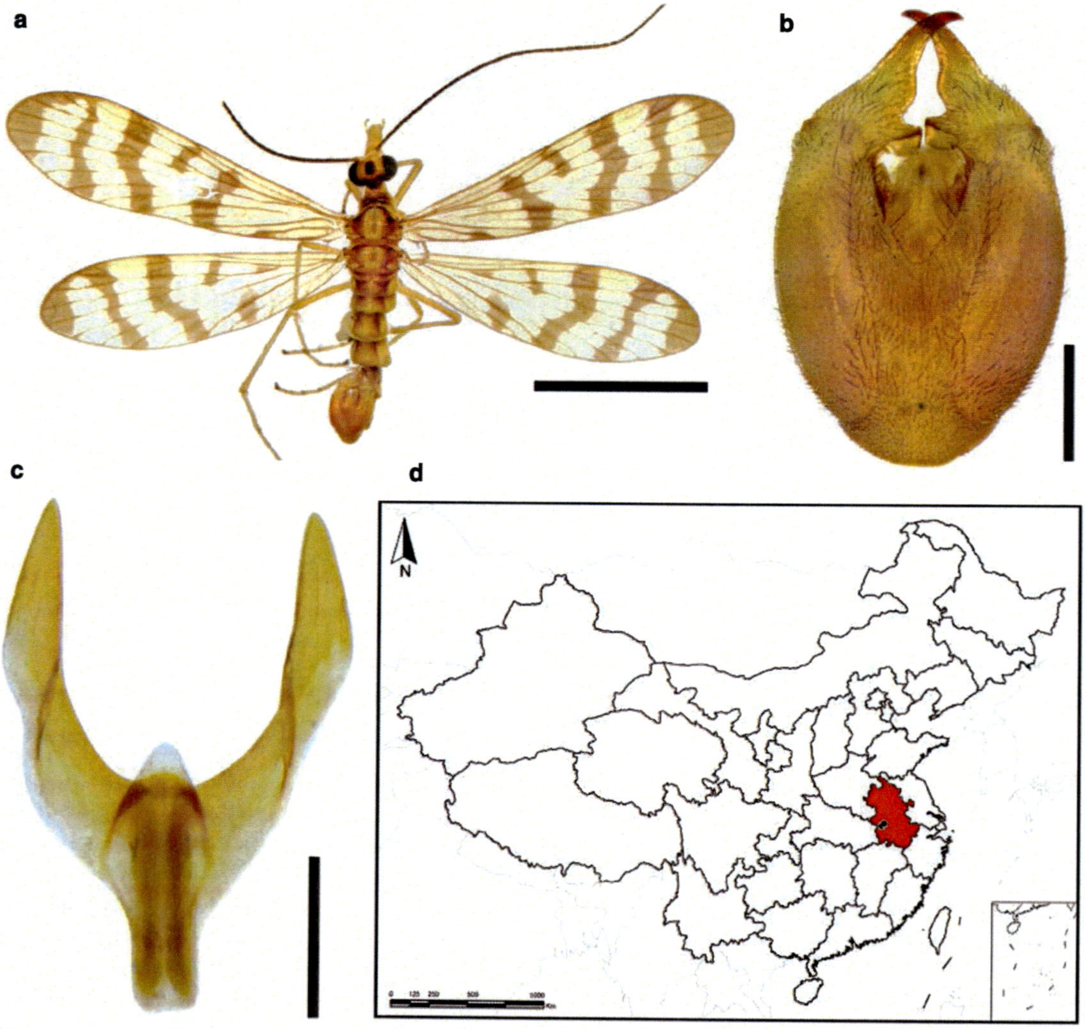

Fig. 10.123 (**a**) Male adult of ***Panorpa menqiuleii***, dorsal view (Yaoluoping, Yuexi, Anhui) (*Scale bar*: 5.0 mm) **[NWAU]**. (**b**) Male genital bulb of ***Panorpa menqiuleii***, ventral view (*Scale bar*: 0.5 mm). (**c**) Female medigynium of ***Panorpa menqiuleii***, ventral view (*Scale bar*: 0.2 mm). (**d**) Distribution of ***Panorpa menqiuleii***

Emergence. August.

Distribution. Anhui (Huoshan, Yuexi).

10.6.5 *Panorpa waongkehzengi* Navás, 1935

(Keh-Zeng Huang's Scorpionfly) [黄氏蝎蛉]

(Fig. 10.124)

Measurements. Forewing length 8.3–10.5 mm.

Morphology. Vertex and rostrum yellowish brown. Ocellar triangle black. Pronotum dark brown. Meso- and metanotum dark brown, with broad yellow median stripe. Pleura and legs yellowish brown. Wing membrane hyaline, slightly tinged with yellow. Markings yellowish brown. Pterostigma yellow and conspicuous. In forewings, apical band split into two slender transverse bands; pterostigmal band complete, with broad basal branch and slender apical branch curving abruptly on posterior portion; marginal spot irregular; basal band split into two spots; basal spot absent. Rs quadrifurcated with unbranched R_2 (most species in Panorpidae have Rs five- or six-branched with R_2 bifurcated or trifurcated). Male T1–T5 yellowish brown. Notal organ flat and covering acute postnotal organ. A6 yellowish brown, short cylindrical. A7 and A8 yellowish brown, short cylindrical, and not constricted basally.

Male genital bulb oval. Epandrium broad and tapering towards slightly emarginate apex. Hypandrium with long and broad basal stalk. Hypovalves pincer-like, bearing obtuse projection on inner margin. Gonostylus half-length of gonocoxites, and bearing obtuse triangular median tooth and columnar basal process. Paramere simple and slender, and bearing row of spines on inner margin. Aedeagus with ventral valves membranous and short, and dorsal valves greatly elongated and exceeding apex of gonocoxites, and diverged apically.

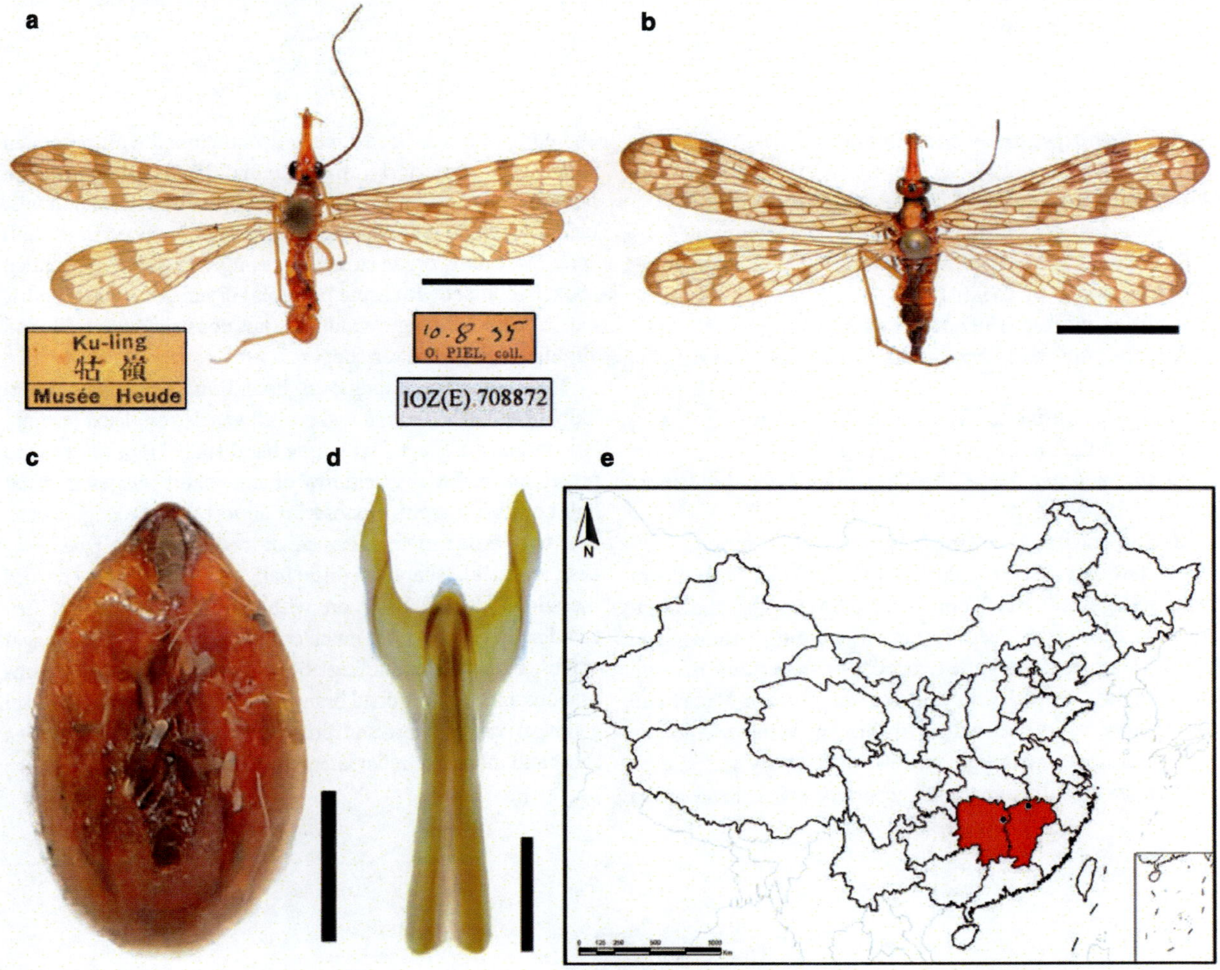

Fig. 10.124 (**a**) Male adult of ***Panorpa waongkehzengi***, dorsal view (Lushan Mountain, Jiangxi) (*Scale bar*: 5.0 mm) **[IZAS]**. (**b**) Female adult of ***Panorpa waongkehzengi***, dorsal view (Liuyang, Hunan) (*Scale bar*: 5.0 mm) **[NWAU]**. (**c**) Male genital bulb of ***Panorpa waongkehzengi***, ventral view (*Scale bar*: 0.5 mm). (**d**) Female medigynium of ***Panorpa waongkehzengi***, ventral view (*Scale bar*: 0.2 mm). (**e**) Distribution of ***Panorpa waongkehzengi***

Female subgenital plate broad bear-shaped, tapering towards slightly emarginated apex. Medigynium with undeveloped main plate; posterior arms slightly shorter than main plate and greatly twisted in distal half; axis three times longer than posterior arms with slightly diverged apodemes; posterior apex of axis subtriangular and slightly extending beyond main plate.

Emergence. August.

Distribution. Hunan (Liuyang), Jiangxi (Lushan Mountain).

Panorpa amurensis Group

This group of species can be recognized by the relatively large body size, the greatly elongated and robust male abdomen, the greatly elongated male gonostylus, and the basally crossed parameres in the male genitalia.

Six species are known from northeastern China, Russian Far East, and North and South Korea. Two of them are included in this atlas.

Key to Species of the Panorpa amurensis Group in This Atlas
(Two species, males)

1. Hypovalves short and narrow, not exceeding apex of gonocoxites; gonostylus slightly curved .... ***Panorpa amurensis*** MacLachlan
– Hypovalves long and broad, exceeding apex of gonocoxites; gonostylus greatly curved in basal 1/3 .... ***Panorpa baohwashana*** Cheng

10.6.6 *Panorpa amurensis* MacLachlan, 1872

(Amur Scorpionfly) [阿穆尔蝎蛉] (Fig. 10.125)

Panorpa orientalis MacLachlan, 1887, **new synonym**.

Panorpa sachalinensis Matsumura, 1911, synonymized with *P. orientalis* by Issiki, 1933

Panorpa liui Hua, 1997, **new synonym**.

Panorpa jilinensis Zhou, 2000, synonymized with *P. liui* by Li et al., 2021

Panorpa neojilinensis Li and Lin in Li et al., 2012, synonymized with *P. liui* by Li et al., 2021

Aulops melania Navás, 1912, synonymized by Esben-Petersen, 1915

Measurements. Forewing length 14.0–17.5 mm.

Morphology. Body color highly variable. Two major forms recognized: Black form (including spring and high-elevation forms) and yellow form (including summer and low-elevation forms). Vertex, ocellar triangle, and rostrum yellow to dark brown. Nota yellow to dark brown. Pleura and legs creamy yellow to yellowish brown. Wing membrane hyaline, or tinged with light yellow, with highly varied dark brown markings. In forewings, generally apical band broad and complete, and occasionally accompanied with separated small spot posteriorly; pterostigmal band complete with basal branch and apical branch; marginal spot conspicuous; basal band split into two spots; basal band reduced into small spot. Male notal organ stout and strongly raised with rounded apex, in contact with acute postnotal organ. A6 long cylindrical. A7 and A8 long cylindrical, not constricted basally and slightly enlarged towards apex.

Male genital bulb long oval. Epandrium emarginate terminally in shallow V-shape and curved ventrad on distal portion. Hypandrium Y-shaped with long basal stalk. Hypovalves with projection subbasally, and row of stout short bristles in distal half on inner margin. Gonostylus almost as long as gonocoxites, with broad triangular median tooth and stout basal process, and sickle-shaped in distal half, which intensively curved mesad and dorsad. Paramere with long basal stalk and bifurcated medially: ventral branch crossed below ventral aedeagal valves, with apex extending above aedeagus, enlarged, membranous and setose; dorsal branch fused with aedeagus. Ventral aedeagal valves short and protruding ventrad; dorsal valves short and inconspicuous; lateral processes flat and rounded.

Female subgenital plate long oval with V-shaped terminal emargination. Medigynium with undeveloped main plate; posterior arms slender; axis twice longer than posterior arms with straight, slightly diverged apodemes.

Emergence. May to September.

Distribution. Liaoning (Huanren, Kuandian, Qianshan, Shenyang), Jilin (Changbai Mountains), Heilongjiang (Jingpo Lake, Mudan Peak), Inner Mongolia (Arxan, Bugt); North & South Korea; Russian Far East (Vladivostock).

a

b

Fig. 10.125 (**a**) Adults of ***Panorpa amurensis*** in the field, yellow form (Jingpohu, Heilongjiang). (**b**) Adults of ***Panorpa amurensis*** in the field, black form (Changbai Mountains, Jilin). (**c**) Male adult of ***Panorpa amurensis***, black-tailed form (Changbai Mountains, Jilin). (**d**) Male genital bulb of ***Panorpa amurensis***, yellow form, ventral view (*Scale bar*: 0.5 mm). (**e**) Male genital bulb of ***Panorpa amurensis***, black form, ventral view (*Scale bar*: 0.5 mm). (**f**) Female medigynium of ***Panorpa amurensis***, ventral view (*Scale bar*: 0.2 mm). (**g**) Distribution of ***Panorpa amurensis***

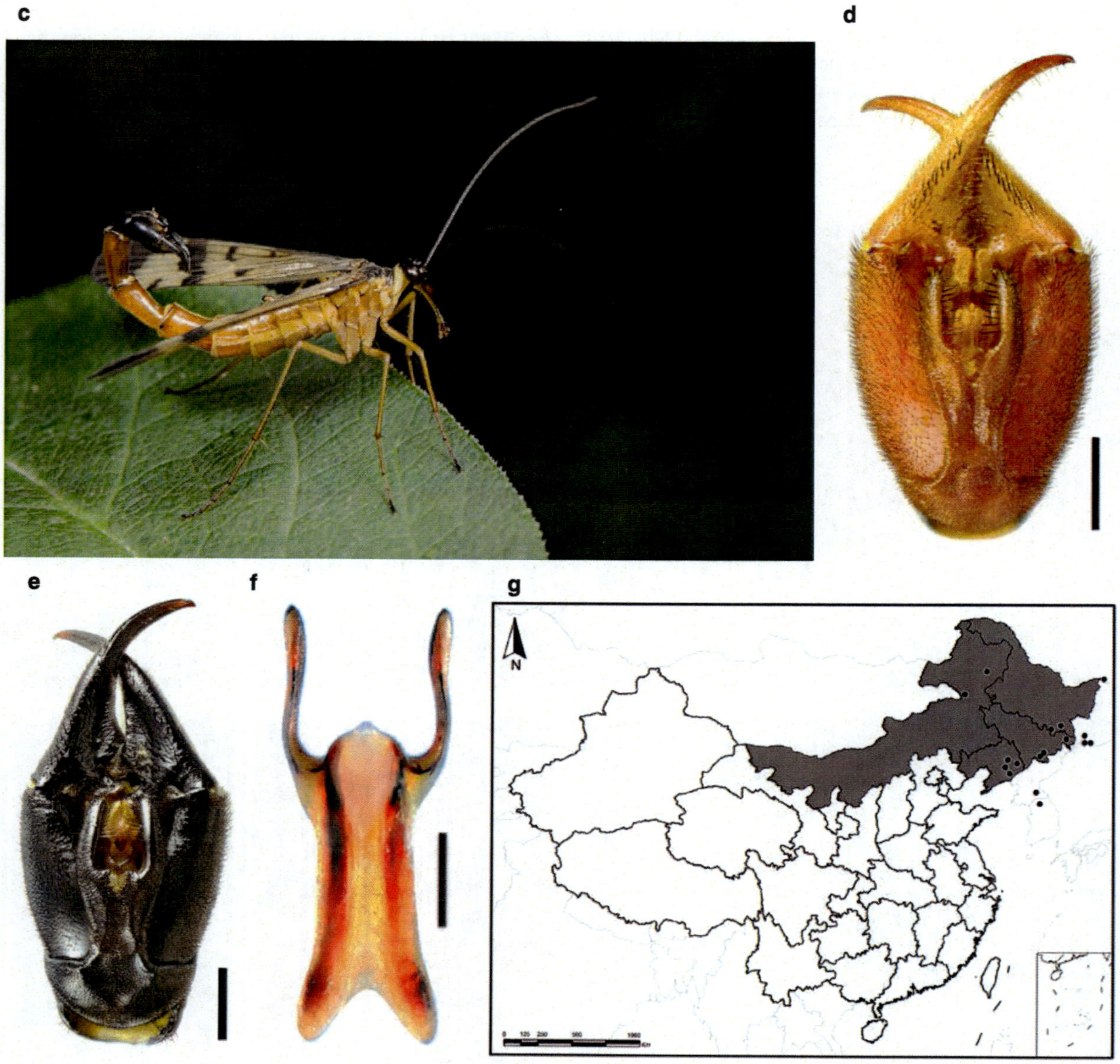

Fig. 10.125 (continued)

10.6.7 *Panorpa baohwashana* Cheng, 1957

(Baohuashan Scorpionfly) [宝华山蝎蛉] (Fig. 10.126)
Measurements. Forewing length 18.0–20.0 mm.

Morphology. Vertex dark brown. Ocellar triangle black. Rostrum dark brown, with yellowish brown flanks. Nota dark brown. Pleura and legs yellowish. Wing membrane hyaline with markings dark brown. In forewings, apical band broad; pterostigmal band with broad basal branch and separated small triangular apical branch; marginal spot small; basal band split into two spots; basal spot absent. Male abdomen dark brown. Notal organ greatly raised and in contact apically with acute postnotal organ. A6 cylindrical, longer than A5. A7 and A8 cylindrical and inconspicuously constricted basally.

Male genital bulb black long oval. Epandrium long and broad, with deep U-shaped terminal emargination, and bended ventrad on apical portion. Hypandrium with greatly shortened basal stalk. Hypovalve very long and broad, wave-like on inner margin, and curved mesad on apical portion. Gonostylus slightly longer than gonocoxites, slender, intensively concaved along outer margin on basal third, arched on distal 2/3, and bearing obtuse triangle median tooth and large oblong basal process.

Parameres bifurcated, with ventral branches crossed near base of aedeagus. Aedeagus acute and long, exceeding apex of gonocoxites; lateral processes finger-like and stout.

Female subgenital plate long oval with V-shaped terminal emargination. Medigynium with undeveloped main plate; posterior arms slender; axis three times longer than posterior arms with sinuate apodemes.

Emergence. July.

Distribution. Jiangsu (Baohua Mountain).

Panorpa guttata Group

This group of species can be recognized by the relatively small body size, the roof-like wing posture, the short male abdomen, and the female medigynium lacking a dorsal plate.

Twenty-three species are known from central and southeastern China. Twenty-two of them are included in this atlas.

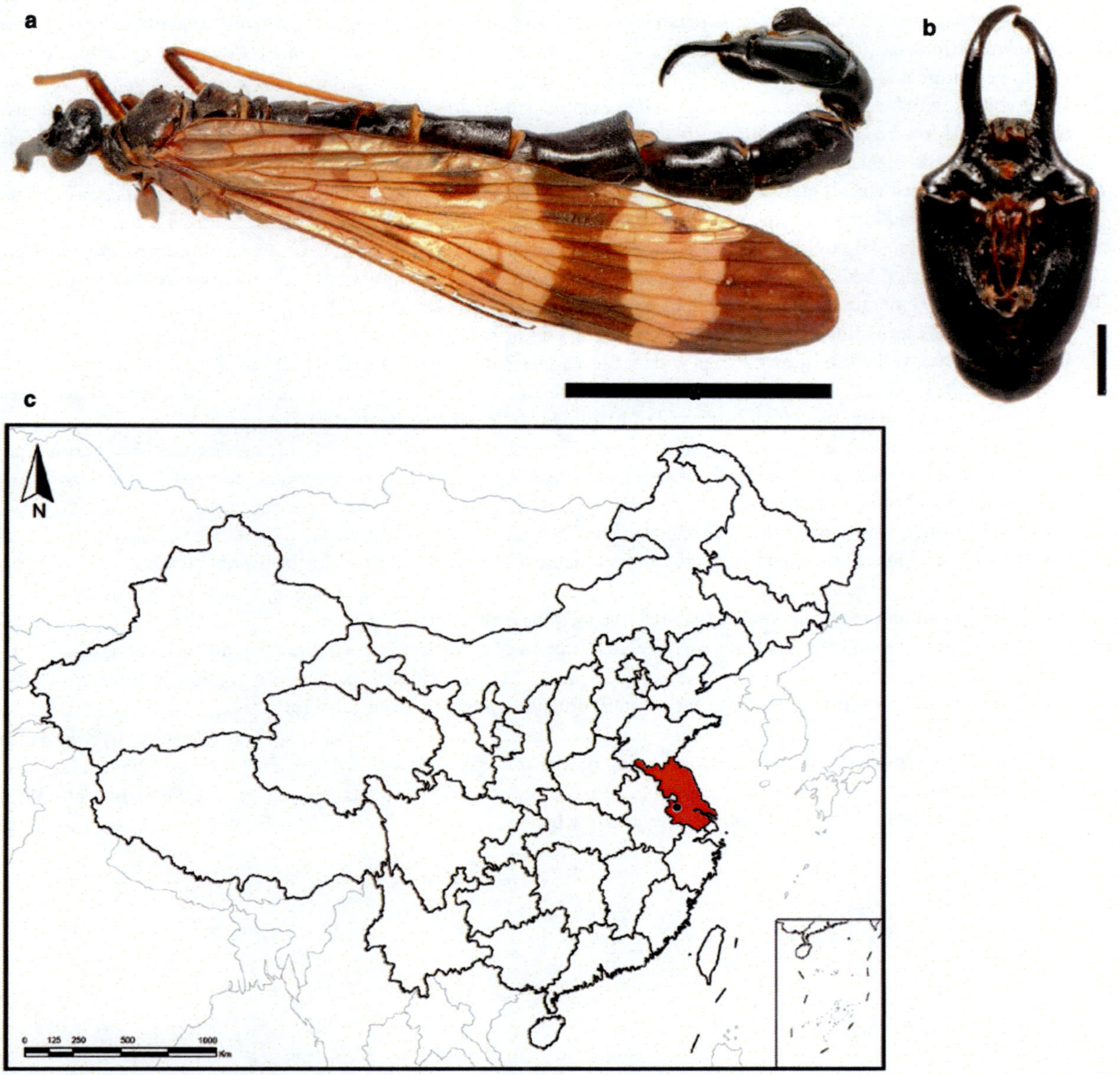

Fig. 10.126 (a) Male adult of ***Panorpa baohwashana***, lateral view (Baohua Mountain, Nanjing, Jiangsu) (*Scale bar*: 5.0 mm) [NWAU]. (b) Male genital bulb of ***Panorpa baohwashana***, ventral view (*Scale bar*: 0.5 mm). (c) Distribution of ***Panorpa baohwashana***

Key to Species of the Panorpa guttata Group in This Atlas
(22 species, males)

1. Wings with greatly developed markings, apical, pterostigmal, basal bands, and marginal and basal spots complete and broad .... ***Panorpa furcata*** Zhou & Zhou
– Wings with markings more or less reduced 2
2. A6 simple on dorsal apex; A7 simple 3
– A6 emarginated on dorsal apex; A7 greatly humped on dorsal apex 10
3. Paramere unfurcated 4
– Paramere bifurcated of trifurcated 6
4. Parameres crossed .... ***Panorpa duanyu*** Wang & Gong
– Parameres uncrossed 5
5. Wings with reduced markings; male paramere slightly enlarged subapically .... ***Panorpa decolorata*** Chou & Wang
– Wings nearly devoid of markings; male paramere simple subapically .... ***Panorpa parallela*** Wang & Hua
6. Parameres trifurcated .... ***Panorpa liaoi*** Zhou & Zhou
– Parameres bifurcated 7
7. Head yellow with black pattern enclosing ocellar triangle and another black patter on occiput; parameres with ventral branch curved towards base of genital bulb .... ***Panorpa reflexa*** Wang & Hua
– Head yellow with ocellar triangle black; parameres with both branches directing caudad 8
8. Wings with reduced apical, pterostigmal, and basal bands .... ***Panorpa caoweii*** Wang
– Wings devoid of markings or with greatly reduced apical and pterostigmal bands 9
9. A6–A8 black; basal stalk of hypandrium shorter than hypovalves .... ***Panorpa nanzhao*** Wang
– A6–A8 yellow; basal stalk of hypandrium approximately as long as hypovalves .... ***Panorpa xuzhu*** Wang & Gong
10. A6 with shallow emargination on dorsal apex and forming pair of inconspicuous lobes 11
– A6 with deep emargination on dorsal apex and forming pair of conspicuous lobes 12
11. Head with vertex yellowish brown; wings with scattered markings; male parameres bifurcated .... ***Panorpa dali*** Wang
– Head with vertex shining black; wings devoid of markings; male parameres trifurcated .... ***Panorpa jiangrixini*** Wang
12. A6 with one or few small and acute teeth on each lateral side below dorsal lobes; gonostylus covered with dozens of long setae on basal half of ventral surface 13
– A6 simple on lateral sides; gonostylus lacking such setae 15
13. A6 with truncated lobes on dorsal apex, and only one acute tooth on each lateral side below dorsal lobes .... ***Panorpa guttata*** Navás
– A6 with subtriangular lobes on dorsal apex, and five teeth on each lateral side 14
14. A6 with dorsal lobes not exceeding middle of A7, and one large tooth and four smaller teeth on each lateral side .... ***Panorpa jinfoshana*** Wang
– A6 with dorsal lobes exceeding middle of A7, and five same-sized teeth on each lateral side .... ***Panorpa latiloba*** Wang
15. Head with large star-shaped black pattern enclosing ocellar triangle, and another black pattern on occiput .... ***Panorpa stella*** Wang
– Head mostly yellowish brown with only ocellar triangle black 16

16. A6 with pair of very small dorsal lobes; each lobe narrower than half width of A6 at base .... ***Panorpa substricta*** Wang
– A6 with pair of broad dorsal lobes; each lobe approximately half as wide as half A6 at base 17
17. Parameres with ventral branch greatly enlarged subapically with pointed apex .... ***Panorpa hirundo*** Wang
– Parameres with ventral branch simple 18
18. Parameres with dorsal branch greatly curved outwards at sub-basal portion .... ***Panorpa filina*** Chou & Wang
– Parameres with dorsal branch relatively straight 19
19. Parameres with dorsal branch slightly lobate and notched on inner side subapically .... ***Panorpa hani*** Wang
– Parameres with dorsal branch simple 20
20. Hypandrium with basal stalk shorter than hypovalves; dorsal branch of paramere convergent apically .... ***Panorpa huangguiqiangi*** Wang
– Hypandrium with basal stalk approximately as long as or longer than hypovalves; dorsal branch of paramere divergent apically 21
21. Hypandrium with hypovalves slightly convergent terminally .... ***Panorpa kunmingensis*** Fu & Hua
– Hypovalves slightly divergent terminally .... ***Panorpa xiaofeng*** Wang & Gong

10.6.8 *Panorpa caoweii* Wang, 2021

(Wei Cao's Scorpionfly) [曹威蝎蛉] (Fig. 10.127)
Measurements. Forewing length 12.0–13.0 mm.

Morphology. Vertex, occiput, and rostrum yellowish brown, ocellar triangle black. Pronotum brown with anterior and posterior margins blackish. Meso- and metanotum brown laterally with pale median stripe. Wings membrane hyaline and tinged with light yellowish brown, markings black. Pterostigma brown. In forewings, apical band greatly reduced, vague, and detached with both anterior and posterior margins; pterostigmal band triangular and reduced at posterior half, with detached spot along hind margin as basal branch and lacking apical branch; basal band reduced to small spot above ending of CuP; marginal and basal spots absent. Male T1–T5 dark brown, with pale, discontinuous median stripe extending from T1 to T3. Notal organ slightly developed and covering acute postnotal organ. A6 blackish brown at basal half and gradually yellowing towards apex, cylindrical. A7 and A8 yellowish orange, A7 nearly cylindrical, A8 constricted basally and beveled apically.

Male genital bulb yellowish orange, long oval. Epandrium extending beyond middle of gonostylus, slightly tapering towards apex, with deep U-shaped terminal emargination and forming two parallel finger-like processes. Hypandrium Y-shaped and approximately 3/4 as long as gonocoxites, with long basal stalk and splitting into pair of hypovalves in distal half. Hypovalves narrow, divergent, and tapering towards apex, with sparse long bristles along inner margin. Gonostyli with disc-shaped basal process and subtriangular median tooth. Parameres bifurcated, with both branches slender and acute at apex; ventral branch approximately half-length of dorsal branch and bending inward; dorsal branch slightly bending outward and extending to apex of gonocoxites. Ventral aedeagal valves membranous and rounded; dorsal processes slender, greatly elongated, nearly parallel, and tapering in to subacute apex in distal 1/3.

Female subgenital plate oblong with rounded apex. Medigynium slender, with main plate approximately half as long as axis; posterior arms short, acute and approximately 1/3 as long as main plate; apodemes of axis slightly divergent.

Emergence. August.

Distribution. Guangxi (Kitten Mountain).

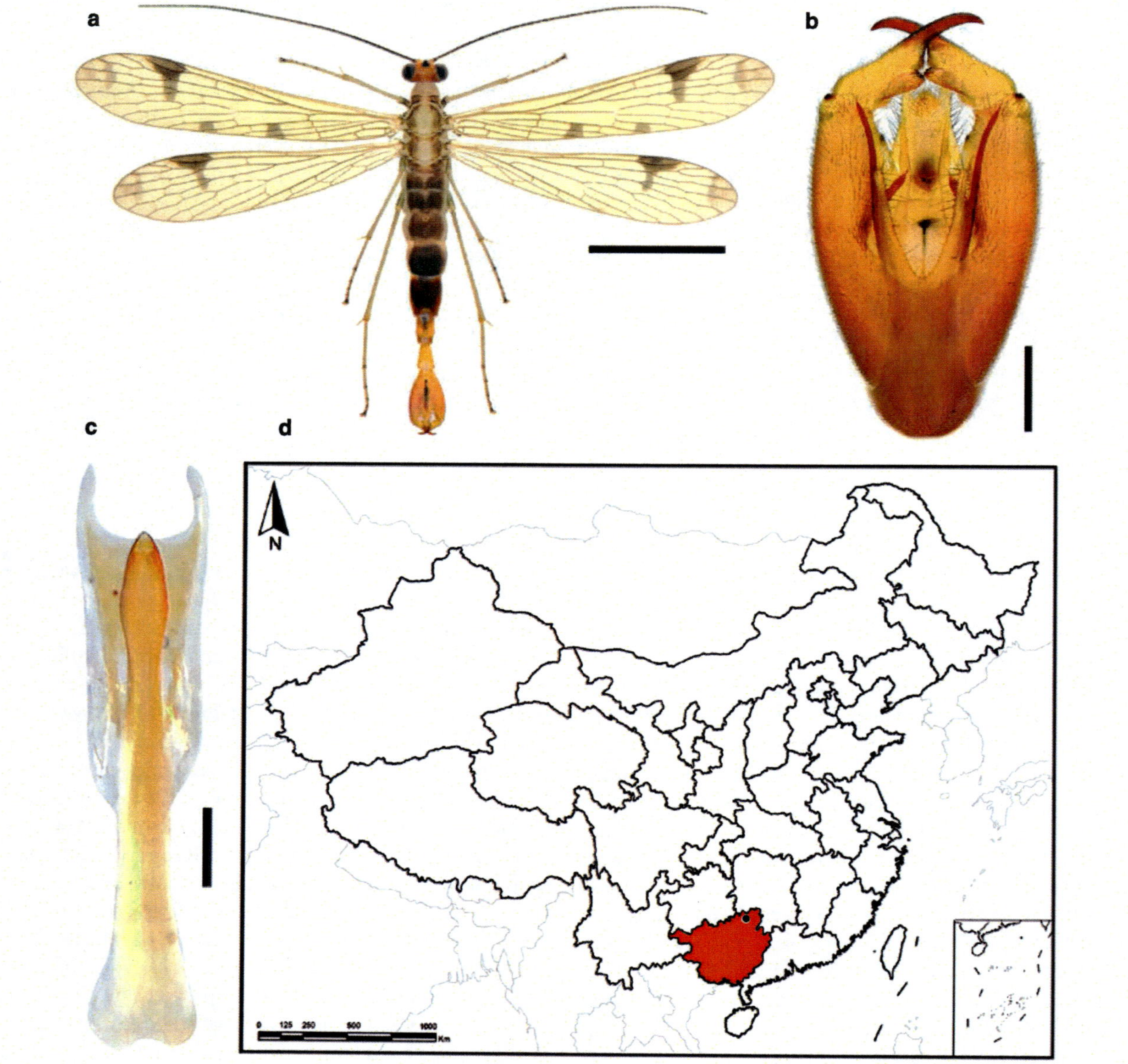

Fig. 10.127 (**a**) Male adult of ***Panorpa caoweii***, dorsal view (Kitten Mountain, Guangxi) (*Scale bar*: 5.0 mm) [NWAU]. (**b**) Male genital bulb of ***Panorpa caoweii***, ventral view (*Scale bar*: 0.5 mm). (**c**) Female medigynium of ***Panorpa caoweii***, ventral view (*Scale bar*: 0.2 mm). (**d**) Distribution of ***Panorpa caoweii***

10.6.9 *Panorpa dali* Wang, 2021

(Dali Scorpionfly) [大理蝎蛉] (Fig. 10.128)

Measurements. Forewing length 8.0–9.0 mm.

Morphology. Vertex, occiput, and rostrum yellowish brown. Ocellar triangle enclosed by blurred dark spot. Nota yellowish brown laterally with broad, light yellow median stripe. Wings narrow in shape, membrane hyaline and tinged with light yellowish brown, markings brown. In forewings, apical and pterostigmal bands greatly reduced into several small spots; basal band reduced to small spot above CuP; marginal and basal spots absent. Hindwings devoid of markings. Male T1–T5 dark brown, with pale, thin, and continuous median stripe extending from T1 to T4. A6 yellowish brown, cylindrical, with very shallow emargination at dorsal apex. A7 and A8 yellowish brown, A7 nearly cylindrical, A8 slightly longer than A7, constricted basally and beveled apically.

Male genital bulb yellowish brown, oval. Epandrium with deep U-shaped terminal emargination and forming two parallel finger-like processes. Hypandrium Y-shaped and approximately 3/4 as long as gonocoxites, with long basal stalk. Hypovalves narrow, divergent at base, and bending inwards at apex, with sparse long bristles along inner margin. Gonostylus greatly incurvated on outer margin, with disc-shaped basal process and subtriangular median tooth. Parameres bifurcated, with ventral branch slender and approximately 3/4 as long as lobate dorsal branch; dorsal branch bearing row of long stout bristles on inner margin. Ventral aedeagal valves membranous; dorsal processes slender, greatly elongated with a longitudinal groove.

Female subgenital plate oblong with rounded apex. Medigynium slender, with main plate broadest at basal 1/3 and approximately half as long as axis; posterior arms short, acute and approximately half as long as main plate; apodemes of axis greatly divergent at base.

Emergence. June to October.

Distribution. Yunnan (Cangshan Mountain).

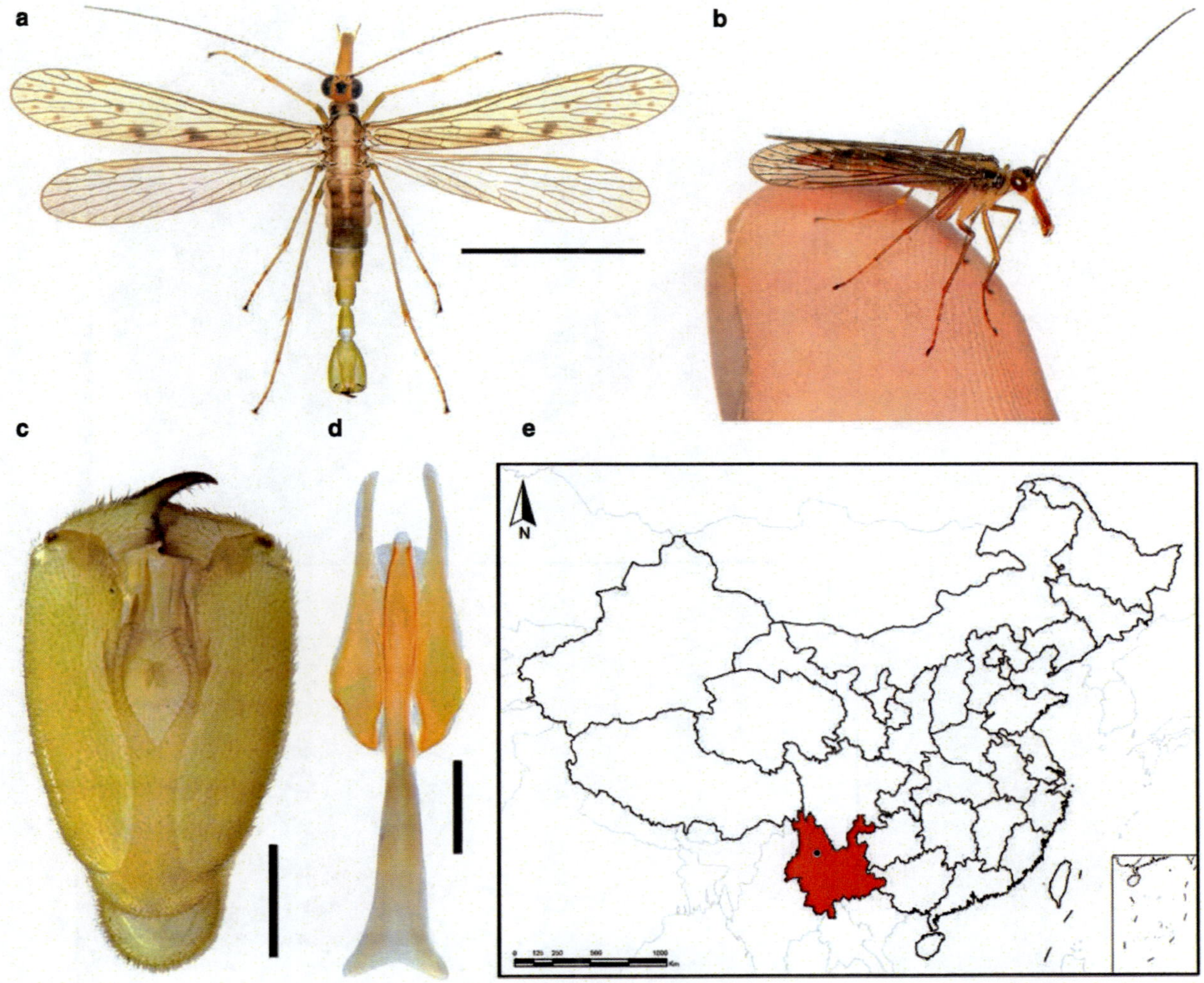

Fig. 10.128 (**a**) Male adult of ***Panorpa dali***, dorsal view (Cangshan Mountain, Dali, Yunnan) (*Scale bar*: 5.0 mm) **[DALU]**. (**b**) Female adult of ***Panorpa dali*** resting on the collector's finger (Cangshan Mountain, Dali, Yunnan). (**c**) Male genital bulb of ***Panorpa dali***, ventral view (*Scale bar*: 0.5 mm). (**d**) Female medigynium of ***Panorpa dali***, ventral view (*Scale bar*: 0.2 mm). (**e**) Distribution of ***Panorpa dali***

10.6.10 *Panorpa decolorata* Chou & Wang, 1981

(Decolorated Scorpionfly) (Fig. 10.129)

Measurements. Forewing length 14.0–15.0 mm.

Morphology. Vertex and rostrum yellowish brown. Ocellar triangle dark brown. Nota light yellowish brown with darkened flanks. Pleura and legs creamy yellow. Wing membrane hyaline with markings brown. Wings held roof-like over abdomen at repose. In forewings, apical band greatly reduced into two faint stripes; pterostigmal band broad anteriorly and narrow posteriorly, with greatly reduced basal and apical branches; basal band reduced into large circular spot near hind margin; marginal and basal spots absent. Male T1–T5 dark brown. Notal organ flat and covering acute postnotal organ. A6 short and cylindrical. A7 and A8 shorter than A6 and cylindrical, not constricted at base.

Male genital bulb elliptical. Epandrium emarginates terminally in deep U-shape and forms pair of finger-like processes laterally. Hypandrium with long broad basal stalk. Hypovalves short, with rounded apexes. Paramere stout and finger-like, with apical portion hooked and bearing dense short spines. Aedeagus with short membranous ventral valves, and foliate separated dorsal valves.

Female subgenital plate oval with shallow terminal emargination. Medigynium with rounded main plate; posterior

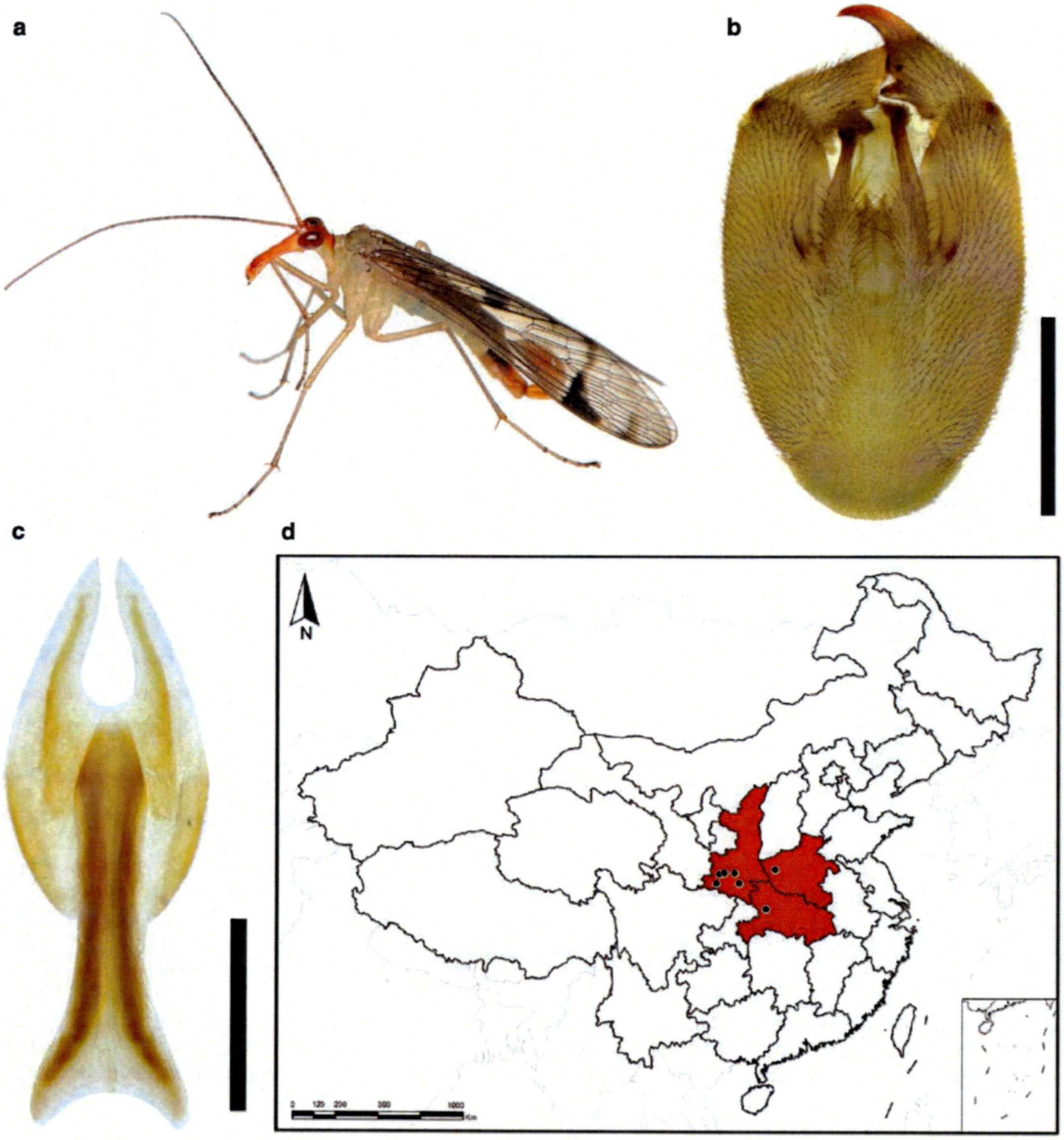

Fig. 10.129 (**a**) Male adult of ***Panorpa decolorata***, lateral view (Tiantaishan National Forest Park, Baoji, Shaanxi). (**b**) Male genital bulb of ***Panorpa decolorata***, ventral view (*Scale bar*: 0.5 mm). (**c**) Female medigynium of ***Panorpa decolorata***, ventral view (*Scale bar*: 0.2 mm). (**d**) Distribution of ***Panorpa decolorata***

arms slender; axis twice as long as posterior arms with greatly diverged apodemes.

Emergence. July to August.

Distribution. Henan (Songxian), Hubei (Shennongjia), Shaanxi (Baoji, Nanzheng, Ningshan, Taibai, Zhouzhi).

10.6.11 *Panorpa duanyu* Wang & Gong, 2021

(Yu Duan's Scorpionfly) [段誉蝎蛉] (Fig. 10.130)
Measurements. Forewing length 9.5–12.0 mm.

Morphology. Vertex and rostrum yellow. Ocellar triangle black. Nota yellowish brown, with broad yellow median stripe. Pleura and coxae creamy yellow; legs mostly light brown. Wing membrane hyaline, lacking distinct markings except greatly reduced apical and pterostigmal bands. Male T1–T5 black. Notal organ flat and covering acute postnotal organ. A6 black, cylindrical. A7 and A8 black, short cylindrical.

Male genital bulb yellowish brown, long oval. Epandrium emarginated terminally in deep U-shape and forming pair of finger-like processes laterally. Hypandrium Y-shaped, with long basal stalk. Hypovalves slender, divergent, with

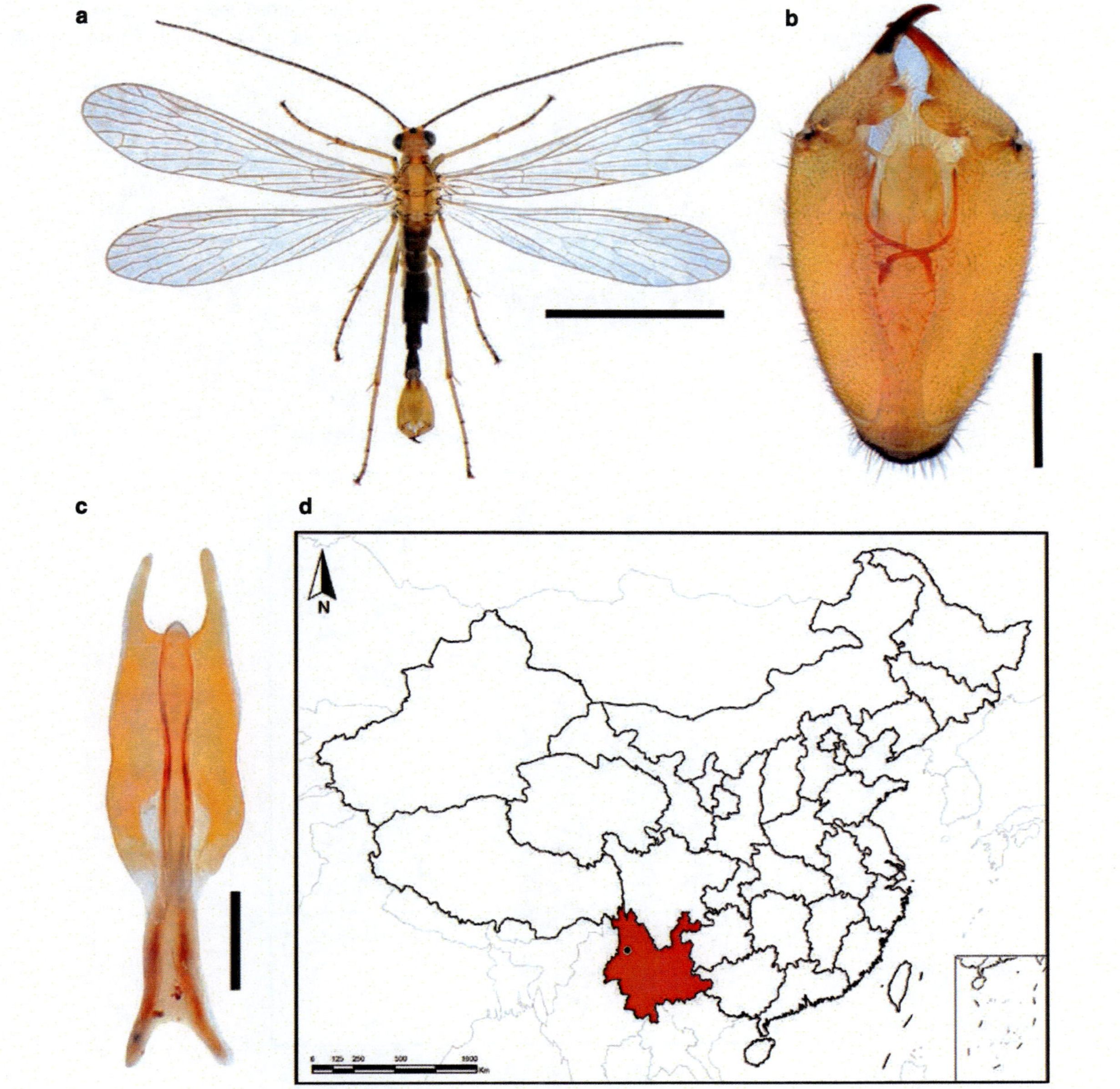

Fig. 10.130 (**a**) Male adult of ***Panorpa duanyu***, dorsal view (Sky Lake, Yunlong, Yunnan) (*Scale bar*: 5.0 mm) **[NWAU]**. (**b**) Male genital bulb of ***Panorpa duanyu***, ventral view (*Scale bar*: 0.5 mm). (**c**) Male genital bulb of ***Panorpa duanyu***, ventral view (*Scale bar*: 0.2 mm). (**d**) Distribution of ***Panorpa duanyu***

sparse long bristles along inner margin. Gonostyli approximately half as long as gonocoxites, with broad earlobe-shaped basal process and rounded median tooth. Parameres slender and covered with numerous microtrichia on inner margin; two parameres crossed at middle portion, with distal portions approximately parallel and directing caudad. Ventral aedeagal valves membranous and covered with numerous microtrichia; dorsal processes stout, greatly elongated, nearly parallel, neck-like in basal half and subacute in distal half.

Female subgenital plate oval with slightly pointed apex, and sparse long setae on distal margin. Medigynium with broad main plate approximately 3/5 as long as axis; posterior arms short and thin, and approximately 1/3 as long as main plate; apodemes greatly divergent basally.

Emergence. July to August.
Distribution. Yunnan (Yunlong).

10.6.12 *Panorpa filina* Chou & Wang, 1987

(Daughter's Scorpionfly) [女儿蝎蛉] (Fig. 10.131)
Measurements. Forewing length 12 mm.

Morphology. Vertex yellow. Ocellar triangle black. Rostrum yellow. Pronotum dark brown. Meso- and metanotum brown with distinct pale yellow median stripe. Pleura and legs sordid yellow. Wing membrane hyaline with markings brown. In forewings, apical band with three small hyaline spots; pterostigmal band unbranched, broad anteriorly and narrow posteriorly; two small spots on end of CuA and 1A,

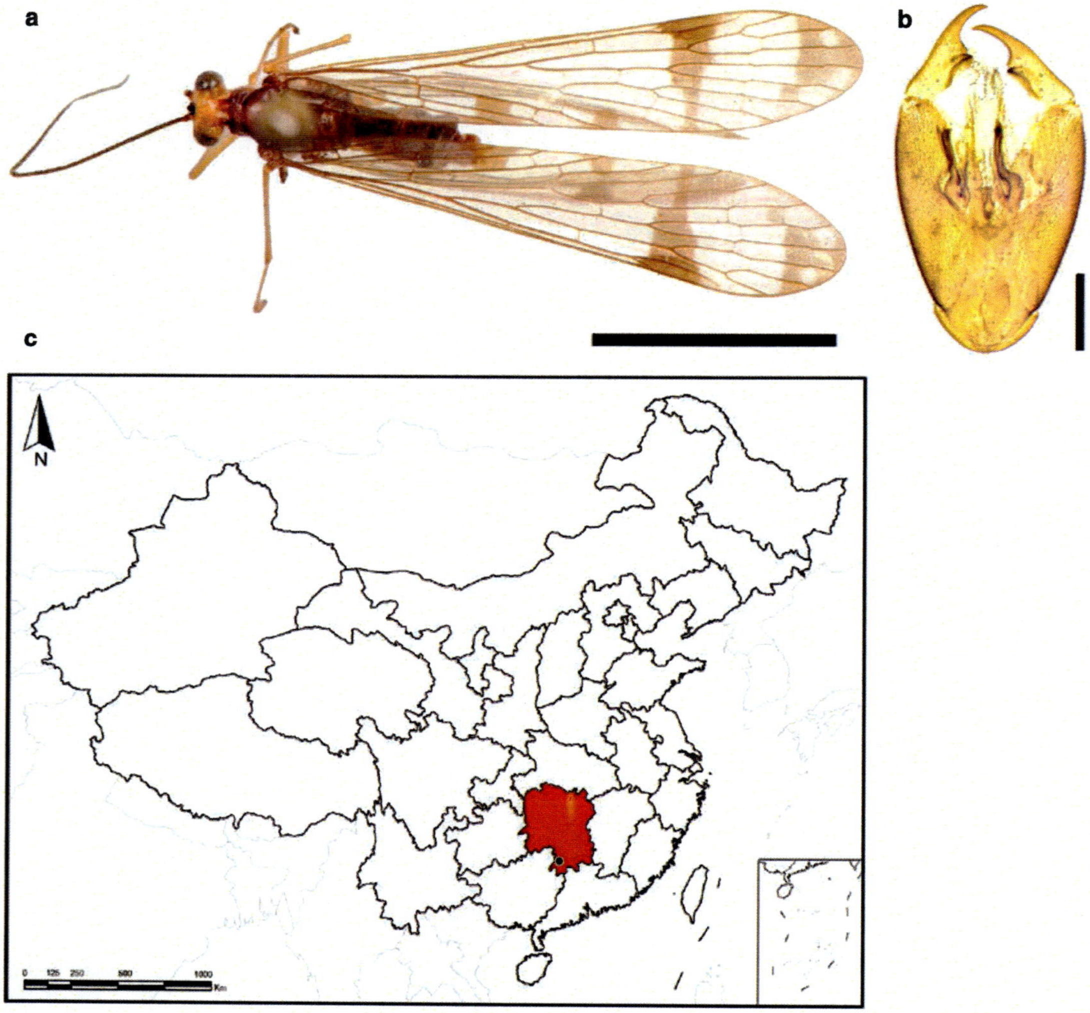

Fig. 10.131 (**a**) Male adult of ***Panorpa filina*** with genitalia removed, dorsal view (Daoxian, Hunan) (*Scale bar*: 0.5 mm) **[NWAU]**. (**b**) Male genital bulb of ***Panorpa filina***, ventral view (*Scale bar*: 0.5 mm). (**c**) Distribution of ***Panorpa filina***

respectively. Male T1–T5 dark yellowish brown. Notal organ flat, semicircular, and covering acute postnotal organ. A6 short cylindrical. A7 and A8 yellow and cylindrical, shorter than A6.

Male genital bulb long oval. Epandrium emarginates terminally in U-shape and forms pair of finger-like processes laterally. Hypandrium Y-shaped with long broad basal stalk; hypovalves finger-like and not exceeding apex of gonocoxites. Gonostylus half-length of gonocoxites, and bearing obtuse triangular median tooth and large disc-like basal process. Paramere bifurcated subbasally, with ventral branch short, and dorsal branch abruptly curved subbasally and bearing two acute teeth apically. Aedeagus with finger-like dorsal valves.

Female unknown.

Emergence. August.

Distribution. Hunan (Daoxian).

10.6.13 *Panorpa furcata* Zhou & Zhou, 2007

(Furcated Scorpionfly) [叉形蝎蛉] (Fig. 10.132)

Measurements. Forewing length 12.0–14.5 mm.

Morphology. Vertex, occiput, and rostrum yellowish brown. Ocellar triangle enclosed by black spot. Nota unevenly dark yellowish brown. Pleura and legs yellowish

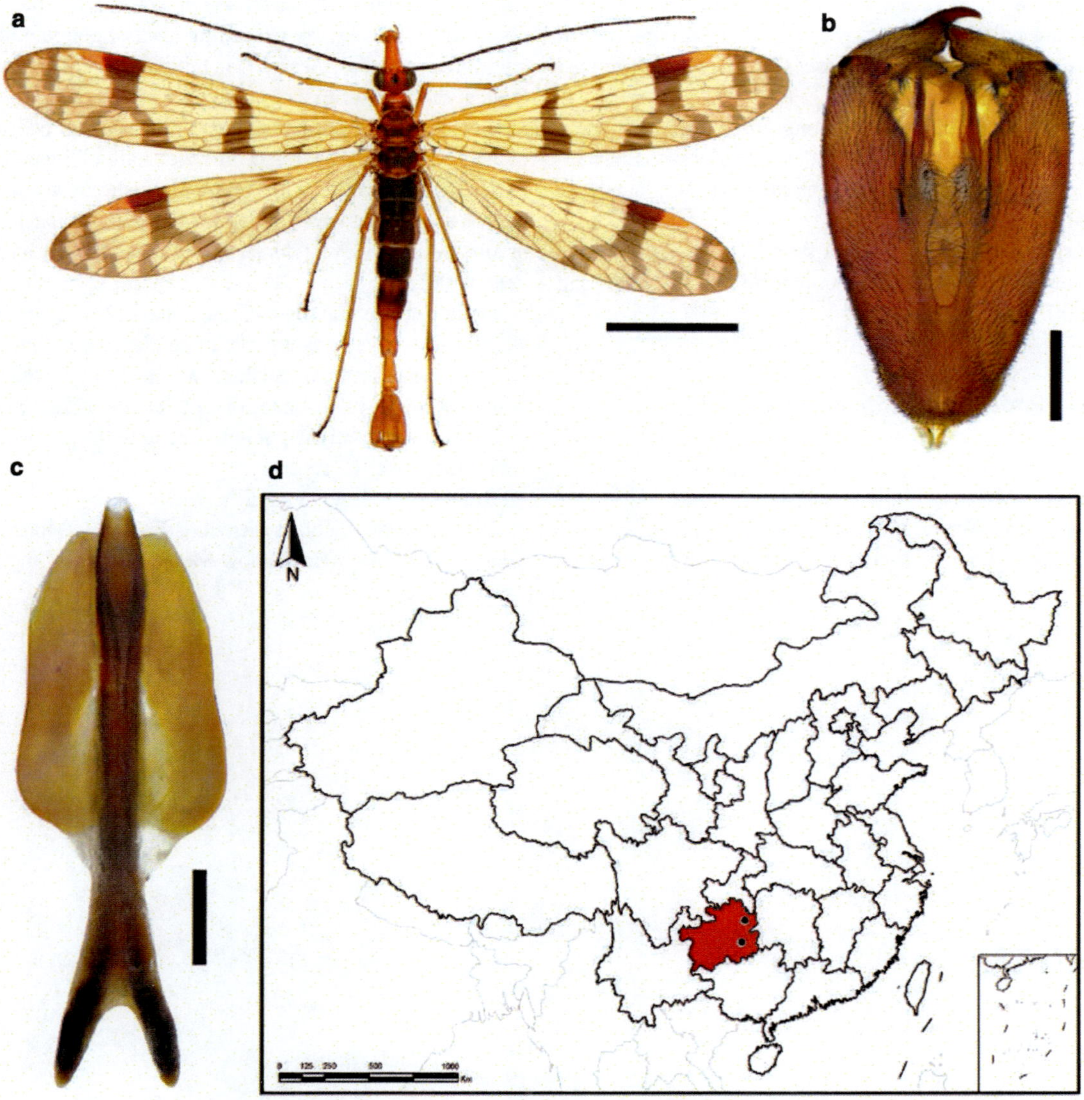

Fig. 10.132 (a) Male adult of ***Panorpa furcata***, dorsal view (Dewang, Jiangkou, Guizhou) (*Scale bar*: 5.0 mm) **[DALU]**. (**b**) Male genital bulb of ***Panorpa furcata***, ventral view (*Scale bar*: 0.5 mm). (**c**) Female medigynium of ***Panorpa furcata***, ventral view (*Scale bar*: 0.2 mm). (**d**) Distribution of ***Panorpa furcata***

brown. Wing membrane strongly tinged with yellow, markings dark brown, pterostigma reddish brown. In forewings, apical band with large hyaline window; pterostigmal band broad with broad basal branch and slightly reduced apical branch; marginal spot small; basal band broad; basal spot shifted posteriorly to ending of 2A. Male T1–T5 black. A6 dark yellowish brown, cylindrical. A7 and A8 yellowish brown, cylindrical.

Male genital bulb long oval, yellowish brown. Epandrium emarginates terminally in U-shape and forms pair of finger-like processes laterally. Hypandrium Y-shaped with long broad basal stalk; hypovalves slender with distal 1/3 slightly broaden, extending to 2/3 of gonocoxites. Gonostylus short, with stout subtriangular median tooth and subacute basal process. Parameres slender, straight, and bearing numerous short spines. Aedeagus with ventral valves membranous; dorsal processes greatly elongated with middle processes.

Female subgenital plate oblong with rounded apex. Genital plate with broad main plate; posterior arms absent; axis slightly shorter than twice length of main plate, with diverged apodemes.

Emergence. October to November.

Distribution. Guizhou (Fanjing Mountain, Leigong Mountain).

10.6.14 *Panorpa guttata* Navás, 1908

(Spotted Scorpionfly) [滴蝎蛉] (Fig. 10.133)

Measurements. Forewing length 11.0–11.5 mm.

Morphology. Vertex and rostrum yellow. Ocellar triangle black. Nota yellowish green, with broad yellow median stripe. Pleura and legs yellow. Wing membrane hyaline, slightly tinged with yellow. Pterostigma yellow and distinct. Markings very faint. In forewings, apical band greatly reduced into slender stripe or absent; pterostigmal band reduced into slender stripe below pterostigma. Other markings absent. Male T1–T5 yellowish green. Notal organ flat, with dense short setae posteriorly, and covering acute post-notal organ. A6 cylindrical, slightly enlarged towards apex, with narrow emargination dorso-apically and forming pair of plates laterally; below each plate are U-shaped emargination and acute tooth. A7 projected dorsad on dorsal apex. A8 short, cylindrical.

Male genital bulb long elliptical. Epandrium emarginates deeply in U-shape terminally and forms pair of finger-like processes laterally. Hypandrium with short broad basal stalk. Hypovalves slender, not exceeding apex of gonocoxites, and slightly projected mesad on middle of inner margins. Gonostylus shorter than gonocoxites, with obtuse triangular median tooth and auriculate basal process, and bearing tuft of dense curved long setae on basal half of inner margin. Paramere bifurcated, with ventral branch short and dorsal branch exceeding apex of gonocoxites. Aedeagus with membranous ventral valves; dorsal valves fused basally and divergent distally.

Female subgenital plate oblong, setose. Medigynium with slightly sclerotized basal plate; main plate narrowing cephalad and expanding caudad; axis slightly longer than twice length of main plate, anterior apodemes widely divergent and connected by slightly sclerotized lamella; posterior arms very short, parallel, acute.

Emergence. June to August.

Distribution. Sichuan (Anzihe, Baoxing, Mount Emei, Shimian, Wolong, Xiling Snow Mountain).

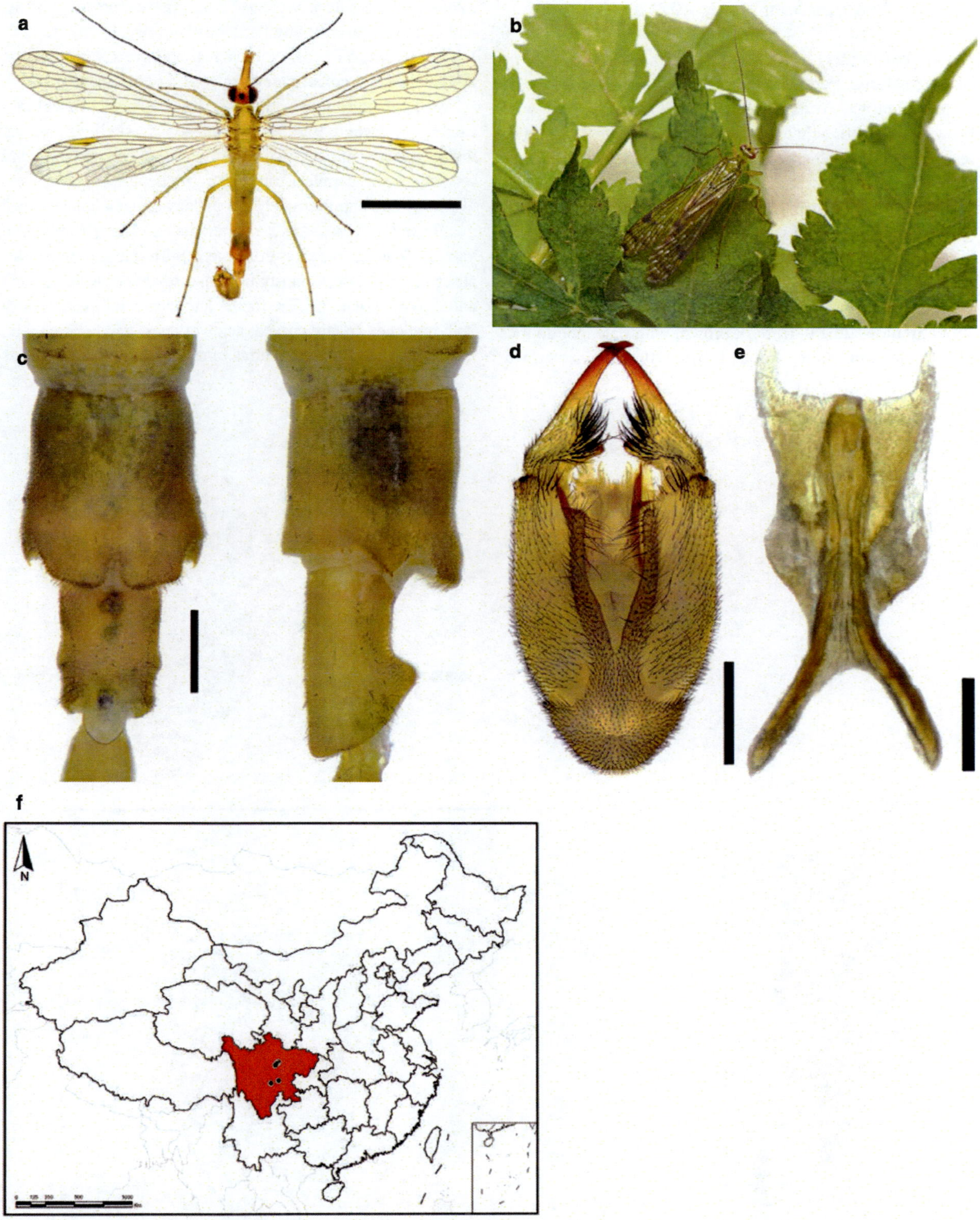

Fig. 10.133 (**a**) Male adult of ***Panorpa guttata***, dorsal view (Gongyihai, Shimian, Sichuan) (*Scale bar*: 5.0 mm) **[NWAU]**. (**b**) Female adult of ***Panorpa guttata*** in the field (Jiguan Mountain, Chongzhou, Sichuan). (**c**) Male A6 and A7 of ***Panorpa guttata*** (*Scale bar*: 1.0 mm). Left. Dorsal view; Right. Lateral view. (**d**) Male genital bulb of ***Panorpa guttata***, ventral view (*Scale bar*: 0.5 mm). (**e**) Female mediygnium of ***Panorpa guttata***, ventral view (*Scale bar*: 0.2 mm). (**f**) Distribution of ***Panorpa guttata***

10.6.15 *Panorpa hani* Wang, 2021

(Hani Scorpionfly) [哈尼蝎蛉] (Fig. 10.134)
Measurements. Forewing length

Morphology. Vertex, occiput, and rostrum yellowish brown, ocellar triangle enclosed by subpentagonal black pattern. Nota yellowish green, with broad yellow median stripe. Wing membrane hyaline and tinged with yellowish brown, markings dark brown. In forewings, apical band broad, vague, and interrupted by two or three hyaline spots; pterostigmal band reduced at posterior half, with two detached spots along hind margin as basal and apical branches; basal band reduced; marginal spot small; basal spot absent. Male T1–T5 dark brown, with pale, continuous median stripe extending from T1 to T5. A6 yellowish brown, cylindrical, with deep V-shaped emargination on dorsal apex, and forming pair of triangular lobes laterally. A7 and A8 yellowish brown, A7 nearly cylindrical and greatly humped on dorsal apex, A8 constricted basally and beveled apically.

Male genital bulb yellowish brown, oval. Epandrium extending beyond middle of gonostylus, slightly tapering towards apex, with deep U-shaped terminal emargination and forming two parallel finger-like processes. Hypandrium Y-shaped and approximately 3/4 as long as gonocoxites, with greatly reduced basal stalk shorter than half-length of hypovalves. Hypovalves narrow, divergent, with sparse long bristles along inner margin. Gonostylus with disc-shaped basal process and subtriangular median tooth. Parameres bifurcated, with both branches greatly curved, acute at apex, and covered with numerous microtrichia; ventral branch approximately 1/4 as

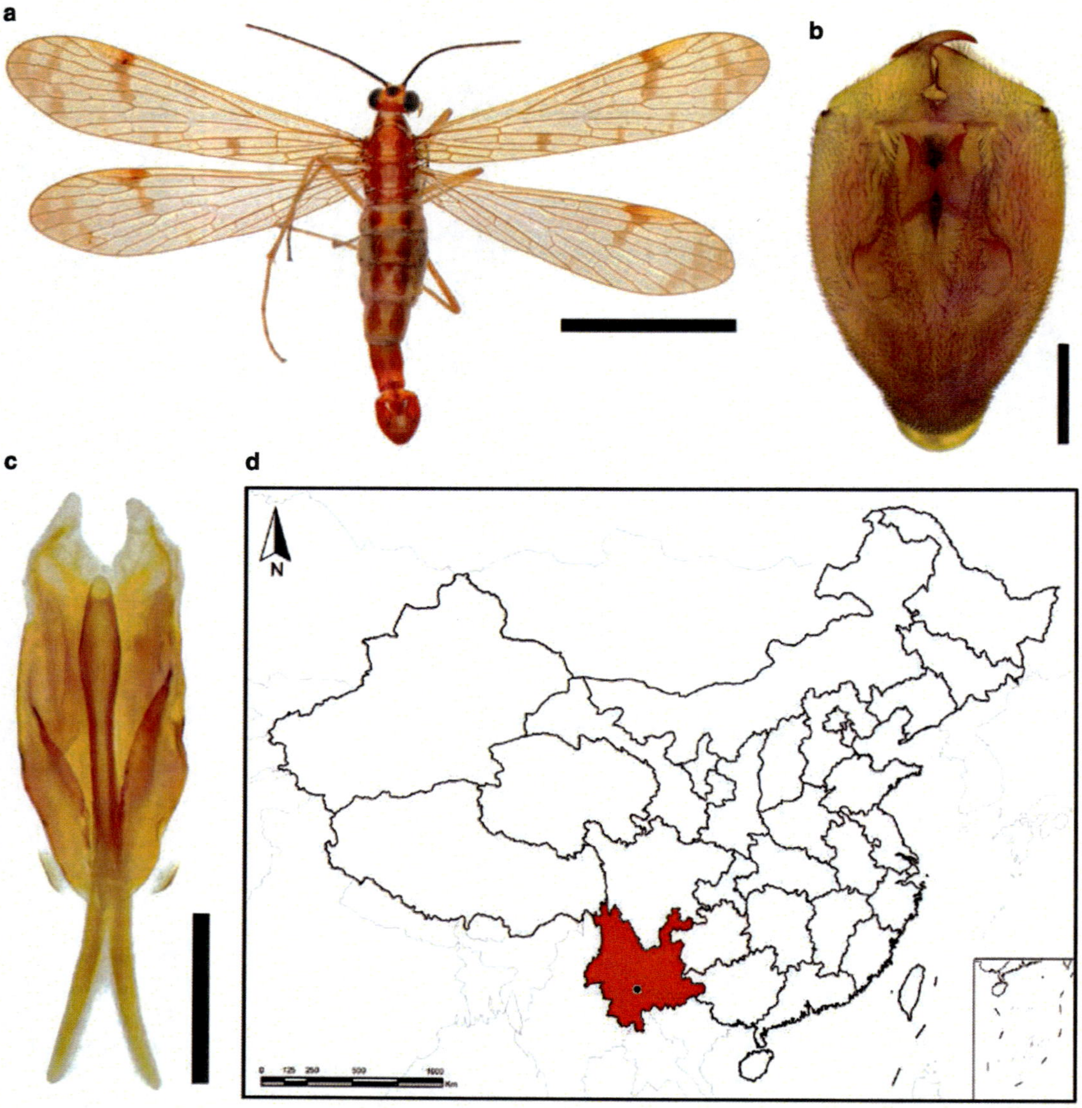

Fig. 10.134 (a) Male adult of ***Panorpa hani***, dorsal view (Zhenyuan, Yunnan) (*Scale bar*: 5.0 mm) **[DALU]**. (**b**) Male genital bulb of ***Panorpa hani***, ventral view (*Scale bar*: 0.5 mm). (**c**) Female medigynium of ***Panorpa hani***, ventral view (*Scale bar*: 0.2 mm). (**d**) Distribution of ***Panorpa hani***

long as dorsal branch and with subapical thorn on outer margin; dorsal branch slightly lobate with subapical notch on inner side, and bending outwards apically. Ventral aedeagal valves membranous and rounded; dorsal processes slender, greatly elongated, nearly parallel, and slightly divergent apically.

Female subgenital plate oblong with nearly truncated apex and sparse long setae on distal margin. Medigynium with broad main plate approximately 2/3 as long as axis; posterior arms very short, stout and approximately 1/5 as long as main plate; apodemes of axis greatly divergent.

Emergence. August.

Distribution. Yunnan (Zhenyuan).

10.6.16 *Panorpa hirundo* Wang, 2021

(Swallow-Tailed Scorpionfly) [燕尾蝎蛉] (Fig. 10.135)

Measurements. Forewing length 11.0–12.0 mm.

Morphology. Vertex, occiput, and rostrum yellowish brown, ocellar triangle black. Nota yellowish green, with broad yellow median stripe. Wing membrane hyaline, tinged with yellowish brown, and devoid of markings. Male A1–A5 dark yellowish green, A6–A9 yellowish brown. A6 yellowish brown, cylindrical, with deep arcuate emargination on dorsal apex, and forming pair of triangular lobes laterally. A7 and A8 yellowish brown, A7 nearly cylindrical and slightly

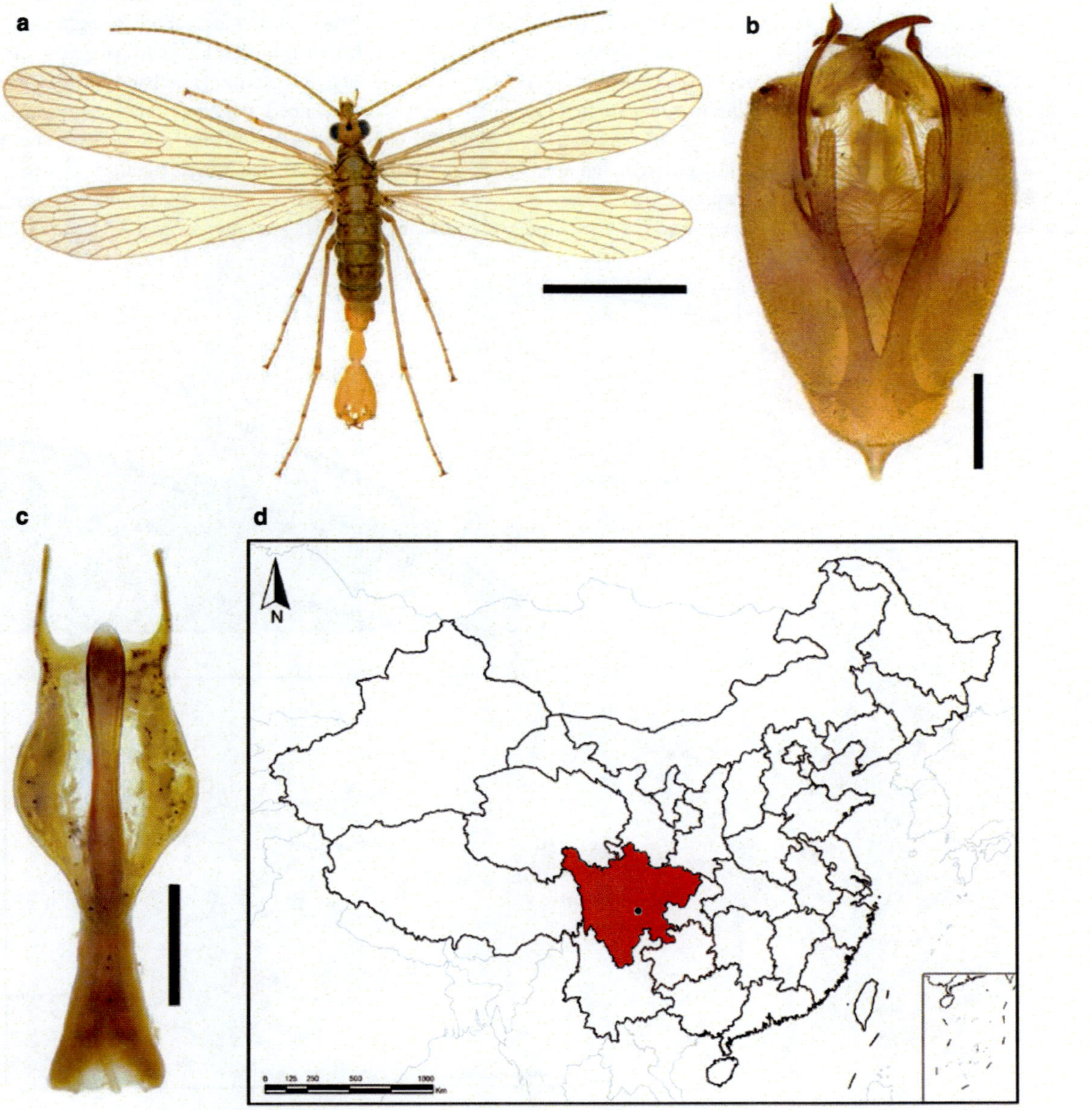

Fig. 10.135 (a) Male adult of ***Panorpa hirundo***, dorsal view (Mount Emei, Sichuan) (*Scale bar*: 5.0 mm) [NWAU]. (b) Male genital bulb of ***Panorpa hirundo***, ventral view (*Scale bar*: 0.5 mm). (c) Female medigynium of ***Panorpa hirundo***, ventral view (*Scale bar*: 0.2 mm). (d) Distribution of ***Panorpa hirundo***

humped on dorsal apex, A8 constricted basally and beveled apically.

Male genital bulb oval. Epandrium slightly tapering towards apex, with deep U-shaped terminal emargination and forming two parallel finger-like processes. Hypandrium Y-shaped and approximately 7/8 as long as gonocoxites, with greatly reduced basal stalk shorter than 1/5 length of hypovalves. Hypovalves narrow, divergent in basal half and nearly parallel in distal half, with sparse long bristles along inner margin. Gonostylus with broad disc-shaped basal process and subtriangular median tooth. Parameres furcated into two slender and glabrous branches; ventral branch long and stout, slightly curved inwards apically, greatly enlarged subapically with pointed apex; dorsal branch thinner and shorter, approximately 4/5 as long as ventral branch and lesser enlarged subapically. Ventral aedeagal valves membranous, rounded, and covered with numerous microtrichia; dorsal processes slender, greatly elongated, nearly parallel, and tapering to subacute apex in distal 1/4.

Female mediygnium oval with slightly emarginated apex. Medigynium with broad main plate approximately half as long as axis and widest at middle; posterior arms very thin, approximately 1/2 as long as main plate; apodemes of axis thick, greatly divergent at base.

Emergence. June to August.

Distribution. Sichuan (Mount Emei).

10.6.17 *Panorpa huangguiqiangi* Wang, 2021

(Gui-Qiang Huang's Scorpionfly) [黄氏蝎蛉]
(Fig. 10.136)

Measurements. Forewing length 9.5–11.0 mm.

Morphology. Vertex, occiput, and rostrum light yellow, ocellar triangle black. Thorax light greenish yellow laterally with light yellow median stripe. Wing membrane hyaline, tinged with yellowish brown, markings dark brown. Pterostigma yellow. In forewings, markings greatly reduced; apical band splitting into two longitudinal parts; inner portion of apical band and pterostigmal band forming two vertically inversed Y-shaped patterns; other markings absent. Male T1–T5 yellowish brown, with yellow, continuous median stripe. A6 yellowish brown, cylindrical, with shallow arcuate emargination on dorsal apex, and forming pair of broad triangular lobes laterally. A7 and A8 yellowish brown,

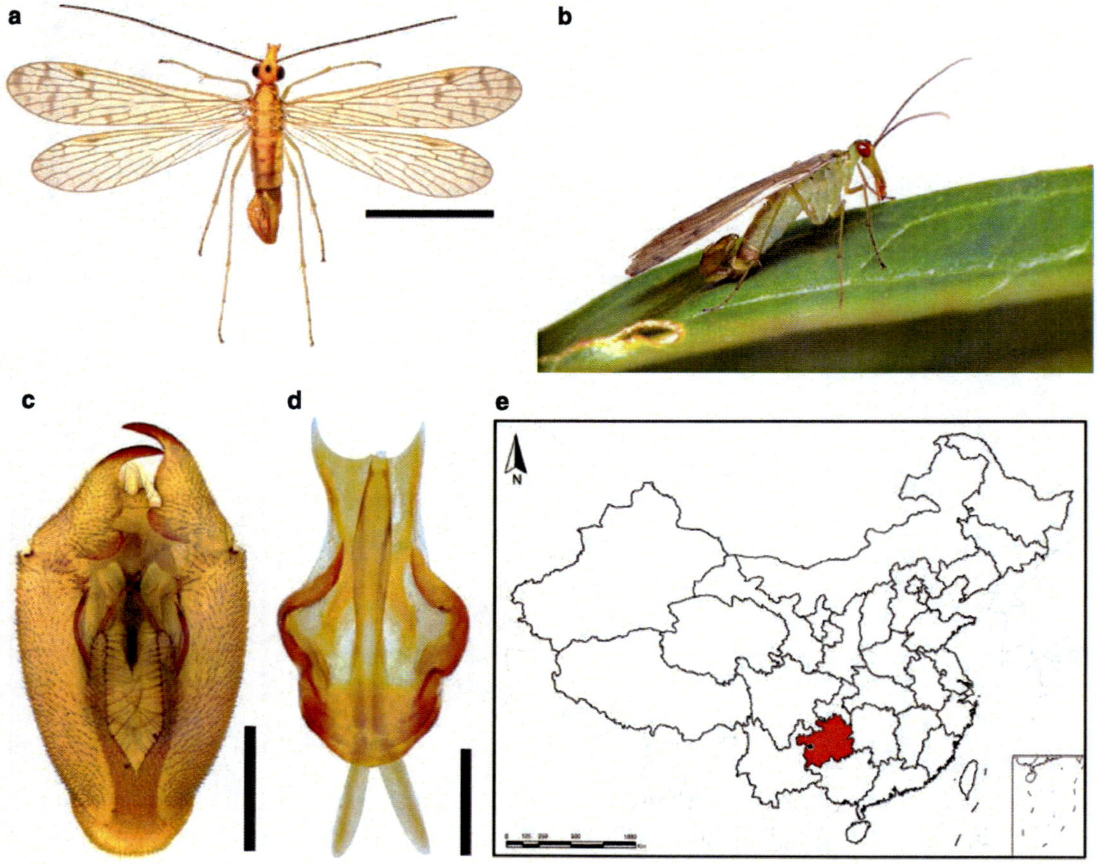

Fig. 10.136 (a) Male adult of ***Panorpa huangguiqiangi***, dorsal view (Yushe National Forest Park, Liupanshui, Guizhou) (*Scale bar*: 5.0 mm) [NWAU]. (b) Male adult of ***Panorpa huangguiqiangi*** in the field (Yushe Forest Park, Liupanshui, Guizhou). (c) Male genital bulb of ***Panorpa huangguiqiangi***, ventral view (*Scale bar*: 0.5 mm). (d) Female medigynium of ***Panorpa huangguiqiangi***, ventral view (*Scale bar*: 0.2 mm). (e) Distribution of ***Panorpa huangguiqiangi***

A7 nearly cylindrical and greatly humped on dorsal apex, A8 constricted basally and beveled apically.

Male genital bulb yellowish brown, oval. Epandrium tapering towards apex, with U-shaped terminal emargination and forming two short finger-like processes. Hypandrium Y-shaped, with basal stalk approximately half as long as hypovalves. Hypovalves narrow, divergent in basal half and nearly parallel in distal half. Gonostylus with broad earlobe-shaped basal process and rounded median tooth. Parameres furcated into two slender branches; ventral branch approximately 4/5 as long as dorsal branch, and stout, curved and pointed apically; dorsal branch similar and crossed subapically. Ventral aedeagal valves membranous, elongated, and covered with numerous microtrichia; dorsal processes stout, greatly elongated, nearly parallel but slightly divergent apically, and with shallow emargination at apex.

Female subgenital plate subtrapezoidal with truncated apex. Medigynium with main plate approximately 3/4 as long as axis, broadest at middle, and greatly curled marginally; posterior arms very short and stout; apodemes divergent basally.

Emergence. September.

Distribution. Guihzhou (Liupanshui).

10.6.18 *Panorpa jiangrixini* Wang, 2021

(Ri-Xin Jiang's Scorpionfly) [姜氏蝎蛉] (Fig. 10.137)

Measurements. Forewing length 12.0–13.0 mm.

Morphology. Vertex, occiput, and ocellar triangle shining black. Rostrum mostly shining black with yellowish lateral margin. Pronotum mostly black with yellowish median spot

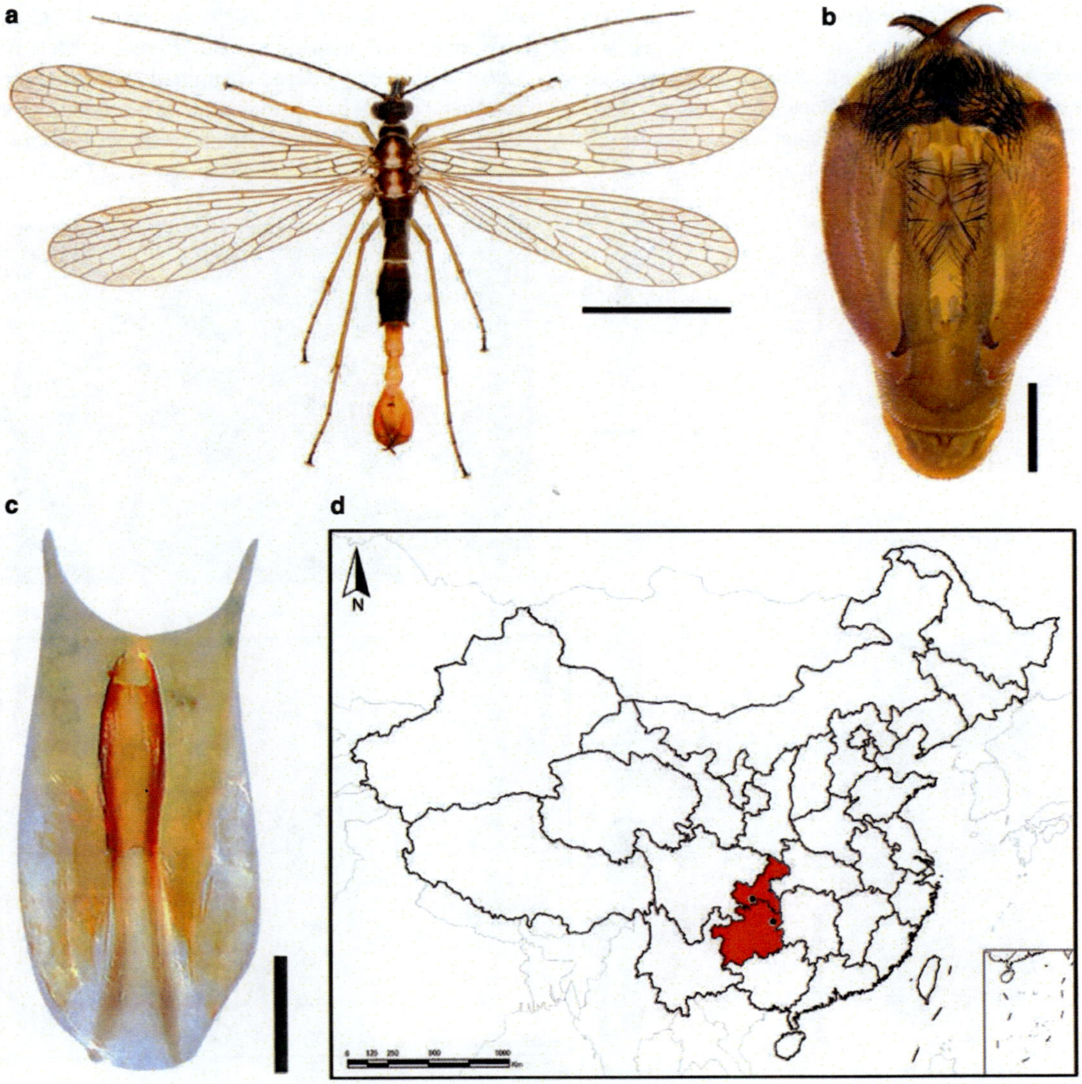

Fig. 10.137 (**a**) Male adult of ***Panorpa jiangrixini***, dorsal view (Fanjing Mountain, Guizhou) (*Scale bar*: 5.0 mm) **[NWAU]**. (**b**) Male genital bulb of ***Panorpa jiangrixini***, ventral view (*Scale bar*: 0.5 mm). (**c**) Female medigynium of ***Panorpa jiangrixini***, ventral view (*Scale bar*: 0.2 mm). (**d**) Distribution of ***Panorpa jiangrixini***

on posterior margin. Meso- and metanotum mostly dark brown with discontinuous yellow median stripe forming two bottle-shaped patterns. Wing membrane hyaline, tinged with light grayish brown, devoid of markings. Male T1–T5 dark brown to black. A6 dark brown at base and gradually yellowing towards apex, cylindrical, with shallow arcuate emargination on dorsal apex. A7 and A8 yellowish brown, A7 nearly cylindrical and slightly humped on dorsal apex, A8 constricted basally and beveled apically.

Male genital bulb yellowish brown, oval. Epandrium broad with U-shaped terminal emargination and forming two short stout processes. Hypandrium Y-shaped with basal stalk approximately 1/3 as long as hypovalves. Hypovalves narrow, nearly parallel, with distal half approximately half as wide as basal half. Gonostylus with broad rounded basal process and subtriangular median tooth, and covered with numerous long stout setae on basal half of ventral surface. Parameres trifurcated: basal branch stout and shortest, stretching towards base of genital bulb, greatly curved outward and hook-shaped, with apexes exposing from lateral margin of hypovalves; ventral and dorsal branches slender and straight, stretching caudad. Ventral aedeagal valves membranous, bottle-shaped, and covered with numerous long microtrichia; dorsal processes stout, greatly elongated, slightly divergent apically.

Female subgenital plate oblong with rounded apex. Medigynium with main plate broad and suboval; axis entirely concealed in main plate with slightly divergent apodemes; posterior arms thin and approximately 1/4 as long as main plate.

Emergence. June to August.

Distribution. Chongqing (Mount Jinfo), Guizhou (Fanjing Mountain).

10.6.19 *Panorpa jinfoshana* Wang, 2021

(Jinfoshan Scorpionfly) [金佛山蝎蛉] (Fig. 10.138)

Measurements. Forewing length 10.5–11.5 mm.

Morphology. Vertex, occiput, and rostrum yellow. Ocellar triangle black. Body mostly greenish yellow. Wing membrane slightly tinged with yellowish brown, with greatly reduced brown markings. Pterostigma yellow and conspicuous. In forewings, markings greatly reduced and scattered; only apical and pterostigmal bands present. Male A6 cylindrical, with deep V-shaped emargination on dorsal apex, and forming pair of broad triangular lobes laterally, and five small pointed teeth (dorsal one largest) below lobes on each lateral side. A7 and A8 yellowish brown, A7 nearly cylindri-

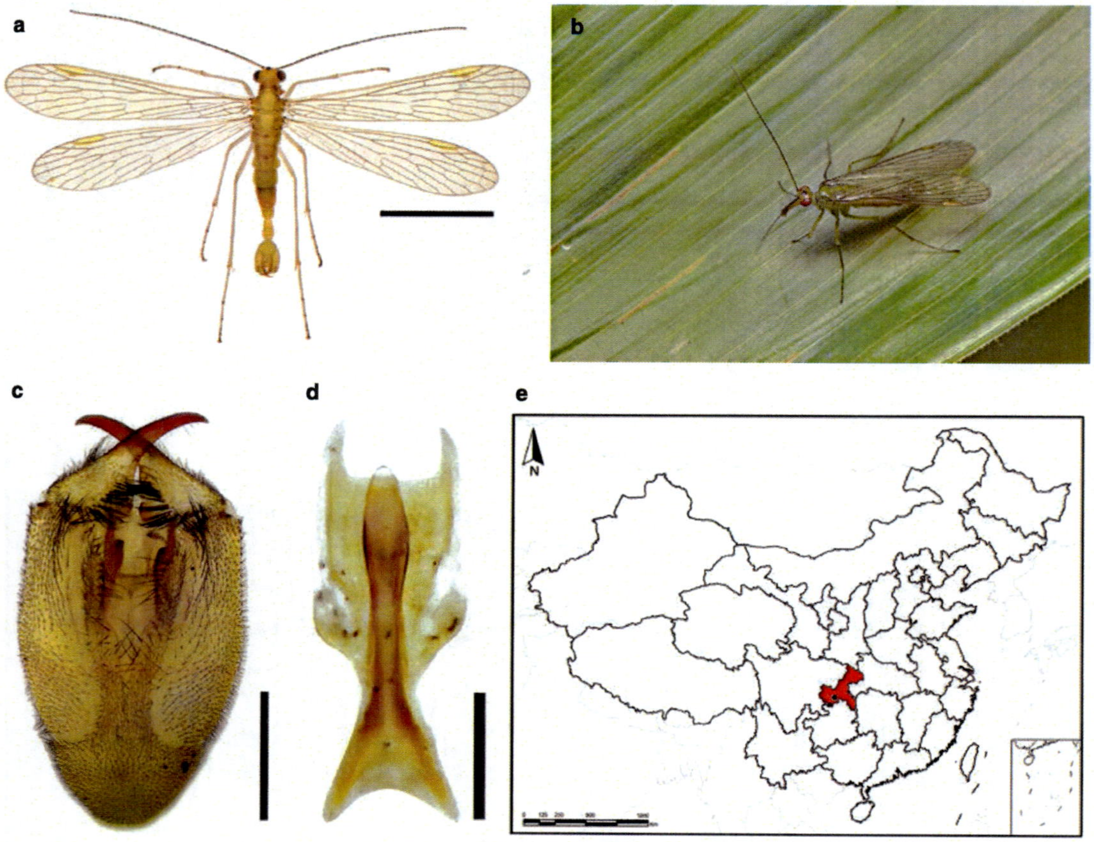

Fig. 10.138 (**a**) Male adult of ***Panorpa jinfoshana***, dorsal view (Mount Jinfo, Nanchuan, Chongqing) (*Scale bar*: 5.0 mm) **[DALU]**. (**b**) Male adult of ***Panorpa jinfoshana*** in the field (Mount Jinfo, Nanchuan, Chongqing). (**c**) Male genital bulb of ***Panorpa jinfoshana***, ventral view (*Scale bar*: 0.5 mm). (**d**) Female medigynium of ***Panorpa jinfoshana***, ventral view (*Scale bar*: 0.2 mm). (**e**) Distribution of ***Panorpa jinfoshana***

cal and greatly humped on dorsal apex, A8 constricted basally and beveled apically.

Male genital bulb oval. Epandrium with U-shaped terminal emargination and forming pair of short finger-like processes. Hypandrium Y-shaped; hypovalves narrow, divergent in basal half and nearly parallel in distal half. Gonostylus with broad earlobe-shaped basal process and rounded median tooth, and long stout setae on basal half of ventral surface. Parameres bifurcated; ventral branch long, stout and pointed apically; dorsal branch approximately 4/5 as long as ventral branch. Ventral aedeagal valves membranous and covered with numerous microtrichia; dorsal processes stout, greatly elongated, nearly parallel, and with greatly constricted apex.

Female subgenital plate oval with rounded apex, and sparse long setae on distal margin. Medigynium with basally constricted basal plate; main plate approximately 1/2 as long as axis; posterior arms short and stout; apodemes greatly divergent basally.

Emergence. August to September.

Distribution. Chongqing (Mount Jinfo).

10.6.20 *Panorpa kunmingensis* Fu & Hua, 2009

(Kunming Scorpionfly) [昆明蝎蛉] (Fig. 10.139)

Measurements. Forewing length 10.5–12.0 mm.

Morphology. Vertex yellow. Ocellar triangle black. Rostrum yellow. Nota brown, with broad pale yellow median stripe. Pleura and legs light brown. Wing membrane hyaline. In forewings, apical band complete and distinct; pterostigmal band scattered into several small spots; basal band greatly reduced into irregular spot on CuA; marginal and basal spots absent. Wings held roof-like at repose. Male T1–T5 brown, bearing yellow median stripe tapering caudad. Notal organ flat and covering acute postnotal organ. A6 deeply emarginate on dorsal apex in V-shape, and forming pair of triangular lobes laterally. A7 very short and projected on dorsal apex, which can fit V-shaped emargination between triangular lobes on A6.

Male genital bulb long oval. Epandrium emarginates terminally in deep U-shape and forms pair of finger-like processes laterally. Hypandrium fork-shaped with long basal

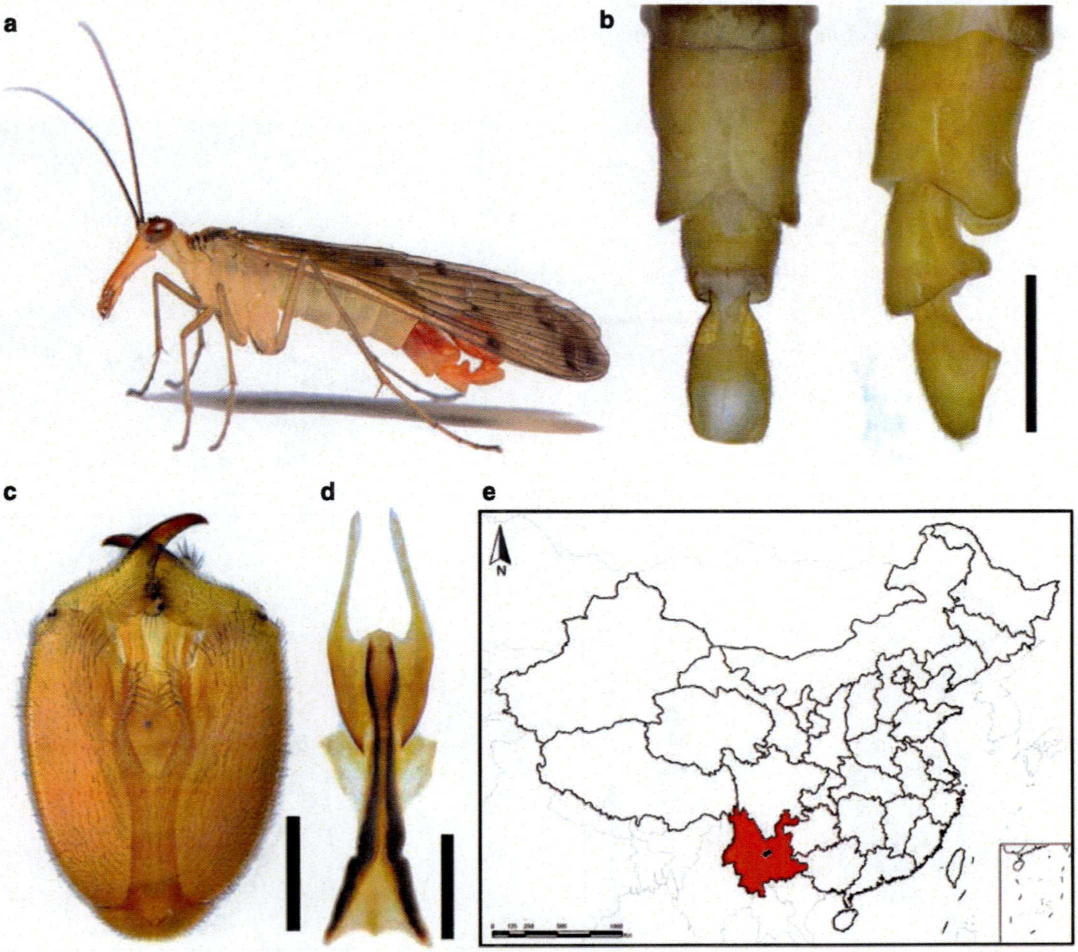

Fig. 10.139 (**a**) Male adult of ***Panorpa kunmingensis***, lateral view (West Hill, Kunming, Yunnan). (**b**) Male A6 and A7 of ***Panorpa kunmingensis*** (*Scale bar*: 1.0 mm). Left. Dorsal view; Right. Lateral view. (**c**) Male genital bulb of ***Panorpa kunmingensis,*** ventral view (*Scale bar*: 0.5 mm). (**d**) Female medigynium of ***Panorpa kunmingensis***, ventral view (*Scale bar*: 0.2 mm). (**e**) Distribution of ***Panorpa kunmingensis***

stalk. Hypovalves relatively short. Gonostylus intensively concaved along outer margin, bearing obtuse triangular median tooth and large disc-like basal process. Paramere bifurcated with thin and short ventral branch, and foliate dorsal branch, which curved laterad and exceeding apex of gonocoxites. Aedeagus with membranous ventral valves and strongly sclerotized and long dorsal valves.

Female subgenital plate oblong with shallow V-shaped emargination terminally. Medigynium with broad main plate approximately half as long as axis; posterior arms slender and slightly longer than main plate; axis greatly elongated, with apodemes greatly diverged.

Emergence. August to October.

Distribution. Yunnan (Kunming, Anning).

10.6.21 *Panorpa latiloba* Wang, 2021

(Broad-lobed Scorpionfly) [宽瓣蝎蛉] (Fig. 10.140)

Measurements. Forewing length 10.0–12.0 mm.

Morphology. Vertex, occiput, and rostrum yellow. Ocellar triangle black. Body mostly greenish yellow. Wing membrane hyaline, devoid of markings. Pterostigma yellow and conspicuous. In forewings, markings greatly reduced and scattered; only apical and pterostigmal bands present. Male A6 cylindrical, with deep V-shaped emargination on dorsal apex, and forming pair of very long and broad triangular lobes laterally. A7 and A8 yellowish brown, A7 nearly cylindrical and greatly humped on dorsal apex, A8 constricted basally and beveled apically.

Male genital bulb oval. Epandrium with U-shaped terminal emargination and forming two finger-like processes. Hypandrium Y-shaped. Hypovalves narrow, broadest at middle, divergent. Gonostylus with earlobe-shaped basal process and rounded median tooth. Parameres bifurcated; ventral branch long, slightly foliate, and three times as wide as dorsal branch; dorsal branch slender and approximately half as long as ventral branch. Ventral aedeagal valves membranous and covered with numerous microtrichia; dorsal processes stout, greatly elongated, nearly parallel, and with beveled apex.

Female subgenital plate oval with rounded apex. Medigynium with basally constricted basal plate; main plate approximately 1/2 as long as axis; posterior arms short and stout, and approximately half as long as main plate; apodemes greatly divergent basally.

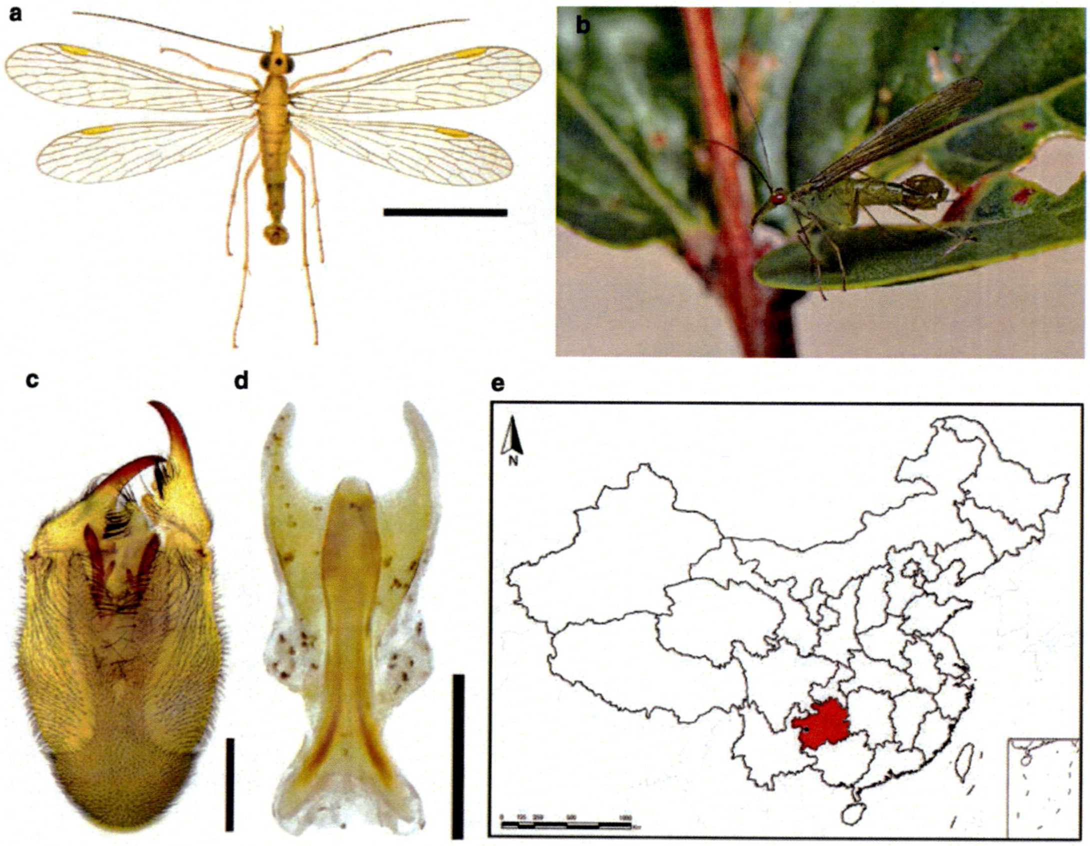

Fig. 10.140 (**a**) Male adult of ***Panorpa latiloba***, dorsal view (Yushe National Forest Park, Liupanshui, Guizhou) (*Scale bar*: 5.0 mm) [NWAU]. (**b**) Male adult of ***Panorpa latiloba*** in the field (Yushe National Forest Park, Liupanshui, Guizhou). (**c**) Male genital bulb of ***Panorpa latiloba***, ventral view (*Scale bar*: 0.5 mm). (**d**) Female medigynium of ***Panorpa latiloba***, ventral view (*Scale bar*: 0.2 mm). (**e**) Distribution of ***Panorpa latiloba***

Emergence. August to September.
Distribution. Guizhou (Liupanshui).

10.6.22 *Panorpa liaoi* Zhou & Zhou, 2007

(Liao's Scorpionfly) [廖氏蝎蛉] (Fig. 10.141)
Measurements. Forewing length 10.0–12.0 mm.

Morphology. Vertex and rostrum yellow. Ocellar triangle black. Nota dark brown, with broad yellow median stripe. Pleura and coxae creamy yellow; legs mostly light brown. Wing membrane hyaline, devoid of markings. Pterostigma light brown. Wings held roof-like over abdomen at repose. Male T1–T5 dark brown. Notal organ flat and covering acute postnotal organ. A6 cylindrical. A7 and A8 short cylindrical.

Male genital bulb long oval. Epandrium emarginated terminally in deep U-shape and forming pair of finger-like processes laterally. Hypandrium Y-shaped, with long basal stalk. Hypovalves slender with rounded projection medially on inner margin. Gonostylus slightly concaved along outer margin, bearing stout median tooth and large disclike basal process. Paramere strongly sclerotized and trifurcated, with median branch bifurcated subapically. Ventral aedeagal

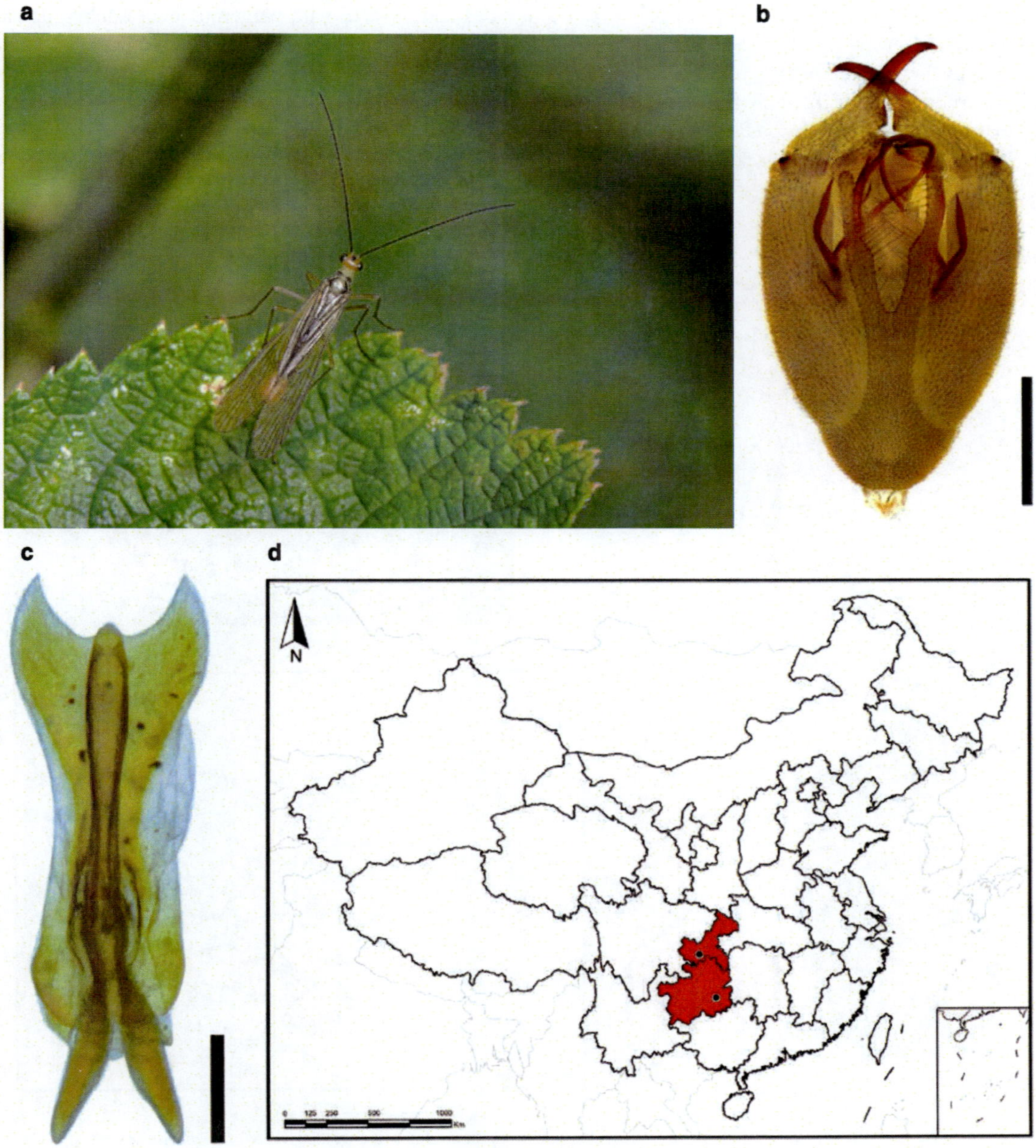

Fig. 10.141 (**a**) Male adult of ***Panorpa liaoi*** in the field (Mount Jinfo, Nanchuan, Chongqing). (**b**) Male genital bulb of ***Panorpa liaoi***, ventral view (*Scale bar*: 0.5 mm). (**c**) Female medigynium of ***Panorpa liaoi***, ventral view (*Scale bar*: 0.2 mm). (**d**) Distribution of ***Panorpa liaoi***

valves membranous and vase-shaped; dorsal valves intensively sclerotized with hook-like lateral processes.

Female subgenital plate broad, subtrapezoidal, with shallow V-shaped emargination terminally. Medigynium with long main plate greatly incurved along lateral margins; posterior arms short and subtriangular; axis longer than main plate with stout apodemes diverged.

Emergence. June to July.

Distribution. Chongqing (Mount Jinfo), Guizhou (Leigong Mountain).

10.6.23 *Panorpa nanzhao* Wang, 2021

(Nanzhao Scorpionfly) [南诏蝎蛉] (Fig. 10.142)

Measurements. Forewing length 10.5–12.0 mm.

Morphology. Vertex and rostrum yellow. Ocellar triangle black. Nota yellowish brown, with broad yellow median stripe. Pleura and coxae creamy yellow; legs mostly light brown. Wing membrane hyaline, lacking distinct markings except greatly reduced apical and pterostigmal bands. Male T1–T5 black. Notal organ flat and covering acute postnotal organ. A6 black, cylindrical. A7 and A8 black, short cylindrical.

Male genital bulb yellowish brown, long oval. Epandrium emarginated terminally in deep U-shape and forming pair of finger-like processes laterally. Hypandrium Y-shaped, with long basal stalk. Hypovalves slender, divergent, with sparse long bristles along inner margin. Gonostyli approximately half as long as gonocoxites, with broad earlobe-shaped basal process and rounded median tooth. Parameres bifurcated: ventral branch extremely short and hook-like; dorsal branch slender, slightly convergent apically, arcuate in basal half and straight in distal half. Ventral aedeagal valves membranous and covered with numerous microtrichia; dorsal processes stout, greatly elongated, nearly parallel, neck-like in basal half and foliate in distal half, and slightly curved inward apically.

Female Subgenital plate oval with slightly pointed apex. Medigynium with broad main plate approximately 3/5 as

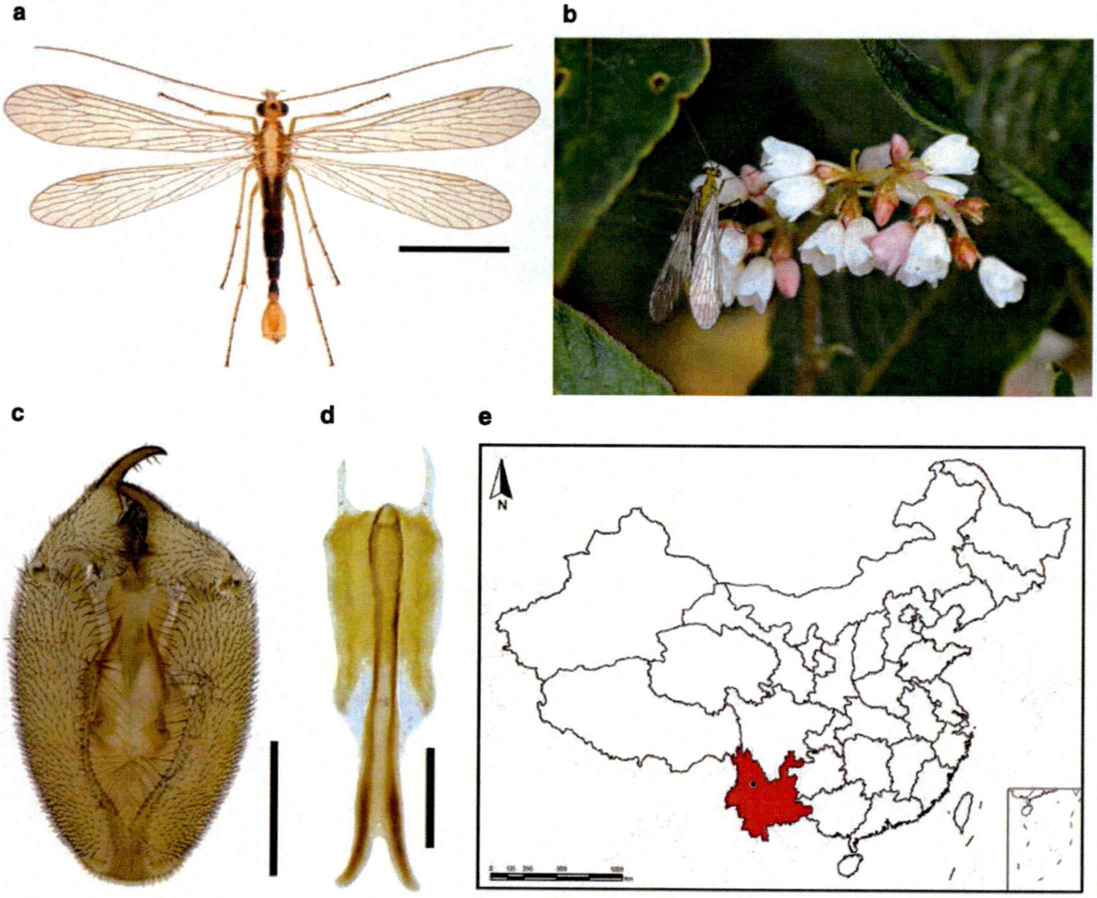

Fig. 10.142 (**a**) Male adult of ***Panorpa nanzhao***, dorsal view (Cangshan Mountain, Dali, Yunnan) (*Scale bar*: 5.0 mm) **[DALU]**. (**b**) Male adult of ***Panorpa nanzhao*** in the field (Cangshan Mountain, Dali, Yunnan). (**c**) Male genital bulb of ***Panorpa nanzhao***, ventral view (*Scale bar*: 0.5 mm). (**d**) Female medigynium of ***Panorpa nanzhao***, ventral view (*Scale bar*: 0.2 mm). (**e**) Distribution of ***Panorpa nanzhao***

long as axis; posterior arms short and thin, and approximately 1/3 as long as main plate; apodemes greatly divergent basally.

Emergence. June to September.

Distribution. Yunnan (Cangshan Mountain).

10.6.24 *Panorpa parallela* Wang & Hua, 2016

(Parallel Scorpionfly) [平枝蝎蛉] (Fig. 10.143)

Measurements. Forewing length 10.0–11.3 mm.

Morphology. Vertex and rostrum yellow. Ocellar triangle black. Nota brown, with broad yellow median stripe. Pleura and legs yellowish brown. Wing membrane hyaline, and slightly tinged with yellow. Pterostigma reddish brown. In forewings, apical band greatly reduced into inconspicuous short stripe; pterostigmal band reduced into irregular spot with acute hind apex. Wings held in close contact along midline over abdomen at repose. Male T1–T5 sordidly yellowish brown, with pale median stripe. Notal organ flat and covering postnotal organ. A6 dark brown dorsally and yellowish brown ventrally, cylindrical. A7 and A8 yellowish brown. A7 short cylindrical, A8 constricted basally.

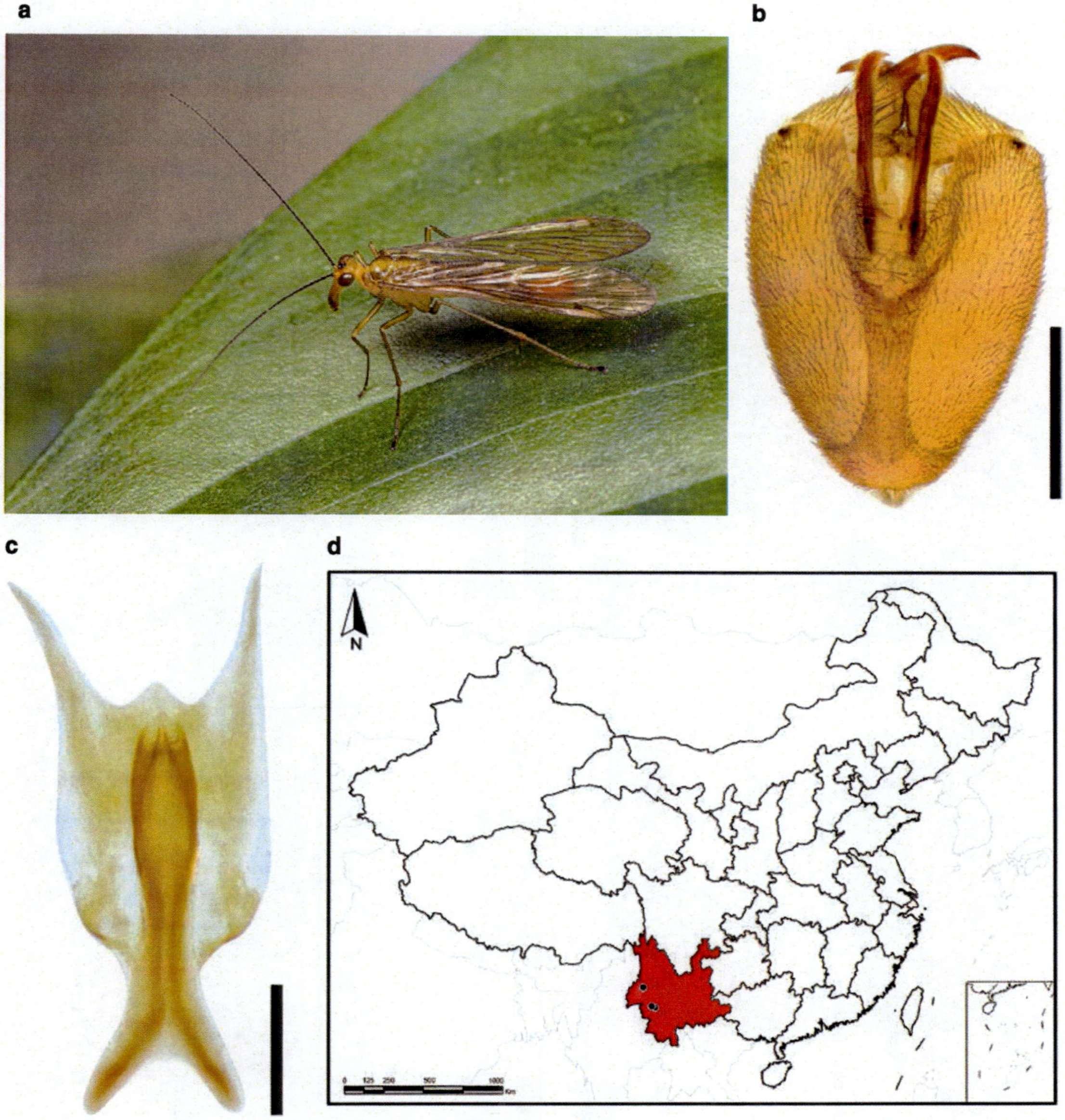

Fig. 10.143 (a) Male adult of ***Panorpa parallela*** in the field (Daxue Mountain, Yongde, Yunnan). (**b**) Male genital bulb of ***Panorpa parallela***, ventral view (*Scale bar*: 0.5 mm). (**c**) Female medigynium of ***Panorpa parallela***, ventral view (*Scale bar*: 0.2 mm). (**d**) Distribution of ***Panorpa parallela***

Male genital bulb oval. Epandrium emarginates terminally in U-shape and forms pair of finger-like processes laterally. Hypandrium with long narrow basal stalk and Y-shaped. Hypovalves relatively short, not extending to apex of gonocoxites. Gonostylus with obtuse triangular median tooth and large disc-like basal process. Paramere stick-like, extending to median tooth of gonostylus, acute and hooked on apex, and bearing numerous short spines on distal third. Ventral aedeagal valves short and membranous; dorsal valves intensively sclerotized, enlarged and flatiron-shaped distally.

Female subgenital plate long elliptic. Medigynium with main plate oblong and intensely constricted at base, posterior margin with triangular mesal prominence; basal plate slightly wider than axis but narrower than main plate; axis elongated beyond main plate with apodemes widely divergent; posterior arms slender.

Emergence. July to August.

Distribution. Yunnan (Gaoligong Mountains, Yongde).

10.6.25 *Panorpa reflexa* Wang & Hua, 2016

(Reflexed Scorpionfly) [反曲蝎蛉] (Fig. 10.144)

Measurements. Forewing length 10.0–11.0 mm.

Morphology. Vertex yellow. Ocellar triangle enclosed by large black pentagonal spot. Another black spot on occiput. Rostrum light yellowish brown. Pronotum dark brown. Meso- and metanotum dark brown, with broad yellow median stripe. Pleura and legs creamy yellow. Wing membrane hyaline. In forewings, pterostigmal band greatly reduced into small triangular spot. Other markings absent. Wings held in close contact along mid-line over abdomen at repose. Male T1–T5 dark brown, with light yellow median stripe tapering caudad. Notal organ flat and covering acute postnotal organ. A6 almost dark brown, but yellowish brown on apex, short cylindrical. A7 and A8 yellowish brown and short cylindrical, and not constricted basally.

Male genital bulb elliptical. Epandrium emarginates terminally in U-shape and forms pair of finger-like processes

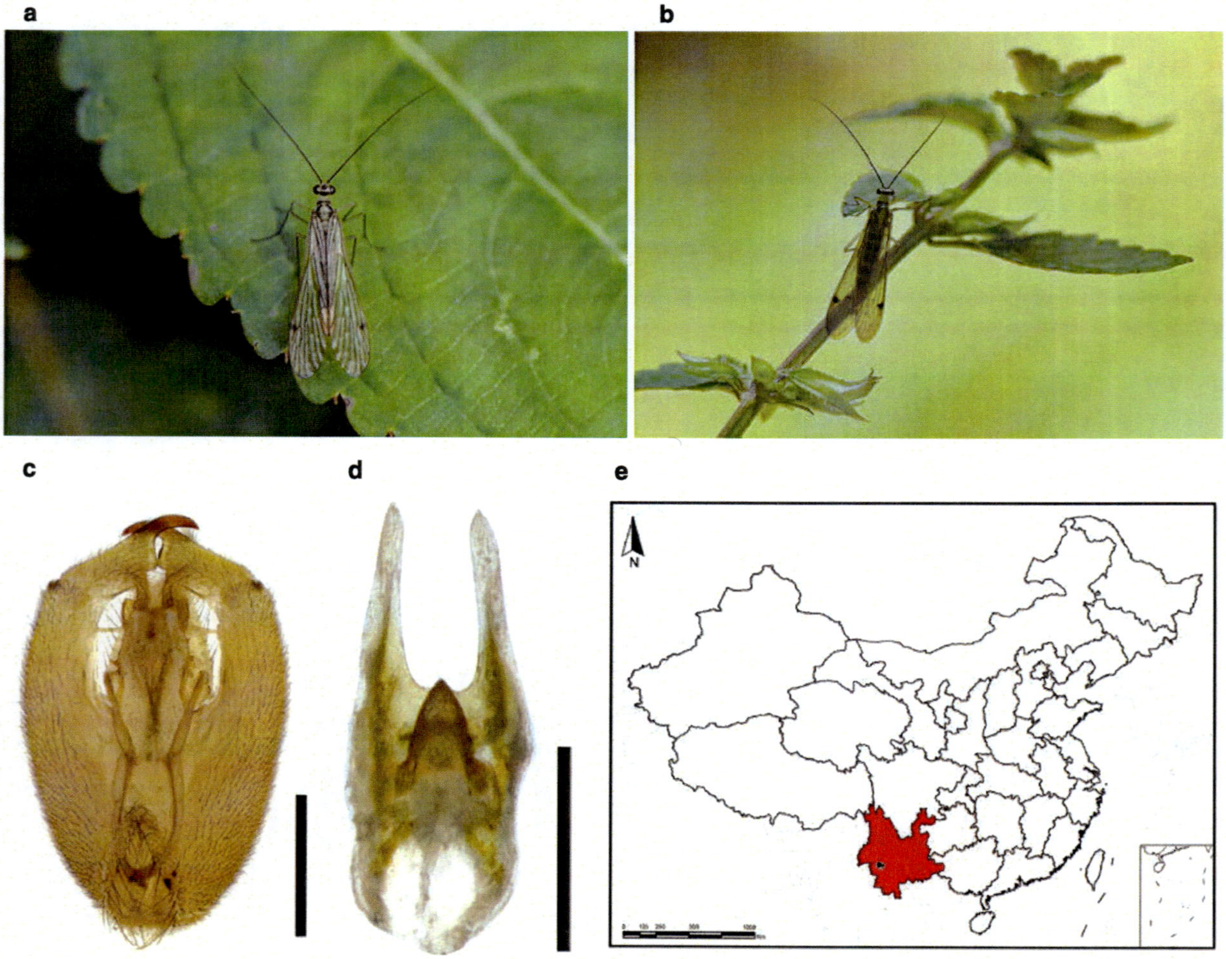

Fig. 10.144 (a) Male adult of ***Panorpa reflexa***, dorsal view (Daxue Mountain, Yongde, Yunnan). (b) Female adult of ***Panorpa reflexa*** in the field (Tangli Mountain, Yongde, Yunnan). (c) Male genital bulb of ***Panorpa reflexa***, ventral view (*Scale bar*: 0.5 mm). (d) Female medigynium of ***Panorpa reflexa***, ventral view (*Scale bar*: 0.2 mm). (e) Distribution of ***Panorpa reflexa***

laterally. Hypandrium greatly shortened, with hypovalves extending to basal third of gonocoxites and bearing row of long bristles on inner margin. Gonostylus short, with broad median tooth and curved basal process truncated apically. Paramere 3-shaped, with ventral branch stretched under hypovalves, and dorsal branch aside aedeagus. Columnar aedeagus with membranous ventral valves and greatly sclerotized dorsal valves bearing pair of auriculate processes apically. Lateral margin of dorsal valves rolled ventrad wave-like, and bearing 3–4 acute teeth. Lateral processes long and acute.

Female subgenital plate greatly elongated, accompanied by well-developed laterotergites, distal half vase-shaped; marginally. Medigynium small, shorter than half length of subgenital plate; axis short, entirely concealed, and lacking apodemes; anterior arms shorter than main plate, curved inward; posterior arms longer than main plate, straight, slightly convergent.

Emergence. July to August.

Distribution. Yunnan (Yongde).

10.6.26 *Panorpa stella* Wang, 2021

(Stellar Scorpionfly) [星斑蝎蛉] (Fig. 10.145)

Measurements. Forewing length 11.0–12.5 mm.

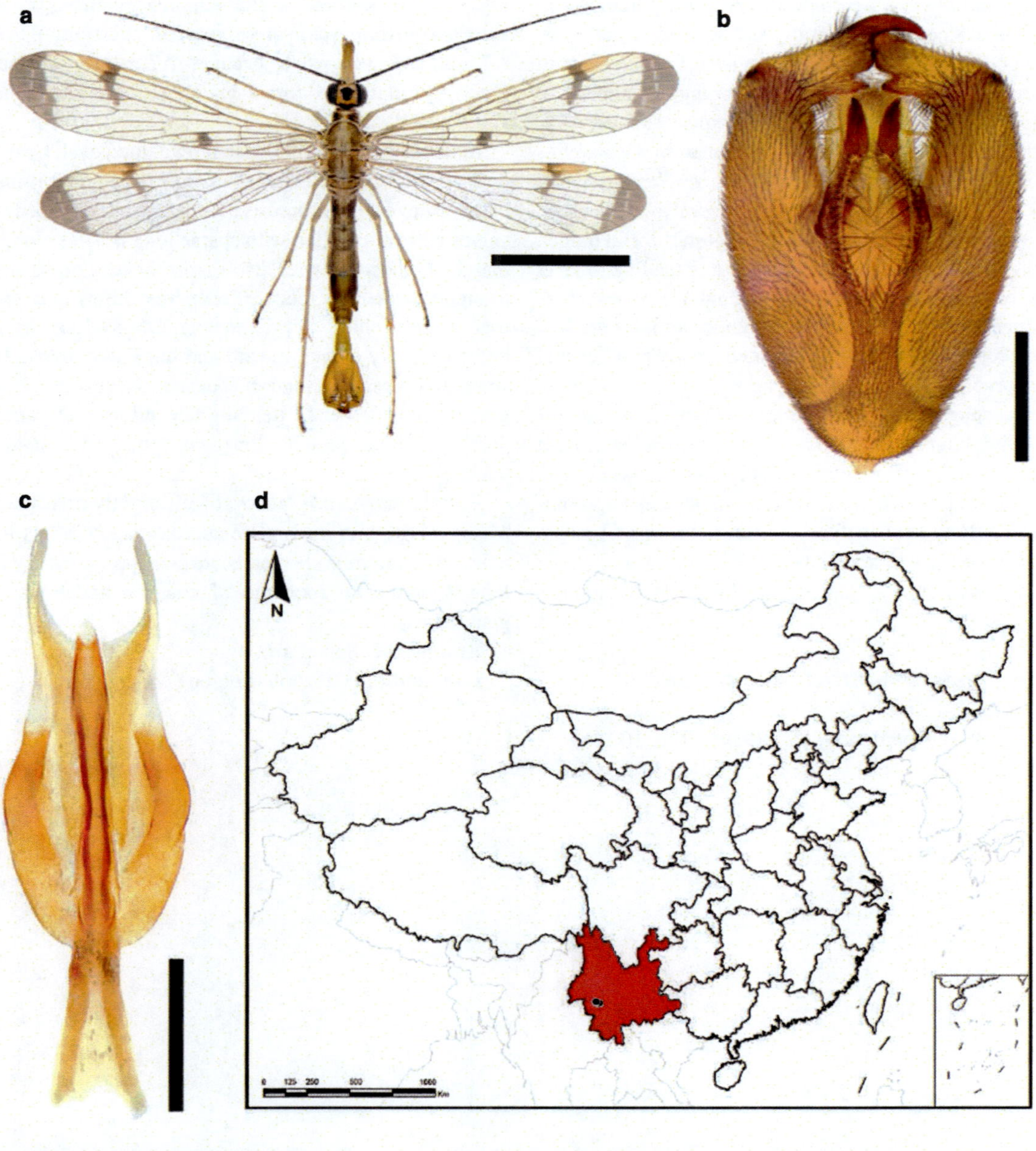

Fig. 10.145 (**a**) Male adult of ***Panorpa stella***, dorsal view (Daxue Mountain, Yongde, Yunnan) (*Scale bar*: 5.0 mm) **[NWAU]**. (**b**) Male genital bulb of ***Panorpa stella***, ventral view (*Scale bar*: 0.5 mm). (**c**) Female medigynium of ***Panorpa stella***, ventral view (*Scale bar*: 0.2 mm). (**d**) Distribution of ***Panorpa stella***

Morphology. Vertex yellow, rostrum yellowish brown. Star-shaped black patter enclosing ocellar triangle, and another black pattern on occiput. Nota grayish brown with broad light brown median stripe. Pleura and legs light yellowish brown. Wing membrane hyaline, slightly tinged with grayish brown, markings blackish brown. Pterostigma reddish brown and conspicuous. In forewings, apical band broad with oblique and irregular inner margin; pterostigma band with only anterior half present; basal band reduced to small spot above ending of 1A; marginal and basal spots absent. Male T1–T5 brown, continuous pale median stripe extending from T1 to T6. A6 yellowish brown and gradually lightening towards apex, cylindrical, with deep U-shaped emargination on dorsal apex, and forming pair of triangular lobes. A7 and A8 yellowish brown; A7 cylindrical and greatly humped on dorsal apex, A8 constricted basally and beveled apically.

Male genital bulb yellowish brown, oval. Epandrium with U-shaped terminal emargination and forming two finger-like processes. Hypandrium Y-shaped.; hypovalves slender, arcuate, with sparse long bristles along inner margin. Gonostylus with subacute earlobe-shaped basal process and rounded median tooth. Parameres bifurcated: ventral branch short, bifurcated subapically into two pointed thorns; dorsal branch stout, slightly convergent apically, almost extending to level of apex of gonocoxites. Ventral aedeagal valves membranous, semicircular, and covered with numerous microtrichia; dorsal processes stout, greatly elongated, nearly parallel, and slightly curved inward at subacute apex.

Female subgenital subtrapezoidal and truncated terminally. Medigynium with broad main plate approximately 3/5 as long as axis; lateral margin of main plate thickened than median region; posterior arms slender, arcuate, and approximately 1/3 as long as main plate; apodemes divergent basally.

Emergence. July to September.

Distribution. Yunnan (Gaoligong Mountains, Yongde).

10.6.27 *Panorpa substricta* Wang, 2021

(Narrow-Lobed Scorpionfly) [窄瓣蝎蛉] (Fig. 10.146)

Measurements. Forewing length 13.0–14.0 mm.

Morphology. Vertex, occiput, and rostrum yellowish brown, ocellar triangle black. Pronotum dark yellowish brown. Meso- and metanotum dark brown laterally with broad yellow median stripe. Pleura and legs yellowish brown. Wing membrane hyaline and tinged with yellowish brown, markings black. In forewings, apical band broad and enclosing two large hyaline spots along apical cross-veins; pterostigmal band triangular with detached basal and apical branches posteriorly; marginal spot small and below R_1; basal band represented by wide spot at ending of 1A; basal spot at inner side of ending of 2A. Male T1–T5 dark brown, with pale, discontinuous median stripe extending from T1 to T4. A6 blackish brown, cylindrical, with shallow V-shaped emargination on dorsal apex and forming pair of very small, subtriangular lobes. A7 and A8 yellowish orange, A7 nearly cylindrical, greatly humped on dorsal apex; A8 constricted basally and beveled apically.

Male genital bulb yellowish brown, long oval. Epandrium with deep U-shaped terminal emargination and forming two parallel finger-like processes. Hypandrium Y-shaped; hypovalves narrow, parallel, with sparse long bristles along inner margin. Gonostylus with disc-shaped basal process and subtriangular median tooth. Parameres bifurcated: ventral branch very short, approximately 1/5 as long as dorsal branch, greatly curved inward and hook-like; dorsal branch stout and straight. Ventral aedeagal valves membranous, greatly elongated and extending beyond apex of parameres; dorsal processes greatly elongated, broad foliate with apex beveled.

Female subgenital plate oval with slightly truncated apex. Medigynium with main plate approximately 2/3 as long as axis; posterior arms slender, acute and approximately half as long as main plate; apodemes of axis stout and greatly divergent basally.

Emergence. September.

Distribution. Guizhou (Leigong Mountain).

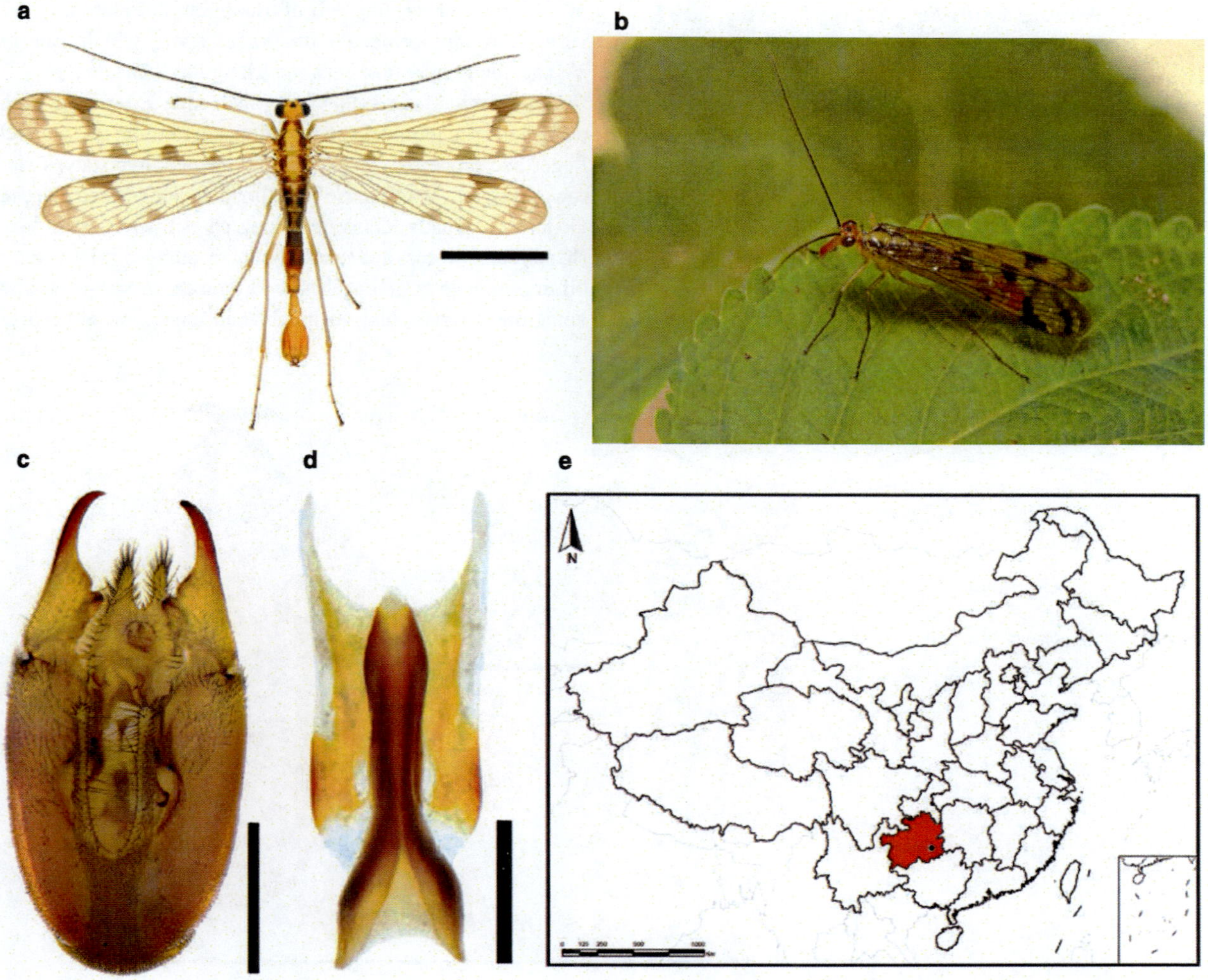

Fig. 10.146 (**a**) Male adult of ***Panorpa substricta***, dorsal view (Leigong Mountain, Guizhou) (*Scale bar*: 5.0 mm) [**DALU**]. (**b**) Male adult of ***Panorpa substricta*** in the field (Leigong Mountain, Guizhou). (**c**) Male genital bulb of ***Panorpa substricta***, ventral view (*Scale bar*: 0.5 mm). (**d**) Female medigynium of ***Panorpa substricta***, ventral view (*Scale bar*: 0.2 mm). (**e**) Distribution of ***Panorpa substricta***

10.6.28 *Panorpa xiaofeng* Wang & Gong, 2021

(Feng Xiao's Scorpionfly) [萧峰蝎蛉] (Fig. 10.147)
Measurements. Forewing length 10.0–12.0 mm.

Morphology. Vertex yellow. Ocellar triangle black. Rostrum yellow. Nota brown, with broad pale yellow median stripe. Pleura and legs light brown. Wing membrane hyaline. In forewings, apical and pterostigmal bands greatly reduced into several small spots. Male T1–T5 brown, bearing yellow median stripe tapering caudad. Notal organ flat and covering acute postnotal organ. A6 deeply emarginate on dorsal apex in V-shape, and forming pair of triangular lobes laterally. A7 very short and projected on dorsal apex, which can fit V-shaped emargination between triangular lobes on A6.

Male genital bulb yellowish brown, oval. Epandrium with deep U-shaped terminal emargination and forming two parallel finger-like processes. Hypandrium Y-shaped; hypovalves narrow, divergent, with sparse long bristles along inner margin. Gonostylus greatly incurvated along outer margin, with disc-shaped basal process and subtriangular median tooth. Parameres bifurcated, with both branches acute at apex, and covered with numerous microtrichia; ventral branch approximately 2/5 as

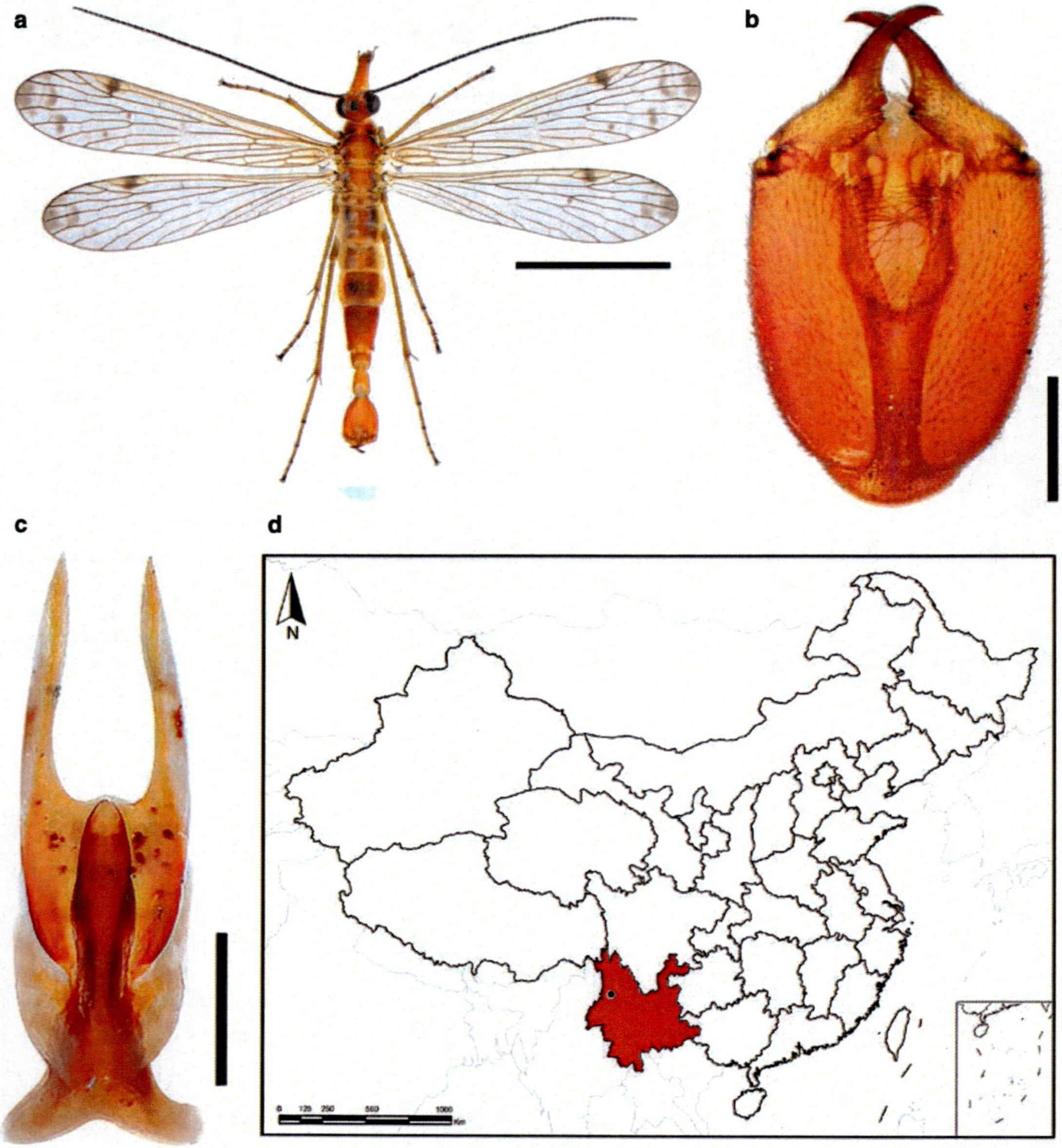

Fig. 10.147 (**a**) Male adult of ***Panorpa xiaofeng***, dorsal view (Sky Lake, Yunlong, Yunnan) (*Scale bar*: 5.0 mm) **[DALU]**. (**b**) Male genital bulb of ***Panorpa xiaofeng***, ventral view (*Scale bar*: 0.5 mm). (**c**) Female medigynium of ***Panorpa xiaofeng***, ventral view (*Scale bar*: 0.2 mm). (**d**) Distribution of ***Panorpa xiaofeng***

long as dorsal branch; dorsal branch greatly elongated and bending outwards apically. Ventral aedeagal valves membranous and rounded; dorsal processes stout, greatly elongated, nearly parallel, and with finger-like projection apically.

Female subgenital plate oblong with slightly emarginated apex. Medigynium with broad main plate approximately half as long as axis; posterior arms slender and slightly longer than main plate; apodemes of axis greatly divergent basally.

Emergence. August to October.

Distribution. Yunnan (Yunlong).

10.6.29 *Panorpa xuzhu* Wang & Gong, 2021

(Xuzhu's Scorpionfly) [虚竹蝎蛉] (Fig. 10.148)

Measurements. Forewing length 9.5–11.5 mm.

Morphology. Vertex and rostrum yellow. Ocellar triangle black. Body mostly yellow. Wing membrane hyaline, lacking distinct markings. Male notal organ flat and covering acute postnotal organ. A6, cylindrical. A7 and A8 short cylindrical.

Male genital bulb yellow, long oval. Epandrium with deep U-shaped terminal emargination and forming two parallel fin-

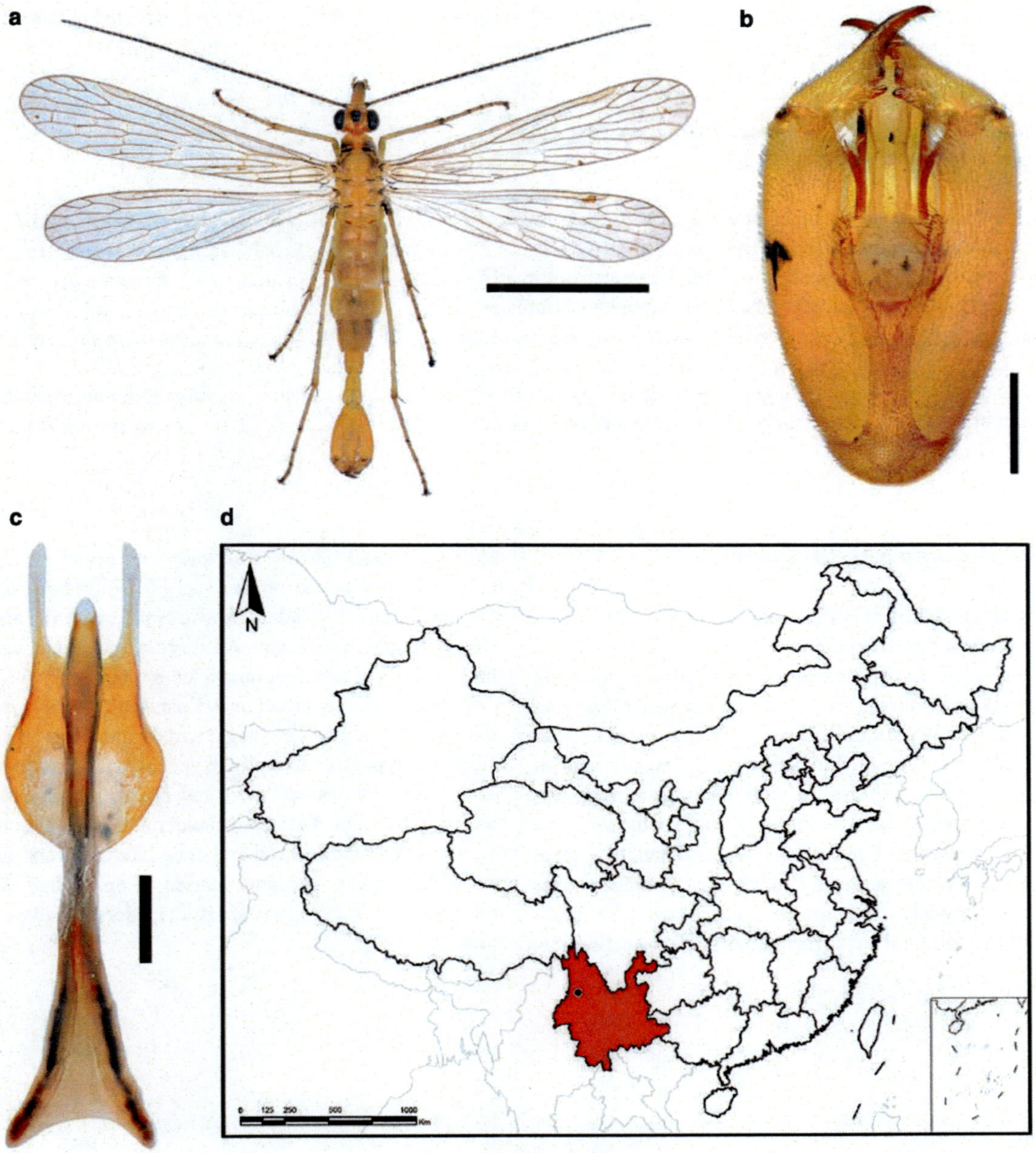

Fig. 10.148 (a) Male adult of ***Panorpa xuzhu***, dorsal view (Sky Lake, Yunlong, Yunnan) (*Scale bar*: 5.0 mm) **[DALU]**. (b) Male genital bulb of ***Panorpa xuzhu***, ventral view (*Scale bar*: 0.5 mm). (c) Female medigynium of ***Panorpa xuzhu***, ventral view (*Scale bar*: 0.2 mm). (d) Distribution of ***Panorpa xuzhu***

ger-like processes. Hypandrium Y-shaped; hypovalves narrow, divergent, with sparse long bristles along inner margin. Gonostylus with disc-shaped basal process and subtriangular median tooth. Parameres bifurcated: ventral branch approximately 4/5 as long as dorsal branch, and greatly bending outwards apically; dorsal branch relatively straight. Ventral aedeagal valves membranous, rounded, and covered with numerous microtrichia; dorsal processes slender, greatly elongated and enlarged towards apex, and with longitudinal groove.

Female subgenital plate oblong with slightly emarginated apex and sparse long setae on distal margin. Medigynium slender, with main plate broadest at basal 2/5 and approximately 1/3 as long as axis; posterior arms slender and approximately 2/3 as long as main plate; apodemes of axis greatly divergent at base.

Emergence. July to August.

Distribution. Yunnan (Yunlong).

Panorpa wormaldi Group

This group of species can be recognized by the relatively small body size, the roof-like wing posture, the short male abdomen, and the female medigynium bearing a dorsal plate.

Nineteen species are known from southeastern China and Japan. Nine species are recorded from China (Wang and Suzuki 2022), and five of them are included in this atlas.

Key to Species of the Panorpa wormaldi Group in This Atlas
(Five species, males)

1. Wing markings greatly reduced or absent; paramere trifurcated .... ***Panorpa longiramina*** Issiki & Cheng
– Wing markings well-developed; paramere bifurcated 2
2. Basal process of gonocoxites bifurcated; parameres unfurcated .... ***Panorpa gressitti*** Byers
– Basal process of gonocoxites unfurcated; parameres bifurcated 3
3. Head yellowish brown with trident-shaped pattern on occiput .... ***Panorpa peterseana*** Issiki
– Head black and lacking pattern 4
4. Ventral branch of paramere longer than 1/3 length of dorsal branch .... ***Panorpa obliquifasciata*** Chou & Wang
– Ventral branch of paramere shorter than 1/3 length of dorsal branch .... ***Panorpa implicata*** Cheng

10.6.30 *Panorpa gressitti* Byers, 1970

(Gressitt's Scorpionfly) [嘉理斯蝎蛉] (Fig. 10.149)

Measurements. Forewing length 11.2 mm.

Morphology. Vertex dark brown, ocellar triangle black. Rostrum yellowish brown. Pronotum dark brown. Meso- and metanotum mostly dark brown, but yellowish white on scutella. Pleura and coxae creamy yellow. Legs mostly reddish brown. Wing membrane hyaline with markings dark brown. In forewings, apical band with 3–4 hyaline window spots innerly; pterostigmal band broad, with separated apical branch fused with apical band, and broad basal branch; marginal spot obliquely elongates and connected with basal branch of pterostigmal band, forming distinct V-shaped pattern; basal band narrow anteriorly and broad posteriorly; series of irregular spots on hind margin. Wings held roof-like at repose. Male T1–T5 dark brown. Notal organ flat and covering acute postnotal organ. A6 short cylindrical. A7 and A8 short cylindrical and not constricted basally.

Male genital bulb broad oval. Epandrium deeply emarginate terminally into subrectangle and forming pair of slender finger-like processes laterally. Hypandrium with long broad basal stalk. Hypovalves short and broad, not extending to base of gonostylus. Paramere slender and exceeding middle of gonostylus, with curved and claw-like apex, and bearing row of long spines on inner margin of apical half. Ventral aedeagal valves mostly trapezoidal; dorsal valves acute apically.

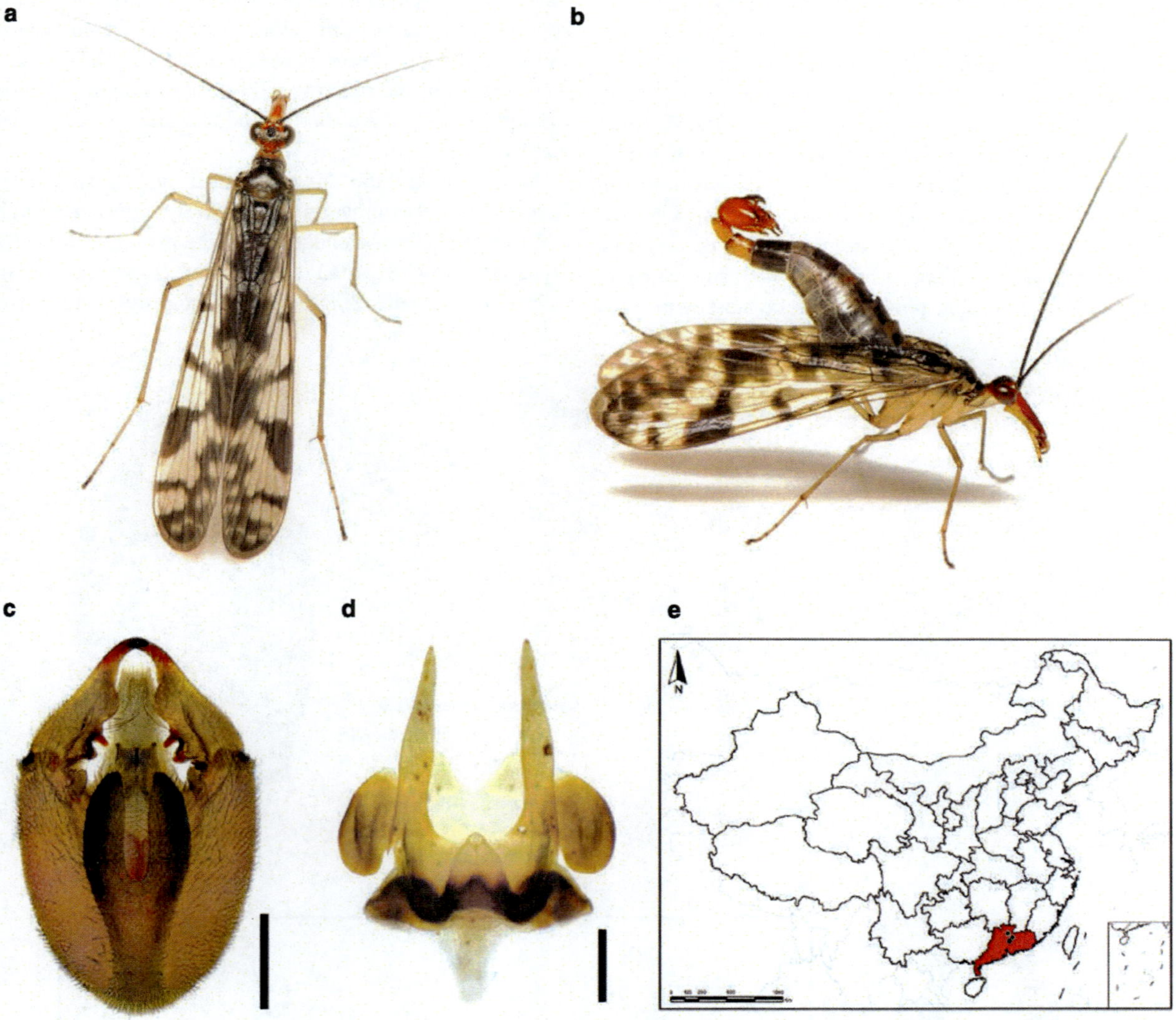

Fig. 10.149 (**a**) Male adult of ***Panorpa gressitti***, dorsal view (Liuxihe, Conghua, Guangdong). (**b**) Male adult of ***Panorpa gressitti***, lateral view (Liuxihe, Conghua, Guangdong). (**c**) Male genital bulb of ***Panorpa gressitti***, ventral view (*Scale bar*: 0.5 mm). (**d**) Female medigynium of ***Panorpa gressitti***, ventral view (*Scale bar*: 0.2 mm). (**e**) Distribution of ***Panorpa gressitti***

Female subgenital plate subtriangular with shallow terminal emargination. Medigynium with subtrapezoidal main plate; broad dorsal plate emarginated in V-shape terminally, and enclosing posterior arms earlobe-like laterally; posterior arms slender; axis short and lacking distinct apodemes; decorated area of axis greatly elongated posteriorly, subtriangular.

Emergence. March to June.

Distribution. Guangdong (Conghua, Longdou Peak, Nankun Mountains).

10.6.31 *Panorpa implicata* Cheng, 1957

(Implicated Scorpionfly) [织纹蝎蛉] (Fig. 10.150)
Measurements. Forewing length 13.5 mm.

Morphology. Vertex and ocellar triangle black. Rostrum reddish brown and relatively stout. Nota dark brown. Pleura and legs yellowish brown. Wing membrane hyaline with markings brown. In forewings, apical band bearing 4–5 large hyaline windows innerly; and forming series of reticulate patterns with separated apical branch of pterostigmal band; marginal spot elongated posteriorly to connect with basal branch of pterostigmal band, forming V-shaped pattern; series of irregular markings along basal half of hind margin; hyaline circular spot near ending of CuP. Male T1–T5 black. Notal organ flat and covering acute postnotal organ. A6 short cylindrical. A7 and A8 short cylindrical and not constricted basally.

Male genital bulb long elliptical, yellowish brown. Epandrium deeply emarginate terminally into subrectangle and forming pair of slender finger-like processes laterally. Hypandrium with broad basal stalk. Hypovalves long, stripe-like, not extending to base of gonostylus. Paramere

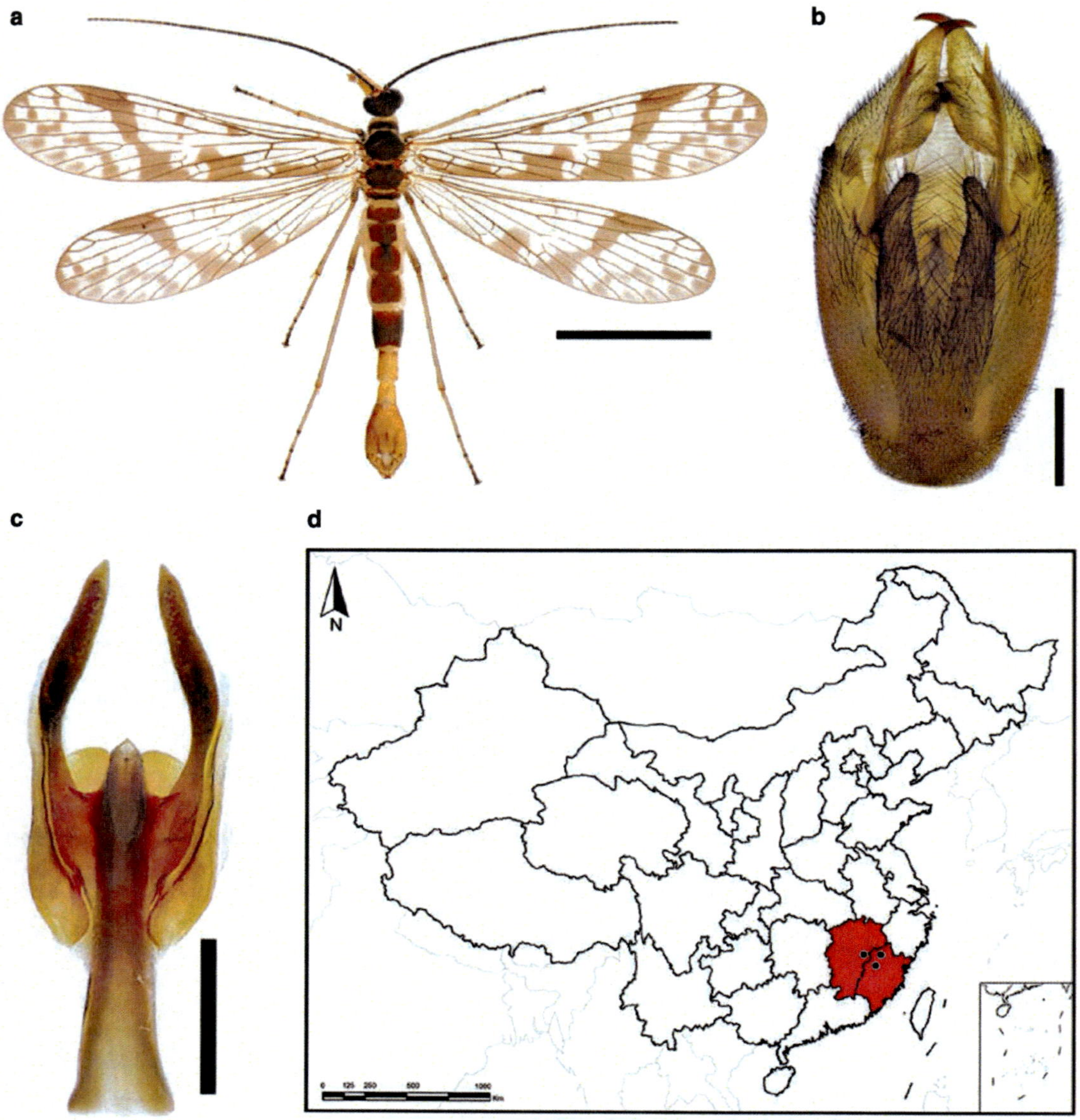

Fig. 10.150 (a) Male adult of ***Panorpa implicata***, dorsal view (Zixi, Jiangxi) (*Scale bar*: 5.0 mm) **[DALU]**. (b) Male genital bulb of ***Panorpa implicata***, ventral view (*Scale bar*: 0.5 mm). (c) Female medigynium of ***Panorpa implicata***, ventral view (*Scale bar*: 0.2 mm). (d) Distribution of ***Panorpa implicata***

bifurcated with delicate and short ventral branch, and strong dorsal branch reaching middle of gonostylus. Ventral aedeagal valves mostly trapezoidal; dorsal valves acute apically.

Female subgenital plate subtriangular with indistinct emargination terminally. Medigynium with subtrapezoidal main plate; broad dorsal plate enclosing posterior arms ear-lobe-like laterally; posterior arms slender and bending near middle; axis longer than posterior arms with slightly diverged apodemes; decorated area of axis greatly elongated posteriorly, subtriangular.

Emergence. April to June.

Distribution. Fujian (Wuyi Mountains), Jiangxi (Zixi).

10.6.32 *Panorpa longiramina* Issiki & Cheng, 1947

(Long-Paramere Scorpionfly) [长枝蝎蛉] (Fig. 10.151)

Measurements. Forewing length 12.5 mm.

Morphology. Vertex dark brown. Ocellar triangle black. Rostrum yellowish brown. Nota black. Pleura and coxae creamy yellow, and legs mostly light yellowish brown. Wing membrane hyaline. Pterostigma brown. Wing markings only represented by small spot below pterostigma. Male T1–T5 black. Notal organ flat and covering acute postnotal organ. A6 black and cylindrical. A7 and A8 yellowish brown, short cylindrical and not constricted basally.

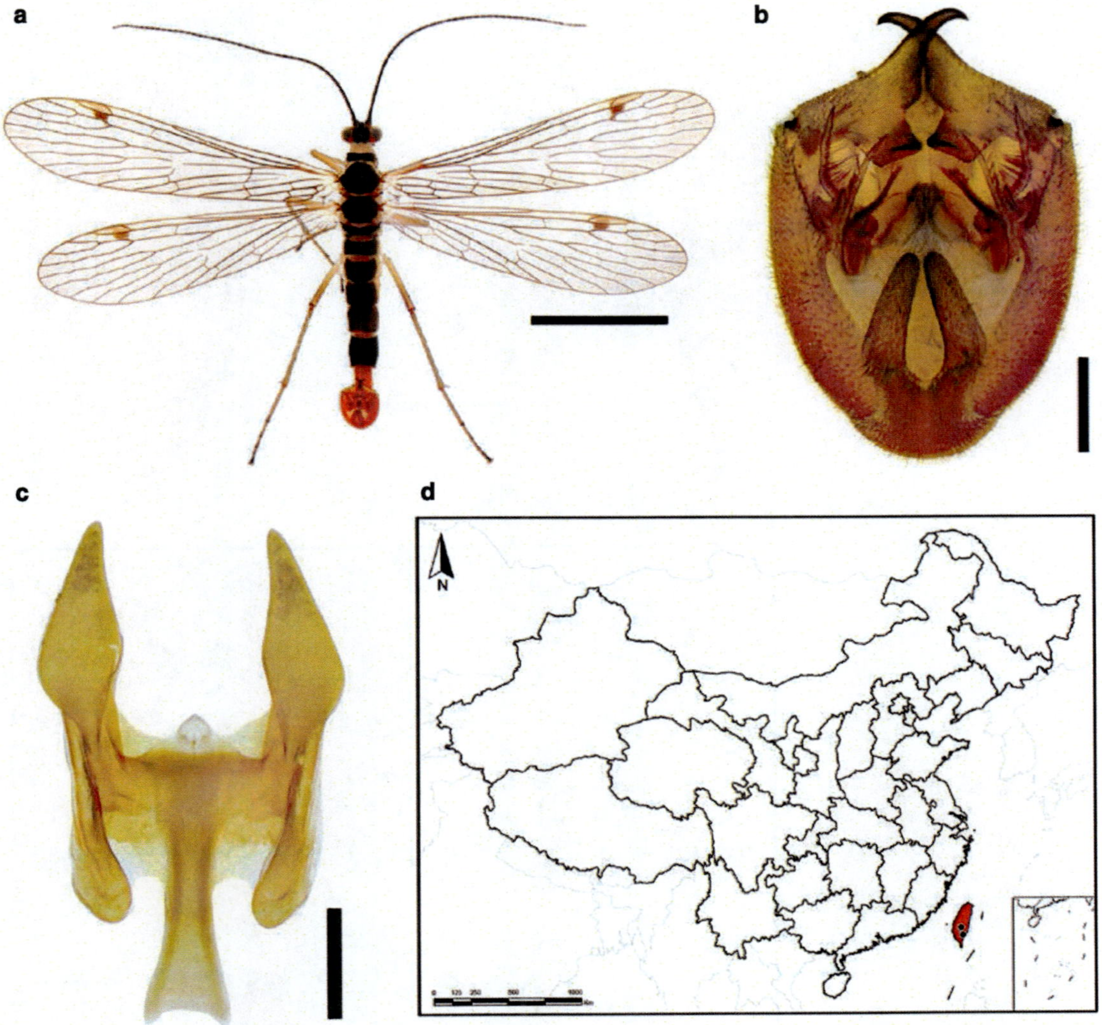

Fig. 10.151 (a) Male adult of ***Panorpa longiramina***, dorsal view (Haituan, Taitung, Taiwan) (*Scale bar*: 5.0 mm) [NWAU]. (b) Male genital bulb of ***Panorpa longiramina***, ventral view (*Scale bar*: 0.5 mm). (c) Female medigynium of ***Panorpa longiramina***, ventral view (*Scale bar*: 0.2 mm). (d) Distribution of ***Panorpa longiramina***

Male genital bulb oval. Epandrium emarginates terminally in deep U-shape and forms pair of finger-like processes laterally. Hypandrium Y-shaped, with short broad basal stalk. Hypovalves pincer-like, very short, only extending to middle of gonocoxites. Gonostylus almost half-length of gonocoxites, concaved on outer margin, bearing obtuse triangular median tooth and stout subtrapezoidal basal process. Paramere trifurcated, with row of comb-like long spines on inner margin of each branch. Ventral aedeagal valves short; dorsal valves broad and finger-like, exceeding apex of gonocoxites.

Female subgenital plate broad subtrapezoidal with indistinct emargination terminally. Medigynium with subtrapezoidal main plate; broad dorsal plate enclosing posterior arms earlobe-like laterally; posterior arms spatulate and broadest near middle; axis nearly as long as posterior arms with slightly diverged apodemes; decorated area of axis elongated posteriorly, subtriangular.

Emergence. May.

Distribution. Taiwan (Nantou, Taitung).

10.6.33 *Panorpa obliquifascia* Chou & Wang, 1987

(Oblique-Striped Scorpionfly) [斜带蝎蛉] (Fig. 10.152)

Measurements. Forewing length 12.5–13.0 mm.

Morphology. Vertex and ocellar triangle black. Rostrum yellowish brown. Nota dark brown. Pleura and legs creamy yellow. Wing membrane hyaline with markings brown. In forewings, apical band scattered with separated thin stripe near inner margin; pterostigmal band with stout basal branch

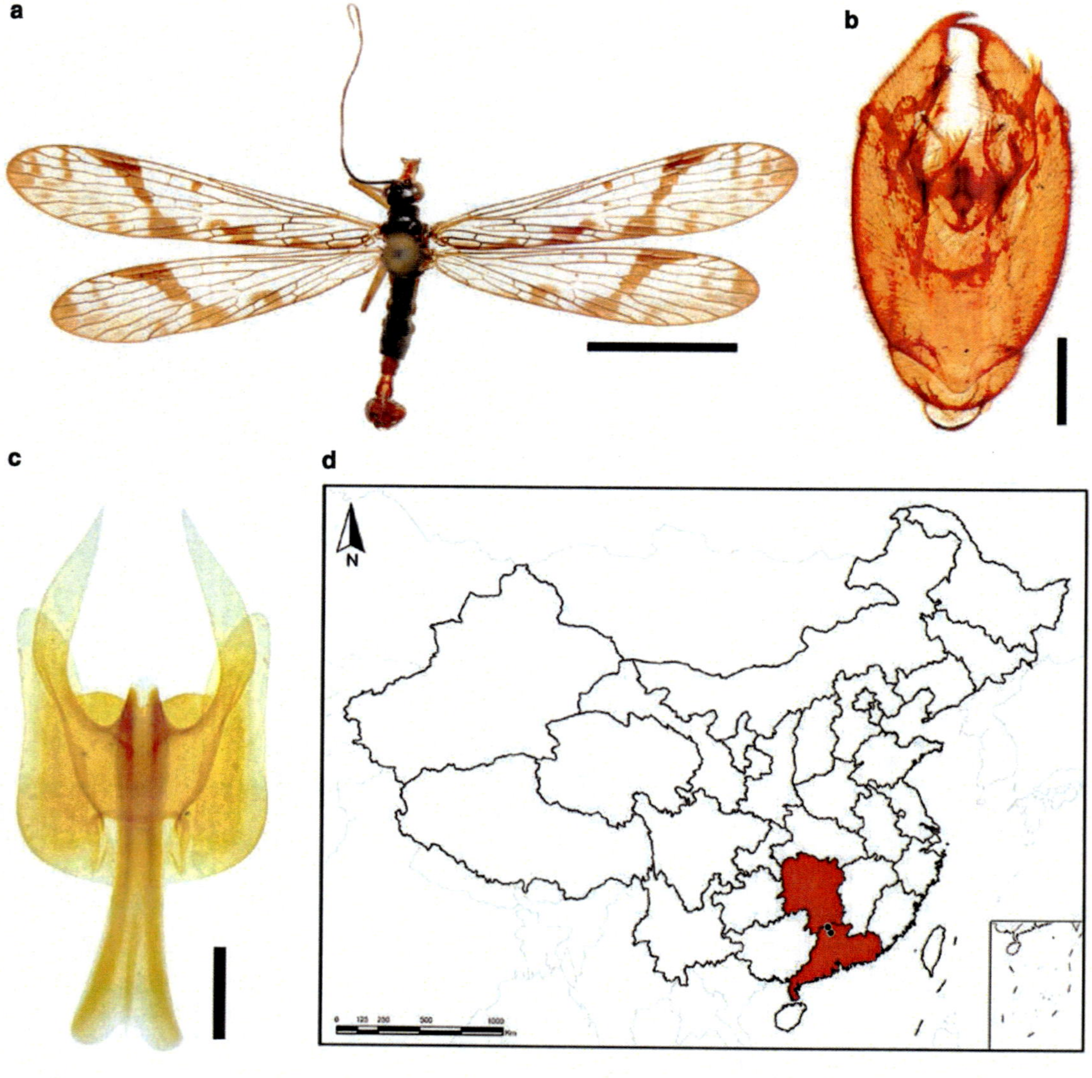

Fig. 10.152 (**a**) Male adult of ***Panorpa obliquifascia***, dorsal view (Nanling Mountains, Guangdong) (*Scale bar*: 5.0 mm) **[NWAU]**. (**b**) Male genital bulb of ***Panorpa obliquifascia***, ventral view (*Scale bar*: 0.5 mm). (**c**) Female medigynium of ***Panorpa obliquifascia***, ventral view (*Scale bar*: 0.2 mm). (**d**) Distribution of ***Panorpa obliquifascia***

and separated faint apical branch, forming conspicuous oblique band; marginal spot small; basal band reduced, forming series of markings along hind margin; basal spot absent. Male T1–T5 dark brown. Notal organ flat and covering acute postnotal organ. A6 dark brown, short cylindrical. A7 and A8 yellowish brown, short cylindrical and not constricted basally.

Male genital bulb long elliptical. Epandrium emarginates terminally in deep U-shape and forms pair of finger-like processes laterally. Hypandrium with long basal stalk and Y-shape. Hypovalves long broad stripe-like, extending beyond apex of gonocoxites, and bearing row of long bristles along inner margin. Gonostylus short, not half as long as gonocoxites, and bearing obtuse triangular median tooth and large bowl-shaped basal process. Paramere bifurcated, with ventral branch short and glabrous; and dorsal branch long and bearing long comb-like spines on inner margin. Aedeagus short and not exceeding apex of gonocoxites.

Female subgenital plate broad oval with indistinct emargination terminally. Medigynium with subtrapezoidal main plate; broad dorsal plate enclosing posterior arms laterally; posterior arms slender near middle, and greatly bending inward; axis longer than posterior arms with slightly diverged apodemes; decorated area of axis elongated posteriorly, subtriangular.

Emergence. May to July.

Distribution. Guangdong (Nanling Mountains), Hunan (Mangshan Mountains).

10.6.34 *Panorpa peterseana* Issiki, 1929

(Petersen's Scorpionfly) [彼得森蝎蛉] (Fig. 10.153)

Measurements. Forewing length 11.5–13.0 mm.

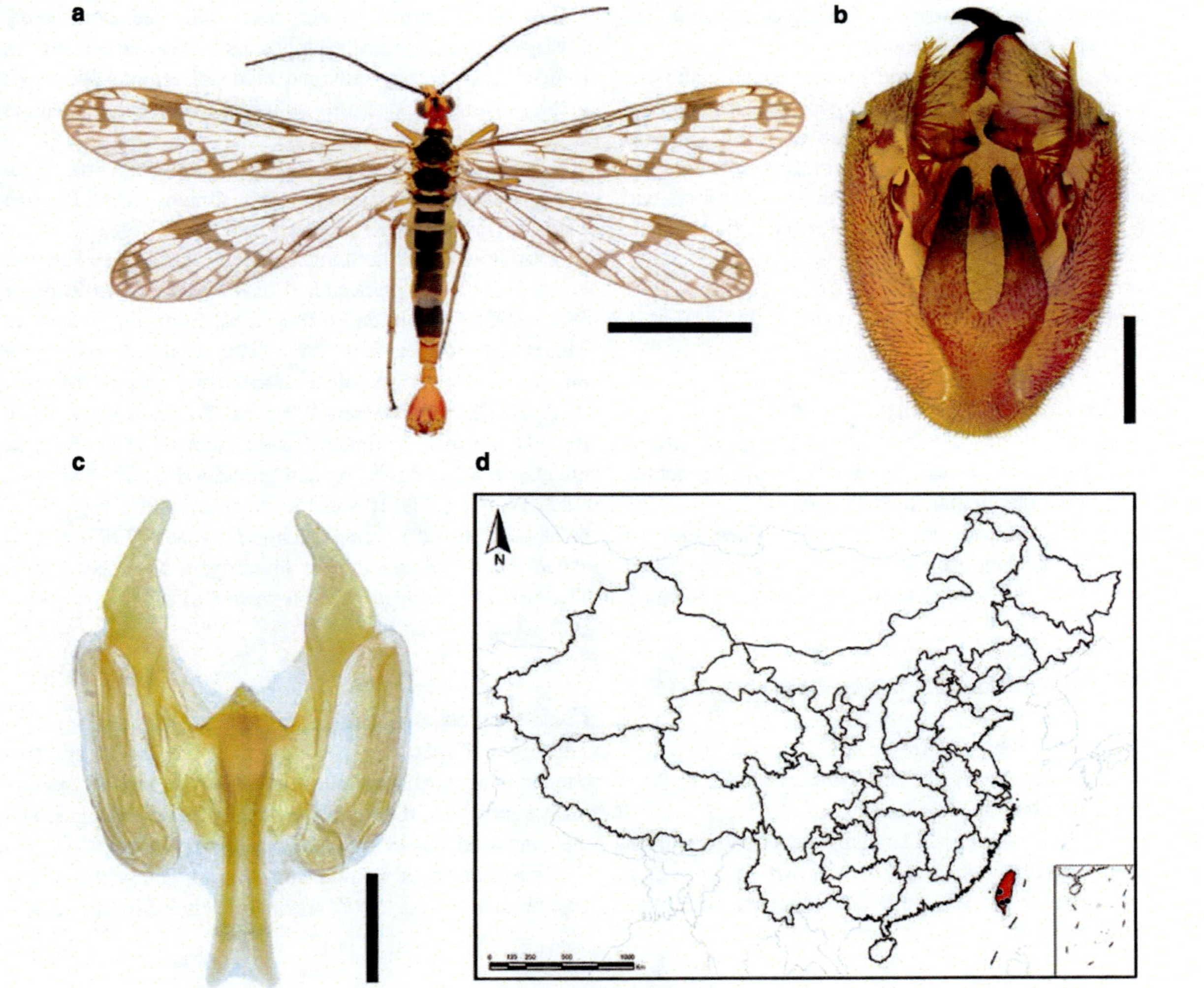

Fig. 10.153 (**a**) Male adult of ***Panorpa peterseana***, dorsal view (Tahan Mountain, Pingtung, Taiwan) (*Scale bar*: 5.0 mm) **[NWAU]**. (**b**) Male genital bulb of ***Panorpa peterseana***, ventral view (*Scale bar*: 0. mm). (**c**) Female medigynium of ***Panorpa peterseana***, ventral view (*Scale bar*: 0.2 mm). (**d**) Distribution of ***Panorpa peterseana***

Morphology. Vertex yellow. Ocellar triangle black. Trident brown pattern on occiput. Rostrum reddish brown. Nota dark brown. Pleura and legs light yellowish brown. Wing membrane hyaline. Markings brown. In forewings, apical band reduced into several scattered spots; pterostigmal band slender, with complete basal branch and inconspicuous apical branch; basal band split into two spots, with posterior spot connected to basal branch of pterostigmal band posteriorly; marginal spot small; basal spot absent; series of irregular markings on hind margin of anal region. Male T1–T5 dark brown. Notal organ flat and covering postnotal organ. A6 dark brown, short cylindrical. A7 and A8 reddish brown, short cylindrical, and not constricted basally.

Male genital bulb oval. Epandrium emarginates terminally in U-shape and forms pair of finger-like processes laterally. Hypandrium with short broad basal stalk. Hypovalves pincer-like, almost extending to apex of gonocoxites. Gonostylus with obtuse triangular median tooth and subtrapezoidal basal process. Paramere bifurcated, with row of long comb-like spines on inner margin, dorsal branch long and extending to middle of gonostylus. Aedeagus with short ventral valves and finger-like dorsal valves.

Female subgenital plate broad subtrapezoidal with indistinct emargination terminally. Medigynium with subtrapezoidal main plate; broad dorsal plate enclosing posterior arms earlobe-like laterally; posterior arms spatulate and broadest near middle; axis shorter than posterior arms with slightly diverged apodemes; decorated area of axis elongated posteriorly, subtriangular.

Emergence. March to May.

Distribution. Taiwan (Nantou, Pingtung, Taitung).

Panorpa communis Group

This group of species can be recognized by the greatly broadened male paramere and the greatly enlarged decorated area in the axis of the female medigynium.

Nineteen species are known from Europe, Western Asia, and northeastern Asia. Only one species, ***Panorpa sibirica*** Esben-Petersen, 1915 has been recorded from northeastern China.

10.6.35 *Panorpa sibirica* Esben-Petersen, 1915

(Siberian Scorpionfly) [西伯利亚蝎蛉]

Panorpa changbaishana Hua, 2008, **new synonym**.

Measurements. Forewing length 12.5–13.0 mm.

Morphology. Vertex, ocellar triangle and rostrum black. Nota black. Pleura and coxae dark brown. Legs mostly deep yellowish brown. Wing membrane hyaline with black markings. In forewings, apical band broad with large hyaline window posteriorly; pterostigmal band broad, with broad basal branch and narrow apical branch; basal band split into two small spots; marginal spot small; basal spot absent. Male T1–T5 dark brown. Notal organ flat and covering acute postnotal organ. A6 dark brown, cylindrical. A7 and A8 black, short, and constricted basally, inverted cone-shaped.

Male genital bulb long oval. Epandrium emarginates terminally in deep U-shape. Hypandrium with long broad basal stalk. Hypovalves slender, slightly expanding laterad at middle, convergent and tapering towards apexes at distal half. Gonocoxites with bifurcated plate on ventral apex. Gonostylus shorter than half-length of gonocoxites, and bearing obtuse triangular median tooth and disc-like basal process. Paramere foliate basally and tapering towards bifurcated acute apex, and bearing row of short spines along inner margin. Aedeagus with short membranous ventral valves; dorsal valves finger-like and bearing membranous enlarged process on dorsal surface apically.

Female subgenital plate oval with subacute apex. Medigynium with well-developed main plate; posterior arms slender; axis greatly enlarged at distal region; apodemes greatly diverged and shortly split into two or three branches basally.

Distribution. Liaoning (Kuandian, Yingkou), Jilin (Changbai Mountains), Heilongjiang (Jingpo Lake); Russian Far East (Primorsky Krai); North and South Korea.

Remarks. *Panorpa changbaishana* Hua, 2008 (= *Panorpa choui* Hua, 1998, preoccupied by *Panorpa choui* Zhou & Wu, 1993) was initially described from the Changbai Mountains, northeastern China (Hua 1998). It shares the identical morphology with *P. sibirica* not only in the wing markings (Esben-Petersen 1921: fig. 24; Martynova 1957: fig. 14) and male abdomen (Esben-Petersen 1921: fig. 26), but also in the male and female genitalia (Fig. 10.154c, d vs. Issiki 1933: figs 7H, 10A & 15; Martynova 1957: figs 36, 58, 66, 82, 95 & 120). In addition, Martynova (1957) clearly stated that *P. sibirica* can be found from Northeast China. Therefore, *P. changbaishana* is treated as a junior synonym of *P. sibirica* in this atlas.

Panorpa cheni Group

This group of species can be recognized by the two- or three-banded wing markings, the greatly enlarged median tooth in male gonostylus, the greatly elongated male aedeagus, and the concealed axis in the female medigynium.

Nine species are known from central, southeastern, and southwestern China. Six of them are included in this atlas.

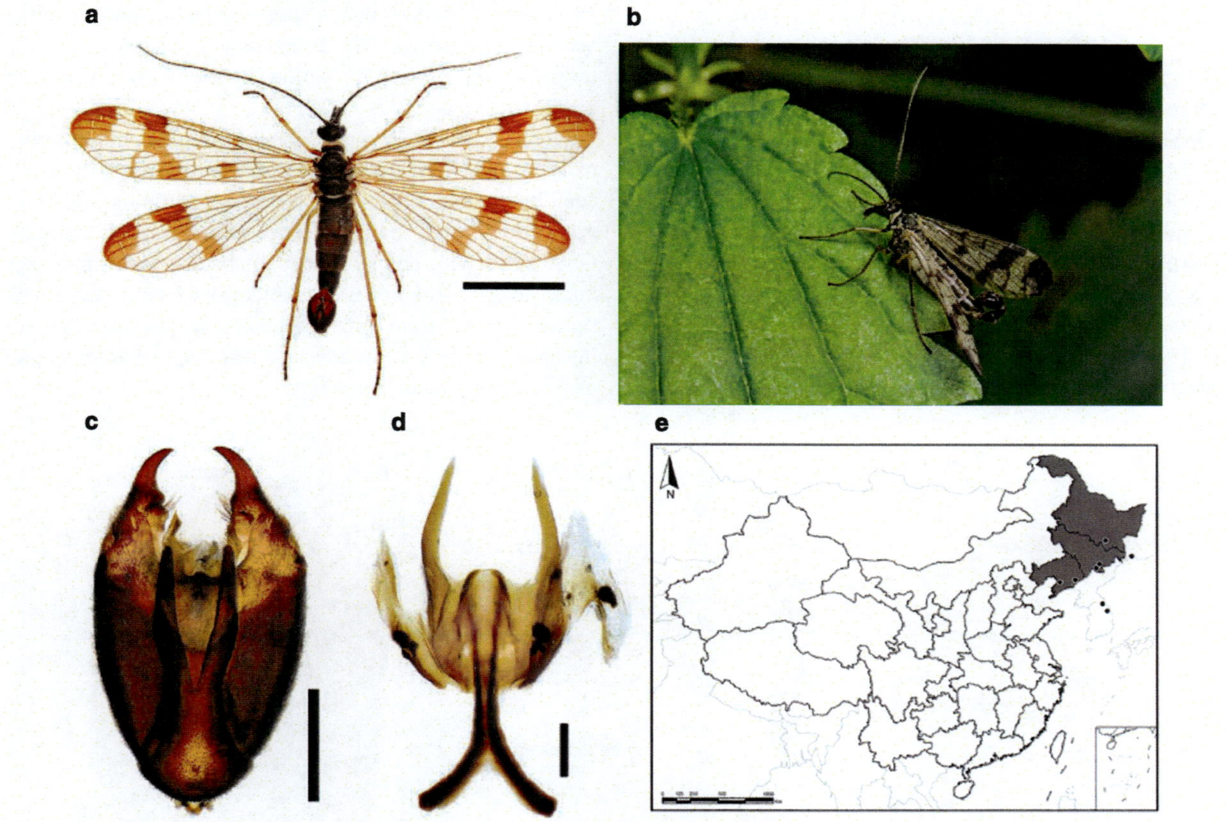

Fig. 10.154 (**a**) Male adult of ***Panorpa sibirica***, dorsal view (Changbai Mountains, Jilin) (*Scale bar*: 5.0 mm) [NWAU]. (**b**) Male adult of ***Panorpa sibirica*** in the field (Changbai Mountains, Jilin). (**c**) Male genital bulb of ***Panorpa sibirica***, ventral view (*Scale bar*: 0.5 mm). (**d**) Female madigynium of ***Panorpa sibirica***, ventral view (*Scale bar*: 0.2 mm). (**e**) Distribution of ***Panorpa sibirica***

Key to Species of the Panorpa cheni Group in This Atlas
(6 species, males)

1. Wings with two transverse bands 2
– Wings with three transverse bands 3
2. Wings with basal band absent; parameres with dorsal branch slender and simple .... ***Panorpa bifasciata*** Chou & Wang
– Wings with basal band present and slightly reduced; parameres with dorsal branch strongly enlarged near apex with short and acute tip .... ***Panorpa cheni*** Cheng
3. Gonostylus greatly elongated, and nearly as long as gonocoxites 4
– Gonostylys shorter than gonocoxites 5
4. Gonostylus with large concaved area on basal half of ventral surface; median tooth acute .... ***Panorpa cladocerca*** Navás
– Gonostylus lacking such concaved area; median tooth rounded .... ***Panorpa choui*** Zhou & Wu
5. Gonocoxites with row of long stout setae on apical half of ventro-inner margin .... ***Panorpa trifasciata*** Cheng
– Gonocoxites lacking such setae .... ***Panorpa tritaenia*** Chou & Wang

10.6.36 *Panorpa bifasciata* Chou & Wang, 1981

(Two-Striped Scorpionfly) [双带蝎蛉] (Fig. 10.155)
Measurements. Forewing length 11.0–13.0 mm.

Morphology. Vertex yellowish brown. Ocellar triangle black. Black transverse stripe passing through ocellar triangle and extending to inner border of compound eyes. Rostrum yellowish brown. Nota dark brown. Pleura and legs yellowish brown. Wing membrane hyaline with dark brown markings. Forewing with broad apical and pterostigmal bands; pterostigmal band lacking apical branch; basal band reduced into small spot or absent; no other markings. Male T1–T5 dark brown. Notal organ flat and covering acute postnotal organ. A6 dark brown, cylindrical, tapering towards apex. A7 and A8 yellowish brown and constricted basally.

Male genital bulb oval. Epandrium emarginated terminally in deep U-shape. Hypandrium with extremely shortened basal stalk. Hypovalves slender, exceeding middle of gonocoxites, and bearing slender setae on inner margins. Gonostylus with obtuse triangular median tooth and large hook-shaped basal process. Paramere slender and acute, with several long and stout spines basally. Aedeagus stout, exceeding apex of gonocoxites; dorsal aedeagal valves forming pair of auriculate processes apically.

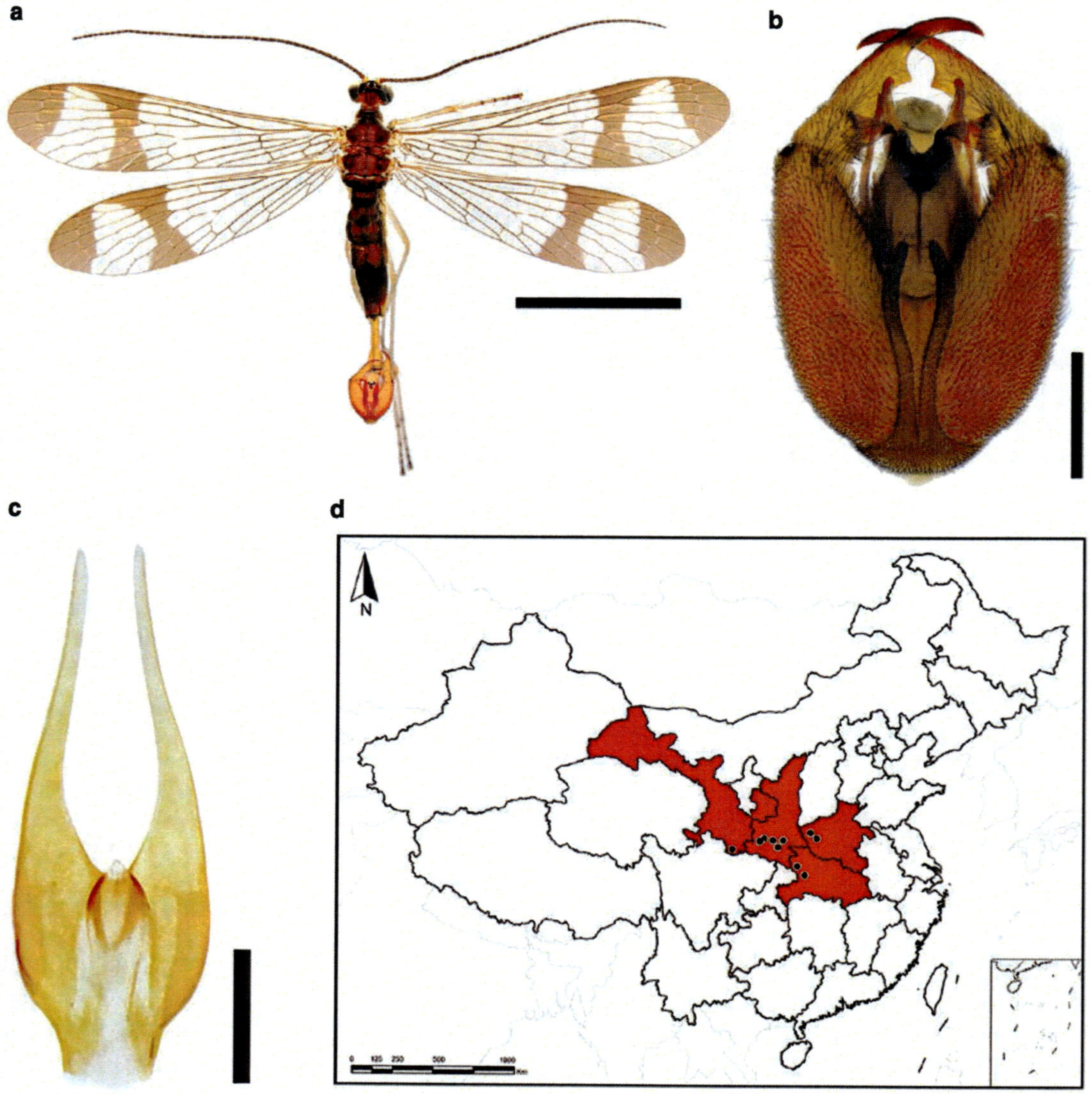

Fig. 10.155 (a) Male adult of ***Panorpa bifasciata***, dorsal view (Huoditang, Ningshan, Shaanxi) (*Scale bar*: 5.0 mm) [NWAU]. (b) Male genital bulb of ***Panorpa bifasciata***, ventral view (*Scale bar*: 0.5 mm). (c) Female medigynium of ***Panorpa bifasciata***, ventral view (*Scale bar*: 0.2 mm). (d) Distribution of ***Panorpa bifasciata***

Female subgenital plate drop-shaped with subrounded apex and well-developed laterotergites. Medigynium with rounded main plate; posterior arms slender, twice as long as main plate; axis lacking apodemes.

Emergence. June to August.

Distribution. Gansu (Wenxian), Henan (Lushi, Songxian), Hubei (Shennongjia, Zhuxi), Shaanxi (Chang'an, Liuba, Ningshan, Taibai, Zhouzhi).

10.6.37 *Panorpa cheni* Cheng, 1957

(Chen's Scorpionfly) [陈氏蝎蛉] (Fig. 10.156)
Measurements. Forewing length 11.5–12.0 mm.

Morphology. Vertex dark brown. Ocellar triangle black. Rostrum yellowish brown. Nota dark brown. Pleura and legs yellowish brown. Wing membrane hyaline with dark brown markings. In forewings, apical band broad; pterostigmal band broad, lacking apical branch; basal band narrow; marginal spot small; basal spot absent. Male T1–T5 black. Notal organ flat and covering acute postnotal organ. A6 black and cylindrical. A7 and A8 yellowish brown and constricted basally.

Male genital bulb oval. Epandrium emarginates terminally in deep U-shape and forms pair of finger-like processes laterally. Hypandrium with greatly shortened basal stalk. Hypovalves slender, extending to middle of gonocoxites, bearing row of long bristles on inner margin. Gonostylus with acute median tooth and broad basal process with short acute tip. Paramere bifurcated: dorsal branch short and bearing tuft of long stout spines; ventral branch strongly enlarged near apex with short and acute tip, resembling drumstick. Aedeagus short and stout, not exceeding apex of gonocoxites; dorsal valves with pair of slender finger-like processes laterally.

Female subgenital plate oval with subrounded apex and well-developed laterotergites. Medigynium with rounded main plate; posterior arms slender, twice as long as main plate; axis lacking apodemes, sinuate, and convergent.

Emergence. August to October.

Distribution. Zhejiang (Tianmu Mountains).

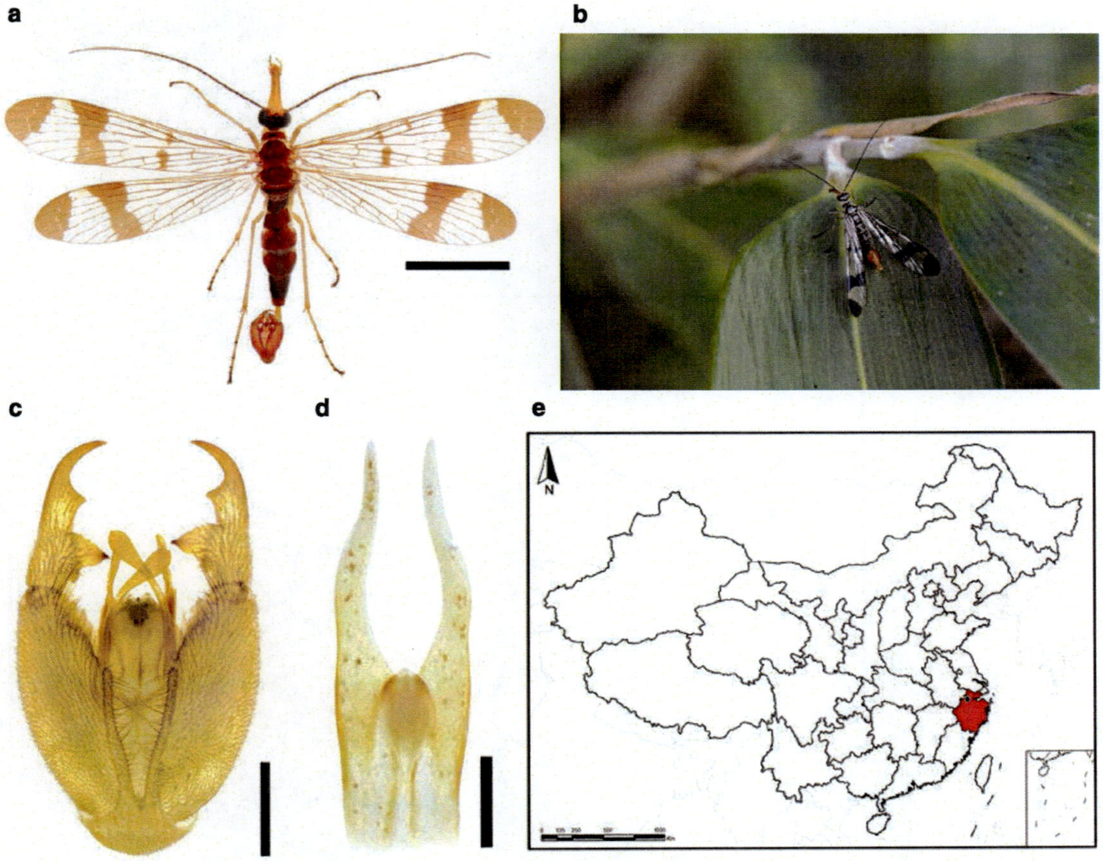

Fig. 10.156 (**a**) Male adult of ***Panorpa cheni***, dorsal view (West Tianmu Mountain, Lin'an, Zhejiang) (*Scale bar*: 5.0 mm) **[NWAU]**. (**b**) Male adult of ***Panorpa cheni*** in the field (West Tianmu Mountain, Lin'an, Zhejiang). (**c**) Male genital bulb of ***Panorpa cheni***, ventral view (*Scale bar*: 0.5 mm). (**d**) Female medigynium of ***Panorpa cheni***, ventral view (*Scale bar*: 0.2 mm). (**e**) Distribution of ***Panorpa cheni***

10.6.38 *Panorpa choui* Zhou & Wu, 1993

(Chou's Scorpionfly) [周氏蝎蛉] (Fig. 10.157)

Measurements. Forewing length 14.0 mm.

Morphology. Vertex black. Ocellar triangle black. Rostrum yellowish brown. Nota dark brown. Pleura and legs yellowish brown. Wing membrane hyaline with brown markings. In forewings, apical band broad; pterostigmal band broad, lacking apical branch and constricted medially; basal band broad; marginal spot and spots absent. Male T1–T5 dark brown. Notal organ flat and covering acute postnotal organ. A6 dark brown and cylindrical. A7 and A8 yellowish brown and constricted basally.

Male genital bulb oval. Epandrium emarginates terminally in deep U-shape and forms pair of finger-like processes laterally. Hypandrium with greatly shortened basal stalk. Hypovalves slender. Gonostylus with obtuse triangular median tooth and broad basal process. The basal process rounded subapically, bearing short acute tip. Paramere sword-like, with row of long spines on inner margin, and extending to median tooth of gonostylus. Aedeagus with ventral valves foliate and rounded apically; dorsal valves with finger-like process laterally.

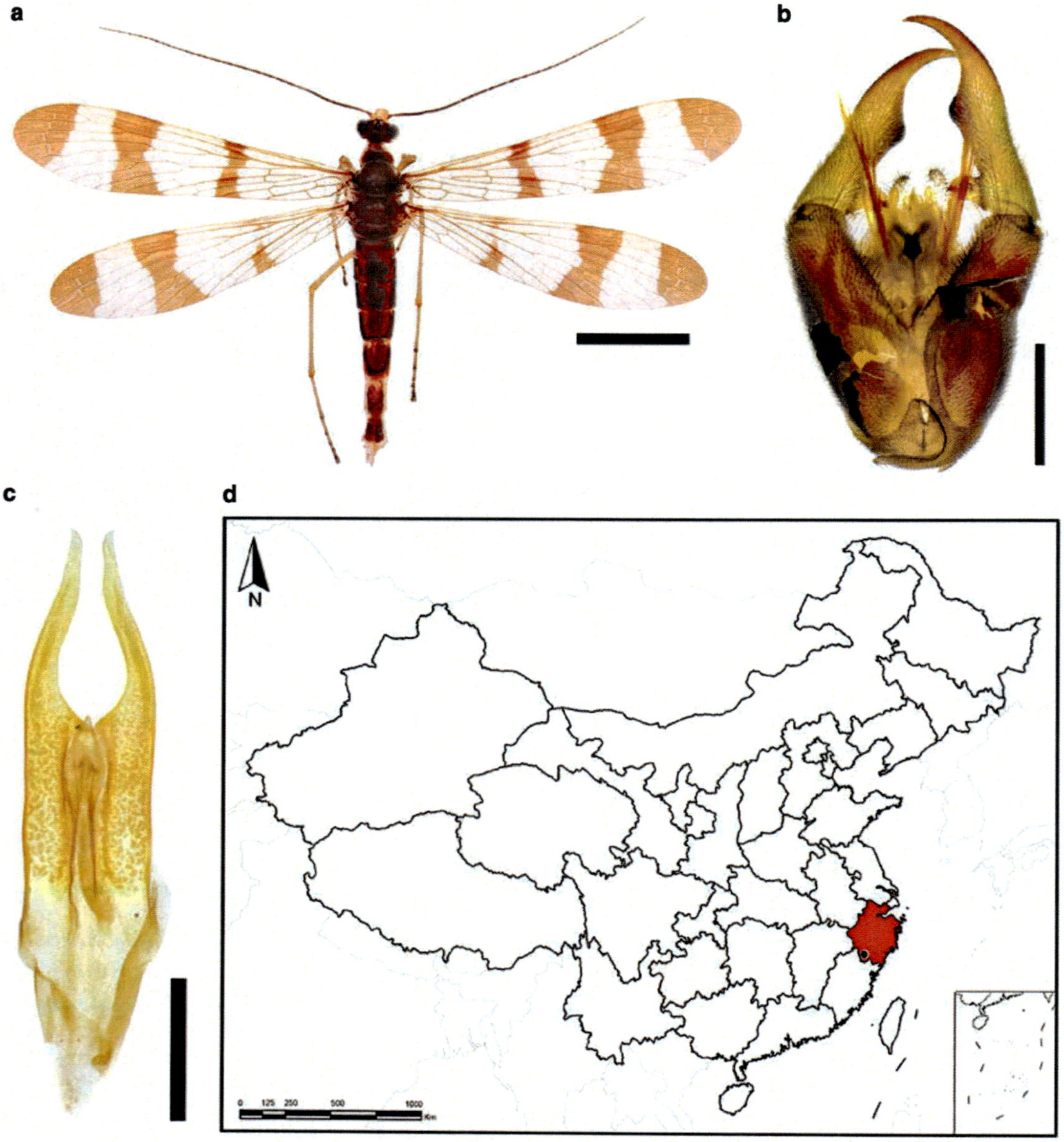

Fig. 10.157 (**a**) Female adult of ***Panorpa choui***, dorsal view (Fengyang Mountain, Zhejiang) (*Scale bar*: 5.0 mm) [**NWAU**]. (**b**) Male genital bulb of ***Panorpa choui***, ventral view (*Scale bar*: 0.5 mm). (**c**) Female medigynium of ***Panorpa choui***, ventral view (*Scale bar*: 0.2 mm). (**d**) Distribution of ***Panorpa choui***

Female subgenital plate oval with subrounded apex and well-developed laterotergites. Medigynium with oblong main plate; posterior arms slender, half as long as main plate; axis lacking apodemes.

Emergence. July to August.

Distribution. Zhejiang (Fengyang Mountain, Mogan Mountain).

10.6.39 *Panorpa cladocerca* Navás, 1935

(Branch-Tailed Scorpionfly) [枝尾蝎蛉] (Fig. 10.158)

Measurements. Forewing length 12.0–14.5 mm.

Morphology. Vertex reddish brown. Ocellar triangle black. Rostrum reddish brown. Nota dark brown. Pleura and coxae creamy yellow. Legs mostly yellowish brown. Wing membrane hyaline, slightly tinged with yellow. Markings dark brown. In forewings, apical band broad; pterostigmal band broad, lacking apical branch and constricted medially; occasionally small triangular spot on hind margin between apical and pterostigmal bands; basal band complete; marginal spot small; basal spot absent. Male T1–T5 dark brown. Notal organ flat and covering acute postnotal organ. A6 dark brown basally and brown distally. A7 and A8 yellowish brown. Base of A7 intensively constricted and stalk-like.

Male genital bulb oval. Epandrium emarginates terminally in deep U-shape and forms pair of slender finger-like processes laterally. Hypandrium with greatly shortened basal stalk. Hypovalves slender and slightly convergent on distal halves. Gonostylus with large concaved area on basal half of ventral surface, and bearing finger-like basal process. Parameres long stout, curved convergently on distal halves, slightly extending into concaved areas of gonostyli, and bearing dense short spines along inner margins. Aedeagus with short ventral valves; dorsal valves slender, finger-like, and separated.

Female subgenital plate subtriangular with subrounded apex and well-developed laterotergites. Medigynium with subtrapezoidal main plate; posterior arms slender, nearly as long as main plate; axis lacking apodemes.

Emergence. August to October.

Distribution. Jiangxi (Lushan Mountain).

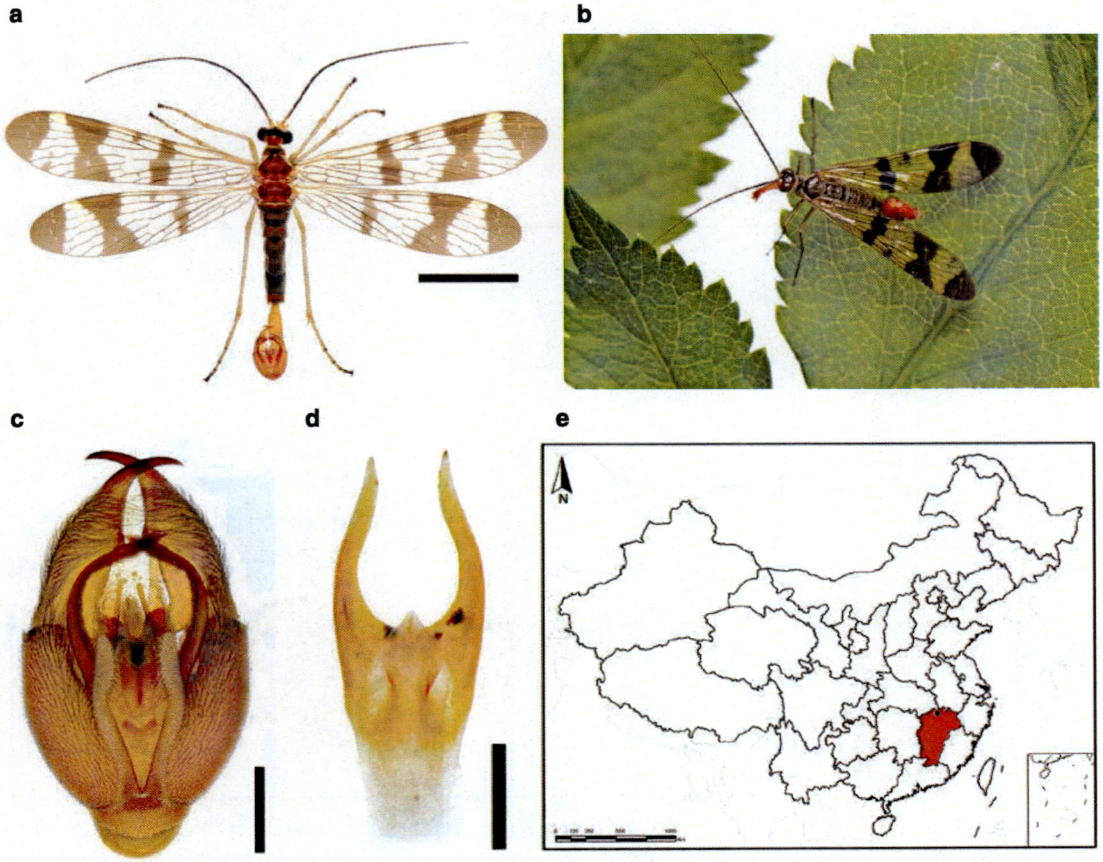

Fig. 10.158 (**a**) Male adult of ***Panorpa cladocerca***, dorsal view (Lushan Mountain, Jiujiang, Jiangxi). (**b**) Male adult of ***Panorpa cladocerca*** in the field (Lushan Mountain, Jiujiang, Jiangxi). (**c**) Male genital bulb of ***Panorpa cladocerca***, ventral view (*Scale bar*: 0.5 mm). (**d**) Female medigyium of ***Panorpa cladocerca***, ventral view (*Scale bar*: 0.2 mm). (**e**) Distribution of ***Panorpa cladocerca***

10.6.40 *Panorpa trifasciata* Cheng, 1957

(Tri-Striped Scorpionfly) [三条蝎蛉] (Fig. 10.159)
Measurements. Forewing length 13.0–14.0 mm.

Morphology. Vertex and ocellar triangle black. Rostrum yellowish brown. Nota black. Pleura and legs light yellowish brown. Wing membrane hyaline. In forewings, apical, pterostigmal, and basal bands distinct into three transverse bands. Pterostigmal band broad and constricted medially, lacking apical branch. Marginal and basal spots absent. Male T1–T5 dark brown. Notal organ flat and covering acute postnotal organ. A6 dark brown, conical. A7 and A8 yellowish brown and constricted basally.

Male genital bulb broad oval. Epandrium emarginates terminally in U-shape and forms pair of finger-like processes laterally. Hypandrium with greatly shortened basal stalk. Hypovalves stripe-like, extending to middle of gonocoxites. Gonocoxites with row of long stout setae on apical half of ventro-inner margin. Gonostylus with acute triangular median tooth and stout basal process. Paramere slender and bearing row of spines on inner margin. Aedeagus with ventral valves short, and dorsal valves stout and divergent.

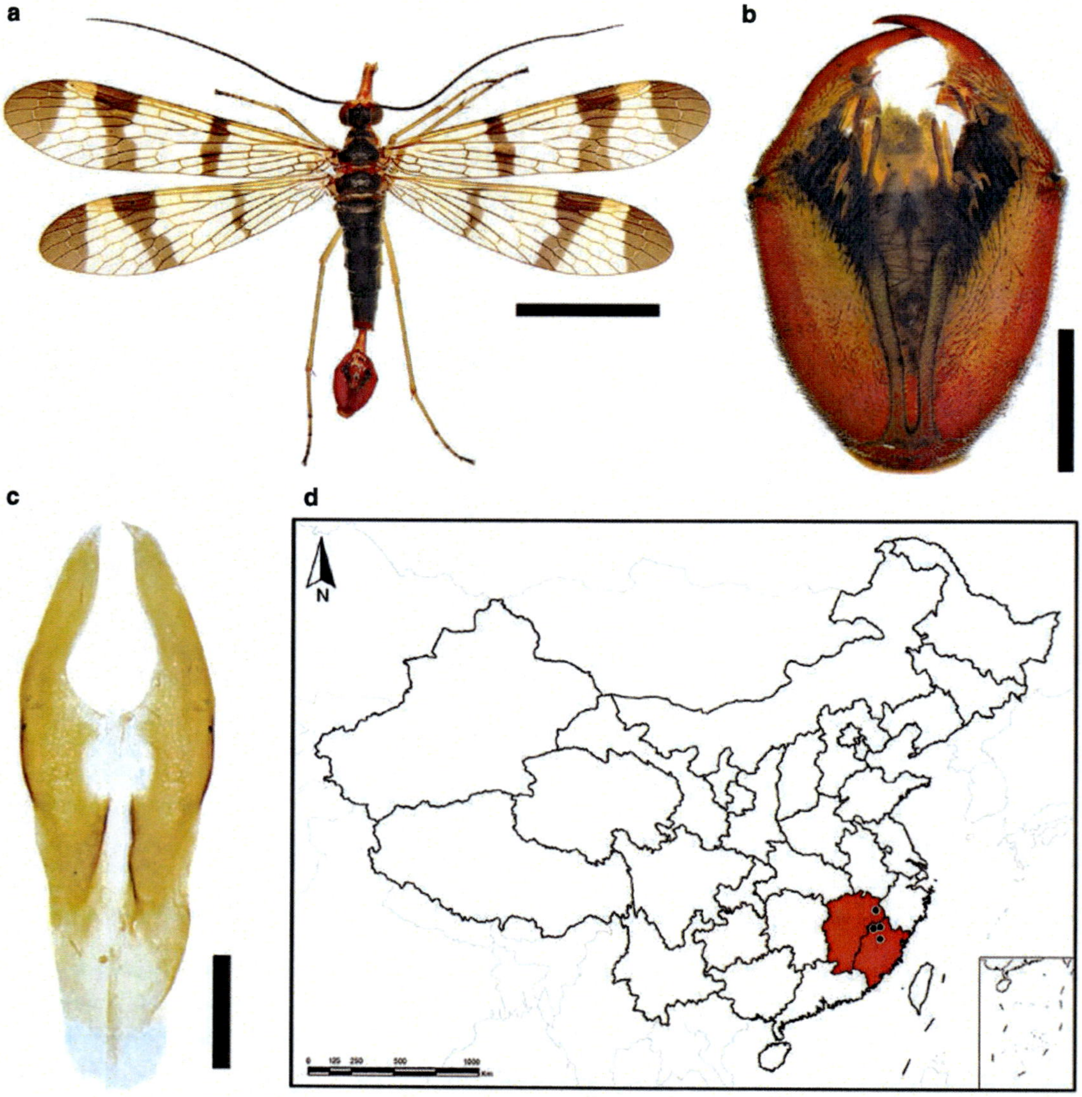

Fig. 10.159 (**a**) Male adult of ***Panorpa trifasciata***, dorsal view (Sanqing Mountain, Jiangxi) (*Scale bar*: 5.0 mm) [NWAU]. (**b**) Male genital bulb of ***Panorpa trifasciata***, ventral view (*Scale bar*: 0.5 mm). (**c**) Female medigynium of ***Panorpa trifasciata***, ventral view (*Scale bar*: 0.2 mm). (**d**) Distribution of ***Panorpa trifasciata***

Female subgenital plate subtriangular with subrounded apex and well-developed laterotergites. Medigynium with narrow main plate; posterior arms broad with constricted and acute apex, shorter than main plate; axis lacking apodemes.

Emergence. June to October.

Distribution. Fujian (Shaowu, Wuyi Mountains).

10.6.41 *Panorpa tritaenia* Chou & Wang, 1987

(Three-Banded Scorpionfly) [三带蝎蛉] (Fig. 10.160)

Measurements. Forewing length 13.0 mm.

Morphology. Vertex and ocellar triangle black. Rostrum yellowish brown. Nota black. Pleura and legs yellowish brown. Wing membrane hyaline. Pterostigma yellow. Markings brown. In forewings, apical, pterostigmal and basal bands arranged into three distinct transverse bands. Pterostigmal band lacking apical branch; and basal band extending from Rs to end of 1A. Other markings absent. Hindwing similar to forewing but lacking basal band. Male T1–T5 dark brown. Notal organ flat and covering acute postnotal organ. A6 dark brown, conical. A7 and A8 yellowish brown and constricted basally.

Male genital bulb broad oval. Epandrium emarginates terminally in U-shape and forms pair of finger-like processes laterally. Hypandrium with short narrow basal stalk. Hypovalves slender, divergent apically, extending to three-fourths of gonocoxites. Gonostylus with broad triangular median tooth and hook-like basal process.

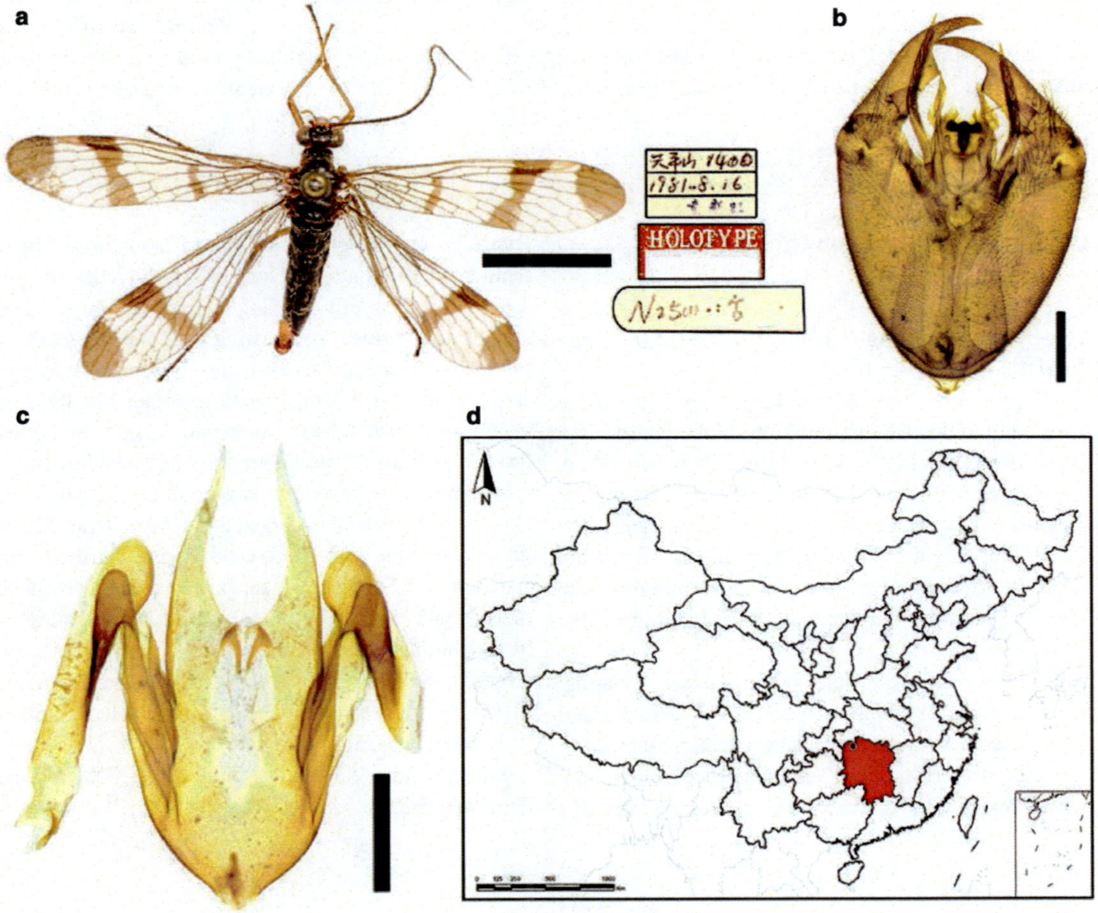

Fig. 10.160 (**a**) Male adult (holotype) of ***Panorpa tritaenia***, dorsal view (Tianping Mountain, Sangzhi, Hunan) (*Scale bar*: 5.0 mm) [NWAU]. (**b**) Male genital bulb of ***Panorpa tritaenia***, ventral view (*Scale bar*: 0.5 mm). (**c**) Female medigynium of ***Panorpa tritaenia***, ventral view (*Scale bar*: 0.2 mm). (**d**) Distribution of ***Panorpa tritaenia***

Paramere slender, distinctly exceeding median tooth of gonostylus, and bearing row of spines along inner margin. Aedeagus stout, with ventral valves membranous and dorsal valves sclerotized. dorsal valve with pair of auriculate processes dorsally and extending to median tooth of gonostylus.

Female subgenital plate subtriangular with subrounded apex and well-developed laterotergites. Medigynium with oblong main plate; posterior arms slender, shorter than main plate; axis lacking apodemes.

Emergence. July to August.

Distribution. Guangxi (Longsheng), Hunan (Sangzhi).

Panorpa semifasciata Group

This group of species can be recognized by the greatly reduced and usually sexually dimorphic wing markings, the greatly elongated median tooth in male gonostylus, the greatly elongated male aedeagus, the broadened female T9, and the concealed axis in the female medigynium.

Four species are known from southwestern China. Two of them are included in this atlas.

Key to Species of the Panorpa semifasciata Group in This Atlas
(Two species, males)

1. Pterostigma reddish brown and distinct; gonostylus longer than gonocoxites; median tooth acute; paramere unbranched .... ***Panorpa semifasciata*** Cheng
– Pterostigma indistinct; gonostylus nearly as long as gonocoxites; median tooth rounded; paramere bifurcated .... ***Panorpa emeishana*** Hua, Sun & Li

10.6.42 *Panorpa emeishana* Hua, Sun & Li, 2001

Measurements. Forewing length 12.0–13.0 mm (Fig. 10.161).

Morphology. Vertex and ocellar triangle black. Rostrum dark brown, with yellowish brown flanks. Nota, pleura, and coxae dark brown. Legs mostly deep yellowish brown. Wing membrane hyaline with markings deep brown. Pterostigma brown and distinct. In males, wing markings usually greatly reduced with only slight tinge on apex; while females with apical band (narrow and irregular on inner margin) and greatly reduced pterostigmal band (a triangular spot below pterostigma).

Male T1–T5 dark brown. Notal organ flat and covering acute postnotal organ. A6 long cylindrical. A7 and A8 dark brown, constricted basally, and truncated apically. Male genital bulb broad oval. Epandrium emarginates terminally in U-shape and forms pair of finger-like processes laterally. Hypandrium with greatly shortened basal stalk. Hypovalves slender and bearing row of long bristles on inner margins, with distal halves abruptly bending laterad. Gonostylus almost as long as gonocoxites, and bearing stout median tooth and thin acute basal process. Paramere closely adjoins aedeagus, slender, and bearing tuft of long stout spines basally. Aedeagus strongly sclerotized with ventral and dorsal valves intensively fused, extending to middle portion of gonostylus, and bearing pair of auriculate processes on apex of dorsal valves.

Female T9 greatly enlarged and curved ventrad, enclosing subgenital plate, with latero-caudal portion greatly elongated and horn-like. Subgenital plate long subtriangular with tongue-shaped and less sclerotized apex, laterotergites narrow. Medigynium with reduced main plate; anterior arms short and diverged; posterior arms broad, tapering towards acute apex, slightly convergent, and twice as long as anterior arms.

Emergence. July to August.

Distribution. Sichuan (Mount Emei, Dawa Mountain, Jiajin Mountain).

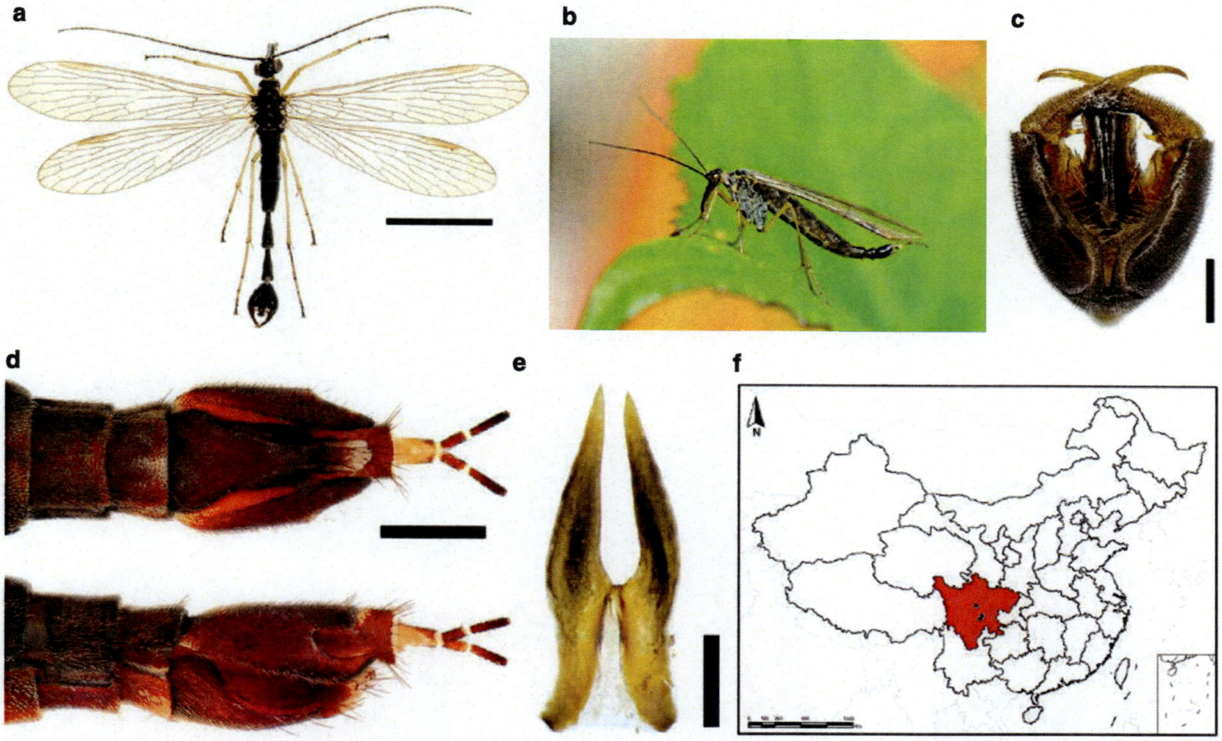

Fig. 10.161 (**a**) Male adult of ***Panorpa emeishana***, dorsal view (Mount Emei, Sichuan) (*Scale bar*: 5.0 mm) [NWAU]. (**b**) Female adult of ***Panorpa emeishana*** in the field (Jajin Mountain, Sichuan). (**c**) Male genital bulb of ***Panorpa emeishana***, ventral view (*Scale bar*: 5.0 mm). (**d**) Female terminal abdomen of ***Panorpa emeishana*** (*Scale bar*: 1.0 mm). Upper, ventral view; Lower, lateral view. (**e**) Female medigynium of ***panorpa emeishana***, ventral view (*Scale bar*: 0.2 mm). (**f**) Distribution of ***Panorpa carpenteri***

10.6.43 *Panorpa semifasciata* Cheng, 1949

Measurements. Forewing length 13.5–14.0 mm (Fig. 10.162).

Morphology. Vertex and ocellar triangle black. Rostrum dark brown. Nota black. Pleura and legs dark brown. Wing membrane hyaline, slightly tinged with yellowish brown and rounded at apex. Pterostigma reddish brown and very conspicuous. Males usually devoid of wing markings. Females with greatly reduced apical and pterostigmal bands: apical band faint with 2–3 hyaline spots innerly; pterostigmal band with disconnected basal branch, which represented by small spot at hind margin; apical branch absent; other markings absent. Male T1–T5 black. Notal organ flat and covering acute postnotal organ. A6 black, long cylindrical and tapering towards apex, without anal horns. A7 and A8 slender and constricted basally.

Male genital bulb broad oval. Epandrium emarginates terminally in U-shape and forms pair of finger-like processes laterally. Hypandrium with greatly shortened basal stalk. Hypovalves stripe-like, extending to 2/3 of gonocoxites and bearing row of long bristles on inner margin, divergent apically. Gonostylus longer than gonocoxites, and with stout and acute median tooth and basal process. Paramere stout, extending to median tooth of gonostylus, with long spines on inner margin and curved mesad on apical portion. Aedeagus with ventral and dorsal valves intensively fused and strongly sclerotized, long columnar and extending to median tooth of gonostylus, with apexes curved dorsad.

Female T9 greatly enlarged and curved ventrad, enclosing subgenital plate. Subgenital plate long subtriangular with tongue-shaped and less sclerotized apex, laterotergites narrow. Medigynium with reduced main plate; anterior arms short and diverged; posterior arms broad, tapering towards acute apex, slightly convergent, and twice as long as anterior arms.

Emergence. July to September.

Distribution. Sichuan (Kangding, Shimian).

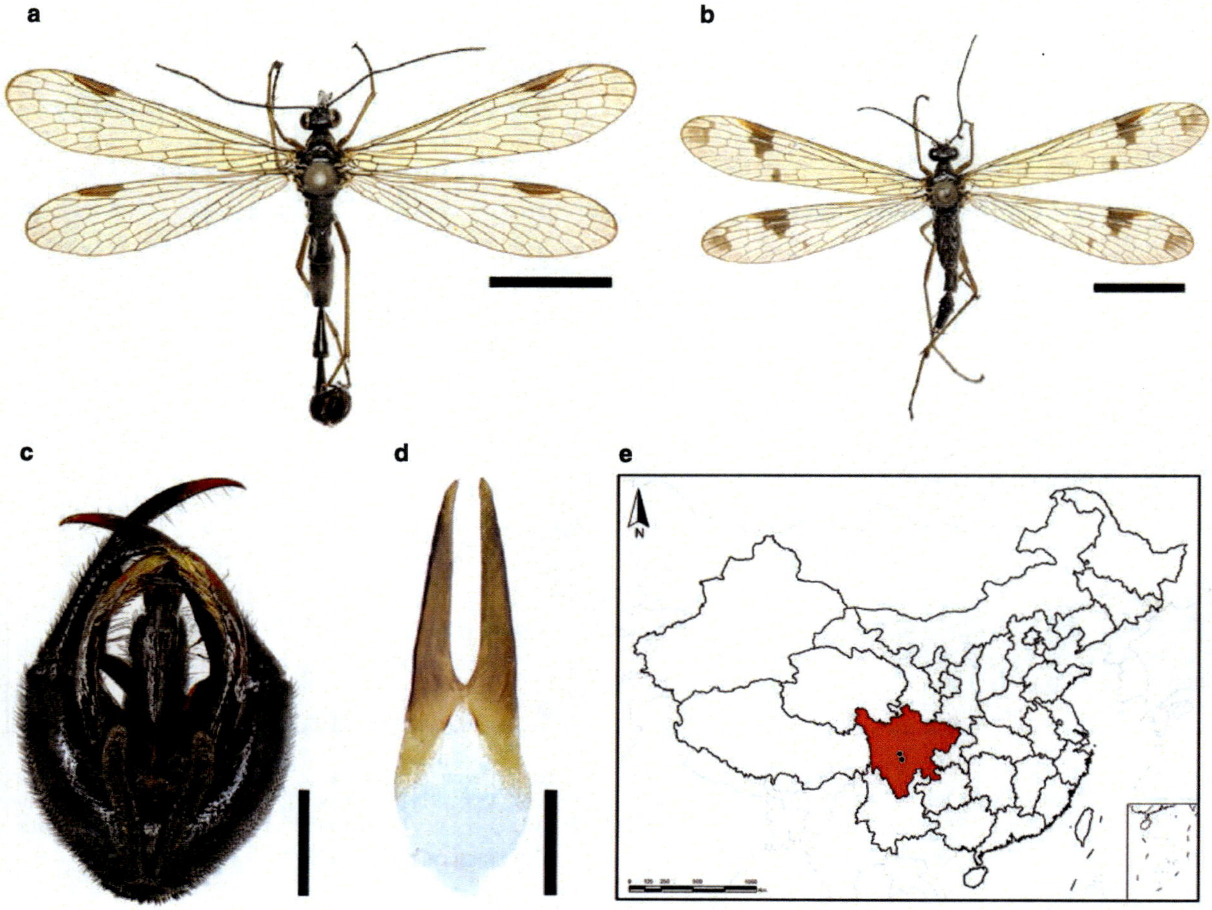

Fig. 10.162 (**a**) Male adult of ***Panorpa semifasciata***, dorsal view (Liziping, Shimian, Sichuan) (*Scale bar*: 5.0 mm) [**NWAU**]. (**b**) Female adult of ***Panorpa semifasciata***, dorsal view (Liziping, Shimina, Sichuan) (*Scale bar*: 5.0 mm) [**NWAU**]. (**c**) Male genital bulb of ***Panorpa semifasciata***, ventral view (*Scale bar*: 0.5 mm). (**d**) Female medigynium of ***Panorpa semifasciata***, ventral view (*Scale bar*: 0.2 mm). (**e**) Distribution of ***Panorpa semifasciata***

Panorpa stigmalis Group

This group of species can be recognized by the large body size, the greatly elongated male abdomen, the greatly developed male notal and postnotal organs, and the greatly elongated and extremely thin axis in the female medigynium.

Three species are known from southwestern China and all of them are included in this atlas.

Key to Species of the Panorpa semifasciata Group in This Atlas
(Three species, males)

1. Wing markings slightly reduced; basal band not extending to anterior margin; basal spot absent; A6 slightly projected dorsad on dorsal apex .... ***Panorpa stigmalis*** Navás
– Wing markings well-developed; basal band extending to anterior margin; basal spot present; A6 lacking projection on dorsal apex 2
2. Wing markings with each band and spot implicated into reticulate patterns; gonocoxites simple; dorsal aedeagal process rounded apically .... ***Panorpa dashahensis*** Zhou & Zhou
– Wing markings with each band and spot separated; gonocoxites bearing lobate process on ventral apex innerly; dorsal aedeagal processes acute apically .... ***Panorpa quadrifasciata*** Chou & Wang

10.6.44 *Panorpa dashahensis* Zhou & Zhou, 2005

(Dashahe Scorpionfly) [大沙河蝎蛉] (Fig. 10.163)

Measurements. Forewing length 16.0–18.0 mm.

Morphology. Vertex yellowish brown with dark brown, transverse band enclosing ocellar triangle. Nota black. Pleura and legs light reddish brown. Wing membrane waxy yellow, translucent. Pterostigma reddish brown. Markings brown, greatly interrupted and reticulate. In forewings, apical band broad with three hyaline windows, and interrupted posteriorly to connect with apical branch of pterostigmal band; pterostigmal band complete, with basal branch and apical branch; marginal spot elongated posteriorly to connect with basal branch of pterostigmal band; basal band complete; marginal spot greatly elongated. Male T1–T5 reddish brown. Notal organ stout clavate, exceeding acute postnotal organ on hind portion of T4. A6 long cylindrical, bearing numerous dense long setae on apical portion of dorsal surface. A7 and A8 elongated and greatly stalked basally.

Male genital bulb long oval. Epandrium emarginate terminally in U-shape and forming pair of finger-like processes laterally. Hypandrium Y-shaped, with long narrow basal stalk. Hypovalves slender, extending to 2/3 of gonocoxites, slightly convergent apically. Gonostylus long, with triangular median tooth and disc-like basal process. Paramere slender and simple, slightly exceeding apex of aedeagus, bearing numerous microtrichia. Ventral aedeagal valves short; dorsal valves elongated, greatly enlarged apically, and exceeding apex of gonocoxites.

Female subgenital plate long oval with subacute apex. Medigynium with broad main plate greatly constricted at basal portion; posterior arms slender, shorter than half-length of main plate; axis thin, nearly twice as long as main plate with parallel apodemes.

Emergence. August to October.

Distribution. Guizhou (Dashahe, Leigong Mountain).

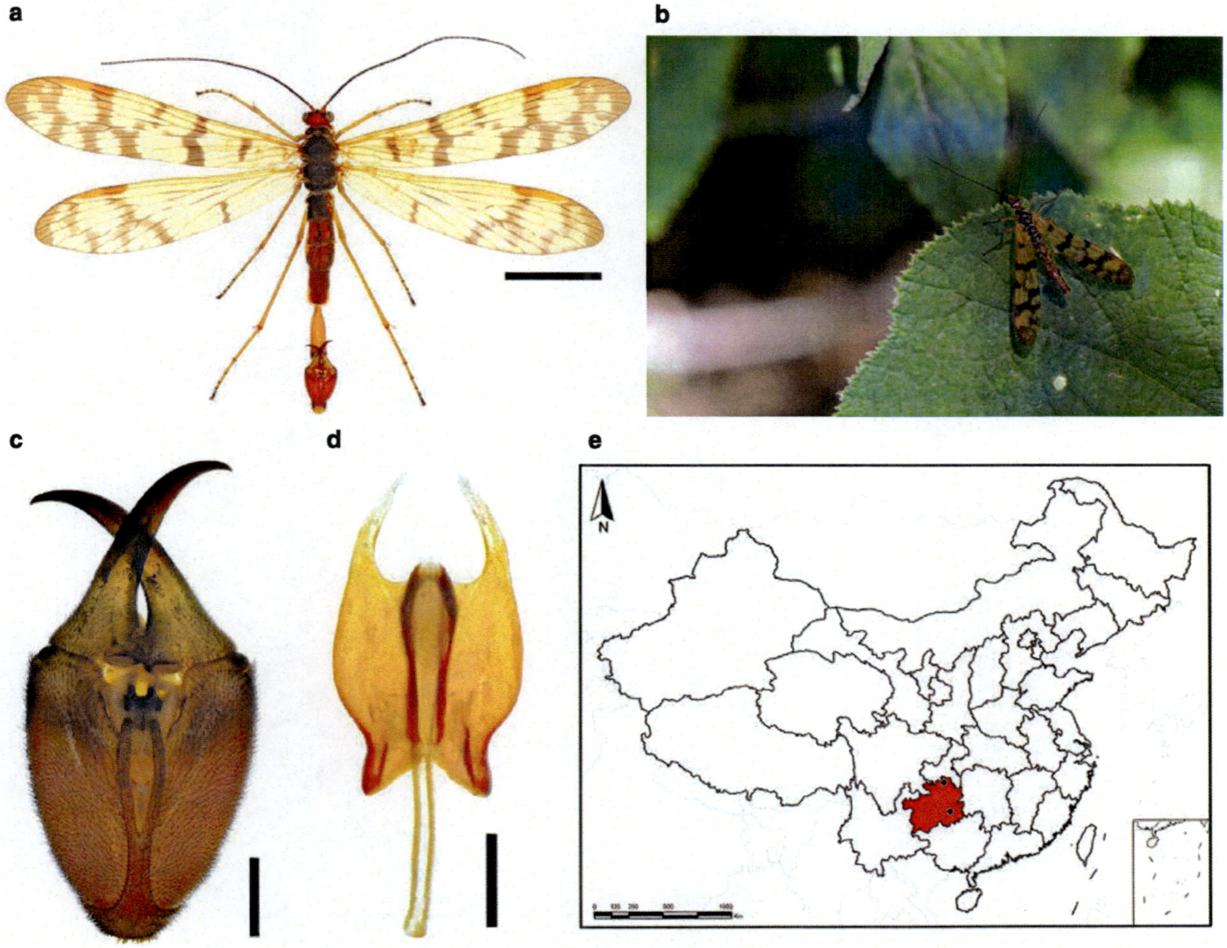

Fig. 10.163 (**a**) Male adult of ***Panorpa dashahensis***, dorsal view (Leigong Mountain, Guizhou) (*Scale bar*: 5.0 mm) **[NWAU]**. (**b**) Male adult of ***Panorpa dashahensis*** in the field (Leigong Mountain, Guizhou). (**c**) Male genital bulb of ***Panorpa dashahensis***, ventral view (*Scale bar*: 0.5 mm). (**d**) Female medigynium of ***Panorpa dashahensis***, ventral view (*Scale bar*: 0.2 mm). (**e**) Distribution of ***Panorpa dashahensis***

10.6.45 *Panorpa quadrifasciata* Chou & Wang, 1987

(Four-Banded Scorpionfly) [四带蝎蛉] (Fig. 10.164)
Panorpa subaurea Chou & Li, 1987, **new synonym.**

Measurements. Forewing length 16.5–18.0 mm.

Morphology. Vertex and rostrum reddish brown. Ocellar triangle black. Nota reddish brown. Pleura and legs light reddish brown. Wing membrane waxy yellow, translucent. Pterostigma reddish brown. Markings brown. In forewings, apical band narrow; pterostigmal band complete, with basal branch and apical branch; small triangular spot on hind margin between apical and pterostigmal bands; basal band complete; marginal and basal spots small and conspicuous. Male T1–T5 reddish brown. Notal organ stout clavate, exceeding acute postnotal organ on middle of T4. A6 long cylindrical, bearing numerous dense long setae on apical portion of dorsal surface. A7 and A8 elongated and greatly stalked basally.

Male genital bulb long oval. Epandrium emarginates terminally in U-shape and forms pair of finger-like processes laterally. Hypandrium Y-shaped, with long narrow basal stalk. Hypovalves narrow and short, extending to 2/3 of gonocoxites, divergent apically. Gonocoxites each with acute triangular plate

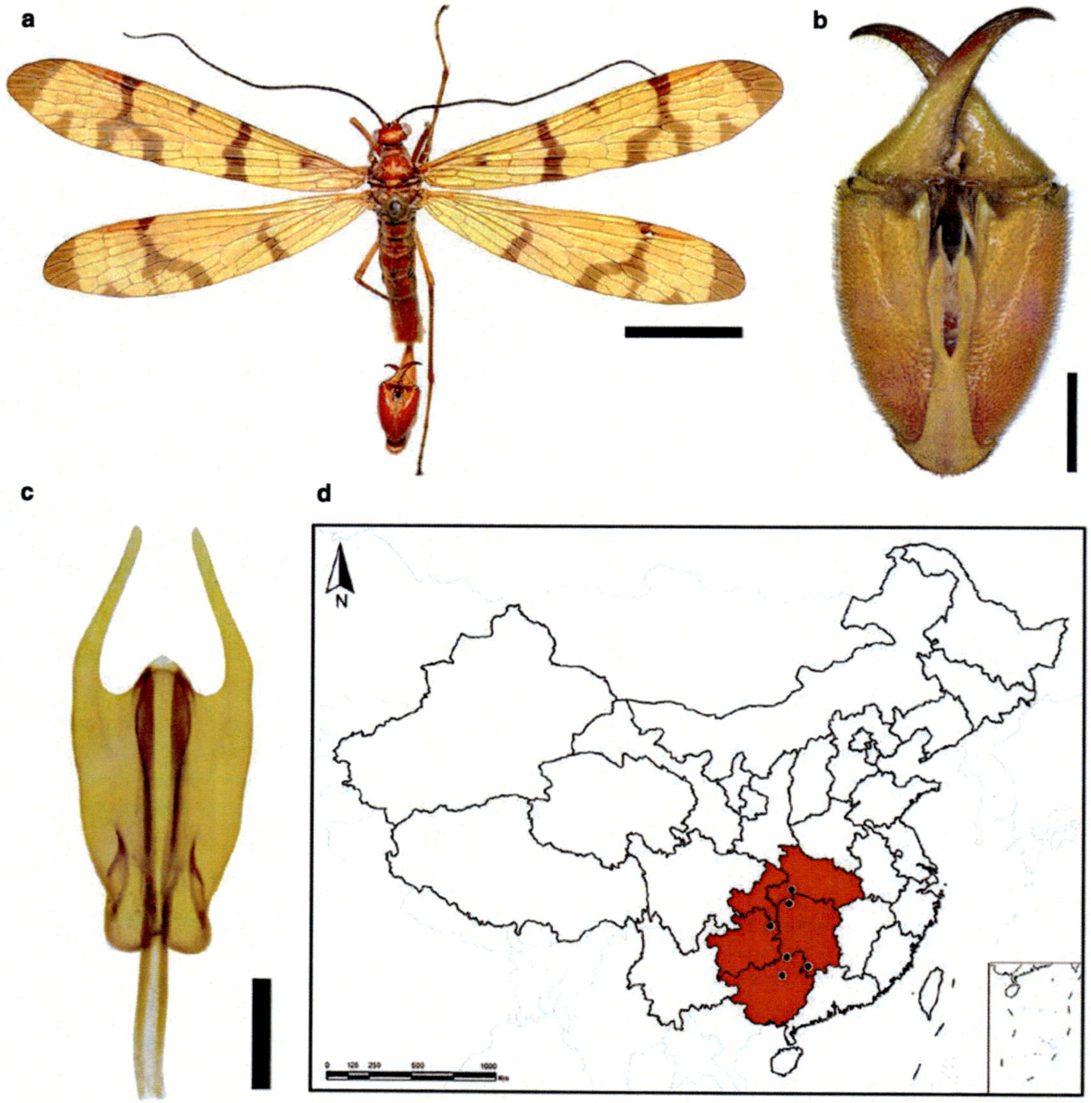

Fig. 10.164 (**a**) Male adult of ***Panorpa quadrifasciata***, dorsal view (Wufeng, Hubei, NWAU). (**b**) Male genital bulb of ***Panorpa quadrifasciata***, ventral view (*Scale bar*: 0.5 mm). (**c**) Female medigyium of ***Panorpa quadrifasciata***, ventral view (*Scale bar*: 0.2 mm). (**d**) Distribution of ***Panorpa quadrifasciata***

directed caudad on ventro-inner apex. Gonostylus long, with triangular median tooth and disc-like basal process. Paramere slender and simple, extending to basal process of gonostylus. Ventral aedeagal valves short; dorsal valves willow leaf-shaped and divergent, exceeding apex of gonocoxites.

Female subgenital plate long oval with rounded apex. Medigynium with broad main plate slightly constricted at basal portion; posterior arms slender, half as long as main plate; axis thin, shorter than twice length of main plate with parallel apodemes.

Emergence. August to October.

Distribution. Hunan (Daoxian, Zhangjiajie), Guangxi (Longsheng, Liuzhou), Guizhou (Fanjing Mountain), and Hubei (Wufeng).

Remarks. *Panorpa subaurea* Chou & Li was described from Mangshan Mountains (southern Hunan) based on a single female holotype. Its morphology properly fits the variation range of *P. quadrifasciata*, especially, the shortened axis in the female medigynium, which can also be observed in a few female specimens of the latter. Therefore, it is treated as a junior synonym of *P. quadrifasciata*.

10.6.46 *Panorpa stigmalis* Navás, 1908

(Stigmal Scorpionfly) [痣带蝎蛉] (Fig. 10.165)

Measurements. Forewing length 16.5–18.0 mm.

Morphology. Vertex yellowish brown. Ocellar triangle dark brown. Rostrum yellowish brown. Nota dark brown. Pleura and legs yellowish brown. Wing membrane translucent, tinged with yellow. Markings brown. Pterostigma reddish brown and conspicuous. In forewings, apical band complete; pterostigmal band narrow, without apical branch; basal band

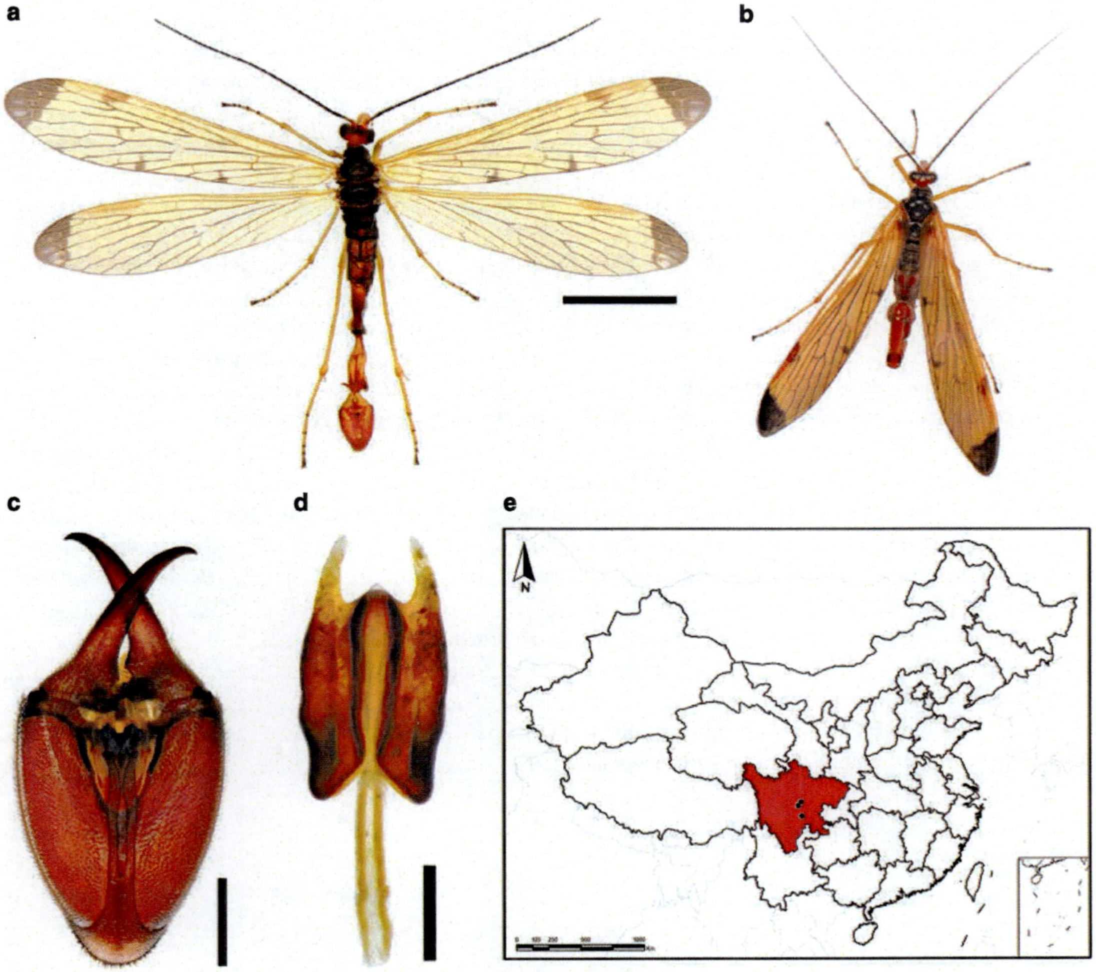

Fig. 10.165 (a) Male adult of ***Panorpa stigmalis***, dorsal view (Jiguan Mountain, Sichuan) (*Scale bar*: 5.0 mm) **[NWAU]**. (**b**) Male adult of ***Panorpa stigmalis***, dorsal view (Baoxing, Sichuan). (**c**) Male genital bulb of ***Panorpa stigmalis***, ventral view (*Scale bar*: 0.5 mm). (**d**) Female medigynium of ***Panorpa stigmalis***, ventral view (*Scale bar*: 0.2 mm). (**e**) Distribution of ***Panorpa stigmalis***

reduced into small stripe-like spot on hind margin. Male T1–T5 yellowish brown. Notal organ strongly developed, clavate, exceeding middle of T4. Postnotal organ on distal third of T4 and strongly projected dorsad. A6 yellowish brown, long cylindrical, slightly projected and bearing tuft of stout setae on dorsal apex, without anal horns. A7 and A8 yellowish brown, with A7 greatly stalked on basal quarter.

Male genital bulb long oval. Epandrium emarginates terminally in U-shape and forms pair of finger-like processes laterally. Hypandrium with long narrow basal stalk. Hypovalves slender, exceeding distal 2/3 of gonocoxites, slightly divergent apically. Gonostylus with obtuse triangular median tooth and trapezoidal basal process. Paramere simple. Aedeagus short and pincer-like, not exceeding apex of gonocoxites.

Female subgenital plate long oval with rounded apex. Medigynium with broad main plate notched in V-shape basally; posterior arms slender, 1/3 as long as main plate; axis thin, twice as long as main plate with parallel apodemes.

Emergence. August to October.

Distribution. Sichuan (Anzihe, Mount Emei, Jiajin Mountain, Qingcheng Mountain).

Panorpa deceptor Group

This group of species can be recognized by the medium to large body size, the conical male A6, the elongated and basally constricted male A7 and A8, the greatly shortened basal stalk of hypandrium, and the stout and elongated female axis, which usually bears basally branched apodemes.

Thirty-eight species are known from southwestern China and northern Thailand. 20 of them are included in this atlas.

Key to Species of the Panorpa deceptor Group in This Atlas

(19 species, males. ***Panorpa insularis*** Hua & Chou is not included because the male is unknown.)

1. A6 bearing conical anal horn on dorsal subapex .... ***Panorpa flavipennis*** Carpenter
– A6 lacking anal horn 2
2. Aedeagus greatly shortened, entirely concealed in aedeagal concavity formed by gonocoxites, with apex far away from apex of gonocoxites 3
– Aedeagus not shortened or only slightly shortened, with apex nearly extending or exceeding apex of gonocoxites 6
3. Hindwing with hind margin expanded between endings of M_1–CuP, and broader than forewing .... ***Panorpa subambra*** Chou & Tong
– Hindwing nearly as wide as or narrower than forewing 4
4. Paramere greatly elongated and wrench-shaped subapically, exceeding apex of gonocoxites .... ***Panorpa sonani*** Issiki
– Paramere extremely short, concealed in aedeagal concavity 5
5. Paramere bearing long hook-like process subbasally, and numerous long spines along inner margin .... ***Panorpa pectinata*** Issiki
– Paramere lacking hook-like process and numerous long spines .... ***Panorpa brevititilana*** Issiki
6. Wing membrane tinged with deep yellow 7
– Wing membrane colorless, light yellowish brown, or greyish brown 10
7. Parameres bifurcated 8
– Parameres unfurcated 9
8. Parameres with ventral branch slightly shorter than dorsal branch .... ***Panorpa aurea*** Cheng
– Parameres with ventral branch twice as long as dorsal branch .... ***Panorpa lutea*** Carpenter

9. Gonostylus bearing large concaved region in ventro-basal portion; parameres bearing long spines along inner margin .... ***Panorpa auripennis*** Bicha
– Gonostylus lacking concaved region; parameres bearing short spines .... ***Panorpa sextaenia*** Zhou & Bao
10. Gonocoxites with broad plate on ventro-inner margin 11
– Gonocoxites lacking such plate 12
11. Gonocoxites with long finger-like process on ventral apex .... ***Panorpa pingjiangensis*** Chou & Wang
– Gonocoxites lacking such process 12
12. Paramere extending to basal process of gonostylus .... ***Panorpa mangshanensis*** Chou & Wang
– Paramere not exceeding apex of gonocoxites 13
13. Wings devoid of marking; pterostigma brown and distinct .... ***Panorpa esakii*** Issiki
– Wings with markings; pterostigma less distinct 14
14. Wing markings faint brown and water mark-like .... ***Panorpa taiwanensis*** Issiki
– Wing markings dense 15
15. Parameres greatly elongated and exceeding middle of gonostylus 16
– Parameres not exceeding middle of gonostylus 17
16. Gonostylus with basal process broad and short; median tooth close to basal process .... ***Panorpa anrenensis*** Chou & Wang
– Gonostylus with basal process elongated; median tooth far away from basal process .... ***Panorpa flavicorporis*** Cheng
17. Gonostylus with basal process greatly elongated; median tooth reduced .... ***Panorpa guidongensis*** Chou & Li
– Gonostylus not modified as above 18
18. Dorsal aedeagal valves finger-like and nearly parallel .... ***Panorpa tetrazonia*** Navás
– Dorsal aedeagal valves slender and greatly curving laterad at apex .... ***Panorpa mokansana*** Cheng

10.6.47 *Panorpa anrenensis* Chou & Wang, 1987

(Anren Scorpionfly) [安仁蝎蛉] (Fig. 10.166)
Measurements. Forewing length 12.0–13.0 mm.

Morphology. Vertex yellowish brown. Ocellar triangle black. Rostrum yellowish brown. Pronotum dark brown. Meso- and metanotum dark brown with greyish yellow median stripe. Pleura and legs yellowish brown. Wing membrane hyaline with brown markings. In forewings, apical band with hyaline spot posteriorly; pterostigmal band complete, with broad basal branch and thin apical branch; basal band complete and broadened towards hind portion; marginal and basal spots conspicuous. Male T1–T5 dark brown. Notal organ slightly raised. A6 dark brown, cylindrical. A7 and A8 yellowish brown and slightly constricted basally.

Male genital bulb long oval. Epandrium with shallow U-shaped terminal emargination. Hypandrium with greatly shortened basal stalk. Hypovalve slender, bearing row of stout bristles on inner margin of apical portion. Gonostylus with obtuse triangular median tooth and pincer-like basal process. Paramere slender, exceeding median tooth of gonostylus and bearing dense short spines on inner surface. Aedeagus with short ventral valves; dorsal valves elongated, finger-like, and divergent.

Female subgenital plate narrow and long with shallow V-shaped emargination terminally. Medigynium with rounded main plate; posterior arms with curved outer margin and straight inner margin, slightly shorter than main plate; slender and acute projection between greatly diverged apodemes of axis.

Emergence. September to October.

Distribution. Hunan (Anren) and Guangxi (Liuzhou).

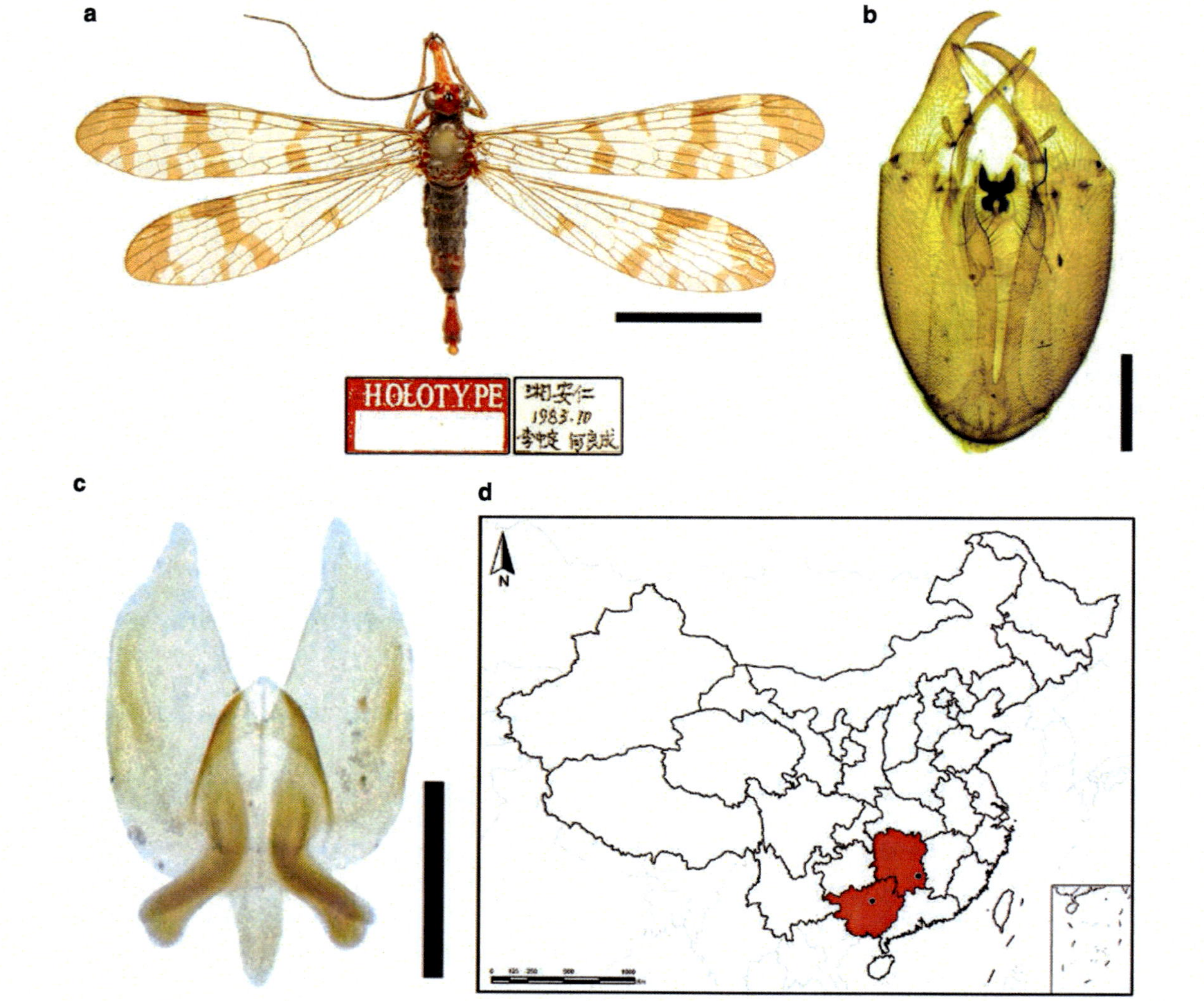

Fig. 10.166 (**a**) Male adult (holotype) of ***Panorpa anrenensis***, dorsal view (Anren, Hunan) (*Scale bar*: 5.0 mm) **[NWAU]**. (**b**) Male genital bulb of ***Panorpa anrenensis***, ventral view (*Scale bar*: 0.5 mm). (**c**) Female medigynium of ***Panorpa anrenensis***, ventral view (*Scale bar*: 0.2 mm). (**d**) Distribution of ***Panorpa anrenensis***

10.6.48 *Panorpa aurea* Cheng, 1957

(Golden Scorpionfly) [金身蝎蛉] (Fig. 10.167)
Measurements. Forewing length 15.5–16.0 mm.

Morphology. Vertex yellowish brown. Ocellar triangle black. Rostrum yellowish brown. Nota dark brown. Pleura and legs yellowish. Wing membrane hyaline, slightly tinged with yellow. Markings yellowish brown. In forewings, apical band broad with large hyaline window posteriorly; pterostigmal and basal bands complete and broad; marginal spot absent; basal spot small. Male T1–T5 dark brown. Notal organ slightly raised and covering acute postnotal organ. A6 yellowish brown, cylindrical and tapering toward apex. A7 and A8 yellowish brown and constricted basally.

Male genital bulb long oval. Epandrium with deep U-shaped terminal emargination, forming pair of finger-like processes laterally. Hypovalve slender, bearing row of long bristles along inner margin, and slightly exceeding middle of gonostylus apically. Gonostylus with obtuse triangular median

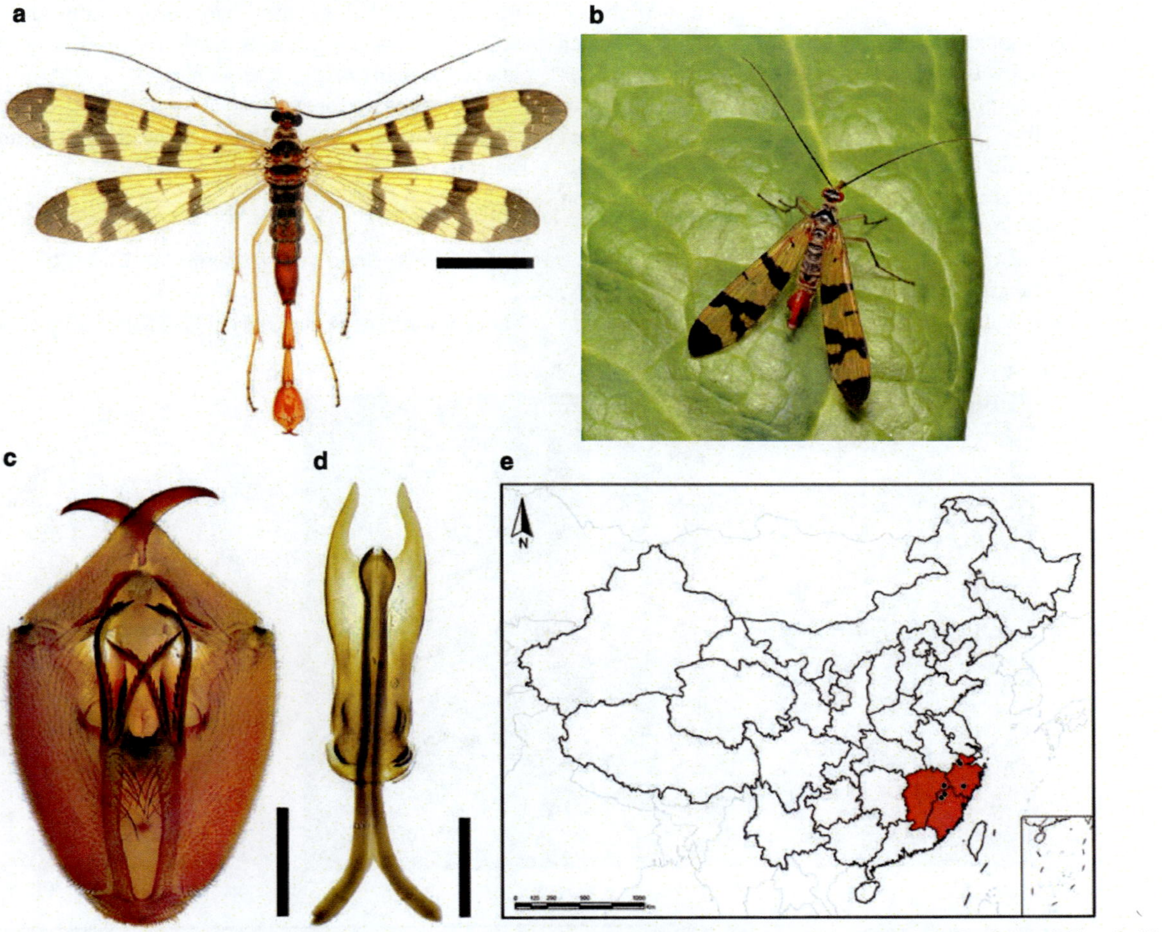

Fig. 10.167 (**a**) Male adult of ***Panorpa aurea***, dorsal view (Sanqing Mountain, Jiangxi) (*Scale bar*: 5.0 mm) [**NWAU**]. (**b**) Male adult of ***Panorpa aurea*** in the field (Sanqing Mountain, Jiangxi). (**c**) Male genital bulb of ***Panorpa aurea***, ventral view (*Scale bar*: 0.5 mm). (**d**) Female medigynium of ***Panorpa aurea***, ventral view (*Scale bar*: 0.2 mm). (**e**) Distribution of ***Panorpa aurea***

tooth and well-developed disc-like basal process. Paramere bifurcated and Y-shaped, and bearing row of stout spines on inner margin. Aedeagus with membranous poorly developed ventral valves; dorsal valves finger-like and divergent.

Female subgenital plate subtriangular with shallow V-shaped emargination terminally. Medigynium with oblong main plate; distal half of main plate with rounded lateral margins and slightly wider than basal half; posterior arms approximately 1/4 as long as main plate; axis slightly shorter than twice length of main plate with greatly diverged apodemes.

Emergence. September to October.

Distribution. Fujian (Wuyi Mountains), Jiangxi (Yanshan), Zhejiang (Fengyang Mountain, Tianmu Mountains).

10.6.49 *Panorpa auripennis* Bicha, 2019 [New Record to China]

(Golden-Winged Scorpionfly) [金翅蝎蛉] (Fig. 10.168)

Measurements. Forewing length 14.5–16.0 mm.

Morphology. Most body parts golden orange. Ocellar triangle black. Wing membrane subhyaline, greatly tinged with golden orange. Markings black. Forewings three-banded, apical band narrow and semicircular; pterostigmal band narrow and unbranched posteriorly; basal band narrow; basal spot small. Male notal organ slightly developed and covering acute postnotal organ. A6 cylindrical and slightly tapering towards apex. A7 and A8 greatly constricted basally.

Male genital bulb oval. Epandrium with deep U-shaped terminal emargination, forming pair of finger-like processes laterally. Hypovalves slender, bearing row of long bristles along inner margin, and slightly exceeding middle of gonostylus apically. Gonostylus with obtuse triangular median tooth and disc-like, concaved basal process. Paramere bifurcated and Y-shaped, and bearing row of long stout spines on inner margin. Aedeagus with membranous ventral valves; dorsal valves finger-like and slightly divergent distally.

Female subgenital plate long oval with subacute apex. Medigynium with main plate oblong; basal plate broad and enclosing basal half of main plate; posterior arms thin, short, and slightly convergent; axis nearly twice as long as main plate with greatly diverged apodemes.

Emergence. October.

Distribution. Yunnan (Pu'er, Xishuangbanna); northern Thailand.

10.6.50 *Panorpa brevititilana* Issiki, 1929

(Short-Paramere Scorpionfly) [短突蝎蛉] (Fig. 10.169)

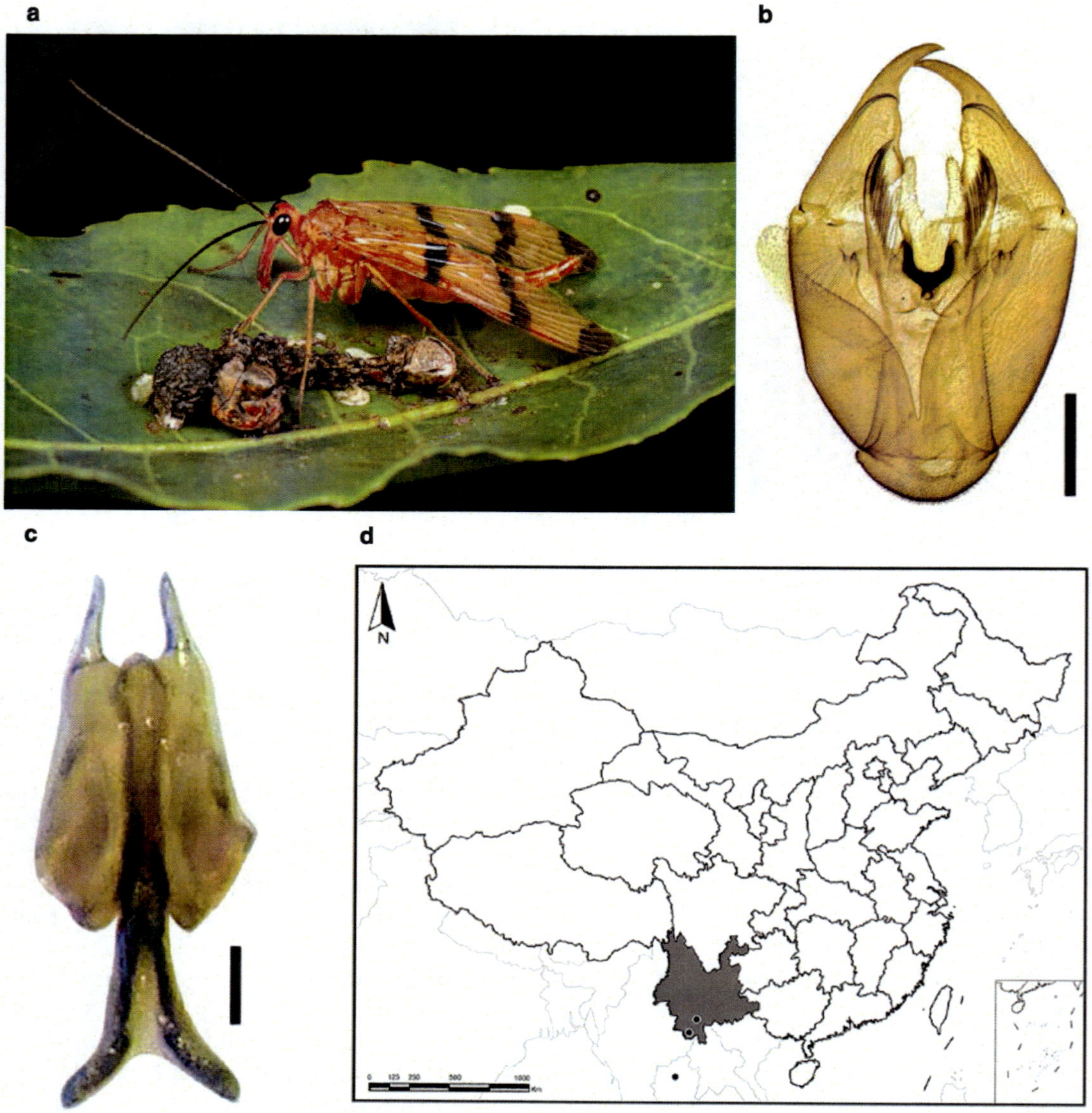

Fig. 10.168 (**a**) Female adult of ***Panorpa auripennis*** in the field (Pu'er, Yunnan, ©John Horstman). (**b**) Male genital bulb of ***Panorpa auripennis***, ventral view (Scale: 0.5 mm). (**c**) Female medigynium of ***Panorpa auripennis***, ventral view (*Scale bar*: 0.2 mm). (**d**) Distribution of ***Panorpa auripennis***

Measurements. Forewing length 11.5–12.5 mm.

Morphology. Vertex dark brown. Ocellar triangle black. Rostrum yellowish brown. Pronotum dark brown. Meso- and metanotum dark brown, with distinct yellow median stripe. Pleura and legs light yellowish brown. Wing membrane hyaline with brown markings. In forewings, apical band broad with irregular inner margin, and with 1 or 2 hyaline window spots innerly; pterostigmal band broad, with thin separated apical branch; marginal spot small; basal band split into two small spots; basal spot small. Male T1–T5 dark brown. Notal organ flat and covering acute postnotal organ. A6 dark brown, cylindrical, tapering towards apex. A7 and A8 yellowish brown. A7 greatly stalked basally.

Male genital bulb long oval. Epandrium emarginated terminally in deep U-shape and forming pair of finger-like processes laterally. Hypandrium with extremely shortened basal stalk. Hypovalves short, only extending to middle of gonocoxites, and bearing row of long bristles on inner margins. Gonostylus with obtuse triangular median tooth and large disc-like basal process. Paramere extremely shortened with acute apex. Aedeagus short and flat.

Female subgenital plate long oval with subacute apex. Medigynium with narrow main plate; posterior arms short; apodemes greatly elongated with widely diverged apodemes.

Emergence. August to October.

Distribution. Taiwan (Taitung, Tainan).

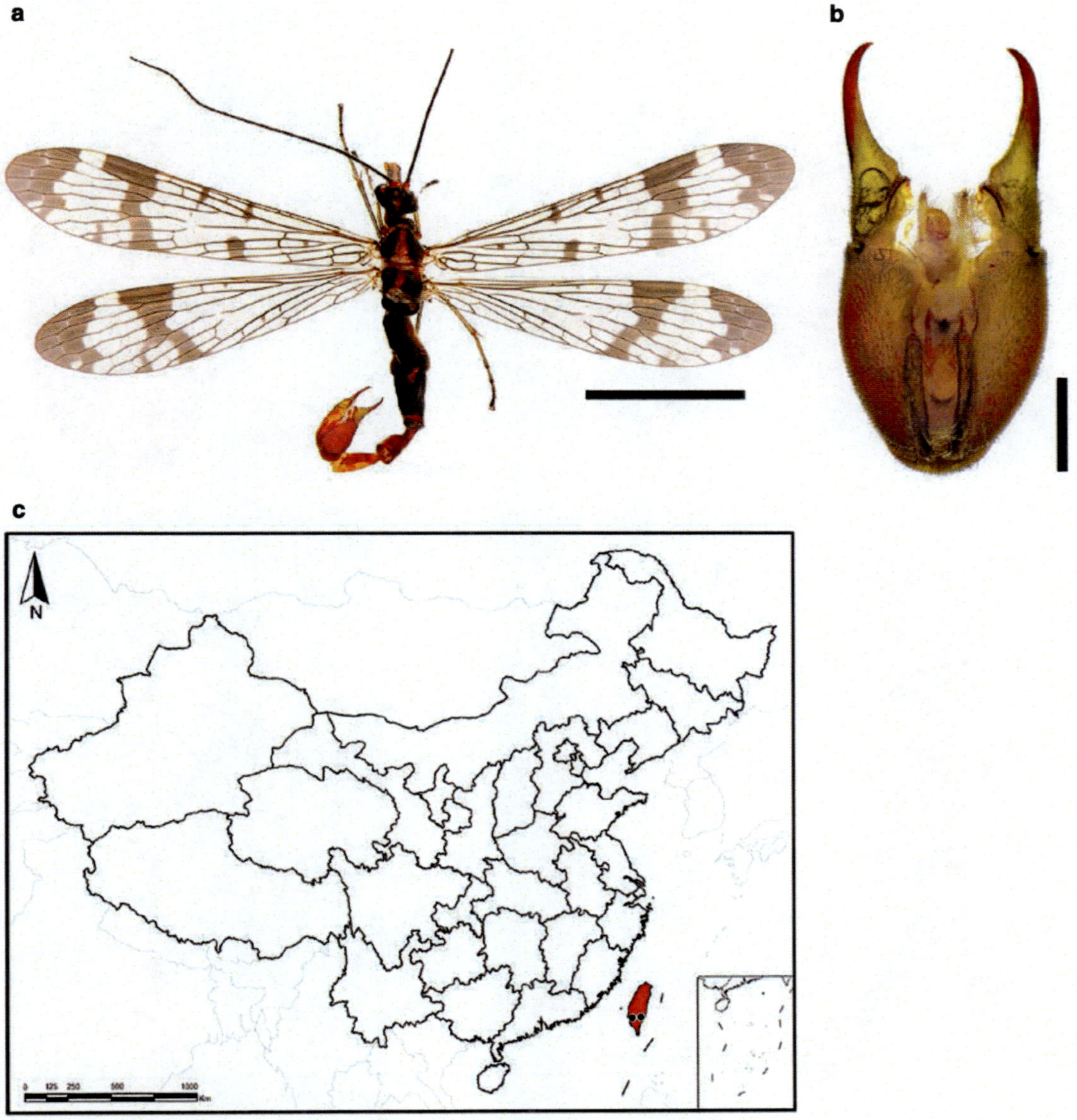

Fig. 10.169 (**a**) Male adult of ***Panorpa brevititilana***, lateral view (Haituan, Taitung, Taiwan) (*Scale bar*: 5.0 mm) **[NWAU]**. (**b**) Male genital bulb of ***Panorpa brevititilana***, ventral view (*Scale bar*: 0.5 mm). (**c**) Distribution of ***Panorpa brevititilana***

10.6.51 *Panorpa esakii* Issiki, 1929

(Esaki's Scorpionfly) [江崎蝎蛉] (Fig. 10.170)
Measurements. Forewing length 14.0 mm.

Morphology. Vertex yellowish brown; broad transverse black stripe passing through ocellar triangle; rostrum; rostrum dark brown and yellowish brown at lateral margins. Pronotum black. Meso- and metanotum mostly black but yellow at scutella; pleura and legs yellowish brown. Wing membrane hyaline, slightly tinged with yellow, lacking distinct markings; pterostigma brown and distinct. Male T1–T5 black. Notal organ flat and covering acute postnotal organ. A6 black, cylindrical, and tapering towards apex. A7 and A8 yellowish brown constricted at base.

Male genital bulb long elliptical. Epandrium emarginates terminally in deep U-shape and forms pair of finger-like processes laterally. Hypandrium with extremely shortened basal stalk. Hypovalves short, only extending to middle of gonocoxites, and bearing row of long bristles on inner margin. Gonostylus with obtuse triangular median tooth and large disc-like basal process. Paramere slender, slightly exceeding apex of gonocoxites; dorsal aedeagal valves slender and foliate.

Female subgenital plate subtrapezoidal in basal half and subtriangular in distal half with subacute apex. Medigynium with main plate oblong and greatly constricted in distal portion; posterior arms short and acute; axis greatly elongated, at least twice as long as main plate; apodemes stout with apex greatly branched.

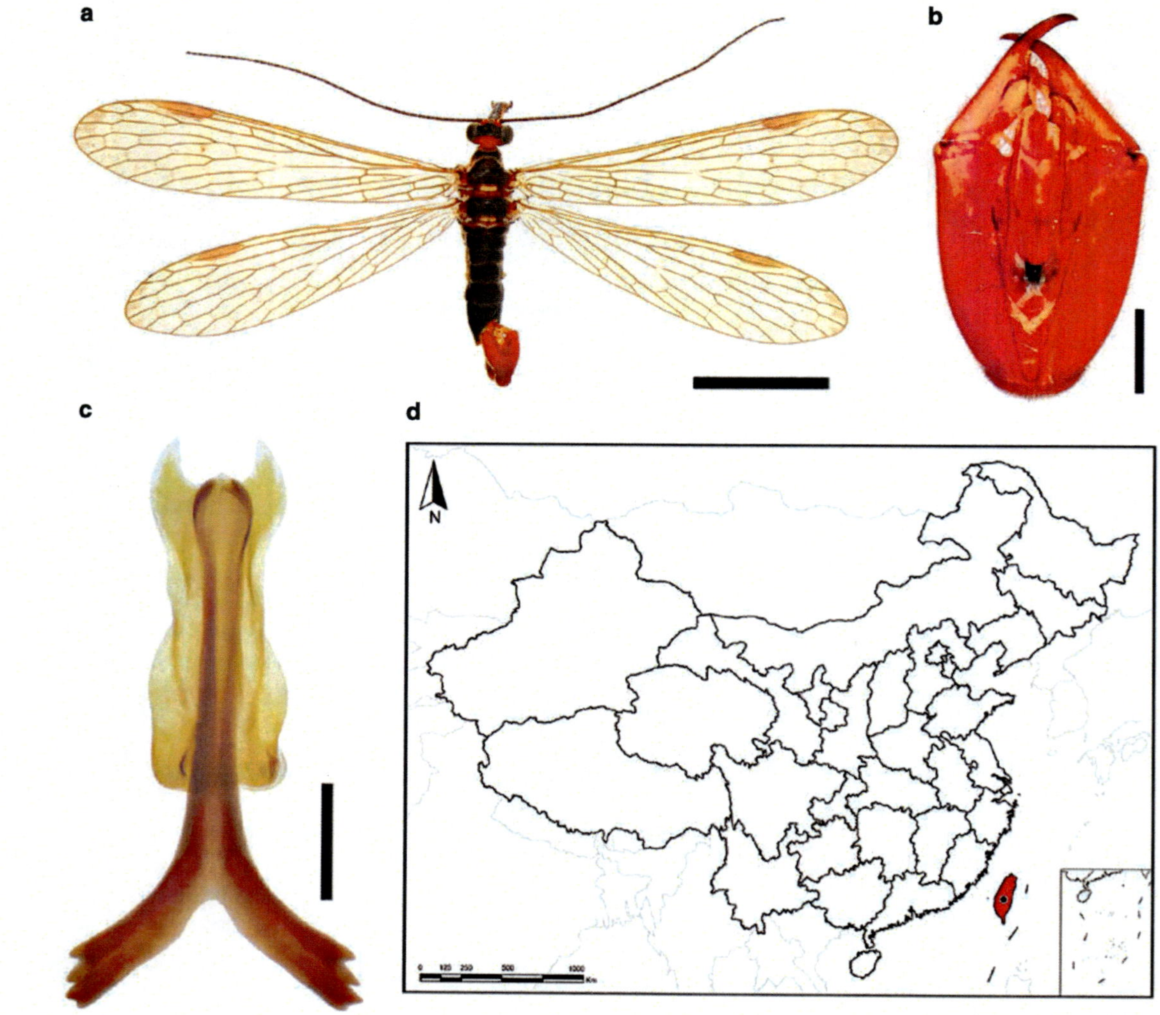

Fig. 10.170 (**a**) Male adult of ***Panorpa esakii***, dorsal view (Ali Mountain, Taiwan) (*Scale bar*: 5.0 mm) [**NWAU**]. (**b**) Male genital bulb of ***Panorpa esakii***, ventral view (*Scale bar*: 0.5 mm). (**c**) Distribution of ***Panorpa esakii***

Emergence. August to October.

Distribution. Taiwan (Ali Mountain).

10.6.52 *Panorpa flavicorporis* Cheng, 1957

(Yellow-Body Scorpionfly) [黄身蝎蛉] (Fig. 10.171)

Measurements. Forewing length 12.0–12.5 mm.

Morphology. Vertex yellow. Ocellar triangle black. Rostrum yellowish brown. Pronotum yellowish brown. Meso- and metanotum yellow, with dark brown flanks. Pleura and legs yellowish brown. Wing membrane hyaline, slightly tinged with yellow. Markings brown. In forewings, apical band broad and complete; pterostigmal band broad, with broad basal branch and thin or separated apical branch; marginal spot distinct and slender; basal band slightly constricted medially; basal spot small. Male T1–T5 deep yellowish brown. Notal organ flat, covering acute postnotal organ. A6 yellowish brown, cylindrical, tapering towards apex. A7 and A8 yellowish brown. A7 greatly stalked basally.

Male genital bulb oval. Epandrium emarginates terminally in U-shape and forms pair of finger-like processes laterally. Hypandrium with greatly shortened basal stalk; hypovalves slender, slightly exceeding middle of gonocoxites. Gonostylus with obtuse triangular median tooth and large basal process. Paramere thread-like and slender, with acute apex, and exceeding median tooth of gonostylus. Aedeagus short with finger-like and divergent dorsal valves.

Female subgenital plate narrow and long with shallow V-shaped emargination terminally. Medigynium with rounded main plate; posterior arms with curved outer margin and straight inner margin, slightly shorter than main plate; triangular projection between greatly diverged apodemes of axis.

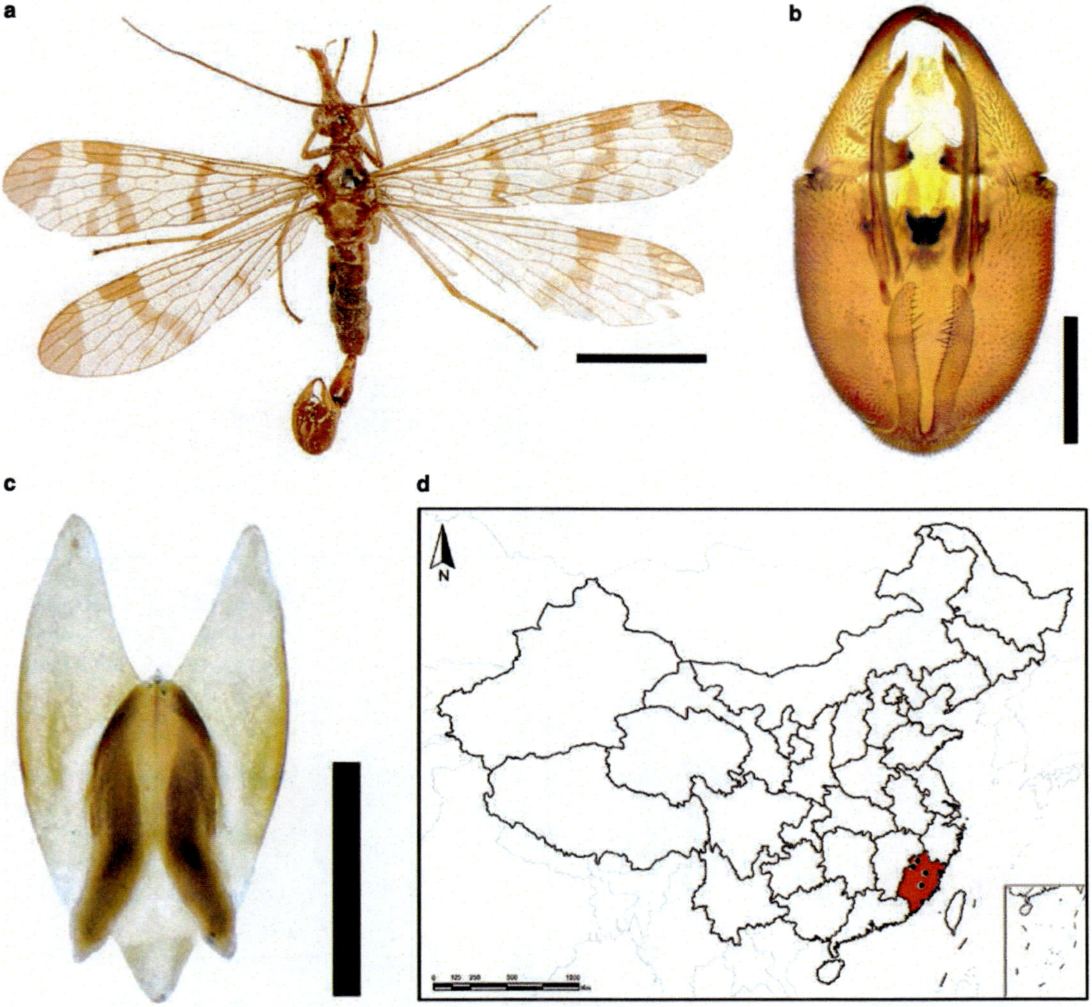

Fig. 10.171 (a) Male adult of ***Panorpa flavicorporis***, dorsal view (Dazhulan, Shaowu, Fujian) (*Scale bar*: 5.0 mm) **[NWAU]**. (b) Male genital bulb of ***Panorpa flavicorporis***, ventral view (*Scale bar*: 0.5 mm). (c) Distribution of ***Panorpa flavicorporis***

Emergence. September to October.

Distribution. Fujian (Dehua, Nanping, Shaowu, Wuyi Mountains).

10.6.53 *Panorpa flavipennis* Carpenter, 1938

(Yellow-Winged Scorpionfly) [黄翅蝎蛉] (Fig. 10.172)
Measurements. Forewing length 13.0–14.5 mm.

Morphology. Vertex and ocellar triangle black. Rostrum reddish brown with two darkened longitudinal stripes laterally. Nota, pleura, and coxae dark brown. Legs mostly deep brown. Wing membrane translucent, deep yellow, markings reddish brown. Pterostigma reddish brown and distinct. In forewings, apical band, and apical branch of pterostigmal band water stain-like; basal band absent in males but present in females; marginal and basal spots absent. Male T1–T5 black. Notal organ flat, and covering acute postnotal organ. A6 cylindrical at basal two-thirds, and conical at distal third, and bearing conical horn directed dorsad on distal third dorsally. A7 and A8 reddish brown and constricted basally, and enlarged towards apexes.

Male genital bulb oval. Epandrium emarginate terminally in shallow V-shape; hypandrium with greatly shortened basal stalk. Hypovalves long stripe-like and bearing row of long bristles on inner margin. Gonostylus with stout median tooth and stout and acute basal process, and slightly concaved along outer margin. Paramere semi-foliate, exceeding apex of gonocoxites and bearing row of comb-like long spines on inner margin. Aedeagus V-shaped, with short ventral valves and finger-like slightly enlarged dorsal valves.

Female subgenital plate long oval with subacute apex. Medigynium with oblong main plate slightly constricted in distal portion; posterior arms slender, approximately 1/3 as long as main plate; two arms usually crossed; axis slightly shorter than twice length of main plate with greatly diverged apodemes.

Emergence. June to October.

Distribution. Sichuan (Anzihe, Bailuding, Wolong).

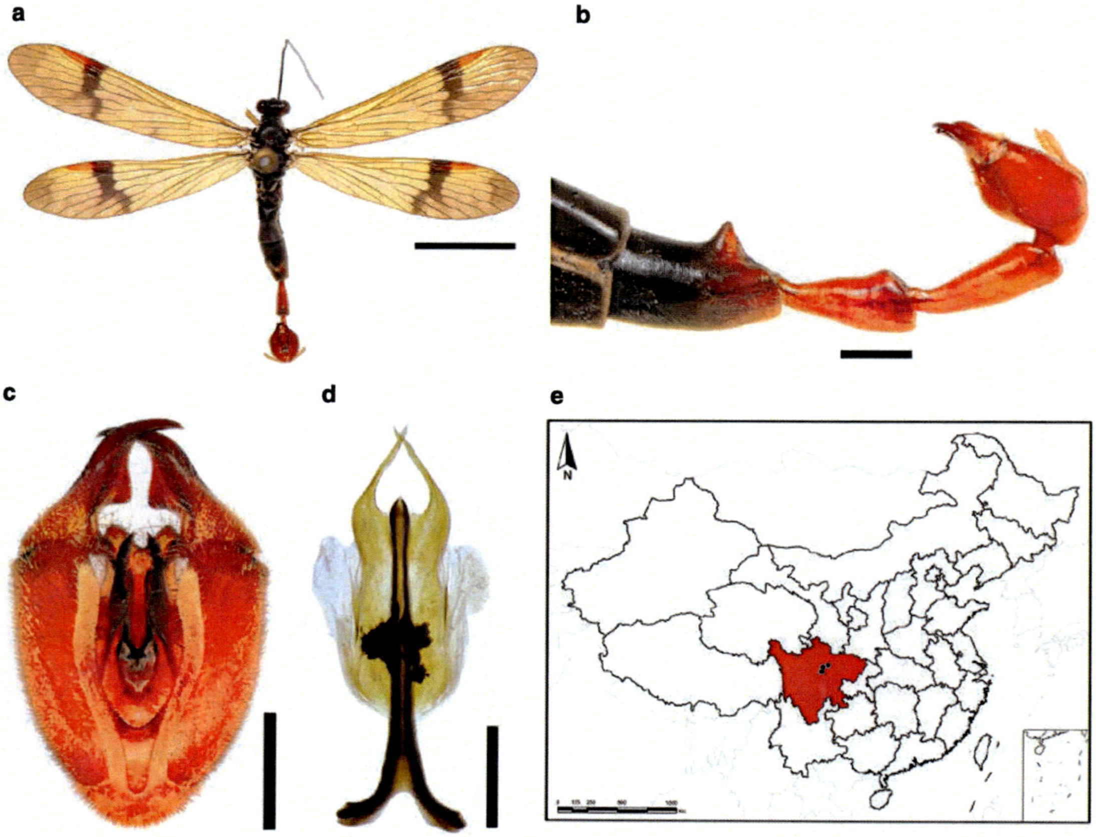

Fig. 10.172 (a) Male adult of ***Panorpa flavipennis***, dorsal view (Anzihe, Chongzhou, Sichuan) (*Scale bar*: 5.0 mm) [NWAU]. (b) Male terminal abdomen of ***Panorpa flavipennis***, lateral view (*Scale bar*: 1.0 mm). (c) Male genital bulb of ***Panorpa flavipennis***, ventral view (*Scale bar*: 0.5 mm). (d) Female medigynium of ***Panorpa flavipennis***, ventral view (*Scale bar*: 0.2 mm). (e) Distribution of ***Panorpa flavipennis***

10.6.54 *Panorpa fukiensis* Tjeder, 1950

(Fujian Scorpionfly) [福建蝎蛉] (Fig. 10.173)
Measurements. Forewing length 14.0 mm.

Morphology. Vertex dark brown. Ocellar triangle black. Rostrum yellowish brown. Nota dark brown. Pleura and legs yellowish brown. Wing membrane hyaline with markings brown. Pterostigma reddish brown. In forewings, apical band broad with incomplete posterior border, accompanied by separated small spot posteriorly, and connected to pterostigmal band anteriorly; pterostigmal band complete, with narrow apical branch; marginal spot long stripe-like; basal band broad; basal spot small. Male T1–T5 blackish brown. A6 devoid of anal horns. A7 and A8 yellowish brown and constricted basally.

Male genital bulb oval. Epandrium emarginates terminally in deep U-shape and forms pair of finger-like processes laterally. Hypandrium with long basal stalk. Hypovalves short and narrow, extending to middle of gonocoxites. Gonostylus with acute median tooth and broad basal process. Paramere with slender basal stalk and enlarged medially, tapering towards apex, and bearing row of long spines on inner margin. Aedeagus with short ventral valves and finger-like divergent dorsal valves, with stout lateral processes.

Female subgenital plate oval with subacute apex. Mediygnium with broad main plate; main plate greatly constricted basally with slightly incurved lateral margins; posterior arms short and thin; axis longer than main plate with widely diverged apodemes.

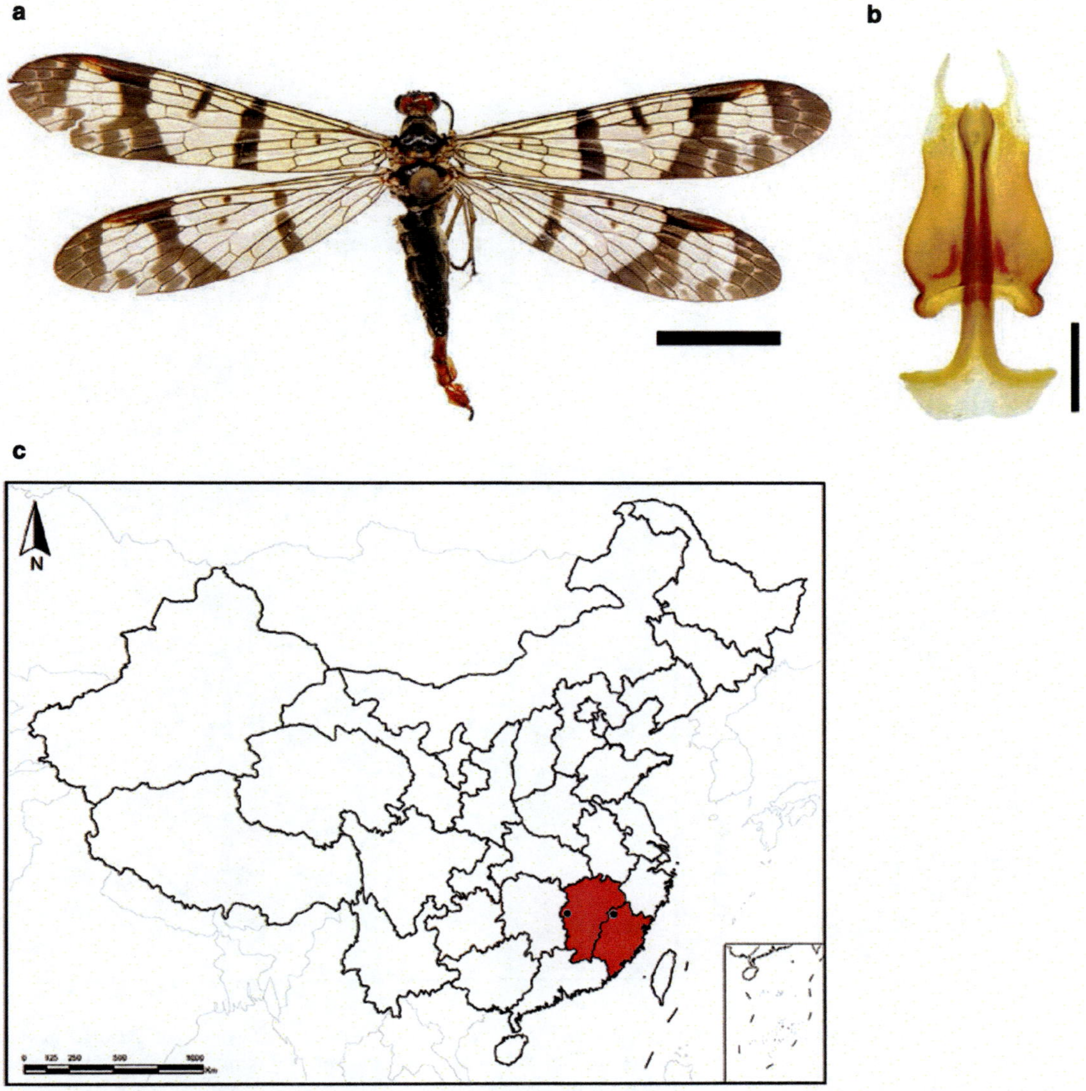

Fig. 10.173 (a) Female adult of ***Panorpa fukiensis***, dorsal view (Wanlong Mountain, Luxi, Jiangxi) (*Scale bar*: 5.0 mm) **[NWAU]. (b)** Distribution of ***Panorpa fukiensis***

Emergence. September to October.
Distribution. Fujian (Guangze), Jiangxi (Luxi).

10.6.55 *Panorpa guidongensis* Chou & Li, 1987

(Guidong Scorpionfly) [桂东蝎蛉] (Fig. 10.174)
Measurements. Forewing length 12.5–13.0 mm.

Morphology. Vertex yellowish brown. Ocellar triangle black. Rostrum yellowish brown. Pronotum dark brown. Meso- and metanotum dark brown with broad yellow median stripe. Pleura and legs yellowish brown. Wing membrane hyaline and slightly tinged with light yellow, markings yellowish brown. Forewing apical band complete and with small elongation on hind portion; pterostigmal band slender with separated apical branch; basal band separated from anterior margin; marginal spot slender; basal spot small. Male T1–T5 dark brown. Notal organ flat and covering acute postnotal organ. A6 dark brown, short conical. A7 and A8 yellowish brown, and constricted basally.

Male genital bulb broad oval. Epandrium deeply emarginate terminally in U-shape and forming pair of finger-like processes laterally. Hypandrium with short basal stalk. Hypovalves slender, exceeding middle of gonocoxites, and bearing row of long bristles on inner margin. Gonostylus with extremely stout basal process and lacking median tooth on inner margin. Paramere trifurcated with numerous spines

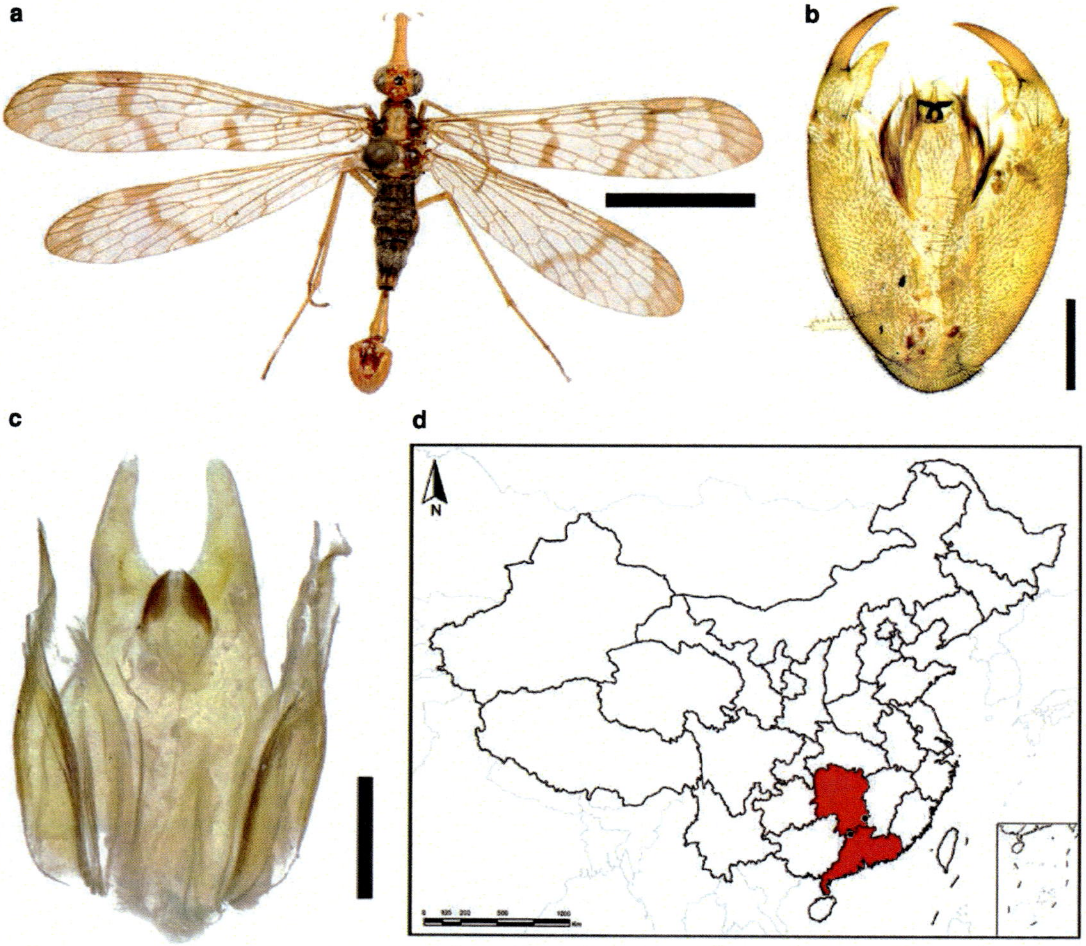

Fig. 10.174 (a) Male adult of ***Panorpa guidongensis***, dorsal view (Bamian Mountain, Guidong, Hunan) (*Scale bar*: 5.0 mm) **[NWAU]**. (b) Male genital bulb of ***Panorpa guidongensis***, ventral view (*Scale bar*: 0.5 mm). (c) Female medigynium of ***Panorpa guidongensis***, ventral view (*Scale bar*: 0.2 mm). (d) Distribution of ***Panorpa guidongensis***

on each branch. Aedeagus strongly sclerotized and exceeding apex of gonocoxites.

Female subgenital plate subtriangular with rounded apex. Medigynium with broad main plate; pair of lateral plates enclosing main plate; posterior arms short and stout; axis concealed and lacking apodemes.

Emergence. September to October.

Distribution. Guangdong (Lianzhou), Hunan (Guidong).

10.6.56 *Panorpa insularis* Hua & Chou, 1998

(Hainan Scorpionfly) [海南蝎蛉] (Fig. 10.175)

Measurements. Forewing length 14.0 mm.

Morphology. Vertex yellowish brown. Ocellar triangle dark brown, accompanied by distinct black V-shaped pattern dorsally. Rostrum yellowish brown, but grayish yellow mesally. Nota dark brown, with distinct pale yellow median

Fig. 10.175 (a) Female adult of *Panorpa insularis* in the field (Jianfengling, Ledong, Hainan, ©Wei-Wei Zhang). (b) Female medigynium of *Panorpa insularis*, ventral view (*Scale bar*: 0.2 mm). (c) Distribution of *Panorpa insularis*

stripe. Pleura and legs light yellowish brown. Wing membrane hyaline, slightly tinged with yellow. Markings dark brown. Pterostigma bright yellow and distinct. In forewings, apical band broad and accompanied by small separated spot posteriorly; pterostigmal band with equally-wide basal and apical branches; marginal spot slender and stripe like; basal band complete, narrow medially and broad posteriorly; basal spot small.

Males unknown.

Female subgenital plate oval with rounded apex. Medigynium with broad main plate and short, rounded basal plate; main plate slightly incurved along lateral margins; posterior arms slender and greatly convergent; axis longer than main plate with greatly diverged apodemes.

Emergence. May, November to December. Possibly bivoltine.

Distribution. Hainan (Jianfengling).

10.6.57 *Panorpa lutea* Carpenter, 1945

(Yellow Scorpionfly) [黄蝎蛉] (Fig. 10.176)
Measurements. Forewing length 17.0–17.5 mm.

Morphology. Vertex dark reddish brown. Ocellar triangle black. Rostrum yellowish brown. Nota dark brown. Pleura and legs yellowish brown. Wing membrane deep wax yellow. Pterostigma reddish brown. Markings blackish brown. In forewings, apical band broad and accompanied by separated small triangular spot posteriorly; pterostigma complete, with broad basal branch and thin apical branch; basal band broad and constricted medially in hourglass-shaped; marginal spot thin and stripe-like; basal spot absent. Male T1–T5 dark brown. Notal organ flat and covering acute postnotal organ. A6 and succeeding segments orange-yellow to yellowish brown. A6 cylindrical. A7 and A8 yellowish brown and constricted basally.

Male genital bulb oval. Epandrium emarginates terminally in U-shape and forms pair of finger-like processes laterally. Hypandrium with greatly shortened basal stalk. Hypovalves slender, extending to middle of gonocoxites, and bearing row of regularly arranged stout bristles in distal half of inner margin. Gonostylus stout, bearing obtuse triangular median tooth and acute and triangular basal process. Paramere bifurcated with short ventral branch, and long dorsal branch. Dorsal branch twice as long as ventral branch, and bearing row of short spines along outer margin. Aedeagus stout, not exceeding apex of gonocoxites.

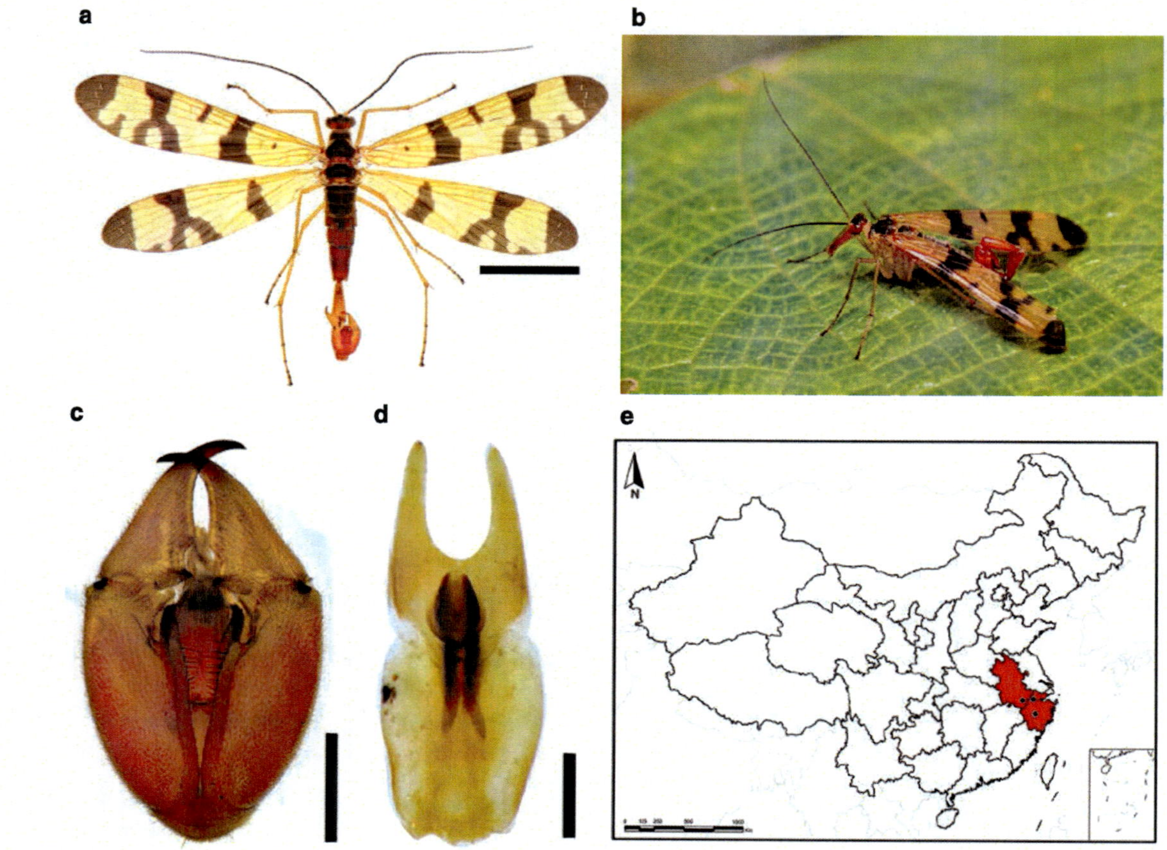

Fig. 10.176 (a) Male adult of ***Panorpa lutea***, dorsal view (West Tianmu Mountain, Lin'an, Zhejiang) (*Scale bar*: 5.0 mm) [NWAU]. (b) Male adult of ***Panorpa lutea*** in the field (West Tianmu Mountain, Lin'an, Zhejiang). (c) Male genital bulb of ***Panorpa lutea***, ventral view (*Scale bar*: 0.5 mm). (d) Female medigynium of ***Panorpa lutea***, ventral view (*Scale bar*: 0.2 mm). (e) Distribution of ***Panorpa lutea***

Female subgenital plate subtriangular with shallow V-shaped emargination terminally. Medigynium with main plate oblong; basal plate broad, 2/3 as long as main plate and folded ventrally to cover main plate; posterior arms half as long as main plate; axis concealed with slightly diverged apodemes.

Emergence. August to October.

Distribution. Anhui (Huangshan Mountain), Zhejiang (Jinhua, Tianmu Mountains).

10.6.58 *Panorpa mangshanensis* Chou & Wang, 1987

(Mangshan Scorpionfly) [莽山蝎蛉] (Fig. 10.177)

Measurements. Forewing length 13.5–14.0 mm.

Morphology. Vertex and ocellar triangle black. Rostrum yellowish brown. Nota black with yellowish brown scutella. Pleura and legs yellowish brown. Wing membrane hyaline with markings brown. In forewings, apical band complete and accompanied by separated faint small spot posteriorly; pterostigmal band connected to apical band anteriorly, and with broad basal branch and thin apical branch; marginal and basal spots very small; basal band complete. Male T1–T5 dark brown. Notal organ flat and covering acute postnotal organ. A6 dark brown, cylindrical. A7 and A8 yellowish brown and constricted basally.

Male genital bulb broad oval. Epandrium emarginates terminally in deep U-shape and forms pair of finger-like processes laterally. Hypandrium with narrow basal stalk. Hypovalves slender and slightly exceeding middle of gonocoxites. Gonocoxites with tuft of long setae on ventral apex, and broad plate directing caudad. Gonostylus with obtuse triangular median tooth and large disc-like basal process. Paramere foliate, extending to middle of gonostylus, and

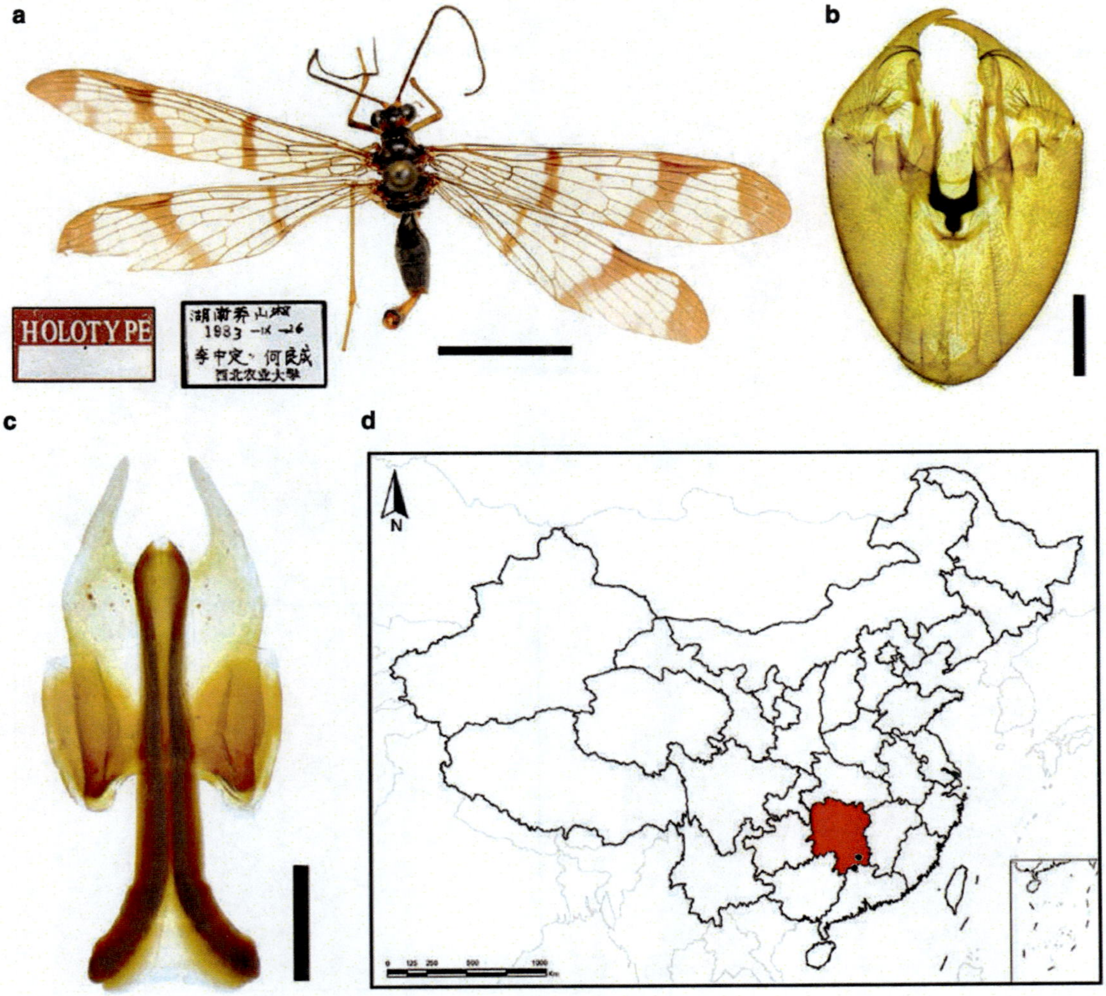

Fig. 10.177 (**a**) Male adult (holotype) of ***Panorpa mangshanensis***, dorsal view (Mangshan Mountains, Hunan) (*Scale bar*: 5.0 mm) [NWAU]. (**b**) Male genital bulb of ***Panorpa mangshanensis***, ventral view (*Scale bar*: 0.5 mm). (**c**) Female medigynium of ***Panorpa mangshanensis***, ventral view (*Scale bar*: 0.2 mm). (**d**) Distribution of ***Panorpa mangshanensis***

bearing row of comb-like spines on inner margin. Aedeagus short; ventral valves inconspicuous; dorsal valves finger-like and divergent.

Female subgenital plate oval with subacute apex; laterotergites well-developed and fused with subgenital plate laterally. Medigynium with broad main plate; pair of basal plates enclosing main plate laterally; posterior arms short, thin, and convergent; axis longer than main plate with greatly diverged apodemes.

Emergence. November.

Distribution. Hunan (Mangshan Mountains, Yongxing).

10.6.59 *Panorpa mokansana* Cheng, 1957

(Moganshan Scorpionfly) [莫干山蝎蛉] (Fig. 10.178)

Measurements. Forewing length 10.5–12.0 mm.

Morphology. Vertex yellow. Large black pattern enclosing ocelli, and extending posteriorly, forming thin black margin aside yellow occiput. Rostrum yellowish brown and deepened toward apex. Pronotum unevenly dark brown. Meso- and metanotum dark brown antero-laterally, with broad yellow mesal stripe. Pleura and legs yellowish brown. Wing membrane hyaline, tinged with yellow; markings black; pterostigma indistinct. In forewings, apical band broad, with large hyaline window in posterior portion; pterostigmal band with basal branch two times as wide as apical branch; marginal spot thick; basal band broad; basal spot large. Male T1–T5 dark brown. Notal organ broad, very short, and covering acute postnotal organ. A6–A8 reddish brown. A6 subcylindrical, slightly tapering from middle toward abruptly beveled apex. A7 greatly constricted stalk-like basally, and greatly enlarged towards truncated apex. A8 similar to A7 but less constricted basally.

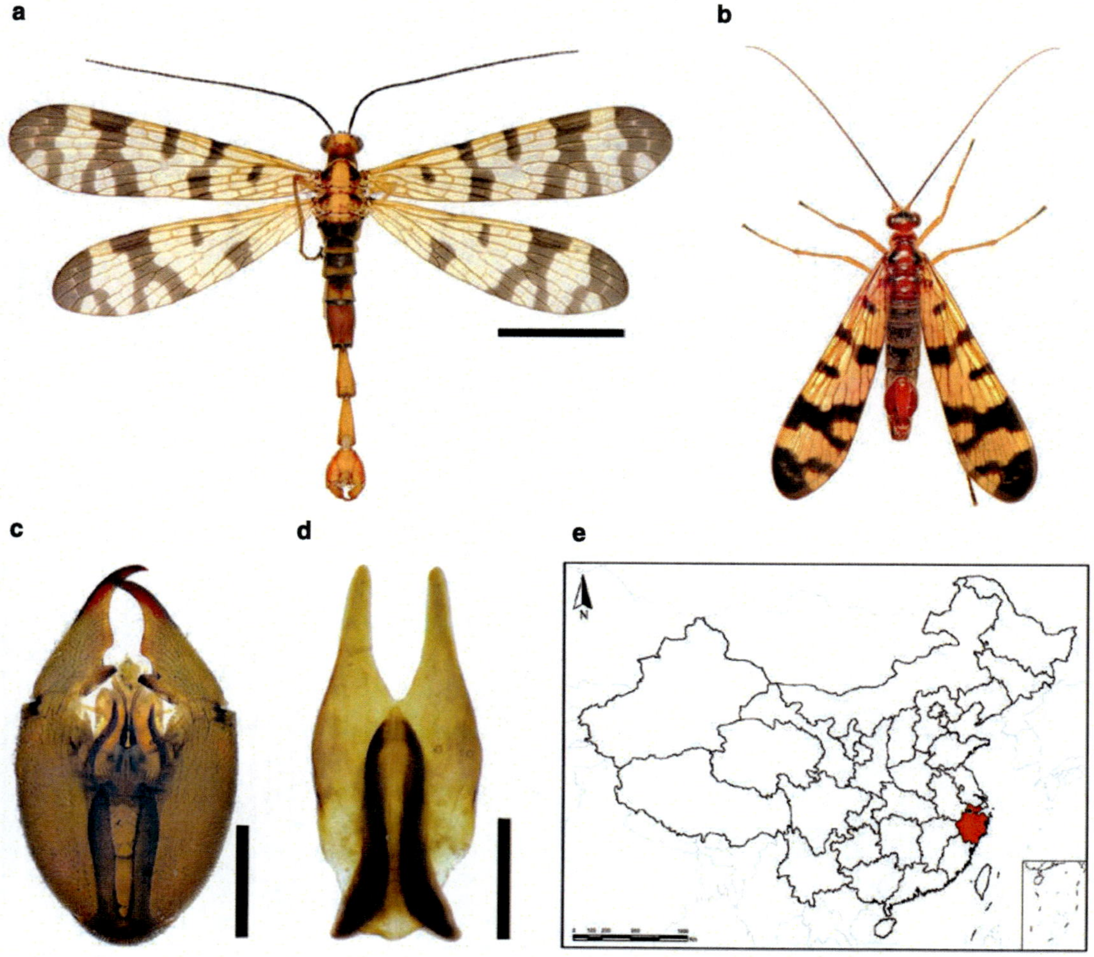

Fig. 10.178 (**a**) Male adult of ***Panorpa mokansana***, dorsal view (Mogan Mountain, Deqing, Zhejiang) (*Scale bar*: 5.0 mm) [**NWAU**]. (**b**) Male adult of ***Panorpa mokansana***, dorsal view (Mogan Mountain, Deqing, Zhejiang). (**c**) Male genital bulb of ***Panorpa mokansana***, ventral view (*Scale bar*: 0.5 mm). (**d**) Female genital bulb of ***Panorpa mokansana***, ventral view (*Scale bar*: 0.2 mm). (**e**) Distribution of ***Panorpa mokansana***

Male genital bulb reddish brown, oval. Epandrium with wide V-shaped emargination terminally and forming pair of stout processes laterally. Hypovalves extending to middle of gonocoxites, and each bearing row of long setae on inner margin of apical half. Gonostylus longer than half length of gonocoxites, with prominent middle tooth and stout basal process. Paramere bifurcated: ventral branch short and stout, curved mesally; dorsal branch long and slender; both branches bearing numerous long spines along posterior margin. Ventral aedeagal valves membranous and inconspicuous; dorsal process broad basally, slender and curved divergently at distal portion.

Female subgenital plate long oval with subacute apex. Mediygnium with broad, oval main plate; posterior arms slender, shorter than main plate; axis nearly as long as main plate; apodemes diverged and slightly exceeding basal border of main plate.

Emergence. September to October.

Distribution. Zhejiang (Mogan Mountain).

10.6.60 *Panorpa pectinata* Issiki, 1929

(Combed Scorpionfly) [栉蝎蛉] (Fig. 10.179)

Measurements. Forewing length 11.5–12.5 mm.

Morphology. Vertex yellowish brown. Ocellar triangle black. Broad black transverse stripe passing through ocellar triangle and extending to inner borders of compound eyes. Rostrum yellowish brown. Pronotum dark brown. Meso- and metanotum brown, with conspicuous broad yel-

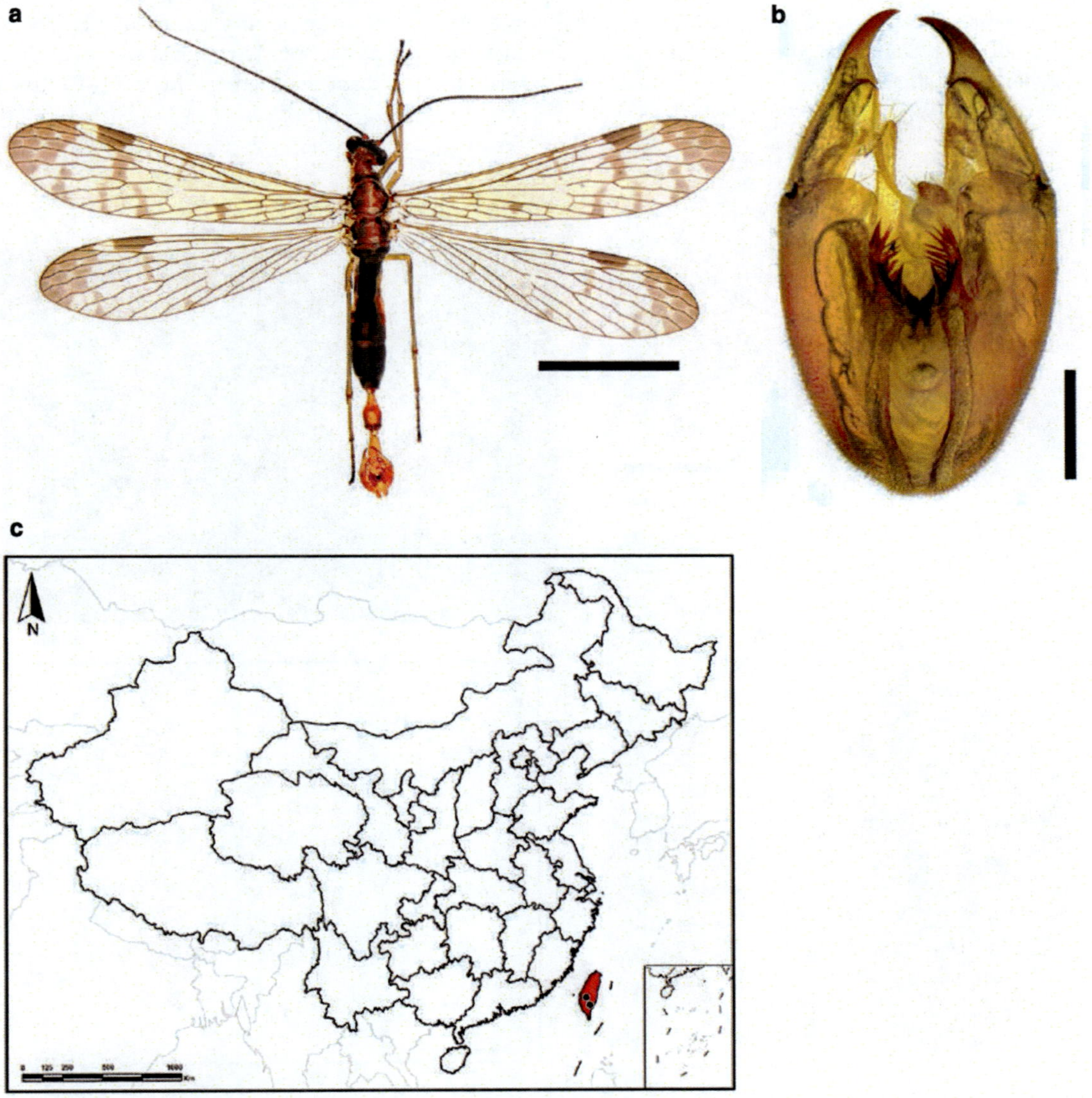

Fig. 10.179 (a) Male adult of ***Panorpa pectinata***, dorsal view (Haituan, Taitungi, Taiwan) (*Scale bar*: 5.0 mm) [NWAU]. (b) Male genital bulb of ***Panorpa pectinata***, ventral view (*Scale bar*: 0.5 mm). (c) Distribution of ***Panorpa pectinata***

low median stripe. Pleura and legs yellowish brown. Wing membrane hyaline and slightly tinged with yellow. Markings brown. In forewings, apical band with 2–3 hyaline spots innerly, and separated stripe posteriorly; pterostigmal band slender with basal branch and apical branch; marginal spot slender; basal band split into two spots; basal spot absent. Male T1–T5 dark brown. Notal organ flat and covering postnotal organ. A6 dark brown, conical. A7 and A8 yellowish brown. A7 greatly stalked basally. A8 slightly constricted basally.

Male genital bulb oval. Epandrium emarginates terminally in deep U-shape and forms pair of finger-like processes laterally. Hypandrium with extremely shortened basal stalk. Hypovalves slender and stripe-like, extending to mid-length of gonocoxites, and bearing row of long bristles on inner margin. Gonostylus with obtuse triangular median tooth and large disclike basal process. Paramere short and stout, with hook-like process subbasally and bifurcated apically, and bearing row of long comb-like spines on inner margin. Aedeagus short and flat.

Female subgenital plate diamond-shaped with subacute apex. Medigynium with broad main plate; posterior arms absent; axis longer than main plate with greatly diverged apodemes.

Emergence. July to August.

Distribution. Taiwan (Ali Mountain, Taitung).

10.6.61 *Panorpa pingjiangensis* Chou & Wang, 1987

(Pingjiang Scorpionfly) [平江蝎蛉] (Fig. 10.180)

Measurements. Forewing length 13.5–15.0 mm.

Morphology. Vertex and ocellar triangle black. Rostrum reddish brown. Nota dark brown. Pleura and legs yellowish brown. Wing membrane hyaline with markings dark brown. Pterostigma reddish brown. In forewings, apical band complete; pterostigmal band connected to apical band anteriorly, and with faint separated apical branch; marginal spot stripe-like; basal band complete; basal spots absent. Male T1–T5 dark brown.

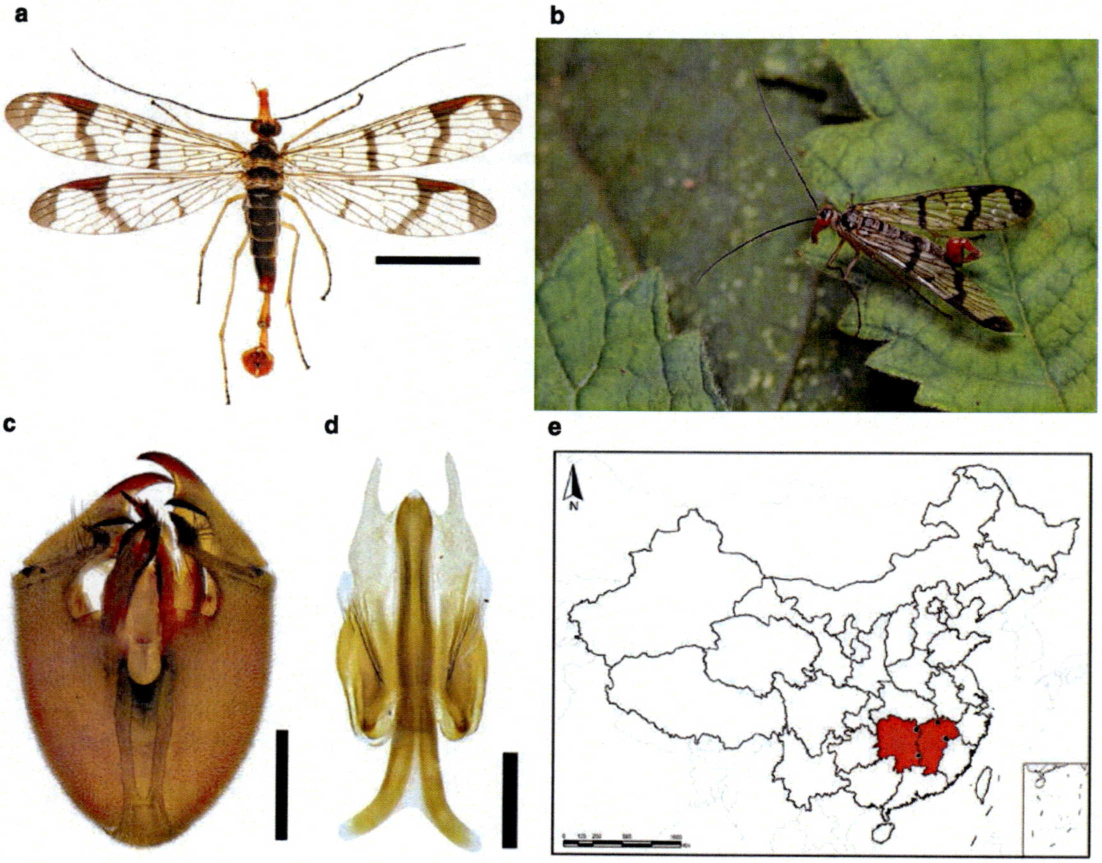

Fig. 10.180 (**a**) Male adult of ***Panorpa pingjiangensis***, dorsal view (Lushan Mountain, Jiangxi) (*Scale bar*: 5.0 mm) [**NWAU**]. (**b**) Male adult of ***Panorpa pingjiangensis*** in the field (Lushan Mountain, Jiangxi). (**c**) Male genital bulb of ***Panorpa pingjiangensis***, ventral view (*Scale bar*: 0.5 mm). (**d**) Female medigynium of ***Panorpa pingjiangensis***, ventral view (*Scale bar*: 0.2 mm). (**e**) Distribution of ***Panorpa pingjiangensis***

Notal organ flat and covering postnotal organ. A6 dark brown, conical. A7 and A8 reddish brown and constricted basally.

Male genital bulb broad oval. Epandrium emarginates terminally in deep U-shape and forms pair of finger-like processes laterally. Hypandrium with short basal stalk. Hypovalves slender, extending to middle of gonocoxites, bearing row of long bristles on inner margin. Gonocoxites with broad plate on middle of ventro-inner margin, and long finger-like process on ventral apex. Gonostylus with inconspicuous triangular median tooth and large disc-like basal process. Paramere stout, bifurcated into two equal-length branches, extending to basal process of gonostylus, branches bearing numerous short spines on inner margins. Aedeagus short with finger-like and divergent dorsal valves.

Female subgenital plate subtriangular with well-developed laterotergites. Medigynium with broad main plate; pair of basal plates enclosing main plate laterally; posterior arms short, thin, and convergent; axis longer than main plate with greatly diverged apodemes.

Emergence. October to November.

Distribution. Hunan (Pingjiang, Yanling), Jiangxi (Lushan Mountain, Zixi).

10.6.62 *Panorpa sextaenia* Zhou & Bao, 2002

(Six-Banded Scorpionfly) [六带蝎蛉] (Fig. 10.181)

Measurements. Forewing length 16.0–19.0 mm.

Morphology. Vertex and rostrum yellow. Ocellar triangle black. Nota, pleura, and legs yellow. Wing membrane waxy yellow. Pterostigma reddish brown and very conspicuous. Markings black and peculiarly arranged. In forewings, apical band absent;

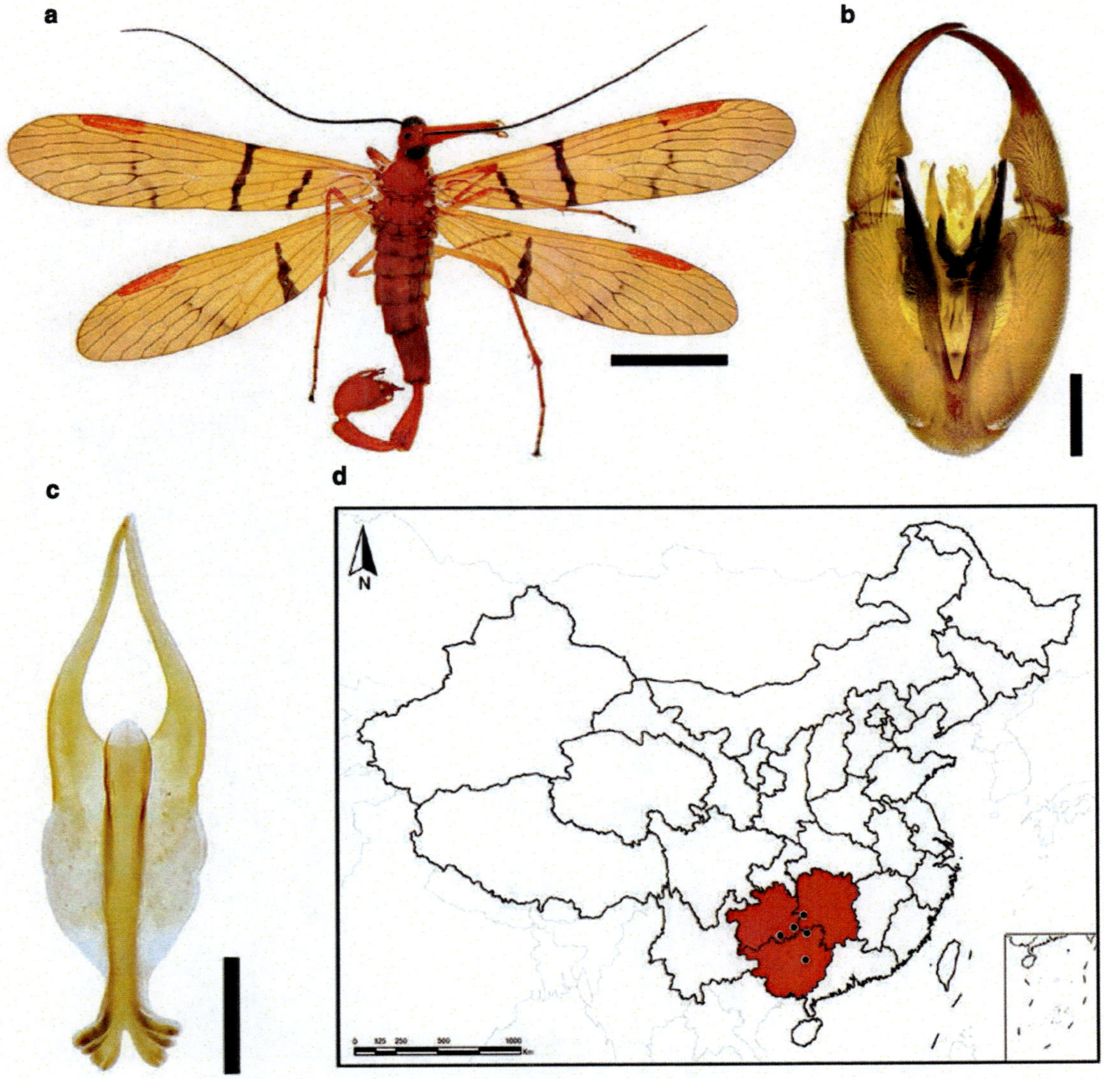

Fig. 10.181 (a) Male adult of ***Panorpa sextaenia***, dorsal view (Liping, Qiandongnan, Guizhou) (*Scale bar*: 5.0 mm) [NWAU]. (b) Male genital bulb of ***Panorpa sextaenia***, ventral view (*Scale bar*: 0.5 mm). (c) Female medigynium of ***Panorpa sextaenia***, ventral view (*Scale bar*: 0.2 mm). (d) Distribution of ***Panorpa sextaenia***

pterostigmal band very faint and without apical branch; marginal spot very small or absent; basal band complete and slender; basal spot elongated into slender stripe and not exceeding M, and occasionally forming small spot on anterior margin. Hindwing only with water stain-like basal branch of pterostigmal band, and posteriorly enlarged basal band. Male abdomen mostly yellow. Notal organ flat and covering acute postnotal organ. A6 long cylindrical, tapering towards apex. A7 and A8 long, slightly curved dorsad and constricted basally.

Male genital bulb broad oval. Epandrium emarginates terminally in U-shape and forms pair of finger-like processes laterally. Hypandrium with short basal stalk. Hypovalves slender, extending to apex of gonocoxites. Gonostylus as long as gonocoxites, and bearing rounded median tooth and small inconspicuous basal process. Paramere stick-like, exceeding basal process of gonostylus, and bearing dense short spines on inner margin. Aedeagus fork-shaped with short membranous ventral valves; dorsal valves sclerotized, slender and foliate with acute apexes.

Female subgenital plate long oval with tongue-shaped apex. Medigynium with broad main plate; posterior arms slender, nearly as long as main plate; axis longer than main plate with greatly diverged apodemes; anterior apex of apodeme stout and four- or more branched.

Emergence. October.

Distribution. Guangxi (Jinxiu, Longsheng), Guizhou (Maolan, Liping), Hunan (Huitong).

10.6.63 *Panorpa sonani* Issiki, 1929

(Sonan's Scorpionfly) [楚南蝎蛉] (Fig. 10.182)
Measurements. Forewing length 14.5–16.0 mm.

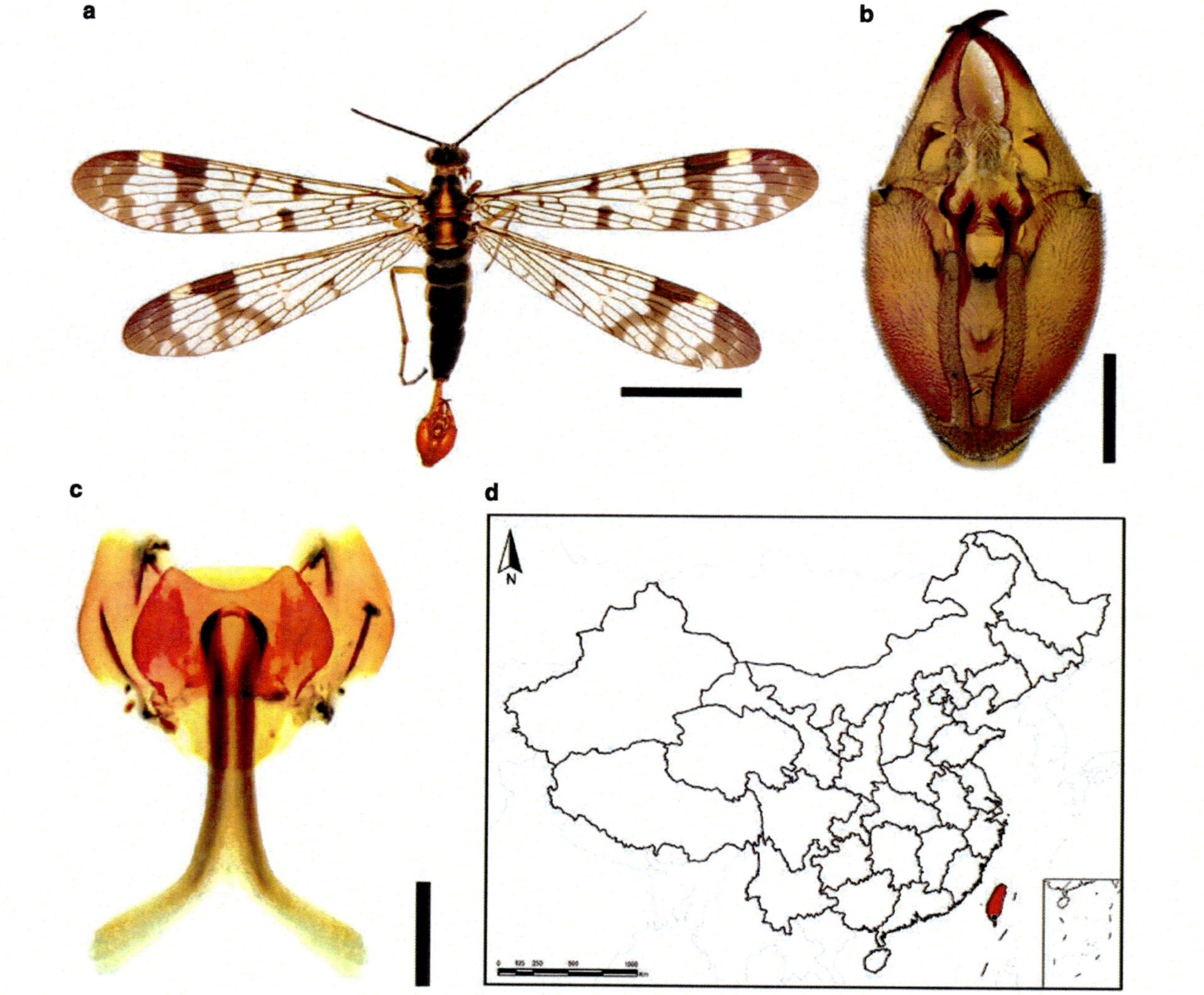

Fig. 10.182 (**a**) Male adult of ***Panorpa sonani***, dorsal view (Tahan Mountain, Pingtung, Taiwan) (*Scale bar*: 5.0 mm) [NWAU]. (**b**) Male genital bulb of ***Panorpa sonani***, ventral view (*Scale bar*: 0.5 mm). (**c**) Female medigynium of ***Panorpa sonani***, ventral view (*Scale bar*: 0.2 mm). (**d**) Distribution of ***Panorpa sonani***

Morphology. Vertex dark brown. Ocellar triangular black. Rostrum yellowish brown, with two darkened longitudinal stripes laterally. Pronotum dark brown. Meso- and metanotum dark brown, with conspicuous yellow median stripe. Pleura and legs yellowish brown. Wing membrane hyaline with markings brown. Pterostigma yellow and conspicuous. In forewings, apical band complete with large hyaline spot posteriorly, and series of hyaline spots along crossveins; pterostigmal band complete with narrow basal and apical branches; marginal spot slender; basal band split into two irregular spots; basal spot small and slightly elongated. Male T1–T5 dark brown. Notal organ flat and covering acute postnotal organ. A6 dark brown, conical. A7 and A8 yellowish brown, and constricted basally.

Male genital bulb long oval. Epandrium emarginates terminally in U-shape and forms pair of finger-like processes laterally. Hypandrium with greatly shortened basal stalk. Hypovalves slender, extending to distal 2/3 of gonocoxites, and bearing row of long bristles on inner margin. Gonostylus with triangular median tooth and broad triangular basal process. Gonostylus also with large concavity on ventral surface subbasally, and acute tooth directed mesad on outer border of concavity. Paramere bifurcated subapically and wrench-like, bearing long comb-like spines on apical half. Aedeagus flat and not exceeding apex of gonocoxites.

Female subgenital plate subtriangular with membranous, concaved region in basal portion. Medigynium with broad main plate; posterior arms nearly reduced; axis longer than three times length of main plate, with apodemes widely diverged.

Emergence. October to November.

Distribution. Taiwan (Pingtung).

10.6.64 *Panorpa subambra* Chou & Tong, 1987

Measurements. Forewing length 12.5–13.0 mm (Fig. 10.183).

Morphology. Vertex yellowish brown. Ocellar triangle black. Rostrum yellowish brown. Nota dark brown. Pleura and legs light yellowish brown. Wing membrane translucent, tinged with waxy yellow. Pterostigma reddish brown. Markings faint yellowish brown. In forewings, pterostigmal band lacking apical branch; other markings inconspicuous.

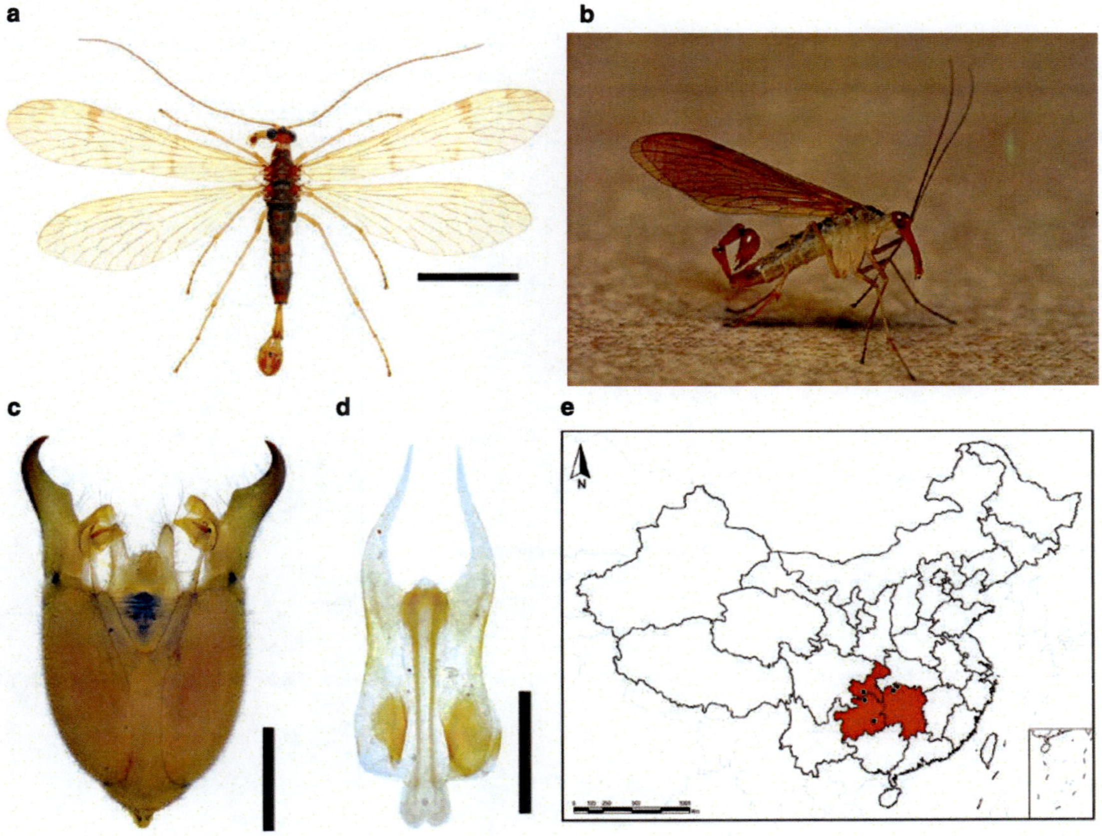

Fig. 10.183 (**a**) Male adult of ***Panorpa subambra***, dorsal view (Huping Mountain, Shimen, Hunan) (*Scale bar*: 5.0 mm) **[NWAU]**. (**b**) Male adult of ***Panorpa subambra*** in the field (Mount Jinfo, Nanchuan, Chongqing). (**c**) Male genital bulb of ***Panorpa subambra***, ventral view (*Scale bar*: 0.5 mm). (**d**) Female medigynium of ***Panorpa subambra***, ventral view (*Scale bar*: 0.2 mm). (**e**) Distribution of ***Panorpa subambra***

Hindwing devoid of markings and distinctly expanded outward from endings of M_3 to CuA, and slightly acute on apex. Male T1–T5 dark brown. Notal organ flat, and covering acute postnotal organ. A6 dark brown, conical and without anal horns. A7 and A8 yellowish brown and constricted basally.

Male genital bulb oval. Epandrium emarginates terminally in deep U-shape and forms pair of finger-like processes laterally; hypandrium Y-shaped, with long narrow basal stalk, and hypovalves extending to distal three-fourths of gonocoxites. Gonocoxites with semicircular plate on ventro-inner margin. Gonostylus concaved on outer margin, and bearing obtuse triangular median tooth and trumpet-shaped basal process. Paramere thread-like, extending beyond basal process of gonocoxites. Aedeagus very short, with membranous ventral valves and sclerotized and divergent dorsal valves.

Female subgenital plate oval with rounded apex. Medigynium with broad main plate slightly incurved along outer margin; posterior arms slender; axis slightly longer than main plate with parallel apodemes.

Emergence. July to August.

Distribution. Chongqing (Mount Jinfo), Guizhou (Kuankuoshui, Leigong Mountain), Hunan (Sangzhi, Shimen).

10.6.65 *Panorpa taiwanensis* Issiki, 1929

(Taiwan Scorpionfly) [台湾蝎蛉] (Fig. 10.184)

Measurements. Forewing length 13.0 mm.

Morphology. Vertex light yellow. Ocellar triangle black. Broad black transverse stripe passing through ocellar triangle and connected to inner borders of compound eyes. Rostrum yellowish brown. Pronotum dark brown. Meso- and metanotum dark brown, with broad light yellow median stripe. Pleura and legs light yellow. Wing membrane hyaline. Markings faint brown. In forewings, apical band scattered into several small spots; pterostigmal band broad near pterostigma, narrowed backward, and lacking apical branch; basal band reduced into irregular spot at CuP; other markings absent. Male T1–T5 dark brown. Notal organ flat and covering acute postnotal organ. A6 mostly dark brown and yellowish brown on apex, conical. A7 and A8 yellowish brown and greatly stalked basally.

Male genital bulb oval. Epandrium emarginates terminally in deep U-shape and forms pair of finger-like processes laterally. Hypandrium with greatly shortened basal stalk. Hypovalves long stripe-like, extending almost to apex of gonocoxites, and bearing row of long bristles along inner margin. Gonostylus with obtuse triangular median tooth and large disc-like basal process. Paramere finger-like and bifurcated subapically with acute apexes, and bearing row of comb-like spines along inner margin. Aedeagus with ventral valves short, dorsal valves broad and divergent.

Female subgenital plate oval with tongue-shaped apex. Medigynium with broad main plate; posterior arms absent; axis longer than main plate with widely diverged apodemes.

Emergence. June to August.

Distribution. Taiwan (Ali Mountain, Hualien, Nantou).

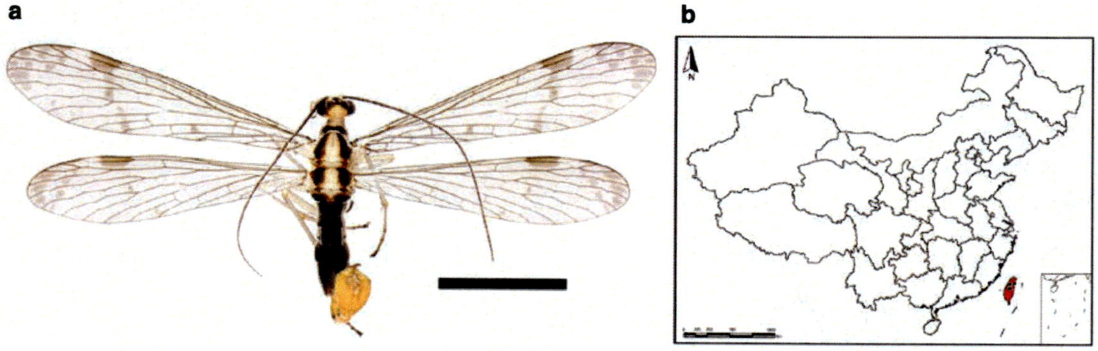

Fig. 10.184 (a) Male adult of ***Panorpa taiwanensis***, dorsal view (Taroko National Park, Hualien, Taiwan) (*Scale bar*: 5.0 mm) **[ECAU]**. **(b)** Distribution of ***Panorpa taiwanensis***

10.6.66 *Panorpa tetrazonia* Navás, 1935

(Four-Banded Lowland Scorpionfly) [四段蝎蛉]
(Fig. 10.185)

Measurements. Forewing length 11.5–12.5 mm.

Morphology. Vertex yellow. Large black pattern enclosing ocelli, and extending posteriorly, forming thin black margin aside yellow occiput. Rostrum yellowish brown and deepened toward apex. Pronotum unevenly dark brown. Meso- and metanotum dark brown anterolaterally, with broad yellow mesal stripe. Pleura and legs yellowish brown. Wing membrane hyaline, tinged with yellow; markings black; pterostigma indistinct. In forewings, apical band broad, with large hyaline window in posterior portion; pterostigmal band with basal branch two times as wide as apical branch; marginal spot thick; basal band broad; basal spot large. Male T1–T5 dark brown. Notal organ broad, very short, and covering acute postnotal organ. A6–A8 reddish brown. A6 subcylindrical, slightly tapering from middle toward abruptly beveled apex. A7 greatly constricted stalk-like basally, and greatly enlarged towards truncated apex. A8 similar to A7 but less constricted basally.

Male genital bulb reddish brown, oval. Epandrium with wide V-shaped emargination terminally and forming pair of stout processes laterally. Hypovalves extending to middle of gonocoxites, and each bearing row of long setae on inner margin of apical half. Gonostylus longer than half length of gonocoxites, with prominent middle tooth and stout basal process. Paramere bifurcated: ventral branch short and stout, curved mesally; dorsal branch long and slender; both branches bearing numerous long spines along posterior margin. Ventral aedeagal valves membranous and inconspicuous; dorsal process broad basally, slender and curved divergently at distal portion.

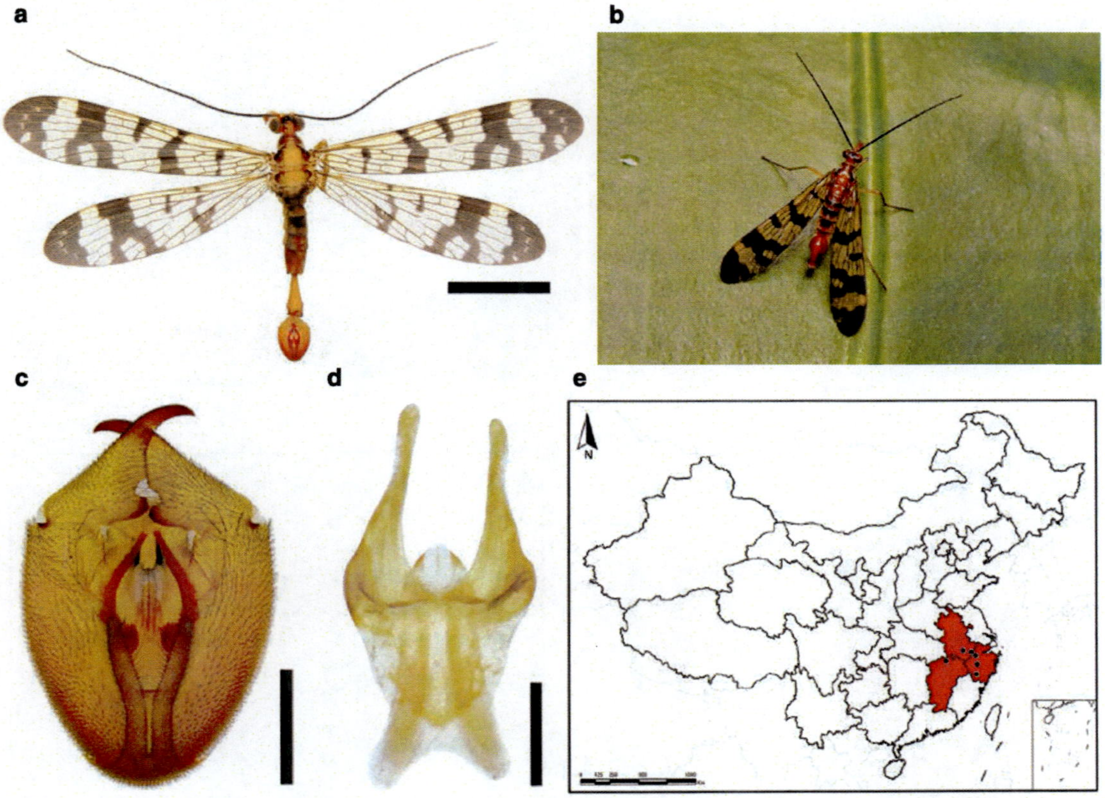

Fig. 10.185 (**a**) Male adult of ***Panorpa tetrazonia*** (Jinhua, Zhejiang) (*Scale bar*: 5.0 mm) [**NWAU**]. (**b**) Male adult of ***Panorpa tetrazonia*** in the field (Jinhua, Zhejiang). (**c**) Male genital bulb of ***Panorpa tetrazonia***, ventral view (*Scale bar*: 0.5 mm). (**d**) Female medigynium of ***Panorpa tetrazonia***, ventral view (*Scale bar*: 0.2 mm). (**e**) Distribution of ***Panorpa tetrazonia***

Female subgenital plate long oval with subacute apex. Mediygnium with broad, triangular main plate; posterior arms slender, nearly as long as main plate; axis longer than main plate with diverged apodemes.

Emergence. September to October.

Distribution. Anhui (Huangshan Mountain, Taiping), Jiangxi (Lushan Mountain), Zhejiang (Jinhua, Lishui, Tianmu Mountains, Tonglu, Qingliang Mountain).

Panorpa cornigera Group

This group of species can be recognized by a single anal horn in the male A6, an aedeagal hamulus in the male genitalia, and the female medigyniu bearing a pair of lateral plates.

Five species are known from northeastern Asia, including northeastern China, Japan, Russian Far East, North and South Korea. Only one species, ***Panorpa cornigera*** MacLachlan, 1887 has been recorded from China.

10.6.67 *Panorpa cornigera* MacLachlan, 1887

(Horned Scorpionfly) [具角蝎蛉] (Fig. 10.186)

Panorpa acutipennis Hua, 1998, **new synonym**.

Panorpa chonmasanus Li and Zhang in Li et al., 2012, **new synonym.**

Measurements. Forewing length 14.0–15.0 mm.

Morphology. Vertex dark brown. Two yellowish brown spots on occiput. Ocellar triangle black. Rostrum reddish brown but paler mesally. Nota black. Pleura yellowish brown to dark brown. Legs mostly yellowish brown. Wing membrane almost opaque and greyish with dark brown markings. Wings slightly acute apically and held roof-like over body at repose. Apical band in forewing broad and forming several separated small spots posteriorly; pterostigmal band with basal branch present and apical branch extremely reduced or absent; marginal spot small; basal band split into two spots; basal spot very small or absent. Male terga I–V dark brown.

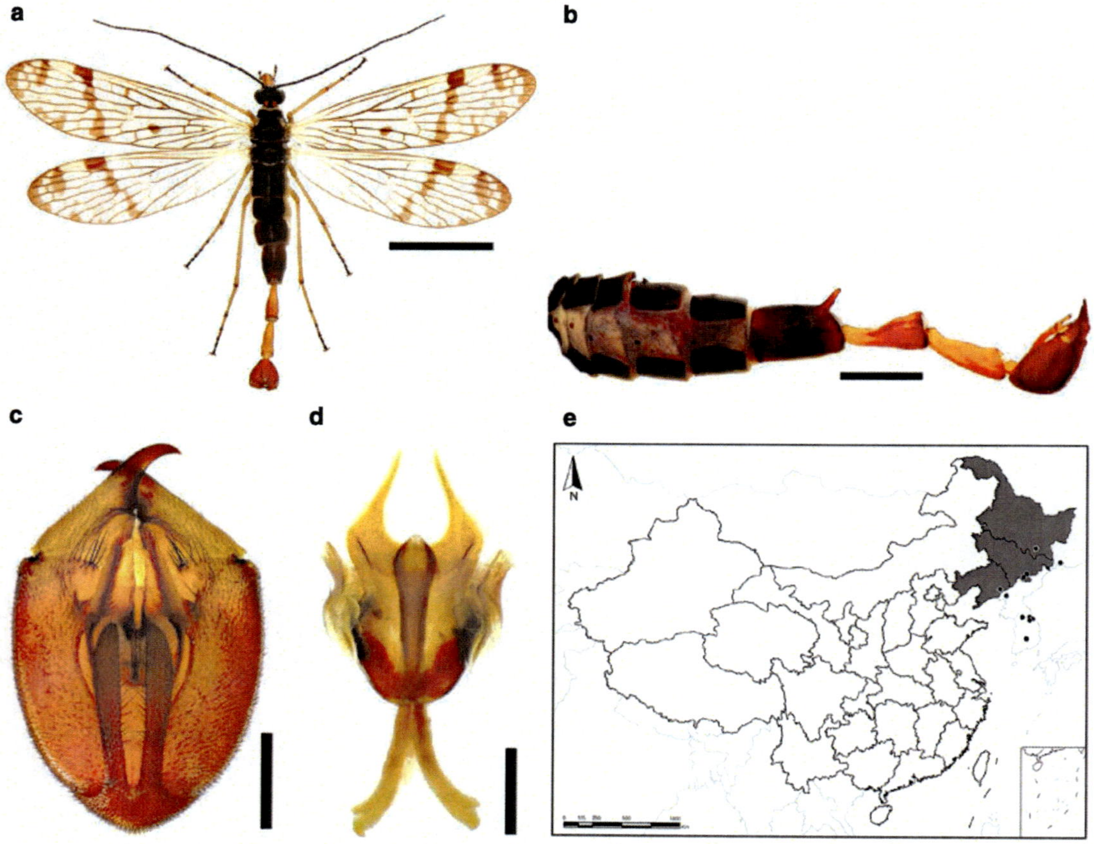

Fig. 10.186 (a) Male adult of ***Panorpa cornigera***, lateral view (Benxi, Liaoning) (*Scale bar*: 5.0 mm) [NWAU]. (b) Male abdomen of ***Panorpa cornigera***, lateral view (*Scale bar*: 2.0 mm). (c) Male genital bulb of ***Panorpa cornigera***, ventral view (*Scale bar*: 0.5 mm). (d) Female medigynium of ***Panorpa cornigera***, ventral view (*Scale bar*: 0.2 mm). (e) Distribution of ***Panorpa cornigera***

Notal organ flat and covering acute postnotal organ. A6 with single brown anal horn on dorsal apex. A7 and A8 constricted basally.

Male genital bulb broad oval. Epandrium emarginate apically and forming pair of finger-like processes laterally. Hypovalves stripe-like and nearly 2/3 length of gonocoxites. Gonocoxites with several long bristles on ventral apex. Gonostylus shorter than gonocoxites and bearing indistinct median tooth, and large disc-like basal process. Paramere curved medially and straight apically, and bearing dense comb-like spines along inner margin. Aedeagus with hamulus basal to ventral valves; dorsal valves bifurcated apically and with stout lateral processes.

Female subgenital plate oval with subacute apex. Medigynium with well-developed main plate; lateral margins of main plate incurved in distal 1/3; posterior arms slender and convergent; axis nearly twice as long as main plate; apodemes greatly diverged and shortly split into two or three branches basally.

Emergence. May to August.

Distribution. Heilongjiang (Mudanjiang), Jilin (Changbai Mountains), Liaoning (Kuandian, Laotudingzi); North & South Korea; Russian Far East (Vladivostok).

Remarks. *Panorpa acutipennis* Hua, 1998 was described from the Changbai Mountains, Northeast China, approximately 340 km southeast from the type locality of *P. cornigera* MacLachlan, Vladivostok, Russian Far East. The type series of *P. acutipennis* are identical to *P. cornigera* not only in the body morphology (Fig. 10.187 vs. Esben-Petersen 1921: figs 61 & 62; Martynova 1957: figs 13 & 25), but also the male and female genitalia, especially the aedeagal hamulus and the bifurcated dorsal aedeagal process (Issiki 1933: fig. 16A; Esben-Petersen 1921: fig. 64; Martynova 1957: fig. 81). In fact, the distribution of *P. cornigera* was reported from northern China by Martynova (1957), but Hua (1998) failed to refer to this record. Therefore, *P. acutipennis* is listed as a junior synonym of *P. cornigera*.

Panorpa chonmasanus Li & Zhang, 2012 was described from Chonmasan, North Pyongan, North Korea (Li et al. 2012), disregarding the fact that *P. cornigera* was previously recorded from the Korean Peninsula by Okamoto (1925) and Martynova (1957). Although a resemblance to *P. acutipennis* was mentioned, Li et al. (2012) did not provide any differentiative characters to separate *P. chonmasanus* from *P. acutipennis*. Therefore, *P. chonmasanus* is also listed as a junior synonym of *P. cornigera*.

Two previously synonymized species, *P. galloisi* Miyaké and *P. glloisi* Navás were described from specimens collected by Gallois from Chuzenji, Nikko, Japan in the adjacent dates (July 27 and July 19, respectively, 1910), and are evidently conspecific (Penny and Byers 1979). According to Miyaké (1911), *P. galloisi* Miyaké differs from the typical *P. cornigera* in the greatly reduced wing markings (cf. more extensive). In addition, early records of *P. cornigera* from Japan (MacLachlan 1887; Miyaké 1908, 1910, 1911, 1913) are likely based on misidentifications, and no further records of *P. cornigera* have been made from Japan since then (Miyamoto 1985; Miyamoto and Nakamura 2008). In our opinion, *P. cornigera* likely confines in the Northeastern Asian mainland and Korean Peninsula, while the Japanese *P. galloisi* Miyaké might represent a distinct species. More material of the nominal *P. galloisi* Miyaké from Nikko, Japan is needed to specify its puzzling status.

Panorpa davidi Group

This group of species can be recognized by the medium body size, the conical male A6, the elongated and basally constricted male A7 and A8, the greatly shortened basal stalk of hypandrium, the usually crossed parameres, and the stout and elongated female axis, which usually bears basally branched apodemes.

Twenty-one species are known from central, southern, and southwestern China. Twenty of them are included in this atlas.

Key to Species of the Panorpa davidi Group in This Atlas
(20 species, males.)

1. Parameres bifurcated .... ***Panorpa biclada*** Zhang & Hua
– Parameres unfurcated 2
2. Forewing with 1A ending far before level of ORs .... ***Panorpa bashanicola*** Hua, Tao & Hua
– Forewing with 1A ending at level of ORs or beyond ORs 3
3. Forewing with 1A ending at level of ORs 4
– Forewing with 1A ending far beyond ORs 5

4. Parameres stout, extending nearly to apex of gonocoxites .... ***Panorpa fulvastra*** Chou
– Parameres slender, intensively curved, and not reaching apex of gonocoxites .... ***Panorpa chengi*** Chou
5. Nota entirely dark brown to black 6
– Nota yellowish brown, or at least with yellow median stripe 9
6. Wing markings speckled .... ***Panorpa fructa*** Cheng
– Wing markings greatly reduced 7
7. Occiput yellowish brown .... ***Panorpa diqingensis*** Li & Hua
– Occiput black 8
8. Gonocoxites bearing several long, curved setae on inner side of ventral apex .... ***Panorpa dispergens*** Li & Hua
– Gonocoxites bearing shorter, straight setae on inner side of ventral apex .... ***Panorpa curvata*** Zhou
9. Wing markings greatly reduced; nota yellowish brown with pale median stripe .... ***Panorpa gaokaii*** Li & Hua
– Wing markings well-developed; nota dark brown to black with yellow median stripe 10
10. Parameres greatly curved, hook-like on apical half, reaching two-thirds of gonocoxites 11
– Parameres thick, greatly twisted in S-shape, reaching, or exceeding apex of gonocoxites 12
11. Meso- and metanotum with extremely broad yellow mesal stripe; basal spot large and conspicuous. .... ***Panorpa yangi*** Chou, 1981
– Meso- and metanotum with narrow yellow mesal stripe; basal spots extremely reduced and faint .... ***Panorpa uncinata*** Li & Hua
12. Gonocoxites with process on inner margin of ventral apex 13
– Gonocoxites without process on inner margin of ventral apex 15
13. Gonocoxites with normal triangular process on inner apex, bearing 1 or 2 long setae on inner margin subapically .... ***Panorpa qinlingensis*** Chou & Ran
– Gonocoxites with elongated process on inner apex, bearing short setae 14
14. Forewings with basal band split into two large spots; basal spot reduced .... ***Panorpa davidi*** Navás
– Forewings with basal band complete; basal spot absent .... ***Panorpa curva*** Carpenter, 1938
15. Forewing with basal band reduced into middle spot, extending from M to anal margin .... ***Panorpa yaoluopingensis*** Li & Hua
– Forewing with broad basal band complete or split into two spots 16
16. Dorsal valves of aedeagus slender and stalked, each with intensively enlarged circular dorsal process on apical portion .... ***Panorpa difficilis*** Carpenter
– Dorsal process of aedeagus not extending to apical portion of dorsal valves 17
17. Forewing with basal band split into two spots 18
– Forewing with basal band complete 19
18. Hypovalves slightly broadened towards rounded apex .... ***Panorpa typicoides*** Cheng
– Hypovalves with apical third widest, then tapered towards apex .... ***Panorpa jinchuana*** Hua, Sun & Li
19. Forewings without basal spot .... ***Panorpa sexspinosa*** Cheng
– Forewings with basal spot .... ***Panorpa neospinosa*** Chou & Wang

10.6.68 *Panorpa bashanicola* Hua, Tao & Hua, 2018

(Bashan Scorpionfly) [巴山蝎蛉]

Measurements. Forewing length 11–12.5 mm.

Morphology. Vertex mostly dark brown. Irregular yellowish brown patterns at occiput. Ocellar triangle black. Nota light yellowish brown. Wing membrane hyaline, devoid of markings. In forewings, 1A ending approximal to level of ORs. Male T1–T5 black with light yellowish brown stripe extending from T1 to T2. Notal organ flat and covering acute postnotal organ. A6 black, slightly tapering towards apex. A7 and A8 yellowish brown and constricted basally.

Male genital bulb long oval, yellowish brown. Epandrium emarginates terminally in deep U-shape and forms pair finger-like processes laterally. Hypandrium with greatly shortened basal stalk. Hypovalves stripe-like, extending beyond

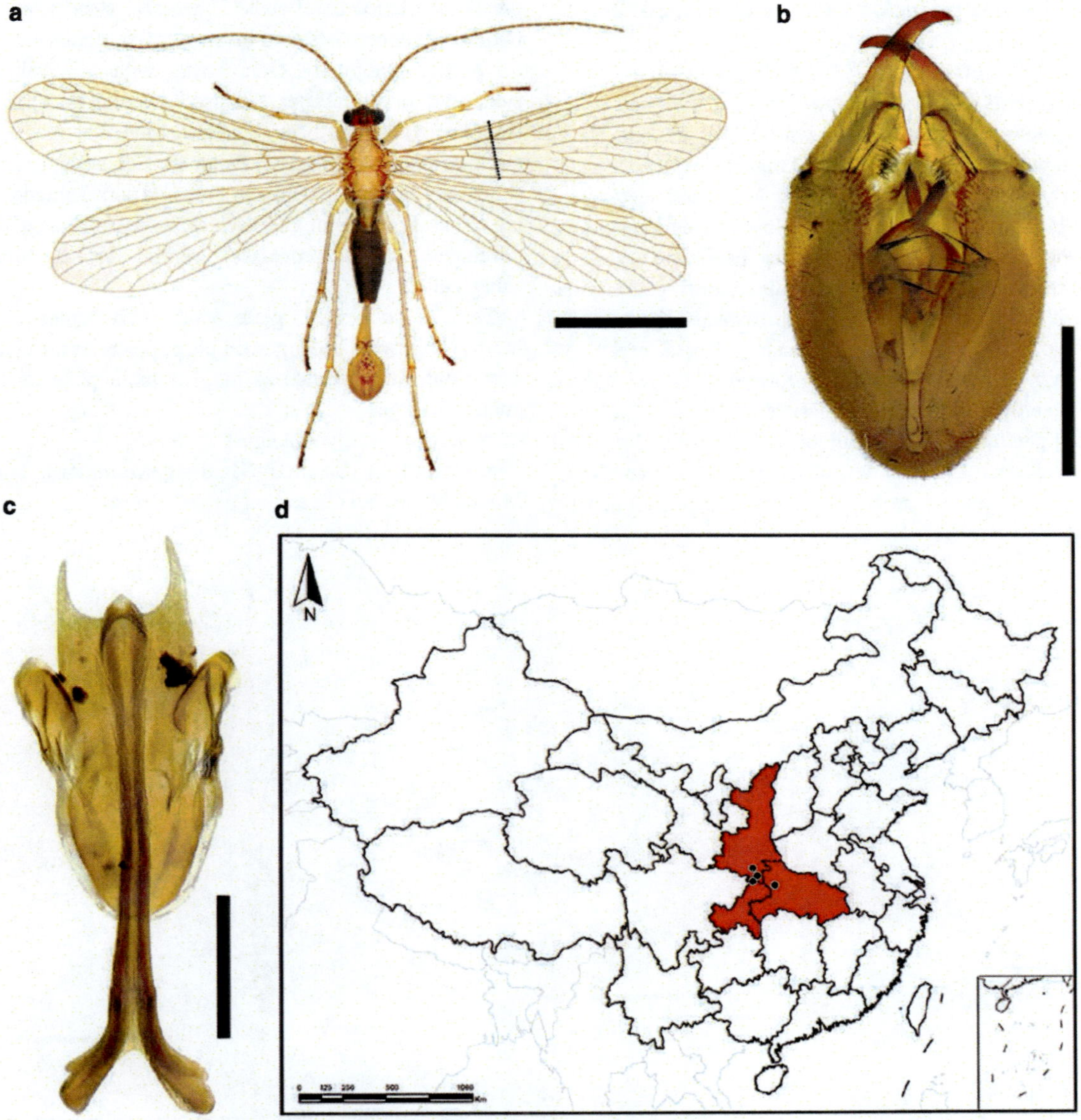

Fig. 10.187 (**a**) Male adult of ***Panorpa bashanicola***, dorsal view (Nangong Mountain, Lan'gao, Shaanxi) (*Scale bar*: 5.0 mm) **[NWAU]**. (**b**) Male genital bulb of ***Panorpa bashanicola***, ventral view (*Scale bar*: 0.5 mm). (**c**) Female medigynium of ***Panorpa bashanicola***, ventral view (*Scale bar*: 0.2 mm). (**d**) Distribution of ***Panorpa bashanicola***

midlength of gonocoxites, bearing several long bristles on apical portion of inner margin. Gonostylus with obtuse triangular median tooth and large disc-like basal process; basal process with triangular and acute inner margin. Parameres spiral, crossed, and bearing row of short microtirchia on inner margin. Aedeagus with short membranous ventral valves; dorsal valves highly sclerotized, finger-like and enlarged on dorsal apex.

Female subgenital oval with subacute apex. Mediygnium with broad main plate; pair of ear-like lateral plates; posterior arms short and acute; axis slightly shorter than twice length of main plate; apodemes greatly diverged basally.

Emergence. June to August.

Distribution. Shaanxi (Lan'gao).

10.6.69 *Panorpa biclada* Zhang & Hua, 2012

(Bicladed Scorpionfly) [双叉蝎蛉] (Fig. 10.188)
Measurements. Forewing length 11.5–12.1 mm.

Morphology. Vertex dark brown. Ocellar triangle black. Rostrum yellowish brown, with two broad deep brown longitudinal stripes laterally. Pronotum black. Meso- and metanotum mostly black, with indistinct light yellow median longitudinal stripe. Pleura and legs light yellow. Wing membrane hyaline. In males, markings usually absent; while distinct pterostigmal band without apical branch present in females. Male T1–T5 black. Notal organ flat, covering acute postnotal organ. A6 black, long cylindrical and tapering towards apex. A7 and A8 yellowish brown and constricted basally.

Male genital bulb long elliptical. Epandrium long and broad, with deep U-shaped terminal emargination, and forming pair of finger-like processes laterally. Hypandrium with greatly shortened basal stalk. Hypovalve slender, exceeding middle of gonocoxites, and bearing sparse stout setae on apical portion marginally. Gonocoxites with 4–5 well-aligned stout setae on ventral apex. Gonostylus shorter than half length of gonocoxites, and bearing obtuse triangular median tooth and disc-like basal process. Paramere slender and bifurcated subbasally, with ventral branch short and dorsal branch extending to basal process of gonostylus. Aedeagus short with well-developed lateral processes; ventral valves bifurcated subapically.

Female subgenital plate oval with subacute apex. Medigynium with oblong main plate; posterior arms slender; axis stout, nearly twice as long as main plate; apodemes greatly diverged.

Emergence. July to August.

Distribution. Shaanxi (Hualong Mountains, Nangong Mountain), Hubei (Shennongjia).

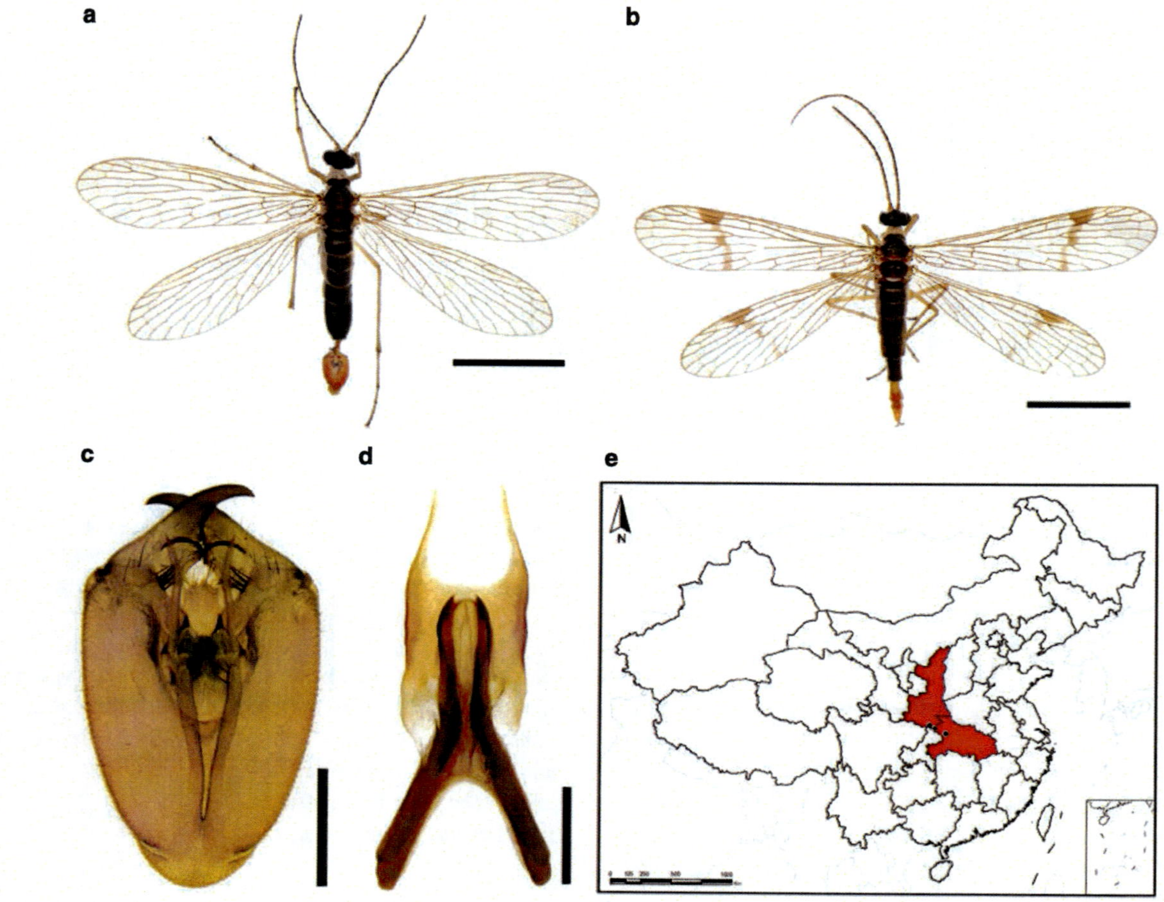

Fig. 10.188 (**a**) Male adult of ***Panorpa biclada***, dorsal view (Nangong Mountain, Lan'gao, Shaanxi) (*Scale bar*: 5.0 mm) [NWAU]. (**b**) Female adult of ***Panorpa biclada***, dorsal view (Nangong Mountain, Lan'gao, Shaanxi). (**c**) Male genital bulb of ***Panorpa biclada***, ventral view (*Scale bar*: 0.5 mm). (**d**) Female medigynium of ***Panorpa biclada***, ventral view (*Scale bar*: 0.2 mm). (**e**) Distribution of ***Panorpa biclada***

10.6.70 *Panorpa chengi* Chou, 1981

(Cheng's Scorpionfly) [郑氏蝎蛉] (Fig. 10.189)
Measurements. Forewing length 11.5–12.0 mm.

Morphology. Vertex yellow. Ocellar triangle black. Brown transversal stripe passing through ocellar triangle. Rostrum reddish brown. Pronotum light yellowish brown. Meso- and metanotum light creamy yellow and darkened on two sides. Pleura creamy yellow. Legs yellowish brown. Wing membrane hyaline, slightly tinged with yellow, devoid of markings or only slightly darkened on apex. Male T1–T5 creamy yellow and darkened laterally. Notal organ flat and covering acute postnotal organ. A6 dark brown, cylindrical, tapering towards apex. A7 and A8 yellowish brown and constricted basally.

Male genital bulb oval. Epandrium emarginated terminally in deep U-shape and forming pair of finger-like processes laterally. Hypandrium with shortened basal stalk. Hypovalves long stripe-like, bearing row of long bristles on inner margin. Gonostylus with broad row of sparse stout setae, and obtuse triangular median tooth and bowl-shaped basal process. Parameres slender, curved, and crossed mesally. Aedeagus short; dorsal aedeagal valves foliate and divergent.

Female subgenital plate oval with subacute apex. Medigynium with oblong main plate; posterior arms slender;

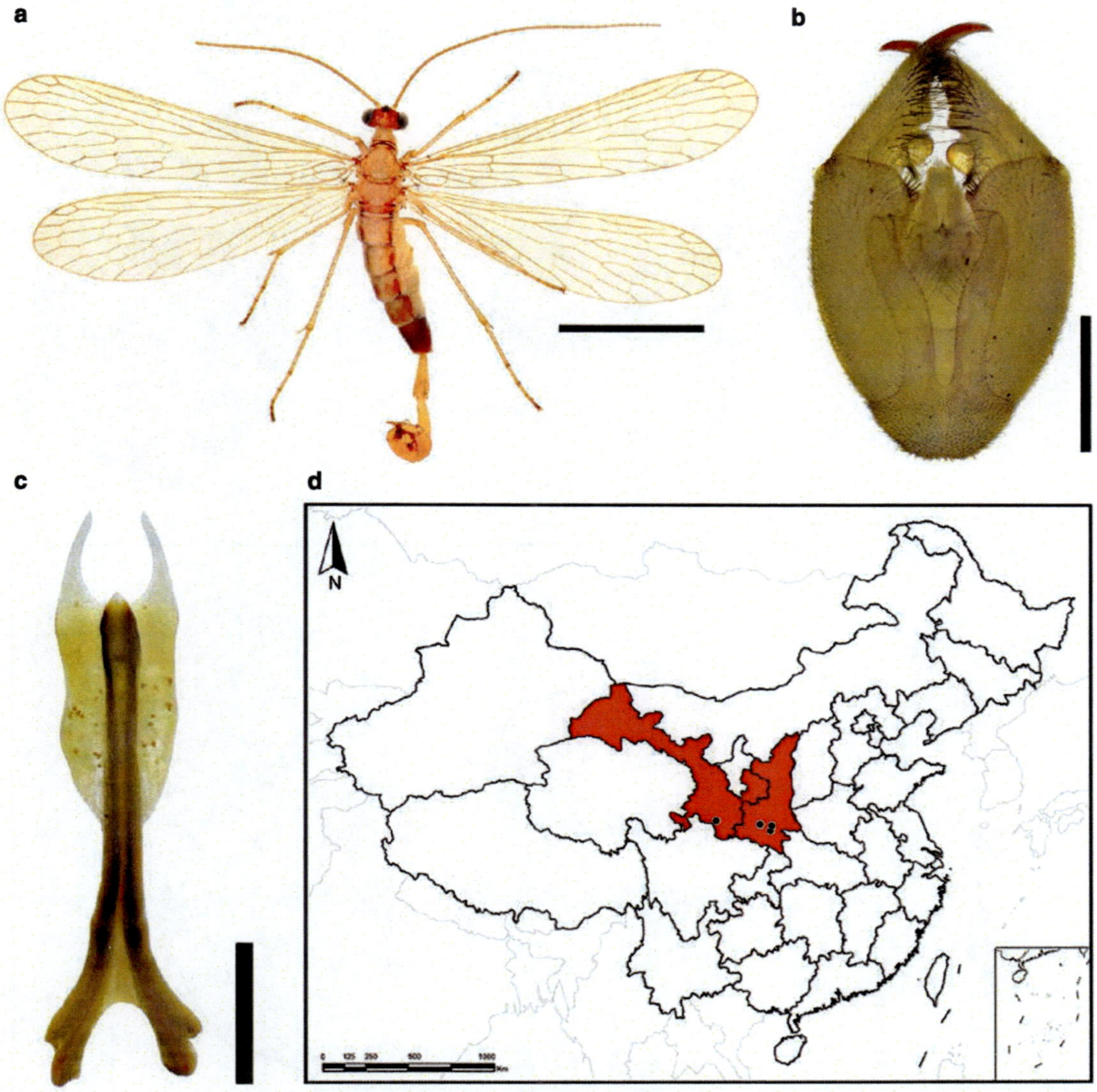

Fig. 10.189 (a) Male adult of ***Panorpa cheng***, dorsal view (Huoditang, Ningshan, Shaanxi) (*Scale bar*: 5.0 mm) **[NWAU]**. **(b)** Male genital bulb of ***Panorpa chengi***, ventral view (*Scale bar*: 0.5 mm). **(c)** Female medigynium of ***Panorpa chengi***, ventral view (*Scale bar*: 0.2 mm). **(d)** Distribution of ***Panorpa chengi***

axis longer than twice length of main plate; apodemes diverged and branched basally.

Emergence. July to August.

Distribution. Gansu (Dangchang), Shaanxi (Huyi, Ningshan, Taibai).

10.6.71 *Panorpa curva* Carpenter, 1938

(Curved Scorpionfly) [弯曲蝎蛉] (Fig. 10.190)

Measurements. Forewing length 12.5–14.0 mm.

Morphology. Vertex reddish brown. Ocellar triangle black. Rostrum yellowish brown to reddish brown. Pronotum dark brown. Meso- and metanotum dark brown, with distinct yellow median stripe. Pleura and legs yellowish brown. Wing membrane hyaline with markings dark brown. Pterostigma reddish brown. In forewings, apical band with hyaline window posteriorly; pterostigmal band lacking apical branch; basal band complete; marginal and basal spots absent. Male T1–T5 dark brown. Notal organ flat and covering acute postnotal organ. A6 dark brown basally and yellowish brown distally, protruding dorsad and bearing long setae on apex. A7 and A8 reddish brown, and constricted basally.

Male genital bulb oval. Epandrium emarginates terminally in deep U-shape and forms pair finger-like processes laterally. Hypandrium with greatly shortened basal stalk. Hypovalves stripe-like, extending beyond midlength of gonocoxites, bearing row of long bristles on inner margin. Gonocoxites with elongated process bearing long dense

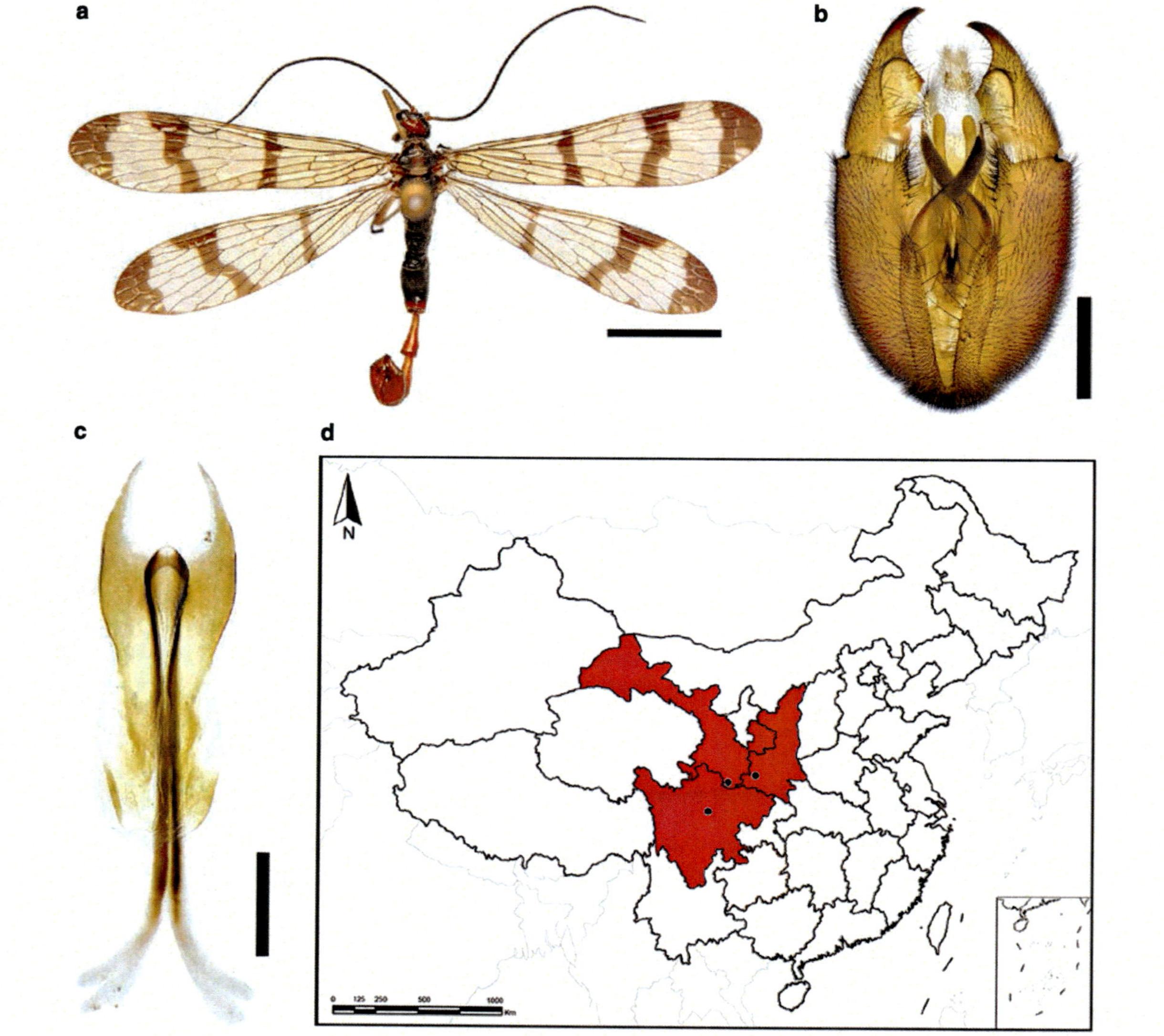

Fig. 10.190 (a) Male adult of ***Panorpa curva***, dorsal view (Wolong, Sichuan) (*Scale*: 5.0 mm) **[NWAU]**. **(b)** Male genital bulb of ***Panorpa curva***, ventral view (Scale bar: 0.5 mm). **(c)** Female medigynium of ***Panorpa curva***, ventral view (*Scale bar*: 0.2 mm). **(d)** Distribution of ***Panorpa curva***

setae on inner margin of ventral apex. Gonostylus with obtuse triangular median tooth and large disc-like basal process. Parameres spiral, crossed, and bearing row of short spines on inner margin. Aedeagus with short membranous ventral valves; dorsal valves highly sclerotized, finger-like and enlarged on dorsal apex.

Female subgenital plate oval with subacute apex. Medigynium with main plate gradually broadening from base to apex; posterior arms short and thin; axis slightly shorter than twice length of main plate with diverged apodemes.

Emergence. June to August.

Distribution. Gansu (Wenxian), Sichuan (Wolong), Shaanxi (Baoji).

10.6.72 *Panorpa curvata* Zhou, 2006

(Alpine Curvated Scorpionfly) [曲杆蝎蛉] (Fig. 10.191)
Panorpa stigmosa Zhou, 2006, **new synonym**.

Measurements. Forewing length 11.0–12.0 mm.

Morphology. Vertex and ocellar triangle black. Rostrum dark brown with yellowish brown genae. Nota black. Pleura and legs light yellow to yellowish brown. Wing membrane hyaline. Pterostigma light yellow. Markings brown and highly variable. In forewings, apical band scattered into series of small spots distributed in apical cells; pterostigmal band broad and complete anteriorly, with separated basal and apical branches posteriorly; basal band reduced into irregular spot between CuA and CuP; marginal spot represented by small rounded spot below R_{4+5}; basal spot absent. In some individuals, markings extremely reduced, with only anterior half of apical band present, or even entirely absent in some males. Male T1–T5 black. Notal organ flat and covering acute postnotal organ. A6 black and conical; A7 and A8 yellowish brown and constricted basally.

Male genital bulb oval. Epandrium emarginates terminally in deep U-shape and forms pair of finger-like processes laterally; hypandrium, with greatly shortened basal stalk. Hypovalve long stripe-like, extending to three-fourths of gonocoxite and slightly projected mesally on middle of inner margin, bearing row of long stout bristles along apical half of inner margin. Gonocoxites slightly projected innerly on ventral apex, with 3–5 long setae. Gonostylus with obtuse triangular median tooth and disc-like basal process; basal process with row of short setae subbasally. Parameres geniculate

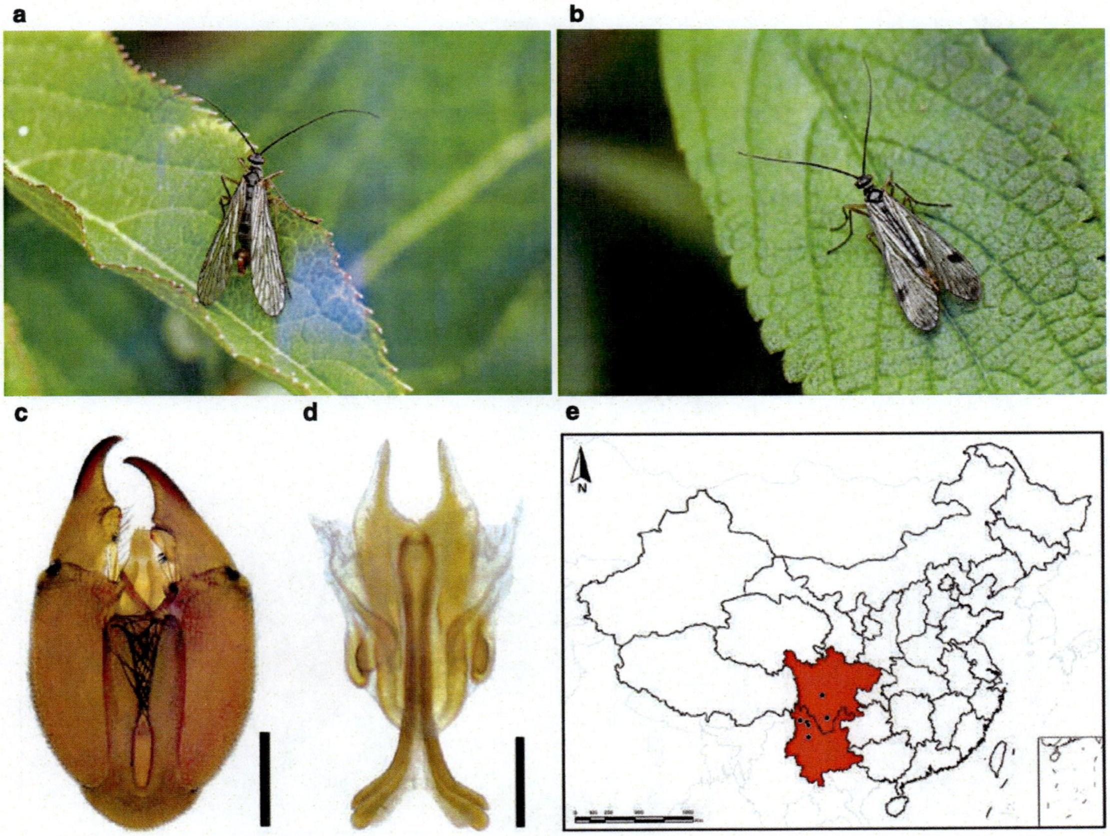

Fig. 10.191 (**a**) Male adult of ***Panorpa curvata*** in the field (Pantiange, Weixi, Yunnan). (**b**) Female adult of ***Panorpa curvata*** in the field (Pantiange, Weixi, Yunnan). (**c**) Male genital bulb of ***Panorpa curvata***, ventral view (*Scale bar*: 0.5 mm). (**d**) Female medigynium of ***Panorpa curvata***, ventral view (*Scale bar*: 0.2 mm). (**e**) Distribution of ***Panorpa curvata***

medially, with spiral apical halves curved mesad and crossed, and each bearing row of short spines on inner margin. Ventral aedeagal valves membranous and short; dorsal valves greatly curved dorsad and bearing membranous enlarged process on dorsal surface subapically.

Female subgenital plate oval with subacute apex. Medigynium with broad main plate; pair of lateral plates ear-like and enclosing main plate; posterior arms short and thin; axis longer than main plate with diverged apodemes.

Emergence. June to July.

Distribution. Sichuan (Luoji Mountain, Mount Gongga), Yunnan (Dali, Yulong Snow Mountain, Weixi).

Remarks. *Panorpa curvata* Zhou, 2006 and *Panorpa stigmosa* Zhou, 2006 were described from Chishui, Guizhou. However, they can only be found from northwestern Yunnan and southern Sichuan in recent decades, in alpine area with altitudes usually above 2500 m. However, in the type locality, the Chishui Tree Fern Nature Reserve, Guizhou, the highest altitude is approximately 1700 m, very unlikely to be suitable for these species. In addition, together with these two, Zhou (2006) also described *Neopanorpa cangshanensis* Zhou, 2006, which is synonymized with *Neopanorpa brisi* (Navás, 1930) in this study. Due to an unknown reason, Zhou's *N. cangshanensis* is likely named after the Cangshan Mountain, Dali, Yunnan, where *N. brisi* is certainly recorded from. Coincidently, the marking-less, male-based *P. curvata* and the multi-markings, female-based *P. stigmosa* were both found from Cangshan Mountain at the same time and localities. By observations in the field, we confirm that these two are conspecific, and misidentified by Zhou (2006) as two species due to the distinct, sexually dimorphic wing markings. Therefore, we herein treat *P. stigmosa* as a junior synonym of *P. curvata*. The holotype of *P. curvata* was probably collected from some alpine area in northwestern Yunnan or southern Sichuan, but erroneously reported from the Chishui Tree Fern Nature Reserve, Guizhou by Zhou (2006).

10.6.73 *Panorpa davidi* Navás, 1908

(Father David's Scorpionfly) [大卫蝎蛉] (Fig. 10.192)
Measurements. Forewing length 13.0 mm.

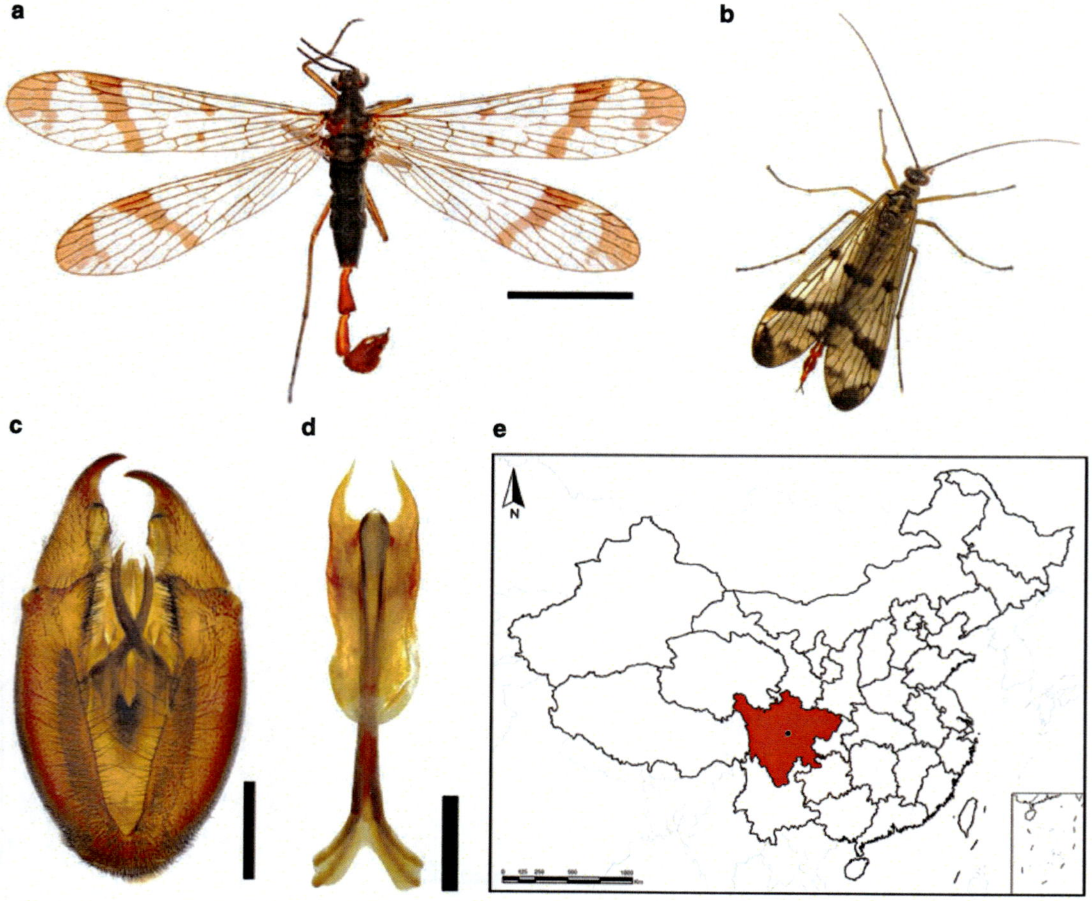

Fig. 10.192 (a) Male adult of ***Panorpa davidi***, dorsal view (Baoxing, Sichuan) (*Scale bar*: 5.0 mm) [NWAU]. (b) Female adult of ***Panorpa davidi***, dorsal view (Baoxing, Sichuan). (c) Male genital bulb of ***Panorpa davidi***, ventral view (*Scale bar*: 0.5 mm). (d) Female medigynium of ***Panorpa davidi***, ventral view (*Scale bar*: 0.2 mm). (e) Distribution of ***Panorpa davidi***

Morphology. Vertex reddish brown. Ocellar triangle black. Rostrum yellowish brown to reddish brown. Pronotum dark brown. Meso- and metanotum dark brown, with distinct yellow median stripe. Pleura and legs yellowish brown. Wing membrane hyaline with markings brown. Pterostigma pale brown. In forewings, apical band with hyaline window posteriorly; pterostigmal band lacking apical branch; basal band split into two spots; marginal and basal spots absent. Male T1–T5 dark brown. Notal organ flat and covering acute postnotal organ. A6 dark brown basally and yellowish brown distally, protruding dorsad and bearing sparse long setae on apex. A7 and A8 reddish brown and constricted basally.

Male genital bulb oval. Epandrium emarginates terminally in deep U-shape and forms pair of finger-like processes laterally. Hypandrium with greatly shortened basal stalk. Hypovalves long stripe-like, extending beyond middle of gonocoxites, and bearing row of long bristles on inner margin. Gonocoxites with protruding process bearing row of stout setae on inner margin of ventral apex. Gonostylus with obtuse triangular median tooth and large disc-like basal process. Parameres slender, crossed, and bearing row of short spines on inner margin. Aedeagus with short membranous ventral valves; dorsal valves highly sclerotized, finger-like and enlarged on dorsal apex.

Female subgenital plate oval with subacute apex. Medigynium with oblong main plate; posterior arms short and slightly convergent; axis twice as long as main plate with greatly diverged apodemes.

Emergence. June to August.

Distribution. Sichuan (Baoxing, Jiajin Mountain).

10.6.74 *Panorpa difficilis* Carpenter, 1938

(Difficult Scorpionfly) [难蝎蛉] (Fig. 10.193)

Measurements. Forewing length 11.0–12.5 mm.

Morphology. Vertex yellowish brown. Ocellar triangle black. Broad black transverse stripe passing through ocellar triangle. Rostrum yellowish brown. Pronotum dark brown. Meso- and metanotum dark brown, with yellow median stripe. Pleura and legs yellowish brown. Wing membrane hyaline with markings brown. In forewings, apical band with two hyaline spots innerly; pterostigmal band broad, with

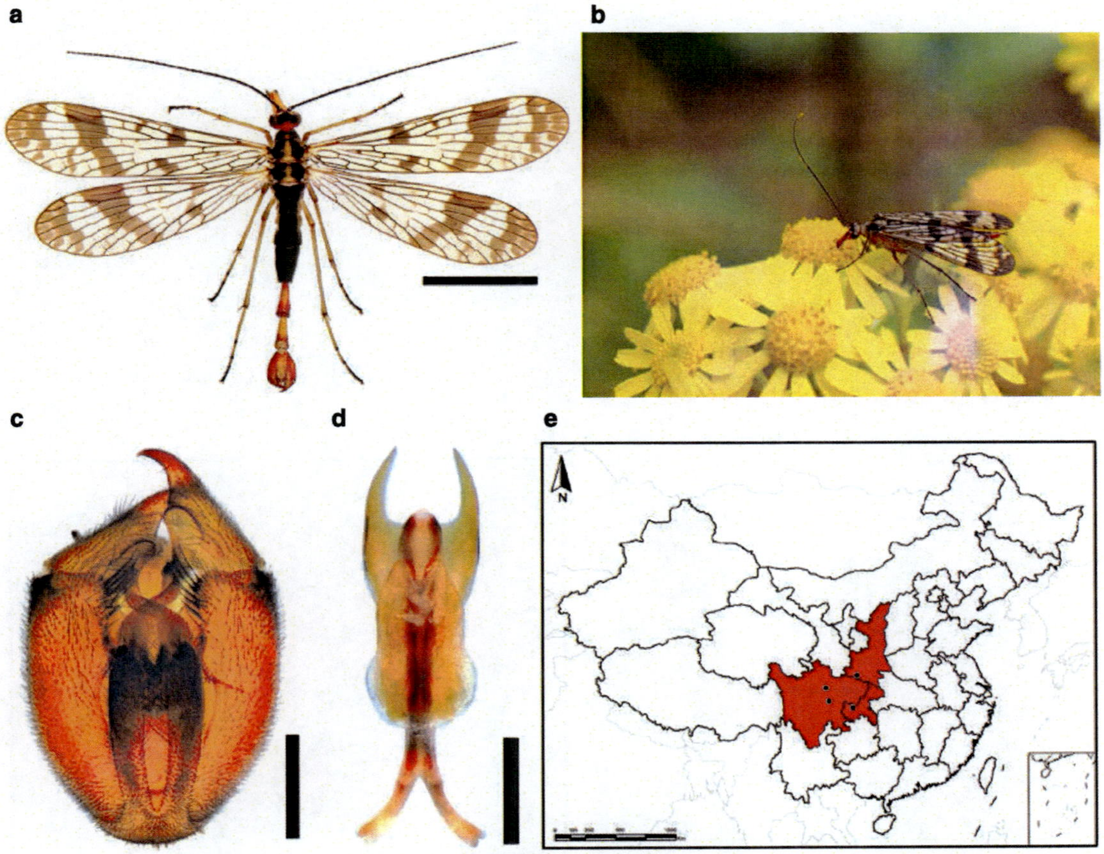

Fig. 10.193 (**a**) Male adult of ***Panopa difficilis***, dorsal view (Tudiling, Maoxian, Sichuan) (*Scale bar*: 5.0 mm) [NWAU]. (**b**) Female adult of ***Panorpa difficilis*** in the field (Xiema, Beibei, Chongqing, ©Hao Xu). (**c**) Male genital bulb of ***Panorpa difficilis***, ventral view (*Scale bar*: 0.5 mm). (**d**) Female medigynium of ***Panorpa difficilis***, ventral view (*Scale bar*: 0.2 mm). (**e**) Distribution of ***Panorpa difficilis***

separated apical branch; marginal spot fusiform; basal band split into two spots; basal spot very small. Male T1–T5 dark brown. Notal organ flat and covering acute postnotal organ. A6 dark brown and slightly projected dorsad on apex. A7 and A8 yellowish brown and constricted basally.

Male genital bulb broad oval. Epandrium emarginated terminally in U-shape and forming pair of finger-like processes laterally. Hypandrium with short basal stalk. Hypovalves long stripe-like, extending to base of gonostylus, and bearing row of long bristles on inner margin. Gonocoxites with tuft of long stout setae on ventral apex. Gonostylus with obtuse triangular median tooth and large disc-like basal process. Paramere spiral in distal half, and bearing row of short spines on inner margin. Aedeagus with ventral valves membranous and short, and dorsal valves finger-like and divergent, bearing membranous enlarged process on dorsal surface subapically.

Female subgenital plate oval with subacute apex. Medigynium with main plate subtrapezoidal in basal half and broader in distal half; posterior arms short and thin; axis slightly shorter than twice length of main plate with diverged apodemes.

Emergence. July to August.

Distribution. Sichuan (Kangding).

10.6.75 *Panorpa diqingensis* Li & Hua, 2020

(Diqing Scorpionfly) [迪庆蝎蛉] (Fig. 10.194)

Measurements. Forewing length 13.0–14.5 mm.

Morphology. Vertex dark brown, occiput yellowish brown. Ocellar triangle black. Rostrum yellowish brown. Pronotum dark brown. Meso- and metanotum dark brown. Pleura and legs yellowish brown. Wing membrane hyaline with markings brown. Male wings devoid of markings; female wings with greatly reduced pterostigmal band. Male T1–T5 black. Notal organ flat and covering acute postnotal organ. A6 dark brown. A7 and A8 yellowish brown and constricted basally.

Male genital bulb broad oval. Epandrium emarginated terminally in U-shape and forming pair of finger-like processes laterally. Hypandrium with short basal stalk. Hypovalves long stripe-like, extending to base of gonostylus, and bearing row of long bristles on inner margin. Gonocoxites with tuft of long stout setae on ventral apex. Gonostylus with obtuse triangular median tooth and large disc-like basal process. Paramere spiral in distal half, and bearing row of short spines on inner margin.

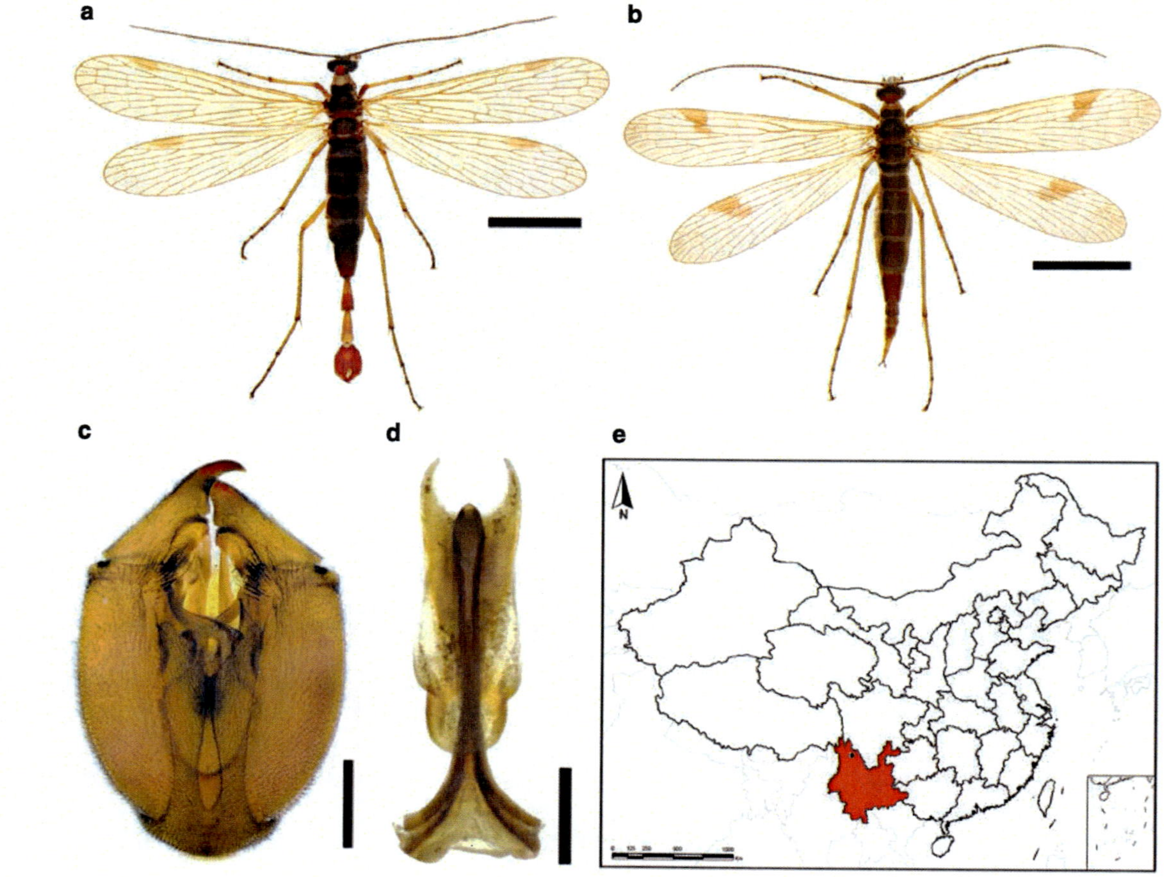

Fig. 10.194 (a) Male adult of ***Panorpa diqingensis***, dorsal view (Haba Snow Mountain, Diqing, Yunnan) (*Scale bar*: 5.0 mm) [NWAU]. (b) Female adult of ***Panorpa diqingensis***, dorsal view (Haba Snow Mountain, Diqing, Yunnan) (*Scale bar*: 5.0 mm) [NWAU]. (c) Male genital bulb of ***Panorpa diqingensis***, ventral view (*Scale bar*: 0.5 mm). (d) Female medigynium of ***Panorpa diqingensis***, ventral view (*Scale bar*: 0.2 mm). (e) Distribution of ***Panorpa diqingensis***

Aedeagus with ventral valves membranous and short, and dorsal valves finger-like and divergent, bearing membranous enlarged process on dorsal surface subapically.

Female subgenital plate oval with subacute apex. Medigynium with main plate oblong, greatly constricted in basal portion; posterior arms short and thin; axis slightly shorter than twice length of main plate with diverged apodemes.

Emergence. July to August.

Distribution. Yunnan (Diqing).

10.6.76 *Panorpa dispergens* Li & Hua, 2020

(Scattered Scorpionfly) [散斑蝎蛉] (Fig. 10.195)

Measurements. Forewing length 12.5–14.0 mm.

Morphology. Vertex dark brown. Ocellar triangle black. Rostrum yellowish brown. Pronotum dark brown. Meso- and metanotum dark brown. Pleura and legs yellowish brown. Wing membrane hyaline with markings brown. In forewings, apical and pterostigmal bands greatly reduced, other markings absent. Male T1–T5 black. Notal organ flat and cover-

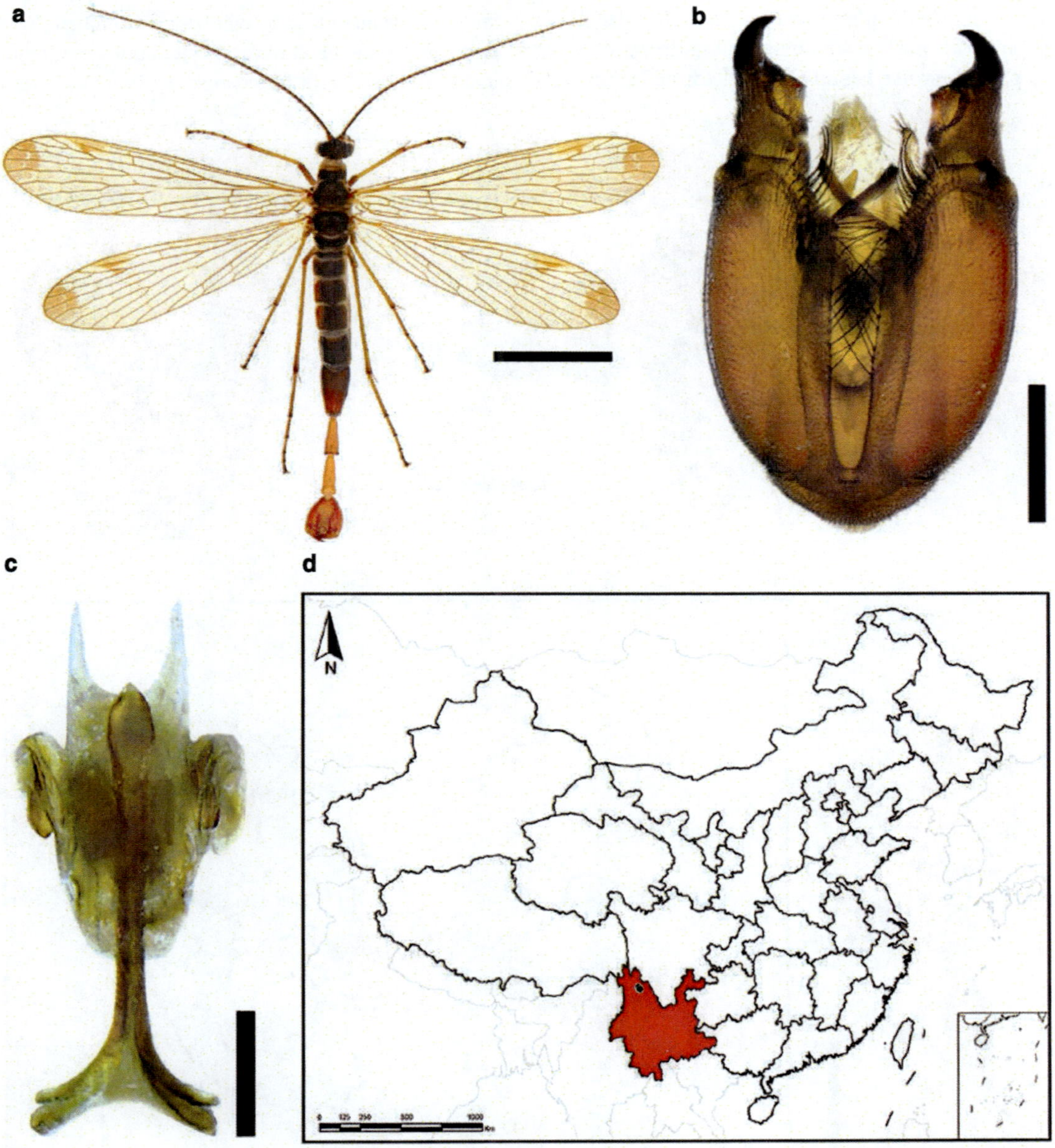

Fig. 10.195 (**a**) Male adult of ***Panorpa dispergens***, dorsal view (Haba Snow Mountain, Diqing, Yunnan) (*Scale bar*: 5.0 mm) **[NWAU]**. (**b**) Male genitalia of ***Panorpa dispergens***, ventral view (*Scale bar*: 0.5 mm). (**c**) Female medigynium of ***Panorpa dispergens***, ventral view (*Scale bar*: 0.2 mm). (**d**) Distribution of ***Panorpa dispergens***

ing acute postnotal organ. A6 dark brown. A7 and A8 yellowish brown and constricted basally.

Male genital bulb broad oval. Epandrium emarginated terminally in U-shape and forming pair of finger-like processes laterally. Hypandrium with short basal stalk. Hypovalves long stripe-like, extending to base of gonostylus, and bearing row of long bristles on inner margin. Gonocoxites with tuft of long stout setae on ventral apex. Gonostylus with obtuse triangular median tooth and large disc-like basal process. Paramere spiral in distal half, and bearing row of short spines on inner margin. Aedeagus with ventral valves membranous and short, and dorsal valves finger-like and divergent, bearing membranous enlarged process on dorsal surface subapically.

Female subgenital plate oval with subacute apex. Medigynium with main plate oblong; posterior arms short and thin; pair of ear-like lateral plates enclosing lateral margin of main plate; axis slightly shorter than twice length of main plate with diverged apodemes.

Emergence. July to August.

Distribution. Yunnan (Diqing).

10.6.77 *Panorpa fructa* Cheng, 1949

(Fruit Scorpionfly) [实蝎蛉] (Fig. 10.196)

Measurements. Forewing length 11.5 mm.

Morphology. Vertex dark brown. Ocellar triangle black. Rostrum reddish brown. Nota dark brown. Pleura and legs yellowish brown. Wing membrane hyaline with scattered brown markings. In forewings, apical band split into series of small faint spots; pterostigmal band incomplete, with separated basal branch and blurred apical branch; marginal spot large; basal band split into two spots; basal spot very small.

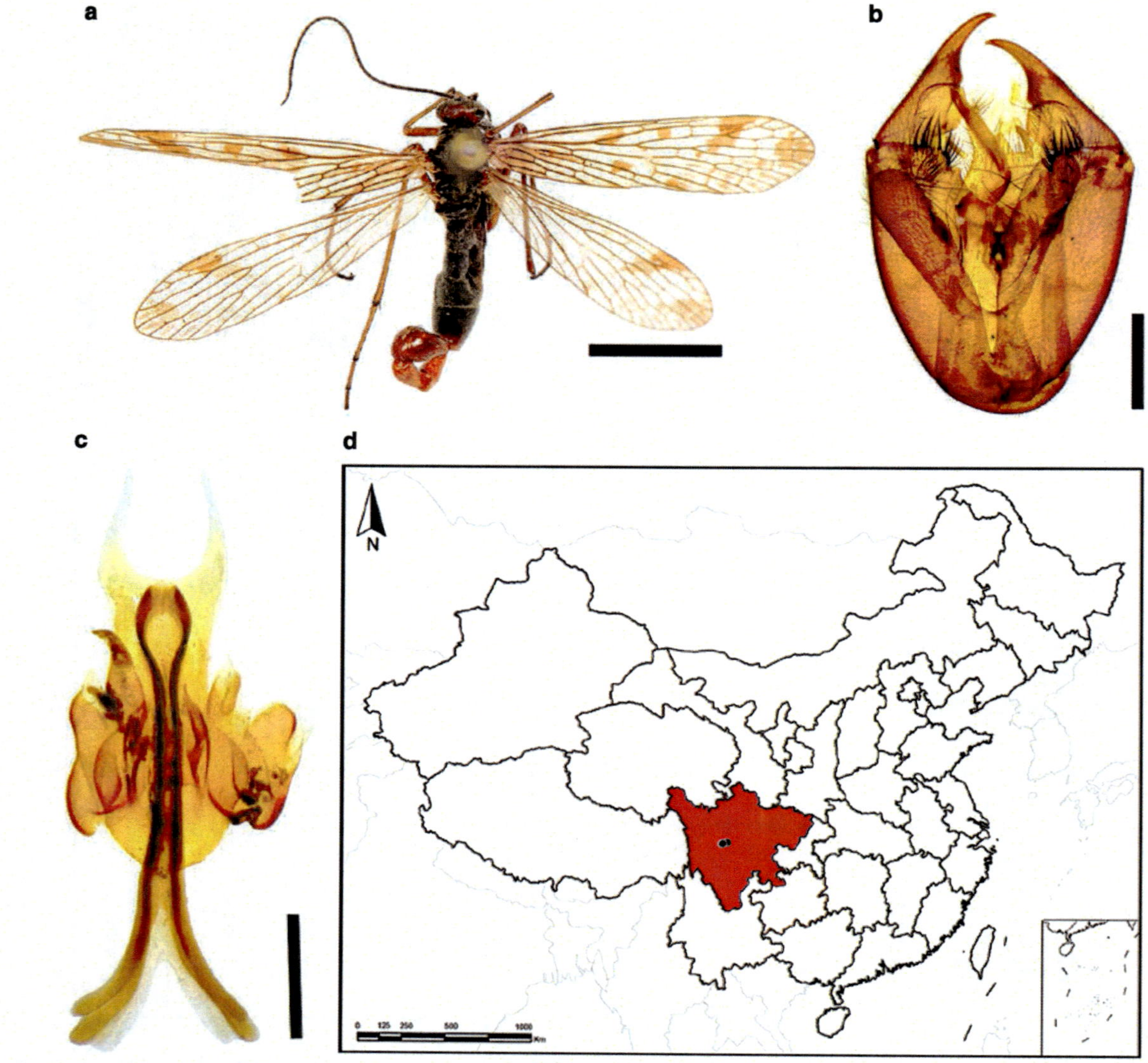

Fig. 10.196 (a) Male adult of ***Panorpa fructa***, dorsal view (Konka Temple, Kangding, Sichuan) (*Scale bar*: 5.0 mm) **[NWAU]**. **(b)** Male genital bulb of ***Panorpa fructa***, ventral view (*Scale bar*: 0.5 mm). **(c)** Female medigynium of ***Panorpa fructa***, ventral view (*Scale bar*: 0.2 mm). **(d)** Distribution of ***Panorpa fructa***

Male T1–T5 dark brown. Notal organ flat, covering acute postnotal organ. A6 conical. A7 and A8 yellowish brown and constricted basally.

Male genital bulb oval. Epandrium emarginates terminally in deep U-shape and forms pair of finger-like processes laterally. Hypandrium with short broad basal stalk. Hypovalves long stripe-like, bearing row of long bristles on inner margins, and almost extending to apex of gonocoxites. Gonocoxites with row of stout setae on ventro-apical margin. Gonostylus with obtuse triangular median tooth and disc-like basal process. Paramere spiral with relatively straight distal portion, bearing row of dense short spines on inner margin. Aedeagus with short ventral valves, and slender divergent dorsal valves.

Female subgenital plate oval with subacute apex. Medigynium with main plate narrow, greatly constricted in middle; pair of lateral plates ear-like and enclosing main plate; posterior arms slender; axis nearly twice as long as main plate with greatly diverged apodemes.

Emergence. August to September.

Distribution. Sichuan (Kangding, Yajiang).

10.6.78 *Panorpa fulvastra* Chou, 1981

(Yellowish Scorpionfly) [淡黄蝎蛉] (Fig. 10.197)

Measurements. Forewing length 10.5–12.0 mm.

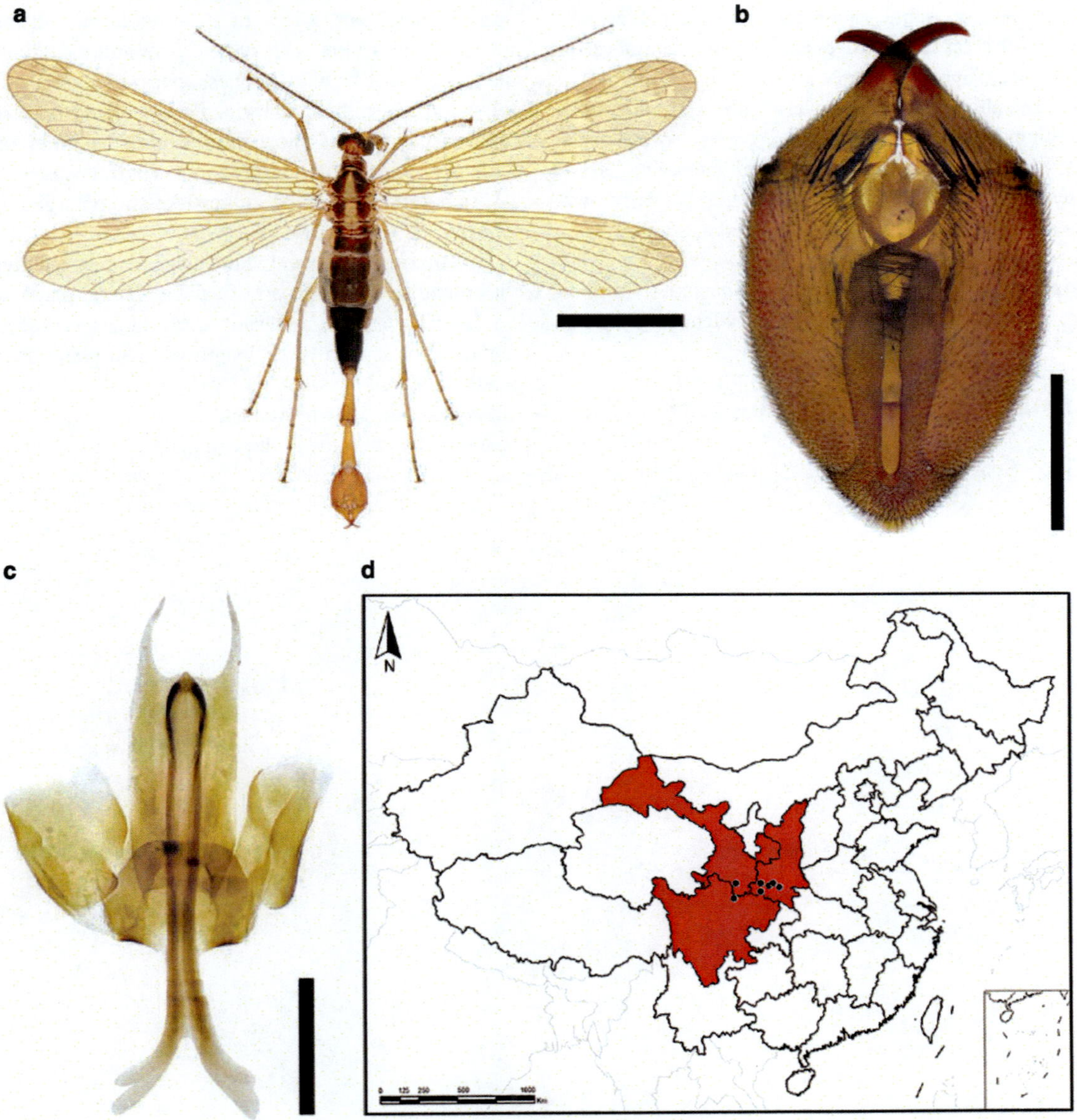

Fig. 10.197 (a) Mal adult of ***Panorpa fulvastra***, lateral view (Qinling Mountains, Shaanxi) (*Scale bar*: 5.0 mm) **[NWAU]**. (**b**) Male genital bulb of ***Panorpa fulvastra***, ventral view (*Scale bar*: 0.5 mm). (**c**) Female medigynium of ***Panorpa fulvastra***, ventral view (*Scale bar*: 0.2 mm). (**d**) Distribution of ***Panorpa fulvastra***

Morphology. Vertex and ocellar triangle black. Rostrum yellowish brown. Pronotum dark brown. Meso- and metanotum dark brown, with distinct pale yellow median stripe. Pleura and legs creamy yellow. Wing membrane hyaline, slightly tinged with yellow. Pterostigma yellow. Markings faint with only apical and pterostigmal bands present. Male T1–T5 blackish brown. Notal organ flat and covering acute postnotal organ. A6 dark brown and cylindrical, tapering towards apex. A7 and A8 yellowish brown and slightly constricted basally.

Male genital bulb oval. Epandrium emarginates terminally in deep U-shape and forms pair of finger-like processes laterally. Hypandrium with extremely shortened basal stalk. Hypovalves stripe-like, extending to middle of gonocoxites. Gonocoxites each bearing 5–6 stout setae on ventral apex innerly. Gonostylus with obtuse triangular median tooth and disc-like basal process. Parameres slender, spiral sigmoidal, and crossed mesally. Aedeagus with short ventral valves and finger-like dorsal valves; dorsal valves enlarged near dorsal apexes.

Female subgenital plate oval with subacute apex. Medigynium with main plate narrow, greatly constricted in middle; pair of lateral plates ear-like and enclosing base of main plate; posterior arms short and thin; axis nearly twice as long as main plate with greatly diverged apodemes.

Emergence. July to August.

Distribution. Gansu (Dangchang), Shaanxi (Ningshan, Foping, Zhouzhi, Nanzheng, Liuba), and Sichuan (Pingwu).

10.6.79 *Panorpa gaokaii* Li & Hua, 2021

(Kai Gao's Scorpionfly) [高氏蝎蛉] (Fig. 10.198)

Measurements. Forewing length 13.0–13.5 mm.

Morphology. Vertex mostly dark brown. Ocellar triangle black. Nota pale brown and dark brown in lateral portions. Wing membrane hyaline, markings indistinct. In forewings, apical and pterostigmal bands greatly reduced, other markings absent. Male T1–T5 black with light yellowish brown stripe extending from T1 to T3. Notal organ flat and covering acute postnotal organ. A6 black, slightly tapering towards apex. A7 and A8 yellowish brown and constricted basally.

Male genital bulb long oval, yellowish brown. Epandrium emarginates terminally in deep U-shape and forms pair finger-like processes laterally. Hypandrium with greatly shortened basal stalk. Hypovalves stripe-like, extending beyond midlength of gonocoxites, bearing several long bristles on apical portion of inner margin. Gonocoxites bearing row of stout setae on ventral apex innerly. Gonostylus with obtuse triangular median tooth and large disc-like basal process; basal process with triangular and acute inner margin. Parameres spiral, crossed, and bearing row of short microtirchia on inner margin. Aedeagus with short membranous ventral valves; dorsal valves highly sclerotized, finger-like and enlarged on dorsal apex.

Female subgenital oval with subacute apex. Mediygnium with broad main plate; pair of ear-like lateral plates enclosing lateral margins; posterior arms thin and acute; axis slightly shorter than twice length of main plate; apodemes greatly diverged basally.

Emergence. June to August.

Distribution. Hubei (Shennongjia).

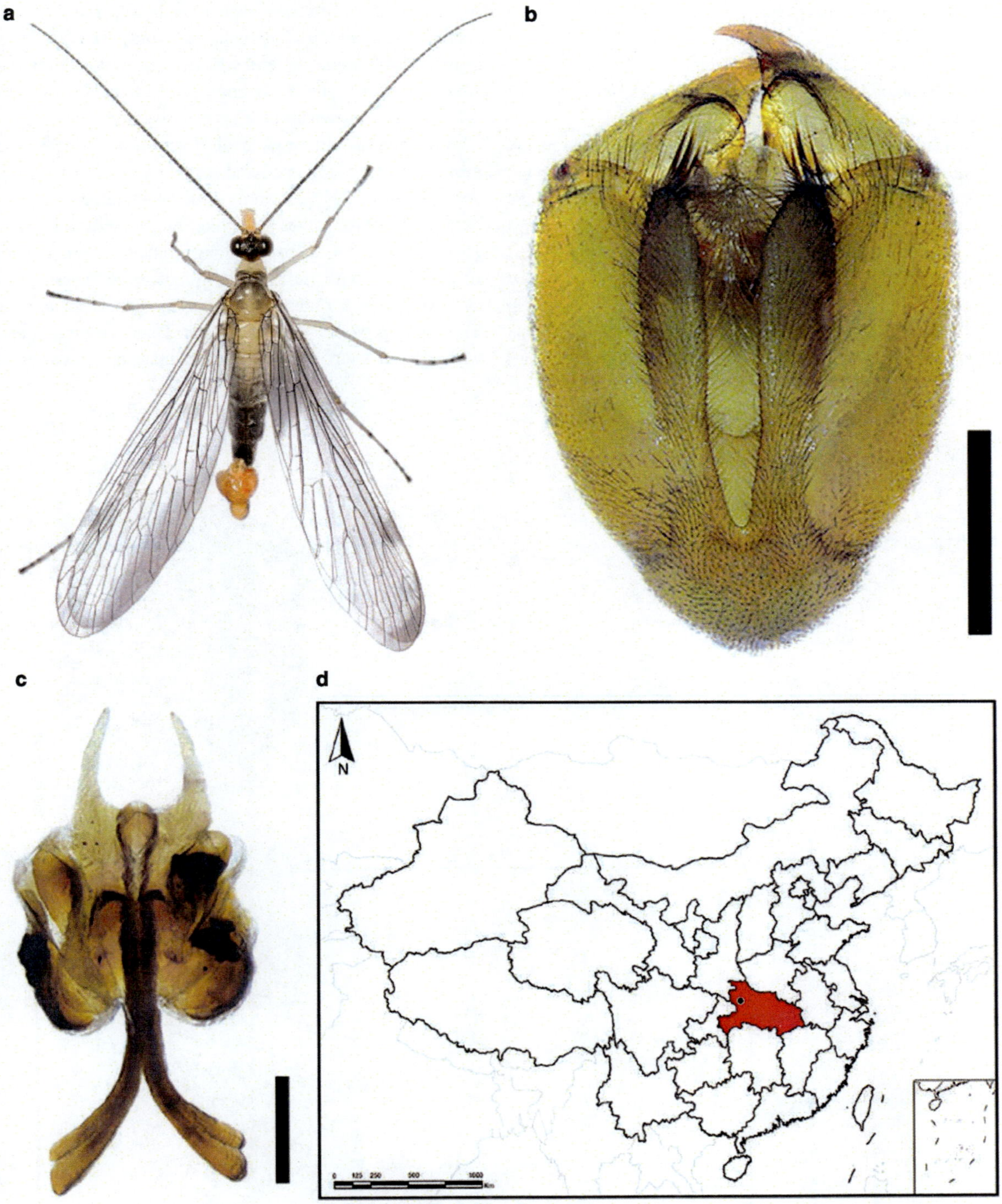

Fig. 10.198 (**a**) Male adult of ***Panorpa gaokaii***, dorsal view (Shennongjia, Hubei). (**b**) Male genitalia of ***Panorpa gaokaii***, ventral view (*Scale bar*: 0.5 mm). (**c**) Female medigynium of ***Panorpa gaokaii***, ventral view (*Scale bar*: 0.2 mm). (**d**) Distribution of ***Panorpa gaokaii***

10.6.80 *Panorpa jinchuana* Hua, Sun & Li, 2001

(Jinchuan Scorpionfly) [金川蝎蛉] (Fig. 10.199)
Measurements. Forewing length 12.5–13.0 mm.

Morphology. Vertex dark brown. Ocellar triangle black. Rostrum yellowish brown. Pronotum dark brown. Mesonotum and metanotum dark brown, with broad yellow mesal stripe. Pleura and legs yellowish brown. Wing membrane hyaline, with markings brown. In forewings, apical band broad with two large hyaline spots innerly; pterostigmal band complete, with broad basal branch and thin apical branch; basal band split into two spots; marginal spot large and conspicuous; basal spot small. Markings in males more reduced than in females, especially in separated or disappeared apical branch of pterostigmal band. Male T1–T5 dark brown. Notal organ flat and covering acute postnotal organ. A6 dark brown, cylindrical, and devoid of anal horns. A7 and A8 yellowish brown and constricted basally.

Male genital bulb broad oval. Epandrium emarginates terminally in deep U-shaped and forms pair of finger-like processes laterally. Hypandrium with an extremely shortened basal stalk. Hypovalves stripelike, not extending to base of gonostylus, and with row of long bristles on inner margin. Gonocoxites with tuft of long setae on ventral apex. Gonostylus with obtuse triangular median tooth and disc-like basal process. Parameres spiral and crossed mesally, with row of short spines on inner margin. Aedeagus with

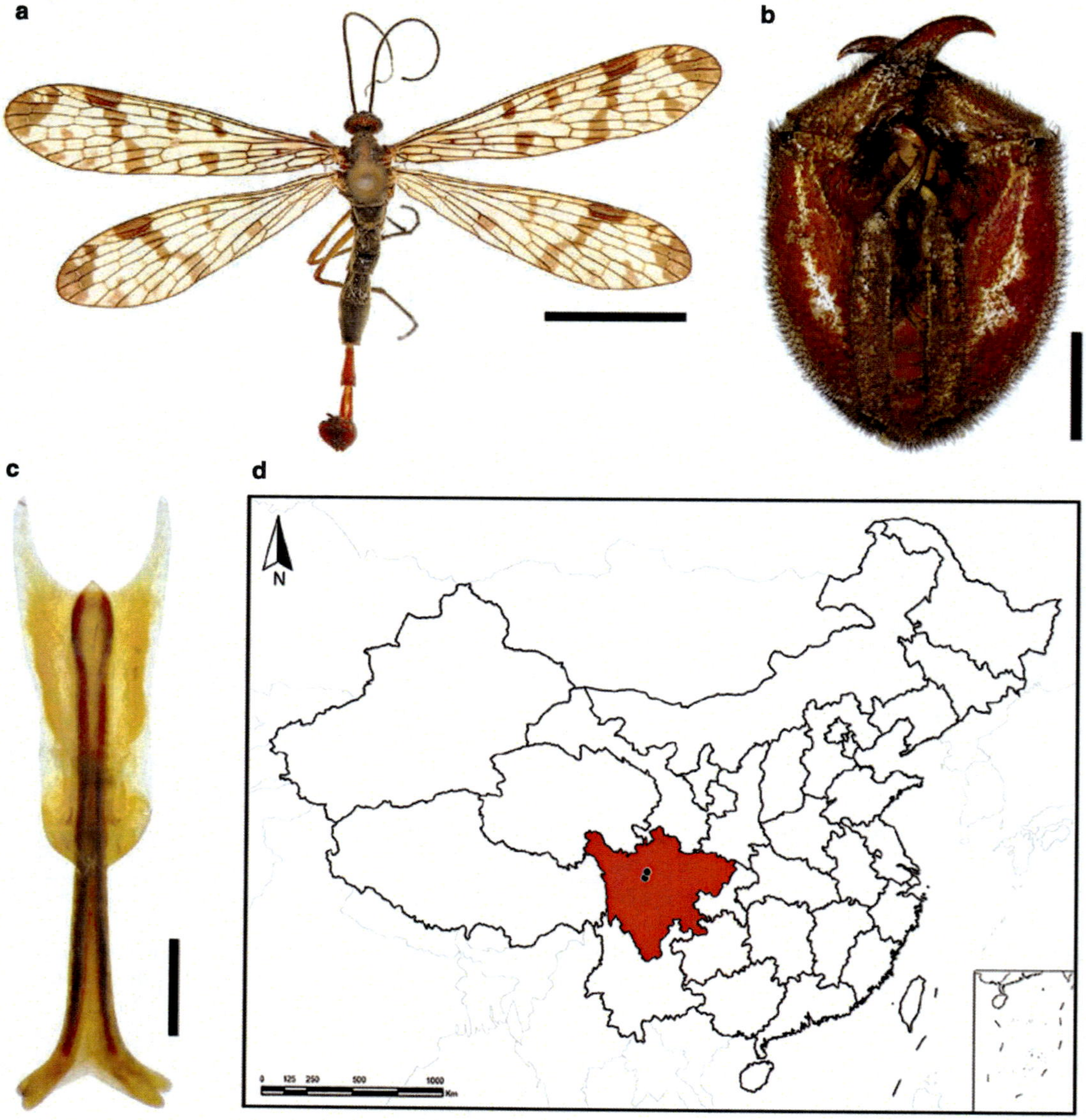

Fig. 10.199 (**a**) Male habitus of ***Panorpa jinchuana***, dorsal view (*Scale bar*: 5.0 mm). (**b**) Male genital bulb of ***Panorpa jinchuana***, ventral view (*Scale bar*: 0.5 mm). (**c**) Female medigynium of ***Panorpa jinchuana***, ventral view (*Scale bar*: 0.2 mm). (**d**) Distribution of ***Panorpa jinchuana***

short membranous ventral valves; dorsal valves fingerlike and enlarged at apex.

Female subgenital plate oval with subacute apex. Medigynium with main plate oblong; posterior arms short and slightly divergent; axis slightly nearly twice as long as main plate with diverged apodemes; anterior apex of apodemes branched.

Emergence. September to October.

Distribution. Sichuan (Danba, Jinchuan).

10.6.81 *Panorpa neospinosa* Chou & Wang, 1981

(New Spined Scorpionfly) [新刺蝎蛉] (Fig. 10.200)

Measurements. Forewing length 9.5–11.0 mm.

Morphology. Vertex and rostrum yellowish brown. Ocellar triangle black. Broad black transverse stripe passing through ocellar triangle. Nota dark brown, with conspicuous yellow median stripe on Meso- and metanotum. Pleura and legs yellowish brown. Wing membrane hyaline with markings brown. In forewings, apical band complete, bearing hyaline spot posteriorly; pterostigmal band complete, with thin apical branch and broad basal branch; basal band broad, complete, or split into two spots; marginal and basal spots large and conspicuous. Male T1–T5 dark brown. Notal organ flat and covering acute postnotal organ. A6 yellowish brown, projected dorsad and setose on apex. A7 and A8 yellowish brown and slightly constricted basally.

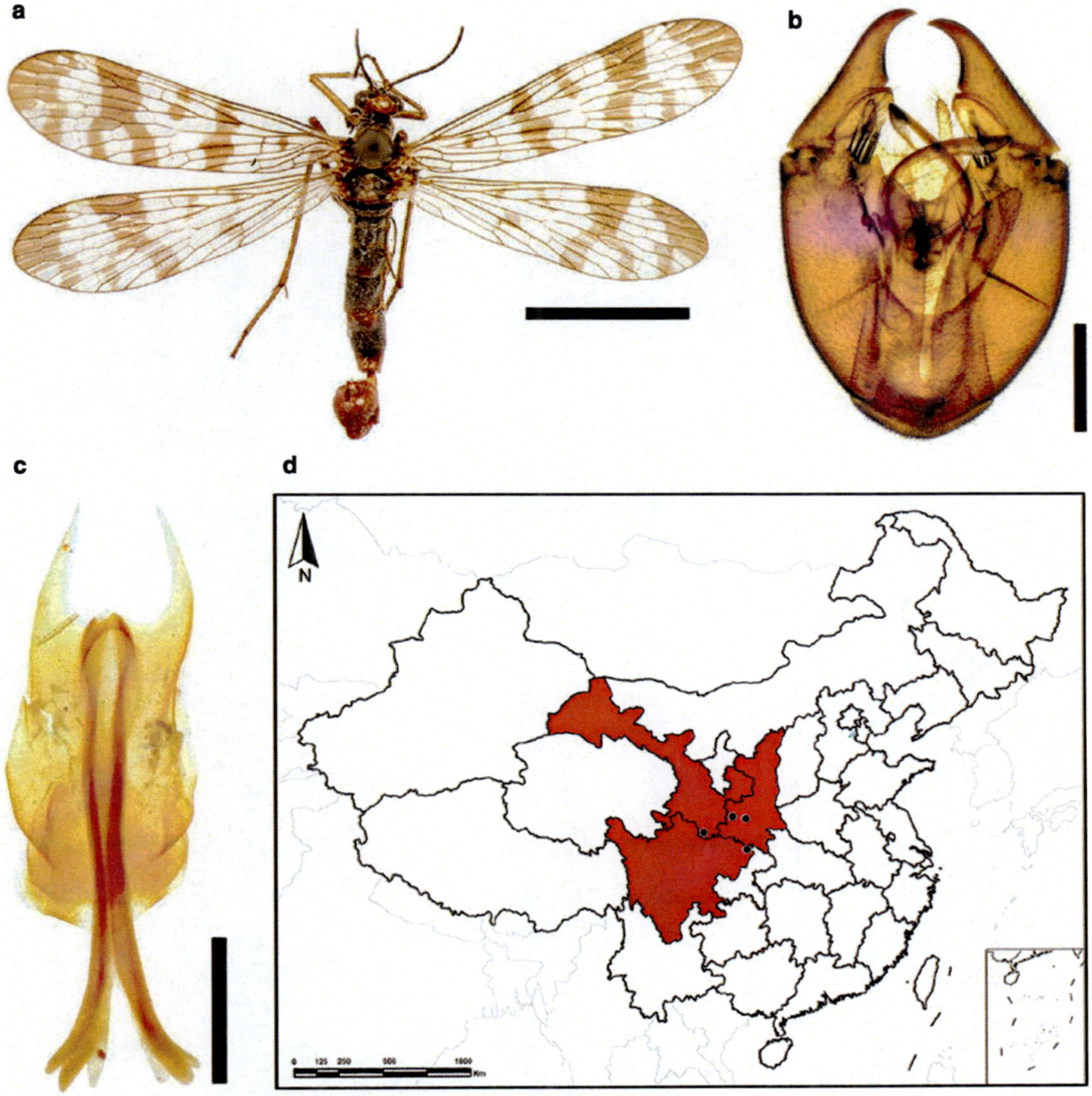

Fig. 10.200 (**a**) Male adult of ***Panorpa neospinosa***, dorsal view (Qinling Mountains, Baoji, Shaanxi) (*Scale bar*: 5.0 mm) **[NWAU]**. (**b**) Male genital bulb of ***Panorpa neospinosa***, ventral view (*Scale bar*: 0.5 mm). (**c**) Female mediygnium of ***Panorpa neospinosa***, ventral view (*Scale bar*: 0.2 mm). (**d**) Distribution of ***Panorpa neospinosa***

Male genital bulb globular. Epandrium emarginates terminally in deep U-shape and forms pair of finger-like processes laterally. Hypandrium with greatly shortened basal stalk. Hypovalves long broad stripe-like, divergent, almost extending to apex of gonocoxites, bearing row of long bristles along inner margin. Gonocoxites with 5–6 stout setae on ventral apex. Gonostylus with obtuse triangular median tooth and large disc-like basal process. Parameres long stripe-like and spiral, crossed mesally, and with dense short spines on inner margins. Ventral aedeagal valves membranous and poorly developed; dorsal valves finger-like, divergent distally, and each with membranous enlarged process on subapical portion dorsally.

Female subgenital plate oval with subacute apex. Medigynium with main plate oblong; posterior arms short and thin; axis slightly shorter than twice length of main plate with diverged apodemes; anterior apex of apodemes branched.

Emergence. July to August.

Distribution. Gansu (Wenxian), Shaanxi (Baoji, Zhouzhi), Sichuan (Wanyuan).

10.6.82 *Panorpa qinlingensis* Chou & Ran, 1981

(Qinling Scorpionfly) [秦岭蝎蛉] (Fig. 10.201)

Measurements. Forewing length 10.5–12.0 mm.

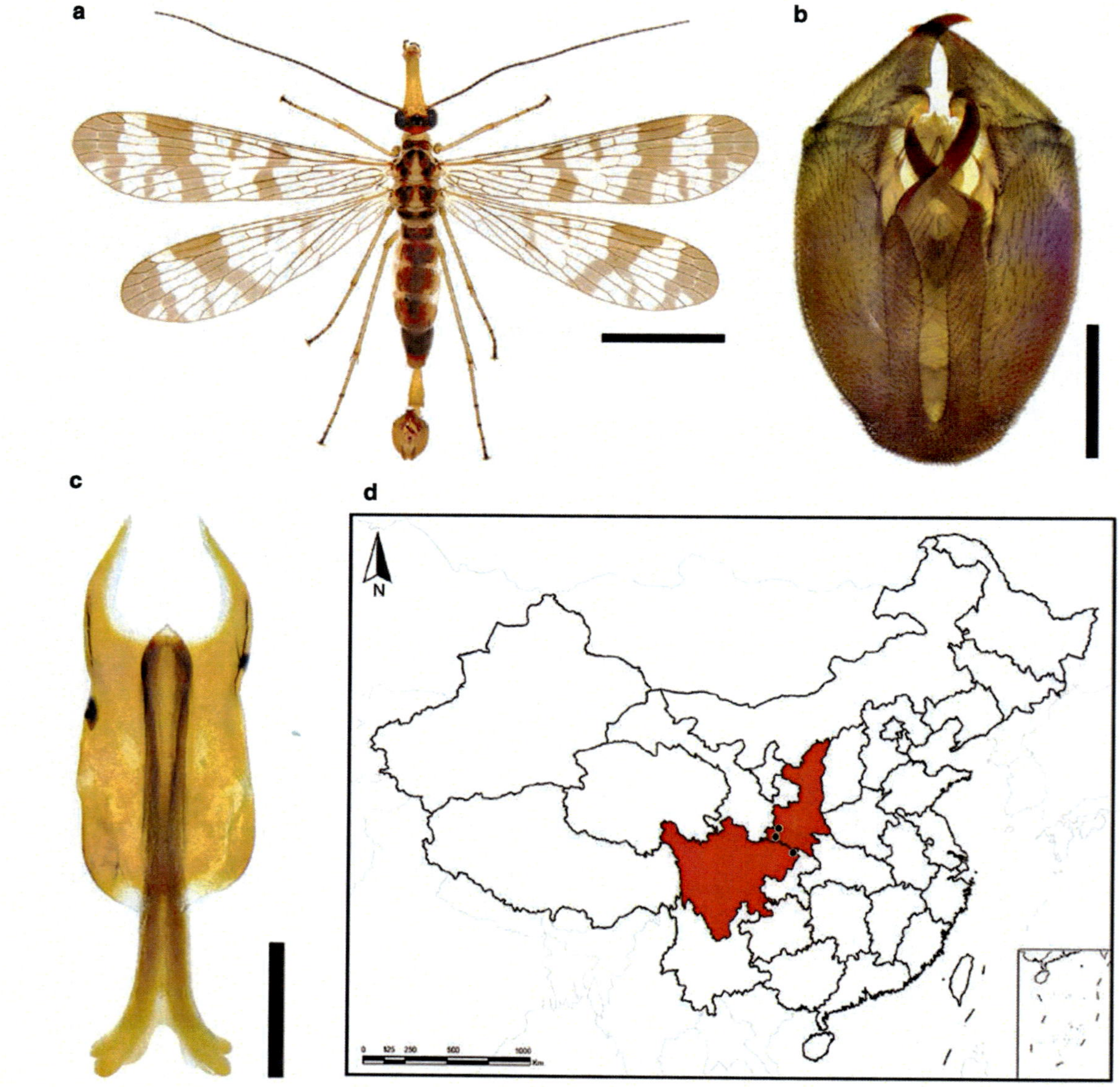

Fig. 10.201 (**a**) Male adult of ***Panorpa qinlingensis***, dorsal view (Liping National Forest Park, Nanzheng, Shaanxi) (*Scale bar*: 5.0 mm) [NWAU]. (**b**) Male genital bulb of ***Panorpa qinlingensis***, ventral view (*Scale bar*: 0.5 mm). (**c**) Female medigynium of ***Panorpa qinlingensis***, ventral view (*Scale bar*: 0.2 mm). (**d**) Distribution of ***Panorpa qinlingensis***

Morphology. Vertex and rostrum yellowish brown. Ocellar triangle black. Nota yellowish brown, with darkened flanks. Pleura and legs creamy yellow. Wing membrane hyaline with markings dark brown. In forewings, apical band complete and bearing large hyaline spot posteriorly; pterostigmal band complete, with thin apical branch and broad basal branch; marginal spot small; basal band split into two small spots; basal spot absent. Male T1–T5 dark brown. Notal organ flat and covering acute postnotal organ. A6 yellowish brown, projected dorsad and setose on apex. A7 and A8 yellowish brown and constricted basally.

Male genital bulb oval. Epandrium emarginates terminally in U-shape and forms pair of finger-like processes laterally. Hypandrium with short basal stalk. Hypovalves stripe-like, extending beyond middle of gonocoxites. Gonocoxites with 1 or 2 long setae on ventral apex. Gonostylus with obtuse triangular median tooth and disc-like basal process. Paramere stout and spiral, bearing dense short spines on inner margin. Ventral aedeagal valves short; dorsal valves finger-like and enlarged and membranous on dorsal apex.

Female subgenital plate oval with subacute apex. Medigynium with main plate oblong; posterior arms short and convergent; axis slightly shorter than twice length of main plate with diverged apodemes; anterior apex of apodemes branched.

Emergence. July to August.

Distribution. Shaanxi (Micang Mountains, Nanzheng, Liuba).

10.6.83 *Panorpa sexspinosa* Cheng, 1949

(Six-Spined Scorpionfly) [六刺蝎蛉] (Fig. 10.202)

Measurements. Forewing length 12.0–15.0 mm.

Morphology. Vertex and rostrum reddish brown. Ocellar triangle black. Pronotum dark brown. Meso- and metanotum dark brown, with pale yellow median stripe. Pleura and legs creamy yellow. Wing membrane hyaline with markings dark brown. Pterostigma brown. In forewings, apical band broad and scattered on posterior margin, and occasionally with hyaline spot innerly; pterostigmal band with broad basal branch and thin or disappeared apical branch; basal band broad; marginal and basal spots absent. Male T1–T5 dark brown. Notal organ flat and covering acute postnotal organ. A6 dark brown in basal half and yellowish brown in distal half, and projected dorsad on dorsal apex with sparse long setae.

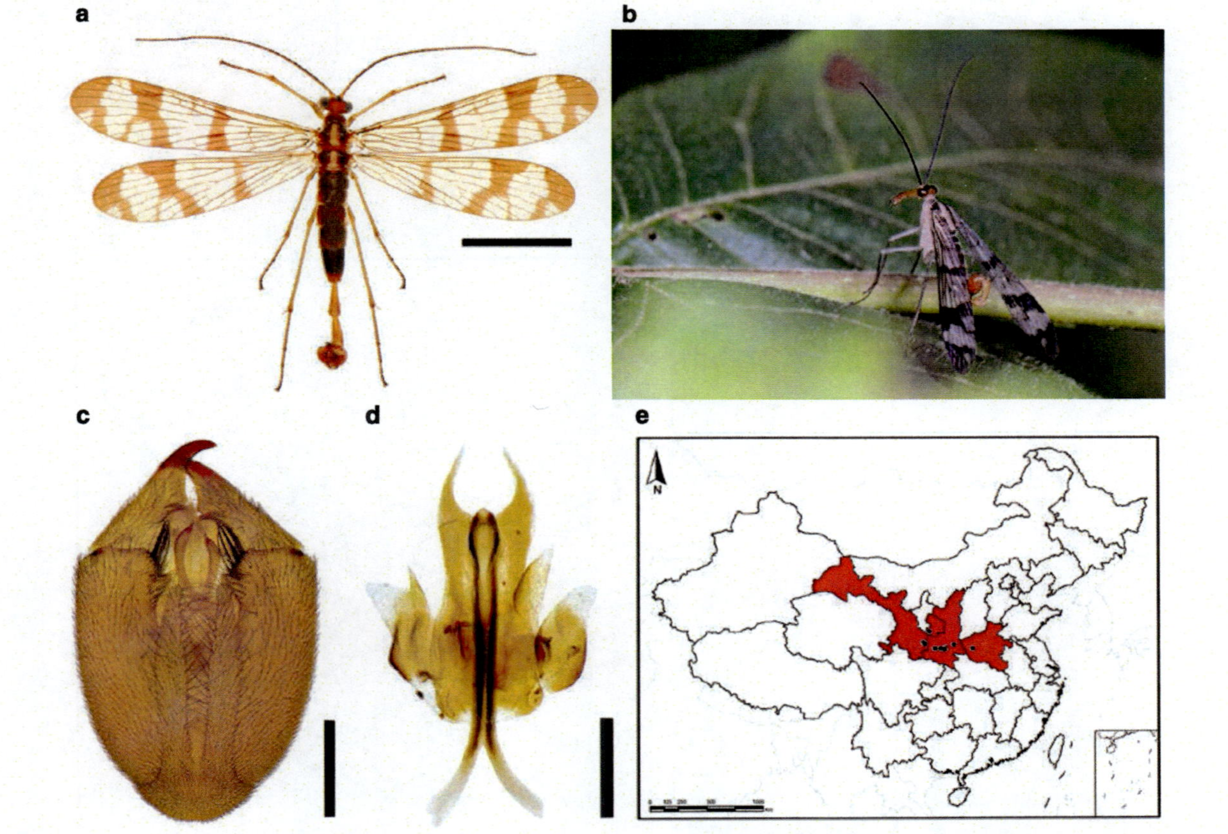

Fig. 10.202 (**a**) Male adult of ***Panorpa sexspinosa***, dorsal view (Qinling Mountains, Shaanxi) (*Scale bar*: 5.0 mm) **[NWAU]**. (**b**) Male adult of ***Panorpa sexspinosa*** in the field (Qinling Mountains, Shaanxi, ©Wei-Wei Zhang). (**c**) Male genital bulb of ***Panorpa sexspinosa***, ventral view (*Scale bar*: 0.5 mm). (**d**) Female medigynium of ***Panorpa sexspinosa***, ventral view (*Scale bar*: 0.2 mm). (**e**) Distribution of ***Panorpa sexspinosa***

Male genital bulb oval. Epandrium emarginates terminally in deep U-shaped and forms pair of finger-like processes laterally. Hypandrium with short basal stalk. Hypovalves stripe-like, not extending to apex of gonocoxites, bearing row of long bristles on inner margin. Gonocoxites with 4–6 stout setae on ventro-apical margin innerly. Gonostylus with obtuse triangular median tooth and large disc-like basal process. Parameres slender and spiral, and crossed mesally. Ventral aedeagal valves short and membranous; dorsal valves sclerotized and elongated, and each with membranous enlarged process on subapical portion dorsally.

Female subgenital plate oval with subacute apex. Medigynium with main plate subtrapezoidal in basal half and broader in distal half; pair of greatly developed lateral plates enclosing base of main plate; posterior arms short and convergent; axis slightly shorter than twice length of main plate with diverged apodemes; anterior apex of apodemes branched.

Emergence. May to August.

Distribution. Gansu (Dangchuan, Pingliang, Tianshui), Henan (Songxian), Shaanxi (Huyi, Lantian, Ningshan, Taibai, Zhouzhi).

10.6.84 *Panorpa typicoides* Cheng, 1949

(Typical-Marking Scorpionfly) [模斑蝎蛉] (Fig. 10.203)

Measurements. Forewing length 12.5 mm.

Morphology. Vertex yellowish brown. Ocellar triangle black. Broad black transverse stripe passing through ocellar triangle. Rostrum yellowish brown. Pronotum dark brown.

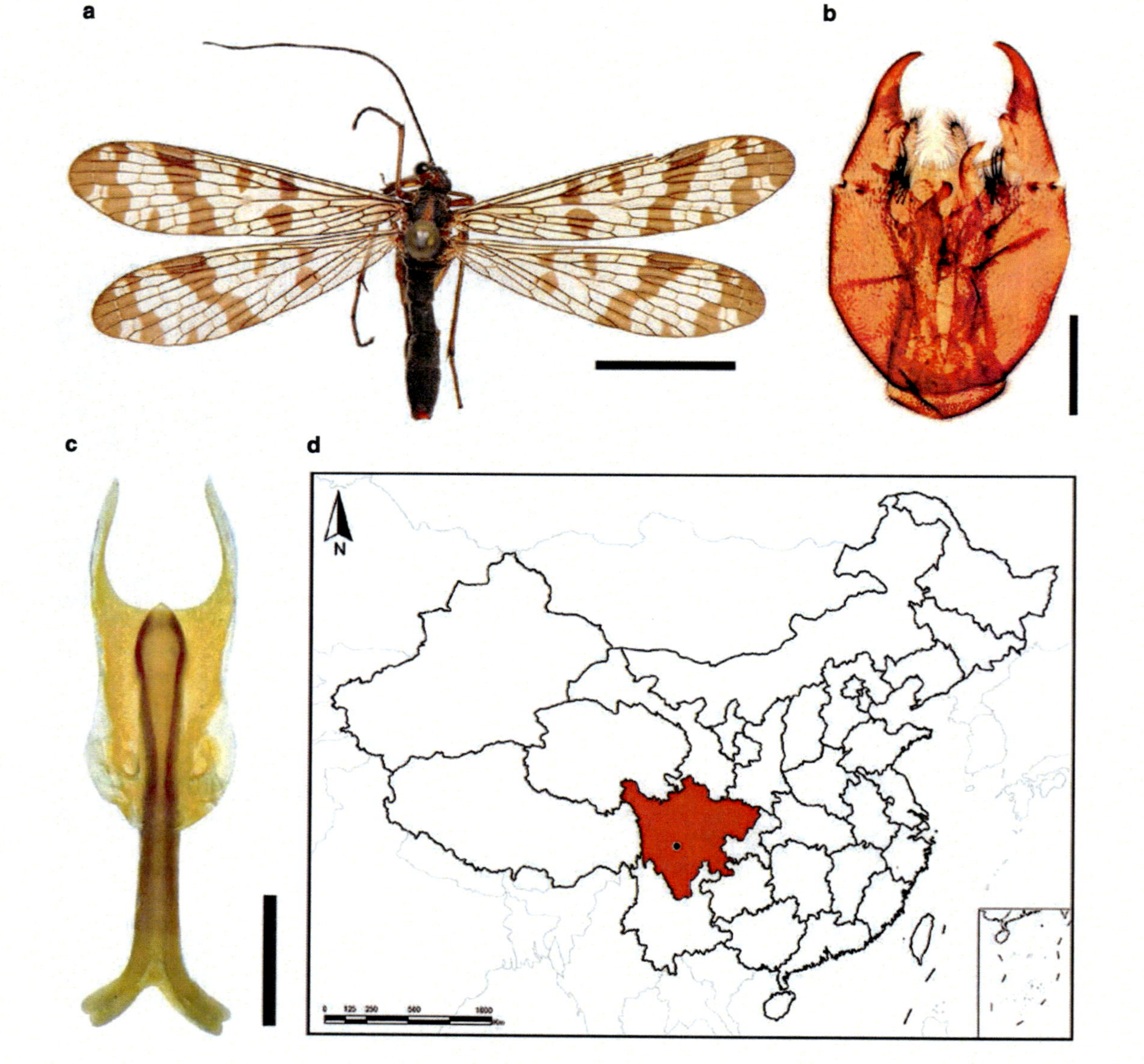

Fig. 10.203 (**a**) Male adult of ***Panorpa typicoides***, dorsal view (Kangding, Sichuan) (*Scale bar*: 5.0 mm) [**NWAU**]. (**b**) Male genital bulb of ***Panorpa typicoides***, ventral view (*Scale bar*: 0.5 mm). (**c**) Female medigynium of ***Panorpa typicoides***, ventral view (*Scale bar*: 0.2 mm). (**d**) Distribution of ***Panorpa typicoides***

Meso- and metanotum dark brown, with yellow median stripe. Pleura and legs yellowish brown. Wing membrane hyaline with markings brown. In forewings, apical band with two hyaline spots innerly; pterostigmal band broad, with separated apical branch; marginal spot fusiform; basal band split into two spots; basal spot very small. Male T1–T5 dark brown. Notal organ flat and covering acute postnotal organ. A6 dark brown and slightly projected dorsad on apex. A7 and A8 yellowish brown and constricted basally.

Male genital bulb broad oval. Epandrium emarginated terminally in U-shape and forming pair of finger-like processes laterally. Hypandrium with short basal stalk. Hypovalves long stripe-like, extending to base of gonostylus, and bearing row of long bristles on inner margin. Gonocoxites with tuft of long stout setae on ventral apex. Gonostylus with obtuse triangular median tooth and large disc-like basal process. Paramere spiral in distal half, and bearing row of short spines on inner margin. Aedeagus with ventral valves membranous and short, and dorsal valves finger-like and divergent, bearing membranous enlarged process on dorsal surface subapically.

Female subgenital plate oval with subacute apex. Medigynium with main plate oblong; posterior arms short and convergent; axis slightly nearly twice as long as main plate with diverged apodemes; anterior apex of apodemes branched.

Emergence. July to August.

Distribution. Sichuan (Kangding).

10.6.85 *Panorpa uncinata* Li & Hua, 2021

(Hooked Scorpionfly) [钩突蝎蛉] (Fig. 10.204)

Measurements. Forewing length 12.0–13.5 mm.

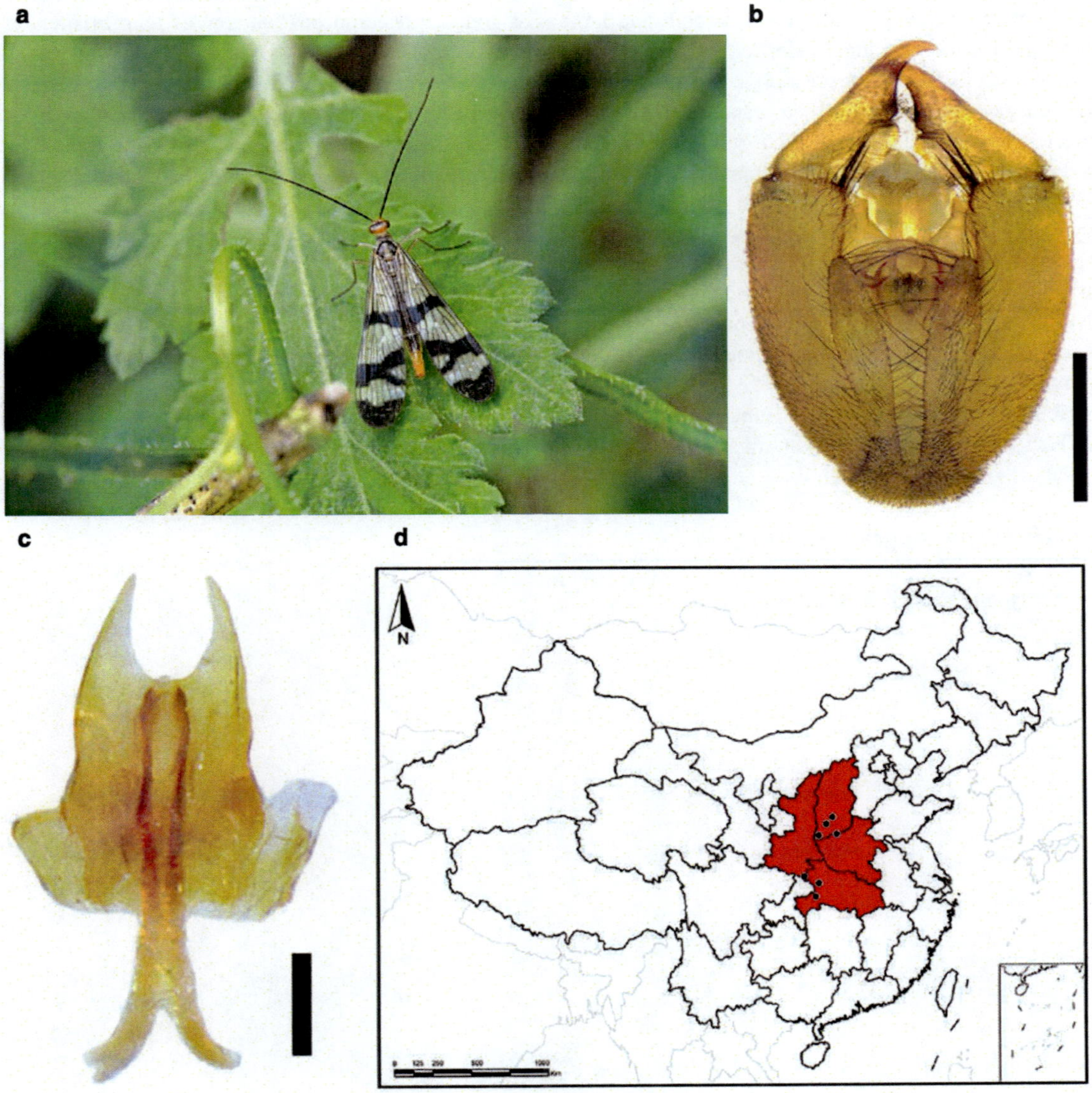

Fig. 10.204 (**a**) Male adult of ***Panorpa uncinata*** in the field (Shennongjia, Hubei). (**b**) Male genital bulb of ***Panorpa uncinata***, ventral view (*Scale bar*: 0.5 mm). (**c**) Female medigynium of ***Panorpa uncinata***, ventral view (*Scale bar*: 0.2 mm). (**d**) Distribution of ***Panorpa uncinata***

Morphology. Vertex and rostrum reddish brown. Ocellar triangle black. Pronotum dark brown. Meso- and metanotum dark brown, with pale yellow median stripe. Pleura and legs creamy yellow. Wing membrane hyaline with markings dark brown. Pterostigma brown. In forewings, apical band broad and scattered on posterior margin, and occasionally with hyaline spot innerly; pterostigmal band with broad basal branch and thin or disappeared apical branch; basal band broad; marginal and basal spots absent. Male T1–T5 dark brown. Notal organ flat and covering acute postnotal organ. A6 dark brown in basal half and yellowish brown in distal half, and projected dorsad on dorsal apex with sparse long setae.

Male genital bulb oval. Epandrium emarginates terminally in deep U-shaped and forms pair of finger-like processes laterally. Hypandrium with short basal stalk. Hypovalves stripe-like, not extending to apex of gonocoxites, bearing row of long bristles on inner margin. Gonocoxites with 4–6 stout setae on ventro-apical margin innerly. Gonostylus with obtuse triangular median tooth and large disc-like basal process. Parameres slender and spiral, hook-like, crossed mesally and not exceeding aedeagus caudally. Ventral aedeagal valves short and membranous; dorsal valves sclerotized and elongated, and each with membranous enlarged process on subapical portion dorsally.

Female subgenital plate oval with subacute apex. Medigynium with main plate subtrapezoidal, nearly as wide as long; pair of greatly developed lateral plates enclosing base of main plate; posterior arms short and convergent; axis slightly shorter than twice length of main plate with diverged apodemes; anterior apex of apodemes branched.

Emergence. May to August.

Distribution. Henan (Jiyuan), Hubei (Shennongjia), Shanxi (Linfen, Yuncheng), Shaanxi (Ankang).

10.6.86 *Panorpa yangi* Chou, 1981

(Yang's Scorpionfly) [杨氏蝎蛉] (Fig. 10.205)

Measurements. Forewing length 9.5–10.5 mm.

Morphology. Vertex yellowish brown. Black transverse stripe passing through black ocellar triangle. Rostrum reddish brown. Nota yellowish brown. Four black spots on latero-anterior corner of Meso- and metanotum. Pleura and legs yellowish brown. Wing membrane hyaline with markings brown. In forewings, apical band broad, with hyaline spot posteriorly; pterostigmal band with narrow apical branch and broad basal branch; basal band complete; marginal and basal spots large and conspicuous. Male T1–T5 yellowish brown. Notal organ flat and covering acute postnotal organ. A6 yellowish brown, cylindrical, and slightly projected on dorsal apex bearing sparse long setae; A7 and A8 yellowish brown and constricted basally.

Male genital bulb broad oval. Epandrium deeply emarginates terminally in deep U-shape and forms pair of finger-like processes laterally. Hypandrium with greatly shortened basal stalk. Hypovalves long and broad, extending to three-fourths of gonocoxites, and bearing row of long bristles on inner margin. Gonocoxites with 4–6 stout setae on ventral apex innerly. Gonostylus with obtuse triangular median tooth and large disclike basal process. Parameres slender, intensively curved and hook-like on apical halves, and crossed mesally. Aedeagus with ventral valves membranous and short; dorsal valves sclerotized and curved dorsad, and each with membranous enlarged process on subapical portion dorsally, and finger-like at apex.

Female subgenital plate oval with subacute apex. Medigynium with main plate oblong; posterior arms very short and slightly convergent; axis slightly shorter than twice length of main plate with diverged apodemes; anterior apex of apodemes branched.

Emergence. July to August.

Distribution. Ningxia (Jingyuan), Shaanxi (Fuxian, Ganquan, Huangling, Xunyi, Ziwuling Mountains).

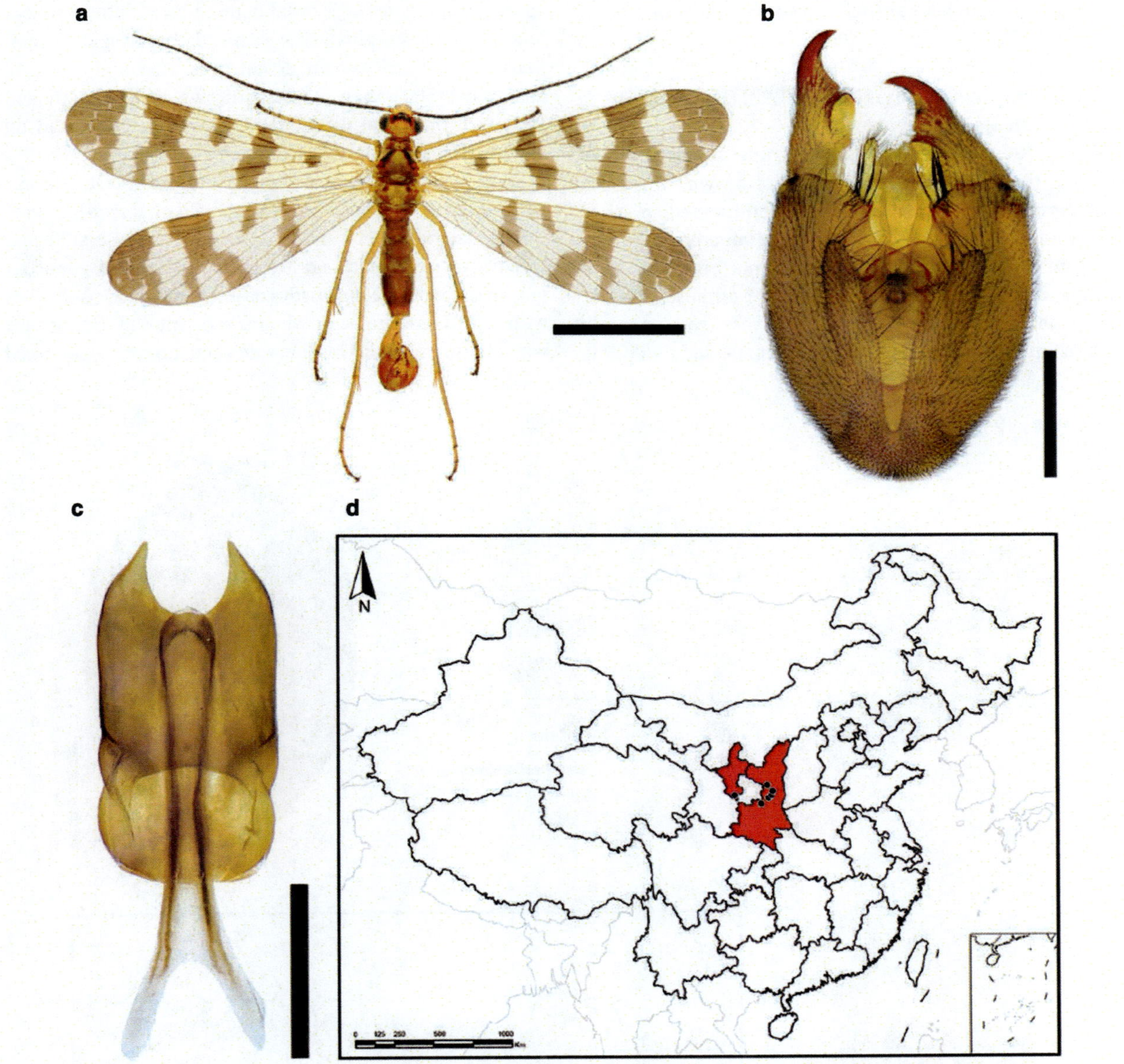

Fig. 10.205 (**a**) Male adult of ***Panorpa yangi***, dorsal view (Huangling, Shaanxi) (*Scale bar*: 5.0 mm) [NWAU]. (**b**) Male genital bulb of ***Panorpa yangi***, ventral view (*Scale bar*: 0.5 mm). (**c**) Female medigynium of ***Panorpa yangi***, ventral view (*Scale bar*: 0.2 mm). (**d**) Distribution of ***Panorpa yangi***

10.6.87 *Panorpa yaoluopingensis* Li & Hua, 2021

(Yaoluoping Scorpionfly) [鹞落坪蝎蛉] (Fig. 10.206)
Measurements. Forewing length 11.0–12.0 mm.

Morphology. Vertex and rostrum reddish brown. Ocellar triangle black. Pronotum dark brown. Meso- and metanotum dark brown, with pale yellow median stripe. Pleura and legs creamy yellow. Wing membrane hyaline with markings dark brown. Pterostigma brown. In forewings, apical band broad and scattered on posterior margin, and occasionally with hyaline spot innerly; pterostigmal band with broad basal branch and thin or disappeared apical branch; basal band reduced into spot near hind margin; marginal spot reduced; basal spot absent. Male T1–T5 dark brown. Notal organ flat and covering acute postnotal organ. A6 dark brown in basal half and yellowish brown in distal half, and projected dorsad on dorsal apex with sparse long setae.

Male genital bulb oval. Epandrium emarginates terminally in deep U-shaped and forms pair of finger-like processes laterally. Hypandrium with short basal stalk. Hypovalves stripe-like, not extending to apex of gonocoxites, bearing row of long bristles on inner margin. Gonocoxites with tuft of stout setae on ventral apex innerly. Gonostylus with obtuse triangular median tooth and large disc-like basal

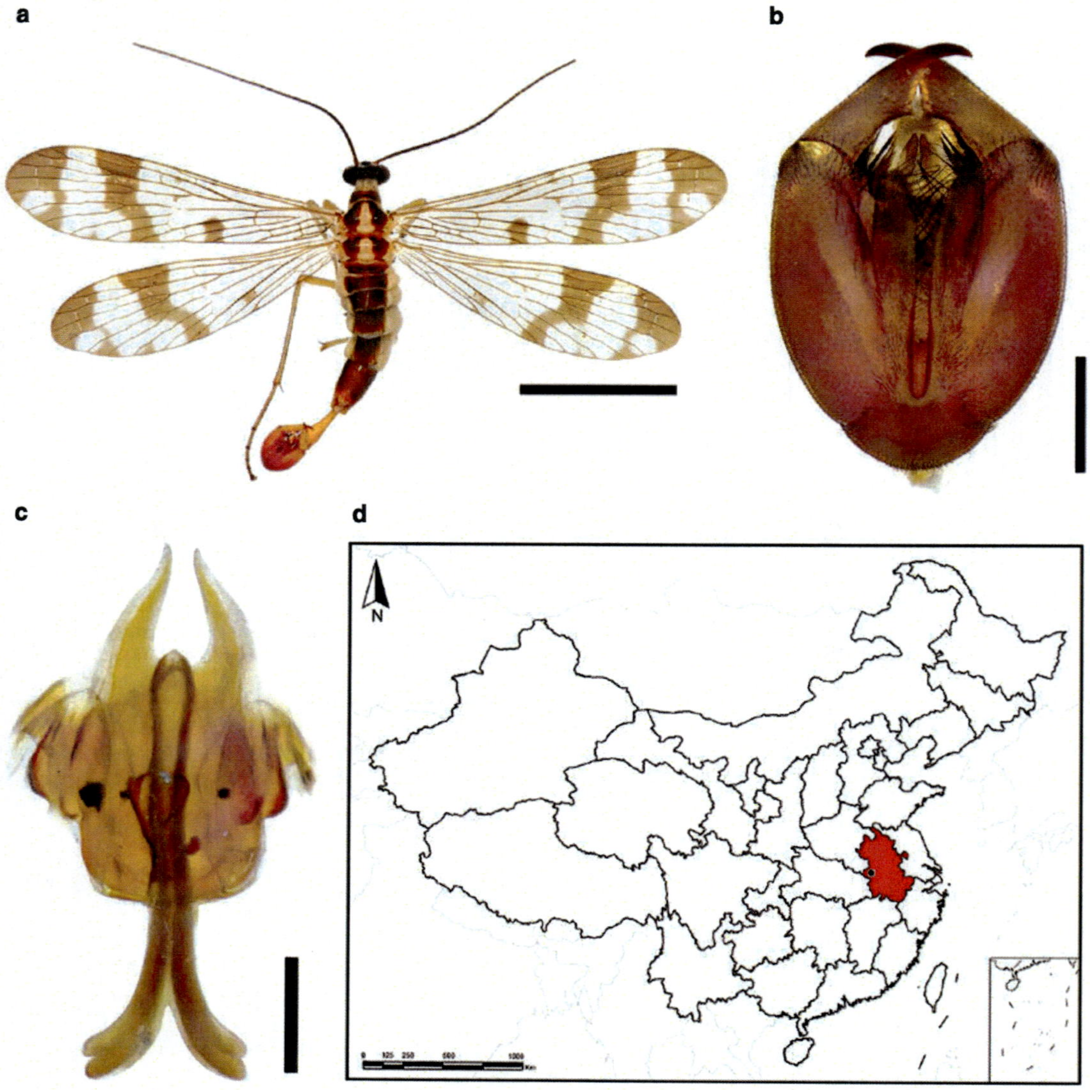

Fig. 10.206 (**a**) Male adult of ***Panorpa yaoluopingensis***, dorsal view (Yaoluoping, Yuexi, Anhui) (*Scale bar*: 5.0 mm) [**NWAU**]. (**b**) Male genital bulb of ***Panorpa yaoluopingensis***, ventral view (*Scale bar*: 0.5 mm). (**c**) Female medigynium of ***Panorpa yaoluopingensis***, ventral view (*Scale bar*: 0.2 mm). (**d**) Distribution of ***Panorpa yaoluopingensis***

process. Parameres slender and spiral, and crossed mesally. Ventral aedeagal valves short and membranous; dorsal valves sclerotized and elongated, and each with membranous enlarged process on subapical portion dorsally.

Female subgenital plate oval with subacute apex. Medigynium with main plate subtrapezoidal; pair of greatly developed lateral plates enclosing lateral margins of main plate; posterior arms short, stout, and greatly convergent; axis slightly shorter than twice length of main plate with diverged apodemes; anterior apex of apodemes branched.

Emergence. May to August.

Distribution. Anhui (Yaoluoping).

Ungrouped Species.

10.6.88 *Panorpa horiensis* Issiki, 1929

(Hori Horned Scorpionfly) [埔里角蝎蛉] (Fig. 10.207)

Measurements. Forewing length 11.5–13.0 mm.

Morphology. Vertex yellow. Ocellar triangle enclosed by large black spot. Rostrum yellowish brown. Nota black. Pleura and legs light yellow. Wing membrane hyaline and slightly tinged with yellow. Apex slightly darkened. Pterostigmal band greatly reduced into series of discontinuous spots. Other markings absent. Male T1–T5 dark brown. Notal organ flat and covering acute postnotal organ. A6 dark brown dorsally and yellowish brown ventrally, and bearing yellowish brown, conical anal horn apically. A7 and A8 yellowish brown and slightly stalked at base, with black lines on lateral surfaces.

Male genital bulb oval. Epandrium emarginates terminally in broad U-shape and forms pair of finger-like processes laterally. Hypovalves short, not extending to middle of gonocoxites, with sparse, stout setae on rounded apexes. Gonostylus shorter than gonocoxites, bearing indistinct median tooth, and large bowl-shaped concavity on basal half of inner margin. Paramere blade-shaped, with base straight, distal portion curved mesad, exceeding apex of gonocoxites, and bearing row of comb-like spines along inner margin. Aedeagus with ventral valves membranous and short; dorsal valves elongated and finger-shaped apically; and lateral processes very stout.

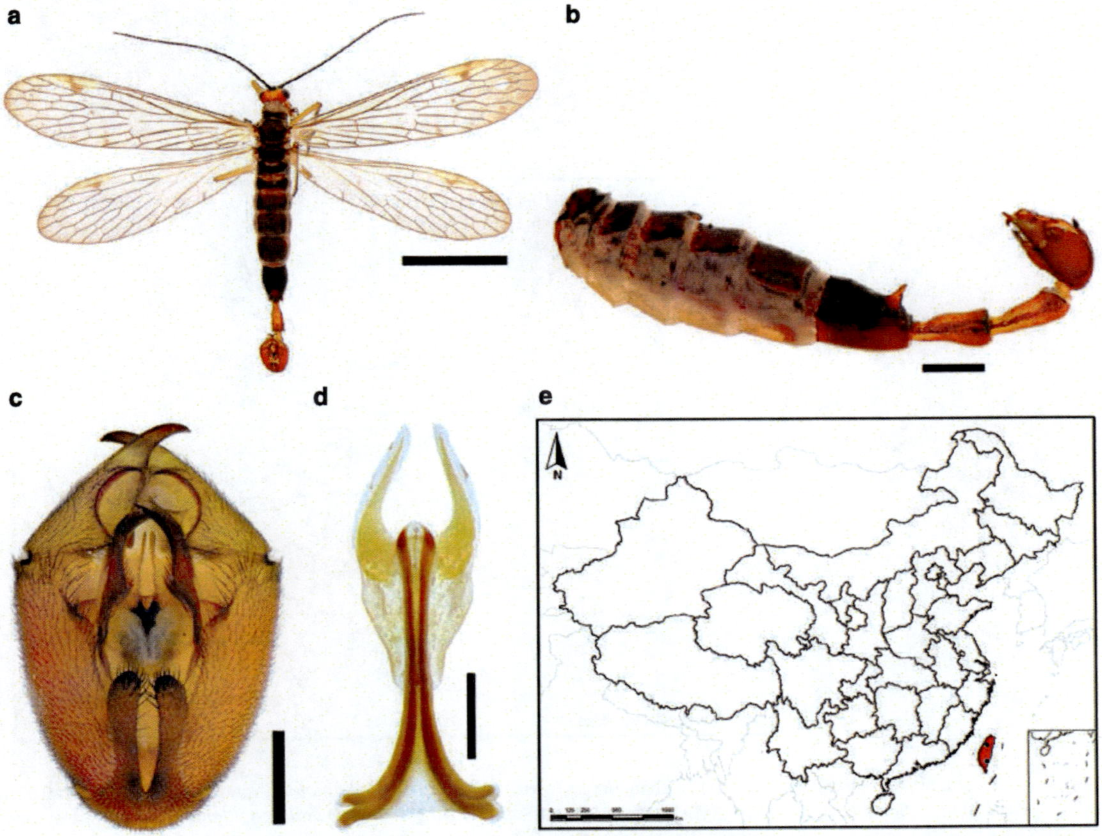

Fig. 10.207 (**a**) Male adult of ***Panorpa horiensis***, dorsal view (Haituan, Taitung, Taiwan) (*Scale bar*: 5.0 mm) **[NWAU]**. (**b**) Male abdomen of ***Panorpa horiensis***, lateral view (*Scale bar*: 2.0 mm). (**c**) Male genital bulb of *Panorpa horiensis*, ventral view (*Scale bar*: 0.5 mm). (**d**) Female medigynium of ***Panorpa horiensis***, ventral view (*Scale bar*: 0.2 mm). (**e**) Distribution of ***Panorpa horiensis***

Female subgenital plate oblong with rounded apex. Medigynium with main plate narow basally and broadening towards apex; posterior arms slender, shorter than main plate, and greatly convergent; axis nearly twice as long as main plate with widely diverged apodemes; anterior apex of apodemes branched.

Emergence. March to May.

Distribution. Taiwan (Nantou, Taitung).

Remarks. Although ***Panorpa horiensis*** bears a single anal horn in the males, it greatly differs from the single horn-bearing *Cerapanorpa* and *Megapanorpa* in the genital morphology. Its systematic position remains unclear.

10.6.89 *Panorpa kiautai* Zhou & Wu, 1993

(Kiauta's Horned Scorpionfly) [尤氏角蝎蛉]
(Fig. 10.208)

Measurements. Forewing length 14.0 mm.

Morphology. Vertex deep yellowish brown. Ocellar triangle black. Rostrum yellowish brown. Pronotum yellowish brown. Meso- and metanotum dark brown, with yellow median stripe. Pleura and legs yellowish brown. Wing membrane hyaline with markings brown. In forewings, apical band with 2–3 large hyaline spots posteriorly; pterostigmal band oblique, without apical branch;

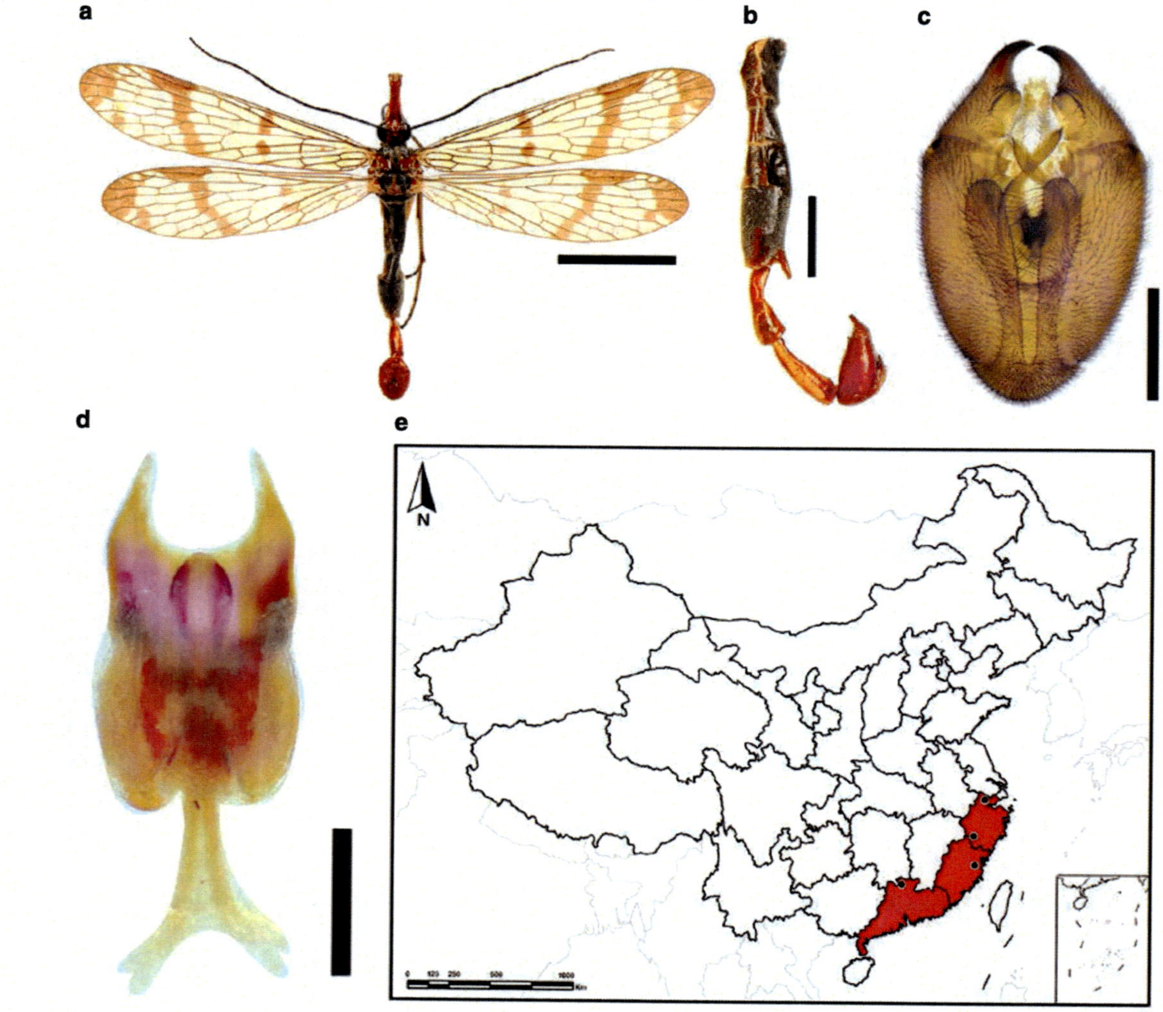

Fig. 10.208 (**a**) Male adult of ***Panorpa kiautai*** (Dongshan Forest Farm, Ruyuan, Guangdong) (*Scale bar*: 5.0 mm) **[NWAU]**. (**b**) Male abdomen of *Panorpa kiautai*, lateral view (*Scale bar*: 2.0 mm). (**c**) Male genital bulb of ***Panorpa kiautai***, ventral view (*Scale bar*: 0.5 mm). (**d**) Female medigynium of ***Panorpa kiautai***, ventral view (*Scale bar*: 0.2 mm). (**e**) Distribution of ***Panorpa kiautai***

basal band broad anteriorly and thin posteriorly; marginal spot absent; basal spot very small. Male T1–T5 dark brown. Notal organ flat and covering acute postnotal organ. A6 with distinct finger-like anal horn on dorsal apex. A7 and A8 yellowish brown and constricted basally.

Male genital bulb oval. Epandrium emarginates terminally in deep U-shape and forms pair of finger-like processes laterally. Hypandrium with extremely shortened basal stalk. Hypovalves stripe-like and distinctly enlarged terminally, with row of long bristles on inner margin. Gonostylus with obtuse triangular median tooth and disc-like basal process. Parameres long stout, spiral and crossed mesally, bearing row of dense short spines on inner margin. Aedeagus with short and membranous ventral valves and finger-like and divergent dorsal valves.

Female subgenital plate oval with subacute apex. Medigynium with main plate subtrapezoidal; pair of greatly developed lateral plates enclosing base of main plate; posterior arms short, stout, and convergent; axis nearly twice as long as main plate with diverged apodemes; anterior apex of apodemes branched.

Emergence. March to April.

Distribution. Guangdong (Ruyuan), Zhejiang (Deqing, Suichang).

Remarks. Although ***Panorpa kiautai*** bears a single anal horn in the males, it greatly differs from the single horn-bearing *Cerapanorpa* and *Megapanorpa* in the genital morphology. The male and female genitalia suggest its possible relationship with the *Panopa davidi* group.

10.7 Genus *Sinopanorpa* Cai & Hua, 2008

The genus *Sinopanorpa* can be recognized mainly by the following diagnostic characters: wing membrane yellow; boundaries of wing markings blurred; A7 extremely constricted and stalk-like at the basal portion in males; and gonostylus longer than gonocoxites.

Four Chinese-endemic species have been described. All of them are included in this atlas (Fig. 10.209).

Fig. 10.209 Male adult of ***Sinopanorpa tincta***, lateral view (©Wen Zhong)

Key to Species of Sinopanorpa in China
(4 species, males)

1. Gonostylus with basal process unfurcated .... ***Sinopanorpa baokangensis*** Wang
– Gonostylus with basal process bifurcated 2
2. Body mostly dark brown; parameres exceeding median tooth of gonostylus .... ***Sinopanorpa nangongshana*** Cai & Hua
– Body mostly yellowish brown; parameres not exceeding median tooth of gonostylus 3
3. Basal process of gonostylus with distal tooth smaller than basal tooth; parameres with long comb-like spines on inner margin .... ***Sinopanorpa tincta*** Navás
– Basal process of gonostylus with distal and basal tooth approximately same sized .... ***Sinopanorpa digitiformis*** Huang & Hua

10.7.1 *Sinopanorpa baokangensis* Wang, 2021

(Baokang Sinoscorpionfly) [保康华蝎蛉] (Fig. 10.210)
Measurements. Forewing length 16.0–17.5 mm.

Morphology. Vertex dark brown. Ocellar triangle black. Rostrum yellowish brown. Nota dark brown. Pleura and legs creamy yellow. Wing membrane translucent, tinged with waxy yellow. Markings yellowish brown. Pterostigma reddish brown. In forewings, apical band broad; pterostigmal band complete, with basal branch and apical branch; basal band split into two spots; marginal spot small; basal spot absent. Male T1–T5 yellowish brown. Notal organ flat and covering the acute postnotal organ. A6 yellowish brown, cylindrical and bearing tuft of long setae on the dorsal surface medially. A7 and A8 yellowish brown. A7 greatly stalked on the basal third; A8 slightly constricted at base.

Male genital bulb ovoid, yellowish brown. Epandrium long and broad, tapering towards apex, and slightly emarginated in V-shape terminally. Hypovalves broad, reaching apex of gonocoxites, and bearing long setae along inner margin. Gonocoxites stout, with apexes nearly truncated. Gonostylus approximately as long as gonostylus, sickle-shaped, with median tooth subtriangular and indistinct, and basal process greatly developed. Tuft of long setae located basal to basal process. Basal process with basal half trapezoidal and distal half thinner and greatly curved. Paramere slender, slightly bent at middle, extending beyond middle of basal process of gonostylus, bearing tuft of long, stout spines at middle of inner margin, and row of short bristles along distal half of inner margin. Aedeagus short, with dorsal processes slightly produced.

Female subgenital plate oval with rounded apex, and bearing several long setae in distal portion. Medigynium with main plate rounded, posterior arms stout and shorter than half length of main plate, and axis slender and longer than twice length of main plate. Apodemes of axis mostly aligned but slightly divergent at base.

Emergence. May to July.

Distribution. Hubei (Baokang).

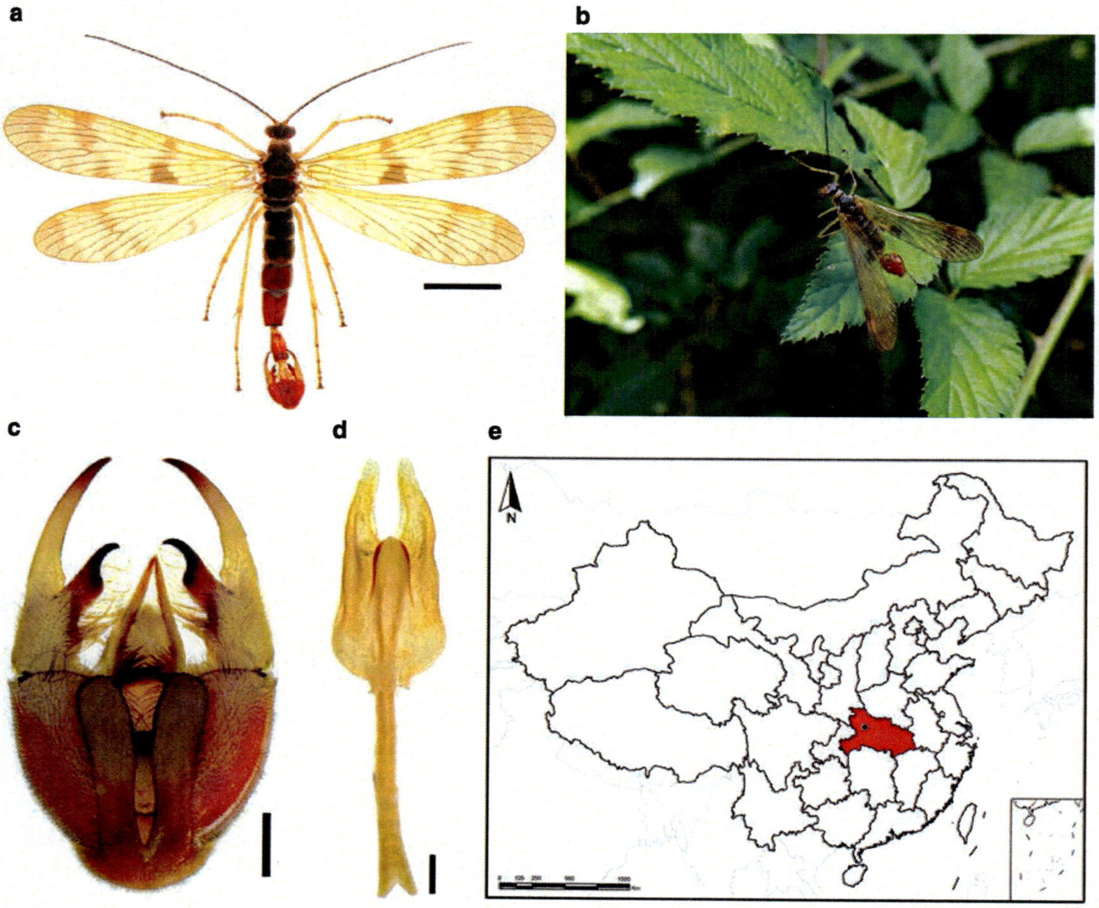

Fig. 10.210 (a) Male adult of ***Sinopanorpa baokangensis***, dorsal view (Baokang, Hubei, ©Mao Ye) (*Scale bar*: 5.0 mm) **[DALU]**. **(b)** Male adult of ***Sinopanorpa baokangensis*** in the field (Baokang, Hubei, ©Mao Ye). **(c)** Male genital bulb of ***Sinopanorpa baokangensis***, ventral view (*Scale bar*: 0.5 mm). **(d)** Female medigynium of ***Sinopanorpa baokangensis***, ventral view (*Scale bar*: 0.2 mm). **(e)** Distribution of ***Sinopanorpa baokangensis***

10.7.2 *Sinopanorpa digitiformis* Huang & Hua, 2008

(Digitate Sinoscorpionfly) [指形华蝎蛉] (Fig. 10.211)

Measurements. Forewing length 17.0–18.0 mm.

Morphology. Vertex dark brown. Ocellar triangle black. Rostrum yellowish brown. Nota dark brown. Pleura and legs light yellowish brown. Wing membrane translucent, tinged with waxy yellow. Markings yellowish brown. Pterostigma reddish brown. In forewings, apical band broad; pterostigmal band complete with basal branch and apical branch; basal band broad; marginal and basal spots absent. Male T1–T5 yellowish brown. Notal organ flat and covering acute postnotal organ. A6 yellowish brown, long cylindrical and bearing tuft of long seta on dorsal surface medially. A7 and A8 yellowish brown. A7 greatly stalked on basal third; A8 slightly constricted at base.

Male genital bulb broad oval. Epandrium long, broad, and slightly emarginate terminally. Hypandrium with greatly shortened basal stalk. Hypovalves long stripe-like with rounded apexes, extending to apex of gonocoxites, and bearing long bristles along inner margin. Gonostylus longer than gonocoxites, and bearing inconspicuous median tooth and stout acute and bifurcated basal process. Paramere finger-like, curved wave-like at middle, and straight in distal half, bearing row of short spines on inner margin. Aedeagus flat and not exceeding apex of gonocoxites.

Female subgenital long oval with subacute apex. Medigynium with broad, subtrapezoidal main plate widest at base; posterior arms short and thin; axis thin, longer than twice length of main plate with slightly diverged apodemes.

Emergence. June to August.

Distribution. Hubei (Shennongjia).

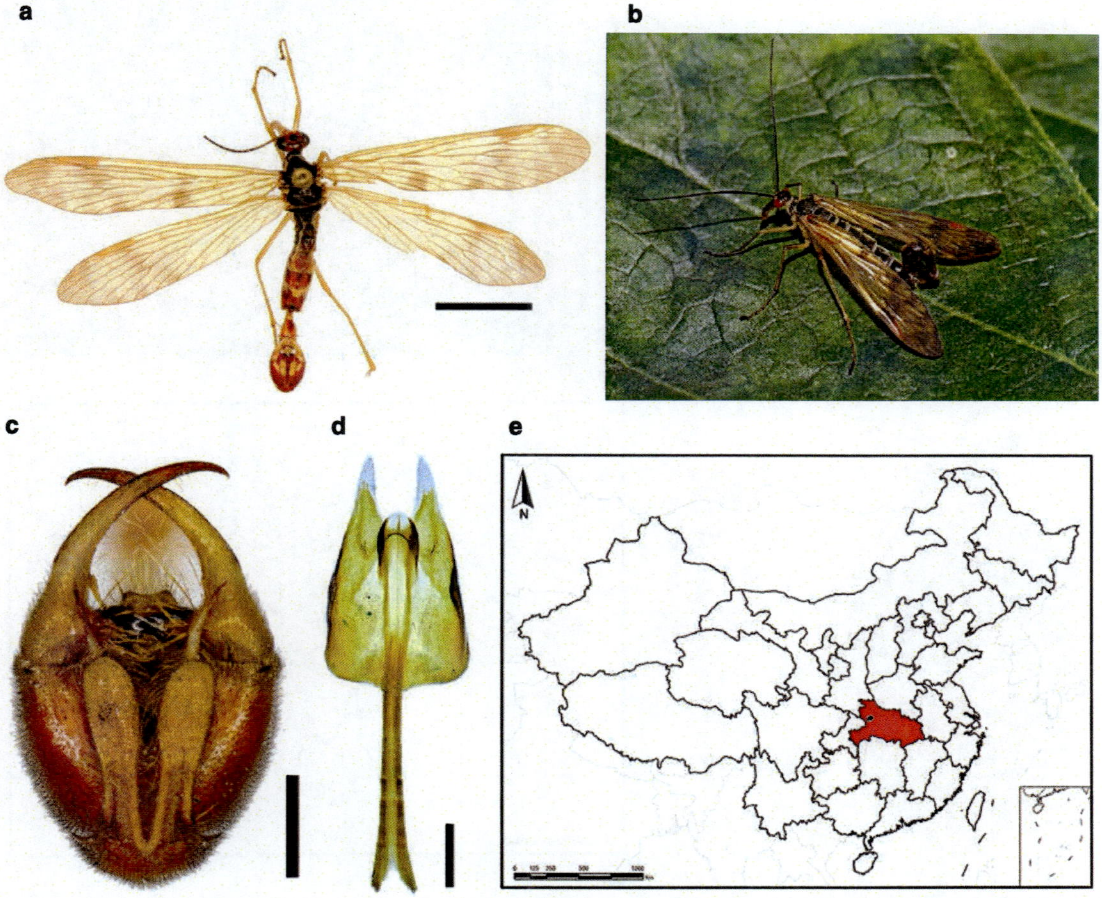

Fig. 10.211 (a) Male adult of ***Sinopanorpa digitiformis***, dorsal view (Shennongjia, Hubei) (*Scale bar*: 5.0 mm) [NWAU]. (b) Male adult of ***Sinopanorpa digitiformis***, dorsal view (Shennongjia, Hubei) (©Kai Gao). (c) Male genital bulb of ***Sinopanorpa digitiformis***, ventral view. (d) Female medigynium of ***Sinopanorpa digitiformis***, ventral view. (e) Distribution of ***Sinopanorpa digitiformis***

10.7.3 *Sinopanorpa nangongshana* Cai & Hua, 2008

(Nangongshan Sinoscorpionfly) [南宫山华蝎蛉]
(Fig. 10.212)

Measurements. Forewing length 16.0–17.0 mm.

Morphology. Vertex reddish brown. Ocellar triangle black. Rostrum yellowish brown. Nota dark brown. Pleura and legs creamy yellow. Wing membrane translucent, tinged with waxy yellow. Markings yellowish brown. Pterostigma reddish brown. In forewings, apical band broad; pterostigmal band complete, with basal branch and apical branch; basal band split into two spots; marginal and basal spots absent. Male T1–T5 dark brown. Notal organ flat and covering acute postnotal organ. A6 dark brown, long cylindrical and bearing tuft of long setae on dorsal surface medially. A7 and A8 dark brown. A7 greatly stalked on basal third; A8 slightly constricted at base.

Male genital bulb broad oval. Epandrium long, broad, and slightly emarginate terminally. Hypandrium with greatly shortened basal stalk. Hypovalves long stripe-like with rounded apexes, extending to apex of gonocoxites, and bearing long bristles along inner margin. Gonostylus longer than gonocoxites, and bearing inconspicuous median tooth and stout acute and bifurcated basal process. Paramere extremely elongated, exceeding median tooth of gonostylus, and bearing dense short spines on apical half. Aedeagus flat and not exceeding apex of gonocoxites.

Female subgenital long oval with subacute apex. Medigynium with broad, oblong main plate incurved along lateral margins; posterior arms short and stout; axis thin, nearly twice as long as main plate with slightly diverged apodemes.

Emergence. July to August.

Distribution. Shaanxi (Hualong Mountain, Nangong Mountain).

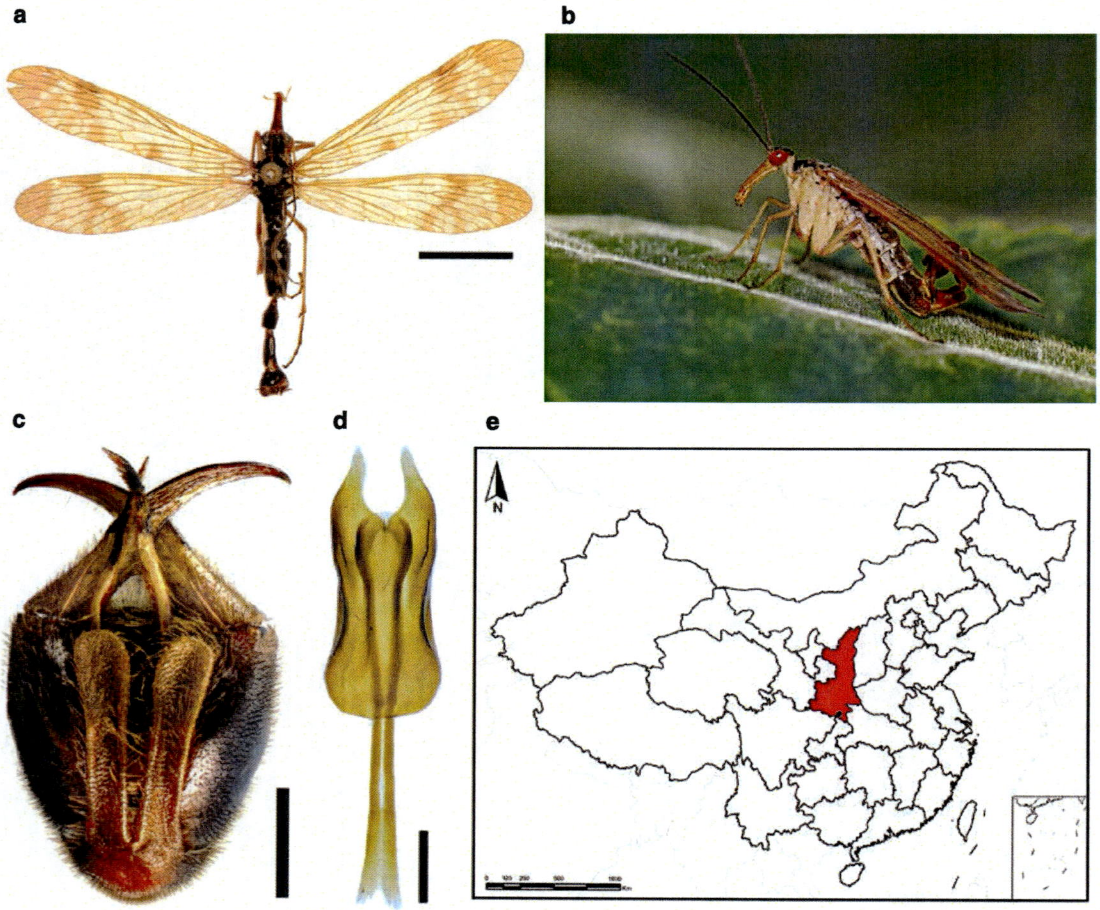

Fig. 10.212 (a) Male adult of ***Sinopanorpa nangongshana***, dorsal view (Nangong Mountain, Lan'gao, Shaanxi) (*Scale bar*: 5.0 mm) [NWAU]. (b) Male adult of ***Sinopanorpa nangongshana*** in the field (Nangong Mountain, Lan'gao, Shaanxi, ©Kai Gao). (c) Male genital bulb of ***Sinopanorpa nangongshana***, ventral view (*Scale bar*: 0.5 mm). (d) Female medigynium of ***Sinopanorpa nangongshana***, ventral view (*Scale bar*: 0.2 mm). (e) Distribution of ***Sinopanorpa nangongshana***

10.7.4 *Sinopanorpa tincta* Navás, 1931

(Tinged Sinoscorpionfly) [染翅华蝎蛉] (Fig. 10.213)
Measurements. Forewing length 15.5–17.0 mm.

Morphology. Vertex dark brown. Ocellar triangle black. Rostrum yellowish brown. Nota dark brown. Pleura and legs creamy yellow. Wing membrane translucent, tinged with waxy yellow. Markings yellowish brown. Pterostigma reddish brown. In forewings, apical band broad; pterostigmal band complete, with basal branch and apical branch; basal band split into two spots; marginal spot small; basal spot absent. Male T1–T5 yellowish brown. Notal organ flat and covering the acute postnotal organ. A6 yellowish brown, cylindrical, and bearing tuft of long setae on the dorsal surface medially. A7 and A8 yellowish brown. A7 greatly stalked on the basal third; A8 slightly constricted at base.

Male genital bulb oval. Epandrium long, and slightly emarginate terminally. Hypandrium with greatly shortened basal stalk. Hypovalves long stripe-like with rounded apexes, extending to the apex of gonocoxites, and bearing long bristles along inner margin. Gonostylus longer than gonocoxites,

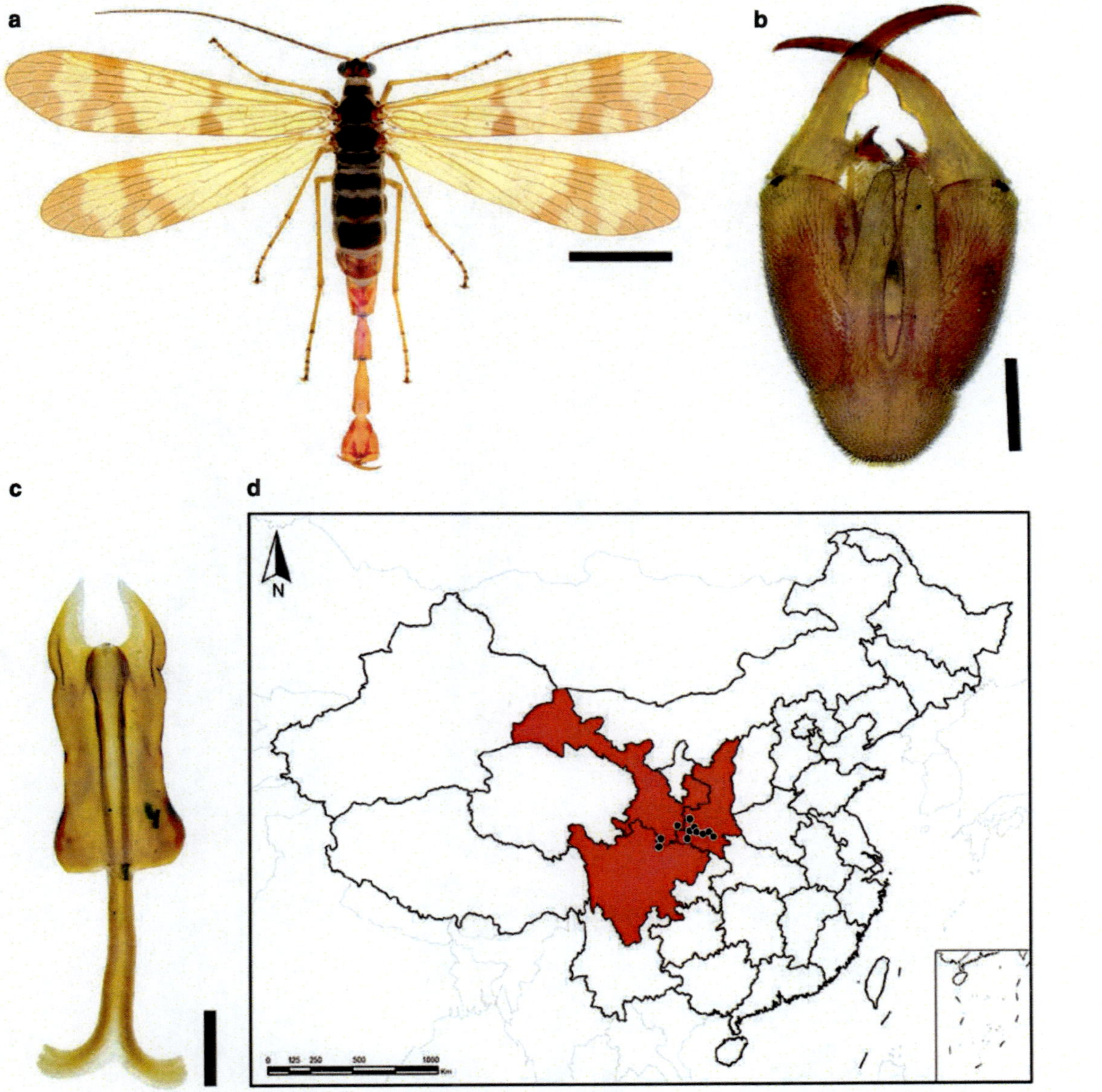

Fig. 10.213 (a) Male adult of ***Sinopanorpa tincta***, dorsal view (Tiantaishan National Forest Park, Baoji, Shaanxi) (*Scale bar*: 5.0 mm) [NWAU]. (b) Male genital bulb of ***Sinopanorpa tincta***, ventral view (*Scale bar*: 0.5 mm). (c) Female medigynium of ***Sinopanorpa tincta***, ventral view (*Scale bar*: 0.2 mm). (d) Distribution of ***Sinopanorpa tincta***

and bearing inconspicuous median tooth and stout acute and bifurcated basal process on the inner margin. Paramere slender, extending to the basal process of gonostylus, and bearing row of comb-like spines along inner margin. Aedeagus flat and not exceeding apex of gonocoxites.

Female subgenital long oval with subacute apex. Medigynium with broad, oblong main plate incurved along lateral margins; posterior arms short and stout; axis thin, nearly twice as long as main plate with apodemes widely diverged at basal portion.

Emergence. July to August.

Distribution. Shaanxi (Baoji, Liuba, Nanzheng, Ningshan, Taibai, Zhouzhi), Gansu (Huixian, Wenxian), Sichuan (Pingwu).

11 Family Panorpodidae Byers, 1965

(Short-Faced Scorpionflies) [拟蝎蛉科]

The short-faced scorpionflies are very similar to the common scorpionflies, but have relatively shorter rostrums and A6–A8. Two genera and more than 20 species have been described worldwide, and only three species in one genus are documented from China. One species is included in this atlas (Fig. 11.1).

Fig. 11.1 Adult of Panorpodidae. Female adult of *Panorpodes paradoxa* in the field (Toyama, Honshu, Japan)

11.1 Genus *Panorpodes* MacLachlan, 1875

Fifteen known species in the genus *Panorpodes* are distributed in the north Pacific Rim, including northeastern China, Japan, Korean Peninsula, and North America. Three species are recorded in China, and only one species is included in this atlas.

11.1.1 *Panorpodes komaensis* Okamoto, 1925

(Korean Short-faced Scorpionfly) [朝鲜拟蝎蛉]

Panorpodes kuandianensis Zhong, Zhang & Hua, 2011, **new synonym** (Figs. 11.2, 11.3, 11.4, and 11.5)

Measurements. Forewing length 12.4–15.2 mm.

Morphology. Vertex yellow. Ocellar triangle yellowish brown, bearing four conspicuous ocellar bristles. Rostrum yellow, short, and stout. Nota yellowish brown and slightly deepened towards lateral portion. Pleura and legs light yellowish brown. Wing membrane translucent, tinged with light yellow. In forewings, apical band irregular with series of large hyaline spots innerly; pterostigmal band greatly reduced; other markings absent. Occasionally, all markings absent in some individuals. Male T1–T5 yellowish brown, with darkened flanks. Notal organ poorly developed and slightly covering acute postnotal organ. A6–A8 similar to the proceeding segments and not elongated as in Panorpidae.

Male genital bulb long elliptical. Epandrium long and broad, with rounded apex. Hypandrium long and broad. Hypovalves truncated terminally, extending beyond the apex of gonocoxites, bearing auriculate process subbasally and 10–16 black short spines on the inner margin; crescent window formed between the two hypovalves basally. Gonostylus shorter than half the length of gonocoxites, and bearing finger-like basal process on the inner margin. Paramere almost fused with aedeagus. Aedeagus with ventral and dorsal valves intensively fused with rounded lateral processes.

Female subgenital plate with unfused gonocoxosternites VIII. Medigynium minute, laterally with two knob-like processes; axis slightly projected caudally and lacking apodemes.

Emergence. June to July.

Distribution. Liaoning (Kuandian); North & South Korea; (possibly) Russian Far East.

Remarks. *Panorpodes kuandianensis* Zhong, Zhang & Hua, 2011 was described from the Huabo Mountain (Kuandian, Liaoning, northeastern China), approximately 50 km north from the Chinese-North Korean border. Zhong et al. (2011) claimed that *Po. kuandianensis* can be differentiated from the Korean species *Panorpodes komaensis* Okamoto, 1925 by "pleura of abdominal segment VII strongly sclerotized and indicated by longitudinal fissures,"

J.-S. Wang, B.-Z. Hua, *A Color Atlas of the Chinese Mecoptera*, https://doi.org/10.1007/978-981-16-9558-2_11

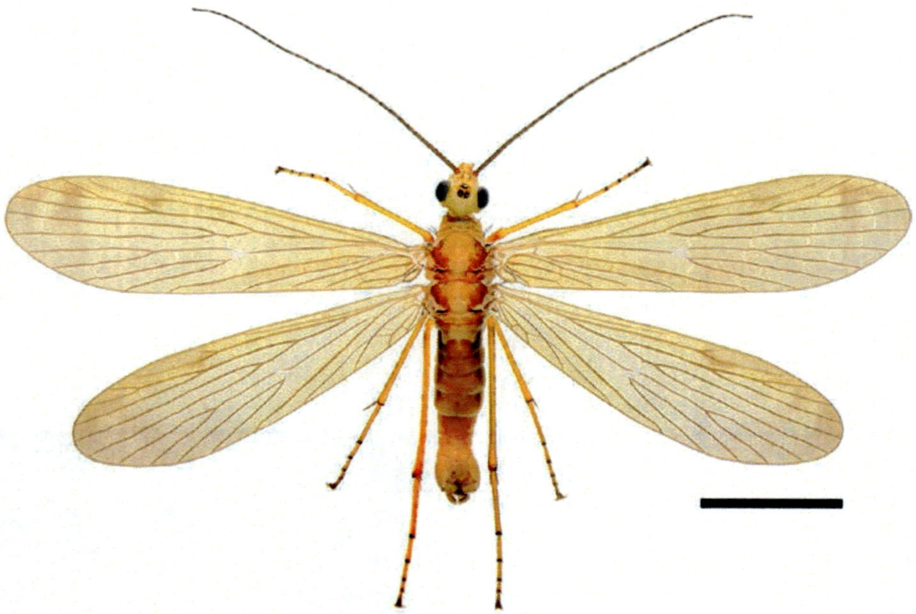

Fig. 11.2 Male adult of *Panorpodes komaensis*, dorsal view (Huabo Mountain, Kuandian, Liaoning) (*Scale bar*: 5.0 mm) [NWAU]

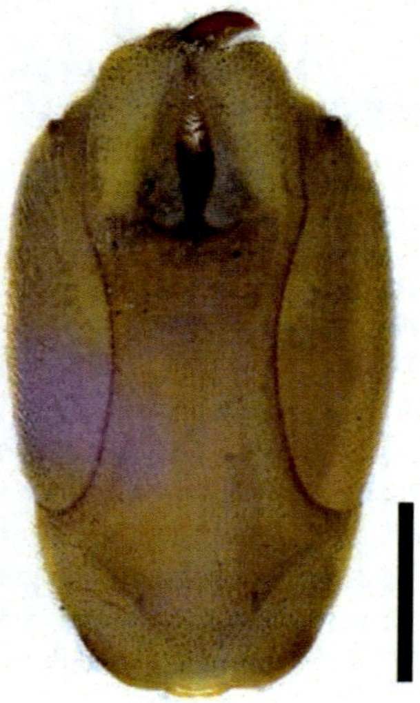

Fig. 11.3 Male genital bulb of *Panorpodes komaensis*, ventral view (*Scale bar*: 0.5 mm)

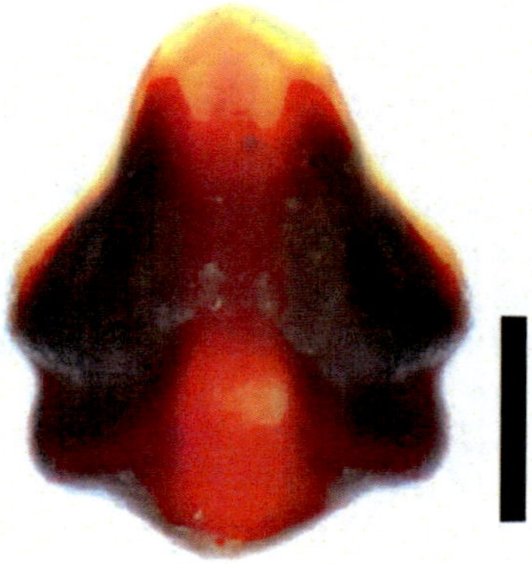

Fig. 11.4 Female medigynium of *Panorpa komaensis*, ventral view (*Scale bar*: 0.2 mm)

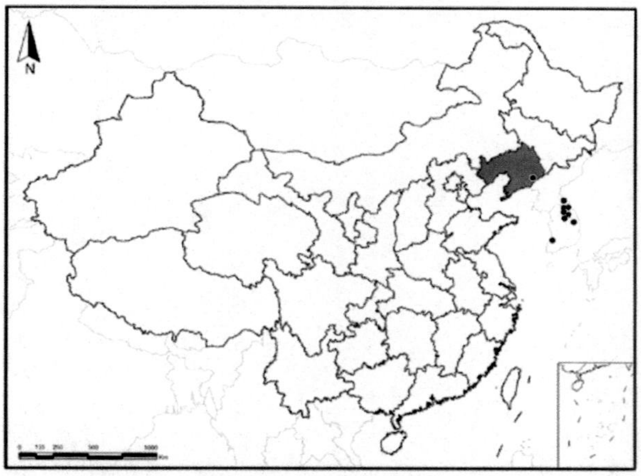

Fig. 11.5 Distribution of *Panorpodes komaensis*

and "hypovalves nearly parallel, shorter than the stalk, with more than 10 black spines along the apical inner margin." However, the "longitudinal fissures" can be observed in all species in *Panorpodes*, the hypovalves of *Po. komaensis* are also shorter than the stalk of hypandrium, and the number of the black spines can be varied in different individuals and by different directions of observation. In addition, by the BLAST analysis in NCBI (National Center for Biological Information), the *COI* (cytochrome oxidase subunit I) genes entirely matched (percent identity: 100%) for three records: *Po. kuandianensis* from China (GenBank Accession Number: KT943364.1), South Korea (LC479905.1), and *Po. komaensis* from South Korea (MN345588.1). It is almost certain that these two names are pertaining to a same species. Therefore, we herein treat *Po. kuandianensis* a junior synonym of *Po. komaensis*.

Checklist of the Chinese Mecoptera

Family Bittacidae Handlirsch, 1906

Genus *Bicaubittacus* Tan & Hua, 2009

1. *Bicaubittacus appendiculatus* (Esben-Petersen, 1927)
2. *Bicaubittacus burmanus* (Tjeder, 1973)
3. *Bicaubittacus longiprocessus* (Huang & Hua, 2005)
4. *Bicaubittapous mengyangicus* Tan & Hua, 2009
5. *Bicaubittacus yangi* Tan & Hua, 2009

Genus *Bittacus* Latreille, 1802

6. *Bittacus acutus* Zhang, Du & Hua, 2020
7. *Bitacus bifurcatus* Hua & Tan in Hua et al., 2008
8. *Bittacus carpenteri* Cheng, 1957
9. *Bittacus choui* Hua & Tan, 2007
10. *Bittacus chujoi* Issiki & Cheng, 1947
11. *Bittacus cirratus* Tjeder, 1956
    Syn. *Bittacus zoensis* Cheng, 1957, synonymized by Byers, 1970
12. *Bittacus contumax* Tjeder, 1956
13. *Bittacus coreanus* Issiki, 1929
14. *Bittacus dashahensis* Zhou & Zhou, 2005
15. *Bittacus diaoluoshamus* Chen & Hua, 2011
16. *Bittacus falcatus* Zhou & Zhou, 2007
17. *Bittacus flavidus* Huang & Hua, 2005
18. *Bittacus formosamus* Issiki, 1927
19. *Bittacus gressitti* Cheng, 1957
20. *Bittacus hainanicus* Tan & Hua, 2008
21. *Bittacus jiulianshanus* Chen, Tan & Hua, 2013
22. *Bittacus kuankuoshuensis* Zhou & Zhou, 2012
23. *Bittacus lii* Zhou, 2003
24. *Bittacus longantennatus* Chen, Tan & Hua, 2013
25. *Bittacus longilobus* Zhang, Du & Hua, 2020
26. *Bittacus maculatus* Issiki, 1927
27. *Bittacus malaisei* Tjeder, 1973
28. *Bittacus monastyrskiyi* Bicha, 2007
29. *Bittacus obscurus* Huang & Hua, 2005
30. *Bittacus pieli* Navás, 1935
31. *Bittacus planus* Cheng, 1949
32. *Bittacus puripennis* Cai & Hua in Cai et al. 2006.
33. *Bittacus setigerus* Chen, Tan & Hua, 2013
34. *Bittacus sinensis* Walker, 1853
    Syn. *Bitacus quaternipunctatus* Enderlein, 1910, synonymized by Esben-Petersen, 1921
    Syn. *Bittacus strategus* Navás, 1913, synonymized by Esben-Petersen, 1921
35. *Bittacus sinicus* Issiki, 1931
36. *Bittacus shaoguanensis* Zhang, Du & Hua, 2020
37. *Bittacus sonani* Issiki, 1929
38. *Bittacus stigmosus* Zhou & Zhou, 2005
39. *Bittacus striatus* Issiki, 1927
40. *Bittacus strigatus* Hua & Chou, 1998
41. *Bittacus tienmushana* Cheng, 1957
42. *Bittacus trapezoideus* Huang & Hua, 2005
43. *Bittacus triangularis* Issiki, 1929
44. *Bittacus ussuriensis* Plutenko, 1985
45. *Bittacus vexiliferus* Byers, 1970
46. *Bittacus zhejiangicus* Tan & Hua, 2008

Genus *Terrobittacus* Tan & Hua, 2009

47. *Terrobittacus angustus* Du & Hua, 2017
48. *Terrobittacus echinatus* (Hua & Huang in Hua et al., 2008)
49. *Terrobittacus implicatus* (Huang & Hua in Cai et al., 2006)
50. *Terrobittacus rostratus* Du & Hua, 2017
51. *Terrobittacus wui* (Zhou, 2001) **[New Combination]**
    Bittacus wui Zhou, 2001
    Terrobittacus longisetus Tan & Hua, 2009, **new synonym**
52. *Terrobittacus xiphicus* Tan & Hua, 2009

Family Panorpidae Latreille, 1802

Subfamily Neopanorpinae Wang & Hua, 2021

Genus *Neopanorpa* van der Weele, 1909

Neopanorpa chillcotti **group**

53. *Neopanorpa chillcotti* Byers, 1971
54. *Neopanorpa hushengchangi* Hua & Chou, 1999
55. *Neopanorpa liuxingyuei* Wang, 2021

J.-S. Wang, B.-Z. Hua, *A Color Atlas of the Chinese Mecoptera*, https://doi.org/10.1007/978-981-16-9558-2

56. *Neopanorpa wuchaoi* Wang, 2021

Neopanorpa pielina **group**

57. *Neopanorpa chaohsiufui* Wang & Hua, 2017
    = *Neopanorpa chaoi* Zhou, 2003, preoccupied by *N. chaoi* Cheng, 1957
58. *Neopanorpa choui* Cheng, 1949
59. *Neopanorpa curva* Zhou, 2000
60. *Neopanorpa diancangshanensis* Wang & Hua, 2018
61. *Neopanorpa dispar* Issiki & Cheng, 1947
62. *Neopanorpa dubis* Chou & Wang in Chou et al., 1988
63. *Neopanorpa forcula* Wang & Hua, 2019
    = *Neopanorpa furcata* Zhou, 2005, preoccupied by *Neopanorpa furcata* (Hardwicke, 1825)
64. *Neopanorpa hainanica* Hua & Chou, 1998
65. *Neopanorpa harmandi* (Navás, 1908)
    Syn. *Neopanorpa harmandi conjuncta* Navás, 1930, synonymized by Byers, 1965
66. *Neopanorpa hualizhongi* Hua & Chou, 1998
67. *Neopanorpa jigongshanensis* Hua, 1998
68. *Neopanorpa kwangtsehi* Cheng, 1957
69. *Neopanorpa leigongshana* Zhou & Zhou, 2007
70. *Neopanorpa lifashengi* Hua & Chou, 1999
71. *Neopanorpa longiprocessa* Hua & Chou, 1997
72. *Neopanorpa longistipitata* Wang & Hua, 2018
73. *Neopanorpa luojishana* Wang & Hua, 2019
74. *Neopanorpa maai* Cheng, 1957
75. *Neopanorpa mangshanensis* Chou & Wang in Chou et al., 1988
76. *Neopanorpa magnatitilana* Wang & Hua, 2018
77. *Neopanorpa makii* Issiki, 1927
78. *Neopanorpa menghaiensis* Zhou & Wu in Zhou et al., 1993a
79. *Neopanorpa minuta* Chou & Wang in Chou et al., 1988
80. *Neopanorpa moganshanensis* Zhou & Wu in Zhou et al., 1993a
81. *Neopanorpa mutabilis* Cheng, 1957
82. *Neopanorpa nigritis* Carpehter, 1938
    Neopanorpa pallivalva Zhou, 2003, **new synonym**
83. *Neopanorpa ovata* Cheng, 1957
84. *Neopanorpa pennyi* Byers, 1999 **[New Record to China]**
    Neopanorpa uncata Zhou, 2000, **new synonym**
85. *Neopanorpa pielina* Navás, 1936
86. *Neopanorpa puripennis* Chou & Wang in Chou et al., 1988
87. *Neopanorpa qudristigma* Wang & Hua, 2018
88. *Neopanorpa retina* Chou & Li in Chou et al., 1988
89. *Neopanorpa semiorbiculata* Wang & Hua, 2018
90. *Neopanorpa setigera* Wang & Hua, 2018
91. *Neopanorpa simulans* Byers, 2008
    = *Neopanorpa similis* Byers, 2002, preoccupied by *N. similis* Byers, 1999
92. *Neopanorpa tengchongensis* Zhou & Wu in Zhou et al., 1993a
93. *Neopanorpa tenuis* Zhou, 2000
94. *Neopanorpa tibetensis* Hua & Chou, 1999
95. *Neopanorpa tienpingshana* Chou & Wang in Chou et al., 1988
96. *Neopanorpa tincta* Wang & Hua, 2018
97. *Neopanorpa triangulata* Wang & Hua, 2018
98. *Neopanorpa uncinella* Qian & Zhou, 2001
99. *Neopanorpa wangcaensis* Zhou & Zhou, 2012
100. *Neopanorpa xingmini* Wang & Hua, 2019
101. *Neapanorpa yingjiangensis* Zhou, 2000
102. *Neopanorpa zhengyucheni* Wang, 2021

Neopanorpa brisi **group**

103. *Neopanorpa apicata* Navás, 1922
104. *Neopanorpa brisi* (Navás, 1930)
    Leptopanorpa brisi Navás, 1930
    Neopanorpa banksi Carpenter, 1938, **new synonym**
    Neopanorpa dimidiata Navás, 1930, synonymized by Wang & Hua, 2018
    Neopanorpa cangshanensis Zhou, 2006, **new synonym**
    Neopanorpa pilosa Carpenter, 1945, **new synonym**
    Neopanorpa taoi Cheng, 1949, **new synonym**
    Neopanorpa validipennis Cheng, 1949, **new synonym**
    Neopanorpa varia Cheng, 1949, **new synonym**
105. *Neopanorpa fangxianga* Zhou & Zhou, 2007
106. *Neopanorpa pendula* Qian & Zhou, 2001

Neopanorpa cavaleriei **group**

107. *Neopanorpa cavaleriei* (Navás, 1908)
    Neopanorpa maolanensis Zhou & Bao, 2002, **new synonym**
108. *Neopanorpa cantonensis* Cheng, 1957
109. *Neopanorpa carpenteri* Cheng, 1957
110. *Neopanorpa chelata* Carpenter, 1938
    Neopanorp parva Carpenter, 1945, **new synonym**
    Neopanorpa latipennis Cheng, 1949, **new synonym**
    Neopanorpa heii Cheng, 1957, **new synonym**
111. *Neopanorpa hunanensis* Hua, 2002
    = *Neopanorpa magna* Chou & Wang in Chou et al., 1988, preoccupied by *N. magna* Isiki, 1927
    Neopanorpa anchoroides Zhou, 2003, **new synonym**
112. *Neopanorpa ocellaris* (Navás, 1908)
113. *Neopanorpa abstrusa* Zhou & Wu in Zhou et al., 1993a
114. *Neopanorpa auriculata* Zhou, 2005

115. *Neopanorpa chaoi* Cheng, 1957
Hom. *Neopanorpa chaoi* Zhou, 2003
116. *Neopanorpa circularis* Ju & Zhou, 2003
117. *Neopanorpa clara* Chou & Wang in Chou et al., 1988
118. *Neopanorpa claripennis* Carpenter, 1938
Neopanorpa lui Chou & Ran in Chou et al., 1981, **new synonym**
119. *Neopanorpa cuspidata* Byers, 1965
120. *Neopanorpa formosana* (Navás, 1911)
121. *Neopanorpa gulinensis* Zhou & Zhou, 2005
Neopanorpa linjiangensis Zhou, 2005, **new synonym**
122. *Neopanorpa hei* Zhou & Fan, 1998
123. *Neopanorpa heii* Cheng, 1949
124. *Neopanorpa k-maculata* Cheng, 1952
125. *Neopanorpa lacunaris* Navás, 1930
126. *Neopanorpa lichuanensis* Cheng, 1957
127. *Neopanorpa lipingensis* Cai & Hua, 2009
128. *Neopanorpa lungtausana* Cheng, 1957
Syn. *Neopanorpa brevivalvae* Chou & Wang in Chou et al., 1988, synonymized by Wang & Hua, 2019b
129. *Neopanorpa magna* Issiki, 1927
Hom. *Neopanorpa magna* Chou & Wang in Chou et al., 1988, replaced by *Neopanorpa hunanensis* Hua, 2002
130. *Neopanorpa mayangensis* Zhou & Zhou, 2010
131. *Neopanorpa nielseni* Byers, 1965
132. *Neopanorpa pulchra* Carpenter, 1945
133. *Neopanorpa parva* Carpenter, 1945
134. *Neopanorpa sheni* Hua & Chou, 1997
135. *Neopanorpa siamensis* Byers, 1965
136. *Neopanorpa spatulata* Byers, 1965
137. *Neopanorpa formosensis* Navás, 1930
138. *Neopanorpa xishuiensis* Zhou, 2005

Neopanorpa tienmushana **group**
139. *Neopanorpa caveata* Cheng, 1957
140. *Neopanorpa gradana* Cheng, 1952
141. *Neopanorpa huangshana* Cheng, 1957
142. *Neopanorpa jiulongensis* Zhou in Zhou et al., 1993b
143. *Neopanorpa sauteri* (Esben-Petersen, 1912)
144. *Neopanorpa mokansana* Cheng, 1957
145. *Neopanorpa montana* Zhou in Zhou et al., 1993b
146. *Neopanorpa ophthalmica* (Navás, 1911)
147. *Neopanorpa tienmushana* Cheng, 1957
148. *Neopanorpa translucida* Cheng, 1957
149. *Neopanorpa youngi* Byers, 1994

Subfamily Panorpinae Latreille, 1802

Genus *Cerapanorpa* Gao, Ma & Hua, 2016
150. *Cerapanorpa alpina* Gao & Hua, 2021
151. *Cerapanorpa baimaensis* Gao & Hua, 2019
152. *Cerapanorpa bicornifera* (Chou & Wang in Chou et al., 1981)
153. *Cerapanorpa brevicornis* (Hua & Li in Li et al., 2007)
154. *Cerapanorpa bonis* (Cheng, 1949)
155. *Cerapanorpa byersi* (Hua & Huang in Li et al., 2007)
156. *Cerapanorpa centralis* (Tjeder, 1936)
157. *Cerapanorpa dubia* (Chou & Wang in Chou et al., 1981)
158. *Cerapanorpa emarginata* (Cheng, 1949)
159. *Cerapanorpa funiushana* (Hua & Chou, 1997)
160. *Cerapanorpa liupanshana* Gao, Ma & Hua, 2016
161. *Cerapanorpa minshana* Gao, Li & Hua, 2020
162. *Cerapanorpa nanwutaina* (Chou in Chou et al., 1981)
163. *Cerapanorpa obtusa* (Cheng, 1949)
Syn. *Panorpa leei* Cheng, 1949, synonymized by Ma et al., 2014
Syn. *Panorpa alticola* Zhou, 2000, synonymized by Gao et al., 2016
164. *Cerapanorpa protrudens* Gao, Ma & Hua, 2016
165. *Cerapanorpa qinlingensis* Gao, Li & Hua, 2020
166. *Cerapanorpa reni* (Chou in Chou et al., 1981)
167. *Cerapanorpa sinuata* Gao, Ma & Hua, 2016
168. *Cerapanorpa taizishana* Gao & Hua, 2019
169. *Cerapanorpa wangwushana* (Huang, Hua & Shen, 2004)
170. *Cerapanorpa xuebaodinga* Gao & Hua, 2019
171. *Cerapanorpa yanggashana* Gao & Hua, 2019

Genus *Dicerapanorpa* Zhong & Hua, 2013

Dicerapanorpa magna **group**
172. *Dicerapanorpa baiyunshana* Zhong & Hua, 2013
173. *Dicerapanorpa hualongshana* Hu & Hua, 2019 in Hu et al., 2019
174. *Dicerapanorpa magna* (Chou in Chou et al., 1981)
175. *Dicerapanorpa minshana* Hu & Hua in Hu et al., 2019
176. *Dicerapanorpa shennongensis* Zhong & Hua, 2013

Dicerapanorpa diceras **group**
177. *Dicerapanorpa bifurcata* Hu & Hua, 2020
178. *Dicerapanorpa diceras* (MacLachlan, 1894)
Syn. *Panorpa grahami* Carpenter, 1938, synonymized by Carpenter, 1948
179. *Dicerapanorpa deqenensis* Hu, Wang & Hua, 2019
180. *Dicerapanorpa harmonia* Wang, 2022
181. *Dicerapanorpa huangguocongi* Wang, 2022
182. *Dicerapanorpa kimminsi* (Carpenter, 1948)
183. *Dicerapanorpa lativalva* Hu & Hua in Hu et al., 2020
184. *Dicerapanorpa luojishana* Hu & Hua, 2019
185. *Dicerapanorpa macula* Hu, Wang & Hua, 2019
186. *Dicerapanorpa nakhi* Wang, 2022
187. *Dicerapanorpa stotzneri* (Esben-Petersen, 1934)
188. *Dicerapanorpa tanae* Hu, Wang & Hua, 2019
189. *Dicerapanorpa tenuis* Hu, Wang & Hua, 2019
190. *Dicerapanorpa tjederi* (Carpenter, 1938)
191. *Dicerapanorpa triclada* (Qian & Zhou, 2001)
192. *Dicerapanorpa yangqichengi* Wang, 2022
193. *Dicerapanorpa yijunae* Hu & Hua, 2019
194. *Dicerapanorpa zhengkuni* Hu & Hua, 2020

195. *Dicerapanorpa zhongdianensis* Hu, Wang & Hua, 2019

Genus *Furcatopanorpa* Ma & Hua, 2011

196. *Furcatopanorpa longihypovalva* (Hua & Cai, 2009)

Genus *Megapanorpa* Wang & Hua, 2019

197. *Megapanorpa absens* Wang & Hua, 2019
198. *Megapanorpa anfracta* (Ju & Zhou, 2003) **[New Combination]**
    Panorpa anfracta Ju & Zhou, 2003
199. *Megapanorpa gaokaii* Wang & Hua, 2019
200. *Megapanorpa grandis* Wang & Hua, 2019
201. *Megapanorpa jiangorum* Wang & Hua, 2019
202. *Megapanorpa wanghongjiani* Wang & Hua, 2019

Genus *Panorpa* Linnaeus, 1758

Panorpa kongosana **group**

203. *Panorpa kongosana* Okamoto, 1925
    Panorpa macrostyla Hua, 1998, **new synonym**
204. *Panorpa okamotona* Issiki, 1927

Panorpa waongkehzengi **group**

205. *Panorpa jinhuaensis* Wang, Gao & Hua, 2019
206. *Panorpa menqiuleii* Wang, Gao & Hua, 2019
207. *Panorpa pieli* Cheng, 1957
208. *Panorpa pusilla* Cheng, 1949
209. *Panorpa waongkehzengi* Navás, 1935

Panorpa amurensis **group**

210. *Panorpa amurensis* MacLaclan, 1872
    Panorpa liui Hua, 1997, **new synonym**
    Syn. *Panorpa jilinensis* Zhou, 2000, synonymized with *P. liui* by Li et al., 2021
    Panorpa orientalis MacLachlan, 1887, **new synonym**
    Syn. *Panorpa sachalinensis* Matsumura, 1911, synonymized with *P. orientalis* by Issiki, 1933
    Panorpa neojilinensis Li & Lin in Li et al., 2012, **new synonym**
    Syn. *Aulops melania* Navás, 1912, synonymized by Esben-Petersen 1915
211. *Panorpa baohwashana* Cheng, 1957

Panorpa japonica **group**

212. *Panorpa kellogi* Cheng, 1957

Panorpa wormaldi **group**

213. *Panorpa fengyanga* Wang & Suzuki, 2022
214. *Panorpa gressitti* Byers, 1970
215. *Panorpa implicata* Cheng, 1957
216. *Panorpa longiramina* Issiki & Cheng, 1947
217. *Panorpa nudiramus* Byers, 2002
218. *Panorpa obliqua* Carpenter, 1945
219. *Panorpa obliquifascia* Chou & Wang in Chou et al., 1987
220. *Panorpa ochraceocauda* Issiki, 1927
221. *Panorpa peterseana* Issiki, 1927
222. *Panorpa zhuohengi* Wang & Suzuki, 2022

Panorpa guttata **group**

223. *Panorpa caoweii* Wang, 2021
224. *Panorpa dali* Wang, 2021
225. *Panorpa decolorata* Chou & Wang in Chou et al., 1981
226. *Panorpa duanyu* Wang & Gong, 2021
227. *Panorpa filina* Chou & Wang in Chou et al., 1987
228. *Panorpa furcata* Zhou & Zhou, 2007
229. *Panorpa guttata* Navás, 1908
230. *Panorpa hani* Wang, 2021
231. *Panorpa hirundo* Wang, 2021
232. *Panorpa huangguiqiangi* Wang, 2021
233. *Panorpa jiangrixini* Wang, 2021
234. *Panorpa jinfoshana* Wang, 2021
235. *Panorpa latiloba* Wang, 2021
236. *Panorpa kunmingensis* Fu & Hua, 2009
237. *Panorpa liaoi* Zhou & Zhou, 2007
    = *Panorpa liaoae* Zhou & Zhou, 2007
238. *Panorpa nanzhao* Wang, 2021
239. *Panorpa parallela* Wang & Hua, 2016
240. *Panorpa qiana* Zhou & Zhou, 2010
241. *Panorpa reflexa* Wang & Hua, 2016
242. *Panorpa stella* Wang, 2021
243. *Panorpa substricta* Wang, 2021
244. *Panorpa xiaofeng* Wang & Gong, 2021
245. *Panorpa xuzhu* Wang & Gong, 2021

Panorpa communis **group**

246. *Panorpa sibirica* Esben-Petersen, 1915
    Panorpa changbaishana Hua in Hou & Hua, 2008, **new synonym**
    = *Panorpa choui* Hua, 1998, preoccupied by *Panorpa choui* Zhou & Wu, 1993

Panorpa cheni **group**

247. *Panorpa alata* Zhou & Zhou, 2005
248. *Panorpa bistriata* Issiki, 1929
249. *Panorpa bifasciata* Chou & Wang in Chou et al., 1981
250. *Panorpa cheni* Cheng, 1957
251. *Panorpa choui* Zhou & Wu in Zhou et al., 1993a
    Hom. *Panorpa choui* Hua, 1998, replaced by *P. changbaishana* Hua in Hou & Hua, 2008
252. *Panorpa cladocerca* Navás, 1935
253. *Panorpa songes* Zhou & Zhou, 2005
254. *Panorpa trifasciata* Cheng, 1957
255. *Panorpa tritaenia* Chou & Wang in Chou et al., 1987

Panorpa semifasciata **group**

256. *Panorpa carpenteri* Cheng, 1957
257. *Panorpa emeishana* Hua, Sun & Li, 2001

258. *Pamorpa grahamana* Cheng, 1957
259. *Panorpa semifasciata* Cheng, 1949

Panorpa stigmalis **group**

260. *Panorpa dashahensis* Zhou & Zhou, 2005
261. *Panorpa quadrifasciata* Chou & Wang in Chou et al., 1987
    Panorpa subaurea Chou & Li in Chou et al., 1987, **new synonym**
262. *Panorpa stigmalis* Navás, 1908

Panorpa deceptor **group**

263. *Panorpa akasakai* Issiki, 1929
264. *Panorpa angustistriata* Issiki, 1929
265. *Panorpa anrenensis* Chou & Wang in Chou et al., 1987
266. *Panorpa apiconebulosa* Issiki, 1929
267. *Panorpa aurea* Cheng, 1957
268. *Panorpa auripennis* Bicha, 2019 [**New Record to China**]
269. *Panorpa brevititilana* Issiki, 1929
270. *Panorpa bunun* Issiki, 1929
271. *Panorpa deceptor* Esben-Petersen, 1913
272. *Panorpa coomani* Cheng, 1957
273. *Panorpa esakii* Issiki, 1929
274. *Panorpa falsa* Issiki & Cheng, 1947
275. *Panorpa flavicorporis* Cheng, 1957
276. *Panorpa flavipennis* Carpenter, 1938
277. *Panorpa fukiensis* Tjeder, 1950
278. *Panorpa hamata* Issiki & Cheng, 1947
279. *Panorpa klapperichi* Tjeder, 1950
280. *Panorpa lintienshana* Cheng, 1952
281. *Panorpa longititilana* Issiki, 1929
282. *Panorpa lutea* Carpenter, 1945
283. *Panorpa mokansana* Cheng, 1957
    Syn. *Panorpa wrighti* Cheng, 1957, synonymized by Wang et al., 2019
284. *Panorpa nokoensis* Issiki, 1929
285. *Panorpa tetrazonia* Navás, 1935
286. *Panorpa insularis* Hua & Chou, 1998
287. *Panorpa issikii* Penny & Byers, 1979
    = *Panorpa acuta* Issiki & Cheng, 1947, preoccupied by *P. acuta* Carpenter, 1931
288. *Panorpa mangshanensis* Chou & Wang in Chou et al., 1987
289. *Panorpa pallidimaculata* Issiki, 1929
290. *Panorpa pectinata* Issiki, 1929
291. *Panorpa pingjiangensis* Chou & Wang in Chou et al., 1987
292. *Panorpa rantaisanensis* Issiki, 1929
293. *Panorpa sextaenia* Zhou & Bao, 2002
294. *Panorpa shibatai* Issiki, 1929
295. *Panorpa sonani* Issiki, 1929
296. *Panorpa subambra* Chou & Tong in Chou et al., 1987
297. *Panorpa taiheisanensis* Issiki, 1929
298. *Panorpa taiwanensis* Issiki, 1929
299. *Panorpa tecta* Byers, 2002
300. *Panorpa yiei* Issiki & Cheng, 1947

Panorpa cornigera **group**

301. *Panorpa cornigera* MacLachlan, 1887
    Panorpa acutipennis Hua, 1998, **new synonym**
    Panorpa chonmasanus Li and Zhang in Li et al., 2012, **new synonym**

Panorpa davidi **group**

302. *Panorpa bashanicola* Hua, Tao & Hua, 2018
303. *Panorpa biclada* Zhang & Hua, 2012
304. *Panorpa chengi* Chou in Chou et al., 1981
305. *Panorpa curva* Carpenter, 1938
306. *Panorpa curvata* Zhou, 2006
    Panorpa stigmosa Zhou, 2006, **new synonym**
307. *Panorpa davidi* Navás, 1908
308. *Panorpa difficilis* Carpenter, 1938
309. *Panorpa diqingensis* Li & Hua, 2020
310. *Panorpa dispergens* Li & Hua, 2020
311. *Panorpa fructa* Cheng, 1949
312. *Panorpa fulvastra* Chou in Chou et al., 1981
313. *Panorpa gaokaii* Li & Hua, 2021
314. *Panorpa huayuani* Li & Hua, 2021
315. *Panorpa jinchuana* Hua, Sun & Li, 2001
316. *Panorpa neospinosa* Chou & Wang in Chou et al., 1981
317. *Panorpa qinlingensis* Chou & Ran in Chou et al., 1981
318. *Panorpa sexspinosa* Cheng, 1949
    Syn. *Panorpa sexspinosa zhongnanensis* Chou & Ran in Chou et al., 1981, synonymized by Li & Hua, 2021
    Syn. *Panorpa shanyangensis* Chou & Wang in Chou et al., 1981, synonymized by Li & Hua, 2021
319. *Panorpa typicoides* Cheng, 1949
320. *Panorpa uncinata* Li & Hua, 2021
321. *Panorpa yangi* Chou in Chou et al., 1981
322. *Panorpa yaoluopingensis* Li & Hua, 2021

Ungrouped species

323. *Panorpa acanthophylla* Zhou, 2006
324. *Panorpa concolor* Esben-Petersen, 1915
325. *Panorpa guidongensis* Chou & Li in Chou et al., 1987
326. *Panorpa horiensis* Issiki, 1929
327. *Panorpa issikiana* Byers, 1970
328. *Panorpa kiautai* Zhou & Wu in Zhou et al., 1993a
329. *Panorpa lachlani* Navás, 1930

Genus *Sinopanorpa* Cai & Hua in Cai et al., 2008

330. *Sinopanorpa baokangensis* Wang, 2021
331. *Sinopanorpa digitiformis* Huang & Hua in Cai et al., 2008
332. *Sinopanorpa nangongshana* Cai & Hua in Cai et al., 2008
333. *Sinopanorpa tincta* (Navás, 1931)
    Syn. *Panorpa statura* Cheng, 1949, synonymized by Hua & Peng, 2005

Family Panorpodidae Byers, 1965

Genus *Panorpodes* MacLachlan, 1875

334. *Panorpodes brachypoda* Tan & Hua, 2008
335. *Panorpodes brevicaudata* (Hua, 1998)
336. *Panorpodes komaensis* Okamoto, 1925
    Panorpodes kuandianensis Zhong, Zhang & Hua, 2011, **new synonym**

References

Benton, M.J. 1995. Diversification and extinction in the history of life. *Science* 268 (5207): 52.

Bicha, W. 1983. A new species of scorpionfly (Mecoptera: Panorpidae) from North America. *Proceedings of the Entomological Society of Washington* 85 (1): 152–155.

———. 2006. New scorpionflies (Mecoptera: Panorpidae) from Jalisco, Michoacán, and Oaxaca, Mexico. *Proceedings of the Entomological Society of Washington* 108 (1): 24–34.

———. 2007. The first hangingfly (Mecoptera: Bittacidae) described from Vietnam. *Russian Entomological Journal* 16: 333–335.

———. 2010. A review of the scorpionflies (Mecoptera) of Indochina with the description of a new species of *Neopanorpa* from northern Thailand. *Zootaxa* 2408 (1): 61–67.

———. 2015a. A nearly cryptic scorpionfly, *Panorpa cryptica* n. sp. (Mecoptera: Panorpidae) from North America. *Zootaxa* 3973 (3): 591–600.

———. 2015b. The scorpionflies (Mecoptera) of Indochina with the description of new species of *Bittacus* and *Neopanorpa*. *Proceedings of the Entomological Society of Washington* 117 (4): 435–451.

———. 2018. Biodiversity of Mecoptera. In *Insect biodiversity: Science and society, II*, ed. R.G. Foottit and P.H. Adler, 705–720. Hoboken: John Wiley & Sons.

Bicha, W., N. Schiff, T.H. Pham, A. Lancaster, and B. Scheffler. 2017. New species of *Neopanorpa* (Mecoptera) from Vietnam, with a key to the species of Mecoptera from Vietnam. *Proceedings of the Entomological Society of Washington* 119 (4): 529–544.

Bicha, W., Y. Chiu, T. Nakamura, and D. Almquist. 2020. Unusual scorpionfly (Mecoptera: Panorpidae) collecting techniques. *Proceedings of the Entomological Society of Washington* 122 (4): 1001–1004.

Bicha, W.J. 2011. A review of the hangingflies (Mecoptera: Bittacidae) of South Asia with the description of a new species of *Bittacus* from Nepal. *Zootaxa*, 3032:60–64.

Bostock, J., and H.T. Riley. 1855. *The natural history of Pliny, volume III*, 29–31. London: H. G. Bohn.

Byers, G.W. 1958. Descriptions and distributional records of American Mecoptera. *Journal of the Kansas Entomological Society* 31: 213–222.

———. 1962. Descriptions and distributional records of American Mecoptera.II. *Journal of the Kansas Entomological Society* 35: 299–307.

———. 1963. The life history of *Panorpa nuptialis* (Mecoptera: Panorpidae). *Annals of the Entomological Society of America* 56 (2): 142–149.

———. 1965. The Mecoptera of Indo-China. *Pacific Insects* 7 (4): 705–748.

———. 1966. Mecoptera from Borneo and Tioman Islands. *Pacific Insects* 8: 885–892.

———. 1970. New and little known Chinese Mecoptera. *Journal of the Kansas Entomological Society* 43 (4): 383–394.

———. 1971. A new *Neopanorpa* from Nepal. *Journal of the Kansas Entomological Society* 44 (4): 534–539.

———. 1973. Descriptions and distributional records of American Mecoptera. III. *Journal of the Kansas Entomological Society* 46: 362–375.

———. 1975. Descriptions and distributional records of African Mecoptera. *Journal of the Entomological Society of Southern Africa* 38: 217–222.

———. 1988. Geographical affinities of the North American Mecoptera. In *Origins of the North American insect Fauna*, Memoirs of the Entomological Society of Canada, ed. J.A. Downes and D.H. Kavanaugh, 25–30. Cambridge: Cambridge University Press.

———. 1989. The Nannochoristidae of South America (Mecoptera). *The University of Kansas Science Bulletin* 54 (2): 25–34.

———. 1993. Autumnal Mecoptera of southeastern United States. *University of Kansas Science Bulletin* 55 (2): 57–96.

———. 1994. Taiwanese species of *Neopanorpa* (Insecta: Mecoptera: Panorpidae). *Annals of Carnegie Museum* 63 (2): 185–192.

———. 1996. Descriptions and distributional records of American Mecoptera IV. *University of Kansas Science Bulletin* 55 (14): 519–547.

———. 1997. Biology of *Brachypanorpa* (Mecoptera: Panorpodidae). *Journal of the Kansas Entomological Society* 70 (4): 313–322.

———. 1999. Thirteen new Panorpidae (Mecoptera) from northern Burma. *Entomologica Scandinavica* 30: 197–218.

———. 2002a. Mecoptera from Taiwan. *Journal of the Kansas Entomological Society* 75 (4): 252–258.

———. 2002b. Scorpionflies, hangingflies, and other Mecoptera. *The Kansas School Naturalist* 48 (1): 1–15.

———. 2008. New names in *Neopanorpa* (Mecoptera: Panorpidae). *Journal of the Kansas Entomological Society* 81 (1): 1–3.

———. 2009. Mecoptera: Scorpionflies, hangingflies. In *Encyclopedia of insects*, ed. V.H. Resh and R.T. Cardé, 2nd ed., 611–614. Cambridge: Academic Press.

Byers, G.W., and R. Thornhill. 1983. Biology of the Mecoptera. *Annual Review of Entomology* 28 (1): 203–228.

Cai, L.-J., and B.-Z. Hua. 2009a. A new *Neopanorpa* (Mecoptera, Panorpidae) from China with notes on its biology. *Deutsche Entomologische Zeitschrift* 56 (1): 93–99.

———. 2009b. Morphology of the immature stages of *Panorpa qinlingensis* (Mecoptera: Panorpidae) with notes on its biology. *Entomologica Fennica* 20 (3): 215–224.

Cai, L.-J., P.-Y. Huang, and B.-Z. Hua. 2006. Two new Chinese *Bittacus* Latreille (Mecoptera: Bittacidae) from Michangshan Mountains. *Entomotaxonomia* 28 (2): 127–130.

———. 2008. *Sinopanorpa*, a new genus of Panorpidae (Mecoptera) from the Oriental China with descriptions of two new species. *Zootaxa* 1941: 43–54.

Carpenter, F.M. 1931. Revision of the Nearctic Mecoptera. *Bulletin of the Museum of Comparative Zoology* 72: 205–277.

J.-S. Wang, B.-Z. Hua, *A Color Atlas of the Chinese Mecoptera*, https://doi.org/10.1007/978-981-16-9558-2

———. 1935. New Nearctic Mecoptera, with notes on other Species. *Psyche* 42: 105–122.
———. 1936. Descriptions and records of Nearctic Mecoptera. *Psyche* 43: 56–64.
———. 1938. Mecoptera from China, with descriptions of new species. *Proceedings of the Entomological Society of Washington* 40 (9): 267–281.
———. 1945. Panorpidae from China (Mecoptera). *Psyche* 52 (1/2): 70–78.
———. 1948. Notes on Chinese Panorpidae (Mecoptera). *Psyche* 55: 28–30.
Chau C.S., Byers G.W. 1978. The Mecoptera of Indonesia: Genus *Neopanorpa*. *The University of Kansas Science Bulletin* 51 (11): 341–405.
Chen, J., and B.-Z. Hua. 2011. A second species of *Bittacus* Latreille, 1805 (Mecoptera: Bittacidae) from Hainan Island, China. *Zootaxa* 3093: 64–68.
Chen, Q.-X., and B.-Z. Hua. 2016. Ultrastructure and morphology of compound eyes of the scorpionfly *Panorpa dubia* (Insecta: Mecoptera: Panorpidae). *PLoS ONE* 11 (6): e0156970.
Chen, J., J.-L. Tan, and B.-Z. Hua. 2013. Review of the Chinese *Bittacus* (Mecoptera: Bittacidae) with descriptions of three new species. *Journal of Natural History* 47 (21/22): 1463–1480.
Cheng, F.Y. 1949. New species of Mecoptera from northwest China. *Psyche* 56 (4): 139–173.
———. 1952. Additions to the Mecopterous fauna of Formosa. *Psyche* 59 (3): 89–95.
———. 1957a. Descriptions of new Panorpidae (Mecoptera) in the collection of the California Academy of Sciences. *Memoirs of the College of Agriculture, National Taiwan University* 5 (1): 27–33.
———. 1957b. Revision of the Chinese Mecoptera. *Bulletin of the Museum of Comparative Zoology at Harvard College* 116 (1): 1–118.
Cheung, D.K., S.A. Marshall, and D.W. Webb. 2006. Mecoptera of Ontario. *Canadian Journal of Arthropod Identification* 1: 1–13.
Chou, I. 1963. Some viewpoints about insect taxonomy. *Acta Entomologica Sinica* 12 (5/6): 586–596.
Chou, I., R.-B. Ran, and S.-M. Wang. 1981. Studies on the classification of Chinese Mecoptera (I, II). *Entomotaxonomia* 3 (1): 1–18.
Chou, I., S.-M. Wang, Z.-D. Li, and X.-W. Tong. 1987. New species of Mecoptera from Hunan Province (I)—Studies on the classification of Chinese Mecoptera (III). *Entomotaxonomia* 9 (3): 201–212.
———. 1988. New species of Mecoptera from Hunan Province (II)—Studies on the classification of Chinese Mecoptera (IV). *Entomotaxonomia* 10 (1/2): 31–42.
Dallai, R., P. Lupetti, B.A. Afzelius, and F. Frati. 2003. Sperm structure of Mecoptera and Siphonaptera (Insecta) and the phylogenetic position of *Boreus hyemalis*. *Zoomorphology* 122 (4): 211–220.
Du, W., and B.-Z. Hua. 2017. Two new species of the genus *Terrobittacus* Tan and Hua, 2009 (Mecoptera: Bittacidae) from southwestern China with a key to species. *European Journal of Taxonomy* 294: 1–13.
Du, X.-L., C. Yue, and B.-Z. Hua. 2009. Embryonic development of the scorpionfly *Panorpa emarginata* Cheng with special reference to external morphology (Mecoptera: Panorpidae). *Journal of Morphology* 270: 984–995.
Du, W., J. Chen, and B.-Z. Hua. 2015. Discovery of *Bittacus ussuriensis* Plutenko (Mecoptera: Bittacidae) from China. *Entomotaxonomia* 37 (4): 1–5.
Dunford, J.C., and L.A. Somma. 2008. Scorpionflies (Mecoptera). In *Encyclopedia of entomology*, ed. J.L. Capinera, 2nd ed., 3304–3310. Berlin: Springer Netherlands.
Enderlein, G. 1910. Über die Phylogenie und Klassifikation der Mecopteren unter Berücksichtigung der fossilen Formen. *Zoologischer Anzeiger* 35: 385–399.
———. 1912. Zur Kenntnis der Mecopteren Javas. *Notes from the Leyden Museum* 34: 235–239.
Erwin, D.H. 1990. Carboniferous-Triassic gastropod diversity patterns and the Permo-Triassic mass extinction. *Paleobiology* 16 (2): 187–203.
Esben-Petersen, P. 1912. H. Sauter's Formosa-Ausbeute: Neuroptera Planipennia. *Entomologische Mitteilungen* 1: 197–198.
———. 1913. H. Sauter's Formosa-Ausbeute: Planipennia II, Megaloptera and Mecoptera. *Entomologische Mitteilungen* 2 (7/8): 222–228.
———. 1915. A synonymic list of the order Mecoptera together with descriptions of new species. *Entomologiska Meddelelser* 10: 216–242.
———. 1921. Mecoptera. Monographic revision: Collections zoologiques du Baron Edm. de Selys Longchamps. *Catalogue Systématique et Descriptif* 5: 1–172.
———. 1927. New and little-known species of Mecoptera and Neuroptera in the Zoological Museum of Helsingfors. *Notulae Entomologicae* 7: 13–18.
———. 1934. Two new species of *Panorpa* Linn. (Mecoptera). *Videnskabelige Meddelelser fra Dansk Naturhistorik Forening, Kjobenhaven* 99: 1–3.
Ferris, G.F., and B.E. Rees. 1939. The morphology of *Panorpa nuptialis* Gerstaecker (Mecoptera: Panorpidae). *Microentomology: Contributions to Entomology from the Natural History Museum of Stanford University* 4 (3): 79–108.
Friedrich, F., and R.G. Beutel. 2009. The thoracic morphology of *Nannochorista* (Nannochoristidae) and its implicationsfor the phylogeny of Mecoptera and Antliophora. *Journal of Zoological Systematics and Evolutionary Research* 48 (1): 50–74.
Friedrich, F., H. Pohl, F. Beckmann, and R.G. Beutel. 2013. The head of *Merope tuber* (Meropeidae) and the phylogeny of Mecoptera (Hexapoda). *Arthropod Structure & Development* 42 (1): 69–88.
Fu, Q., and B.-Z. Hua. 2009. A new species of *Panorpa* (Mecoptera: Panorpidae) from Yunnan, China. *Entomotaxonomia* 31 (3): 201–205.
Fu, Q., S. Xie, and B.-Z. Hua. 2010. Mecoptera: Panorpidae, Bittacidae. In *Insects of Fengyangshan national nature reserve*, ed. H.-C. Xu, 229–230. Beijing: China Forestry Publishing House.
Gao, Q.-H., and B.-Z. Hua. 2013. Co-evolution of the mating position and male genitalia in insects: A case study of a hangingfly. *PLoS ONE* 8 (12): e80651.
Gao, K., and B.-Z. Hua. 2019. Revision of the genus *Cerapanorpa* (Mecoptera: Panorpidae) with descriptions of four new species. *European Journal of Taxonomy* 537: 1–23.
———. 2021. A new species of the genus *Cerapanorpa* (Mecoptera, Panorpidae) from the eastern Bashan Mountains. *Biodiversity Data Journal* 9: e72451.
Gao, C., N. Ma, and B.-Z. Hua. 2016. *Cerapanorpa*, a new genus of Panorpidae (Insecta: Mecoptera) with descriptions of three new species. *Zootaxa* 4158 (1): 93–104.
Gao, K., M.-D. Li, and B.-Z. Hua. 2020. Two new species of *Cerapanorpa* (Mecoptera, Panorpidae) from the Qinling and Minshan mountains. *ZooKeys* 971: 17–30.
Gao, K., Y. Hua, L.-X. Xing, and B.-Z. Hua. 2021. Speciation of the cold-adapted scorpionfly *Cerapanorpa brevicornis* (Mecoptera: Panorpidae) via interglacial refugia. *Insect Conservation and Diversity*. https://doi.org/10.1111/icad.12519.
Gassner, G. 1963. Notes on the biology and immature stages of *Panorpa nuptialis* Gerst. (Mecoptera: Panorpidae). *Texas Journal of Science* 15: 142–154.
Grimaldi, D., and M.S. Engel. 2005. *Evolution of the insects*, 772. Cambridge: Cambridge University Press.
Grimaldi, D.A., A. Rasnitsyn, J.-F. Zhang, and N. Fraser. 2005. Revision of the bizarre Mesozoic scorpionflies in the

Pseudopolycentropodidae (Mecopteroidea). *Insect Systematics & Evolution* 36 (4): 443–458.

Handlirsch, A. 1906. Die Fossilen Insekten und die Phylogenie der Rezenten Formen, parts I–IV. // *Ein Handbuch für Paläontologen und Zoologen*, pp. 1–640, Leipizig: Verlag von Wilhelm Engelmann.

Hennig, W. 1969. *Die Stammesgeschichte der Insekten*, 436. Frankfurt: Waldemar Kramer Verlag.

Hinton, H.E. 2009. On the structure, function, and distribution of the prolegs of the Panorpoidea, with a criticism of the Berlese-Imms Theory. *Transitions of the Royal Entomological Society of London* 106 (13): 455–540.

Hou, X.-Y., and B.-Z. Hua. 2008. Structures of the female reproductive systems in Panorpidae (Mecoptera) with remarks on their taxonomic significance. *Acta Zootaxonomica Sinica* 33 (3): 427–434.

Hsu, T.-C., W.-J. Wu, and L.-Y. Chou. 1999. A check list of Mecoptera (Insecta) of Taiwan. *Journal of Agricultural Research of China* 48 (2): 149–155.

Hu, G.-L., and B.-Z. Hua. 2016. Historical biogeography of the short-faced scorpionflies (Insecta: Mecoptera: Panorpodidae). *Journal of Biogeography* 43 (9): 1762–1772.

Hu, G.-L., G. Yan, H. Xu, and B.-Z. Hua. 2015. Molecular phylogeny of Panorpidae (Insecta: Mecoptera) based on mitochondrial and nuclear genes. *Molecular phylogenetics and Evolution* 85: 22–31.

Hua, B.-Z. 1997. A new species of the genus *Panorpa* (Mecoptera: Panorpidae) from northeast China. *Entomotaxonomia* 19 (3): 213–215.

———. 1998. *Panorpa* in Changbai Mountains with descriptions of four new species (Mecoptera: Panorpidae). *Entomotaxonomia* 20 (1): 37–43.

———. 1999. A new species of Panorpidae of Jigong Mountain in Henan (Mecoptera). In *The Fauna and Taxonomy of Insects in Henan. Vol. 3. Insects of the Jigong Mountain Region*, ed. X.-C. Shen and G.-F. Deng, 14–15. Beijing: China Agricultural Scientech Press.

Hua, L.-Z. 2002. *List of Chinese insects. Vol. II*, 612. Guangzhou: Zhongshan (Sun Yat-sen) University Press.

Hua, B.-Z. 2004. Panorpodidae: A new recorded family of Mecoptera from China. *Entomotaxonomia* 26 (4): 284–287.

Hua, B.-Z., and L.-J. Cai. 2009. A new species of the genus *Panorpa* (Mecoptera: Panorpidae) from China with notes on its biology. *Journal of Natural History* 43 (9/10): 545–552.

Hua, B.-Z., and I. Chou. 1997. The Panorpidae (Mecoptera) of Funiu Mountain in Henan Province. *Entomotaxonomia* 19 (4): 273–278.

———. 1998a. Panorpidae (Mecoptera) in Hainan Island. *Entomotaxonomia* 20 (2): 133–139.

———. 1998b. The Bittacidae of Funiu Mountain in Henan (Mecoptera). In *The Fauna and Taxonomy of Insects in Henan. Vol. 2. Insects of the Funiu Mountains Region*, ed. X.-C. Shen and Z.-Y. Shi, 64–67. Beijing: China Agricultural Scientech Press.

———. 1999. Panorpidae (Mecoptera) in Xizang, China, with descriptions of three new species of the genus *Neopanorpa*. *Entomologia Sinica* 6 (1): 11–17.

Hua, B.-Z., and T. Peng. 2005. A new synonym of *Panorpa tincta* Navas, 1931 (Mecoptera: Panorpidae). *Entomotaxonomia* 27 (2): 123–124.

Hua, B.-Z., and J.-L. Tan. 2007. A new species of *Bittacus* Latreille (Mecoptera, Bittacidae) from Daba Mountain in China. *Acta Zootaxonomica Sinica* 32 (2): 455–458.

Hua, B.-Z., G.-H. Sun, and M.-L. Li. 2001. Sichuan Panorpidae (Mecoptera) kept in the Tianjin Natural History Museum. *Entomotaxonomia* 23 (2): 120–123.

Hua, B.-Z., J.-L. Tan, and P.-Y. Huang. 2008. Two new species of the genus *Bittacus* (Mecoptera: Bittacidae) from China. *Zootaxa* 1749 (1): 62–68.

Hua, Y., S.-H. Tao, and B.-Z. Hua. 2018. An enigmatic new species of *Panorpa* Linnaeus from the Bashan Mountains (Mecoptera, Panorpidae). *ZooKeys* 777: 109–118.

Huang, P.-Y., and B.-Z. Hua. 2005. Four new species of the Chinese *Bittacus* Latreille (Mecoptera: Bittacidae). *Acta Zootaxonomica Sinica* 30 (2): 393–398.

Huang, P.-Y., B.-Z. Hua, and X.-C. Shen. 2004. A new species of the genus *Panorpa* (Mecoptera: Panorpidae) from Wangwu Mountain. *Entomotaxonomia* 26 (1): 29–31.

Hünefeld, F., and R.G. Beutel. 2005. The sperm pumps of Strepsiptera and Antliophora (Hexapoda). *Journal of Zoological Systematics and Evolutionary Research* 43 (4): 297–306.

Issiki, S. 1927. New and rare species of Mecoptera from Corea, Formosa and Japan. *Insecta Matsumurana* 2 (1): 1–12.

———. 1929a. Descriptions of new Mecoptera from the Japanese Empire. *Transactions of the Natural History Society of Formosa* 19: 260–314.

———. 1929b. Descriptions of new species of the genus *Panorpa* from Japan and Formosa. *Journal of the Society of Tropical Agriculture* 1: 182–191.

———. 1931. Two new species of scorpion flies (Insecta, Order Mecoptera). *Annals and Magazine of Natural History* 7: 219–222.

———. 1933. Morphological studies on the Panorpidae of Japan and adjoining countries and comparison with American and European forms. *Japanese Journal of Zoology* 4: 315–416.

———. 1965. Notes on some Mecoptera from Formosa. *Special Bulletin of Lepidopterological Society of Japan* 1: 245.

Issiki, S., and F.Y. Cheng. 1947. Formosan Mecoptera with descriptions of new species. *Memoirs of the College of Agriculture, National Taiwan University* 1 (4): 1–17.

Iwasaki, Y. 1996. Hunting and mating behavior in the Japanese hangingfly *Bittacus mastrillii* (Mecoptera: Bittacidae). *Annals of the Entomological Society of America* 89: 869–874.

Jiang, L., and B.-Z. Hua. 2013. Morphology and chaetotaxy of the immature stages of the scorpionfly *Panorpa liui* Hua (Mecoptera: Panorpidae) with notes on its biology. *Journal of Natural History* 47 (41/42): 2691–2705.

———. 2015. Morphological comparison of the larvae of *Panorpa obtusa* Cheng and Neopanorpa lui Chou & Ran (Mecoptera: Panorpidae). *Zoologischer Anzeiger* 255: 62–70.

Jiang L., C. Yue, and B.-Z. Hua. 2014. Larval morphology of *Panorpodes kuandianensis* (Insecta, Mecoptera, Panorpodidae) and its evolutionary implications. *ZooKeys* 398: 69–82.

Jiang, L., Q.-H. Gao, and B.-Z. Hua. 2015. Larval morphology of the hanging-fly *Bittacus trapezoideus* Huang & Hua (Insecta: Mecoptera: Bittacidae). *Zootaxa* 3957 (3): 324–333.

Jiang L., Y., Hua G.-L. Hu, and B.-Z. Hua. 2019b. Habitat divergence shapes the morphological diversity of larval insects: insights from scorpionflies. *Scientific Reports* 9: 12708.

Ju, J.-S., and W.-B. Zhou. 2003. Two new species of Mecoptera from Jiulongshan. *China. Jounal of Zhejiang Forestry College* 20 (1): 37–40.

Kaltenbach, A. 1978. *Mecoptera (Schnabelhafte, Schnabefliegen)*, Handbuch der Zoologie, 1–111. New York: Walter de Gruyter Berlin.

Kelly, C.A., A.J. Norbutus, A.F. Lagalante, and V.K. Iyengar. 2012. Male courtship pheromones as indicators of genetic quality in an arctiid moth (*Utetheisa ornatrix*). *Behavioral Ecology* 23 (5): 1009–1014.

Kou, L.-X., and B.-Z. Hua. 2016. Comparative embryogenesis of Mecoptera and Lepidoptera with special reference to the abdominal prolegs. *Journal of Morphology* 277 (5): 585–593.

Kristensen, N.P. 1981. Phylogeny of insect orders. *Annual Review of Entomology* 26 (1): 135–157.

———. 1999. Phylogeny of endopterygote insects, the most successful lineage of living organisms. *European Journal of Entomology* 96: 237–254.

Latreille, P.A. 1802. *Histoire Naturelle, Générale et Particuliere des Crustacés et des Insectes, Tome Troisieme*, 1–467. Paris: De l'Imprimerie de F. Dufart.

LeBas, N.R., L.R. Hockham, and M.G. Ritchie. 2004. Sexual selection in the gift-giving dance fly, *Rhamphomyia sulcata*, favors small males carrying small gifts. *Evolution* 58 (8): 1763–1772.

Li, Y.-L., and D. Ren. 2009. History and development of researches on Bittacidae (Insecta: Mecoptera). *Acta Geoscientica Sinica* 30 (4): 554–560.

Li, X., B.-Z. Hua, L.-J. Cai, and P.-Y. Huang. 2007. Two new species of the genus *Panorpa* (Mecoptera: Panorpidae) from Shaanxi, China with notes on their biology. *Zootaxa* 1542: 59–67.

Li, J.-K., X.-P. Zhang, L. Lin, and M.-X. Gao. 2012. The Panorpidae from North Korea and Northeast China (Mecoptera, Panorpidae). *Les Cahiers Magellanes* 35: 1–5.

Lieftinck, M.A. 1933. De in Ned. Indie voorkomende Mecoptera *(Panorpa). Verslagen von de vergaderingen der Afdeling nederlandsch Ost-Indie van de Nederlandsche Entomologische Vereening*, 4: 117–120.

———. 1934. Uit het leven der schorpioenvliegen (Panorpidae). *Tropische Natuur* 23: 64–70.

———. 1936. Studies in Oriental Mecopter. I. The genus *Leptopanorpa* in Malaysia. *Treubia* 15: 271–320.

Lindgren, N.K., M.S. Sisson, A.D. Archambeault, B.C. Rahlwes, J.R. Willett, and S.R. Bucheli. 2015. Four forensic entomology case studies: Records and behavioral observations on seldom reported cadaver fauna with notes on relevant previous occurrences and ecology. *Journal of Medical Entomology* 52 (2): 143–150.

Linnaeus, C. 1758. *Systema Naturae*. 10th ed, 824. Stockholm: Laurentius Salvius.

Ma, N., and B.-Z. Hua. 2011. *Furcatopanorpa*: A new genus of Panorpidae (Mecoptera) from China. *Journal of Natural History* 45 (35/36): 2251–2261.

Ma, N., W. Zhong, Q.-H. Gao, and B.-Z. Hua. 2012. Female genital plate diversity and phylogenetic analyses of East Asian Panorpidae (Mecoptera). *Systematics and Biodiversity* 10 (2): 159–178.

Ma, N., G.-L. Hu, J.-X. Zhang, and B.-Z. Hua. 2014. Morphological variation of the scorpionfly *Panorpa obtusa* Cheng (Mecoptera: Panorpidae) with a new synonym. *PLoS ONE* 9 (9): e108545.

MacLachlan, R. 1875. A sketch of our present knowledge of the neuropterous fauna of Japan (excluding Odonata and Trichoptera). *Transactions of the Entomological Society of London* 23 (2): 167–190.

———. 1887. Description de plusieurs nouvelles espèces de Panorpides provenant du Japon et de la Sibérie orientale. *Mitteilugen der Schweizerischen Entomologischen Gesellschaft* 7: 400–406.

———. 1894. On two small collections of Neuroptera from Tachien-lu, in the province of Szechuen, western China, on the frontier of Thibet. *Annals and Megazine of Natural History* 6 (13): 421–436.

Mader, D. 2020. Kopulation und Sexualethologie von Skorpionsfliegen, anderen Schnabelfliegen, Kamelhalsfliegen, Schlamm-fliegen, Steinfliegen, Köcherfliegen und Eintagsfliegen. *Galathea, Beiträge des Kreises Nürnberger Entomologen* 36: 61–184.

Martynova, O.M. 1954. Skorpionnitzy (Mecoptera) fauny SSSR I. *Trudy zoologicheskogo Instituta, Akademia Nauk SSSR* 15: 54–66.

———. 1957. Skorpionnitzy (Mecoptera) fauny SSSR II. Semejstvo Panorpidae. *Entomologischeskoe Obozreni* 36: 721–747.

Meusemann, K., Trautwein, M., Friedrich F., Beutel, R.G., Wiegmann, B.M., Donath, A., et al. 2020. Are fleas highly modified Mecoptera? Phylogenetic resolution of Antliophora (Insecta: Holometabola). *BioRxiv*. https://doi.org/10.1101/2020.11.19.390666.

Miao, Y., N. Ma, and B.-Z. Hua. 2017. Cytotaxonomy and molecular phylogeny of the genus *Cerapanorpa* Gao, Ma & Hua, 2016 (Mecoptera: Panorpidae). *Scientific Reports* 7: 4493.

Miao, Y., J.-S. Wang, and B.-Z. Hua. 2019. Molecular phylogeny of the scorpionflies Panorpidae (Insecta: Mecoptera) and chromosomal evolution. *Cladistics* 35: 385–400.

Mickoleit, G. 1976. The genital and postgenital segments of the Mecoptera females (Insecta, Holometabola). *Zoomorphologie* 85 (2): 133–156.

Mickoleit, G. 1978. Die phylogenetischen Beziehungen der Schnabelfliegen-Familien aufgrund morphologischer Ausprägungen der weiblichen Genital- und Postgenitalsegmente (Mecoptera). *Entomologica Germanica* 4: 258–271.

Misof, B., S.-L. Liu, K. Meusemann, R.S. Peters, A. Donath, C. Mayer, et al. 2014. Phylogenomics resolves the timing and pattern of insect evolution. *Science* 346 (6210): 763–767.

Miyaké, T. 1908. A list of Panorpidae of Japan, with descriptions of ten new species. *Bulletin of the College of Agriculture, Imperial University of Tokyo* 8: 1–12.

———. 1912. The life-history of *Panorpa klugi* M'Lachlan. *Journal of the College of Agriculture, Imperial University of Tokyo* 4 (2): 118–138.

———. 1913. Studies on the Mecoptera of Japan. *Journal of the College of Agriculture of the Imperial University of Tokyo* 4: 265–400.

Navás, L. 1908. Neurópteros nuevos. *Memorias de la real Academia de Ciencias y Artes de Barcelona* 6 (25): 401–423.

———. 1911. Névroptères nouveaux de l'extreme Orient. *Russkoe Entomologicheskoe Obozreni* 11: 111–117.

———. 1913. Névroptères du Japon recueillis par M. Edme Gallois. *Bulletin du Museue National d'Histoire Naturelle, Paris* 19: 441–451.

———. 1927. Commiones entomológicas 10: Mis excursiones científicas en 1927. *Revista de la Academia de Ciencias Exactas, Físicas Químicas y Naturales de Zaragoza* 11: 79–137.

———. 1930a. Insecta orientalia: VII series. *Memotie dell'Accademia pontifica dei Nuovi Lincei* 14: 409–418.

———. 1930b. Névroptères et insectes voisins: Chine et pays environnants [I] serie. *Notes d'Entomologie Chinoise* 4: 1–12.

———. 1931. Decadas de insectos nuevos. *Revista de la Real Academia de Ciencias Exactas, Fisicasy Naturales de Madrid* 26: 60–86.

———. 1935. Névroptères et insectes voisins: Chine et pays environnants. Huitieme [VII] serie. *Notes d'Entomologie Chinoise* 2: 85–103.

———. 1936. Névroptères et insectes voisins: Chine et pays environnants. Neuvieme [IX] serie. *Notes d'Entomologie Chinoise* 3: 37–62.

Okamoto H. 1925. The Mecoptera of Corea. *Bulletin of the agricultural Experiment Station, Government-General of Chosen* 2:17–24.

Palmer, C.M. 2010. Diversity of feeding strategies in adult Mecoptera. *Terrestrial Arthropod Reviews* 3 (2): 111–128.

Penny, N.D. 1975. Evolution of the extant Mecoptera. *Journal of the Kansas Entomological Society* 48 (3): 331–350.

——— N.D., Byers G.W. 1979. A check-list of the Mecoptera of the world. *Acta Amazonica* 9(2): 365–388.

Plutenko, A.V. 1985. New and little-known species of Mecoptera from the Soviet Far East. *Entomologicheskoye Obozrenie* 64 (1): 171–176.

Qian, Z.-X., and W.-B. Zhou. 2001. Three new species of Mecoptera from Yunnan. *Journal of Zhejiang Forestry College* 18 (3): 297–300.

Ren, D. 1993. First discovery of fossil Bittacids from China. *Acta Geologica Sinica* 67 (4): 376.

Ren, D., C.C. Labandeira, J.A. Santiago-Blay, A. Rasnitsyn, C.-K. Shih, A. Bashkuev, et al. 2009. A probable pollination mode before angiosperms: Eurasian, long-proboscid scorpionflies. *Science* 326 (5954): 840–847.

Riek, E.F. 1954. The Australian Mecoptera or scorpion-flies. *Australian Journal of Zoology* 2: 143–168.

———. 1973. A revision of Australian scorpion flies of the family Choristidae (Mecoptera). *Journal of the Australian Entomological Society* 12: 103–112.

Rust, M.K., and G.W. Byers. 1976. The Mecoptera of India and adjacent regions. *The University of Kansas Science Bulletin* 51 (2): 19–90.

Shorthouse, D.P. 2010. SimpleMappr, an online tool to produce publication-quality point maps. Accessed August 25, 2021 https://www.simplemappr.net.

Soszyńska-Maj, A., E. Krzemińska, R. Pérez-de la Fuente, J.-S. Wang, K. Szpila, K. Skibińska, et al. 2022. Evolution of sexual conflict in scorpionflies. *eLife* 11: e70508.

Tan, J.-L., and B.-Z. Hua. 2006. Behaviors of the hangingfly *Bittacus planus*. *Chinese Bulletin of Entomology* 43 (3): 348–351.

———. 2008a. A new species of the genus *Bittacus* (Mecoptera, Bittacidae) from Zhejiang. *China. Acta Zootaxonomica Sinica* 33 (3): 487–490.

———. 2008b. First discovery of Bittacidae (Mecoptera) in Hainan Island, China, with description of a new species. *Entomological News* 119 (5): 497–500.

———. 2008c. The second species of the Chinese Panorpodidae (Mecoptera), *Panorpodes brachypodus* sp. nov. *Zootaxa* 1751: 59–64.

———. 2009a. *Bicaubittacus*, a new genus of the Oriental Bittacidae (Mecoptera) with descriptions of two new species. *Zootaxa* 2221: 27–40.

———. 2009b. *Terrobittacus*, a new genus of the Chinese Bittacidae (Mecoptera) with descriptions of two new species. *Journal of Natural History* 43 (47/48): 2937–2954.

Thornhill, R. 1975. Scorpionflies as kleptoparasites of web-building spiders. *Nature* 258 (5537): 709–711.

———. 1976. Sexual selection and nuptial feeding behavior in *Bittacus apicalis* (Insecta: Mecoptera). *The American Naturalist* 110 (974): 529–548.

Tihelka, E., M. Giacomelli, D.-Y. Huang, D. Pisani, P.C.J. Donoghue, and C.-Y. Cai. 2020. Fleas are parasitic scorpionflies. *Palaeoentomology* 3: 641–653.

Tillyard, R.J. 1917. Mesozoic insects of Queensland. No. 1. Planipennia, Trichoptera, and the new order Protomecoptera. *Proceedings of the Linnaean Society of New South Wales* 42: 175–200.

Tjeder, B. 1936. Schwedish-chinensische wissenschaftliche Expedition nach den nordwestlichen Provinzen Chinas, under Leitung von Dr. Sven Hedin und Prof. Su Ping-chang. Insekten gesammelt vom schwedischen Arzt der Expedition Dr. David Hummel 1927–1930. *Arkiv for Zoologi* 27A (33): 1–14.

———. 1950. Mecopterens aus Fukien. *Bonner zoologische Beiträge* 1 (2/4): 286–290.

———. 1956. Zwei neue ost-asiatische *Bittacus*-Arten. *Beiträge zur Entomologie* 6: 45–53.

———. 1973. Bittacidae from Burma, collected by R. Malaise (Mecoptera). *Entomologisk Tidskrift* 94 (3/4): 188–193.

Tong, X., L. Jiang, and B.-Z. Hua. 2017. A unique mating pattern of *Panorpodes kuandianensis* (Mecoptera: Panorpodidae). *Contributions to Zoology* 86 (3): 229–237.

Walker, F. 1853. Genus 5: *Bittacus*. In *Catalogue of the Specimens of Neuropterous Insects in the Collection of the British Museum, Part II (Sialides–Nemopterides)*, ed. J.E. Gray, 465–469. London: British Museum [Natural History].

Wang, J.-S. 2022. New and little-known species of the genus *Dicerapanorpa* from northwestern Yunnan, China (Mecoptera: Panorpidae). *Acta Entomologica Musei Nationalis Pragae* 62: 1–13.

Wang, J.-S., and T. Suzuki. 2022. Review of the *Panorpa wormaldi* group (Mecoptera: Panorpidae), with descriptions of two new species. *European Journal of Taxonomy* 794: 18–39.

Wang, J.-S. 2021a. Bittacidae of Yunnan, China and adjacent regions, with first discovery of male *Bittacus malaisei* Tjeder, 1973 (Mecoptera). *Annales de la Société entomologique de France* 57: 77–83.

———. 2021b. *Neopanorpa* (Mecoptera: Panorpidae) from the Himalayas and adjacent regions, with descriptions of three new species. *Acta Entomologica Musei Nationalis Pragae* 61 (1): 203–212.

———. 2021c. A new species of the genus *Sinopanorpa* Cai & Hua, 2008 from Hubei Province, China (Mecoptera: Panorpidae). *Faunitaxys* 9 (29): 1–6.

Wang, J.-S., and Y.-J. Gong. 2021. Taxonomy of the *Panorpa guttata* group (Mecoptera: Panorpidae), with descriptions of fourteen new species from China. *Zootaxa* 4981 (2): 241–274.

Wang, J.-S., and B.-Z. Hua. 2016. Two new species of the genus *Panorpa* Linnaeus (Mecoptera, Panorpidae) from Yunnan, China. *ZooKeys* 587: 151–162.

———. 2017a. An annotated checklist of the Chinese Mecoptera with description of male *Panorpa guttata* Navás, 1908. *Entomotaxonomia* 39 (1): 24–42.

Wang, M., and B.-Z. Hua. 2017b. Discovery of *Neopanorpa chillcotti* Byers (Mecoptera: Panorpidae) from Tibet, China, with discussion of its generic status. *Zootaxa* 4232 (2): 241–250.

———. 2018a. A new species of *Neopanorpa* with an extremely long notal organ from Sichuan, China (Mecoptera, Panorpidae). *ZooKeys* 750: 131–140.

———. 2018b. High species diversity of the genus *Neopanorpa* (Mecoptera: Panorpidae) in Yunnan Province, China. *Zootaxa* 4483 (1): 36–66.

Wang, J.-S., and B.-Z. Hua. 2019a. *Megapanorpa*, a new genus with a single anal horn in males from Oriental China (Mecoptera: Panorpidae). *Entomological Science* 22: 64–79.

Wang, J.-S., and B.-Z. Hua. 2019b. Taxonomy of the genus *Neopanorpa* van der Weele, 1909 (Mecoptera, Panorpidae) from the Oriental Region, with the description of two new species. *European Journal of Taxonomy* 543: 1–17.

———. 2020. Taxonomic revision and phylogenetic analysis of the enigmatic scorpionfly genus *Leptopanorpa* MacLachlan (Mecoptera: Panorpidae). *Journal of Zoological Systematics and Evolutionary Research* 58: 900–928.

———. 2021. Morphological phylogeny of Panorpidae (Mecoptera: Panorpodidae). *Systematic Entomology* 46 (3): 526–557.

Wang, Y., C.C. Labandeira, C.-K. Shih, Q.-L. Ding, C. Wang, Y.-Y. Zhao, et al. 2012. Jurassic mimicry between a hangingfly and a ginkgo from China. *Proceedings of the National Academy of Sciences* 109 (50): 20514–20519.

Wang, J.-S., X.-T. Gao, and B.-Z. Hua. 2019. Two new species of the genus *Panorpa* (Mecoptera, Panorpidae) from eastern China and a new synonym. *ZooKeys* 874: 149–164.

Wedell, N. 1993. Spermatophore size in bushcrickets: Comparative evidence for nuptial gifts as a sperm protection device. *Evolution* 47 (4): 1203–1212.

van der Weele, H.W. 1909. Mecoptera and Planipennia of Insulinde. *Notes from the Leyden Museum* 31 (1): 1–100.

Wei, Z., X. Tong, and B.-Z. Hua. 2020. How does the male penisfilum enter the female copulatory pore in hangingflies? *Insects* 11 (2): 123.

Whiting, M.F. 2002. Mecoptera is paraphyletic: Multiple genes and phylogeny of Mecoptera and Siphonaptera. *Zoologica Scripta* 31 (1): 93–104.

Whiting, M.F., J.M. Carpenter, Q. Wheeler, and W.C. Wheeler. 1997. The Strepsiptera problem: Phylogeny of the holometabolous insect orders inferred from 18S and 28S ribosomal DNA sequences and morphology. *Systematic Biology* 46 (1): 1–68.

Willmann, R. 1987. The phylogenetic system of the Mecoptera. *Systematic Entomology* 12: 519–524.

———. 1989. Evolution und phylogenetisches System der Mecoptera (Insecta: Holometabola). *Abhandlungen der Senckenbergischen Naturforschenden Gesellschaft* 544: 1–153.

Yie, S.T. 1951. The biology of Formosan Panorpidae and morphology of eleven species of their immature stages. *Memoirs of the College of Agriculture, National Taiwan University* 2 (4): 1–111.

Yue, C., and B.-Z. Hua. 2010. Are abdominal prolegs serially homologous with the thoracic legs in Panorpidae (Insecta: Mecoptera): Embryological evidence. *Journal of Morphology* 271: 1366–1373.

Zhang, J.-X., and B.-Z. Hua. 2012. A new species of the genus *Panorpa* L. (Mecoptera: Panorpidae) from the Daba Mountains of central China. *Entomotaxonomia* 34 (3): 541–547.

Zhang, Y.N., W., Du., B.-Z., and Hua. 2020. Three new species of the genus Bittacus Latreille, 1805 (Mecoptera: Bittacidae), with a key to the species of Bittacidae in South China. *Zootaxa* 4718: 381–90.

Zhong, W., and B.-Z. Hua. 2013a. *Dicerapanorpa*, a new genus of East Asian Panorpidae (Insecta: Mecoptera: Panorpidae) with descriptions of two new species. *Journal of Natural History* 47 (13/14): 1019–1046.

———. 2013b. Mating behaviour and copulatory mechanism in the scorpionfly *Neopanorpa longiprocessa* (Mecoptera: Panorpidae). *PLoS ONE* 8 (9): e74781.

Zhong, W., J.-X. Zhang, and B.-Z. Hua. 2011. *Panorpodes kuandianensis*, a new species of short-faced scorpionflies (Mecoptera, Panorpodidae) from Liaoning, China. *Zootaxa* 2921: 47–55.

Zhong, W., G. Ding, and B.-Z. Hua. 2015a. The role of male's anal horns in copulation of a scorpionfly. *Journal of Zoology* 295 (3): 170–177.

Zhong, W., Z.-Y. Qi, and B.-Z. Hua. 2015b. Atypical mating in a scorpionfly without a notal organ. *Contributions to Zoology* 84 (4): 305–315.

Zhou, W.-B. 2000. Six new species of Mecoptera from China. *Journal of Zhejiang Forestry College* 17 (3): 248–254.

———. 2001. Mecoptera. In *Insects of Tianmushan National Nature Reserve*, ed. H. Wu and C.-W. Pan, 393–396. Beijing: Science Press.

———. 2003. Four new species of Mecoptera from China. *Wuyi Science Journal* 19: 88–94.

———. 2005. Mecoptera. In *Insects from Xishui Landscape*, ed. D.-C. Jin and Z.-Z. Li, 383–387. Guiyang: Guizhou Science and Technology Publishing House.

———. 2006. Mecoptera. In *Insects from Chishui Suoluo Landscape*, ed. D.-C. Jin and Z.-Z. Li, 273–375. Guiyang: Guizhou Science and Technology Publishing House.

Zhou, W.-B., and K.-H. Bao. 2002. Mecoptera. In *Insects from Maolan Landscape*, ed. Z.-Z. Li and D.-C. Jin, 409–411. Guiyang: Guizhou Science and Technology Publishing House.

Zhou, W.-B., and Z.-Y. Fan. 1998. Mecoptera. In *Insects of Longwangshan Nature Reserve*, ed. H. Wu, 145–147. Beijing: China Forestry Publishing House.

Zhou, W.-B., and X. Zhou. 2005. Mecoptera. In *Insects from Dashahe nature reserve of Guizhou*, ed. M.-F. Yang and D.-C. Jin, 374–379. Guiyang: Guizhou People's Publishing House.

———. 2006. Mecoptera. In *Insects from Fanjingshan Landscape*, ed. Z.-Z. Li and D.-C. Jin, 432–436. Guiyang: Guizhou Science and Technology Publishing House.

Zhou, X., and W.-B. Zhou. 2007. Mecoptera. In *Insects from Leigongshan Landscape*, ed. Z.-Z. Li and D.-C. Jin, 450–456. Guiyang: Guizhou Science and Technology Publishing House.

———. 2010. Mecoptera. In *Insects from Mayanghe Landscape*, ed. X.-S. Chen, Z.-Z. Li, and D.-C. Jin, 56–58. Guiyang: Guizhou Science and Technology Publishing House.

———. 2012. Mecoptera. In *Insects from Kuankuoshui Landscape*, ed. R.-H. Dai, Z.-Z. Li, and D.-C. Jin, 508–561. Guiyang: Guizhou Science and Technology Publishing House.

———. 2014. Mecoptera. In *Insects in Qingliangfeng, Zhejiang Province*, ed. Y.-P. Wang and C.-L. Tong, 197–199. Beijing: China Forestry Publishing House.

Zhou, W.-B., Y.-X. Hu, X.-P. Wu, and H. Wu. 1993a. Six new species and two new records of Mecoptera from China. *Journal of Zhejiang Forestry College* 10 (2): 189–196.

Zhou, W.-B., Y.-X. Hu, X.-P. Wu, and Q.-Z. Zheng. 1993b. Two new species of *Neopanorpa* from Zhejiang (Mecoptera: Panorpidae). *Journal of Zhejiang Forestry College* 10 (3): 314–317.